Circular Economy of Plastics

Peter Orth · Jürgen Bruder · Ulrich Liman ·
Manfred Rink
Editors

Circular Economy of Plastics

From Plastics Recycling Towards Feedstock Transformation

Springer

Editors
Peter Orth
OPC - Orth Plastic Consulting
Cologne, Germany

Jürgen Bruder
BPC - Bruder Plastic Consulting
Potsdam, Germany

Ulrich Liman
ULC - Ulrich Liman Consulting
Langenfeld, Germany

Manfred Rink
mrink-Consulting
Cologne, Germany

ISBN 978-3-031-78828-4 ISBN 978-3-031-78829-1 (eBook)
https://doi.org/10.1007/978-3-031-78829-1

This Springer imprint is published by the registered company Springer Nature Switzerland AG
The registered company address is: Gewerbestrasse 11, 6330 Cham, Switzerland

We dedicate this book to the late Prof. Dr. Klaus Töpfer, who, as German Federal Minister for the Environment, Nature Conservation and Nuclear Safety and Executive Director of the United Nations Environment Programme (UNEP), was a pioneer of the Circular Economy and a champion of sustainable development.

Foreword by André Bardow, ETH Zurich

"The revolution devours its children" was a common saying during the French Revolution, but the statement might also apply to today's plastics revolution. Plastics have undoubtedly revolutionized our lives since their discovery by Staudinger at ETH Zurich 100 years ago. Unfortunately, this triumph of plastics is increasingly showing its harmful effects through pollution in all areas of the environment. Therefore, we need a fresh start for plastics that addresses potential environmental impacts throughout their life cycle.

Less visible than pollution by plastic waste, but just as central, are greenhouse gas emissions: plastics demand has become a primary driver of global petroleum consumption and thus also of the resulting CO_2 emissions. To achieve our climate targets, we therefore need climate-neutral plastics. For this purpose, the following circular raw materials are available: Recyclates, biomass, and CO_2. Our analyses show that these circular raw materials already allow the climate-neutral production of plastics with today's technologies. However, much remains to be done to make this a reality: in particular, we need more and better recycling on a much larger scale.

To realize circular carbon-neutral plastics, the entire value chain must change, and our perspective must expand: the life cycle of plastics does not end at the factory gate. We must critically examine where the use of plastics makes sense—and where it can be avoided. CO_2 emissions must be transparently recorded and priced throughout the life cycle, including potential end-of-life incineration. In particular, product design must already consider the end of life. Plastic waste must ultimately be valued to motivate consumers to collect and sort it. Such a life cycle perspective on plastics could enable an actual "win-win" situation: fossil feedstock is replaced by circular plastics that no longer end up in the environment. To achieve this vision, we must invest consistently in circular instead of fossil feedstock.

With this book, leading industry experts set out their plan for a circular economy for plastics. I wish the book a broad readership to take us to the second plastics revolution: towards sustainability.

André Bardow

Prof. Dr.-Ing. André Bardow is Professor of Energy and Process Systems Engineering at ETH Zurich since 2020. Previously, he held the Chair of Technical Thermodynamics (LTT) at RWTH Aachen University (2010–2020), was the founding Director of the Institute for Energy and Climate Research (IEK-10) at Forschungszentrum Jülich (2017–2022) and was Associate Professor at TU Delft (2007–2010). He was a visiting professor at the University of California, Santa Barbara (2015/2016). Professor Bardow is a Fellow of the Royal Chemical Society and Chairman of the Thermodynamics Committee of the German Engineers' Association VDI.

Foreword by Martin Faulstich, INZIN, Düsseldorf

Over 50 years ago, the Club of Rome published its epoch-making study "The Limits to Growth". Then, as now, it was argued that insufficient account had been taken of creativity and technological progress. Despite all the justified criticism of the study, this is actually true, but in a different way than the critics probably thought. A number of developments have even been strengthened since then or were only discernible at the time in rudimentary form. For example, research and development have made it possible to use all the elements of the periodic table that can be used for technical purposes. This is particularly true of the metals, which are known to be finite on our planet and cannot be synthesized. In addition, climate change hardly played a role in our consciousness at that time. The limited supplies of coal, oil and gas were in the foreground and the absorption capacity of the atmosphere for carbon dioxide and other greenhouse gases was not yet an issue.

The terms sustainability, Anthropocene and planetary boundaries have since shaped the ecological and economic debate. In its report "Responsibility in a finite world", the German Advisory Council on the Environment made it impressively clear that the natural balance is also the foundation for a prosperous economy and a thriving society. It is more important than ever to decouple welfare and resource use, and the potential for doing so is far from exhausted. However, many impressive efficiency gains are currently being undone by rebound effects.

Plastics are an important part of our civilisation; with their diverse properties, they play an important role in all areas of life. Production and use have therefore risen sharply worldwide in recent decades, but success in keeping them in the cycle is still quite modest. This is where the new book "Circular Economy of Plastics" by the renowned authors' collective comes in and takes a competent look at all stages of the value chain: Production with renewable raw materials instead of crude oil, waste avoidance as a priority, recycling and the necessary use of recyclates, thermal treatment for unavoidable and non-recyclable waste.

The plastics industry will undoubtedly make its contribution to transforming our still predominantly fossil-based industrial society into a sustainable industrial society. A greenhouse gas-neutral and resource-efficient society will then be based on two essential pillars—renewable energy sources and recycled raw materials, so the energy turnaround also needs the raw materials turnaround.

A few years ago, "Kunststoffe im Kreislauf" became the standard German-language work in this field. It is therefore very welcome that this book has now been significantly expanded and is now also available to international experts in an English-language version. We wish this work, which is rich in perspectives, an interested readership and a wide distribution.

Martin Faulstich

Prof. Dr. Martin Faulstich is a Senior Professor at the Chair of Resource and Energy Systems at TU Dortmund University, having previously held chairs at TU Clausthal and TU Munich. He is a board member of the INZIN Institute for the Future of Industrial Society in Düsseldorf. From 2008 to 2016, he was Chairman of the German Advisory Council on the Environment. In addition, he was and is active in renowned committees, such as Co-Chairman of the Resource Commission at the Federal Environment Agency, on the Board of Trustees of the Daimler Benz Foundation and as a trustee of the Ifo Institute for Economic Research.

Foreword by Peter Vanacker, CEO LyondellBasell

As we look toward a future defined by both opportunity and challenge, the need for a sustainable and circular economy for plastics has never been greater and the chemical industry is undergoing a significant transformation. We are actively seeking solutions to mitigate climate change, reduce our environmental footprint, and develop innovative new products that meet the needs of a growing society. This journey requires collaboration across the entire value chain and the circular economy of plastics is a critical contribution to this collective effort.

This book delves into the complexities of creating a closed-loop system for plastics. It explores the challenges and opportunities associated with recycling, feedstock diversification, and technological advancements. By outlining a roadmap for a circular plastics economy, this book serves as a resource for policymakers, industry leaders, and all stakeholders invested in a sustainable future.

As a leader in the global chemical industry, we understand the critical role plastics play in society. They enhance people's lives by enabling many core applications, from healthcare and housing to food packaging and sustainable transportation, as well as renewable energy generation. However, plastic pollution is a critical issue, and we recognize the need for our industry to move faster. This is why we are accelerating our efforts to innovate, scale and deliver solutions to turn post-use plastics into everyday products and enable a circular economy.

Driven by this belief, we are already active in some of the efforts outlined in the "Circular Economy of Plastics" and are already creating solutions and addressing rising demand for more sustainable products from our customers and society. We are investing across the value chain to establish LYB as a full solution provider with the ability to scale critical recycling technologies. We develop advanced and mechanical recycling technologies and explore renewable feedstocks to grow our *Circulen* family of polymers. We believe that a circular economy for plastics is not only achievable, but one of the essential elements for ensuring a sustainable future for our industry and our planet.

This book summarizes the shared commitment to building a future where plastics are no longer a source of waste, but a valuable resource continuously cycled within a sustainable system. For the benefit of all of us.

Peter Vanacker

Peter Vanacker is Chief Executive Officer of LyondellBasell, a leader in the global chemical industry. Prior to joining the company, Vanacker was president, CEO and chair of the executive committee at Neste Corporation, a renewable products company. Vanacker earned his master's degree in chemical engineering with a specialty in polymers engineering from the University of Ghent. Among his leading professional engagements, he sits on the executive committee and board of directors of the American Chemistry Council (ACC), ACC's board strategy and board plastics committees. Vanacker also serves in a similar capacity at the International Consortium of Chemical Associations (ICCA), as well as the Alliance to End Plastic Waste (AEPW).

Foreword by Philipp Lehner, CEO ALPLA Group

Plastics surround and accompany us in our daily lives. They are the basis for our high standard of living and at the same time the material of choice for solving the social challenges of the future.

As a manufacturing company in the FMCG sector with around 200 plants and more than 23,000 employees worldwide, we reach around three billion people every day with our plastic packaging. Our mission is clear and ambitious: we want to provide everyday goods safely, affordably and sustainably.

A central aspect of our corporate philosophy is the circular economy. For us, this means reusing and recycling materials. Through our extensive recycling activities and close cooperation with partners worldwide, we do everything we can to keep plastics in the cycle and minimise the environmental and climate impact.

However, implementing this philosophy brings with it numerous challenges. From the perspective of a processor and recycler, we must ensure that our products are optimised for recycling. Design for recycling is therefore an essential part of our development processes.

Another important point is the use of recyclates in our products. Here we face the challenge of ensuring both the quantity and the quality of the recyclates. Our own recycling activities and the organisation of collection together with partners, particularly in countries without existing collection systems, are crucial to ensuring a constant and high-quality supply of recyclates.

Collaboration with our customers and other stakeholders in the cycle is also of crucial importance. We see ourselves not only as a producer, but also as a consultant and partner in matters of sustainability. In recent years, our customers have been listening to us more and more and are increasingly seeking our advice. This close cooperation enables us to develop innovative and sustainable solutions that meet the requirements of the market.

Despite the numerous challenges, the future offers more opportunities than ever to keep plastics in circulation. Through research and development, we can create products and materials that are more sustainable and efficient than ever. New technologies enable us to produce packaging with a lower weight and higher recycling content. And new business areas open up additional opportunities for us.

The last few years have shown that a rethink is taking place. Sustainable behaviour is becoming increasingly urgent in society and the reduction of CO_2 emissions is gaining in importance. Despite bureaucratic and cost-related challenges, particularly in the EU, we are committed to establishing plastic as an important part of the solution for our society. After all, plastic is in most cases the most climate-friendly option compared to other materials and, as a hidden climate champion, is therefore an important building block in the fight against the climate crisis. Plastic is more indispensable than ever before.

I am sure that this valuable book will not only give a comprehensive insight into the complex challenges and opportunities of our industry, but also inspires the readers to promote circular and therefore sustainable practices in their own environment.

Philipp Lehner

Philipp Lehner, CEO ALPLA Group is Chief Executive Officer (CEO) of the internationally active packaging and recycling specialist ALPLA Group. After working in the finance and consulting industry and completing an MBA at Harvard Business School, Philipp Lehner joined the family business in June 2014. He initially completed a six-month plant trainee programme in Lübeck, Germany, before moving to ALPLA Iowa in the USA for 18 months as Plant Manager. He then took on overall responsibility for business in the USA as Regional Manager for two and a half years before returning to the Austrian headquarters in Hard/Vorarlberg at the end of 2018. As CFO of the ALPLA Group from January 2019 to December 2020, he was responsible for finance, information technology, digitalisation and human resources before taking over as CEO in January 2021.

Preface

Nothing retains its own form; but nature, the greater renewer, ever makes up forms from forms. Believe me, nothing perishes in this whole wide universe; it does but vary and renew its form. What we call birth is but a different beginning; death is but ceasing to be the same. Though one thing may pass into another, yet the sum of all remains unchanged.

—Ovid[1]

The plastics industry must overcome major challenges in this decade of action. Together with its customers and partners, it is transitioning to a circular economy in almost all sectors of the economy. The historically evolved linear economy of producing, using, and discharging waste into the terrestrial, aquatic, or atmospheric environment without taking into account the vulnerability of the earth's ecosystem and the limited resources of our planet is being fundamentally replaced by an economy and way of life whose production and consumption is in harmony with boundaries of our planet and is oriented towards the imperative of climate neutrality, among other things through the continuous recycling of all materials. The role of the organic material plastic in this cycle is ambivalent, on the one hand, the increasing quantities of waste and the greenhouse gas emissions they induce must be managed, and on the other hand, the resources to produce plastics must be secured. We therefore need a strict closure of the carbon cycle, to which all stages of the value chain circle from chemicals to plastics production, plastics processing and product manufacture, trade and consumers, right through to disposal and recycling must contribute.

At the same time, the plastics industry is shaping a large-scale transformation process to replace the fossil feedstocks coal, oil and gas, which have formed the basis of its production for more than 70 years, with those from circular carbon sources, thereby completing a raw materials transition. An important interim goal is to achieve climate neutrality by 2050. This approach reflects the UN Sustainable Development Goals (SDGs) on industry, innovation and infrastructure (SDG 9) and on sustainable

[1] Ovid (43 BC–17 AD), actually Publius Ovidius Naso, Roman epic poet. Metamorphoses, Book 15, Lines 165–172, "Pythagoras' Speech on … the Imperishability of Matter".

consumption and production (SDG 12). Both measures require paradigm shifts in industry, the economy and society; they are far-reaching transformations.

In the first German-language edition of this book, which was published by Springer Vieweg in 2022 under the title "Kunststoffe im Kreislauf—vom Recycling zur Rohstoffwende", we approached this extensive range of topics for the first time. The acceptance of this book among German-speaking readers prompted the publisher to ask us for a second, English-language edition of the book for an international audience, a challenge that we gladly accepted after some clarifying discussions. We repositioned ourselves for this and designed a genuine editorial work, for which we were able to strengthen the team (Peter Orth, Jürgen Bruder and Manfred Rink) with Ulrich Liman, who was most recently a member of senior management at Covestro AG. The work was compiled based on the editors' decades of experience as employees in the plastics industry and as managing directors of German, European and international plastics associations.

We are now presenting the book in English in a significantly expanded form and with a new structure, making it accessible and readable for international business and science on all continents, as English has become the lingua franca or, in other words, the business language. However, we are aware that we can hardly meet the requirements of a European edition, as the majority of the authors we have been able to attract are of German-speaking origin and we have not succeeded in bringing together a truly representative group of European authors to reflect the diversity of our continent and the very different perspectives associated with it. And this is certainly not an international edition, as there are no authors from America or China, for example. A truly global presentation of the situation and treatment of the problems arising from the UN's predicted growth of humanity and the tension between growth and sustainability, which the topic deserves, is therefore a task for the future.

The title of the book "Circular Economy of Plastics—from Recycling to Raw Materials Transition" names the two core themes of the book. The introduction of a circular economy for plastics, in which the resources once introduced into the technosphere are recycled without constantly extracting further resources from the earth, and the substitution of fossil raw materials with circular raw materials.

The book covers five multifaceted subject areas. Part I describes the status quo of the material in four chapters—production, applications, volumes and volume flows, structures of the plastics industry; Part II—one of the central parts—describes the emerging circular economy of plastics in seven chapters using selected examples, focusing on the product, market, waste and raw material; Part III—the other central part—presents the very different circular carbon sources for plastics production: Plastic waste, biomass and carbon dioxide; Part IV deals in four chapters with four cross-cutting topics essential for the metric in circular economy—Life Cycle Assessment, Product Carbon Footprint, Digitization and Standardization; finally, Part V deals in three chapters with the less consistent regulatory apparatus of the European Union relevant to our context and in particular the Single Use Plastics Directive (SUPD) and the Packaging and Packaging Waste Regulation (PPWR). A concluding chapter provides an outlook for the future and summarizes the associated challenges for stakeholders in the circular economy of plastics. The book concludes with a glossary and a list of further readings in nine sections.

The ambition and scope of the book thus go well beyond the German edition, without being able to present the topic in a truly comprehensive manner. In total, the book contains 51 contributions by 69 authors in the Front Matter and in the 22 chapters of the five parts mentioned, they contribute their scientific and industrial perspectives to describe products, processes and services as well as the political framework conditions and their further developmental work in the course of the dynamic transformation taking place. The breadth of the range of topics reflects the complexity and scope of the ongoing transformation to a circular economy and the developing raw materials transition. However, it also gives an idea of the problems, challenges and chances that still lie ahead on the path to climate neutrality.

This book is aimed at decision-makers and experts in the plastics industry and the plastics value chains, at politicians and their staff in all areas of politics and the employees of public authorities involved in the subject area, at employees and members of NGOs involved in the circular economy, at students, lecturers and scholars at universities specializing in macromolecular chemistry, plastics technology, industrial chemistry and engineering, chemical engineering, process engineering and mechanical engineering, and finally at journalists and editors of the media who work in these economic and social sectors and beyond.

We would like to express our sincere thanks to all the authors, whose contributions were generally made alongside their regular professional activities, for their efforts, expenditure and time invested. And, of course, for the patience they have shown with us and our ever-expanding requests. It was only through their commitment that this practice-oriented book, which incorporates their expertise and proposed solutions for the further development of the circular economy, could be produced. We would also like to thank Springer-Verlag, who made it possible for us to publish this work and make it available to a large audience, as well as the editorial team led by Dr. Daniel Fröhlich and Dr. Barbara Haider, who accompanied and supported our work.

This work is a snapshot of a complex of technical, scientific, political and social topics that is developing at a rapid pace. The authors are aware that it contains incompleteness and gaps, that some things will turn out differently than expected and that many new developments will be added. They ask for the reader's understanding in this regard. We wish you, our estimated readers, a helpful and exciting read on this journey into the future of plastics. We look forward to your feedback, suggestions and criticism.

Cologne, Germany — Peter Orth
Langenfeld, Germany — Jürgen Bruder
Potsdam, Germany — Ulrich Liman
August 2024 — Manfred Rink

Contents

Editors and Contributors

About the Editors

Dr. Peter Orth is a chemist by education; with OPC—Orth Plastics Consulting he leads a group of senior consultants with decades of professional experience in leading positions in the German and European plastics industry and its associations. His main focus is on innovation and sustainability in cooperation with science, industry and politics. He is currently working in the fields of CO_2 as a feedstock, resource efficiency, and recycling technologies.

Dr. Jürgen Bruder is a chemist with additional degree in business. He worked in higher education and the chemical industry, before becoming Director General of IK, the German plastic packaging association. He held positions in national, European and global committees including the Global Plastics Alliance. Today he deals with the transformation of the plastics industry to a circular economy focusing on plastic packaging and plastic applications in agriculture.

Dr. Ulrich Liman is a chemist by education, he joined Bayer AG 1989 in marketing of Polyurethane and Polycarbonates and held various managerial positions in Japan, China and Germany. Until 2022 he was responsible for polyurethane R&D and held a seat in the Supervisory Board of Covestro AG.

Manfred Rink is a mechanical engineer; his professional path led him from an automotive supplier to one of the large plastics manufacturers where he held various management positions. Most recently, he dealt with future trends in the plastics industry, risks and opportunities. Today, he works in the fields of sustainability and circular economy in the plastics industry.

Contributors

Najine Ameli phase4:Institut gGmbH i.G., Bochum, Germany

Andreas Bastian Plastship GmbH, Waldems, Germany

Jan Bauer RIGK GmbH, Wiesbaden, Germany

Martin Beeh beeh_innovation, Cologne, Germany

Stefanie Bierwirth DIN e.V., Berlin, Germany

Christian Bonten Institut für Kunststofftechnik, Universität Stuttgart, Stuttgart, Germany

Volker Brennecke Meerbusch, Germany

Jürgen Bruder BPC - Bruder Plastics Consulting, Potsdam, Germany

Dietmar Böhm PreZero Stiftung & Co. KG, Neckarsulm, Germany

Lars Börger nova-Institut GmbH, Hürth, Germany

Michael Carus nova-Institut GmbH, Hürth, Germany

Stephanie Cieplik BKV GmbH, Frankfurt am Main, Germany

Francesco Degli Innocenti Firenze, Italy

Karl Edsjö Saltsjö-Boo, Sweden

Bita Fesidis FOM Hochschule, Düsseldorf, Germany

Silvia Freni-Sterrantino EuPC, Brussels, Belgium

Lorena Fricke Industrievereinigung Kunststoffverpackungen e.V, Bad Homburg vor der Höhe, Germany

Ace Fung Neste Oyj, Porvoo, Finland

Dagmar Glatz dm-Drogeriemarkt GmbH & Co. KG, Karlsruhe, Germany

Sarah Grede GS1 Germany GmbH, Cologne, Germany

Andreas Hanus Sesotec GmbH, Schönberg, Germany

Ronja Hermanns CarbonMinds GmbH, Cologne, Germany

Kathrin Hesse Technische Hochschule Köln, Cologne, Germany

Michael Heyde Alpla Werke Alwin Lehner GmbH & Co KG, Markdorf, Germany

Frauke Hoffmann Hoffmann + Voss GmbH, Viersen, Germany

Thomas Hülsmann VinylPlus Deutschland e.V., Bonn, Germany

Peter Jetzer Recycario Data Science Institute for Economical Plastics Recycling, Erlensee, Germany

Benjamin Kampmann Pöppelmann GmbH & Co. KG, Lohne, Germany

Oona Koski UPM-Kymmene Oyj, Tampere, Finland

Lars Krause nova-Institut GmbH, Hürth, Germany

Günter Krieglstein Tomra Systems ASA c/o Tomra Leergutsysteme Ges.m.b.H, Brunn am Gebirge, Austria

Thomas Krupp Technische Hochschule Köln, Cologne, Germany

Miia Laamanen Neste Corporation, Porvoo, Finland

Reinhold W. Lang Johannes Kepler Universität Linz, Linz, Austria

Ulrike Lange VDI Technologiezentrum GmbH, Berlin, Germany

Petri Lehmus Neste Corporation, Porvoo, Finland

Walter Leitner Max Planck Institute for Chemical Energy Conversion, Mühlheim, Germany

Ulrich Liman ULC - Ulrich Liman Consulting, Langenfeld, Germany

Michael Löwe Tomra Systems ASA c/o Tomra Systems GmbH, Langenfeld, Germany

Anni Luu Technische Hochschule Köln, Cologne, Germany

Raoul Meys CarbonMinds GmbH, Cologne, Germany

Andreas Neumann LyondellBasell Industries, Basell Polyolefine GmbH, Wesseling, Germany

Peter Orth OPC - Orth Plastics Consulting, Cologne, Germany

Nico Pastewski VDI Technologiezentrum GmbH, Berlin, Germany

Michael Perl Sesotec GmbH, Schönberg, Germany

Rudolf Pfaendner Fraunhofer Institute LBF, Darmstadt, Germany

Manfred Renner Fraunhofer Institute UMSICHT, Oberhausen, Germany

Manfred Rink mrink-Consulting, Cologne, Germany

Antonello Romano Stora Enso, Brussels, Belgium

Stefan P. Schleicher Wegener Center for Climate and Global Change, Graz, Austria

Isabell Schmidt Industrievereinigung Kunststoffverpackungen e.V., Bad Homburg vor der Höhe, Germany

Michael J. Schneider Remondis Sustainable Services GmbH, Lünen, Germany

Frank Schockemöhle Pöppelmann GmbH & Co. KG, Lohne, Germany

Guido Schroer RWTH Aachen University—Power2Polymers GmbH, Aachen, Germany

Herbert Snell Snell GmbH, Tittling, Germany

David Stakemeier dm-Drogeriemarkt GmbH & Co. KG, Edewecht, Germany

Oliver Stengel Hochschule Bochum, Bochum, Germany

Michael Vetter Rewindo GmbH, Bonn, Germany

Wei Min Wang VDI Technologiezentrum GmbH, Berlin, Germany

Johannes Wedi Bischof+Klein SE & Co. KG, Lengerich, Germany

Victoria Wessolowski BASF SE, Ludwigshafen am Rhein, Germany

Daniel F. White LyondellBasell Industries, Basell Polyolefine GmbH, Wesseling, Germany

Herwart Wilms Remondis Service International GmbH, Lünen, Germany

Klaus Wittstock BASF SE, Ludwigshafen am Rhein, Germany

Anna Ziessow Reyda GmbH, Freiburg, Germany

Abbreviations

AFNOR	French Standardization Association (Association Française de Normalisation)
APR	Association of Plastics Recyclers (American)
ASF	Additive Sustainability Footprint
ASTM	ASTM International (formerly known as American Society for Testing and Materials)
BDE	Federal Association of the German Waste Management Industry (Bundesverband der Deutschen Entsorgungswirtschaft e.V.)
BIR	Bureau of International Recycling
BKV	Beteiligungs- und Kunststoffverwertungsgesellschaft mbH
BMC	Bulk Molding Compound
BVSE	German Federal Association of Secondary Raw Materials and Disposal (Bundesverband Sekundärrohstoffe und Entsorgung e.V.)
CAGR	Compound Annual Growth Rate
CCE	Circular Carbon Economy
CCS	Carbon Capture and Storage
CDR	Carbon Dioxide Removal
CEAP	EU Circular Economy Action Plan
CEFLEX	Circular Economy for Flexible Packaging
CEN	European Standardization Committee (Comité Européen de Normalisation)
CENELEC	European Committee for Electrotechnical Standardisation
CLP Regulation	Classification, Labelling and Packaging Regulation
CO_2 e	CO_2 equivalent
CPA	Circular Plastics Alliance
CPG	Consumer Packaged Goods
CPR	Construction Products Regulation
CSRD	Corporate Sustainability Reporting Directive
DAC	Direct Air Capture
DPP	Digital Product Passport

DRS	Deposit Return Scheme/System
ECHA	European Chemicals Agency
ELV	End-of-life vehicles
EPD	Environmental Product Declaration
EPPA	European Trade Association of PVC Window System Suppliers
EPR	Extended Producer Responsibility
ESG	Environmental Social Governance
ESPR	Ecodesign for Sustainable Products Regulation
EU ETS	European Emission Trading System
EWC	European Waste Catalogue
FAO	Food and Agriculture Organization of the United Nations
FMCG	Fast Moving Consumer Goods
GHG	Greenhouse Gases
GLBIOM	Global Biosphere Management Model
GPC	Global Product Classification
GPT	Global Plastic Treaty
GTIN	Global Trade Item Number
GWP	Global Warming Potential
HTA	Hard-to-abate
HVO	Hydrogenated Vegetable Oil
IBC	Intermediate Bulk Container
ICIS	Independent Commodity Information Services
ICT	Information, Communication, and Technology
IEA	International Energy Agency
IEC	International Electrotechnical Commission
IGVP	Institut für Grenzflächenverfahren und Plasmatechnologie of Stuttgart University
IK	German Packaging Manufacturers Association (Industrievereinigung Kunststoffverpackungen e.V.)
IPCC	Intergovernmental Panel on Climate Change
ISCC	International Sustainability and Carbon Certification
ISO	International Standardization Organization
JTC	Joint Technical Committee
LCA	Life Cycle Analysis
MBA	Mass Balance Approach
MDO	Machine Direct Oriented
MSWI	Municipal Solid Waste Incineration
Mt	Million tons
MVR	Melt Volume-flow Rate
NBN	Bureau de Normalisation/Bureau voor Normalisatie
NCS	National Collection Scheme
NGO	Non-governmental Organization
NTA	National Technical Agreement (Netherlands)
OECD	Organization for Economic Co-operation and Development
PCF	Product Carbon Footprint

PCR	Post Consumer Recyclate
PEFC	Program for the Endorsement of Forest Certification
PIR	Post Industrial Recyclate
PPWR	Packaging and Packaging Waste Regulation
PRE	Plastics Recyclers Europe Association
PWL	EU Plastic Waste Levy
RCI	Renewable Carbon Initiative
RDF	Refuse-Derived Fuel
REACH	Registration, Evaluation, Authorisation and Restriction of Chemicals
RED	Renewable Energy Directive
RFID	Radio Frequency Identification Device
RSB	Roundtable on Sustainable Biomaterials
SCC	Social Cost of Carbon
SFDR	Sustainable Finance Disclosure Regulation
SFM	Sustainable Forest Management
SME	Small and Medium-sized Enterprises
SPRIND	German Agency for Disruptive Innovation (Bundesagentur für Sprung-Innovationen)
SUPD	Single Use Plastics Directive
SVHC	Substances of Very High Concern
TfS	Together for Sustainability
TRL	Technical Readiness Level
UBA	German Environmental Protection Agency (Umweltbundesamt)
UNEP/EA	UN Environmental Program/Environmental Assembly
UNI	Ente Italiano di Normazione
USP	Unique Selling Point
VCI	German Chemical Industry Association (Verband der Chemischen Industrie e.V.)
VDI	German Engineers Association (Verein Deutscher Ingenieure e.V.)
VDID	German Industrial Designers Association (Verein Deutscher Industriedesigner e.V.)
VOC	Volatile Organic Compound
WBCSD	World Business Council on Sustainable Development
WEEE	Waste from Electrical and Electronic Equipment
WPC	Wood Plastics Composites
WRAP	Waste and Resource Action Program
WRI	World Resource Institute
ZSVR	German Central Agency Packaging Register (Zentrale Stelle Verpackungsregister)

Chapter 1
Introduction

Peter Orth and Jürgen Bruder

For more than 60 years, plastics have been a group of materials that have become an integral part of our modern world, be it in everyday life, industry, food supply, healthcare, public amenities, mobility, building and construction, etc. Enormous annual growth in world plastic production has led to the fact that by 1989, the world production of plastic in terms of volume surpassed steel (Source: Plastics Europe) and has quadrupled again since then. Not only the low density but also many other outstanding and sometimes unique properties have contributed to this astonishing development.

However, the linear production and consumption model of plastics (and the entire economy) over these decades has led to serious environmental problems. Today, this results in a broad societal discussion about plastics, particularly their reduction and avoidance, areas of application, necessary collection and recycling, and thus the prevention of plastics entering the environment with all the negative effects and the regulation of all these issues.

As early as the 1980s, society became aware of waste problems as the first challenge in relation to plastics. Overflowing garbage bins, largely caused by plastic packaging and an extreme shortage of landfill space, led to political measures such as the Packaging Directive in Germany in 1991 and the EU Packaging Directive in 1994. In the late 2000s and early 2010s, the release of plastic waste into the environment, particularly into the oceans, was perceived as a global challenge. Images of marine litter, animals dying from plastic, littered bays, coastlines, and beaches have led to a broad social awareness of this plastic waste problem.

The second challenge that has developed relatively steadily is the realization, especially in European countries with a largely functioning disposal system for plastic

P. Orth (✉)
OPC - Orth Plastics Consulting, Cologne, Germany
e-mail: opc.cologne@t-online.de

J. Bruder
BPC- Bruder Plastics Consulting, Potsdam, Germany
e-mail: plastics@jbruder.de

P. Orth et al. (eds.), *Circular Economy of Plastics*,
https://doi.org/10.1007/978-3-031-78829-1_1

waste, that the resource requirements for plastics, like other materials, can only be secured in the long term through the development of a circular economy of plastics. The need to consider recyclates as valuable sources of raw materials has been and will be reinforced by the necessity to substitute oil and natural gas for climate reasons, not only in the energy sector but also in the materials industry.

This results in a third challenge: the substitution of fossil raw materials with the aim of reducing CO_2 emissions, which are generated during the extraction of raw materials, the production of plastics, and the incineration of plastics. The defossilization of the plastics value chain in the EU must align with the goals of the Green Deal to create a climate-neutral industry by 2050.

All three major challenges:

- Preventing the release of plastic waste into the environment,
- Development of a circular economy of plastics to secure the demand for raw materials,
- Defossilization of the plastics value chain.

Stand alone as individual social, technological, and political/regulatory challenges, but they are also closely interrelated and influence each other. These challenges are addressed in this book with varying degrees of emphasis, focusing on the development of a circular economy and the defossilization of the plastics value chain. They are part of overarching global challenges and must make an important contribution to overcoming them.

The above three major challenges are embedded in the global tension between demographic developments and planetary limits. As the global population is projected to reach 9.7 billion by 2050, the tension becomes ever more acute. Humanity is on a collision course with the natural limits of our planet. This tension manifests in several critical ways:

- Increased carbon emissions (including plastics production and incineration) exacerbate global warming and unprecedented climate change. Urbanization and industrial activities amplify this problem. The issues are rising temperatures, extreme weather events, and sea-level rise and the impacts are threats to agriculture, water supply, and human health (IPCC 2022).
- As populations grow, so does the demand for resources (including plastics), outpacing the planet's capacity to regenerate. The issues are the finite availability of raw materials and fresh water, and the impacts are increased competition for resources, potential conflicts, and economic instability (UNEP 2024).
- Overconsumption and waste generation (including plastics) lead to widespread pollution and ecosystem damage. The issues are pollution from industrial activities, plastic waste, and chemical contaminants and the impacts are health hazards, degradation of ecosystems, and loss of biodiversity (UNEP 2019).
- Human expansion and economic activities endanger biodiversity, destroy natural habitats, and push many species towards extinction. The issues are habitat destruction, overexploitation, and invasive species and the impacts are collapse of ecosystems, extinction of species, and loss of genetic diversity (IPBES 2019).

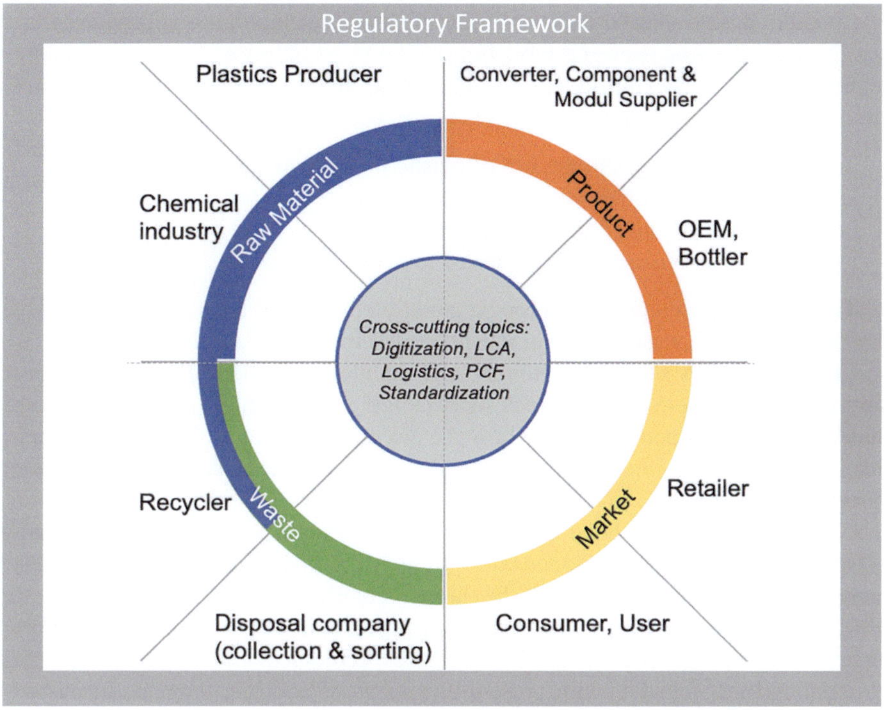

Fig. 1.1 Schematic diagram of the plastics cycle (*Source* VDI/own presentation)

The demographic pressures increase the demand for plastics. The issues of growing global population, urbanization, and increased consumption lead to a strain on infrastructure, food security, and social services (Pew 2021). The reader will find scientific contributions, case studies, and further reading on all three challenges in various chapters of the book.

In Part I, "Plastic the Material," important basics for understanding the challenges are compiled, such as a brief description of the essential properties of the material (Chap. 2), its current and future feedstock base (Chap. 3), and the structure of the complex plastics ecosystem (Chap. 4). The following chapter (Chap. 5) deals with the scope of the regional and global material flows of virgin plastics, plastic waste, and recyclates, and the data situation on the global and European levels.

Part II, "Circular Economy of Plastics," consisting of Chaps. 6, 7, 8, 9, 10, 11 and 12, covers the current central challenge of the plastics value chains in Europe, the establishment of a circular plastics economy. In Chap. 6, the cross-stage challenges in the establishment and development of a circular economy are addressed. Systemic approaches like moderated stakeholder processes and common design concepts contribute to overcoming economic, cultural, and language barriers between the players in the various stages of the plastics cycle (Fig. 1.1).

Chapters 7, 8, 9 and 10 deal with the respective technologies, experiences, and challenges in the processes of establishing a circular economy of plastics with its four segments in the cycle (Introduction to Part II). Concrete examples and experiences are provided by stakeholders:

- in plastics processing and the use of plastic products (Chap. 7),
- in the retail sector (Chap. 8),
- in waste logistics and sorting (Chap. 9),
- in the recycling business as part of raw material production (Chap. 10).

The reader will search in vain for a chapter on the role of the consumer or user as the "source" of the waste; this gap must be filled by the social sciences.

Chapter 11 uses selected voluntary and statutory Extended Producer Responsibility (EPR) systems in the areas of packaging, agricultural plastics, furniture, building, and construction to show how the circular economy is developing in practice in different value chains and what challenges the initiators and operators of such systems are facing.

Chapter 12 addresses the global plastic stream, which not only completely eludes a circular economy to be developed but also enters the environment in an uncontrolled manner. This addresses the first of the three major challenges mentioned above, the prevention of plastic waste entering the environment, with a focus on the global environmental problem of marine litter. Causes, ecological and other consequences are summarized, together with the current status of sources, input pathways and quantities, and existing and planned measures to combat the input of plastics into the environment.

Part III, "Circular Feedstock," deals with the third and most recent major challenge in relation to plastics, the feedstock change from fossil to circular raw materials, also known as defossilization. Chapters 13, 14 and 15 deal in more depth with the three basic resources for circular raw materials: plastic wastes (Chap. 13), biomass (Chap. 14), and CO_2 (Chap. 15). In connection with the successive increase and growth in the use of circular feedstocks, corresponding verification, and certification procedures are necessary, which are discussed as mass balancing in Chap. 16.

Part IV, "Cross-cutting Technologies," acknowledges that a functioning circular economy requires cross-cutting instruments and methods that are accepted and applied by all players in the cycle as well as by regulators. These include Comparative Life Cycle Assessments (LCA) of circular plastics (Chap. 17) as well as the determination of Product Carbon Footprints (Chap. 18) of products in general and plastics in particular. Digitization is seen as key to developing a circular economy in the entire plastics ecosystem but also in the individual stages of the cycle. Its contributions and potential are outlined in Chap. 19. The transformation to a circular economy will only succeed if the technical framework conditions are rapidly defined and continuously developed by standardization (Chap. 20).

The three challenges relating to plastics require a political framework to be solved. Part V, "Regulatory Framework," explains the EU Directives, regulations, strategy documents, and legislative projects that deal with or affect both challenges: preventing the release of plastic waste into the environment and establishing a circular economy of plastics. Specific regulations on the defossilization of plastic production or incineration have not yet been issued.

Chapter 21 provides an overview of the most important European legal regulations related to the markets of plastics such as packaging (draft PPWR), building & construction (CPR), electrical and electronic equipment (WEEE), automotive (ELV), and single-use plastics (SUPD), but also to fiscal instruments like Taxonomy or Plastics Tax. Chapter 22 deals in detail with the new regulatory core elements of SUPD and draft PPWR to enhance the use of recyclates, the recycled content quota, and investigates the polymer-specific substitution quota as an alternative instrument. Outlook and Challenges is the title of the concluding Chap. 23, which summarises the quantitative development up to 2050 and the most important technological problems to be solved along the way. These are technical, economic and ecological in nature and are tasks for industry on the one hand and for individuals and society on the other. Politics plays an important role here, as it must ensure the necessary framework conditions with a series of measures.

Each chapter is provided with an independent bibliography to facilitate the reader's meaningful engagement with subsections of the work. The book ends with a Back Matter containing important sources for further reading and a glossary.

References

IPCC 2022: Intergovernmental Panel on Climate Change (IPCC), "Climate Change 2022: Impacts, Adaptation and Vulnerability", https://www.ipcc.ch/report/ar6/wg2/

IPBES 2019: Intergovernmental Science-Policy Platform on Biodiversity and Ecosystem Services (IPBES), "Global Assessment Report on Biodiversity and Ecosystem Services". https://files.ipbes.net/ipbes-web-prod-public-files/ipbes_global_assessment_report_summary_for_policymakers.pdf

Pew 2021: Pew Research Center, "Global Population Growth and Demographic Projections". https://www.pewresearch.org/short-reads/2019/06/17/worlds-population-is-projected-to-nearly-stop-growing-by-the-end-of-the-century/

UNEP 2019: United Nations Environment Programme (UNEP), "Global Chemicals Outlook II". https://www.unep.org/resources/report/global-chemicals-outlook-ii-legacies-innovative-solutions

UNEP 2024: "Global Resources Outlook 2024". https://www.unep.org/resources/Global-Resource-Outlook-2024

Dr. Peter Orth is a chemist by education; with OPC—Orth Plastics Consulting, he leads a group of senior consultants with decades of professional experience in leading positions in the German and European plastics industry and its associations. His main focus is on innovation and sustainability in cooperation with science, industry, and politics. He is currently working in the fields of CO_2 as a feedstock, resource efficiency, and recycling technologies.

Dr. Jürgen Bruder is a chemist with an additional degree in business. He worked in higher education and the chemical industry, before becoming Director General of IK, the German plastic packaging association. He held positions in national, European, and global committees including the Global Plastics Alliance. Today he deals with the transformation of the plastics industry to a circular economy focusing on plastic packaging and plastic applications in agriculture.

Part I
Plastic—The Material

Part I of the book explains the status quo of today's, predominantly linearly organized plastics industry. The properties of plastics, their applications, and life cycles, as well as important economic aspects of the material group, are comprehensively presented. The success of plastics as a material is essentially based on efficient linear value chains, which begin with a cost-effective feedstock base.

The unique properties of the material (Chap. 2), the predominantly fossil feedstock base (Chap. 3), the plastics ecosystem with its main players (Chap. 4), and the current plastic flows at global and European levels (Chap. 5) provided convey an overall picture of the initial situation for the transformation to a circular plastics economy representing a paradigm shift in the economy and society. The inevitability of the transformation to a circular economy and the simultaneous switch to renewable carbon sources is substantiated by facts about the current treatment of plastic waste and its impact on the earth's climate.

Part I of the book provides the reader with a fundamental basis for understanding and classifying the subsequent parts and thus for the challenges that the plastics industry is going to face in designing sustainable solutions and structures in the next decades.

Chapter 2
The Material

Manfred Rink and Ulrich Liman

Contents

Summary

Since 1950, when the first million tons of plastic were produced worldwide and used in the market, the success story of this carbon-based class of materials has continued unabated to this day, based on its almost infinite variability ("made-to-measure material") and high functionality. Minerals (glass, concrete), metals (steel and aluminium) and wood (furniture, paper) have a common acceptance as engineering materials in society. Polymers (plastic) are often considered to be easily replaceable by such classical materials. Since the industrial production of polymers plastics have proven to be indispensable and irreplaceable in their application possibilities for social, economic and ecological reasons, i.e. as a sustainable material. An understanding of the basic physical material and service properties in the most important areas of application and of the life cycles of the material in the various applications is necessary to give the often polemically conducted discussion about the advantages and disadvantages of this important class of materials an objective and thus target-oriented direction

M. Rink (✉)
mrink-consulting, Cologne, Germany
e-mail: manfred@rink-consulting.de

U. Liman
ULC—Ulrich Liman Consulting, Langenfeld, Germany

P. Orth et al. (eds.), *Circular Economy of Plastics*,
https://doi.org/10.1007/978-3-031-78829-1_2

in the sense of the circular economy. The chapter is supplemented by aspects of economic efficiency, the nature of plastics and a brief description of the types of plastics produced today and available on the market.

2.1 Definitions

This chapter covers thermoplastics and thermosets used mainly in three-dimensional shaped parts. Polymers used for textiles, paints, varnishes, textiles waterproofing, or within the production of cosmetics, medicines or chemical processes are not included. PVC-, PO- and PU-fibres are included, whereas PA-fibres, PET-, PBT-, or Acrylic polyesters are not included.

2.2 Physical Properties

Density: Compared to metals or glass, plastics have a low density: polyolefins < 1 g/cm^3, reinforced plastics (GRP, CFRP) > 1 g/cm^3. With the combination of low density and high strength, plastic is a lightweight construction material par excellence compared with metals or materials based on the lithosphere.

Shape and Form: With manufacturing processes such as injection moulding, deep drawing, and blow moulding, plastics can be shaped with relatively little effort at comparatively low temperatures, and a complex design can be achieved in a single operation. “Additive manufacturing” (3D printing), which is still in its infancy, further expands the range of shaping possibilities without the need for a tool.

Elasticity and Toughness: Plastics are generally elastic, i.e. they reverse an externally induced change in shape under their own power, and tough, they are resistant to fracture or crack propagation.

Corrosion and Resistance: Many plastics especially polymers in the chemical sense (PE, PP, PVC, rubbers) are long-term and weather-resistant, i.e. they do not corrode; chemical degradation takes decades or even longer. This apparent advantage of longevity is reversed in the case of plastic waste that is released into the environment in an unordered manner, as it can hardly be degraded there.

Colour: Thermoplastics in particular are dyed by compounding with appropriate pigments, the range of which is constantly being expanded. Practically every conceivable colour can be adjusted in this way.

Optical Properties: Amorphous thermoplastics such as PC or PMMA are transparent, i.e. translucent, making them suitable for a wide range of optical applications (light guides, etc.) as well as for “glazing”.

Electromagnetic Properties: Most plastics are insulators, i.e. they have only low electrical conductivity but can be modified to become electrical conductors to a certain extent by incorporating conductive materials.

Surface: Injection-molded components precisely reproduce the surface of the mold from high-gloss to almost any desired texture.

2.3 Usage Properties and Applications

Protective functions: In terms of volume, the most important area of application for plastics is packaging and thus the protection of goods of all kinds, especially foodstuffs. Relevant properties for this are freedom of shaping, low density, easy surface modification, (partial) impermeability to light, gases (oxygen in particular), germs, etc.

A WWF study (WWF 2015) states that food losses along the production chain, including consumers, are still between 30 and 40% worldwide. These losses can only be significantly reduced further through improved packaging, transport and storage in and with plastics (denkstatt 2020; OFI et al. 2022).

Transport and distribution: Plastic sacks and bottles/canisters, containers and boxes enable easy, safe and robust portioning, storage and transport of goods of all types and quantities in the commercial and private sectors. They are the basis for efficient and economical distribution of goods—from production to the end consumer. This also includes the cold chains, which are becoming more important due to the further increase in urbanisation and global warming effects.

Housing: Devices or appliances are given protection and form by plastic housings. Colour, surface and design are equally brand ambassadors and design elements of our living environment.

Insulation and comfort: Soft and rigid foams made of PE and PS, PUR or rubbers can be found in countless construction applications, where they are used, for example, for heat and cold insulation, cavity sealing or shaping, as well as for comfort reasons in furniture upholstery, mattresses or car seats.

Durability: Chemical and weathering resistance predestine PVC and the polyolefins—in particular for use as dissolving or operating agent containers (fuel tanks), as buoys, ropes or nets in maritime environments or as covering sheets and tarpaulins for construction and civil engineering industry or commercial vehicles.

Structural properties: After reinforcement with glass or carbon fibres for selective adjustment of higher elastic moduli, many plastics are also used as materials for structural components. Thermoplastics reinforced with short glass fibres can be injection moulded, resins impregnated with glass fibre braids are moulded and cured in the SMC or BMC process. Carbon fibers are embedded in thermoplastic matrices, among other things, and enter the market as semi-finished products. Such materials achieve Young's moduli that exceed those of metals and have very favourable strength-to-weight ratios. However, they are rather expensive and are therefore preferably used as lightweight materials in aircraft construction. Not all of the properties described can be realized with a single material and not all of them are required for a specific application. To meet these and other requirements, an enormous variety of different types of plastics has been developed and combined as composites a development that continues unabated. What is initially a great advantage turns out to be complicated when it comes to recycling: we should not speak of "plastics recycling", but of "recycling of plastics". Nevertheless, plastics as an engineering material have the same importance in the future as other classical materials from minerals or wood or other biotic resources. All materials encounter environmental challenges in emissions of GHG like CO_2. Plastic in its use keeps actually carbon away from emission into the atmosphere.

2.4 Applications and Life Cycles

Applications

The following is a list of typical material applications for the six largest global plastics markets. The largest polymer grades used in these markets in terms of volume are also listed:

Packaging

- Extruded films (LDPE, LLDPE, PP)
- Blow-moulded hollow bodies, bottles, canisters, barrels, etc. (HDPE, PET)
- Thermoformed cups, trays, blisters, etc. (PP, PET, PS)
- Foams (EPS, XPS, EPP, etc.)

Building and Construction

- Hoses, pipes, fittings (PE, PP, PVC)
- Foams (EPS, XPS, PU, PE)
- Foils (PE, PP, PVC)
- Sheets (PVC, PMMA, PC)

Automotive/Vehicles

- Bumper (PP)
- Interior parts (PP, ABS, PC/ABS, PU)
- Lightning (PC, PMMA)
- Engine compartment (PA, PP)
- Electrical (PVC, PP, PA)
- Tyres (NR, SBR)

Electrical and Electronic

- Cable (PVC, PP, TPE)
- Housing (PP, ABS, PC/ABS)
- Fastening elements (PE, PP, PA, PBT)
- Printed circuit boards (EP, UP, PI)

Household/Leisure/Sport

- Toys (PE, PP, ABS)
- Kitchen utensils (PE, PP, PA)
- Mattresses (NR, PUR)
- Shoes (PUR)

Agriculture

- Films (LDPE, LLDPE)
- Container (HDPE)
- Greenhouses (LDPE, PMMA, PC)
- Nets (HDPE)
- Yarns (PP)
- Hoses (PE)

Depending on the market and application, plastics have very different life cycles, which are determined less by the plastic itself than by the service life of the respective input product. The following are some typical examples:

- Packaging—a few days (food) to several years (transport containers);
- Technical applications (vehicles, electrical appliances)—several years to several decades;
- Construction—applications (buildings, roads, gas and water networks) decades to centuries;
- Medical articles—often disposable.

Product life cycles vary by orders of magnitude depending on the applications in which plastics are used (cf. Fig. 2.1). Typical consumer packaging is only in use for weeks or months. The same is true for many applications in agriculture. Sports and leisure articles are subject to fashion trends that artificially limit their service life. Electrical items have been in use for decades, and many electronic products are used for only a few years due to the pace of innovation in their markets. Automobiles, which are typically reused (sold and bought used), may well last more than 20 years. Finally, construction products disappear for 100 years in the ground or in buildings and often only go out of service when they are demolished. In a study from Geyer (2017) it is stipulated that 30% of the total amount of produced plastic is still in use. Due to the long lifetime of plastic applications plastics are bound in an "anthropogenic storage" before it is available for a recycling process. Similar issues are also known in metal recycling (Umweltbundesamt 2023).

Fig. 2.1 Life cycles and business aspects of plastics applications (*Source* own representation)

2.5 Economic Aspects

The majority of plastics come onto the market at prices around and around 2–5 €/kg; the thermoplastics that dominate in terms of volume are in the form of free-flowing granulates that are easy to transport, store and handle. In addition, compared to metals, they can be formed into complex shapes at comparatively low temperatures and with relatively little effort. Snap-fit connections, for example, can be formed in the same manufacturing step without much additional effort which later enables cost-effective assembly and disassembly.

Plastic processing machines impress with their moderate investment costs and rather low energy consumption. This has promoted the emergence of a plastics industry with a strong structure of small and medium-sized enterprises, which today is at home in almost all countries of the world and provides jobs for millions of people in the proverbial garage as well as in large-scale operations.

When considering economic efficiency, the entire life cycle (cf. Chap. 17) of a product must be taken into account. In addition to the material price and the investment costs as part of the manufacturing costs, ease of repair, costs during use and costs for disposal or recycling must be considered as well. As plastics are produced from fossil raw materials so far, as well as using fossil-based energy plastics are currently debated as inefficient for climate change purposes. Nevertheless, the picture will change if recycling and green energy foster the use of plastic for sustainable applications including packaging. Processing plastics for finished goods and application will not harm greenhouse gas emissions, as no carbon will enter the atmosphere.

Finally, and increasingly, the costs of disposing plastic waste have to be controlled, if it is used for incineration, land or maritime storage (PEW 2020).

Similar to other biotic or abiotic raw materials for engineering purposes recycling is increasingly economical, similar to metals (Umweltbundesamt 2023).

2.6 On the Nature of Plastics

The term polymers is used here to refer to all macromolecular substances—plastics, rubbers, varnishes, fibres, adhesives, etc. which are produced synthetically or are obtained from natural substances by chemical or biological modification, and which are used as materials, insulating materials, surface coatings, woven fabrics, textiles, etc. In the following, however, we will refer to plastics for the sake of simplicity, as they account for by far the largest share of the volume.

Approximately 70% of plastics are thermoplastic in nature; here the macromolecules can be remelted, extruded or shaped for the purpose of processing or forming mainly by applying heat. Approximately 30% of plastics are thermoset or elastomeric in nature; here, shaping is usually accompanied by crosslinking, so that they cannot be remelted and reshaped.

Chemists also distinguish between "polymers" (e.g. polyethylene), "polycondensates" (e.g. polyester) and "polyadducts" (e.g. polyurethane), which describes the nature of the bond in the chemical reaction of building up polymers from monomers. This is important for the stability of plastics in their applications (corrosion behaviour) and for any degradation possibilities, i.e. for chemical recycling (cf. Sub-Sect. 10.2.3).

Plastics have an average specific weight (density) of approx 1 g/cm^3; one tonne therefore has a volume of 1000 L. By comparison, steel has an average density of approx 8 g/cm^3. A ton of steel therefore has a volume of 125 L and thus only 1/8 of the volume of a ton of plastic. Plastics are therefore more "visible" and "present" than steel by a factor of 8 (cf. Fig. 2.2). This makes them the material par excellence for lightweight construction and energy-efficient solutions for sustainable applications in our living standards (BMUV 2023; Conversio 2019). With the increasing challenges of population increase, sustainable living standard conditions and climate change reduction plastics will further grow and play an important role in the contribution.

2.7 Plastic Types

With the focus on the main growing plastic types of the future we describe here only plastics with an annual production of > 1 Mt should be mentioned here (cf. Fig. 5.4).

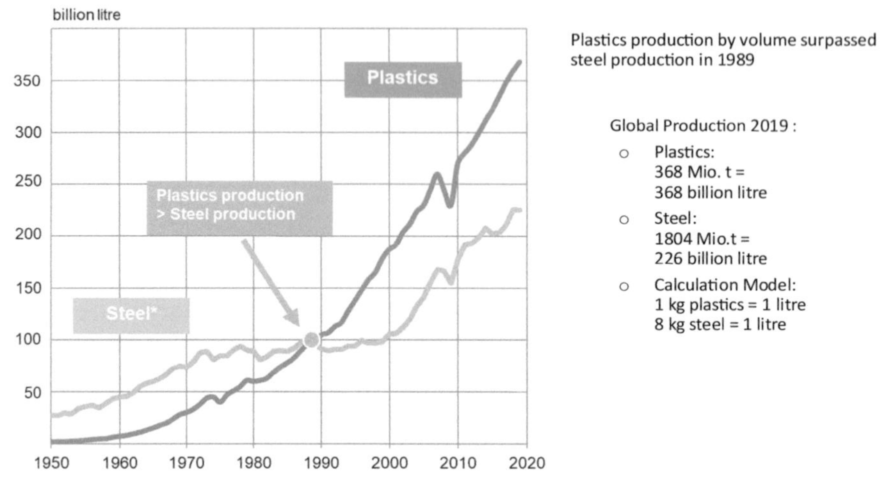

Fig. 2.2 Comparison of density and volume of plastic/steel (*Source* PlasticsEurope)

- The largest single group among plastics are the thermoplastic polyolefins polyethylene (HDPE, LDPE, LLDPE) and polypropylene (PP), which today already account for more than 60% of the total volume of plastics produced;
- The next largest group are the thermoplastics polystyrene (PS) and polyvinyl chloride (PVC) as well as polyamide (PA) and polyethylene terephthalate (PET);
- These are followed by the engineering thermoplastics acrylonitrile butadiene styrene (ABS), polycarbonate (PC) and polyoxymethylene (POM);
- Then there is the group of polyurethanes (PUR) and thermosets, of which the styrene and epoxy resins should be mentioned;
- Finally, rubbers (abbreviation: R), of which natural rubber (NR), butadiene (BR) and styrene-butadiene rubbers (SBR) as well as chlorinated (CR) and nitrile rubbers (NBR) contribute significantly.

References

BMUV: https://www.bmuv.de/themen/nachhaltigkeit/ueberblick-nachhaltigkeit (last access: Aug. 21, 2024).

Conversio 2019: https://www.bkv-gmbh.de/5075-study-global-plastics-flow-2018.html (last access: Aug. 21, 2024).

denkstatt 2020: Food, Packaging, Sustainability. Update 2020. https://denkstatt.at/wp-content/uploads/2023/05/guideline_stopwaste_e_082020_web.pdf (last access: Aug. 21, 2024).

Geyer et al. 2017: R. Geyer, J. R. Jambeck, K. L. Law, Production, use, and fate of all plastics ever made. https://advances.sciencemag.org/content/3/7/e1700782 (last access: Aug. 21, 2024).

ISO 2018: https://www.iso.org/standard/71206.html (last access: Aug. 21, 2024).

OFI 2022, denkstatt and BOKU: STOP WASTE—SAVE FOOD. Guide for packaging manufacturers, food processors, trade, politics and NGOs. Vienna, February 2020. https://denkstatt.eu/publications/?lang=de (last access: Nov. 11, 2023).

PEW 2020: https://www.pewtrusts.org/en/research-and-analysis/articles/2020/07/23/breaking-the-plastic-wave-top-findings (last access: Aug. 21, 2024).

Umweltbundesamt 2023: bit.ly/2dowYYI, p. 10.

UNO 2019: https://population.un.org/wpp/Download/Probabilistic/Population/ (last access: Aug. 21, 2024).

WWF 2015: https://www.wwf.de/themen-projekte/landwirtschaft/ernaehrung-konsum/lebensmittelverschwendung/das-grosse-wegschmeissen (last access: Aug. 21, 2024).

Manfred Rink is a mechanical engineer; his professional path led him from an automotive supplier to one of the large plastics manufacturers where he held various management positions. Most recently, he dealt with future trends in the plastics industry, risks and opportunities. Today, he works in the fields of sustainability and circular economy in the plastics industry.

Dr. Ulrich Liman is a chemist by education, he joined Bayer AG in 1989 in the marketing of Polyurethane and Polycarbonates and held various managerial positions in Japan, China and Germany. Until 2022, he was responsible for polyurethane R&D and held a seat on the Supervisory Board of Covestro AG.

Chapter 3
Today's Feedstock Base

Ulrich Liman and Peter Orth

Contents

Summary

Around 90% of polymers are currently made from fossil feedstocks oil and gas. To conserve resources, recycle waste, and avoid CO_2 emissions, the fossil feedstock for plastic production must be replaced step by step by circular raw materials over the next two decades—the core idea of the circular economy.

There are basically three different carbon sources available instead:

U. Liman (✉)
ULC - Ulrich Liman Consulting, Langenfeld, Germany
e-mail: druliman2@gmail.com

P. Orth
OPC - Orth Plastics Consulting, Cologne, Germany
e-mail: opc.cologne@t-online.de

P. Orth et al. (eds.), *Circular Economy of Plastics*,
https://doi.org/10.1007/978-3-031-78829-1_3

- Recycled materials made from plastic waste using mechanical or chemical recycling processes;
- Carbon dioxide (CO_2), which is released into the atmosphere as gaseous waste from the cement industry and incineration processes;
- renewable raw materials from different forms and sources of biomass.

The individual sources, including the fossil ones, are explained in basic terms regarding their conception. In a separate scenario, the further development of emissions is shown based on the growth forecast for plastic production and the possible reduction in emissions when using circular raw materials in the context of the German and European climate goals.

3.1 Feedstock

3.1.1 Fossil Feedstock

Around 90% of all polymers are currently made from fossil raw materials oil and gas. The use of coal as a raw material is practically a thing of the past.

Oil is by far the most important raw material for the production of plastics. The raw or light gasoline naphtha as well as the hydrocarbons methane, ethane and liquid petrol gas (LPG—a mixture of predominantly propane and butane) are obtained in refineries through fractional distillation, from which in turn they are produced through steam cracking (cf. Sects. 10.4 and 13.3). The important basic chemicals[1] ethylene, propylene, and the so-called C_4 fraction are produced. The aromatic hydrocarbons benzene, toluene, and the xylene isomers (BTX) are usually produced from naphtha by catalytic reforming. Methanol is produced directly from methane.

Naturally occurring ethane is also of great importance, which is obtained as shale gas through fracking, but which is also produced as a by-product (strand gas) during oil production and was usually burned (flared) in the past. The ethane is dehydrogenated to ethylene (ethene) on an industrial scale in energy-consuming gas crackers (Weissermel and Arpe 2010).

The basic chemicals ethylene and propylene are also the most important monomers in terms of quantity. Polyethylene and polypropylene made from them already account for almost 2/3 of world plastic production. They are also starting products for other monomers, such as vinyl chloride (PVC), styrene (polystyrene), acrylic acid (polyacrylate), or acrylonitrile (polyacrylonitrile).

The C_4 fraction, consisting of butadiene, isobutylene, and butenes, is, among other things, the basis of synthetic rubbers.

The production sites for these plastics are usually immediately adjacent to the refineries; the most important interfaces between petrochemicals and chemistry are

---

[1] The list of ten basic chemicals: the olefins ethylene and propylene, the aromatics benzene, toluene and xylene (BTX), the group of C_4 chemicals, methanol, chlorine, caustic soda and ammonia.

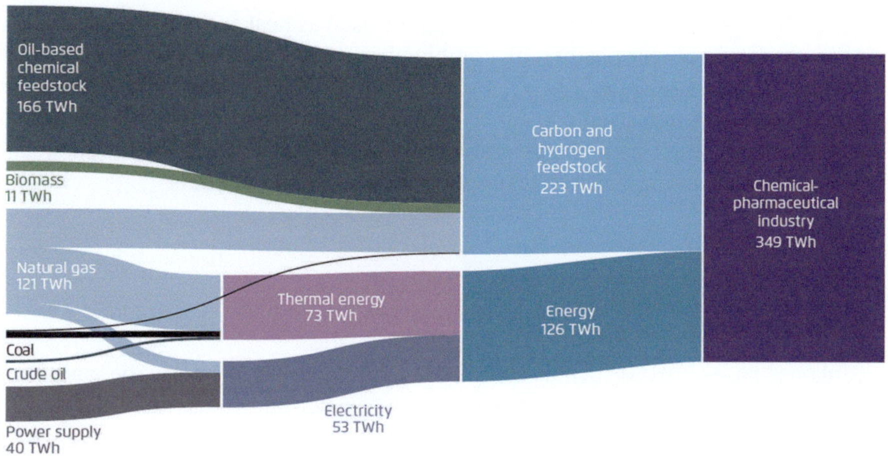

Fig. 3.1 Raw material and energy consumption in the chemical industry in Germany 2019 (*Source* Agora Industrie 2023)

the steam crackers. In recent decades, the construction of refineries has moved ever closer to the oil sources for logistical reasons; the same applies to crackers and plastic production as well as the associated added value. As a result, the production of polyolefins in particular is shifting to a large extent to the countries bordering the Persian-Arab Gulf (including Saudi Arabia, Kuwait, U.A.E.) and the Gulf of Mexico (including the US states of Texas and Louisiana). The same holds true for gas-based production. Conversely, production—including upstream petrochemicals—is continuously decreasing in the classic target markets, corresponding plants are being shut down, and new construction is only taking place sporadically.

3.1.2 Resource Requirements[2]

In 2019, the chemical industry in Germany consumed 349 terawatt hours (based on the lower heat value, TWh_{LHV}), mainly fossil energy and raw materials to produce chemicals. The chemical industry is also a large electricity consumer with 53 TWh, with around 25% produced locally and 75% purchased via the power grid. The raw material and energy base of the chemical industry is shown in Fig. 3.1: 36% of the resources used are used as fuel for the supply of process energy (electricity and heat) to produce chemicals from 64% of the raw materials used (Agora Industrie 2023).

[2] The following statements are based on the publication "Chemistry in Transition" by Agora Industrie (2023). The data refers to the (entire) chemical production in Germany in 2019. They are an example of the situation in Europe. These data are discussed as representative of plastic production.

58% of the process energy of 126 TWh_{LHV} is used as heat and 42% as electricity. Natural gas-based power plants with combined heat and power (CHP) are the most important suppliers of locally generated heat and electricity, making the chemical industry the largest industrial consumer of natural gas (Agora Energiewende 2023). Refinery products and natural gas make up around 83% of the German chemical industry's energy and raw material supply as fuel and raw materials.

3.2 CO_2 Emissions

Greenhouse gases (GHG) such as carbon dioxide, methane, nitrous oxide, and fluorinated gases—emitted, among other things, when burning fossil fuels—enter the earth's atmosphere via different processes, where they accumulate and are broken down with different half-lives (cf. Fig. 3.2). The greenhouse effect raises the average temperature in the earth's atmosphere, resulting in undesirable changes in the climate (Häckel 2021). The different effectiveness of greenhouse gases is taken into account using a so-called CO_2 equivalent (CO_2e).

The GHG emissions generated in the production of the chemical industry cover the entire life cycle of the products—from raw materials through production and use to disposal at the end of the life cycle.

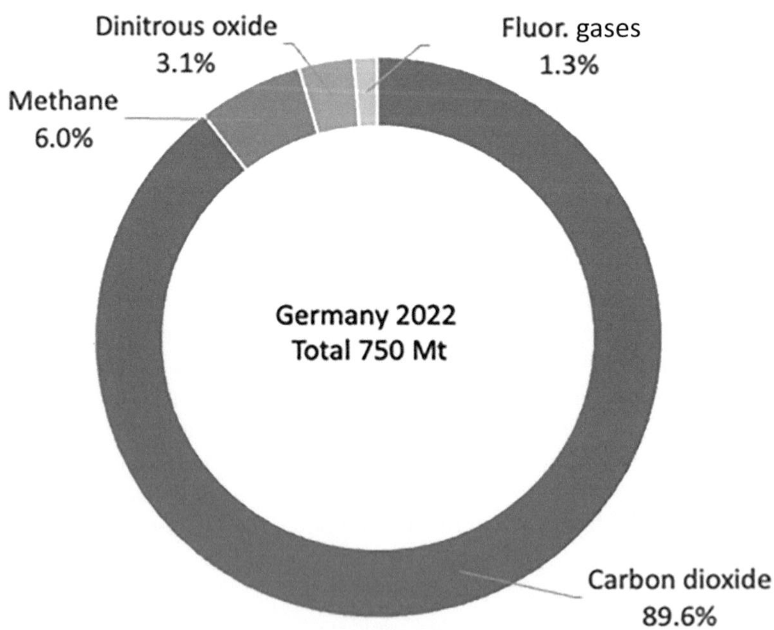

Fig. 3.2 Shares of various greenhouse gases in total emissions in Germany in CO_2 equivalents (*Source* Own representation based on UBA 2024)

In 2019, direct emissions from the production of chemicals amounted to 39 Mt CO_2 (VCI 2022a), approx. 21% of all industrial emissions in Germany, with ammonia production alone accounting for 6 Mt CO_2e (approx. 15%). Although reducing direct production-related emissions is of great importance, they are only a fraction of the total life cycle emissions of chemical products. However, due to the lack of uniform standards and specifications for emissions accounting, reliable data is hardly available. Based on the following estimates, it is assumed that indirect upstream and downstream emissions are approximately three times as high as direct emissions.

Upstream emissions. Based on statistical emission[3] factors for fossil resources and the quantities of fossil resources required for the production of the ten basic chemicals, the extraction and provision of fossil raw materials account for around 6 Mt CO_2e. These are largely volatile methane emissions—a particularly climate-damaging greenhouse gas.

Downstream emissions. The emissions at the end of the life cycle are also difficult to estimate—more than 60% of the products are exported and are therefore not available for recycling in Germany. However, products remaining in Germany are often inadequately disposed of—due to the use (cosmetics, detergents, tire abrasion, etc.) or due to insufficient waste collection. Since recycling rates are low in relation to the production volumes of all basic chemicals, incineration is the most common method of waste treatment. If all basic chemicals and their derivatives were burned at the end of their life cycle and the carbon they contained entered the atmosphere as CO_2, the emissions at the end of the life cycle would be 42 Mt CO_2. However, the actual value differs from this estimate—due, among other things, to methane emissions from landfills or due to the accumulation of waste in nature that does not end up as CO_2 in the atmosphere (e.g. microplastics). The service life of the products also varies, so end-of-life emissions only occur with a long delay. Figure 3.3 illustrates the life cycle emissions of basic chemicals broken down by type.

- upstream emissions (provision of fossil resources),
- indirect emissions from the energy supply (fossil energies),
- direct process-related emissions from the production of basic chemicals and
- End-of-life emissions under the described assumptions.

Process emissions from the processing of basic chemicals into intermediate and end products are not shown.

[3] Statistical emission factors for fossil resources based on the cm.chemicals database (Carbon Minds 2020).

Fig. 3.3 Estimation of the life cycle emissions of basic chemicals produced in Germany in 2019 (*Source* Agora Industrie 2023)

3.3 Reduction Strategies

For climate protection reasons, the emissions associated with the production of plastics under today's conditions must be drastically reduced and ideally reduced to zero—the goal is climate neutrality (net zero emissions).

In addition, in consideration of planetary boundaries, fossil carbon must be replaced by renewable carbon.

Based on the above considerations, the following strategies can be formulated:

(1) Reducing plastic production
(2) Increasing efficiency in the supply chains of fossil resources
(3) Reduction in the use of fossil fuels
(4) Alternative carbon development
(5) Transformation of today's linear economy into a circular economy

These strategies are briefly discussed here.

3.3.1 Production Reduction

Reducing plastic consumption and, as a result, reducing plastic production is a demand that is repeatedly raised both in science and by various NGOs (Vidal et al. 2024; Exit Plastics 2024). This is understandable and, in some cases, justified. However, what is usually ignored is that there are good reasons for using plastics in all areas of application and that there are either no alternative solutions or that they are ecologically less favorable (Meng et al. 2024).

Such demands mostly come from the countries of the wealthy global north, where the plastics markets are hardly growing anymore. However, the growth is taking place in the global south (cf. Chapt. 5.8), where plastic applications are essential and cannot be substituted for, among other things, food (protection, durability, distribution, cooling, etc.), housing (building materials, utilities, household appliances, living environment, etc.), Mobility (lightweight materials, tires, batteries, cables, etc.), hygiene and medical care (water treatment and distribution, blood supply, medical materials and devices, etc.). What do you want to explain to the people there and how are the relevant Sustainable Development Goals of the United Nations, which aim at a more equitable distribution of security and prosperity, to be achieved?

There is no space here to elaborate on this further.

3.3.2 Increased Efficiency

Although renewable resources will take their place in the long term, fossil resources will still be needed for many years to come. Any increase in efficiency aimed at avoiding emissions in the production of oil or gas, in transport and storage, in preparation and processing (refining and cracking), in the construction and maintenance of plants, etc. is therefore necessary and sensible. The same applies to the pricing of such emissions within the framework of emissions trading systems to promote efficiency and reduce the attractiveness of fossil resources. In this context, the plastics manufacturing industry must contribute to the industrial heating transition by increasing efficiency. IN4Climate. NRW has developed a four-stage model for climate-neutral heat supply (IN4Climate 2021).

3.3.3 Fossil Fuel Replacement

The replacement of fossil fuels, i.e., hydrocarbons, whose chemically stored energy is used to generate the majority of the process energy during combustion (oxidation to CO_2) with the development of heat (see above), must be done through the successive use of electrical energy, for which this must be available. This requires the rapid expansion of climate-neutral electricity generation, especially alternative energies—especially offshore wind farms—and the expansion of the networks required for transport and distribution. Chemicals will require very large amounts of electricity at prices <5 cents/kWh. Due to the uncertainty about the availability of these amounts of electricity from public networks, individual companies are taking matters into their own hands. Covestro concluded a supply contract for a capacity of 100 MW of offshore wind turbines with the Danish electricity supplier Ørsted for delivery to begin in 2025 (Ørsted 2020). BASF, together with RWE, is planning an offshore wind farm with a capacity of 2 GW, which is to be delivered in 2030 (BASF 2021), and, together with Vattenfall and Allianz Capital Partners as financial investors, is taking the Hollandse Kust Zuid offshore wind farm with an output of 1.5 from GW in operation (BASF 2023). The electricity from these plants will be used, among others, to supply the BASF locations in Antwerp and Ludwigshafen.

At the same time, companies in the chemical industry and plant engineering are working on the electrification of crackers, the largest individual energy-consuming systems in the chemical industry. LyondellBasell (LYB), CPChem, and Technip Energies are collaborating on the electrification of a cracker at the LYB site in Channelview, Texas (LyondellBasell 2023), with the goal of ultimately reducing scope 1 and 2 GHG emissions by 90% (LyondellBasell 2023).

BASF, Sabic, and Linde Engineering report on the commissioning of a demonstration plant in the cracker operation of the BASF Ludwigshafen plant, which is the final step to gather the necessary experience for the industrial application of this technology (BASF 2024). It is emphasized that the furnace's technology holds huge potential for the sustainability of the global petrochemical industry and that it can demonstrate the role that renewable electricity can play in higher efficiency and low-emission chemical processing.

3.3.4 Renewable Carbon—Circular Raw Materials

Basically, three different carbon sources can be considered as substitutes for fossil raw materials:

- plastic waste, i.e., the carbon found in the technosphere;
- renewable raw materials from the biosphere, i.e. different forms and sources of biomass;
- Carbon dioxide (CO_2), which has so far been released into the atmosphere as gaseous waste from various technical processes.

3.3.4.1 Plastic Waste

In 2019, around 30 Mt of recyclates were produced through mechanical recycling from processed plastic waste using mechanical recycling processes, i.e., around 8% of world plastic production (cf. Chap. 5) (Conversio 2020a). Chemical recycling processes are being developed with great intensity worldwide, but there are currently only small quantities of such origin on the market. Chemical recycling products will have to be appropriately incorporated into established large-scale processes, for example, to be further processed as pyrolysis oil in a cracker (cf. Sect. 13.3).

The term recyclate is currently (still) mainly used for the target product of plastic recycling processes, but other industries are also using it more and more. Examples include reprocessed/recycled non-ferrous metals for batteries. In the discussion about the EU battery regulation, there is talk of a minimum content of recyclates of cobalt, lead, nickel, and lithium for the production of batteries (Halleux 2021).

The cross-material term "secondary raw material", which, in the spirit of the circular economy, emphasizes the raw material character associated with value retention, is more likely to be found in scientific literature but is rarely used in everyday language. It remains to be seen whether plastics that have been partially or completely produced via chemical recycling (see Chap. 10.3) will be referred to as "recycled materials" in the future, even though they are of virgin quality. In the future, plastics will predominantly and ultimately exclusively be made from circular raw materials. A linguistic differentiation would not be effective.

3.3.4.2 Biomass

There are great hopes associated with the extraction of raw materials from biomass.

Biogenic raw materials are currently only used to produce two large-volume products: the latex precursor for natural rubber comes from rubber tree plantations in Southeast Asia (Hofmann et al. 2001), and polyethylene of Brazilian origin (Braskem 2021) is produced on an industrial scale based on sugar cane.

In our context, the term biogenic raw materials refers to raw materials that are plant-based, i.e. renewable. The result so far has been the state-supported cultivation of so-called "energy crops" (oil palms, corn, sugar cane, rapeseed, wheat), which compete with food and fodder crops for cultivation space, especially to produce (bio)ethanol.

The focus is currently more on renewable raw materials of the 2nd generation (from wood and plant residues such as leaves, stems, and roots) and the 3rd generation (from algae and enzymatic processes) (cf. Chap. 14). An example is the synthesis of polyethylene furanoate (PEF), for which 5-hydromethylfurfural (HMF) is first produced from the vegetable sugar fructose (from straw) by dehydrogenation, which is further oxidized to 2,5-furandicarboxylic acid (FDCA). PEF is produced from the dicarboxylic acid by polycondensation with ethylene glycol. PEF is an alternative to the established, fossil-based PET with interesting properties (PEF 2021).

In recent years, the concept of the bioeconomy has developed from a niche topic to the political mainstream worldwide (cf. Chap. 14). Numerous countries have published bioeconomy strategies. The EU Commission emphasizes that the bioeconomy is key to a climate-neutral economy and can contribute to all dimensions and goals of the "European Green Deal" (VDI 2021b).

3.3.4.3 CO_2—Carbon Dioxide

CO_2 (carbon dioxide) is the highest oxidation product of carbon, the most important product of any combustion process, an inert, chemically stable molecule that, with a proportion of approx. 400 ppm in the earth's atmosphere, is considered a trace gas, but has a significant greenhouse effect. The reduction of anthropogenic CO_2 emissions is one of the main current challenges for the earth's population. The Paris

Climate Protection Agreement of 2015 set goals for this, which the vast majority of signatory states committed to complying with.

Anyone who generates CO_2 in the future will not simply be able to release it into the atmosphere but will have to pay for it. Alternatively, it can be intercepted and used as a material.

The lignite-burning energy producers have demonstrated this by installing flue gas desulfurization systems (FGD) in their power plants as an example of the capture and utilization of SO_2; this gave rise to a veritable industry that revolutionized parts of the construction industry (drywall construction with plasterboard).

The captured CO_2 must be reduced in order to be used again in chemistry and plastic production. As a result, the carbon is fed back into the cycle. This means that combustion can ultimately lead to material use (cf. Chap. 15).

However, the reduction of CO_2 is energy-intensive, so technical use has so far only taken place in a few places (Covestro 2020). Corresponding research and development work is underway (Kopernikus 2021), ranging from the mapping of industrial CO_2 sources (VDI 2021a) to collection and purification techniques, to the development of catalyst systems to reduce the activation energy (Dechema 2021). However, the basic requirement will be the availability of hydrogen and large quantities of inexpensive renewable energy. The scientific and technical solutions for reducing CO_2, summarized under the term Carbon Capture and Usage (CCU), are hopeful but have hardly progressed beyond the pilot plant stage due to the high energy requirements (VDI 2021a).

Figure 3.4 shows an estimate by the Nova Institute for the carbon sources mentioned from which circular raw materials are produced for the chemical and plastics manufacturing industry. As defossilization gets underway, the proportion of fossil raw materials is expected to fall to zero by 2050. In Part III of this work, circular raw materials are discussed in detail.

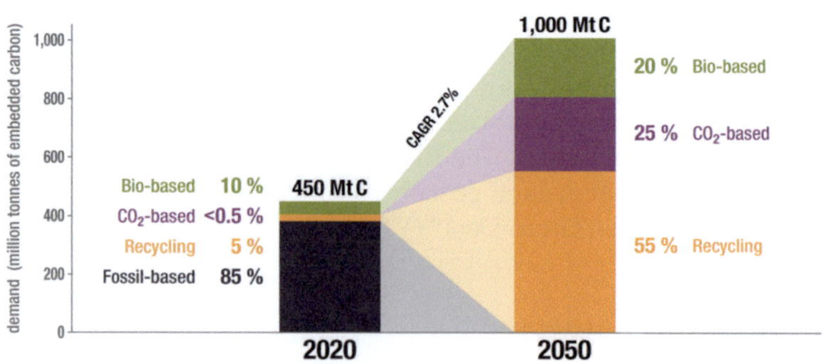

Fig. 3.4 Global carbon demand for chemicals and derived materials (*Source* Nova-Institute 2021)

3.3.5 *Circular Economy*

However, the path described here is not just a question of chemistry, but of the entire economy and society. To be able to quantitatively use plastic in the technosphere[4] as a source of raw materials in the future, the following requirements, divided into four segments, must be met:

- Product manufacturers (brand owners, OEMs, etc.) are willing and able to use plastics that are made entirely or partially from circular raw materials and supplement the design of their products with circular suitability.
- Providers on the market—distributors and retailers—assume their product responsibility, organize, and support the return of their products; their customers—consumers and users—decide the fate of the products they own and hand over the waste to the resource economy.
- Waste disposal becomes a resource economy—from the collection of the waste from the consumer and the processing to the handover of the secondary raw material of a defined quality to the raw material producers.
- Raw material producers (chemical and plastics manufacturing industries) modify their processes so that they can safely and verifiably produce the basic chemicals and building blocks required for the production of their products from circular raw materials and can make the expected offer to product manufacturers (cf. Chap. 16).

We represent the 4 segments in the circuit diagram as in the introduction and will discuss them in detail in Part II of this work.

3.4 Perspectives

One of the key measures against climate change is the decarbonization and defossilization, resp., of fuels and raw materials, i.e. the move away from oil and gas. For chemistry and plastics production, this will result in a complete transformation of the raw material base in the coming decades. Basic chemicals made from circular and renewable raw materials ("Renewable Carbon") (Nova 2021) will become the mass basis for plastic production in the next two decades. To conserve resources, recycle waste, and minimize CO_2 emissions, fossil raw materials for future plastic production must be gradually replaced by circular raw materials. Assuming that by 2050 the use of fossil raw materials will decrease by 15% every five years to a remaining 10% (the Nova Institute anticipates complete defossilization, see Fig. 3.4), we get a global 4% growth scenario, the result shown in Fig. 3.5. The need for circular raw materials and the carbon embedded, resp., will, therefore, increase to significantly more than 1,000 Mt by 2050 (cf. Chap. 5). The prerequisite for this is the industrial transformation from the linear to the circular economy ("circular economy"), in which plastic waste remains a valuable product that is processed and recycled to flow

[4] In a future circular economy, this applies equally to all materials.

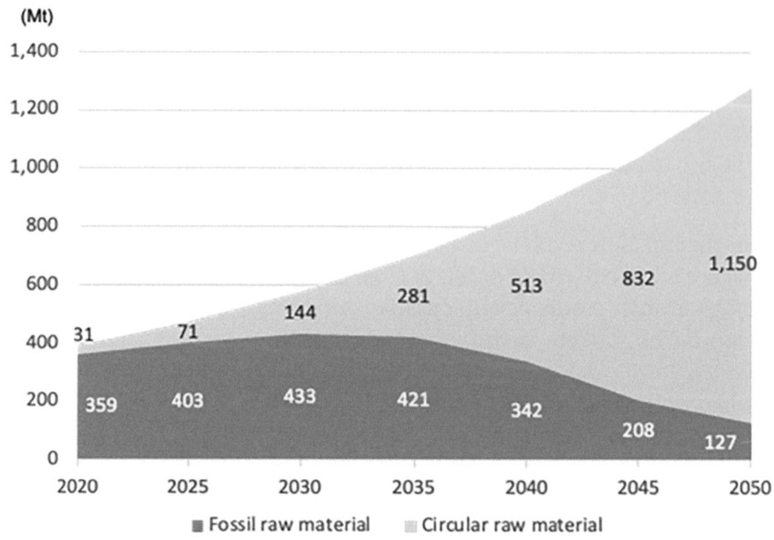

Fig. 3.5 Development of global fossil and circular raw material demand (*Source* Own illustration)

back into the cycle as a circular raw material, in which everyone involved ensures the suitability and management of the cycle of its products.

All exploitation technologies are needed for this and only with technological openness will we be able to overcome the challenges that lie ahead. The processes for mechanical recycling must be further developed and transferred to other scales. The development of chemical recycling processes must be promoted and implemented in large-scale plants to recycle waste that cannot be mechanically recycled with reasonable ecological and economic effort. Various waste fractions must continue to be burned in the future; the resulting CO_2 must be captured and fed back into the cycle using reductive, energy-intensive processes (CCU). Second- and third-generation renewable raw materials will also play a role as circular raw materials. The incineration of plastic waste will become a transition technology for leftovers that cannot otherwise be used. Waste management and redistribution logistics will face significantly more demanding tasks.

References

Agora 2020: https://www.agora-energiewende.org/publications (last access: Aug. 21, 2024)

Agora 2021: https://static.agora-energiewende.de/fileadmin/Projekte/2020/2020-09_DE-Call_for_Action_Industry/A-EW_204_Klimaneutralitaet-2050_Was-Industrie-von-politik-bedarf_WEB.pdf

Agora Industry 2023: Chemistry in transition. The three fundamental pillars for the transformation of chemical value chains. Berlin 2023.

BASF 2021: https://www.basf.com/global/de/media/news-releases/2021/05/p-21-217.html

BASF 2023: https://www.basf.com/global/de/media/news-releases/2023/09/p-23-318.html
BASF 2024: https://www.basf.com/global/en/media/news-releases/2024/04/p-24-177.html
BMU 2018: https://www.bmuv.de/fileadmin/Daten_BMU/Pools/Broschueren/klimaschutz_in_zahlen_2018_bf. pdf. 05.2018
Braskem 2021: https://www.braskem.com.br/portal/imgreen/arquivos/catalogos/Catalogo_Im_green_ING_AF_BX-.pdf. 04.2021
Conversio 2020: https://www.bkv-gmbh.de/studien/studie-global-plastics-flow-2018-conversio.html
Covestro 2020: https://www.covestro.com/de/sustainability/flagship-solutions/CO2-as-a-raw-material. (last access: Feb. 15, 2022)
Dechema 2021: https://dechema.de/EVT21.html
DIN EN ISO 14021: Environmental labels and declarations - Environmental provider declarations. Edition 2016–07
EDGAR 2021: https://edgar.jrc.ec.europa.eu/report_2021. (last access: Feb. 15, 2022)
Exit Plastik 2024: https://exit-plastik.de/ueber-uns/. (last access: Aug. 21, 2024)
FNR 2022: https://www.fnr.de/nachwachsende-rohstoffe/nachwachsende-rohstoffe-im-ueberblick. (last access: Feb. 15, 2022)
Häckel 2021: Häckel, H. Meteorology. (9th ed.). utb GmbH. https://doi.org/ https://doi.org/10.36198/9783838555041
Halleux 2021: New EU regulatory framework for batteries - Setting sustainability requirements. EPRS_BRI (2021) 689337_EN.pdf. (last access: Dec. 13, 2021)
Hofmann et al. 2001: Hofmann, W., Gupta, H., Burger, M.: Handbook of rubber technology. Gupta, Ratingen 2001
IN4Climate 2021: Climate-neutral industrial heat: Strategies and prerequisites for transformation. A discussion paper from the Heat Working Group, Gelsenkirchen, 2021
IPCC 2018: https://www.ipcc.ch/site/assets/uploads/2020/07/SR1.5-SPM_de_sperrfrei.pdf
Kopernikus 2021: https://www.kopernikus-projekte.de/projekte/p2x. (last access: Feb. 15, 2022)
LyondellBasell 2023: https://www.lyondellbasell.com/en/news-events/corporate--financial-news/lyondellbasell-pursues-new-electric-furnace-technology-collaborates-with-technip-energies-and-chevronphillips/
Meng et al. 2024 : Meng, F., Brandão, M., Cullen, J.M., Replacing Plastics with Alternatives Is Worse for Greenhouse Gas Emissions in Most Cases. Environ. Sci. Technol. 2024, 58, 2716−2727. https://doi.org/10.1021/acs.est.3c05191
Nova 2021: https://renewable-carbon.eu/publications/?search=1&publication-type=graphics
Ørsted 2020: https://orsted.com/en/media/newsroom/news/2019/12/799146415974816
PEF 2021: https://www.chemie.de/news/thema/polyethylenefuranoat/. (last access: Nov. 25, 2021)
UBA 2024: https://www.umweltbundesamt.de/daten/klima/treibhausgas-emissionen-in-deutschland#emissionsentwicklung
VDI 2021a: https://www.vdi.de/ueber-uns/presse/publikationen/details/industrielle-CO2-kreislaeufe. (last access: Feb. 15, 2022)
VDI 2021b: Bioeconomy - together for a sustainable economy. VDI fields of action November 2021
Vidal et al. 2024: Vidal, F., van der Marel, E.R., Kerr, R.W.F. et al. Designing a circular carbon and plastics economy for a sustainable future. Nature 626, 45–57 (2024). https://doi.org/10.1038/s41586-023-06939-z
Weissermel and Arpe 2010: Weissermel, K, Arpe, HJ. Industrial Organic Chemistry. Wiley-VCH, Weinheim 2010. https://doi.org/10.1002/9783527619191.fmatter

Dr. Ulrich Liman is a chemist by education, he joined Bayer AG in 1989 in the marketing of Polyurethane and Polycarbonates and held various managerial positions in Japan, China, and Germany. Until 2022, he was responsible for polyurethane R&D and held a seat on the Supervisory Board of Covestro AG.

Dr. Peter Orth is a chemist by education; with OPC—Orth Plastics Consulting he leads a group of senior consultants with decades of professional experience in leading positions in the German and European plastics industry and its associations. His main focus is on innovation and sustainability in cooperation with science, industry, and politics. He is currently working in the fields of CO_2 as a feedstock, resource efficiency, and recycling technologies.

Chapter 4
The Plastics Ecosystem

Peter Orth and Jürgen Bruder

Contents

Summary

The plastics industry is an important part of the European and global economy. Beyond the traditional definition of the plastics industry with the three domains of plastics production, plastics conversion, and plastics machinery, there are important other areas belonging to the plastics ecosystem namely recyclers, service providers, science, associations, trade fairs and many others.

All participating institutions work closely together in complex, mostly international networks, often within the framework of customer–supplier relationships that have been built up and proven over decades.

The necessary transformation of today's linear production system is dominated by linear supply chains to a circular economy, with its demanding requirements, deeply intervenes in these structures and workflows. The phasing out of fossil raw materials while waste is becoming an important feedstock resource is not only changing the world in plastics production but also in all other areas of the plastics ecosystem.

P. Orth (✉) · J. Bruder
OPC - Orth Plastics Consulting, Cologne, Germany
e-mail: opc.cologne@t-online.de

P. Orth et al. (eds.), *Circular Economy of Plastics*,
https://doi.org/10.1007/978-3-031-78829-1_4

To assess the impact of this change on established processes and companies, it is necessary to understand the structures existing today.

In this chapter, the sectors and sub-sectors are presented in detail. In doing so, the essential characteristics of each sector are dealt with in a structured manner, the important actors are named, and their roles are described. Regional aspects and key figures complete the picture.

4.1 Plastics Industry and the Plastics Ecosystem

For about 20 years, the plastics industry in the narrower sense has been defined as the totality of the three very different sectors of plastics production (part of the chemical industry), plastics machinery manufacturing (part of mechanical engineering), and plastics processing. The plastics industry in Europe has >1.5 million employees, >53,000 companies (mostly SMEs, see Sect. 4.3), and generates a turnover of >400 billion euros (PlasticsEurope 2024).

The plastics ecosystem as defined in this chapter also includes, among others, compounders, distributors (often with their own compounding operations) and additive/auxiliary manufacturers; specialised logistics companies; sorting companies and recyclers (traditionally regarded as part of the waste management industry); trade fair companies, publishers and many other service providers; and finally the relevant scientific community. In addition, there are the associations of the sectors, working and interest groups, think tanks (Agora 2021), market researchers (Conversio 2020) and the like.

4.2 Plastics Production

General

Companies in the plastics manufacturing industry are usually either part of the chemical or petrochemical industry. Today, most of them are globally positioned and supply their customers, being also globally active, with products of the same specification from regional production. Production is location-sensitive and capital-intensive; world-scale units typically have >500 kt annual production (e.g., polyolefins or isocyanates). They are integrated into complex global supply chains and depend on their reliable functioning. Consolidation of this industry is well advanced but ongoing. The range of annual sales in plastics production extends from approx. 1 billion to approx. 25 billion euros. As a rule, these are publicly listed companies.

Regional aspects

The focus of this paragraph is on Europe as a plastics region, even though this is becoming relatively less important due to various developments. Currently, about

14% of the world's production is still produced in existing plants; new plants are only being built to a minor extent (replacement investments). A stagnating plastics market does not indicate particular growth prospects.

The most important oil-producing region is the Middle East, namely Saudi Arabia and the Emirates in the Persian-Arabian Gulf; due to recent political developments, Iraq and Iran currently play only a minor role. For about 20 years, the governments there have been pursuing a policy of reducing dependence on pure oil exports and keeping a larger share of the value added in the country, i.e. producing plastics locally. Currently, the region (incl. Africa) is home to approx. 9% of global production.

China is by far the largest plastics production location and currently accounts for approx. 32% of global production, for which large quantities of raw materials are imported. In addition, enormous quantities of plastics are imported from other regions and countries (USA, Persian-Arabian Gulf) to cover the constantly growing demand of the Chinese plastics processing industry. A further 22% is produced in Japan, India, Malaysia, Singapore, South Korea, Taiwan, and Thailand. Today, Asia and Oceania account for appr. 54% of global production capacity.

Plastics production in North America has experienced a renaissance in recent years due to the massive availability of cheap shale gas, which has been effective like a drug for the plastics industry there. The Americas (including South America) account for around 21% of global production. In addition, shale gas (ethane) is liquefied and exported to Europe, e.g., where it is reduced in gas crackers. The remaining 2% is produced in the CIS and Russia respectively.

All volumes mentioned are from PlasticsEurope (2024).

Companies

The main polyolefin producers can be grouped into two groups.

Plastics/chemical subsidiaries of major oil companies such as ChevronPhilips (CPC), ENI (Versalis), ExxonMobil (EMC), OMV (Borealis), Petrobras (Braskem), Repsol, Saudi Aramco (Sabic) and Total; petrochemical/chemical companies such as Dow, LyondellBasell, and Ineos in North America and Europe; ChemChina, Petrochina, and Sinopec in China; Asahi, Mitsubishi, Mitsui, and Sumitomo in Japan; LG and SK in South Korea, Reliance (India), PTT and SCG (Thailand) and Formosa Plastics (Taiwan).

Chemical companies such as BASF (polyamides and PUR) and Covestro (polycarbonates and PUR), which emerged from the Bayer Group, have become specialists—but they are among the world market leaders with their respective products.

This market is in a constant state of flux, with acquisitions, divestments, and start-ups taking place all the time. Any overview is therefore only just a snapshot.

The defossilisation aimed at the goal of climate neutrality means that many of these companies are revising and modifying their business models. This is leading to restructuring, consolidation, and massive changes in the market. The Abu Dhabi National Oil Company (ADNOC) seemingly is amongst the forerunners in that very process.

4.3 Plastics Conversion

Plastics conversion is at the heart of the plastics industry. Here, the plastic pellets and liquids delivered by plastics manufacturers are formed on plastics converting machines into the shape in which they meet private or commercial consumers in the form of semi-finished or finished parts such as packaging, technical parts, insulating materials and other building products like pipes or fittings, consumer goods, or plastic products for agriculture and fisheries. The plastics converter uses various processing methods such as injection moulding, reaction injection moulding, extrusion, blow moulding, thermoforming, rotational sintering, etc. Larger plastics converters may often be recognized "from the outside" by the silos in which the granulates, usually delivered by tanker truck, are kept for production.

According to the European Plastics Converters Association (2024) the industry, which is made up of predominantly small and medium-sized companies, comprises over 50,000 companies across Europe, makes an annual turnover of around 260 billion euros and employs around 1,470,000 people. 54.1 Mt of plastics were converted in Europe in 2022 (PlasticsEurope 2024), of which 8.7 Mt were plastic recyclates or over 16%. The plastics processing volume in Europe (see Table 5.2) is distributed as follows.

- 39% plastic packaging;
- 23% construction products;
- 8% vehicle construction.

Plastics conversion also takes place in sectors that do not belong to the plastics industry and thus to the plastics ecosystem, such as the beverage industry, the furniture industry, the automotive industry, the toy industry, and many others. For example, PET bottles and their preforms are now mainly produced by mineral water and beverage bottlers themselves and are only supplied to a lesser extent by plastics converters. Most companies in the automotive and electrical industries also have large in-house plastics converting capacities (in-house production).

Small converters focusing on specialties have sales in the single-digit millions, while large packaging manufacturers (films, industrial packaging) generate sales of 1 billion euros and more. Many of the companies, which operate nationally and increasingly on a European scale, are family-run. Some of the large companies are also family-run, while others belong to international groups that are often listed on the stock exchange.

During the Corona pandemic, state economic ministries classified a number of manufacturers of plastic packaging (packaging for disinfectants, pharmaceuticals, foodstuffs, etc.) as systemically relevant.

4.4 Plastics Machinery

Special machines are necessary for processing and shaping the plastics, usually delivered in the form of pellets or liquids. The most important types are extruders for melting and compounding as well as for the extrusion of semi-finished products (films and sheets, hoses, pipes, profiles), thermoforming machines for the production of open hollow bodies (cups, bowls, pots, blisters, etc.), injection moulding machines for the production of moulded parts with diameters from the mm to the m range as well as RIM machines for the production of soft and hard foam parts based on polyurethane. Moulds, presses, calendars, depositing and winding equipment, autoclaves, blow moulding machines for bottles, canisters, drums, intermediate bulk containers (IBC) or fuel tanks are further moulding units. In addition, there are now numerous manufacturers of equipment for 3D printing (additive manufacturing), which is evolving from a special process to a standard process for prototypes, small series, complicated geometries, and remote processing. The list of peripheral devices and the corresponding manufacturers is almost endless. Austria, Germany, Italy, and Switzerland are the technological centres of plastics machinery manufacturing and the largest exporters, but the largest manufacturer globally has been China for years. The industry has a rather medium-sized structure; a few companies (mostly injection moulding machine manufacturers) have annual sales in the order of 1 billion euros. The typical mechanical engineering company is family-owned and operates in Europe or even worldwide.

According to Euromap (2024), the plastics machinery industry comprises about 1,000 companies with a workforce of some 57,000 people creating a turnover of € 14.3 billion. In 2021 machines for the most important processes have the following market shares in Europe (VDMA 2021):

- Injection moulding 27.3%
- Extrusion 49.1%
- Blow molding 7.0%
- EPS Foaming 3.2%
- PET Preforms and Stretch Blow Moulds 5.9%

The development towards a recycling economy brings new requirements and opportunities for plastics machinery and plant engineering in the field of plastics recycling. This initially involves the processing of used, mixed and contaminated waste by washing, separating, and drying to produce clean production waste. After shredding, fine grinding and compacting (agglomeration), the waste is transferred to the classic manufacturing processes (Herbold 2021). Detailed figures are not available yet from the industry.

4.5 Recyclers

In the 1980s, the recycling of plastic waste was only weakly developed; it was a niche business for the recycling of production waste and selected separately generated commercial waste. Since then, many recycling companies have been established in Europe and all over the world. It is currently estimated that there are around 2,500 companies and projects active in mechanical recycling worldwide (ICIS 2024), producing 35.5 Mt of recyclate in 2022 (cf. Table 5.1). Plastics recyclers in Europe run around 850 facilities with 12.5 Mt of installed capacity. Some 30.000 employees create over € 10.4 billion in turnover (PRE 2024).

In Europe, many of these companies are medium-sized enterprises, each of which produces 10–50 kt of recyclates, markets them independently and achieves a turnover of < € 20 million.

As indicated in Sect. 4.4 already, recycling begins with classification, screening, sorting, and separation processes followed by homogenization at the heart of which are specialized extruders (cf. Chap. 10.1). The plants are geared to the special characteristics of the input waste material. The plastics recycling industry has been undergoing a (cultural) change from a waste management orientation to an integrated part of the plastics value chain for about ten years. This change is embedded in that of the waste industry towards a secondary raw materials economy and is accelerated by ongoing consolidation. Thus, medium-sized plastics recycling companies are being taken over by various players in the market. Not only large, internationally positioned waste management companies such as Remondis or Veolia are active, but also plastics manufacturers such as Borealis, LyondellBasell, or Total, plastics distributors or compounders respectively such as Mocom, large processors such as Alpla, Berry Global, or Trioworld, dual systems (in Germany) such as DSD or retail chains such as the Schwarz Group (Lidl/Kaufland) with its independent legal entity PreZero being an integrated environmental service provider.

4.6 Value Chains

A typical feature of the plastics industry is its organisation in linear value chains—as is the case in the automotive industry, the packaging manufacturing and filling/packing industry, the E & E industry, the medical technology industry, etc. Such chains are organized from both ends, live from direct, often long-standing contacts and the contractual obligation of almost all of their members to each other, the cooperation with experienced logistics companies (cf. Chap. 9), being sometimes IT companies themselves or integrate them. Such chains are rarely organized on a national or regional basis; in many cases, they are now global organizational structures whose complexity poses major challenges, especially for SMEs.

In addition to the manufacture and supply of plastics by the plastics manufacturing industry, there is close cooperation between the plastics manufacturing and plastics processing industries right from the development stage and during production. Only by bundling and utilizing all competencies along the entire value chain can the challenges posed by competitive products be met. This cooperation is becoming even more important as further and new challenges have arisen and will continue to arise with the circular economy (cf. Sect. 6.1).

Some value-added cycles/networks have been successfully established, e.g. for the PET value chain (PETCore 2024), or are being established, for example in some areas of the construction industry (window frames, flooring, etc.), driven forward by the PVC industry (VinylPlus 2024), but are far from being the rule. (cf. Sect. 7.4).

4.7 Service Providers

Trade fairs are important service providers for the industry. They fulfil important functions as marketplaces and information exchanges, are meeting points and communication centres and, of course, and above all, they show innovation in competition. They enable the exhibitors' customers and their competitors to make direct comparisons, pick up on trends and reinforce them.

The two important leading global trade fairs for the plastics industry are the "K" in Düsseldorf, Germany, and "Chinaplas" in Shanghai and Shenzhen, China. Due to the Corona pandemic, almost all trade fairs planned for 2020 and 2021 were cancelled or postponed. Where trade fairs did take place, it was in experimental virtual formats—visitor presence was impossible. Structures and processes of future so-called hybrid trade fairs will be significantly changed after the pandemic.

The International Plastics Fair "K" takes place every three years in Düsseldorf—most recently in 2022 for the first time after the pandemic. Approx. 3,000 companies (mainly machine builders and raw material manufacturers) exhibited on 175,000 m^2 of hall space and welcomed approx. 180,000 visitors. The "K" is still *the* meeting place for the industry, traditionally based in Germany, the technological centre of the plastics industry. The organiser, Messe Duesseldorf, has a portfolio of a total of 9 plastics trade fairs worldwide, making it probably the world market leader in this sector.

The International Trade Fair "Chinaplas" is held annually, alternating between Shenzhen or Shanghai, with the following figures for Shanghai 2018: exhibition area: 320,000 m^2, exhibitors: >3,400, trade visitors: >150,000. Chinaplas represents the largest and most dynamic market in the plastics industry. Chinaplas's sequence was interrupted by the pandemic, too, and it will after all be important to see how it will come back.

Of regional importance are the NPE in Orlando/FL for North America and Fakuma in Friedrichshafen, Germany, for Europe. Mention should also be made of the JEC in Paris, France, the world's largest trade fair for composite materials, the PlastPol fair in Kielce, Poland, and PLAST in Milano, Italy.

4.8 Science

Across Europe, there is a large and diverse scientific landscape dealing with almost all of the above-mentioned aspects of plastics as a material.

Universities

Polymer sciences with a focus on macromolecular chemistry and plastics engineering are represented at more than 50 universities in AT/CH/DE with about 150 professorships in research and teaching. At many of these universities, there are independent degree programmes in the disciplines mentioned (and beyond), whose graduates find work in the landscape of relevant commercial enterprises, public authorities, and science. The proportion of students from other European and international countries is growing steadily; when they return to their home countries, they become part of a sometimes global network. However, the number of students is currently declining, partly due to the diminished image of plastics as a material because of the unresolved waste problem.

The authors don't have respective figures for all countries in Europe but estimate that the abovementioned numbers stand for almost 40% of European scientific capacities in macromolecular chemistry and plastics engineering.

Non-university Research

Closely linked to the universities are the institutes and working groups of national science foundations, scientific societies, etc. like Christian Doppler Gesellschaft in Austria, Centre National de la Recherche Scientifique (CNRS) in France, Max Planck Society or Fraunhofer Society in Germany, or the Netherlands Organisation for Applied Scientific Research (TNO). Depending on the character of the respective organisation, they are engaged in basic or applied research. In most cases, they are linked to the plastics industry through projects or committee memberships. Of particular importance in the context of this book is the Fraunhofer Cluster of Excellence "Circular Plastics Economy" (FhG 2024), which as a virtual institute (out of 6 real institutes) deals with questions of the plastics circular economy (cf. Back matter). Aspects of the circular economy are now being worked on intensively in almost all plastics technology institutes in Europe.

Of course, no conclusive consideration can be made here. In summary, however, the authors note that the scientific potential available in Central Europe is (still) almost unique worldwide. It contributes significantly to the technological leadership of the plastics industry in Europe.

4.9 European Associations

The plastics industry in Europe is organised by numerous regional and national associations and similar organisations. Within the context of this book, we can only mention on the European level.

PlasticsEurope (2024) is the pan-European association of the plastics manufacturing industry. Centered in Brussels, Belgium, it has national/regional centres in Den Haag, The Netherlands; Frankfurt, Germany; London, U.K.; Madrid, Spain; Paris, France; Vienna, Austria; and Warsaw, Poland.

The European Plastics Converters Association (2024) also centered in Brussels is a European umbrella organization representing 28 national associations from across Europe and 18 sectoral organisations mostly located in Brussels, too.

EUROMAP (2024) is the European umbrella organisation of national machine builders member associations located in Austria, France, Germany, Italy, Luxembourg, Spain, Switzerland, Turkey, and the United Kingdom. Euromap is a resident of the German Plastics Machinery Association in Frankfurt, Germany. EUROMAP collects information about all associations within the plastics and rubber industry. The list with actually 289 items provided on their website should cover most of the plastics industry's associations worldwide.

Plastics Recyclers Europe (2024) also located in Brussels represents plastics recyclers in Europe. The association lists 165 member companies, 10 national member associations, 7 working groups, and 8 task forces.

4.10 Other Parts of the Plastics Ecosystem

(Chemical) trading houses and logistics companies (cf. Chap. 9) play an important role in the distribution and marketing of plastics. The large plastics producers generally concentrate on compounding and packaging a few types of their products, which they have transported to their large customers in the largest possible containers (containers, silo vehicles). They usually sell smaller containers via their distribution partners, who often also act as compounders of smaller types and specialities in terms of volume.

Other ingredients (besides plastic) of the compounds are supplied by specialists who very often exclusively run a plastics-related business: Reinforcing fibers (glass and carbon fibers), fillers ($CaCO_3$, talc, silica gel, carbon black, etc.), pigments, stabilizers, flame retardants, processing aids, and much more.

System houses support the business of PUR raw material producers in formulation, distribution, customer advisory services, etc.

In the circular economy environment, start-ups such as Cirplus (2022) or Plastship (2022), which provide their customers with improved access to recyclates, among other things, or software houses such as Simcon (2022), which offers software and services for parts and tool development and for optimising the production process, are establishing themselves.

The Ellen MacArthur Foundation is a UK charity which aims to speed up the transition to the circular economy. Since it was set up, the charity has emerged as a global thought leader, putting the circular economy on the agenda of decision-makers across business, government, and academia.

The BKV, a company run by the German plastics industry, focuses on the generation of market data, technology assessment, the promotion of the plastics recycling industry, and the treatment of the marine litter problem (BKV 2024).

The Nova Institute for Political and Ecological Innovation is a private enterprise engaged in research (desk-top research) and consultancy in the field of circular raw materials (renewable carbon) (Nova 2022).

References

Agora 2021: https://www.agora-energiewende.de (last access: Oct. 21, 2021)
BKV 2024: https://www.bkv-gmbh.de/home-en.html (last access: Apr. 9, 2024)
Cirplus 2022: https://www.cirplus.com/de. (last access: Feb. 16, 2022
Conversio 2020: https://www.bkv-gmbh.de/studien/studie-stoffstrombild-kunststoffe-in-deutschland-2019-conversio.html (last access: Oct.28, 2021)
Euromap 2024: https://www.euromap.org/markets/the-industry-at-a-glance (last access: Apr. 9, 2024)
FhG 2024: https://www.ccpe.fraunhofer.de/ (last access: Apr. 9, 2024)
EuPC 2024: https://www.plasticsconverters.eu/ (last access: Apr. 9, 2024)
Herbold 2021: https://www.herbold.com (last access: Oct. 30, 2021)
ICIS 2024: https://www.icis.com/explore/sectors/recycling/ (last access: Apr. 9, 2024)
Nova 2022: https://nova-institut.de/ (last access: Oct. 24, 2022)
PETCore 2024: https://www.petcore-europe.org/ (last access: Apr. 9, 2024)
PlasticsEurope 2024: https://plasticseurope.org/media/plastics-europe-launches-the-plastics-the-fast-facts-2023/ (last access: Apr. 9, 2024)
Plastship 2022: https://plastship.com. (last access: Feb. 16, 2022)
PRE 2024: https://www.plasticsrecyclers.eu/about/ (last access: Apr. 9, 2024)
Simcon 2022: https://www.simcon.com. (last access: Feb. 16, 2022
VDMA 2021: Personal communication
VinylPlus 2024: https://www.vinylplus.eu/ (last access: Apr. 9, 2024)

Dr. Peter Orth is a chemist by education; with OPC—Orth Plastics Consulting, he leads a group of senior consultants with decades of professional experience in leading positions in the German and European plastics industry and its associations. His main focus is on innovation and sustainability in cooperation with science, industry, and politics. He is currently working in the fields of CO_2 as a feedstock, resource efficiency, and recycling technologies.

Dr. Jürgen Bruder is a chemist with an additional degree in business. He worked in higher education and the chemical industry, before becoming Director General of IK, the German plastic packaging association. He held positions in national, European, and global committees including the Global Plastics Alliance. Today he deals with the transformation of the plastics industry to a circular economy focusing on plastic packaging and plastic applications in agriculture.

Chapter 5
Material Flows

Jürgen Bruder, Manfred Rink, and Peter Orth

Contents

Summary

This chapter describes the paths that plastics provided and accompanied by the plastics industry taken during their life cycles. Experts refer to these as material flows.

To clarify the dimensions of the material flows, the quantitative side, i.e. volumes produced and processed, wastes generated and their fate, regionally differentiated for Europe (EU27+3), China, and North America are compared with global figures. Eight European countries with highly developed plastics ecosystems (Belgium, France, Germany, Italy, Netherlands, Poland, Spain, and the U.K.) are compared to each

J. Bruder (✉)
BPC—Bruder Plastics Consulting, Potsdam, Germany
e-mail: plastics@jbruder.de

M. Rink
mrink-Consulting, Cologne, Germany
e-mail: manfred@rink-consulting.de

P. Orth
OPC—Orth Plastics Consulting, Cologne, Germany

P. Orth et al. (eds.), *Circular Economy of Plastics*,
https://doi.org/10.1007/978-3-031-78829-1_5

other and with EU figures. The latest available data for production, processing, and waste treatment in Europe refer to 2022. The last available data for processing and waste treatment in North America and China refer to 2018.

Regional distribution of plastics processed in the largest markets and respective applications in Europe are then dealt with in more detail. Detailed information on the production and use of recyclates in the EU 27+3 and in Germany are presented. Trends in energy recovery, landfill and exports of plastic wastes are shown. The possible development of demand and required production volumes is presented and discussed in various scenarios.

In addition to Conversio/PlasticsEurope data, data and forecasts from OECD's Global Plastic Outlook study are presented taking into account polymers for tires and fibers, too.

References to the most important sources of market data on plastics and the plastics industry, supplemented by literature references, conclude the chapter.

5.1 Global and European Plastic Volume

To make the (quantitative) dimensions of the challenges the plastics industry is faced visible, the material flows, i.e., the paths from the production of plastics, their processing and use, as well as disposal and recycling, are shown here in a structured manner according to region and time allocation (cf. Tables 5.1 and 5.2). In the following, the direct material flows, i.e., the volumes of plastics not yet maintained in the cycle, are considered.

The data presented cover thermoplastics and thermosets. Polymers used for textiles, paints, varnishes, textiles waterproofing, or for cosmetics, medicines, or chemical processes are not included in the scope of this data. PVC-, PO-, and PU-fibres are included, whereas PA-fibres, PET-, PBT-, or Acrylic polyesters are not included (PlasticsEurope 2022).

5.2 Production of Plastics

5.2.1 Virgin Material

In 2022, 400 Mt of plastics (fossil-based, PCR, and bio-based) were produced globally, of which Asia accounted for 54% (with an upward trend), North America for 17% (almost constant), and Europe for 14% (with a downward trend) (cf. Fig. 5.1). In China alone, 32% of world production took place. Included in the worldwide plastic production are approx. 36 Mt of post-consumer recyclates and 2.3 Mt of bio-based plastic, including bio-attributed plastics (PlasticsEurope 2024a).

Table 5.1 Overview of plastic production, processing, and waste—World, Europe, China, and North America (Data in Mt)

Region	World	EU27 + UK/CH/NO	China	North America
Production 2022	**400.3**	**58.8**[1]	**128**	**68.5**
Fossil based	362.3	47.2	118.8	65.3
PCR	35.5	7.6	8.5	2.8
Biobased	2.3	0.4	0.7	0.4
Processing Plastics converter demand (virgin + recycled)	**390.7** (2021)	**54.1** (2022)	**75.2** (2018)	**38.6** (2018)
Packaging	172	21.1	32	20
Building and construction	70	12.4	N/A	N/A
Automotive	31	4.5	N/A	N/A
Electrical and electronic	27	3.1	N/A	N/A
Waste collected	**175** (2018)	**32.3** (2022)	**40.1** (2018)	**37** (2018)
Recycling	50	8.7	12.6	4
Energy recovery	50	16.0	19.2	5.2
Sanitary landfill	75	7.6	8.3	27.8
Unmanaged waste 2018	75	1.0	19	0.7

The above data are rounded estimations
Sources PlasticsEurope (2022), (2024a, b), Conversio (2020)
[1]Including 3.8 Mt PIR, not shown separately for all regions

In Europe (EU27+UK/CH/NO), 58.8 Mt of plastics were produced in 2022 continuing the decline in European plastics production compared to 2018 with 62.8 Mt. The above quantities include fossil-based thermoplastics, thermosets, and PUR used for plastic parts and products and PCR, PIR and bio-based plastic, but no coatings, fibers or rubbers. Concerning the plastics trade of the EU27 plastics industry in 2022, the US, South Korea, and China were the top importers, and the UK, the US, and Turkey the main exporters (PlasticsEurope 2024a).

In Germany, 13 Mt of plastics were produced in 2022, of which 11,5 Mt were fossil based and 1.7 Mt were recyclates for material applications (PlasticsEurope 2024a). Italy with 1.1 Mt or 30% has a relatively high share of recyclate with a total plastics production volume of 3.7 Mt.

OECD's Global Plastic Outlook (cf. Section 5.8) mentions a global total plastic use by polymer of slightly more than 460 Mt for 2022. The difference is due to the inclusion of fibres (60 Mt) and elastomers for tyres (8 Mt) not considered by PlasticsEurope data (OECD 2023).

Table 5.2 Overview of plastic production, processing, and waste—Europe, Belgium, France, Germany, Italy, Netherlands, Poland, Spain, UK

Data in Mt	EU27+UK/ CH/NO	Belgium	France	Germany	Italy	Netherland	Poland	Spain	UK
Production 2022	**58.8**[1]	**7.3**	**5.5**	**13,0**	**3.7**	**6,2**	**2.4**	**5.3**	**2.6**
Fossil based	47.2	6.9	4.5	11,5	2	5.5	1.7	4.1	1.8
PCR	7.6	0.2	0.7	1.7	1.1	0.4	0.5	1,1	0.6
Biobased	0.4	<<0.1	<<0.1	0.3	0.1	<<0.1	<<0.1	<<0.1	<<0.1
Processing 2022 Plastics converter demand (virgin + recycled)	**54.1**	**2.4**	**5.0**	**12.7**	**7.6**	**2.3**	**3.9**	**4.4**	**3.7**
Packaging	21.1	1.1	2.2	4.4	3.0	0.9	1.4	1.9	1.5
Building and construction	12.4	0.5	1.1	3.1	1.4	0.6	1.1	0.8	1.0
Automotive	4.5	0.2	0.4	1.3	0.5	0.1	0.3	0.3	0.2
Electro and electronic	3.1	0.1	0.3	0.8	0.5	0.1	0.3	0.2	0.2
Waste collected	**32.3**	**0.6**	**4.1**	**5.6**	**3.9**	**1.1**	**2.1**	**2.9**	**4.2**
Recycling	8.7	0.3	0.8	2.0	1.2	0.4	0.5	1.2	1.2
Energy recovery	16.0	0.4	2.1	3.5	1.8	0.7	0.8	0.7	2.3
Sanitary landfill	7.6	0.01	1.1	0.03	1.0	0.002	0.9	1.1	0.7

The above data are rounded estimations

Source PlasticsEurope (2024a, b, c)

[1] Including 3,8 Mt PIR, not shown separately for all countries

5.2.2 *Recyclates and Their Use*

In Europe (EU27+3) 7.6 Mt of post-consumer recyclates were produced in 2022 (PlasticsEurope 2024b). The largest PCR producers in the EU are, in this sequence: Germany, Italy, Spain, France, and the UK, (cf. Table 5.2), consequently 12.6% of the plastics produced in Europe are PCR recyclates. The inclusion of post-industrial recyclates (PIR) as a non-fossil raw material base is controversial and is not pursued here. Nevertheless, post-industrial recyclates (PIR) are an important partially existential basis for recyclers; they are generally not a topic in the political discussion on increasing the use of recyclates since there is hardly any potential for an increase here—and an increase is of course not intended. The potential for increase therefore lies exclusively with post-consumer recyclates.

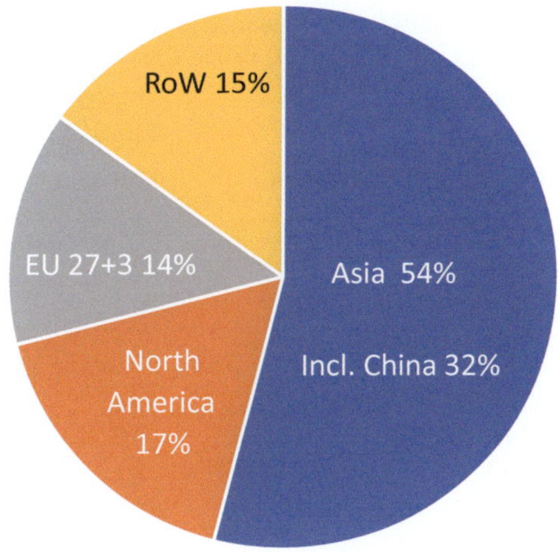

Fig. 5.1 Global plastics production (400 Mt in 2022) differentiated by region Asia, North America, and EU in %. (*Source* Own representation based on PlasticsEurope data)

Progress in the use of circular raw materials for plastic products is crucial for gradual defossilization. Looking at the global substitution of fossil-based plastics by circular raw materials such as PCR and bio-based plastics, a share of around 9% (400 kt in 2022) can be determined. In the EU 27+3 5.5 Mt of post-consumer recyclates were used for internal EU conversion in 2021, corresponding to a substitution rate of 9.9% with PCR. That is to say that 9.9% overall recycled content was used to produce new products in the EU 27+3. Compared to 2018 (4 Mt) and 2020 (4.6 Mt) a significant increase in the use of recyclates is to be stated. The 5.5 Mt achieved in 2021 and the PCR production of 6.8 Mt in 2022 makes the ambitious target of the EU Commission—10 Mt of PCR recyclates in 2025—achievable.

Table 5.3 shows the use of PCR recyclates in new products in the EU 27+3 and in Germany, broken down by the most important plastic markets. The highest proportion of recyclates in the production of new plastic products is found for agriculture, farming, and gardening applications, while in absolute terms most recyclates are used in building & construction products, followed by packaging and agriculture.

It is remarkable that the three markets of building & construction, packaging, and agriculture/farming/gardening take up more than 80% of the PCR recyclates used in absolute terms, with a share of more than 60% of the plastics processed in the EU 27 + 3. The remaining approx. 20% of PCR recyclates are distributed over the other markets, representing less than 40% of volume-based plastics processing.

The above-mentioned 6.8 Mt PCR recyclates are mainly used in products for the following markets:

- Building and construction: road construction products, door and window profiles, cable protection and drainage pipes, construction foils and profiles, flooring;
- Packaging: PET bottles, packaging and transport films, bubble films, trays, bottles for household cleaners, buckets, pallets, boxes, large containers;
- Agriculture and gardening: agricultural films, flower and plant pots, rain barrels, composters.

Table 5.3 Share of PCR content in new products in the EU 27+3 in 2022 and in Germany in 2021. (*Source* PlasticsEurope 2024b and Conversio 2022)

Application	EU 27+3 2022 (Mt)	EU 27+3 2022 (%)	Germany 2021 (Mt)	Germany 2021 (%)
Building and construction	2.8	22.7	0.53	14.5
Packaging	2.0	9.7	0.37	8.4
Agriculture/Farming/Gardening	0.9	37.5	0.15	25.2
Automotive	0.2	4.6	0.03	2.7
Electrical and electronics	0.1	3.2	0.02	1.8
Houseware/Leisure/Sports	0.1	4.5	0.01	2.8
Others	0.6	7.2	0.16	1.4
Total	6.8	12.6	1.27	9.1

5.3 Plastic Applications and Markets

The regional distribution of global converter demand is shown in Fig. 5.2. Worldwide, the demand for plastics is distributed more evenly among the different regions and countries, but China dominates, too.

The regional distribution of plastics converter demands within the EU 27+3 (54 Mt in 2022), is shown in Fig. 5.3. According to this, almost a quarter of plastics processed in Europe were sold in Germany with 23%, Italy was the next largest market with 14%, Great Britain, Poland, Spain, the BeNeLux countries, and France come with 7–10% each to a total of 41%, the 21 other national markets take up a total of 22%.

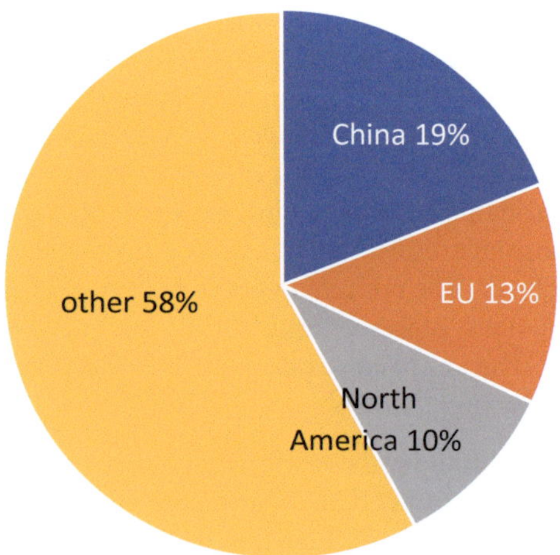

Fig. 5.2 Regional distribution of plastics converter demand of 400 Mt in 2022 in %. (*Source* Own representation based on PlasticsEurope)

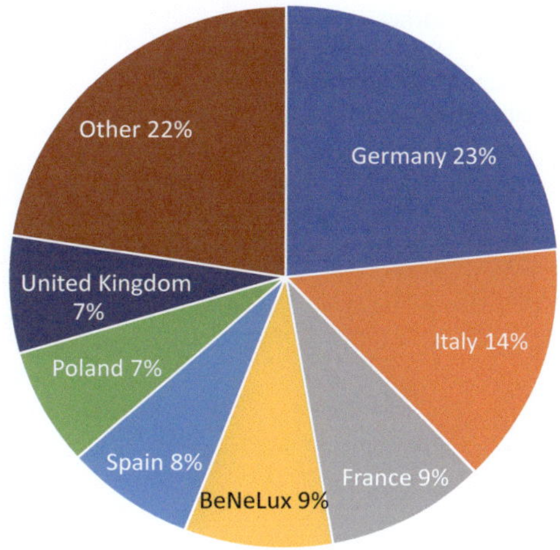

Fig. 5.3 Regional distribution of EU plastics converter demand of 54 Mt in 2022 in %. (*Source* Own representation based on PlasticsEurope)

Market shares of the European plastics converters demand by type and application with the volume in 2021 (50.3 Mt) are shown in Fig. 5.4 (PlasticsEurope 2022).

- Polyolefins (PE and PP) have a market share of around 50%, mainly for packaging applications, and PP also for a wide range of more technical applications;
- PVC, with a 10% market share, is the material of choice for many construction applications;
- PET is one of the important packaging materials, accounting for about 8%;
- Polyurethanes (PUR) account for about 8% of the construction market (mostly rigid foams) and the automotive and household markets (mostly flexible foams);
- PS and EPS, with a market share of around 6%, are used in a wide range of applications;

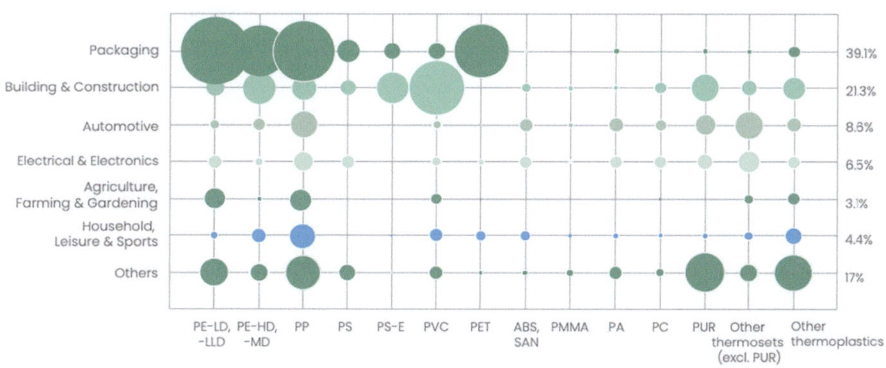

Fig. 5.4 Plastics converter demand in the EU by type and application in 2021. (*Source* PlasticsEurope)

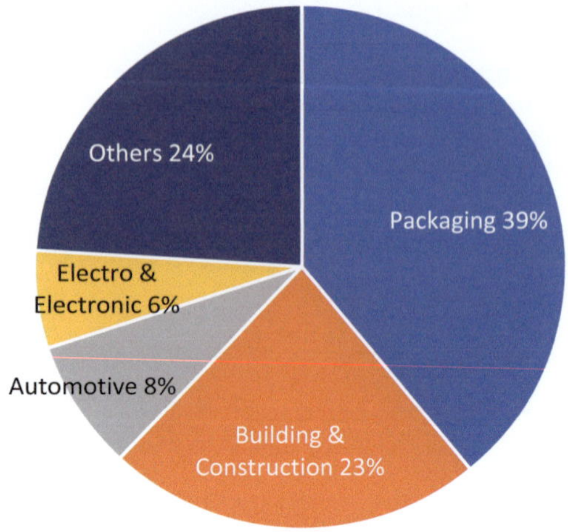

Fig. 5.5 Market sizes in the EU in %. (*Source* Own representation based on PlasticsEurope)

- All other plastics account for a cumulative market share of around 18%.

Broken down to applications or corresponding market sectors, the picture shown in Fig. 5.5 results for the 54 Mt of plastics processed in the EU 27+ 3 in 2022 on average in Europe. 39% of the materials are used in consumer packaging, durables, and industrial goods markets—i.e. to manufacture private consumer packaging or industrial/commercial packaging; 21% are used for construction and infrastructure applications, i.e. in the building and civil engineering industry; 8% are used in the automotive industry; almost 6% are used for manufacturing electrical and electronic equipment; the sports and leisure market, and the agricultural market count for 4% and 3%, respectively; 17% is distributed into a variety of other markets.

For the European countries with higher plastics production, there are comparable data available (cf. Table 5.2).

For the global market, PlasticsEurope (2022) and Conversio (2020) provide figures for the packaging market accounting for 44% of all plastics, having a 42% market share in China and 51% in North America, thus being significantly larger than the respective European market.

5.4 Energy Recovery and Landfill

Tables 5.1 and 5.2 also provide an overview of the plastic waste collected in the world and in the EU and to which of the three treatment processes these wastes are directed to: recycling, energy recovery, or landfill.

The topic of recycling is examined from different perspectives in many chapters of this book, so only a brief classification of energy recovery and landfill will be given here.

Energy recovery of plastic waste takes place primarily in municipal waste incineration plants with the recovery of useful energy (district heating, industrial heat, industrial steam, steam for electricity generation) and in industrial plants such as the cement industry, where plastic waste substitutes fossil raw materials. The calorific value of plastic waste is sometimes higher than that of fossil fuels such as coal or heating oil. The combustion product is CO_2, which acts as a greenhouse gas.

Of the 32 Mt of plastic wastes collected in the EU 27+3 in 2022, 50% was recovered as energy, i.e. incinerated with energy recovery. In the EU countries with a complete or partial landfill ban for high-calorific waste, this proportion is significantly higher, thus in Belgium (59%), Germany (63%), or the Netherlands (62%) (cf. Table 5.2). Finland has the highest share of energy recovery from plastic waste within the EU (77%) (PlasticsEurope 2022).

Globally, the share of energy recovery from collected waste (not from waste as a whole!) was just under 30% in 2018, and around 20% in relation to all wastes, including wastes entering the environment as improper disposal or leakage (cf. Chap. 12) (Conversio 2020).

When it comes to landfilling, a fundamental distinction must be made between sanitary or managed landfill and unmanaged landfill. We deal with the latter in Chap. 12, as waste is released uncontrolled into the environment. However, the transitions are fluid.

In the EU 27+3, around 24% of collected plastic wastes were landfilled in 2022, with a decreasing trend in recent years (PlasticsEurope 2022). The share of plastic wastes landfilled shows a clear gradient within the EU, from central and northern Europe (landfill rate close to 0) to the east, south, and west, where up to 75% are still landfilled (ibid).

Globally, around 43% of collected plastic waste was landfilled in 2018, with around 30% of all plastic waste ending up in sanitary landfills. With almost 75 Mt, the amount of plastic waste landfilled in 2018 was roughly the same as the amount of uncontrolled waste released into the environment (Conversio 2020).

The development of a circular plastics economy in the EU requires a continuous increase in the proportion of recycled plastic waste and a corresponding reduction in the proportion of energy recovery and landfilling, especially. The introduction of landfill bans for high-calorific waste, including plastic waste, in Northern and Central European countries in the 2000s and 2010s has led to a significant increase in recycling in these countries, but also to an increase in energy recovery and a temporary increase in the export of plastic waste. The theory that a restriction on landfilling is a prerequisite for an increase in plastic recycling cannot be empirically proven. Spain, for example, has one of the highest recycling rates in the EU at 40%, but also still has a high proportion of landfilling in plastic waste treatment of 38% (cf. Table 5.2).

With its regulations, the EU Commission is pushing for the sorting of all recyclable or other recoverable components from municipal waste by 2030 and a reduction in landfilling of municipal waste to a maximum of 10% in 2035, considering the very different levels of development of waste management systems across the EU (EU 2018).

A reduction in the proportion of energy recovery from plastic wastes in favor of recycling should also be aimed. A significant part of wastes in the energy recovery stream are difficult or impossible to be recycled mechanically, chemical recycling should be further developed and established.

The future position of energy recovery in a plastics and carbon circular economy can be described as follows: "Incineration with energy recovery is best applied to inseparable or contaminated wastes that cannot be treated through other forms of recycling. Although energy is recovered, this option sits at the bottom of the recirculation pathway hierarchy because it requires the capture of carbon emissions and destroys the material structure and value" (Vidal et al. 2024, p. 46).

In the future, thermal recovery plants will increasingly become the conversion stage for the destruction of environmentally hazardous substances and the removal of pollutants. For example, medical products to be disposed of must be incinerated in a controlled manner for hygienic reasons, as must contaminated pesticide containers and sorting residues for which there is no economically or ecologically viable recycling route. Such non-recyclable quantities of plastic waste must be safely removed from the product cycle and inorganic incineration residues must be safely deposited. The authors estimate that around 10% of all plastic waste will have to be energetically recovered in the future if it is not to be landfilled.

5.5 Export and Import of Plastic Waste

Export of Plastic Waste to Other EU Countries

Not all plastic waste generated domestically can be recovered in the particular EU Member State, whether for capacity reasons, competition reasons, or because special recycling plants, especially for pre-treatment, are required for certain waste fractions. In the latter case, recyclers source their input from several Member States to reach an economic scale in capacity. The latter is the case, for example, for HDPE bale nets from agriculture, whose collection volumes even in large EU agricultural countries are not sufficient to economically utilise a dedicated national recycling plant. Furthermore, an input mix from several EU countries can be useful for the recycler to achieve certain recyclate qualities. Waste exports and imports within the EU internal market, being monitored in accordance with uniform EU law in each Member State, are therefore useful for the development of a European circular economy by securing the corresponding raw materials base. A negative connotation—as sometimes happens in the media—is therefore not appropriate for the intra-EU shipment of plastic waste. The political agreement achieved between the European Parliament and the Council (EU 2023a) on stronger control of waste exports in November 2023 covers both, the intra-EU waste shipment and the export to non-EU countries. This agreement is to be transformed into a new Regulation on waste shipments. This Regulation also aims at a better "circulation of waste for recycling and re-use between

EU Member States, for the EU's transition to a circular economy and the security of supply of raw materials" (ibid).

Export of Plastic Wastes to Non-EU Countries

The situation is different for waste exports to non-EU countries, especially to non-OECD countries. In recent years, reports of improperly recycled plastic waste that ended up in the environment have repeatedly made headlines, especially in Asian recipient countries such as Malaysia, Vietnam, Indonesia, and India. In this regard, the Interpol report of August 2020 discloses the causes, modes of action, and facilitating circumstances of crime around illegal waste trade and treatment (Interpol 2020). Despite the significant decline in exports of plastic waste to non-EU countries since the middle/end of the 2010s, illegal waste exports to non-OECD countries and inadequate treatment of waste there could not be completely prevented and are a cause for concern. In addition to the planned establishment of an EU Circular Economy based on EU waste, this is another reason for preparing a new EU Regulation on waste shipments.

Until 2017 China was the main buyer of plastic waste exports from EU countries. Since the Green Fence measures consistently pursued by China as of the beginning of 2018, preventing the import of non-pretreated waste, the export of plastic waste from EU Member States has decreased significantly. Other Asian countries have also introduced stricter import bans, and EU regulation has also contributed to the reduction in EU exports (EU 2020). The export of plastic waste from the EU to non-EU countries fell from 2.61 Mt in 2016 to 1.14 Mt in 2022 (cf. Table 5.5).

The most important target countries for plastic waste exports in 2021 were Turkey, Malaysia, and Vietnam (cf. Table 5.4).

According to the drafted Regulation on waste shipments, the export of plastic wastes will continue to decrease significantly over the next few years until the export to non-OECD countries is finished. It is planned that two and a half years after adoption, an export ban to non-OECD countries will be imposed on all plastic wastes, including non-hazardous plastic wastes. Only after a further two and a half years

Table 5.4 EU plastic waste exports in 2021. (*Source* Statista 2023)

Target countries	Quantities (kt)
Turkey	394.5
Malaysia	133.4
Vietnam	121.4
UK	101.4
Indonesia	94.3
USA	37.8
South Korea	35.6
India	24.7
China (incl. Hong Kong)	0.7
Total	1.140

Table 5.5 Total quantities of plastic waste exported from Germany to EU and non-EU countries compared with EU exports (*Sources* Destatis 2023; Statista 2023; FAZ 2024)

Year	Exported quantity from Germany to EU and non-EU countries in Mt	Exported quantity from EU to non-EU countries in Mt
2016	1.46	2.61
2017	1.22	2.16
2018	1.07	1.59
2019	1.09	1.52
2020	1.02	1.59
2021	0.82	1.14
2022	0.74	n.a
2023	0.69	n.a

have elapsed interested countries apply to import plastic waste. To do so, they must provide appropriate evidence that they fulfil EU standards for environmental and health protection, to be checked by accredited institutions. The EU Commission is to pay particular attention to the export ban to ensure that plastic waste that has been exported to date is not diverted to EU member states that do not have sufficient capacity to treat it in a sustainable manner (EU 2023b).

Case Study for Germany

Like the overall EU trend, the member state with the largest plastics processing share in the EU (23% of all plastics in the EU are processed in Germany) has also seen a sharp decline in exports. However, the export figures here include both intra-EU waste shipments and exports to non-EU countries.

Exports of plastic waste from Germany halved from about 1.5 Mt/a to about 0.7 Mt/a between 2016 and 2022. The full picture of foreign trade in plastic waste also includes the fact that 0.446 Mt of plastic waste was imported to Germany in 2021, 9% less than during the previous year (DPA 2022).

The main destinations for exports of plastic waste from Germany in 2021 were the Netherlands (0.153 Mt) with Rotterdam as a transshipment centre for sea transport, Turkey (0.092 Mt), and Poland (0,082 Mt) (Destatis 2023). The main exports were industrial films, production waste, and food packaging, whereby food packaging was exported almost exclusively to EU countries.

According to an overview for 2022, 99.9% of plastic packaging waste from the dual systems (separate household collection) is recycled in the EU (84.4% in Germany, 14.5% in other EU countries) with 0.1% going to third countries, mainly Turkey, Switzerland and Serbia (IK 2023).

5.6 Forecasts

For 2050, the UN (2019) expects a world population of almost 10 billion people, all striving for a safe and good life (food, housing, mobility, …). Plastics are indispensable for this, so volumes will increase. In the future, too, the high performance and favourable price of plastics will meet the requirements of consumers and users even more, e.g. for packaging (protection of contents, light, break-proof, transparent, …) or building applications (durables, insulation, corrosion-resistance, …), like no other material does. A linear extrapolation of the current per capita consumption in the developed countries of approx. 150 kg/a to the expected world population would require a production of approx. 1,500 Mt in 2050—i.e. almost a quadrupling of current production volume.

Various scenarios are being discussed for the future development of demand and the required production volumes. These range from an expected or hoped-for decrease in consumption (PEW 2020) to an annual global growth of 5% (Geyer 2017). For Europe and especially for Germany, we expect significantly lower growth rates of 1–2% only, since these markets are saturated. The annual world production of plastics increased from 2 Mt in 1950 to about 400 Mt in 2022 (cf. Table 5.1). This corresponds to an average growth rate of 8.4% per year (Geyer 2017). If a further annual growth rate of 4% is assumed, annual production in 2050 would be at 1.200 Mt.

The cumulative global production of plastics (thermosets, thermoplastics, polyurethanes, elastomers, paints, adhesives, sealants) from 1950 to 2015 was 8,900 Mt (8,300 Mt virgin and 600 Mt recycled). By 2020, about 10,000 Mt were produced in this way. Geyer et al. (2017) expect a cumulative production of about 43,000 Mt by the end of 2050, of which 34,000 Mt virgin and 9,000 Mt recyclates. Of this, 10,000 Mt will still be in use (23%) in 2050, while 33,000 Mt (77%) will have become waste. From today's perspective without large technological and financial efforts, 12,000 Mt of this waste will be released into the environment, a further 12,000 Mt will be incinerated, and 9,000 Mt will be recycled (cf. Tables 5.6 and 5.7).

Plastic Waste in the Environment

The assumptions (see above) and results of Geyer et al. are quite controversial, but they represent the first published approach of this kind and in any case give an idea of the dimension of the challenges society, economy, and industry are facing—and

Table 5.6 Estimated and extrapolated cumulative plastic production and its fate from 1950 to 2020 and 1950 to 2050. (*Source* Own presentation based on Geyer et al. 2017)

Figures in Mt	1950–2020	1950–2050
Cumulative production of which	10,000	43,000
In use	3,000	10,000
Recycled	1,000	9,000
Incinerated	1,000	12,000
In the environment	5,000	12,000

each individual citizen, too. We are running into a gigantic environmental problem, constantly increasing due to our actions, which we will only solve if we no longer see the dissipated plastic in all its manifestations as waste—i.e. as a problem—but as a raw material—i.e. as an opportunity—and make every effort to use this raw material again (KI 2021).

5.7 Data and Sources

The most important source of market data on plastics and the plastics industry is the regular publications of PlasticsEurope (together with EPRO), collected by the PlasticsEurope Market Research Group (PEMRG) in cooperation with market research company Conversio. The results for Europe are published appr. every year as "Plastics the Facts", most recently for 2023 (PlasticsEurope 2024a). A disadvantage here, however, is the consistently varying forms of presentation and the ever-changing range of data presented chosen by the editors of this publication. In addition, PlasticsEurope published to date three in-depth Circular Economy of Plastics reports with slightly changing titles making the orientation difficult. The last one was published in March 2024 with reference year 2022 (PlasticsEurope 2024b).

Market data for Germany—published by actually 15 associations and other organizations of the German plastics industry under the leadership of BKV GmbH in cooperation with Conversio—are published every two years in August/September, most recently in August 2022 (Conversio 2022).

Global data were published for the first time at K 2019; again, the publishers were several plastics industry associations having commissioned Conversio (Conversio 2020). Data was compiled from a total of 44 countries, covering appr. 60% of world population and around 80% of global GDP. Data collection and discussion of results took place in close cooperation with national plastics associations in Asia, Africa, and the Americas. It is hoped that this activity will be continued, as there is no known data collection that is even roughly comparable.

These studies have in common that, in addition to information on the production and use of plastics, they provide clear information on the recycling of plastic waste in a hitherto unique way. A perspective that is becoming increasingly significant in view of globalization and climate change: a globally active plastics industry would do well to provide the public and politicians with comprehensive data on its activities and their effects to accompany and promote the necessary industrial transformation. It is hoped that the PRISM (*Plastics Recovery Insights and Steering Model*) project run by the Alliance to End Plastics Waste (AEPW 2021) will help to generate and make available more and more reliable global data, even though there was little communication only in recent years.

In 2021, the association-like European initiative PolyREC was founded, a non-profit organisation to monitor, verify, and document information on recycled plastics (PolyREC 2022). PolyREC thus is also the official rapporteur of the Circular Plastics Alliance (CPA).

Currently, there is no comparable data available for the American and Asian markets, becoming larger and larger and lagging behind in terms of plastic waste recovery. China, in particular, the largest plastics producer, processor, consumer, and exporter, still is a blank spot on the map in this respect. However, the appearances of Chinese associations at K 2019 (K-Aktuell 2019) suggest that data and strategies are available in China but are not made available internationally to the extent required. At least the China Association of Circular Economy—CACE should be mentioned here, which communicates comprehensive goals and declarations of intent, but little concrete information (CACE 2021). Caused by the COVID-19 pandemic this starting development was sharply interrupted.

Finally, the Polyglobe database (Polyglobe 2021) should be mentioned, an online database published by KI-Verlag, Bad Homburg, Germany (KI 2021), providing comprehensive information on worldwide production capacities of plastics and their most important precursors. It contains daily updated details of around 5,500 plants from more than 2,000 producers at almost 1,100 locations in 79 countries, representing a total capacity of more than 1 Gt of monomers and polymers. The data goes back to 2011 and the outlook covers five years. For each plant, nominal capacity, product, producer, location, and expansion/shutdown plans as well as maintenance data and status (force majeure, curtailment, etc.) are given. The data can be selected, combined, and aggregated interactively via products or product groups, producers, products, regions, or countries. Unfortunately, this database is dealing with virgin plastics only up to now.

5.8 OECD Data

About OECD

The Organization for Economic Co-operation and Development (OECD) is an international organization with 38 member states. Most members belong to high per capita income countries and are considered developed countries. In 2022, the OECD published its report "Global Plastics Outlook: Policy Scenarios to 2060" (OECD 2022b) with data and forecasts on global plastic consumption and environmental pollution from plastic waste. The report outlines how plastic production, use, and pollution will evolve over the forecast period, comparing a business-as-usual scenario with one of ambitious global action. The report looks at the sources and impacts of plastic waste pollution and suggests measures to control or prevent the release of plastic waste into the environment.

The report is tied in with the previous report "Global Plastics Outlook: Economic Drivers, Environmental Impacts and Policy Options" (OECD 2022a), published in February 2022, stating that plastic waste has doubled in the past two decades, mostly due to waste in landfills, burned, or otherwise released into the environment. Based on this report, the United Nations is expected to deliver a legally binding agreement on plastics pollution by the end of 2024 (UNDP 2024). The OECD Global Plastic

Outlook database comprises tires (rubbers) and fibers (plastics), too (cf. Table 5.1), in so far, the term plastic has a slightly different meaning in that context resulting in larger volumes and different conditions than in other sections of this book.

Plastic and Waste Volume Projections

According to OECD (cf. Fig. 5.6) global plastic consumption and thus the market for plastic manufacturers and processors will almost triple from 460 Mt in 2019 to 1,231 Mt in 2060 (Baseline scenario), if current economic practices are maintained. The reasons for this lie in population growth and in global economic growth, quantified as gross domestic product (GDP). The biggest growth is expected in the emerging countries of Africa (Sub-Saharan) and Asia. The use of recycled (secondary) plastics is growing faster than that of virgin material (fossil-based) but will only reach a share of around 12% in 2060 (2019: 6%). In absolute figures, this means that the need or consumption of secondary materials will increase fivefold from 30 Mt in 2019 to 150 Mt in 2060.

The largest consumers of plastics are the 38 OECD countries of the "Global North" and China, with a combined 66% of global consumption, with the packaging (31%), construction (17%) and transport (12%) markets being the main buyers.

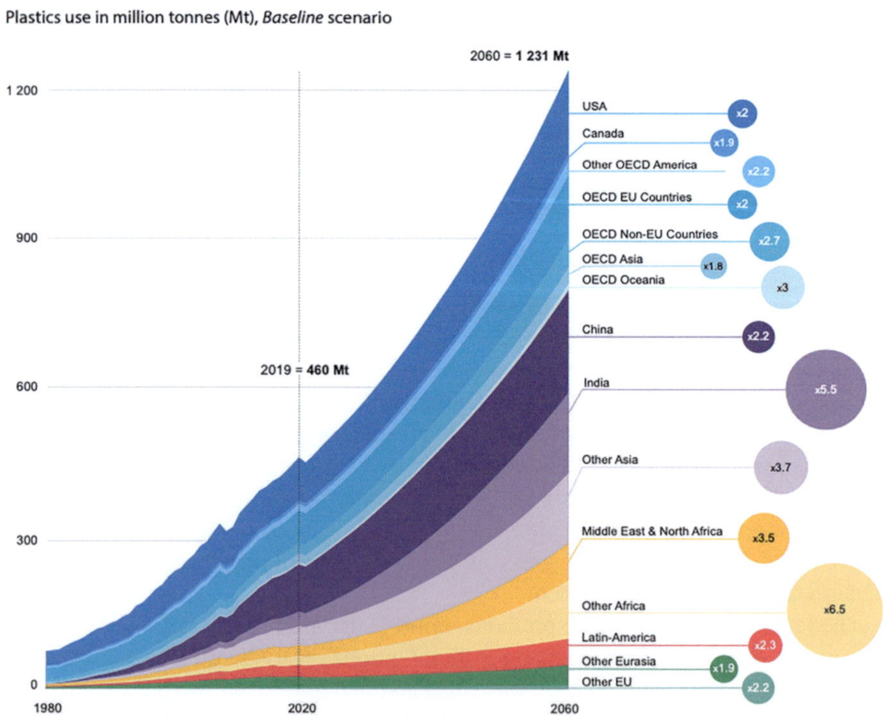

Fig. 5.6 OECD global plastics outlook 2060 (Plastics use in Mt, Baseline scenario). *Source* OECD, Global plastic outlook, Policy scenarios to 2060)

OECD data show a significant correlation between GDP (value of goods and services produced in a region), plastic demand, and plastic waste. Economic growth and prosperity development in a region go hand in hand with the need for plastic. In the OECD regions EUR and USA, the percentage share of world GDP will fall from 2019 to 2060 (forecast for the baseline scenario), with the demand for plastic "only" doubling. India will more than double its share of GDP over the same period, while demand for plastics will increase fivefold. By 2060, India will catch up with the US and China in terms of volume and overtake the OECD EUR countries. Also noteworthy is the "Other Africa" region, which is not listed in the table, with plastic demand growing by a factor of 6.5 in the reporting period. The volume of plastic demand in this region will grow to a level comparable to the OECD-EUR-region.

OECD characterizes this business-as-usual outlook as unsustainable. Recycled (secondary) plastics are projected to grow quicker than virgin (primary) plastics and double their market share to 12% of total plastics use in 2060, indeed. But plastic waste is expected to almost triple, with half of all plastic waste produced still going to landfill and <20% be recycled. Improvements in waste management partially mitigate increases in the amount of mismanaged waste, which however almost doubles. Plastic leakage to the environment is going to double to 44 Mt a year and plastic accumulated in maritime environment will more than triple, from 140 Mt in 2019 to 493 Mt in 2060 (cf. Fig. 5.7). Microplastic leakage is projected to increase in all regions, highlighting the need for better mitigation solutions.

Under these conditions, Greenhouse Gas emissions (GHG) from plastics lifecycle are projected to more than double, to 4.3 Gt CO_2eq, together with other lifecycle impacts.

OECD therefore asks for a comprehensive policy considering the whole plastics lifecycle, comprising shared objectives and co-ordinated efforts at the global level together with policy measures targeting all phases of the plastic lifecycle on the countries' level, with the following pillars:

(1) **Reduce the quantity of new plastic** by restraining plastic demand and enhancing circularity. Relevant policies should assess plastic taxes to reduce the most polluting types of plastic, incentivise circularity by decreasing the relative

Table 5.7 GDP, plastic consumption, and plastic waste in selected regions and countries 2019/2060. (*Source* OECD 2022a)

		World	EUR[1]	USA	China	India
GDP	2019	100%	13%	15%	20%	8%
	2060	100%	9%	11%	18%	18%
Market (Mt)	2019	460	64	84	94	29
	2060	1,231	125	166	203	160
Wastes (Mt)	2019	353	51	73	65	19
	2060	1,014	100	142	183	128

[1] The OECD-EUR region comprises the EU member states (except BG, CY, HR, LU, MT, and RO) and CH, GB, IL, IS, NO, and TR

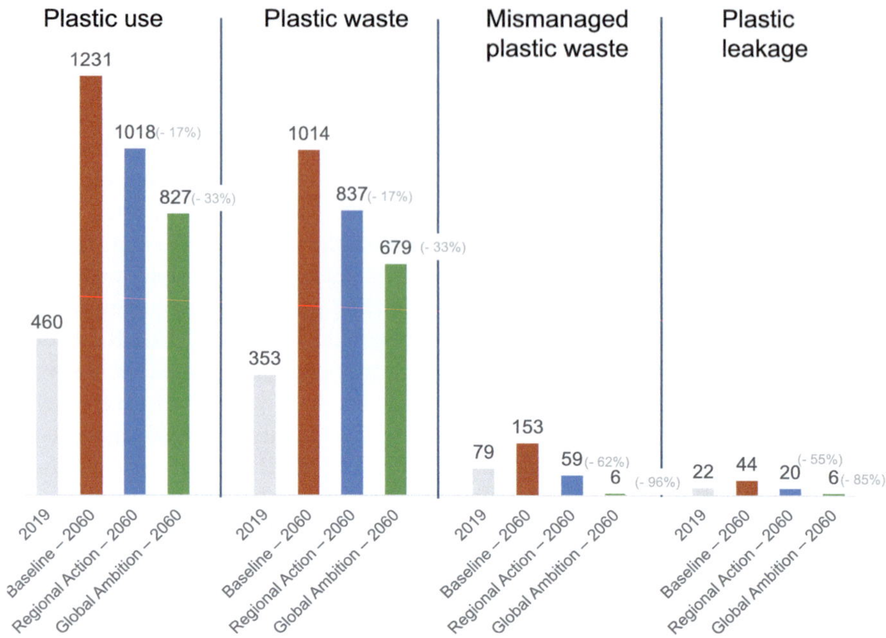

Fig. 5.7 Combining policies that target different lifecycle stages can drastically reduce plastics leakage to the environment. Yearly value in Mt, percentage change compared to Baseline. (*Source* OECD 2022b)

price of secondary plastics, and foster the design of plastics more durable and repairable to extend product lifespans (*Plastics use* cf. Fig. 5.7).

(2) **Enhance recycling** to ensure that unavoidable increase in plastic demand is met by inter alia higher recycling rates through recycled content targets, Extended Producer Responsibility (EPR) schemes for packaging and products such as electronics, vehicles, and clothing, or recycling rate targets (*Plastics waste* cf. Fig. 5.7).

(3) **Close leakage pathways** through investments in waste collection, sanitary landfills, and improved collection of litter to adequately manage generated plastic waste and prevent it from ending up in the environment (*Plastics leakage* cf. Fig. 5.7).

(4) **Regional Action** should target at halving plastic leakage by 2060 (compared to *baseline*), to increase the market-share of recycled plastics to 29%, and to dip mismanaged waste below 2019 levels (largely in non-OECD countries) (cf. Table 5.8).

(5) **Global Ambition** should aim at reducing plastic leakage to near zero by 2060, to have nearly 60% of waste recycled and the market share of recycled plastics surged to 41% (cf. Table 5.8).

Table 5.8 Share of recycled plastics in 2060 for three scenarios as compared to the situation in 2019 (*Source* Own presentation based on OECD data)

Year/Scenarios	Baseline	Regional action	Global ambition
2019	6	–	–
2060	12	29	41

References

AEPW 2021: https://endplasticwaste.org/-/media/Project/AEPW/Alliance/Media-Page/AllianceInAction_ProgressReport_2021.pdf?rev=b6abf6408ca84fcdb03d1ffdb51cfa85 (last access: Nov. 12, 2021)
CACE 2021 : http://en.chinacace.or (last access: Nov. 30, 2021)
Conversio 2020: https://www.bkv-gmbh.de/studien/studie-global-plastics-flow-2018-conversio.html (last access: Nov. 18, 2023)
Conversio 2022: https://www.bvse.de/dateien2020/2-PDF/01-Nachrichten/03-Kunststoff/2022/Kurzfassung_Stoffstrombild_2021_13102022_1_.pdf (last access: Nov. 18, 2023)
DESTATIS 2023: https://www.destatis.de/DE/Presse/Pressemitteilungen/Zahl-der-Woche/2023/PD23_23_p002.html (last access: Dec. 8, 2023)
DPA 2022: dpa-infocom, dpa:220109–99–640345/6
EU 2018: Directive (EU) 2018/850 of 30 May 2018 amending Directive 1999/31/EC on the landfill of waste. https://eur-lex.europa.eu/legal-content/EN/TXT/HTML/?uri=CELEX:32018L0850
EU 2020: Commission Delegated Regulation (EU) 2020/2174 of 19 October 2020 amending Annexes IC, III, IIIA, IV, V, VII and VIII to Regulation (EC) No 1013/2006 of the European Parliament and of the Council on shipments of waste. https://eur-lex.europa.eu/legal-content/DE/ALL/?uri=CELEX%3A32020R2174
EU 2023a: https://www.consilium.europa.eu/en/press/press-releases/2023/11/17/waste-shipments-council-and-parliament-reach-agreement-on-more-efficient-and-updated-rules/
EU 2023b: Council of the European Union: Proposal for a Regulation of the European Parliament and of the Council on shipments of waste and amending Regulations (EU) No 1257/2013 and (EU) No 2020/1056. ST-16528–2023-INIT_en.pdf
FAZ 2024: Deutsche Exporte von Plastikmüll auf dem Tiefstand. (German exports of plastic waste at an all-time low). FAZ 15.02.2024
Geyer 2017: R. Geyer, J. R. Jambeck, K. L. Law, Production, use, and fate of all plastics ever made. Sci. Adv. 3, e1700782 (2017) (last access: Nov. 18, 2023)
IK 2023: https://newsroom.kunststoffverpackungen.de/2023/12/04/recycling-rekord-kunststoffverpackungen/
Interpol 2020: Strategic Analysis Report. Emerging criminal trends in the global waste market since January 2018. https://www.interpol.int/en/News-and-Events/News/2020/INTERPOL-report-alerts-to-sharp-rise-in-plastic-waste-crime
K-Aktuell 2019: https://www.k-profi.de/ausgaben/pdf/k2019/K-AKTUELL_2019_AUSGABE_4.pdf (last access: Aug. 06, 2021)
KI 2021: https://www.kiweb.de/ (last access: Feb. 03, 2021)
Plastics Country 2021: https://www.kunststoffland-nrw.de/fileadmin/Downloads/NRW-Report/NRW_report_2021_01_HP_ds_1.pdf, p.40
OECD 2022a: Organization for Economic Co-operation and Development: Global Plastics Outlook: Economic Drivers, Environmental Impacts and Policy Options. https://www.oecd-ilibrary.org/environment/global-plastics-outlook_de747aef-en (last access: Feb. 26, 2024)
OECD 2022b: Organization for Economic Co-operation and Development: Global Plastics Outlook: Policy Scenarios to 2060

PEW 2020: https://www.pewtrusts.org/en/research-and-analysis/articles/2020/07/23/breaking-the-plastic-wave-top-findings (last access: Nov. 18, 2023)
PlasticsEurope 2022: https://plasticseurope.org/knowledge-hub/plastics-the-facts-2022/
PlasticsEurope 2024a: Plastics - the fast facts 2023. https://plasticseurope.org/de/resources/market-data/ (last access: Apr. 15, 2024)
PlasticsEurope 2024b: The Circular Economy for Plastics. A European analysis. March 2024. https://plasticseurope.org/knowledge-hub/the-circular-economy-for-plastics-a-european-analysis-2024/
PlasticsEurope 2024c: National Infographics: Circular Economy of Plastics 2024. plasticseurope/resources/publications
Polyglobe 2021: www.polyglobe.net (last access: Jan. 29, 2022)
PolyREC 2022: https://www.polyrec.eu/ (last access: Mar. 31, 2022)
Statista 2023: https://www.statista.com/statistics/1235938/annual-plastic-waste-exports-destinations-european-union/ (last access: Dec. 8, 2023)
UNDP 2024: https://www.undp.org/blog/beginning-end-plastics-pollution
UNO 2019: https://population.un.org/wpp/Download/Probabilistic/Population/ (last access: Nov. 18, 2023)
Vidal et al 2024: Fernando Vidal, Eva R. van der Marel, Ryan W. F. Kerr, Caitlin McElroy, Nadia Schroeder, Celia Mitchell, Gloria Rosetto, Thomas T. D. Chen, Richard M. Bailey, Cameron Hepburn, Catherine Redgwell & Charlotte K. Williams: Designing a circular carbon and plastics economy for a sustainable future. *Nature* **626**, 45–57 (2024). https://doi.org/10.1038/s41586-023-06939-z

Dr. Jürgen Bruder is a chemist with an additional degree in business. He worked in higher education and the chemical industry, before becoming Director General of IK, the German plastic packaging association. He held positions in national, European, and global committees including the Global Plastics Alliance. Today he deals with the transformation of the plastics industry to a circular economy focusing on plastic packaging and plastic applications in agriculture.

Manfred Rink is a mechanical engineer; his professional path led him from an automotive supplier to one of the large plastics manufacturers where he held various management positions. Most recently, he dealt with future trends in the plastics industry, risks, and opportunities. Today, he works in the fields of sustainability and circular economy in the plastics industry.

Dr. Peter Orth is a chemist by education; with OPC—Orth Plastics Consulting he leads a group of senior consultants with decades of professional experience in leading positions in the German and European plastics industry and its associations. His main focus is on innovation and sustainability in cooperation with science, industry, and politics. He is currently working in the fields of CO_2 as a feedstock, resource efficiency, and recycling technologies.

Part II
Plastics Circular Economy

In a circular economy, all actors in the circular value chain must master their specific challenges resulting from their position in the cycle and communicate and cooperate with upstream and downstream stages at the interfaces in such a way that their material, product, or service meets all the requirements of the cycle. The structure of Part II of this book is shaped by this principle as presented in the circuit diagram.

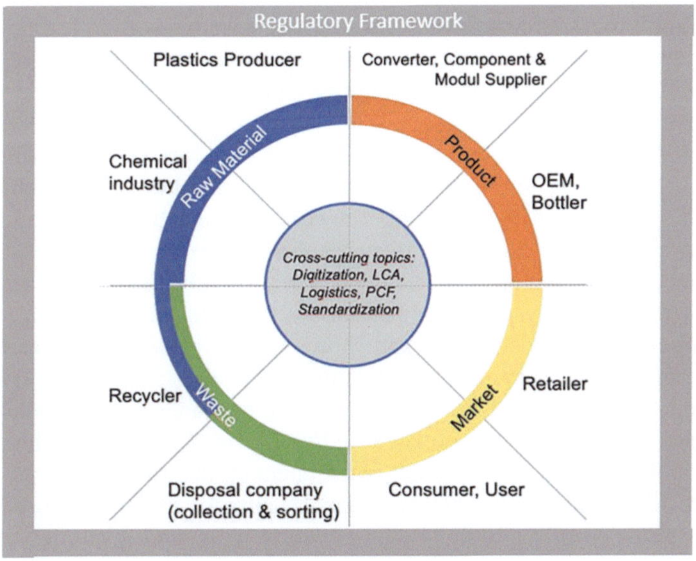

Chapter 6 is dedicated to systemic approaches that examine the collaboration of various stakeholders at the interfaces of the cycle segments and beyond and examine it from different perspectives. The results and experiences of a moderated 3-year dialogue process on the transformation from a linear to a circular plastics economy are included, as is the development of design guidelines, which represent an indispensable common platform for the stakeholders. Since the sustainability and recyclability of a product are largely determined by its design, the content and the future relevance of various design concepts are presented and discussed.

Chapters 7 (product), 8 (market), 9 (waste) and 10 (raw material) deal with the specific challenges for the players in the respective segments of the plastics cycle. The schematic representation of the cycle illustrates the sequence of the mentioned segments from product to market, from market to waste, from waste to raw material, and finally from raw material back to product.

Chapter 7 on the product segment, in contrast to other common representations, intentionally opens up the representation of the cycle, since the product is the essential object of all economic activities and since its design by processors, OEMs or bottlers—expressed in Design for Circularity or the use of recyclates—significantly determines the structure and processes of the value chain circle.

Chapter 8 deals with the market segment based on the extensive experiences of trading companies with packaging circuits for fast-moving consumer goods (FMCG). Of particular importance is environmental communication between retailers and their customers, the consumers.

Chapter 9 on the waste segment deals with waste logistics and sorting. Extensive pre-sorting by the user who determines the end of the product's life, efficient structures and processes in waste logistics, and innovative sorting technologies are the prerequisites to produce recyclates/secondary raw materials that can replace new materials of defined quality and availability and thus contribute to the feedstock change.

Chapter 10 on the raw materials segment closes the loop. The technological principles and challenges for producing specified plastic recyclates from mechanical and chemical recycling are presented here. The case study of the corresponding operations of a well-known plastics manufacturer shows how the gradual substitution of fossil-based plastics with recyclates and other circular plastics is implemented.

Chapter 11 deals with specific collaborations between stakeholders from various circular segments discussed above and shows how Expanded Producer Responsibility systems (EPR) work. Examples from various plastics markets (packaging, agriculture, construction and furniture industry) show concretely how plastic cycles are successfully closed, often initiated and/or financed by the stakeholders on a voluntary and collective basis without government regulation.

Chapter 12 is dedicated to plastics leakage into the environment, particularly the marine litter problem. It shows causes, entry paths and quantities, as well as ecological and other impacts, and finally measures to reduce pollution, including global regulation within the framework of a UN Plastics Treaty. The quantities of plastic not recycled today represent enormous potential for a future global circular economy of plastics.

Chapter 6
Systemic Approaches

Bita Fesidis, Volker Brennecke, Klaus Wittstock, Victoria Wessolowski, Martin Beeh, Najine Ameli, Oliver Stengel, and Jürgen Bruder

Contents

B. Fesidis (✉)
FOM Hochschule, Düsseldorf, Germany
e-mail: Fesidis@vdi.de

V. Brennecke
Meerbusch, Germany
e-mail: volker.brennecke@gmx.de

K. Wittstock (✉) · V. Wessolowski
BASF SE, Ludwigshafen am Rhein, Germany
e-mail: klaus.wittstock@basf.com

V. Wessolowski
e-mail: victoria.wessolowski@basf.com

M. Beeh (✉)
beeh_innovation, Cologne, Germany
e-mail: martin.beeh@beeh-innovation.com

N. Ameli
phase4:Institut gGmbH i.G., Bochum, Germany
e-mail: n.ameli@phase4-institut.de

O. Stengel
Hochschule Bochum, Bochum, Germany
e-mail: oliver.stengel@bochum.de

J. Bruder (✉)
BPC- Bruder Plastics Consulting, Potsdam, Germany
e-mail: plastics@jbruder.de

P. Orth et al. (eds.), *Circular Economy of Plastics*,
https://doi.org/10.1007/978-3-031-78829-1_6

Summary

The opening chapter of Part II "Circular Economy" focuses on new forms of cooperation and communication between the players at different stages of the circular economy, including policymakers. Solving cross-stage challenges in the establishment and development of a circular economy requires overcoming economic, cultural, and language barriers between the players in the various stages. Suitable tools for this include moderated stakeholder processes or joint design concepts underpinned by parameters or instructions for action. Both tools are presented here based on concrete experience reports or existing and practiced design concepts. The case studies and design concepts presented in the articles each use a systemic approach to discuss or solve the respective challenges.

Section 6.1 focuses on the case study of a three-year moderated stakeholder process in Germany on the transformation from a linear to a circular plastics industry. The experiences and methodology of the multi-stage dialogue process with representatives from industry, science, politics, and society provide important insights into how hurdles and conflicting goals can be minimized to close the loop.

Section 6.2 reflects the design possibilities of plastics manufacturers in the context of the overarching design system. The comprehensive overview of different design concepts, their advantages and disadvantages, including their increasing complexity, leads to the conclusion of the need for their harmonization and institutionalization. The analysis of the relationship between design and regulation identifies an increasing trend away from the previous focus on Design for Recycling (DfR) towards more comprehensive design concepts.

The following Sects. 6.3 and 6.4 deal with design concepts in different plastic application areas from different perspectives.

Section 6.3 discusses a number of issues relating to greater consideration of design for circularity from the perspective of an industrial designer for a wide range of consumer goods such as household appliances, furniture and cars. Using concrete product examples, questions are discussed such as: What role can industrial design play in the context of sustainability? What is design for circularity with plastics? What is good design (for circularity)? How can leap innovations be achieved in the field of design?

Section 6.4 takes a closer look at two design concepts that play an important role in the cycle stages of plastic packaging, Eco-design and Design for Recycling (DfR). The management guide on Eco-design of Plastics Packaging, developed as a result of a multi-year moderated stakeholder process with representatives from the packaging chain (packaging manufacturers, brand owners, retailers, recyclers), academia and society, defines Eco-design of plastics packaging not only as a design concept for reducing all environmental impacts but also as a management tool. As an overarching design concept focused on plastic packaging, the comprehensive Eco-design integrates here four design concepts geared towards sustainability, one of which is the Design for Recycling (DfR). The concept of DfR is explained with references to definitions, and many existing guidelines up to the draft of the PPWR, whereby the duality of packaging design and the existence of a sorting and recycling infrastructure is discussed.

6.1 Circular Economy—Dialogue and Collaboration as a Solution Approach[1]

Bita Fesidis and Volker Brennecke

6.1.1 Introduction

In terms of the circular economy, the waste concept is a thing of the past. A circular economy is an economy in which waste becomes resources. The main driver of the transformation—away from the linear model of the "throwaway economy" towards the circular economy—is the European Commission's Green Deal, with the goal of complete climate neutrality by 2050. Crises such as the pandemic and the Russian war against Ukraine have an industrial policy relevance for European energy and raw material independence. A circular economy can make an enormous contribution to becoming more independent of resources and reducing CO_2 emissions.

Plastics are one of the most relevant material flows for circular value creation. Because of their wide range of functions, they are relevant for almost all industries (e.g. medicine, construction, automotive, packaging industry, electrical engineering). Plastics also play an important role in innovative sustainable solutions when it comes

[1] Modified version of Chap. "12. Circular Economy in der Kunststoffindustrie—Dialog und Kollaboration als Lösungsansatz" in Fesidis, B., Röss, S.A., Rummel, S. (editors) "Mit Digitalisierung und Nachhaltigkeit zum klimaneutralen Unternehmen". Springer 2023.

to materials for solar modules, insulation, or lightweight construction. Plastics are important as "enablers" for sustainable solutions and must themselves be sustainable.

Awareness of the necessary transformation of the plastics industry towards a circular economy is increasing. The German chemical industry has set itself the goal of completely recycling the carbon used and being greenhouse gas neutral by 2050 (VCI 2020). However, there is still a long way to go, and fundamental changes are required, as around 400 Mt of plastics were produced worldwide in 2022, most of which have not yet found their way back into the material cycle. Even if recycling rates are high in Germany, where, for example, approx. 99% of plastic waste is recovered, of which around 47% is mechanically recycled and almost 53% is incinerated with energy recovery (Conversio 2022), the actual closure of the cycle is lagging. Only around 12% of new plastic is replaced by recycled materials (cf. Chap. 5).

One reason for the low substitution rates is the classic orientation of political regulation towards waste management. The term circular economy was first defined in the German Circular Economy Act 2012 as "avoidance and recycling of waste" (Section 3 No. 19 KrWG). In practice, this developed into a recycling-oriented waste management system that focused on defined recycling quotas and thus on end-of-life solutions. Aspects such as extending the use phase or increasing the use of carbon stocks in a circular economy are ignored with such a narrow understanding (Acatech 2021). The literature describes, although not always homogeneously, various R strategies that support a broader understanding of the circular economy. The 10 Rs defined by Potting can be summarized in three pillars (Potting et al. 2017). The first pillar concerns "smart product use and production" (Refuse, Rethink, Reduce), the second concerns "increasing the lifespan of products or product components" (Reuse, Repair, Refurbish, Remanufacture, Repurpose), the third pillar concerns, if possible, "long-term use of materials" (Recycle, Recover). However, the plastics industry has some catching up to do when it comes to contributing to these principles.

6.1.2 Challenges for the Transformation to a Plastic Circular Economy

The plastics industry's current efforts to create a circular economy and the speed of action are not sufficient to achieve the goals of the European Green Deal or the Paris and Glasgow climate agreements. A study by SYSTEMIQ shows that if the measures decided by around 2020/2021 are implemented, the European plastics system will only achieve a recyclability of 33% by 2050 (SYSTEMIQ/Plastics Europe 2022). At the same time, several scenarios are drawn up illustrating the potential of the circular economy. In a SYSTEMIQ scenario in which all levers of the circular economy are used, the European plastics economy will achieve a circularity of 78% by 2050. Greenhouse gas emissions are reduced by over 60% in this scenario. The levers include different measures, such as elimination, reuse, substitution, design for recycling as well as investments in collection and recycling and exploiting the potential of mechanical and chemical recycling to better close cycles. The scenario assumes investments of approximately € 160–180 billion between 2020 and 2050, significant changes in consumer behavior, and political framework conditions that offer incentives for investments and innovations.

To follow the path of such a positive scenario, one must understand the status quo and the current challenges in the plastics industry. Some particularly relevant and cross-cycle hurdles are addressed below. They were worked out in the VDI's Round Table Circular Economy for Plastics, in which the authors of the article played a key role (VDI 2022a).

Product design

The transformation to a circular economy requires a rethink in design and thinking in material flows instead of individual products; this affects not only the product manufacturers (distributors) but all circular partners. A holistic view is required in which the origin, properties, and possible uses of the material are considered as well as information and awareness of the consumer right through to the recycling options (cf. Sect. 6.2.4) (VDI 2022a).

Competitiveness of recyclates compared to virgin materials

Recycled materials are often not competitive with virgin materials in terms of price, quality, and availability. In relation to the price, the effort that must be made for the use of recyclate—e.g. in the form of collection, sorting, recycling, adjustments in plastic processing—is high, especially in comparison to virgin material based on oil (with volatile crude oil prices). In addition, supply and demand often do not match regarding the quality requirements for recyclates and their availability. Recycled materials suitable for downcycling, i.e. for applications with lower quality requirements, are usually available in sufficient quantities. However, this is rarely the case for high-quality recyclates suitable for applications with high-quality requirements. However, their constant availability is usually a requirement for the customer. Losses in quality are rarely accepted—in some cases, this is not compatible with legal regulations (cf. Chap. 13).

Technological challenges and conflicting goals

Much plastic waste still cannot be technically recycled, or current technologies are reaching their limits when it comes to developing high-quality recyclates. Technological challenges can be overcome, for example, through changes in product design, improved sorting and collection systems, or new recycling techniques. However, ecological, and economic goal conflicts often arise, e.g. due to increased energy and raw material consumption or changed product performance. In this context, the availability of cost-effective renewable energies is a prerequisite for exploiting technology potential (cf. Chap. 10).

Existing specifications and regulations on product safety as well as approval procedures for new materials (e.g. specifications from the European Food Safety Authority EFSA) currently make the use of recyclates more difficult. From the perspective of those involved in the cycle, adjustments are required in this regard. At the same time, standardization and standard setting are a prerequisite for optimizing the information and data flows in the cycle (cf. Chap. 19).

Public communication/change in social values

To understand waste as a raw material, a changed awareness in society is also needed. This includes awareness of how to use products responsibly (e.g. repairability, upgrading), empowering consumers to make the right sustainable product decisions or responsibly dealing with their waste (cf. Chap. 8).

The challenges described make it clear that the transformation to circular value creation is a complex task. We have widespread agreement that there is a need for political action to set the right framework conditions for the transformation (VDI 2022a, Acatech 2021, PF I4.0 2021). In addition, it is often emphasized in the literature that transformation can only be successful if it is systemically linked to digitalization (RNE 2021, Wilts and Berg 2017, VDMA 2019).

The need for political action will be discussed in more detail below and in Chaps. 21 and 22, on digitalization please see Chap. 20.

6.1.3 A Systemic Perspective

When developing solutions for the circular economy, the different cycle stages—from raw material production to recycling—must be considered. The effects of political measures, for example, must be considered holistically, i.e. with a view to the entire cycle. The circular economy should therefore be understood as a system in which the various actors, industries, and their value chains operate and interact. Activities at one level can affect other levels and the overall system.

From such a systemic perspective, the VDI, together with representatives from all stages of the cycle (chemical industry and plastics manufacturers, plastics processors, OEMs and other plastics users, trade, consumers, logistics and disposal, and recyclers eventually developed a common image that depicts the circular economy for plastics. The circuit diagram developed jointly serves to provide clarity and reduce complexity. In practice, these are often highly interconnected, global supply networks across and at all levels (cf. Fig. 1.1).

The systemic view of the actors and their actions in the cycle requires a rethinking of the parties involved. Traditionally, each actor optimizes its own value chain, interacts with directly upstream and downstream actors, and acts primarily based on (business) economic interests.

However, new circular solutions must be developed in collaboration with all relevant actors. This requires exchange with value creation stages that are further away. For example, product designers (segment: OEM/user) must consider the value of the product at the end of its life during product development and think less in terms of products and more about material flows.

For this purpose, the exchange with the stages of disposal and recycling is essential. Recyclers become raw material suppliers with whom specifications must be agreed upon. When it comes to exchanges with more distant levels of the value cycle, there is often a lack of common language between the actors or an understanding of the logic of action in the other industry.

In searching for solutions, the actors have different interests. (Business) economic interests usually have priority and must be served. Conflicts of interest arise that inhibit the implementation of a circular economy. These are mostly conflicts between economic and ecological goals.

The different interests and perspectives are also described in the context of innovation research. Design thinking emphasizes three components that must be combined for successful innovations and valuable problem solutions (Brown 2016):

- (technological) feasibility,
- (economic) marketability and
- human desires.

This concept is also useful for circular economy (and sustainability) solutions but must be expanded to include the perspective of society. The starting point should not just be people and their benefits—beyond the individual, the view of social, ecological, and social impact must be added (cf. Fig. 6.1).

Bringing a social and user-centered perspective into harmony with requirements for technological feasibility and economic viability is already challenging from the individual perspective of a company. From a system perspective, it becomes even more complex: measures that seem sensible from the perspective of a company

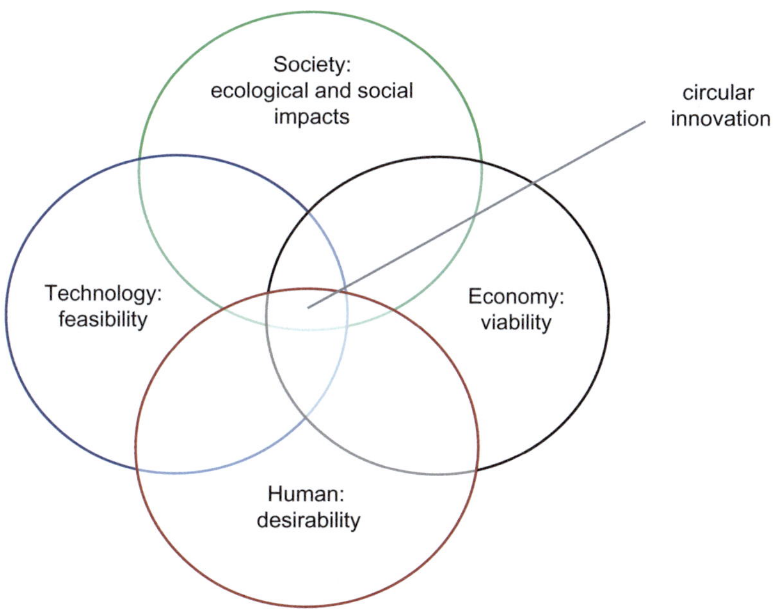

Fig. 6.1 Circular (sustainable) innovation (*Source* Own presentation)

or a cycle stage can have negative effects on the entire system. For example, the ecologically best solution from the point of view of a cycle stage does not necessarily have to be the best solution from the point of view of the entire system. When finding a solution, a holistic perspective on system effects must therefore be taken, requiring all relevant stakeholders to be included.

6.1.4 The VDI Round Table—A Case Study

With the Round Table "Circular Economy in the Plastics Industry", the VDI conducted a dialogue with politics, business, science, and civil society and discussed how the systemic change from the linear to the circular economy in the field of plastics can succeed (VDI 2022a). The dialogue process lasted almost 3 years (2019–2022). Methodologically, the process was based on agile approaches guided by overarching goals. At the same time, the dialogues provided space for new insights that pointed the way for the further process. The dialogues were result-oriented and exploratory. For example, new insights into barriers to the circular economy or necessary areas of action were gained. In contrast to agile approaches, which are purely user-centered (in the sense of a "user-centered" design), the focus in this process was on the selected stakeholders from industry, politics, science, and society (NGOs). The first phase of the project was initially about building a common understanding of the problem from the perspective of the various stakeholders, sharing this understanding, and determining goals together. In the subsequent second Phase, solution ideas were developed.

In Phase 1, the right participants were identified before the dialogues began, for which the VDI brought together representatives from all cycle stages of the plastics industry, NGOs, and science. Pioneers with established market relevance were selected from the industry for each level. Guests from relevant Federal Ministries and the Federal Environment Agency were also included. First, a broad understanding of the problem was built up—in this Phase 1.1 "Understanding", the focus was on divergence (cf. Fig. 6.2).

For this purpose, the dialogue was moderated over several round table sessions, the problems and needs of the participants regarding the transformation to a circular economy were identified, shared, and discussed in the group. The result was a common understanding of the problem across the board and an understanding of the individual interests and logic of action at individual stages. Building on this, an in-depth understanding of the problem was developed in group discussions on individual aspects, for which the potential for closing the cycle and existing hurdles were discussed with the respective cycle stages.

Results were then condensed, consolidated, and shared with all those involved—In Phase 1.2 "Focusing", the focus was on convergence (cf. Fig. 6.2).

In this way, the participants' mutual understanding could be deepened and knowledge deficits at the individual levels about value creation processes and action patterns at other levels could be uncovered. The result of the first Phase (understanding and

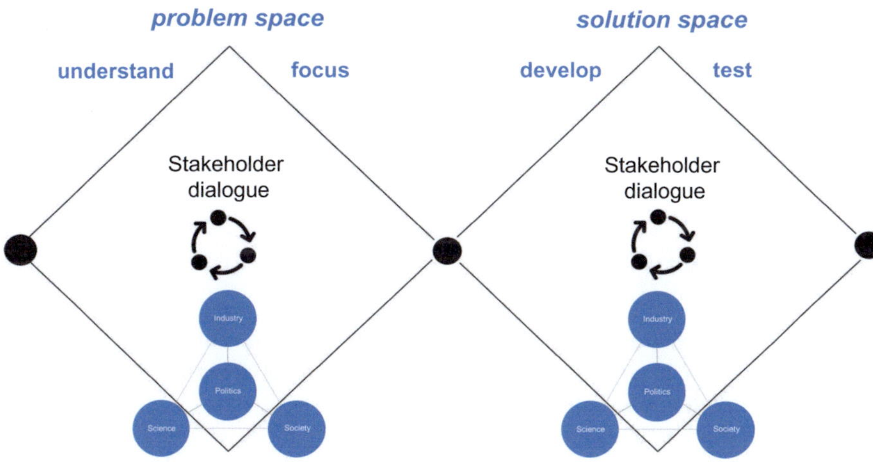

Fig. 6.2 Dialogue process for social transformation projects (*Source* Own presentation)

focusing) was, among other things, agreement on common goals and fields of action and a mutual understanding of the roles and value-creation logic of the circular actors. The approach helped to build trust between participants.

In Phase 2, the solution space was considered. Based on the results in Phase 1, one focus here was on the necessary regulatory incentives. Possible political steering options were initially identified in the dialogue. In addition to the recycled content rate for product groups, which was also discussed at the EU level, the polymer-specific substitution rate, which had previously been less discussed, was also considered. This type of quota does not apply to a product group, but rather to the material plastic. Plastic producers would be obliged to generate a certain minimum proportion of the plastics they sell on the market from non-fossil raw materials (VDI 2022a, AGVU/IK 2021). Further control options were considered in the form of economic control instruments (e.g. CO_2 pricing, new product pricing, fund models).

The selected steering instruments were then discussed in depth from a systemic perspective. The advantages and disadvantages of each steering instrument were discussed from the perspective of the representatives of the cycle stages and the overall effects on the system were discussed in dialogue with everyone involved. To conduct the discussion about the respective steering options objectively and based on facts, a catalog of evaluation criteria was developed. This was based on ecological and economic as well as technical and regulatory criteria and underlying key questions (VDI 2022a). This approach was helpful in identifying different objectives and conflicting goals from a systemic perspective.

In addition to the need for regulatory action, further recommendations for action were developed in Phase 2. For this purpose, the work results of the areas of action identified in Phase 1 were summarized in writing, iteratively refined together with the participants, and conclusions and recommendations were discussed in dialogue.

The design of the solution approaches would be conceivable by continuing the methodology using various methods, be it simulation games, scenario techniques or road mapping. Testing, e.g. in real laboratories, in which regulatory adjustments can also be tested regarding their feasibility, would also be conceivable and recommended.

6.1.5 Reflection of the Methodology

The procedure described above in the case study illustrates how dialogue can be used as a productive tool to solve transformation challenges collaboratively. It enables a new culture of dialogue to solve complex problems.

When reflecting on the approach with a view to innovation research on the circular economy and the stakeholder integration discussed there, it is noticeable that there are hardly any approaches that see dialogues as tools and deepen their design. Nevertheless, there are points of connection. For example, Eisenreich et al. (2021) developed a network approach for circular innovations as a result of a study on open innovation and circular economy. Regarding the involvement of politics in the innovation process, a mixed picture emerges from the companies' perspective: on the one hand, companies support direct cooperation with government and politics, as these are crucial for systemic solutions. On the other hand, the involvement of politicians and regulators in open innovation processes is considered difficult because they have a different way of thinking and cooperation requires a longer time horizon. This view illustrates the current challenges and the need for a new culture of dialogue that starts early and is productive.

Other approaches that deal with collaboration in the circular economy come from research on innovation ecosystems. Konietzko et al. (2021) derive initial principles for innovations in circular ecosystems from a study. The selection of partners in the ecosystem is seen as critical to the success of the collaboration. What is crucial is bringing together new and previously unconnected actors as well as a mix of actors from business, research, politics, and civil society. This combination also proved to be useful in the case study presented. In addition, it illustrates the relevance of pioneers (regarding companies) and thought leaders (regarding participating individuals). In relation to the individual characteristics of the participants, openness to dialogue as well as exploratory and systemic thinking skills were crucial for success.

Other aspects critical to success emphasized by the authors of the above-mentioned study are building and maintaining trust (against the background of differences of opinion between the actors) and the reconciliation of individual and common interests. The development of a common vision, strategy, and goal setting is considered useful—assumptions that were confirmed in the case study. It also became clear how dialogue can serve as a means of developing common goals, building trust, and dealing with conflicts of interest or goals through a fact-based discourse.

In the context of circular product design, IDEO and Ellen MacArthur Foundation offer assistance and tools for implementing circular design thinking and takes a systemic perspective (IDEO and EMF 2022). Tools such as Materials and Product Journey Mapping are helpful for designers to think holistically in the cycle and design products accordingly. The establishment of new partnerships and stakeholder integration are mentioned within the approach, but dialogues are not seen as a tool for implementation. The addition of a dialogue-oriented development approach can help here.

6.1.6 Conclusion

The contribution shows that the transformation from a linear to a circular plastics economy can only succeed if it is approached together, in collaboration between the players in the plastics industry, politics, and society. It is about minimizing hurdles to closing the cycle in the industry (such as conflicting objectives or price or quality-related competitive disadvantages of recyclates) and exploiting digitalization potential. To achieve the transformation goals at the necessary speed, political framework conditions and steering are required, both nationally and at the EU level.

As a solution, the contribution proposes a methodology that was applied to a case study. The basic idea is to solve complex social problems (such as the transformation to a circular economy) through a dialogue-oriented, multi-stage process. Stakeholder groups—consisting of relevant industry representatives, politics, science, and society—are integrated in the early phases of the problem-solving process. Following a common vision, the first step is to build an understanding of the problem through dialogue and to define socially desired target states and evaluation criteria, which represent important guidelines in the further development process. The integration and comparison of perspectives and needs take place continuously in the process, right up to finding a solution.

Dialogue is a productive development tool in that it is results oriented. The approach goes beyond pure (individual) user focus. The focus is on social benefits and the design of desirable futures in the dialogue between business/industry, politics, science, and society (Schütz 2020).

The challenges of transformation described can only be overcome if such a dialogue-oriented and collaborative path is taken. This applies to political measures that need to be developed, such as the implementation and specification of regulatory control options or the design and implementation of open and cross-sector data spaces (e.g. digital product passport). The methodology presented does not contradict the requirement for speed. On the contrary: innovative methods are needed that help deal with typical conflicting goals when it comes to sustainability issues and bring together different ways of thinking and working (e.g. between politics and business or between sectors) productively. Against this background, further research and funding for technology development, collaboration, and dialogue formats are also required.

6.2 Plastics Design for Sustainability

Klaus Wittstock and Victoria Wessolowski

The following contribution investigates the topic of the design of plastics seen from the perspective of the chemical industry respectively plastic manufacturers. Hence, the statements are made only with a view on plastics as basic materials in their primary production to serve diverse applications in subsequent use. The way how a final product containing plastic is devised depends mainly on the expertise and decision-making of the final product manufacturer ultimately combining the entire spectrum of materials and design factors. The first part of this section reflects the design possibilities of plastics manufacturers in the context of the overarching design system. The second part describes the rising complexity of design concepts finally culminating in the need for their harmonization and institutionalization.

6.2.1 Design on Product versus Material Level

In general, design is never a detached theoretical concept if it needs to materialize. It operates in a system where the intended product, available or to-be-developed material(s), and relevant process(es) must be accordingly adapted to each other.

Product design

Product design is about planning of the composition of a final product which can be a relatively simple structure with one component consisting of only one material (e.g. plastic bag, plastic wash basket) or a complex structure with several components made of different materials (e.g. cars, electronic devices). Products should be ideally designed to meet their intended purpose during their use, integrating various kinds of functionality such as modularity allowing easy disassembly or strong inflexibility, durability or short-lived nature, reusability, or single use, and/or recyclability. This means using materials that are, e.g. strong or flexible, heavy or lightweight, resistant to wear and tear, or easily biodegradable. By considering these factors, products can be designed to minimize material use and energy consumption, reduce the need for replacement and disposal, and allow for parts to be reused or recycled. Often use-phase effects are dominant over the impact of the production phase or end-of-life management of the product.

Material design

Material design is about planning of a class, type, and grade of plastic polymer and its additives, fillers, and blends having various features like polypropylene or polyamide with or without flame retardants or UV stabilizers, glass fiber-reinforced or flexible, etc. For example, the design for the recycling of molecules involves the deliberate molecular engineering of polymers to facilitate recycling through re-monomerization, transitioning thermosets into thermoplastics, and building on existing natural biogenic polymers by modification. This approach focuses on interventions at the molecular level to reduce material complexity and enable the recycling of polymers, thereby contributing to the circularity of the materials (von Vacano et al. 2022). From the large variety of available materials and polymers, based on fossil or renewable feedstocks, designers narrow the selection to the specific choice for the to-be-designed product (OECD 2021). Where possible, materials should be recyclable, enabling them to be reused in new products. However, this should be done without compromising the safety or performance in the subsequent use. Where there intended purpose for plastics to biodegrade, they need to break down naturally and safely in a variety of environments, such as home composting, industrial composting facilities, or in soil on fields. The choice of the appropriate material depends on many factors, among others, how it fits the purpose of the final product as well as what impacts it bears along the entire lifecycle to be accepted by the final product design on the application level. Lifecycle analysis can help identify materials with the lowest environmental impact.

Process design

Process design is about planning of manufacturing and/or set-up processes of materials and products. Process design is an important tool for identifying efficiencies and continually improving (production) processes. It includes, for example, finding areas to reduce resource and energy consumption, implementing pollution and emission control measures, regularly reviewing, and evaluating the production process as well as using new and sometimes not obvious collaboration mechanisms (Van den Beuken et al. 2023).

Plastics are frequently criticized due to their sheer amount, with renewed calls to reduce this abundance and even create a "moon shot" or "super polymer with the functionality of today's polymers but with superior" environmental performance (EMAF 2016). In a way, a polymer that could replace all existing polymers in all applications. This would elegantly solve the issue of design at the material level since there would be only one material. Manufacturing processes could be simplified, recycling processes could be optimized to this one type and upscaled easily as well as uptake of recycled material would become very easy. Unfortunately, this magical polymer remains just a thought experiment, not to mention the numerous and intensive attempts by various researchers to realize this idea.

The key challenge to resolve here is the fact that, for different applications, different polymer types are needed to fulfill diverse requirements such as durability or short life span, transparency or opacity, heat or cold resistance, elasticity or rigidness, electrical conductivity, thermal insulation properties, etc. (cf. Fig. 6.3). Not to be neglected are also such aspects that low-cost mass-produced plastics first give many people access to modern products, while high-performance plastics in smaller niche applications first make renewable energies or medical products and other demanding applications possible.

In summary, the design of plastics as such, detached from any particular use in a final product, is sometimes possible but needed only to a certain extent. As the

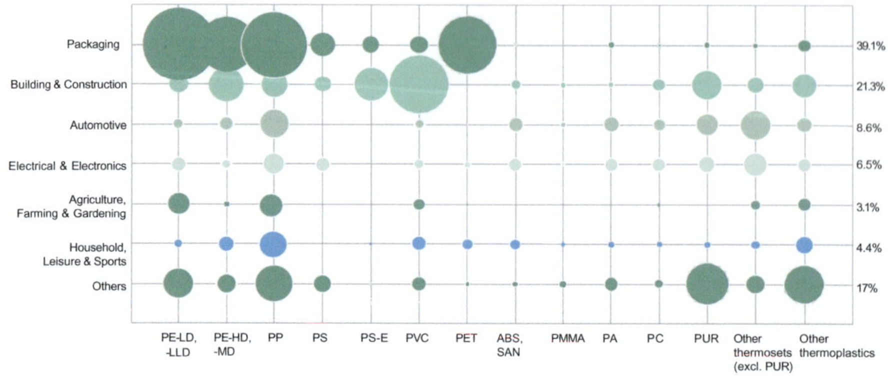

Fig. 6.3 Breakdown of polymer types by application in Europe (*Source* Plastics Europe 2022)

technical properties of polymers vary greatly, the selection of a suitable polymer for a respective application should be done on the final product level also considering polymer fitness along the entire lifecycle. Hereby, one rather speaks of "design with" and not "design of" plastics.

6.2.2 Design Concepts and Parameters

During the last decades, design concepts and included parameters are continuously being further developed, starting with the somewhat "classical" Design for Performance (DfP), going, among others, via Design for Recycling (DfR or D4R) and Design from Recycling (Tan et al. 2023) or—if incorporating both aspects—rather Design for Circularity (DfC) or Design for Circular Economy (DfCE) into Design for Environment (DfE) to a broader and more inclusive Design for Sustainability (DfS) (cf. Fig. 6.4). All these design for X (DfX) terms are distinct concepts often used interchangeably having different meanings and objectives, with "X" representing a central theme to the design (Sassanelli et al. 2020).

Design for Performance (DfP)

For decades, the DfP was the usual approach—products containing plastics were optimized solely for their performance, rarely for resource-efficient manufacturing, and almost never for their end-of-life recovery when they became waste. The waste was collected, disassembled, sorted, partly landfilled, partly used for energy recovery, and partly recycled. The end-of-life management was not part of the design process but completely decoupled from it.

Design for Recycling (DfR)

DfR refers to the process of designing products in a way that makes them easy to recycle. The objective of DfR is mainly to reduce the barriers to and increase the

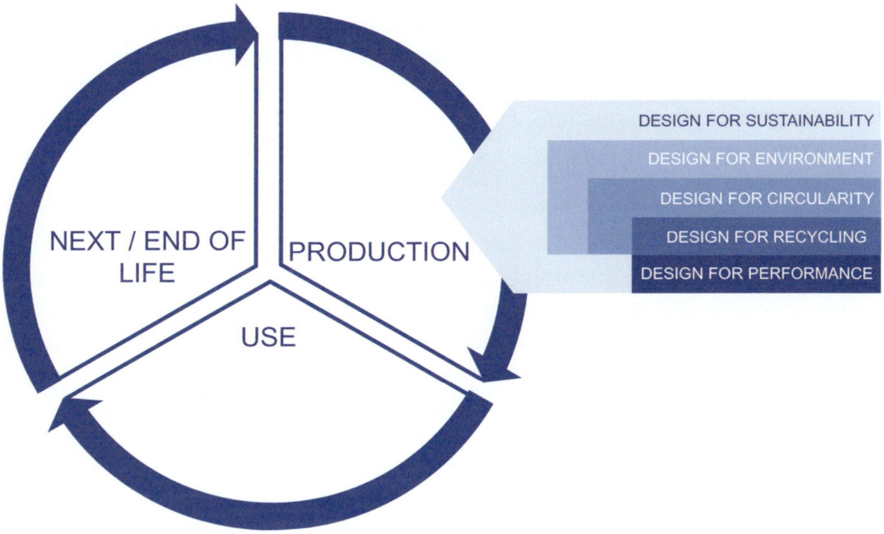

Fig. 6.4 Build-up of some design concepts in the product's lifecycle (*Source* BASF SE)

efficiency of the mechanical recycling of post-industrial and post-consumer plastic-containing waste streams. This concept emerged when it became clear that a traditional DfP concept often leads to waste streams which are difficult to recycle or even not recyclable at all. DfR involves selecting polymers, additives, colors, adhesives, materials, etc. that are easily recyclable respectively can be recycled with the existing mechanical recycling infrastructure. Current approaches to DfR often favor the reduction in the number of polymer types, avoiding the use of mixed materials and contaminants, and designing products that can be easily disassembled. This might be helpful to deal with certain types of relatively simple applications but not with the general diversity of products respectively the wish to use the recycling output in multiple new products. With mechanical recycling, the final properties of recyclates are heavily dictated by the properties of the waste in the recycling stream, leading to the addition of additives, fillers, etc. to maintain similar quality in renewed applications (Billiet and Trenor 2020).

Mechanical recycling

As for mechanical recycling (cf. Sect. 10.1), almost clean and homogenous waste streams as feedstock are a prerequisite, many DfR guidelines reward the use of mono-materials such as PE, PP, PET. Such an approach does not assess the environmental impact of alternative (multi-) material solutions or questions whether the chosen material is indeed mechanically recycled at the end of the day. The knowledge of waste management processes, especially collection and sorting, helps to define future end-of-life options during the design phase. If the designer knows that due to the size (too small) or to other reasons (e.g. material combination, missing sorting line), the product will end in a mixed waste stream destined for chemical recycling, incineration with or without energy recovery or even landfilling, the focus on mono-materials might be counterproductive. Good examples of such cases are very small plastic parts in electronic devices or cars, which will never be dismantled, collected, and sorted into homogenous plastic fractions. Here it is obvious that Design for Environment or Design for Sustainability approaches lead to overall better products than DfR (Ding and Zhu 2023).

Chemical recycling

A somewhat compelling question is also whether chemical recycling might hinder design for mechanical recycling, potentially allowing companies not to improve the design of their products at all, and whether chemical recycling might need a dedicated design for chemical recycling, as some technologies have certain constraints on the processed materials either way. The chemical recycling technologies (like solvolysis or depolymerization, pyrolysis, and gasification) on average have a broader waste input specification than mechanical recycling and generate "virgin"-like quality of output materials. Depending on the technology, they can (cf. Sect. 10.3) accept varying ranges of impurities or homogeneity of plastic types. For example, gasification can accept a broad mixture of organic waste feedstock, showing successful results even in processing automotive shredder residue (Buttker et al. 2005). For this recycling technology, a dedicated design is not necessary as it is the methodology of choice for mixed plastics which cannot be sorted for economic or technical reasons. On the opposite, most depolymerization processes can recycle plastics waste again to virgin-plastics quality but require narrow waste feedstock specifications like those of mechanical recycling.

Biopolymers

When designing biopolymers for recycling, it is important to consider the type of polymer being used. Bio-based and non-biodegradable polymers can be recycled via the same recycling routes as fossil-based polymers, therefore the DfR approach for them is the same as for fossil-based polymers (cf. Sect. 14.1). In this case, the design decision is rather based on the possibility to reduce the product's carbon footprint than to make the product recyclable. On the other hand, biodegradable polymers (independent whether bio-based or fossil-based) are designed to be collected with bio-waste and food waste (such as food packaging) and rather recycled organically (i.e. aerobically by industrial and home composting or anaerobically by bio-gasification with subsequent composting) or biodegrade into soil (such as agricultural mulch films) (Bartolucci et al. 2023). It is important to note that biodegradation is not a license to litter. Biodegradable materials still need to be disposed of properly and should not be discarded in the environment.

Labeling

For DfR, labeling plays a crucial role in designing products made of plastics to guide various stakeholders on how to deal with plastic products at the end of life (Niessner 2022):

- Identification: Labels help identify the plastic type using resin codes (e.g. PET #1, HDPE #2, etc.). For consumers, they provide clear instructions for proper disposal thus generally ensuring that end-of-life plastic products are sorted into the appropriate waste bin and collected for further waste treatment in the first place.
- To make sure that biopolymers can be recycled or biodegraded properly, it is important to use appropriate and certified labeling (such as "Seedling") to ensure that they are not mixed with other materials that may contaminate the conventional recycling or composting processes.
- Sorting and separation: Clear labels accurately assist also in the sorting process of plastics, which is vital during disassembling and partly in automated recycling facilities to direct them to the right recycling streams.
- Material information: Labels disclose material composition, such as additives, fillers, and recycled content in the product.

As new waste treatment technologies emerge, it is thus crucial to keep DfR up to date to enable the recovery of more waste than is currently handled. This includes advancements in collection, sorting ((water digital) marking, mobile near-infrared spectroscopy, etc.), as well as mechanical (solvent-based purification, magnetic density separation, fluorescent imaging, etc.), chemical, and organic recycling technologies.

Design for Circularity (DfC)

DfC is a term that is often used to consider the so-called closed- or open-loop biological or technical circularity of plastics from the very beginning (Brouwer et al. 2020). It differs from the so-called end-of-pipe approaches, where one tries to recycle given end-of-life products and waste streams without being able to influence their design or composition. Under DfC, products should be designed to fit within circular systems, enabling them to be reused, upgraded, refurbished, remanufactured, mechanically, chemically, or organically recycled. In the case of recycling, one tries to combine the recycling of waste and the use of recycled output material in applications being able to be recycled anew (EMAF 2013). However, to require plastics to be designed only for infinite and absolute circularity is limited as plastics physically degrade over time thus forcing the replacement of the worn-out products or the integration of primary resources to fulfill the required functions. It is obvious that effective collection and processing systems should be established for DfR and DfC to materialize at all, which means that the infrastructure for collecting and processing waste must be broadly and technologically expanded, and consumers must be given convenient access to it (Billiet and Trenor 2020).

Design for Environment (DfE)

DfE on the other hand, is a broader concept that encompasses all environmental aspects of the product lifecycle. The objective is to reduce the environmental footprint of products from raw material extraction via the use phase to disposal. This includes selecting materials that are environmentally benign, i.e. minimizing resource and energy consumption and emissions during production, reducing the use of hazardous chemicals, and designing products that can be reused, recycled, or repurposed.

Another term used in this context is eco-design (cf. Sect. 6.4), which was for a long time the same as DfE. In Europe, however, eco-design has become part of comprehensive legislation with a broader range of requirements that products must meet to make sustainable products the standard in Europe. To this end, the European institutions use a triad of tools to develop more and more products to a higher standard over time: Eco-design sets minimum requirements for products that are placed on the market, the Organizational and Product Environmental Footprints provide a tool to measure and compare environmental performance of organizations and products, and eco-labels identify the best-performing products.

Design for Sustainability (DfS)

Further development of the more advanced design represents the DfS which is the most comprehensive of the above-mentioned design concepts and combines somewhat all of them (cf. Fig. 6.5). It integrates sustainability principles into the creation of materials, products, services, and systems, aiming to minimize negative environmental, economic, and social impacts and balancing their needs to ensure the well-being of current and future generations. The entire lifecycle is applied as fundamental thinking across these three impact categories (Sild 2022). UNEP (2009) and OECD (2021) provide good overviews and tools for step-by-step DfS considerations that should be taken during sourcing, manufacturing, use, and end-of-use phases as well as trade-offs that may arise during these stages. With respect to the environment, the aim of DfS is to increase resource efficiency and reduce waste generation thus conserving and reusing resources and eliminating pollution. Economic aspects define the way the business will run and stay successful in the long term. Social safety, equity, or affordability are starting to rightly come to the fore in recent times. Section 6.4 shows exemplary what criteria can be considered in designing products with sustainability considerations in mind. The key challenge is to integrate and balance the three impact categories and not to limit one or other (OECD 2021).

In this regard, DfS tries to counterbalance potential target conflicts between the concepts, as DfR, DfC, and DfE may have trade-offs when optimizing products on a specific level. Designers will need to transparently determine how these conflicts can be resolved by sometimes defining what is non-negotiable and how the aspects can be ranked (OECD 2021). Exemplary,

- if one designs a plastic packaging product that is easy to recycle, one may use a thick-wall mono-material that is easy to sort and to recycle mechanically but is not light-weight (with negative environmental impact during transports) and increases the overall material consumption, compared to a thin-wall multi-layer plastic packaging that might currently not be recycled. In this case, there is a target conflict between the objective of recycling and the objective of reducing the overall environmental impact.
- a target conflict between sustainability and solely environmental criteria occurs if a product requires a material composition for optimal health or safety reasons

Fig. 6.5 Environmental, economic, and social criteria in the design for sustainability approach (*Source* BASF SE)

having a higher carbon footprint than alternative but less safe materials (e.g. in medical applications or for safety parts in automotive).

It is yet worthwhile to mention that sustainability criteria are not purely scientific ones, because they consider the specific regional or local context and different understanding of sustainability by stakeholders applying it. Each player has one's own definition of sustainability and set of criteria to assess whether a design is sustainable or not. For example, a player in one region may prioritize environmental sustainability when making decisions, while a player in another region may prioritize economic sustainability. Players may also prioritize sustainability criteria differently based on their current needs and goals. Sustainability as a fixed state is not given, because sustainability criteria evolve over time changing the understanding of sustainability as such. DfS constantly implies innovation and requires a high degree of flexibility to embrace disruption. All in all, sustainability assessments of designs and decisions about products differ greatly depending on the set of chosen criteria and over time.

In a nutshell, DfP strives to deliver the best performance of products in production and use. DfR focuses on improving the efficiency of recycling processes, whereas DfC also focuses on integrating the recycling output back into the manufacturing of new products and looking at how to uphold the longevity of products by reuse or repurposing. DfE aims to reduce the entire environmental impact of products, whereas DfS takes a more holistic and inclusive approach by balancing economic, environmental, and social considerations.

6.2.3 From Individual Tools to a Mandatory Policy Instrument

Recently, the way design is done has been evolving from a more "classical" way of developing plastics over many years in a more closed environment such as university or corporate research to a more flexible multi-stakeholder approach in an open environment. Today, stakeholders with different kinds of expertise may contribute to various design steps on definition, plan, research, concept, protype, and implementation, as is, for example, practiced in BASF's Creation Centers in China, Germany, India, and Japan.

As the demands for how a product is designed, what kind of expertise is pulled together, and how design parameters are interpreted by various stakeholders tremendously differ, industry players recognize the need for harmonization and mutual recognition of design approaches. Frequently, there seems also to be a certain development path in how design ultimately manifests itself.

Due to increasingly complex external requirements, established individual design approaches currently often lack recognition power, even by those wanting to implement them. In the first step, industry players try to define design guidelines and assessments under industry initiatives. For instance, there is an abundance of guidelines for DfR and DfE of plastic packaging developed by various players from the packaging value chain, think tanks, and technical organizations on national, regional, and international levels, e.g.

- National: COTREP (France), cyclos, Zentrale Stelle Verpackungsregister, Der Runde Tisch Eco Design of IK (Germany), APR (USA);
- Europe: RecyClass, Petcore Europe, CEFLEX; Globally: UNEP, The Consumer Goods Forum.

A recent overview by cyclos-HTP for BKV GmbH shows similarities, differences, gaps, and inconsistencies between the most prominent DfR guidelines and assessments on plastic packaging. The major gaps are the insufficient reproduction of physical, chemical, and technical rules of recycling, missing or even non-transparency, and a lot of room for interpretation in implementing DfR. Among others, the authors recommend using at minimum a 4-class categorization (e.g. full compatibility, tolerated compatibility, limited compatibility, and incompatibility/disruptiveness) with clear demarcation rules for, e.g. separability in the recycling process. As a reference for the assessment of formulations and components of a particular packaging, one should rather use recyclate application and not virgin polymer, because the properties of packaging are not only defined by the polymer classes, but rather polymer types, additives, blends, and copolymers that ensure the processing properties and functional requirements of the individual packaging (BKV 2024).

Meanwhile, design guidelines on product level exist for all major plastic applications and product passports are on uprise, however, the trend is to gain "ultimate" acceptance across all stakeholder groups. When voluntary approaches do not suffice, one goes for an established recognition route of standardization that often requires a third-party certification as a proof. In the case of packaging, DfR standards for different types of plastic packaging are currently in development by CEN, the European Standardization Committee (CEN 2023), with different certification bodies competing for the recognition of one's own approach as well as industry players protecting plastic materials from unfavorable classification.

Design and regulation

This almost inevitably leads to a binding implementation of design requirements that are regulated by policy, with standards that can be referred to in legislation. In recent years, the design of products is seen by policy makers as a political instrument to prescribe the requirements otherwise not being achieved by voluntary commitments of the industry. In this area, there is also a shift from DfR to DfS in recent times as well as an increased push to not only regulate but specially to mandate a more benign design of not only final products but also of components, materials, and even substances, to be exemplarily seen in Europe:

- The German Packaging Ordinance (Bundesregierung 1991) and the subsequent European Packaging and Packaging Waste Directive 94/62/EC from 1994 (EU 1994) were kind of frontrunners in defining DfR as one of the essential requirements for packaging to be placed on the German or European market, respectively. This includes the use of materials that are easily recyclable or recoverable and avoiding the use of mixed materials and hazardous substances that could hinder the recycling process.
- The European End-of-Life-Vehicles Directive 2000/53/EC (EU 2000) was next in requesting from producers the integration of DfR in new vehicles, components, and materials, by pertaining the same DfR logic as in the packaging sector.
- The European Waste Framework Directive 2008/98/EC (EU 2008) expanded further the scope by not only referring to "environmental impacts" to be reduced by design but also introducing extended producer responsibility as "one of the means to support the design … which take into full account and facilitate the efficient use of resources during their whole life cycle including their repair, re-use, disassembly and recycling." The recent revisions of this so-called "EU Waste Package" are seeing the uptake of further design requirements than the one on recycling only.
- For the first time in 2019, the so-called single-use plastics product Directive 2019/904 (EU 2019), combined the targets for recycling and the use of recycled content for plastic beverage bottles.
- The need for a more benign design is taken up by sectoral approaches also on construction products, electronic and electrical equipment, textiles, fishing gear, etc. with the Ecodesign for Sustainable Product Regulation (ESPR) (COM 2022a) aiming to develop a horizontal design approach for almost all physical goods placed on the EU market. Products should hereby become more durable, reusable, upgradable, reparable, easier to maintain, refurbish, and recycle, and have lower energy and resource consumption throughout their lifecycle as well as contain recycled content.

- ESPR builds upon the framework of the so-called Safe and Sustainable by Design criteria (SSbD) for chemicals and materials to be defined by the European Commission under the EU Chemical Strategy for Sustainability. SSbD puts together safety, environmental sustainability, and socio-economic dimensions to be considered along the entire lifecycle for the first time by proposing a hierarchical approach in which the safety aspects are considered first. It also includes substitution and minimization of the negative impacts of substances of concern as well as those hindering recycling processes in line with existing and beyond upcoming regulations (COM 2022b).

Internationally, more benign design is now in scope, among others, by the OECD which developed a common reference for governments on how to make design more impactful for implementing the UN Sustainable Development Goals (OECD 2023). The UN Treaty to end plastic pollution is expected to cover the whole lifecycle of plastics, from design, extraction, and production to consumption and disposal, covering materials, products, chemicals, additives, and microplastics (UNEP 2023).

Technological and digital innovation

Innovation in sorting and recycling processing technology is a challenge. Improving the quality of the collected waste streams is therefore crucial for the recycling of plastics (Knappe et al. 2021). To optimize quality, information about the plastics used in an application is key. There is a growing trend towards (digital) product passports that serve as a documentation system to track and communicate critical information throughout the lifecycle of a product, with Europe aiming to use them for the products regulated under ESPR. Product passports are physical or digital documents that accompany a product, providing detailed information about manufacturing processes and the type of plastics used, their composition, and the presence of additives or fillers. Additionally, product passports may contain information about the bio-based and recycled content of the product, indicating the percentage of alternative materials used in its production. They can also propose recycling and disposal routes for the products (Jensen et al. 2023). Digital product passports take the concept of physical passports further by leveraging technology to store and access product information digitally. They can be accessed through QR codes, RFID tags, or other digital identifiers. Digital passports offer the advantage of real-time updates and accessibility, allowing consumers, recyclers, and manufacturers to access the most up-to-date information on a product's material composition and recycling options (Adisorn et al. 2021). For example, Catena-X (2024) is the first open-source and interoperable data ecosystem aiming to provide an environment for the creation, operation, and collaborative use of data along the automotive value chain. These details are essential for ensuring compliance with legislation or meeting the requirements of downstream industries if a product passport is required.

6.2.4 Conclusion

The often cited, but not really proven, estimate that 80% of the product's lifecycle environmental impact is determined at the design phase (EMAF 2022) depends on the factual treatment of the product by society. Many lifecycle analyses of plastics in various applications show that often the main impact occurs during the use-phase. Ensuring the required performance in production, use, and recycling at their end of life or considering the overarching environmental lifecycle impacts is crucial to be considered in designing products but all these approaches cover only single aspects. Design for Sustainability (DfS) should be used as the main design concept to avoid unintended negative consequences in environmental, social, and economic areas and fulfill the United Nations Sustainable Development Goals (SDGs) holistically and in a balanced way. Today, sustainability is not yet automatically or systematically considered in its entirety when designing products and applications and the plastics they contain (OECD 2021). Incorporating these design principles into plastics manufacturing will help to create a more sustainable economy for plastics and move society towards a more sustainable future. It is not surprising that the currently developed UN legally binding instrument against plastic pollution is expected to cover the whole lifecycle of plastics accepting the urgent need to tackle various "plastics issues" globally (UNEP 2023). It goes without saying that DfS is not used alone. Ongoing innovation and collaboration across the different stakeholders will be critical to continue improving the sustainability of plastics, making the right tradeoffs, and ensuring that plastics can continue to be used in a responsible way.

6.3 Industrial Design for Circularity

Martin Beeh, Najine Ameli and Oliver Stengel

6.3.1 Overview

The design discipline, like the corporate functions that interact with it, is subject to constant change, but it also actively shapes this change. From the design of functional, high-quality industrial products designed with pleasing aesthetics, to responsibility for the economic, social, and sustainable effects of mass products, the discipline took a long, ambiguous, and sometimes controversial path. Today, the focus is on the designer's integral responsibility for the product, service, and to a certain extent also for the business model, which helps shape it and can only be experienced by its users through the design work. This section shows to what extent design and circular economy fit together and what opportunities arise from this.

The term "circular economy" has gone from being a fringe topic "for the distant future" to a core term of the paradigm shift for designers and engineers. Traditional product development processes, established material applications and ways of production as well as linear business models quickly reach their limits when it comes to circular economy and circular design. Companies large and small have recognized that "circular" business models are needed. In addition, the "participation society" today requires companies, brands, and retailers to responsibly develop new products, product cycles, and service/product systems in collaborative value chains. What effects this has and what new thinking and new actions this makes necessary for industrial designers in the design process is the subject of this section.

Pioneers of Design for Circularity are introduced, and their work is recognized. Based on the principles formulated by the designer Dieter Rams, the authors draft a current requirements profile for a design for circularity and the work of industrial designers, product managers, and designers. The guideline "Development of technical products and systems (product development model)" (VDI TZ 2021) and the new guideline "Industrial design—user-centered design in the product development process" (VDI-ZRE 2424) of VDI and VDID are discussed in the context of a design for circularity. Finally, the authors provide an outlook on how Design for Circularity can be implemented in a targeted, concrete, and promising manner in companies and design studios.

6.3.2 Design in the Context of Sustainability: Origin, Present, and Necessary Perspectives

Why design?

The beginning of design as a separate discipline can be dated to the nineteenth century. With the beginning of industrialization, the production process was separated into two sections that had previously always been connected: the design of objects (intellectual performance) was separated from their production (mass production, technical performance) because both became increasingly complex (Schneider 2005).

While engineers took on machine production, designers were defined according to this division of labor as "one who is qualified by training, technical knowledge, experience, and visual sensibility to determine the materials, mechanisms, shape, color, surface finishes, and decoration of objects which are reproduced in quantity by industrial processes" (ICSID 1959).

This division of labor also occurred because of the growing economic importance of design. As early as the second half of the nineteenth century, the qualitative product differences between competing manufacturers became increasingly smaller. The design thus became "the final and most important differentiating factor when competing with the competition" (Hauffe 2000). The design was consequently used to achieve increasing sales rates and economic growth. With mass production, items could be produced more cheaply. Both developments drove consumption, which in turn caused energy and resource consumption to rise.

Which design?

In the twentieth century, the population of most countries grew and with it the number of consumers. After 1970, the threat of overloading global ecological borders showed that business and design as usual were becoming increasingly problematic.

As early as 1969, the architect Buckminster Fuller formulated the message in his book "Instructions for Spaceship Earth" that, despite worsening ecological problems, human intelligence can provide solutions that will continue to enable life on this planet in the future. One of these solutions that is still valid today is the 3R concept: Reduce, Repair, Recycle (Fuller 1969).

Reduce refers to reducing the number of products. Until now, it has never really been reflected on how many products are needed to enable as many people in a society as possible to enjoy their benefits. Does every household need its own freezer, its own cordless drill, its own car with its own parking space, etc.? Less-produced products require less energy and fewer raw materials to produce. Many everyday objects should be designed so that they can be used together—and developing solutions for the shared use of things (sharing by social design) has also become a task for designers. Extending the useful life also allows the number of products consumed to be reduced. This also includes the improved repairability of many products.

The EU Commission and EU Parliament proposed the right to repair in 2023 (EP 2023): According to the EU, around 35 Mt of waste are generated every year simply because products are thrown away or disposed of far too early instead of being repaired. Design also plays a part in this because many products are designed in such a way that they are almost impossible to repair. That should change from now on. Manufacturers are obliged to offer a free repair within the legal (and extended) guarantee period unless this is more expensive than a replacement or impossible. If a product cannot be repaired, the manufacturer should offer a refurbished device. If repairs take a long time, a replacement device should be made available. For many manufacturers, a more robust product design may therefore be the more economically sensible option. Such an approach is ecologically necessary anyway.

At the end of an extended useful life, the product should be recycled as best as possible in the spirit of zero waste. For this purpose, the (composite) materials must be selected and processed accordingly. In addition, there is the EU's intention to accelerate the transition to a circular economy for the furniture industry, among other things, as shown by the timeline for the implementation of the ESPR Regulation (European Eco-design for Sustainable Products Regulation) ESPR 2022), which the circular economy expert Rainer Strack from the Association of the German Furniture Industry shows in an article explained on "Möbel Digital" (Möbel 2023).

Since EU standards are usually very demanding and apply to an economically important market, they often become a global benchmark: companies from Asia or North America that want to import their products into the EU usually adapt their product lines to the highest standards, as it is often more economical to produce just one line for all markets.

3R consequently places new demands on design worldwide: Producing things in an increasingly restrictive environment requires more and increasing skills. In this field, designers play the role of mediator between new technologies, user needs, and sustainable solutions (Esslinger 2011).

Contrary to the previous maxim of promoting consumption figures, designers are increasingly bearing moral, ecological, and social responsibility for the products they design. The designer Victor Papanek called on his guild to do this back in the 1980s (Papanek 1985). The crisis-like deterioration of global environmental conditions is currently being followed by tighter and stricter political regulation of processes, products, and markets at national and supra-national levels. While in the nineteenth and twentieth centuries, it was the designers who had to work on generating a mass market (Haug 2009), in the twenty-first century they are now faced with new ecological and legal framework conditions that force them to do so becoming generalists and working on solutions in an interdisciplinary manner. Reducing consumption is the top priority, as there is no need to (laboriously) recycle what was not produced.

At the same time, this expansion of the discipline allows the designer to question and change many things that previously seemed set in stone.

Design: Quo vadis?

Since its creation, the scope of design has changed because it has to adapt to constantly changing conditions: "Obviously the boundaries of design are not clear. Indeed, they ought not to be crisp to ensure that there is enough flexibility to allow design as a discipline and a body of knowledge to evolve with humanity's understanding of the universe" (Salustri and Nathan 2007).

Consequently, over the last few decades, more and more design trends have emerged that are important for future-oriented design professionals, like the authors and their international networks: Green Design, Eco Design (Tischler and Moser 2023), Design for Sustainable Behavior (Selvefors and Renström 2018), service design (Mager 2020), design thinking and service design thinking (Stickdorn and Schneider 2011), product service systems (VDI-ZRE 2024).

These terms indicate where design will develop in the future: Design should contribute to the realization of a sustainable society. Economist Jeffrey Sachs has declared the twenty-first century to be the "Age of Sustainable Development"—and that is what it will have to be (Sachs 2015). Since the discipline of design has always had a creative component, it should succeed in this turn.

6.3.3 Design for Circularity with Plastics

The design industry has already been part of the exploration of what is technically and creatively possible with early mental models and experimental projects. Student work has already researched products and cycles based on sustainable principles. This experience through very concrete experimentation with separable and recyclable materials has seen its first implementation in low-complexity products. An example of this is the work according to the Cradle-to-Cradle principle, which was described by Michael Braungart and William McDonough in the "Hannover Principles" (Braungart and McDonough 1992).

Pioneers of eco-design

The term eco-design encompasses all design activities that make products and services more environmentally friendly. Questions about the responsibility of designers were already discussed at the Hochschule für Gestaltung Ulm (Ulm School of Design, 1953 to 1968). Dieter Rams, Braun's long-time chief designer, has developed the "10 theses for good design" since the 1970s (Rams 2017).

Thought leaders like Ezio Manzini (1990) and the designer group O2 Ecodesign were already working on fundamentally new approaches to "eco-design" in the 1980s. The exhibition "Sustainable Everyday-Scenarios of urban life" developed by Ezio Manzini and François Jégou for "La Triennale di Milano" (Manzini and Jégou 2003) was sensational. For a long time, they were considered "exotics", but in recent years they have become intensively involved in working with the manufacturing industry and trade. Also mentioned are the eco-design pioneers Albin Kälin from EPEA Switzerland, Ursula Tischner from Cologne, and Ernst-Jan van Hattum from Switzerland (O2 Ecodesign 2022).

6.3.4 Examples of Sustainable Design with Design for Circularity

Climatex® (JAB Anstoetz)

The furniture and interior design upholstery materials are based on the so-called "textile screw", which connects natural and synthetic fibers to optimally use the advantages of each fiber. The textiles can be sorted according to wear and can be returned to biological or technical cycles. The natural fibers are made from high-quality farm wool or flame-retardant European hornbeam wood; the synthetic fiber Cradura™ can be repeatedly recycled (Climatex 2022).

Mirra Chair (Herman Miller)

This office chair from the American office furniture manufacturer and industry pioneer was designed in 2003 by the Berlin design practice Studio 7.5 (2003) and has been continually developed further. In 2019, all of Herman Miller's European production was certified "Cradle-to-Cradle". The company provides an example of how a complex product can be consistently designed to be recyclable. The well-known photography of all parts of the Mirra Chair also visually represents the circular ideas in industrial design very well (Miller 2019).

Organic Chair (Kartell)

In 2016, the Italian furniture manufacturer Kartell presented the new recycled material "Biodura". The Italian designer Antonio Citterio created the "Organic Chair", which is elegant, durable, and also biodegradable. The architecture and design platform "Stylepark" (Stylepark 2023) reports on the choice of materials for Kartell's Organic Chair. In a biological process, the waste material is combined with microorganisms. This creates a biomass that forms the basis for further processing. For the first time in the furniture industry, Kartell is using the injection and pressing process for production (Kartell 2016).

Vytal

This Cologne-based company sells high-quality plastic reusable containers for "food-to-go". The use of the containers is organized with their own app: an exemplary product-service system. The Dutch company MEPAL manufactures these multiple containers from polypropylene, the lid is made from a thermoplastic elastomer (TPE). The plastics are easy to recycle, and their ecological footprint is smaller than that of stainless-steel containers (Vytal 2022).

Electrolux

The Swedish household appliance manufacturer has set itself ambitious goals for 2030: in certain product families, at least 50% of the plastic should consist of recycled material. Suppliers should use scrap-based steel. Circular-based, scalable business models should be developed.

The group wants to use recycled materials in production, as well as sustainable packaging. Spare parts should be permanently available. Electrolux is one of the signatories of the EU Plastic Pledge to promote the use of recycled plastics (EU 2020).

In 2020, together with the Swedish recycling company Stena Recycling, they presented a vacuum cleaner prototype made from 100% recycled materials and components.

Electrolux has made circular not only the production of the vacuum cleaner but also the business model. The company describes its approach as follows (Electrolux 2020):

"Over the last two years, a pilot of a circular subscription business model for the (…) robot vacuum cleaner in Sweden has proven popular with consumers. Since the end of 2018, the Pure i9 has been offered as part of a trial subscription on a pay-per-use basis, with delivery, consumables, service, and maintenance included in the fee. (…).

The team also examined the lifespan of recycled products and components to gain insight into both the recyclability of the selected materials and customer attitudes towards it. Key components such as motors from the recycled vacuum cleaners used in the project still had an average of 83% of their lifespan. (…).

If a customer returns the product to Electrolux and decides to terminate the subscription, it will be refurbished and made available to the next customer. This ensures that every device is used to its full potential and promotes the circular economy throughout the life of the product."

The above-mentioned vacuum cleaner, but also a steam laundry cleaner, an unpainted stick vacuum cleaner, packaging for small appliances made from recycled material, an inner liner of a refrigerator with 75% recycled material, and the GRO concept for plant-based nutrition are part of the Electrolux Better Living Action Plan. This is intended to show how circular suitability can be concretely implemented and at the same time become part of a new circular business model. Here, Design for Circularity can be experienced specifically in the context of a large company that is subject to complex challenges between profitability and sustainability (see Sect. 7.5).

BMW Circular Lab (Circular Design/Reduce, Reuse, Recycle)

The ambitious study is based on fundamentally new materials, new technologies, new processes, and new ideas. The comprehensively communicated project is intended to make BMW fit for the circular economy. The circular processes presented in the project are not revolutionary per se. However, the project is unique to date in terms of its complexity and its aim of organizing the automotive industry in a circular economy (BMW 2022a). Many materials in the BMW i3 electric car have already been selected according to strict sustainability criteria. However, in contradiction to the Design for Circularity, the passenger cell was developed from CFRP. According to BMW, the i3 is 95% recyclable (BMW 2022b).

6.3.5 From Dieter Rams to a Design for Circularity

> Exemplary designs and high functionality have in the past made Braun products distinctive and in a class of their own. Prototypical examples are the coffee grinders. They are characterized by clarity, balance and honesty of form. This also results in the shape consistency, which is unusual for industrial products. Only a few companies manage to design products in such a way that they remain on the market unchanged or with little modification for such a long time
>
> Günther Staeffler. In "Design + Text" (Staeffler 2004)

The industrial designer Dieter Rams represented here by one of his products (cf. Fig. 6.6) also took important sustainability aspects into account in his "10 design theses". How can this "ethics of design" be transferred to the design of and with plastics? Here are some suggestions for five selected theses from Rams (Vitsoe 2024):

Good design is innovative

Use circular plastics, use current results in plastics and processing development, even if the price of the product increases. Direct contact with plastic producers and processors is required to select the best possible material in the spirit of Design for Circularity. Through thoughtful design, possible new manufacturing methods, or optimized processes, the additional material costs can be offset if necessary.

Fig. 6.6 Coffee grinder KSM 1 from Braun. Designed by Dieter Rams. (*Image* Braun Collection, Kronberg)

Good design is honest

Use plastic authentically, do not mix it with metal particles, chrome it, or paint it. The surface quality and mechanical strength of the selected plastic must be durable. This throws some of the successful recipes of industrial design for surface design in products such as automobiles, small electrical appliances, computers, or writing instruments overboard.

Good design is long-lasting

The plastic must fulfill its task over the long term and be selected and designed accordingly. A film hinge experiencing "controlled" wear is unsuitable in the long term. All parts of a product must be repairable or replaceable.

Long-term storage of spare parts and effective repair options (DIY or by specialist personnel) are better than planned obsolescence. Increased production costs must be compensated for through strategic product management, functional optimization, or better user-friendliness.

Good design is environmentally friendly

When selecting materials, it is important to ensure that they are of high quality and functional and that the components are replaceable. Recyclability and the use of recyclates is a must. The plastics must be of the same type and connections must be removable. The composite materials often valued by designers do not meet these requirements. This applies to packaging as well as mobility and sports products or ICT products such as smartphones, laptops, and tablets.

Good design is as little design as possible

Once the right plastic has been chosen in the spirit of a Design for Circularity, the use of plastic should be as low as possible. Long-term user-friendliness through ergonomic design as well as mechanical and chemical resistance of the components—together with processing requirements—should be the only criteria that determine the size of the component and the amount of material used.

It should be noted that the sustainability-oriented use of suitable plastics usually contradicts the principle of the "cheapest price": everyday objects are often designed in such a way that they are damaged and unusable after just a short time. This applies to "Fast Moving Consumer Goods" (FMCG) as well as "Consumer Durables", e.g. ICT devices, household appliances, or furniture. Here the designers are called upon to find economically sustainable solutions together with their clients from the companies. Away from "single-use use" towards other models of permanent or shared use in the sense of a "shared economy". How nice would it be to be able to pass on a classic chair made of plastic (e.g. Panton chair, Bofinger chair, "Universale" by Joe Colombo) or to be able to view it permanently in a museum?

6.3.6 *Design and Product Development with Design for Circularity*

Guideline 2424 "Industrial design—user-centered design in the product development process" from VDI and VDID (VDI 2022b) lists the following activities that are based on the technical product life cycle according to guideline 2221 "Development of technical products and systems (model of product development)" of VDI (cf. Fig. 6.7):

1. Corporate strategy
2. Product creation with strategic planning, product development, realization, and production
3. Product usage
4. End of product life

According to guideline VDI 2221 (2019), the following work stages are mentioned in product development (see Fig. 6.8):

1. Planning
2. Conceptualize
3. Design
4. Elaborate.

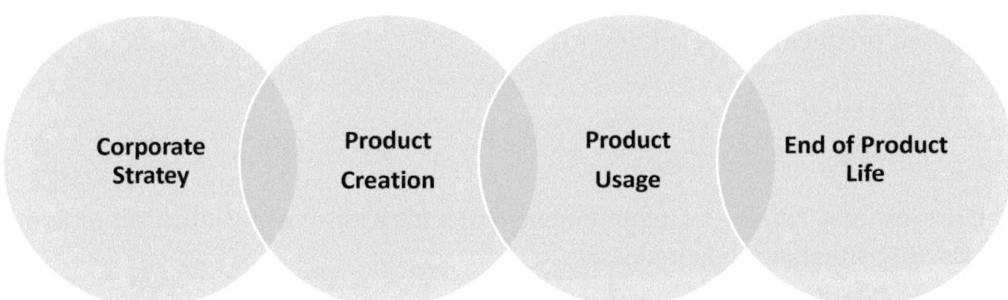

Fig. 6.7 Activities in technical product life cycle according to guideline VDI 2221 (*Source* Own presentation based on guideline VDI 2221)

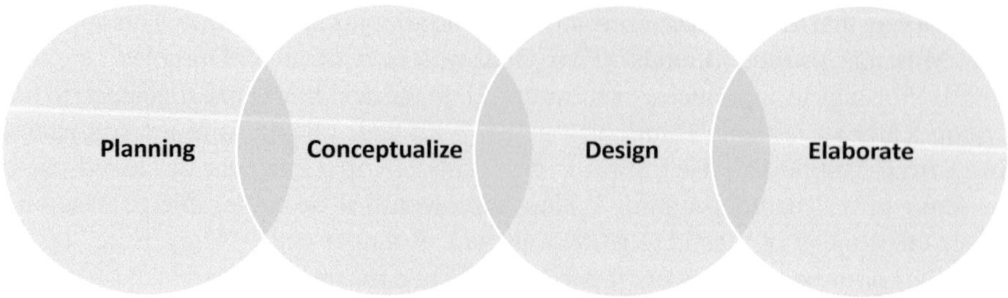

Fig. 6.8 Product development work stages (*Source* Own presentation based on guideline VDI 2221)

The results of the work sections mentioned are evaluated and corrected if necessary.

In the new guideline 2424, the definition of the design and the corresponding selection of materials are strongly associated with engineering activities. To also take the necessary aspects of Design for Circularity into account in the design work of industrial designers based on the technical product life cycle described in guideline VDI 2221, the following additions—divided into activities and results—are listed here.

(1) **Strategic planning (product planning)**

Design for Circularity activities

i. Acquiring knowledge about plastics and other suitable materials through dialogue with research institutions, manufacturers, and processors.
ii. Creating or using a knowledge database on circular materials and their appropriate use in design, product development, manufacturing, recycling.
iii. Design of the complete product life cycle, including maintenance, repair options, resale or rental (product service system).
iv. Reuse, recycle, reprocess.

Design for Circularity result:

i. Documentation of the acquired knowledge on material selection.
ii. Closed product life cycle (circuit).

(2) **Pre-development**

Design for Circularity activities

i. Creation of a catalog of requirements (specifications) and evaluation tools for selecting suitable product ideas.

Design for Circularity results

i. Catalog of requirements and assessment tools for circular industrial design with plastics.

ii. Results of the requirements analysis and the assessment tool.
iii. Recommendation for the best possible use of materials, optimal processing processes, and an integral cycle concept.

(3) **Product Usage**

Design for Circularity activities

i. Usage analysis and user surveys for the continuous development of efficient material applications.
ii. Design of the complete product life cycle, including maintenance, repair options, reuse, recycling, reprocessing.

Design for Circularity result

i. Product life cycle
ii. Continuous improvement process (CIP).

Leap innovation for the circular economy

The steps of a continuous improvement process (CIP) are not enough to implement the necessary transformation from a linear product life cycle to a closed cycle with the necessary urgency. In many cases, this requires so-called leap innovations (Federal Agency for Leap Innovations SPRIND). American companies (Apple, Tesla, Amazon, Alphabet, Microsoft) or innovative German companies like Siemens (WiWo 2023) use innovation to take bold, decisive steps towards new business ideas that are distinctive. However, it remains to be proven how such innovative leaps can be successfully implemented in a circular economy. Great times for explorers, pioneers, and managers with long breaths!

6.3.7 Outlook

One way to implement leap innovation in the circular economy can be a so-called "Design for Circularity Accelerator". How could one be defined, financed, start quickly, and soon bring success? The following approaches show possible perspectives and refer to successful attempts.

Knowledge Pool

In today's multi-crisis times, change and innovation no longer happen gradually, but rather in leaps and bounds. Rapid change and well-founded success criteria must be anticipated, imagined in scenarios, and tested. A knowledge pool that is created through interdisciplinary cooperation between designers, companies, universities, and research institutes helps. More knowledge about product-service systems, new product usage concepts, and sustainable value creation models need to be developed.

Lab

"The proof of the pudding is to eat it" or in the words of Donald Norman (Norman 2022), the pioneer of user-centered design: "Observe—Test—Iterate—Learn". Knowledge pools should carry out "laboratory experiments" to test new products, product-service systems, and material applications in the sense of circular design, to test these new approaches, and to learn new things quickly. Failure, as is unfortunately the case with many pioneering projects (Smart Mercedes-Benz, BMW i3, Audi A2), should be avoided. The economic criteria must also be defined more sustainably. What is pleasing about BMW's "Circular Lab" project is that Milan's star designer Patricia Urquiola has designed so-called "Circular Heroes" that explain the principle of circular design in a playful, intuitive way. Designers can demonstrate interface competence as proven "project integrators", knowledge collectors, and knowledge brokers. Design is known to make ideas tangible.

Lighthouse Projects

Pilot projects are needed to quickly gain valuable experience with prototype products and services in an economically protected area. The Electrolux vacuum cleaner mentioned in Sect. 7.5 was a valuable test of the possibility. Companies and designers are advised to carry out pilot projects with novel Design for Circularity approaches. In the automotive sector, these are traditionally concept studies. BMW is going further with its "Circular Lab" and is developing an entire "world of experience". The focus should be on recyclable plastics and their further development. mentioned Electrolux vacuum cleaner was a valuable test of the possibility. Companies and designers are advised to carry out pilot projects with novel Design for Circularity approaches. In the automotive sector, these are traditionally concept studies. BMW is going further with its "Circular Lab" and is developing an entire "World of Experience". The focus should be on recyclable plastics and their further development.

Process

Traditional product development processes, established material applications and production forms as well as linear business models ("fast moving consumer goods") quickly reach their limits when it comes to the circular economy and circular design. Large companies like Electrolux and BMW have recognized that "circular" business models are also needed. However, these will only have lasting success if non-linear, loop-like, almost "biologically" developing, and constantly adapting processes and transformations are strategically planned, consistently implemented, and constantly optimized, changed, or even destroyed, in the sense of "creative destruction " based on Pablo Picasso and Joseph Schumpeter (HBM 2013). Harvard Professor Clayton Christensen took this idea further in "The Innovator's Dilemma" (Christensen 1997) with "disruptive innovation". "Wirtschaftswoche" writes about this in 2023: "Accordingly, it is often new providers or newcomers to an industry who revolutionize a market with groundbreaking innovations - not established corporations." (WiWo 2023).

It will be interesting to see which industries and companies benefit from the circular economy in the coming years. Will these be established industry giants, or rather the waste management companies, the powerful IT companies, or new market players? In the end, the "lateral thinking" of the design thinking process also helps in the transformation from a throw-away economy with endless amounts of plastic waste (keyword "marine littering") to a circular economy in which plastic waste is a raw material and is almost endlessly used via a recycling process? Let's recall, nature makes use of every natural "waste".

6.4 Eco-Design and Design for Recycling of Plastic Packaging

Jürgen Bruder

6.4.1 Design Terms in Plastic Packaging Supply Chains

The development of guidelines and recommendations for environmentally sound packaging design and, above all, their implementation in concrete product and packaging developments has gathered enormous momentum in Europe since the early/mid-2010s, (cf. Sect. 6.2). In particular, the focus on improving the ecological footprint of plastic packaging, as well as the conflicting goals that arise in the process, have led to several ecological design concepts. They are not always clearly distinguished from one another in terms of content and are sometimes used synonymously in language use. Besides Design for Sustainability[2] or Design for Circular Economy (CEFLEX 2020) the terms Eco-design and Design for Recycling or Design for Recyclability appear most frequently in discussions and in documents.

6.4.2 Eco-Design of Plastic Packaging

Design principle

The term Eco-design was used in the past with different emphases, depending on sectors and products. According to the EU Eco-design Directive (2009), eco-design

[2] The term is sometimes used when it is a question of emphasising the unity of the economic efficiency and functionality of a packaging with ecological criteria such as recyclability or recycled content. In a sense, Design for Sustainability is intended to protect against the absolutisation of ecological aspects in relation to packaging functions. Design for Sustainability is then emphasized as the decisive, higher goal in discussions on concrete conflicting goals.

was linked with the environmental impacts of energy consumption of relevant products primarily during their use phase. The EU proposal for an Eco-design for Sustainable Products Regulation (Eco-design 2022) when the EU harmonized the successor regulation of the EU Eco-design Directive defines eco-design as more comprehensive as follows:" 'Eco-design' means the integration of environmental sustainability considerations into the characteristics of a product and the processes taking place throughout the product's value chain" (Eco-design 2022, Art. 2).

This may include a wide range of product-related environmental requirements (e.g. durability, reusability, reparability, energy and resource efficiency, minimum recycled content), requirements for the upstream and downstream supply chain (e.g. presence of substances that inhibit circularity, remanufacturing, and recycling) but also information requirements. It seems that this very comprehensive approach to eco-design is becoming increasingly prevalent in all sectors, including the plastics industry. In the plastic packaging and filling industries, which were under political and public pressure to improve the environmental performance of their products earlier and more strongly than other sectors, eco-design was interpreted very broadly from the outset. In addition, eco-design was understood not only as a set of requirements and criteria but also as a design principle that aims to reduce all environmental impacts emanating from packaging (from the use of materials or recyclates to design for recycling and the prevention of littering) and relates to the entire life cycle. The packaging functionality is defined in advance and thus set.

Management guide on eco-design of plastic packaging

The management guide of the "Round Table on Eco-design of Plastic Packaging" (IK 2019) is probably the most comprehensive approach in Europe to deal with the Eco-design of plastic packaging. The Round Table's guide considers Design for Recycling (DfR) as an important sub-strategy but goes significantly beyond this in accordance with the Eco-design approach. In this context, Eco-design is defined not only as a design principle for reducing all environmental impacts, but also as a management tool. The experts of the Round Table from various stages of the supply chain, science, and NGOs assign the following functions to the management tool (IK 2019). "Eco-design of packaging

- aims to minimize the environmental impact of packaging and packaged goods throughout the product life cycle;
- is part of the decision-making process on the development and marketing of a packaged product;
- includes a holistic view of the entire packaging system (primary, secondary, and tertiary packaging);
- ranges from incremental product improvements (e.g. material savings) to system innovations (e.g. optimized goods logistics and waste collection);
- deals with conflicts of objectives at different levels between different environmental aspects (e.g. material savings vs. recycled content) and environmental and other aspects (e.g. functionality, consumer protection, cost).

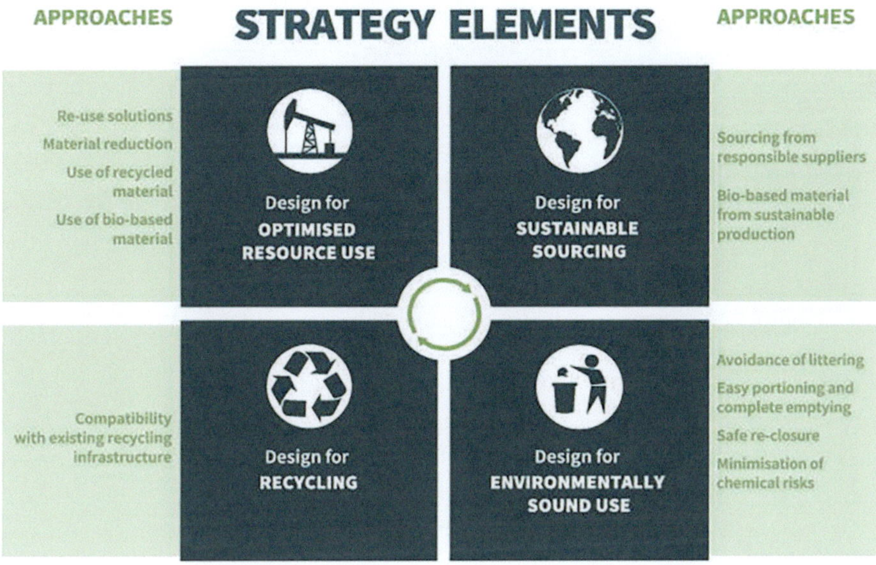

Fig. 6.9 Eco-design strategies (*Source* IK)

Regarding the application of Eco-design as a design principle, the Round Table's guidelines summarize all environmental impacts of plastic packaging in four strategy elements (cf. Fig. 6.9) to minimize the specific environmental impacts. In this way, the current requirements for packaging such as

- improving Design for Recycling,
- increasing the use of recycled material,
- reduction of resource consumption,
- prevention of littering,

are integrated into the strategy elements as optimization approaches.

Based on the following strategy elements of Eco-design interactions and trade-offs can be identified:

- Design for Optimised Resource Use
- Design for Sustainable Sourcing (Design for Sustainable Material Procurement)
- Design for Recycling
- Design for Environmentally Sound Use

An example is the (possible) trade-off between increasing recyclate use and reducing overall resource consumption while ensuring the core functions of a packaging.

Checklists present these interrelationships in a comprehensive manner. At the same time, the guide provides instructions for various management levels, from packaging designers to marketing, and top management, whose decisions ultimately create the conditions for a sustainable packaging solution (cf. Fig. 6.10) (IK 2019).

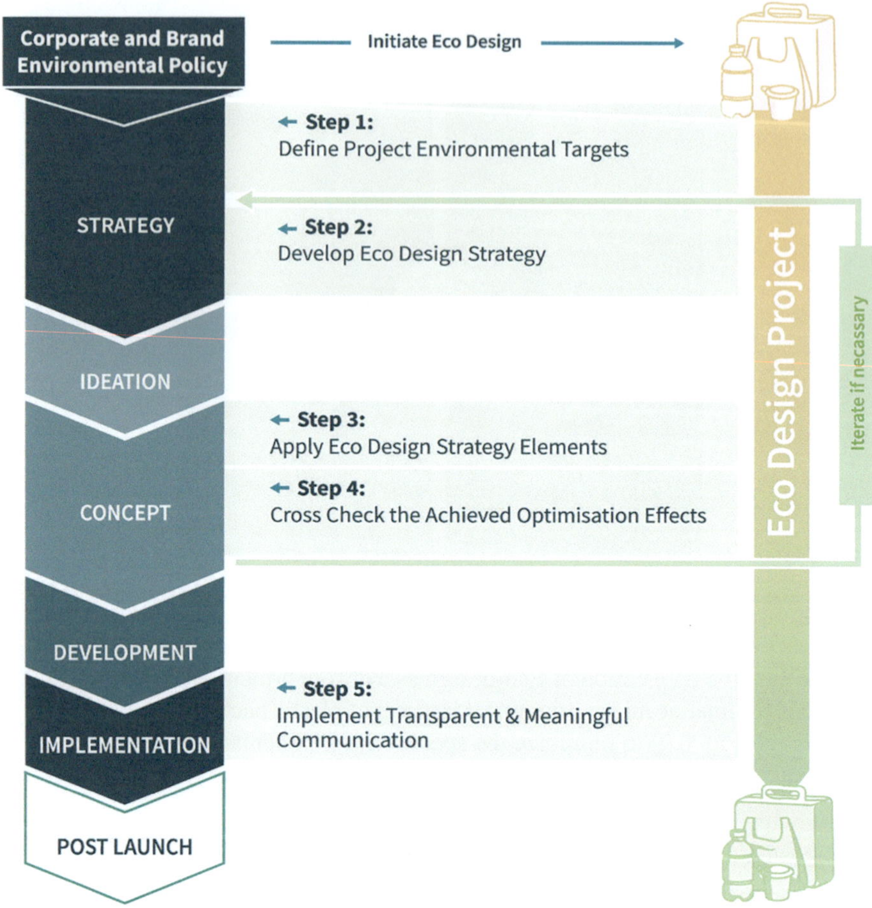

Fig. 6.10 Integration of Eco-design into the management process (*Source* IK)

The Joint Research Centre of the European Commission (JRC) assessed the Round Table guide as one of the top guides in Europe (JRC 2020), because it considers existing tools tried and tested in large companies and integrates the improvement of DfR into the management process of a packaging's design development. A toolbox bundles guidelines and other tools on the categories of Eco-design also from other countries, whereby not surprisingly DfR tools are most frequently represented.

6.4.3 *Design for Recycling of Plastic Packaging*

Design Principle

Design for Recycling (DfR) is a design principle that aims to achieve high recyclability of a product or packaging at an early stage of product or packaging development. As an example of a very advanced and demanding definition of DfR, the Minimum Standard for Recyclability of the Zentrale Stelle Verpackungsregister (ZSVR) is quoted: "Recyclability encompasses the fundamental and gradual suitability of any given packaging to substitute virgin material in applications typical for that material after undergoing recycling processes available on an industrial scale" (ZSVR 2023).

With reference to material-typical applications in this definition, plastic packaging which, after use, can only be used as wood or concrete substitutes (park benches, lawn edging stones, etc.) is classified a priori as non-recyclable. The definition also shows that recyclability is not an immutable, universally applicable packaging property, but must always be related to the specific recycling infrastructure in place. This means that the recyclability of packaging can basically be achieved in two ways:

- Adaptation of the packaging design to the existing recycling infrastructure
- Development of a recycling infrastructure that enables existing design variants of packaging to be recycled to a greater extent.

From this perspective, a lack of recyclability is a consequence of delays in adaptation at both levels (GVM 2022).

Design guidelines

The design for recycling of a product or packaging is an essential prerequisite for meeting legally prescribed recycling input quotas and recycled content requirements and increasing the availability of recyclates in the required volume and quality. It is therefore logical that in recent years many guidelines and evaluation procedures have been developed in Europe for Design for Recycling, particularly for plastic packaging, which is the focus of political attention.[3]

One of the best-known and most widely used DfR guidelines in Europe is that of Recyclass (Recyclass 2024). With its traffic light tables (green, yellow, red), it provides specific information on what can be used in terms of recyclability of the specific packaging group and what should (rather) be avoided. Guidelines with individual traffic light tables are now available for the following packaging groups:

- PET bottles (clear/light blue and transparent)
- PET trays (transparent clear)
- HDPE containers and tubes (natural and coloured)
- PP containers (natural and coloured)

[3] In a study commissioned by BKV nine European and American DfR guidelines and assessments were compared. Based on the comparative studies, recommendations and proposals were developed in preparation of the delegated acts on EU recycling criteria and recycling performance grades for packaging by the EU Commission (BKV 2024).

- EPS containers
- PE films (natural and coloured)
- PP films (natural and coloured)
- PS container dyed
- HDPE and PP crates, boxes, and pallets.

The tables for the packaging groups are similar in structure and indicate what is desirable for the packaging body (subdivided, for example, by material, color, barrier, additives), for the accessories (closure system, seals, etc.) and for the decoration (printing ink, label, adhesive for the label) in each case for recycling, still tolerable and finally to be avoided at all costs.

Recyclass not only includes a design guide, but also offers an evaluation procedure for individual packaging, the result of which is a classification from A (very good) to F (for "fire"—only energetically recoverable). Design guidelines and evaluation procedures are two different things, although they can be closely related.

Regardless of the assessment procedure, each assessment can be used as a yardstick for the recyclability achieved. However, an assessment is also necessary to implement a financial incentive system for the improvement of recyclability envisaged by the legislator in a practicable manner.

Among the assessment procedures, the Minimum Standard of ZSVR deserves special mention, as it had an enormous influence on discussions and measures to improve the recyclability of plastic packaging in Germany and has now also attracted considerable European attention in conjunction with the preparation of the PPWR. The minimum standard is adapted annually to both the state of development of sorting and recycling technology and the development of packaging design by a group of experts from the packaging industry and published by ZSVR in agreement with the Federal Agency of Environment (UBA). The minimum standard is based on three minimum criteria when assessing the recyclability of specific packaging (ZSVR 2023),

- the existence of a sorting and recycling infrastructure for this packaging,
- the sortability and separability of the packaging,
- the absence of recycling incompatibilities.

The result of the evaluation is the available relative recyclable material content in relation to the total packaging, with which its recyclability can be quantitatively assessed.

Concerning the criterion of absence of recycling incompatibilities the Minimum Standard is one of the few in Europe for assessing recyclability that carries out a binary assessment of incompatibilities (BKV 2024).[4]

[4] The annual revision of the Minimum Standard increasingly takes into account incompatible substances that have been proven to hinder recycling and reduce the quality of recyclates. For example, in the 2023 version nitrocellulose (NC) based inks used in adhesive side printing of PE films > A4 size were assessed as incompatible for the first time with the corresponding consequences for the assessment of relevant packaging (ZSVR 2023, Annex 3).

The reference to the (national) sorting and recycling infrastructure in several guidelines may lead to the fact that packaging may be classified as recyclable in one country but not in another. Both the EU Commission and the EU-wide distributing industry are interested in introducing a uniform EU definition of recyclability for individual packaging and a uniform assessment.

Design for recycling criteria and recycling performance grades in the draft of the PPWR

With the PPWR the EU Commission shall be empowered to adopt delegated acts "to establish design for recycling criteria and recycling performance grades" (PPWR 2024, Art. 6). Categories and parameters for the assessment of recyclability shall be established for 22 packaging categories, 13 of them relate to plastic packaging (three based on PET, five on polyolefines, two on PS, three on others). The PPWR draft from March 2024 (PPWR 2024) provides three recyclability performance grades from A to C, whereby grade A represents a recyclability per unit of 95% or higher, grade B of 80% or higher, grade C of 70% or higher. The aim of this fundamental EU regulation is that from 2030 on all packaging placed on the EU market shall be recyclable.

References

Acatech 2021: Circular Economy Roadmap für Deutschland. https://www.circular-economy-initiative.de/circular-economy-roadmap-fr-deutschland. (Last access: Dec. 12, 2022)

Adirson et al. 2021: Adisorn, T; Tholen, L, and Götz, T. Towards a Digital Product Passport Fit for Contributing to a Circular Economy. Energies 2021, 14 (8), 2289. https://doi.org/10.3390/en14082289

AGVU and IK 2021: Arbeitsgemeinschaft Verpackung und Umwelt and IK Industrievereinigung Kunststoffverpackungen e.V., Instrumente zur Steigerung des Rezyklatanteils in Kunststoffverpackungen. https://www.agvu.de/de/instrumente-zur-steigerung-des-rezyklatanteils-in-kunststoffverpackungen-2627/ (Last access: Dec. 12, 2022)

Bartolucci et al. 2023: Bartolucci, L; Cordiner, S; De Maina, E; Kumar, G; Mele, P; Mulone, Vi; Igliński, B; Piechota, G. Sustainable Valorization of Bioplastic Waste: A Review on Effective Recycling Routes for the Most Widely Used Biopolymers. International Journal of Molecular Sciences 2023, 24 (9), 7696. https://doi.org/10.3390/ijms24097696.

Billiet and Trenor 2020: Billiet, S; Trenor, SR. 100th Anniversary of Macromolecular Science Viewpoint: Need for Plastics Packaging Circularity. ACS Macro Letters 2020, 9, 1376–1390. https://doi.org/10.1021/acsmacrolett.0c00437?ref=pdf.

BKV 2024: Vergleich unterschiedlicher Standards zur Bewertung der Recyclingfähigkeit von Kunststoffverpackungen. (Comparison of different standards for assessing the recyclability of plastic packaging). https://www.bkv-gmbh.de/5125-study-comparison-of-different-standards-for-assessing-the-recyclability-of-plastic-packaging.html (Last access: Aug. 20,2024)

BMW 2022a: https://www.bmw.com/en/magazine/sustainability/circular-lab.html

BMW 2022b: https://www.bmw.de/de/home.html (Last access: May 1, 2025)

Braungart and McDonough 1992: Cradle-to-Cradle Center http://www.c2c-center.com/sites/default/files/Casestudy%20Climatex%20IEHN.pdf (Last access: Feb. 1, 2022)

Brouwer et al. 2020: Brouwer, MT; Thoden van Velzen, EU; Ragaert, K; ten Klooster, R. Technical Limits in Circularity for Plastic Packages. Sustainability 2020, 12 (23), 10021. https://doi.org/10.3390/su122310021.

Brown, T. 2016: Change by Design. Wie Design Thinking Organisationen verändert und zu mehr Innovationen führt. Vahlen.

Bundesregierung 1991: Bundesgesetzblatt, Jahrgang 1991, Teil I: Verordnung über die Vermeidung von Verpackungsabfällen. https://www.bgbl.de/xaver/bgbl/start.xav?start=%2F%2F*%5B%40attr_id%3D%27bgbl191s1234.pdf%27%5D#/switch/tocPane?_ts=1746523047072

Buttker et al. 2005: Buttker, B; Giering, R; Schlotter, U; Himmelreich, B; Wittstock, K. Full Scale Industrial Recovery Trials of Shredder Residue in a High Temperature Slagging-Bed-Gasifier in Germany. Sekundärrohstoff Verwertungszentrum Schwarze Pumpe (SVZ), TechnologieentwicklungsGmbH für ökoeffiziente Polymerverwertung (Tecpol), PlasticsEurope. 2005. https://docplayer.net/59757558-Full-scale-industrial-recovery-trials-of-shredder-residue-in-a-high-temperature-slagging-bed-gasifier-in-germany.html. (Last access: Aug. 20,2024)

Catena-X 2024: https://catena-x.net/en/. (Last access: Aug. 20,2024)

CEFLEX 2020: Designing for a Circular Economy, Guidelines. https://guidelines.ceflex.eu/guidelines/ (Last access: Apr. 18, 2024)

CEN 2022: CEN/CENELEC. European Standards support the European Strategy for Plastics in a Circular Economy. https://www.cencenelec.eu/news-and-events/news/2022/brief-news/2022-09-07-plastics-in-circular-economy/. (Last access: Aug. 20, 2024)

Climatex 2022: https://www.climatex.com/nachhaltigkeit/materials/ (Last access: Feb. 1, 2022)

COM 2022a: Ecodesign for Sustainable Products Regulation (ESPR), published on 30 March 2022. https://commission.europa.eu/energy-climate-change-environment/standards-tools-and-labels/products-labelling-rules-and-requirements/sustainable-products/ecodesign-sustainable-products-regulation_en. (Last access: Aug. 20, 2024)

COM 2022b: European Commission & Joint Research Centre: Caldeira, C; Farcal, LR; Garmendia, AI; Mancini, L; Tosches, D. Safe and sustainable by design chemicals and materials—Framework for the definition of criteria and evaluation procedure for chemicals and materials. Publications Office of the European Union, 2022. https://data.europa.eu/doi/https://doi.org/10.2760/487955

Conversio 2021: Stoffstrombild Kunststoffe in Deutschland 2021. Kurzfassung der Studie. https://www.bkv-gmbh.de/files/bkv-neu/studien/Kurzfassung_Stoffstrombild_2021_13102022_1%20.pdf. (Last access: Dec. 12, 2022)

Ding and Zhu 2023: Ding, Q; Zhu, H. The Key to Solving Plastic Packaging Wastes: Design for Recycling and Recycling Technology. Polymers 2023, 15 (6), 1485. https://doi.org/10.3390/polym15061485.

Eco-design 2009: Directive 2009/125/EC of the European Parliament and of the Council of 21 October 2009 establishing a framework for the setting of ecodesign requirements for energy-related products. https://eur-lex.europa.eu/legal-content/EN/ALL/?uri=CELEX:32009L0125 (Last access: Feb. 24, 2024)

Eco-design 2022: Proposal for a Regulation of the European Parliament and of the Council establishing a framework for setting ecodesign requirements for sustainable products. https://eur-lex.europa.eu/legalcontent/EN/TXT/?uri=CELEX:52022PC0142 (Last access: Feb. 24, 2024)

Eisenreich et al. 2021: Eisenreich, A., Füller, J., Stuchtey, M. Open Circular Innovation: How Companies Can Develop Circular Innovations in Collaboration with Stakeholders. Sustainability 2021, 13, 13456. https://doi.org/10.3390/su132313456

Electrolux 2020: https://www.electroluxgroup.com/sustainabilityreports/2020/case-studies/toward-more-circular-products-and-services/ (Last access: Dec. 8, 2023) https://www.electroluxgroup.com/sustainabilityreports/2020/key-priorities-and-progress-2020/our-nine-promises/better-solutions/offer-circular-products-and-business-solutions/#key_priorities_progress (Last access: Jan. 2, 2022)

EMAF 2015: Ellen MacArthur Foundation. Towards the circular economy Volume 1: Economic and business rationale for an accelerated transition. https://www.ellenmacarthurfoundation.org/towards-a-circular-economy-business-rationale-for-an-accelerated-transition. (Last access: Aug. 20, 2024)
EMAF 2022: Ellen MacArthur Foundation. An Introduction to Circular Design. https://www.ellenmacarthurfoundation.org/news/an-introduction-to-circular-design. (Last access: Aug. 20, 2024)
EP 2023: European Parliament position. https://www.europarl.europa.eu/news/en/press-room/20231117IPR12211/new-eu-rules-encouraging-consumers-to-repair-devices-over-replacing-them (Last access: Feb. 22, 24)
ESPR 2022: European Commission. European Eco-design for Sustainable Products Regulation 2022. European Eco-design for Sustainable Products Regulation. https://commission.europa.eu/energy-climate-change-environment/standards-tools-and-labels/products-labelling-rules-and-requirements/sustainable-products/ecodesign-sustainable-products-regulation_en (Last access: Aug. 20, 2024)
Esslinger 2011: Esslinger, H.: Sustainable Design: Beyond the Innovation-Driven Business Model. In: Journal of Product Innovation Management 28, pp. 401-404.
EU 1994: European Parliament and Council Directive 94/62/EC of 20 December 1994 on packaging and packaging waste. https://eur-lex.europa.eu/legal-content/EN/TXT/?uri=celex%3A31994L0062. (Last access: Aug. 20, 2024)
EU 2000: Directive 2000/53/EC of the European Parliament and of the Council of 18 September 2000 on end-of life vehicles. https://eur-lex.europa.eu/legal-content/EN/ALL/?uri=celex%3A32000L0053. (Last access: Aug. 20, 2024)
EU 2008: Directive 2008/98/EC of the European Parliament and of the Council of 19 November 2008 on waste and repealing certain Directives. https://eur-lex.europa.eu/legal-content/EN/TXT/?uri=celex%3A32008L0098. (Last access: Aug. 20, 2024)
EU 2019: Directive (EU) 2019/904 of the European Parliament and of the Council of 5 June 2019 on the reduction of the impact of certain plastic products on the environment. https://eur-lex.europa.eu/eli/dir/2019/904/oj. (Last access: Aug. 20, 2024)
EU 2020: EU Circular Economy Action Plan. https://eur-lex.europa.eu/legal-content/EN/TXT/?qid=1583933814386&uri=COM:2020:98:FIN (Last access: Feb. 1, 2022)
Fuller 1969: Buckminster Fuller, R.: Operating Manual for Spaceship Earth. London: Simon & Schuster 1969.
GVM 2022: Presentation charts from IK General Meeting on Feb. 17, 2022
Hauffe 2000: Hauffe, T, design crash course. Cologne: Dumont 2000.
Haug 2009: Haug, F.: Critique of commodity aesthetics. Berlin: Suhrkamp, 2009.
ICSID 1959: International Center for Settlement of Investment Disputes. History of ICSID. Online at: icsid.org
IDEO & EMF 2022: The Circular Design Guide. https://www.ellenmacarthurfoundation.org/circular-design-guide/methods (Last access: Apr. 29, 2024)
IK 2019: https://ecodesign-packaging.org (Last access: Feb. 24, 2024)
Jensen et al. 2023: Jensen, SF; Kristensen, JH; Adamsen, S; Christensen, A; Waehrens, BV. Digital Product Passports for a Circular Economy: Data Needs for Product Life Cycle Decision-Making. Sustainable Production and Consumption. Vol. 37, May 2023. PP. 242–255. https://doi.org/10.1016/j.spc.2023.02.021.
JRC 2020: https://www.researchgate.net/publication/348481213_Support_to_the_Circular_Plastics_Alliance_in_establishing_a_work_plan_to_develop_guidelines_and_standards_on_design-for-_recycling_of_plastic_products_Final_report. (Last access: Aug. 20, 2024)
Kartell 2016: On Italian Bark interior design blog) https://www.italianbark.com/design-trends-what-is-circular-design/ (Last access: Feb. 1, 2022)
Knappe et al. 2021: Knappe, F; Reinhardt, J; Kauertz, B. Technische Potenzialanalyse zur Steigerung des Kunststoffrecyclings und des Rezyklateinsatzes. Umweltbundesamt. Ressortforschungsplan des Bundesministeriums für Umwelt, Naturschutz und nukleare Sicherheit. Texte 92, 2021. https://www.umweltbundesamt.de/sites/default/files/medien/1410/publikationen/2021-12-10_texte_92-2021_potenzialanalyse-kunststoffrecycling.pdf. (Last access: Aug. 20,2024)

Konietzko et al. 2021: Konietzko, J., Bocken, N.M.P., Hultink, E.J. Circular ecosystem innovation: An initial set of principles. Journal of Cleaner Production. https://doi.org/10.1016/j.jclepro.2019.119942

Mager 2020: Mager, B.: The Future of Service Design (TH Cologne)

Manzini 1990: Manzini, E.: Artifacts: Towards a New Ecology of the Artificial Environment. Domus Academy, 1990

Manzini and Jégou 2003: Manzini, E., Jégou, F. https://www.strategicdesignscenarios.net/wp-content/uploads/2012/05/SUSTAINABLE-EVERYDAY_-Scenarios-of-urban-life.pdf (Last access: Feb. 1, 2022)

Miller 2019: Herman Miller https://www.hermanmiller.com/better-world/sustainability/ (Last access: Feb. 1, 2022)

Möbel 2023: Furniture Digital. https://www.moebeldigital.de/de/kreis-training-fuer-den-greendeal/ (Last access Feb. 21, 2024)

Niessner 2022: Niessner, N. (Ed.): Plastics Recycling. Carl Hanser Verlag. 2022.

Norman 2022: Norman,D., Nielsen Norman Group. https://www.nngroup.com/videos/observe-test-iterate-and-learn-don-norman/ (Last access: Feb. 4, 2022)

O2 2022: Global Network. https://www.o2.org/about-o2 (Last access: Feb. 1, 2022)

OECD 2021: A Chemicals Perspective on Designing with Sustainable Plastics: Goals, Considerations and Trade-offs. https://doi.org/10.1787/f2ba8ff3-en.

OECD 2023: Impact by Design. Effective Results Frameworks for Sustainable Development. https://one.oecd.org/document/DCD(2023)17/en/pdf. (Last access: Aug. 20, 2024)

Papanek 1985: Papanek, V.: Design for the Real World: Human Ecology and Social Change. New York: Van Nostrand Reinhold, 1985.

PF I4.0 2021: Plattform Industrie 4.0. Thesenpapier Industrie 4.0 und Nachhaltigkeit. Zehn Thesen, wie digitale Geschäftsmodelle Nachhaltigkeit in der Industrie 4.0 fördern. https://www.plattform-i40.de/IP/Redaktion/DE/Downloads/Publikation/Thesen-Nachhaltigkeit-Geschaeftsmodelle.html. (Last access: Dec. 12, 2022)

Plastics Europe 2022: Plastics—the Facts 2022. https://plasticseurope.org/knowledge-hub/plastics-the-facts-2022/. (Last access: Aug. 20, 2024)

PlasticsEurope 2022: Plastics—the Facts 2022: An analysis of European plastics production, demand and waste data. PlasticsEurope. Association of Plastics Manufacturers. https://plasticseurope.org/knowledge-hub/plastics-the-facts-2022/ (Last access: Dec. 20, 2022)

Potting et al. 2017: Potting, J., Worrell, E., Hekkert, M. P. Circular Economy: Measuring innovation in the product chain. PBL Netherlands Environmental Assessment Agency. The Hague.

PPWR 2024: Council of the European Union, Mar. 15, 2024. Outcome of proceedings. https://data.consilium.europa.eu/doc/document/ST-7859-2024-INIT/en/pdf (Last access: May 20, 2024)

Rams 2017: Rams, D. "10 theses for good design". Munich: Prestel, 2017

Recyclass 2024: https://recyclass.eu/recyclability/design-for-recycling-guidelines/ (Last access: Feb. 24, 2024)

RNE 2021: Rat für nachhaltige Entwicklung. Statement. Circular Economy: Leveraging a Sustainable Transformation. https://www.nachhaltigkeitsrat.de/wp-content/uploads/2022/02/20211005_RNE-Statement_Circular-Economy-1.pdf. (Last access: Dec. 12, 2022)

Sachs 2015: Sachs, Jeffrey. The Age of Sustainable Development. New York: Columbia University Press, 2015.

Salustri and Nathan 2007: Salustri, F.A.; Nathan, E.L.: Design as…: Thinking of what Design might be. In: Journal of Design Principles and Practices. Vol. 1, No. 1. 2007, p. 20.

Sassanelli et al. 2020: Sassanelli, C; Urbinati, A; Rosa, P; Chiaroni, D; Terzi, S. Addressing circular economy through design for X approaches: A systematic literature review. Computers in industry. Vol. 120. September 2020, 103245. https://doi.org/10.1016/j.compind.2020.103245.

Schneider 2005: Schneider, B: Design—an introduction. Design in a social, cultural, and economic context. Berlin: Birkhäuser 2005.

Schumpeter 2013: Schumpeter, J. Adapted from: Harvard Business Manager. https://www.manager-magazin.de/harvard/management/joseph-schumpeter-innovation-und-schoepferische-zerstoerung-a-00000000-0002-0001-0000-000091405742 (Last access: Dec. 8, 2023)

Schütz 2020: Schütz, F. Das Geschäftsmodell kollaborativer Innovation. Eine empirische Analyse zu funktionalen Rollen in Quadruple-Helix-Innovationsprozessen. Dissertation. Technische Universität Berlin.

Selvefors and Renström 2018: Selvefors, A.; Renström, S.: Sustainability Guide (Hållbarhetsguiden, © SVID, Stiftelse Svensk Industridesign, 2018) https://sustainabilityguide.eu/methods/design-sustainable-behaviour (Last access: Feb. 21, 2024)

Sild 2022: Sild, S. Design for Sustainability. https://fractory.com/design-for-sustainability/#:~:text=Design%20for%20Sustainability%20(also%20sometimes,generation's%20ability%20to%20use%20them. (Last access: Aug. 20, 2024)

SPRIND 2022: Federal Agency for Jump Innovations—SPRIND. https://www.sprind.org/de/ (Last access: Feb. 1, 2022)

Staeffler 2024: Staeffler, G.: Interpretation of long-lasting design. Braun coffee grinders. In: Design+Design 67, March-May 2004, 3–7. https://www.designundtext.com/pdf/design-und-design/2.1.68.pdf. (Last access Feb. 1, 2022)

Stickdorn and Schneider 2011: Stickdorn, M.; Schneider, J.: This is Service Design Thinking. Amsterdam: BIS publishers, 2011.

Studio 7.5 2003: Studio 7.5 Berlin. https://www.seven5.com/pages/en.php (Last access: Feb. 1, 2022)

Stylepark 2023: https://www.stylepark.com/de/kartell/organic-chair (Last access: Dec. 9, 2023)

SYSTEMIQ and PlasticsEurope 2022: ReShaping Plastics: Pathways to a Circular, Climate Neutral Plastics System in Europe. https://plasticseurope.org/wp-content/uploads/2022/04/SYSTEMIQ-ReShapingPlastics-April2022.pdf. (Last access: Dec. 22, 2022)

Tan 2023: Tan, J; Jia, S; Ramakrishna; S. Accelerating Plastic Circularity: A Critical Assessment of the Pathways and Processes to Circular Plastics. Processes 11 (5) 2023, 1457. https://doi.org/10.3390/pr11051457

Tischler and Moser 2023: Tischler, U.; Moser, H.: How to do Ecodesign? UBA 2023. https://www.umweltbundesamt.de/publikationen/how-to-do-ecodesign (Last access: Aug. 20, 2024)

UNEP 2009: United Nation Environmental Programme; Delft University of Technology. Design for Sustainability. A Step-by-Step Approach. https://wedocs.unep.org/bitstream/handle/20.500.11822/8742/DesignforSustainability.pdf?sequence=3&amp%3BisAllowed=. (Last access: Aug. 20, 2024)

UNEP 2023: United Nations Environment Programme. Synthesis report on the submissions received on elements not discussed at the second session, such as principles and scope of the instrument. https://wedocs.unep.org/bitstream/handle/20.500.11822/43802/SynthesisPaper.pdf?sequence=1&isAllowed=y. (Last access: Aug. 20, 2024)

Van den Beuken et al. 2023: Van den Beuken, E; Bertling, J; Kabasci, S; Ligthart, T; Schulte, A; Stegmann, P; Urbanus, JH; van Harmelen, T. TNO innovation for life; Fraunhofer UMSICHT. White paper. From #plasticfree to future-proof plastics. How to use plastics in a circular economy. https://www.umsicht.fraunhofer.de/content/dam/umsicht/en/documents/publications/2023/White-paper-From-plasticfree-to-future-proof-plastics.pdf. (Last access: Aug. 20, 2024)

VCI 2020: Branchenstrategie zur zirkulären Wirtschaft der deutschen chemischen Industrie. https://www.vci.de/ergaenzende-downloads/2020-03-11-branchenstrategie-zirkulaere-wirtschaft-vci.pdf. (Last access: Dec. 12, 2022)

VDI 2019: VDI Directive 2221: Design of technical products and systems - Model of product design. https://www.vdi.de/fileadmin/pages/vdi_de/editors/guidelines/content-directory/3091310.pdf (Last access: Feb. 1, 2022)

VDI 2022a: Circular Economy für Kunststoffe neu denken. Wie die Transformation zur zirkulären Wertschöpfung gelingen kann. https://www.vdi.de/themen/politischer-dialog/dialog-events/vdi-round-table-circular-economy. (Last access: Dec. 12, 2022)

VDI 2022b: VDI/VDID Guideline 2424: https://www.vdi.de/linien/details/vdivdid-2424-industrie design-nutzerzentre-gestalt-im-produktentwicklungverfahren (Last access: Feb. 1, 2022)
VDI TZ 2021: VDI Research. Digitale Zwillinge—heute schon weit verbreitet, morgen Allerorten. https://www.vditz.de/service/publikationen/details/digitale-zwillinge-heute-schon-weit-verbreitet-morgen-allerorten. (Last access: Dec. 12, 2022)
VDI-ZRE 2024: VDI Center for Resource Efficiency. Product service systems. https://www.ressou rce-deutschland.de/werkzeuge/loesungsentwicklung/strategies-massnahme/produkt-service-systeme/ (Last access: Feb. 21, 2024)
VDMA 2019: Kreislaufwirtschaft 4.0: Wie die Digitalisierung die Kreislaufwirtschaft ankurbeln kann—Positionspapier. VDMA European Office.
Vitsoe 2024: https://www.vitsoe.com/de/ueber-vitsoe/gutes-design (Last access: Feb. 24, 2024)
Von Vacano et al. 2022: Von Vacano, B; Mangold, H; Vandermeulen, GWM; Battagliarin, G; Hofmann, M; Bean, J; Künkel, A. Sustainable Design of Structural and Functional Polymers for a Circular Economy. Angewandte Chemie: International Edition 2022. https://doi.org/10.1002/anie.202210823.
Vytal 2022: https://www.vytal.org/faq (last access: Feb. 1, 2022)
Wilts and Berg 2017: Wilts, H., Berg, H. Digitale Kreislaufwirtschaft: Die Digitale Transformation als Wegbereiter ressourcenschonender Stoffkreisläufe. Wuppertal Institut. https://nbn-resolving.org/urn:nbn:de:bsz:wup4-opus-69775. (Last access: Dec. 12, 2022)
WiWo 2023: Wirtschaftswoche: The 50 most innovative companies in the world (June 2, 2023) https://www.wiwo.de/technologie/forschung/fuenf-deutsche-unternehmen-verstellen-das-sind-die-50-innovativsten-unternehmen-der-welt/29183742.html (Last access: Dec. 9, 2023)
ZSVR 2023: https://www.verpackungsregister.org/fileadmin/files/Mindeststandard/Minimum_standard_Packaging-Act_Edition_2023.pdf (Last access: Feb. 24, 2024)

Prof. Dr. Bita Fesidis is a full-time lecturer at FOM University in Düsseldorf. She holds a professorship for innovation management and digitalization and is a co-opted scientist at the Competence Center for Technology & Innovation Management. She also works as a project manager at VDI. Bita Fesidis studied technical communication at RWTH Aachen University and earned her doctorate while working. Through various professional positions, including in the packaging industry, she has relevant practical experience in the areas of digitalization and sustainability.

Dr. Volker Brennecke was head of the Politics and Society Department of the VDI and Director of Technology in Dialogue. He studied political and social sciences and history in Hanover and Hamburg and completed a technical vocational training. Among others, he deals with dialogic processes in technology design between industry, society, and politics, especially in large industrial infrastructure projects. He initiated VDI's round table on the plastics industry and led the process.

Dr. Klaus Wittstock is Director Industry Affairs with BASF SE. He studied chemical engineering at the Technical University of Karlsruhe, where he also obtained his doctorate. He has been working in the field of plastics recycling since 2001, as an expert in the plastics application technology group, then for plastics and recycling sustainability. He is Deputy Chairman PlasticsEurope's Deutschland board, a member of BKV's advisory board, and chairman of working groups on sustainable consumption and production with VCI and BDI.

Victoria Wessolowski heads the Global Advocacy for Plastics project at BASF SE. Together with an international network of colleagues, she develops advocacy processes and corporate positions. She did English and American Studies as well as Business Administration with a focus on International and Strategic Management at the University of Mannheim.

Martin Beeh studied Industrial Design and Economics at Darmstadt University of Applied Sciences and ENSCI-Les Ateliers, Paris. He worked at Décathlon International in France and with Electrolux as Industrial Designer and Design Manager. He then became General Manager of Material ConneXion in Cologne, founded beeh_innovation, focussing on design management and materials expertise, and was Professor for Design Management at the University of Applied Sciences in Lemgo. In 2018, he created the platform materials.cologne. He is a red dot product design juror and a member of VDID.

Dr. Najine Ameli studied integrated product design at Coburg University of Applied Sciences and art and design sciences at the University Duisburg-Essen. She worked for the Wuppertal Institute for Climate, Environment, and Energy and later on at Bochum University of Applied Sciences where she also received her Ph.D. In 2021 she co-founded the "Library of Things Bochum". Since then, she works as Managing Director of this project.

Prof. Dr. Oliver Stengel has an education in ethnology, philosophy, and educational science from Heidelberg University and sociology, psychology, and political science with Universität Jena where he received his Ph.D. He then worked as a research assistant with Jena Universität, the Wuppertal Institute, Ruhr University Bochum, and the University of Duisburg-Essen. Since 2021, he is Professor for Sustainable Development at Bochum University of Applied Sciences.

Dr. Jürgen Bruder is a chemist with an additional degree in business. He worked in higher education and the chemical industry, before becoming Director General of IK, the German plastic packaging association. He held positions in national, European, and global committees including the Global Plastics Alliance. Today he deals with the transformation of the plastics industry to a circular economy focusing on plastic packaging and plastic applications in agriculture.

Chapter 7
Product—Converters and Applications

Michael Heyde, Benjamin Kampmann, Frank Schockemöhle, Johannes Wedi, Thomas Hülsmann, and Karl Edsjö

Contents

M. Heyde (✉)
Alpla Werke Alwin Lehner GmbH & Co KG, Markdorf, Germany
e-mail: Michael.Heyde@alpla.com

B. Kampmann (✉) · F. Schockemöhle
Pöppelmann GmbH & Co KG, Lohne, Germany
e-mail: benjaminkampmann@poeppelmann.com

F. Schockemöhle
e-mail: FrankSchockemoehle@poeppelmann.com

J. Wedi (✉)
Bischof+Klein SE &Co KG, Lengerich, Germany
e-mail: Johannes.Wedi@bk-international.com

T. Hülsmann (✉)
VinylPlus Deutschland e.V., Bonn, Germany
e-mail: thomas.huelsmann@vinylplus.de

K. Edsjö (✉)
Saltsjö-Boo, Sweden
e-mail: karl.edsjo@cirkelar.com

P. Orth et al. (eds.), *Circular Economy of Plastics*,
https://doi.org/10.1007/978-3-031-78829-1_7

Summary

Chapter 7 focuses on the contribution of plastics converters and product manufacturers to the development of a circular plastics economy as presented in the product segment of the circuit diagram (cf. Fig. 7.1).

Concrete practical experiences from renowned companies show the challenges that these actors in the cycle have to overcome in terms of design for recycling, own production or sourcing of recyclates, usage of recyclates, and conformity of products with recycled content, and which solutions they have already achieved. The examples from the areas of non-food packaging, both rigid and flexible, technical parts for the automotive and construction/building industries, and electrical household appliances provide a broad overview, in particular of the issues and problems of continuous and safe use of recyclates as HDPE, PP, LDPE, PVC, or ABS. Readers from this area of the plastics cycle will be particularly interested in how these companies have gradually acquired know-how on recyclates and their use, sometimes in the process from simple to very sophisticated applications. Conclusions and outlooks are given by the authors among others for the recyclability of flexible multi-layer packaging.

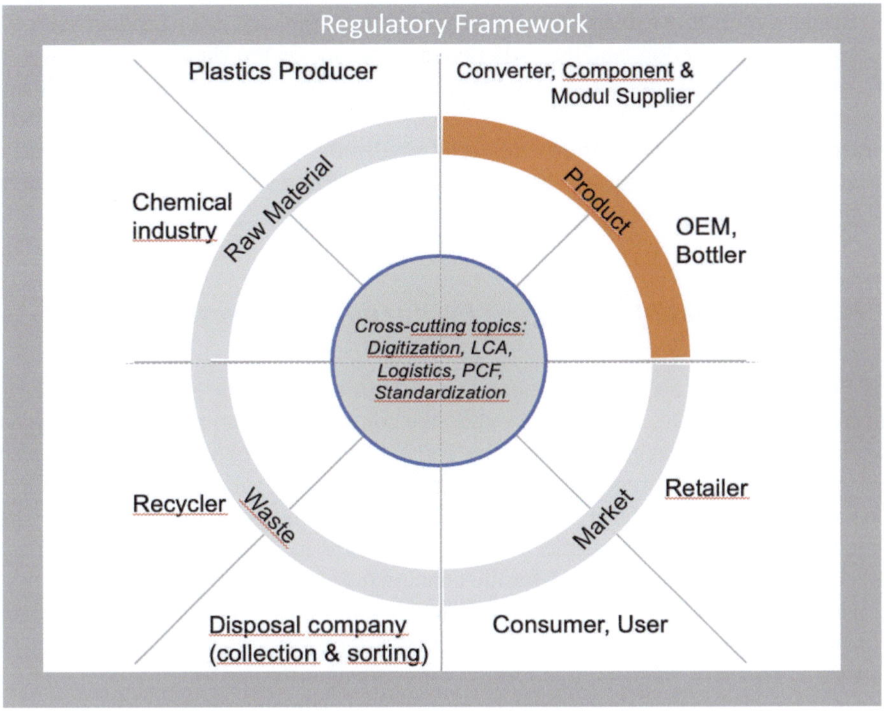

Fig. 7.1 Schematic diagram of the plastics cycle. (*Source* Own presentation)

7.1 Closing the Loop for Non-Food Plastic Bottles

Michael Heyde

7.1.1 Introduction

The use of recycled plastics is beneficial for the environment in all applications when reducing virgin plastics consumption. This applies if the amount discussed is small compared to the total material consumption. With increased recycled content the match between individual material properties and specific requirements of the application will gain importance in terms of material efficiency. Closing the loop into similar or equivalent applications is one option to meet this requirement as it sustains property profiles specific to the respective application. Impurities and detrimental blends caused by product design are becoming transparent immediately. They can be controlled and compensated if all relevant players in the value chain are willing to contribute:

- Brand owners make decisions about packaging design, driven by product aspects, consumer protection, and the esthetical appearance on the shelf;
- Packaging producers must match the requirements of brand owners with the material they get;
- Recyclers must cope with raw material grades they get from the market (baled and/or presorted plastics waste) using available processing technology to meet producers' requirements.

7.1.2 Synergies of Packaging Production and Recycling

ALPLA Group as a major producer of rigid packaging with global presence, is strongly committed to circularity. By 2025 the complete packaging volume produced worldwide by ALPLA is planned to be 100% recyclable and to contain a recycled content of 25% on average. ALPLA's Recycling Division runs recycling capacities on four continents for HDPE and PET with an output capacity of 350 Mt in 2024. The objective is to use post-consumer packaging waste as input material for mechanical recycling to produce new bottles for the consumer goods industry. Close-to-the-end-consumer applications have demanding quality requirements: package function and consumer safety are not negotiable and esthetical appearance is hard to negotiate with most brand owners. ALPLA copes with these challenges through internal cooperation in packaging design, packaging production, and recycling technology.

7.1.3 Historical Background and Challenges for HDPE Bottle Recycling

Mechanical recycling of HDPE containers was one of the very early activities in plastics recycling already performed 40 years ago when rigid HDPE from consumer packaging was identified as an interesting raw material source. HDPE is mostly used in blow-molded hollow bodies made from mono-material. Depending on the intended application and on typical initial contamination the material can be cut into flakes and pelletized even without upfront washing. rHDPE is produced across Europe from used bottles stemming from private households. Thick-walled products for the construction industry have been the typical application of recycled material for a long time. The re-introduction of rHDPE into bottles has failed for many years for three reasons:

(1) The mixture of colors used for consumer packaging results in an unattractive grey tone for products using rHDPE. Despite an intensive use of additional pigments via adding masterbatches product design was restricted to a narrow range of dark colors with no chance for a bright and attractive appearance;
(2) Cross-contamination even with small amounts of PP deteriorated properties;

(3) The typical odor of recycled HDPE was inacceptable for consumer applications.

Technical solutions developed to solve these problems are as follows:

(1) Color Separation

The color split of HDPE bottles coming from collection schemes is attractive for color sorting in many cases: in many regional markets the share of white and "natural" bottles is 40—60 % and two more dominating colors stand for 10—15 % of the total. Hence, a sorting concept focusing on 4 color groups has a yield potential of appr. 70 %. The remaining material may still be used for conventional applications in grey color.

(2) Cross-Contamination

Caps and closures have always been looked at as a quality challenge for rHDPE from household bottles, because of different colors and resin types. Caps for HDPE bottles are mostly made of PP and many bottles enter the recycling process with their caps. But PE and PP are not compatible in compounding and melt flow behavior differences of injection grade PP and extrusion blow moulding grade HDPE result in an inhomogeneous material distribution (PP zones in a continuous HDPE phase) and mechanical instability, e.g. in weld lines. The technical solution is flake sorting. Flake sorters are excellent in removing residual "foreign" colors from a main material and hence, removing the cap PP which as a flake isn't attached to the bottle HDPE any longer.

(3) Odor

Smell of recycled HDPE is caused by volatile substances left in the resin during the use phase of its second life. Most of those substances have been captured by the polymer during its first use phase, as they migrate from the filling into the bottle wall. The dominant applications of rHDPE bottles in end-consumer markets are home care, body care, and cosmetics. In all cases, bottles contain a lot of aromas and perfume substances not welcome in the polymer's next application. State-of-the-art approach to removing these substances is a modified drying process offered by different suppliers for this application.

7.1.4 Standard Process for HDPE Bottle Recycling

The schematic structure of a recycling process for post-consumer rHPDE is depicted in Fig. 7.2.

Presorted bales of post-consumer HDPE bottles are available in many countries all over the world. They may be purchased from interfaces provided by producer responsibility schemes (PRS) in Europe (Grüner Punkt 2023) or from private collectors in countries with limited formal waste management structure. The quality and value of those bales depend on.

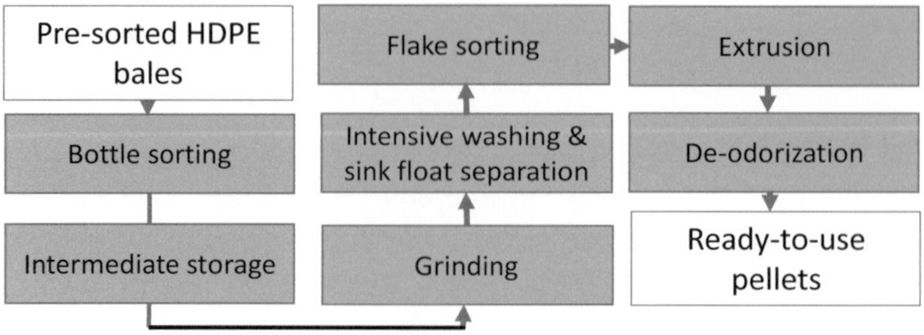

Fig. 7.2 Schematic recycling process for post-consumer HDPE bottles (*Source* Own representation)

- HDPE content, content of packaging items;
- Consistency of previous application (household packaging, industrial packaging);
- Size of packaging items;
- Type of impurities (rigid plastics, film, paper, organics, other mixed waste, fines);
- Surface contamination.

7.1.4.1 Color Separation

Bottle Sorting

During bottle sorting all non-HDPE items are removed using NIR-sorting technology (cf. Sects. 9.3 and 10.1) and the target material is separated into color groups. Target color groups are defined following the local distribution of colors in the input material and market demand for "mono color" rHDPE pellets. Separation by color and material is performed in state-of-the-art installations with the same sorting machines thus increasing efficiency. After conventional mechanical pretreatment with bale opening, removal of fines, oversized items, and "2-dimensional" material like paper and films, the target color groups can be sorted positively in one step each (cf. Figs. 7.3 and 7.4).

Sorted products are stored in bunkers and/or bales. Residues are separated and prepared for disposal as bales or in containers. Each color fraction factually represents a variety of colors defined by sorting parameters. The bandwidth of colors in one fraction affects the yield but not necessarily the color stability, as far as the composition remains stable. Figure 7.5 shows examples of color variations from Spanish household material.

Washing

Most washing installations use multi-step washing processes preferably equipped with individual water circuits for each step following the counter current principle for washing water and product stream handling. At best the product is de-watered after each washing step. Most impurities are typically removed during the first step

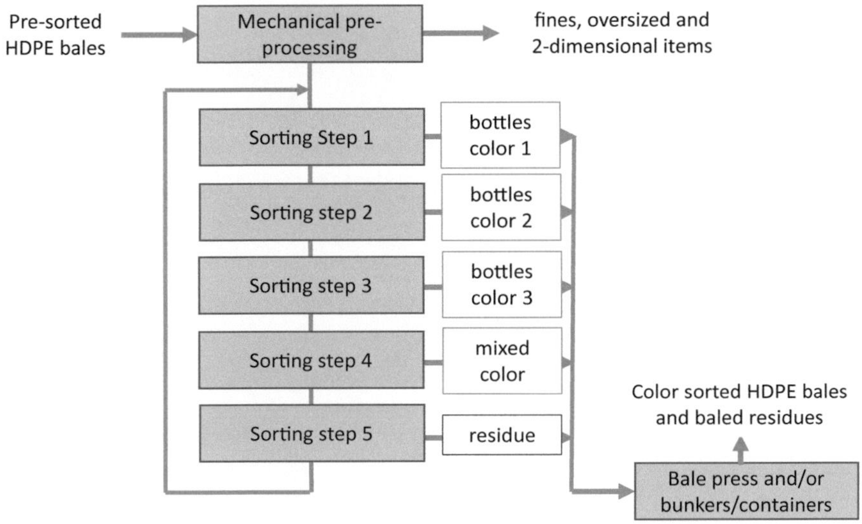

Fig. 7.3 Schematic bottle sorting process. (*Source* Own representation)

be it a bottle prewashing or a wet grinding. Material grinding can be done in one or two steps with a slowly moving shredder in the first step, and a fast-moving grinder in the second step. Separation of non-floatable material is mostly applied in more than one of the washing steps. "Two-dimensional" materials like labels not being removed via sink fractionation consist of polyolefins and can contribute to the overall yield of the process. If color and odor are relevant thin flexibles are removed and valorized as a separate stream. Some machine manufacturers apply this process step as close as possible to the position where the flexibles are removed from the surface to prevent them from releasing soluble or semi-soluble contaminants (colorants, inks, glues, etc.) into the process water (cf. Fig. 7.6).

The addition of washing agents and increasing process temperatures necessary to activate the washing agents can be helpful. The variety is huge but published technical knowledge is limited. The influence of washing agents on quality parameters like detectable residues of critical substances must be examined to improve the washing technology.

7.1.4.2 Cross Contamination

Flake Sorting

Efficient flake sorting systems (cf. Sects. 9.3 and 10.1) are available from different suppliers facilitating separation setups with several purifying and re-sort steps securing reasonable purity and yield (cf. Fig. 7.7). Color-sorted bottles still contain

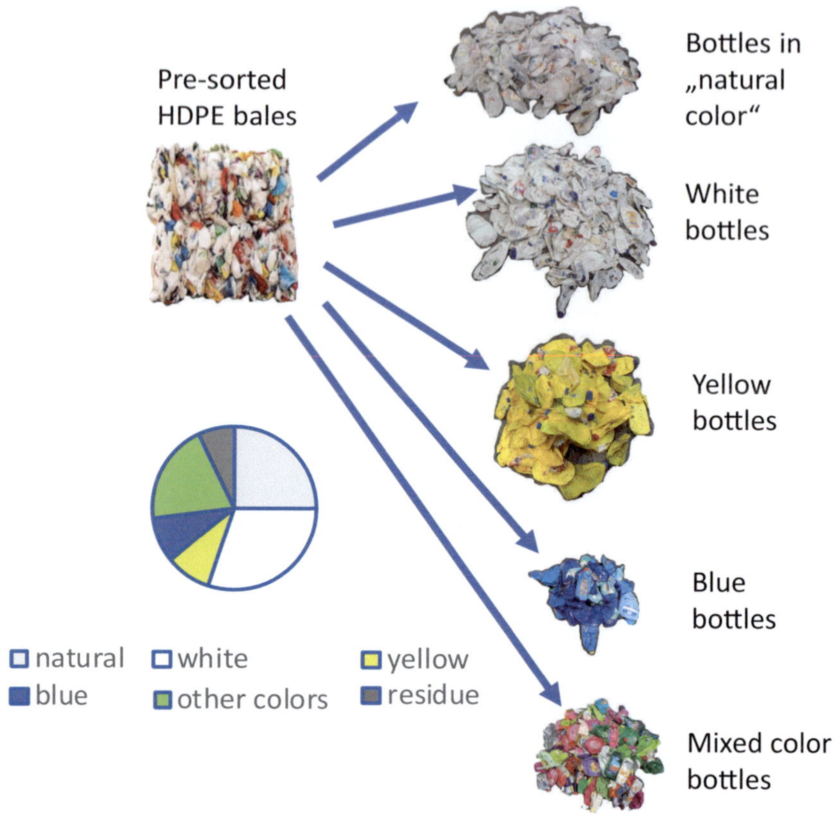

Fig. 7.4 Bottle sorting according to colors. (*Source* Own representation)

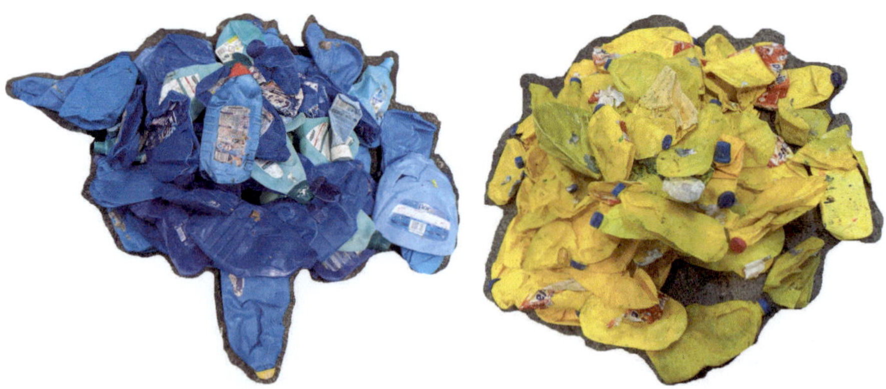

Fig. 7.5 Color variation in blue and yellow color groups. (*Source* Own representation)

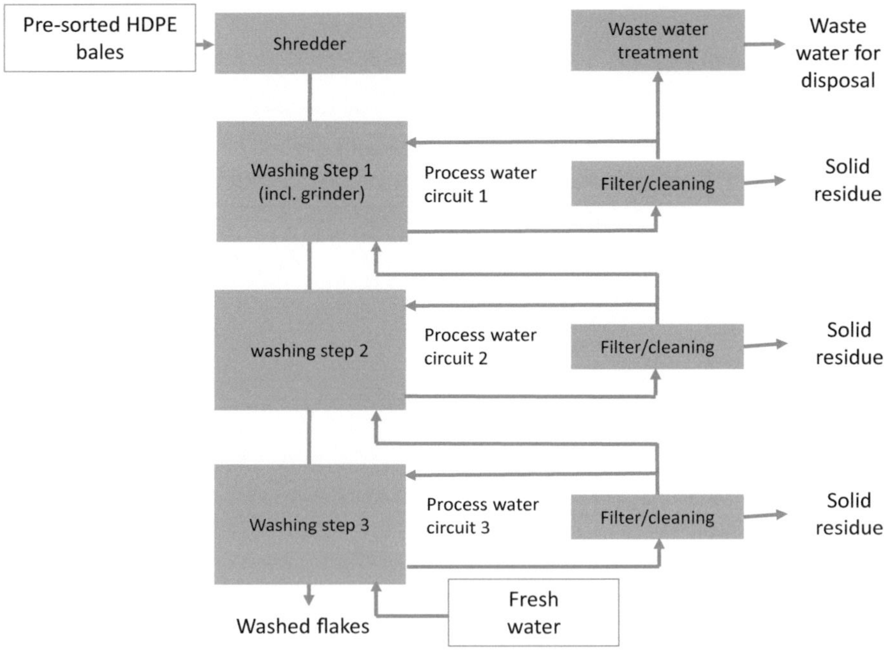

Fig. 7.6 Schematic counter current process. (*Source* Own representation)

items of non-targeted color and the color of the closure may deviate from the main color as shown in Fig. 7.5.

Figure 7.8 visualizes how the flakes look like after washing: the caps now cut into pieces can easily be separated by flake sorting. The separation of "foreign" colors works to a satisfying extent with two sorting steps in line ("purifiers"). The rejects of both purifiers are transferred to a third sorting step separating the final reject from residual flakes of target color. They are recirculated to the front end. This recirculation step reduces the overall capacity, but it reduces losses of target color flakes significantly.

Extrusion

HDPE flakes are mostly processed in single-screw extrusion systems with vacuum degassing, melt filtration, and pelletizing. Pre-heating units are recommended for removing some volatiles (VOCs) upfront, whereas others are removed in the degassing zone of the extruder. Melt filtration can be applied before or after degassing. For high contaminated grades two filtration steps can be applied but, in most cases, the second filter is omitted while increasing the cleaning effort in the washing process. Filtration mesh sizes vary between 80 and 150 µm. The dosing of additives is not very widespread for rHDPE. Adding a masterbatch is recommended to adjust the color directly during the production of the final product. For pelletizing the water ring technology is most common.

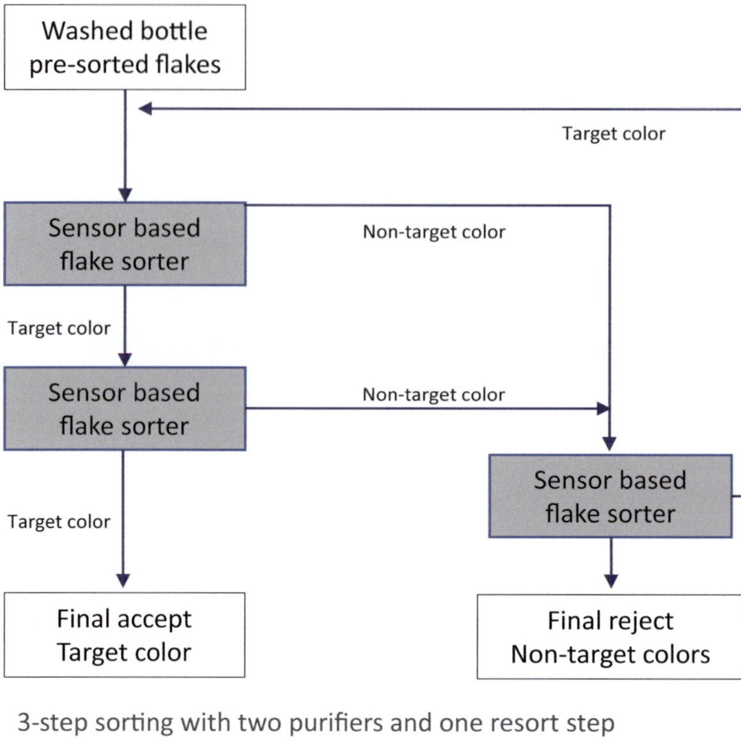

Fig. 7.7 Schematic flake sorting process. (*Source* Own representation)

Fig. 7.8 Flakes from presorted yellow bottles, final reject and final accept of process. (*Source* Own representation)

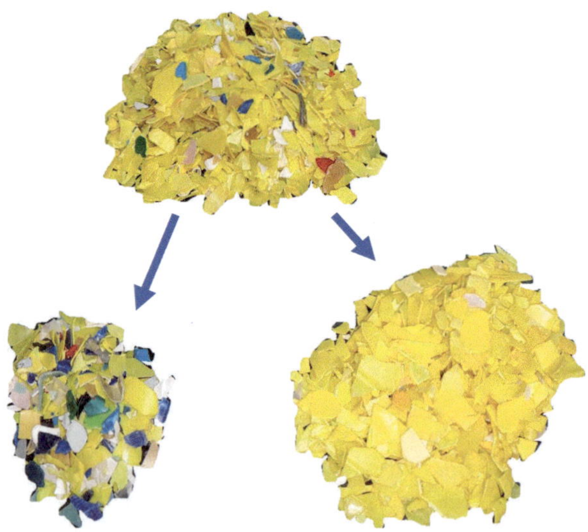

7.1.4.3 Odor Removal

Powerful equipment for the removal of volatile substances from rHDPE and other resins mostly based on modified drying equipment is available in the market. Technology is described in patents (e.g. Hallaji 2017) and works for many applications and input materials quite successful. With a given set of process parameters (temperature, air flow, and residence time) it is obvious that odor can be reduced by increasing the duration of the treatment in drying equipment as mentioned above. Evaluation/approval of odor reduction can be provided by sensory panels (Heyde 2016) Reduction of odor goes along with the reduction of VOCs (Heyde 2016; Kitzberger 2017).

Further improvement in terms of better results and/or better efficiency struggles with the lack of technically detectable target parameters. The measurement of volatiles by headspace GC-FID has already been applied in an early phase of the development and the limonene peak in a typical chromatogram was chosen as an indicator for deodorization performance due to its reliable occurrence in most European post-consumer HDPE samples (cf. Fig. 7.9). But it became obvious that the peak size did not represent the sensory significance of the substances behind. While combining a GC with a sniffing port (GC-O) allows panelists to notice intensity and kind of odor related to a specific retention time and hence to a specific substance, it could be shown that substances with large peaks are not significant in terms of odor and others with strong odor are found at low concentration or were even not detectable with the detector in place (Kitzberger 2017).

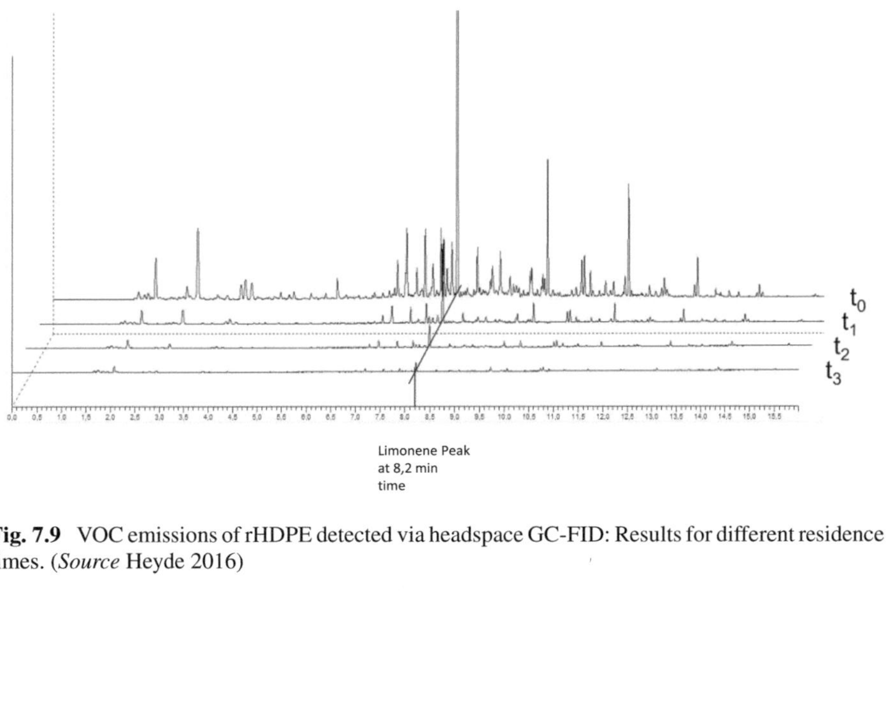

Fig. 7.9 VOC emissions of rHDPE detected via headspace GC-FID: Results for different residence times. (*Source* Heyde 2016)

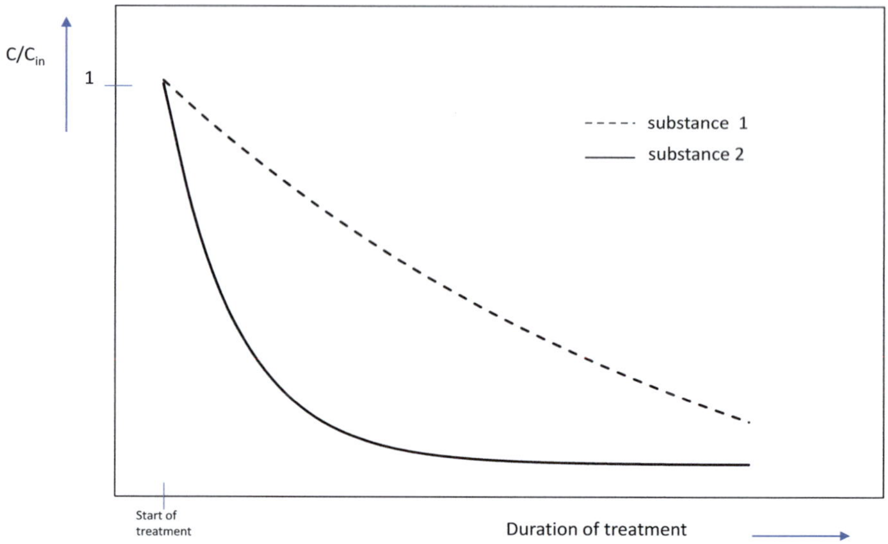

Fig. 7.10 Concentration of volatile substances C emitted by rHDPE samples depending on the duration of odor treatment related to initial concentration C_{in}. (*Source* Own representation)

The removal of volatiles with increasing residence time in the drying unit depends on the molecular weight and molecular structure of the substance to be removed: while substances with low mass transfer resistance are almost completely removed after relevant treatment time (substance 2 in Fig. 7.10) others remain at 50% or more of their initial concentration independent of its absolute value.

Strangl et al. (2019) improved the detectability of substances relevant to odor formation by concentrating them via dichloromethane extraction and distillation. A two-dimensional (2D) GC–MS/O was applied to the stepwise re-diluted concentrate, using one dimension for the sniffing port and the other for a mass spectrometer. The dilution ratio which decreases the concentration below the level of recognition for a specific substance, represents the quantification of the sensory significance of this substance in its environment. The authors applied that procedure to rHDPE samples taken from a deodorization process with different treatment times. It could be demonstrated that most of the relevant substances studied can be reduced after 7 h down to less than 5 percent of the initial level (cf. Table 7.1).

Migration Potential

Interaction between filled products and packaging made of recycled plastics was discussed predominantly for food packaging. HDPE may absorb a lot of substances from its environment during the first use phase and will release them during the next one. If the specific application implies a long-term full surface contact as for liquids in bottles the probability is high of finding substances in the product originating from the plastic's first use phase. First use includes both intended use and un-intended use

Table 7.1 Removal of sensory-relevant substances from post-consumer rHDPE after 7 h of thermal treatment (*Source* Strangl et al. 2019)

Odorant	Decontamination efficiency (%)	Molecular weight (g/mo)	Boiling point at 760 mm Hg (°C)
(*Trans*)-anethole	99.3	148.2	234
(*RS*)-(±)-linalool	97.1	154.3	198
2-methoxynaphthalene	97.1	158.2	274
γ-undecalactone	94.7	184.3	258
Verdyl acetate	94.5	192.3	297
β-Ionone	90.6	192.3	271
Octanal	85.1	128.2	171
Dodecanoic acid	64.0	200.3	299
(E)-oct-2-enal	51.0	126.3	206–208

including misuse like storage of toxic substances. EU Reg 1616/2023 (EU 2023) therefore requires.

- $>99\%$ of food packaging in the input of the recycling process and
- Proof of cleaning efficiency via challenge test.

Challenge tests are performed with artificially contaminated virgin plastics using a defined set of surrogates with different properties. The cleaning efficiency required depends.

- On the probability of toxic contamination caused, e.g. by misuse of the European Food Safety Authority (EFSA) derives from published data (cf. Franz et al. 2022) and
- On exposure scenarios, depending on the type of food, consumption scenario, and packaging size.

The latest published EFSA opinion about an available process already introduced to the market is dated January 2022 (EFSA 2022). The publication states the "…limited efficiency of the technology to decontaminate contaminants with molecular weight above 400 Da which would migrate into water and other beverages." This statement makes clear somehow where a focus of technology development needs to be.

7.1.5 Some Market Aspects

The discussion about recycled HDPE for direct food contact somehow passes by market reality in Europe. Food packaging using HDPE plays a significant role in milk and dairy products national markets in France and in the U.K. and to some extent for the Iberian Peninsula and Poland. For the rest of Europe, there are no significant market volumes. U.K. authorities have accepted state-of-the-art technology to recycle British milk bottles back into milk bottles with recycled content of max. 40%. For price reasons, the British dairy industry by far did not use the available amounts for many years. The "run" on "food grade" rHDPE in the U.K. was not driven by food packaging for several years but by cosmetic packaging. Lacking a product-specific regulation for cosmetic packaging, the industry tried to generate an approach by using food-grade material. Obviously, all goals and claims published by the industry and supposed to come from the regulatory side are not achievable without increasing the sourcing options. This is why some relevant players in the cosmetic industry and partners in the value chain started a project in 2020 to define quality standards for recycled polyolefins in their fields of application (COSPATOX 2023).

7.1.6 Technical Perspectives

Closing the loop for household packaging will lead to applications being "close to the end consumer" for food contact and other applications where consumer protection has to be considered. Technical purification is only one option to follow, the other one is to control input contamination. Besides incidental contamination, the previous application of the packaging triggers the type and level of impurities the material catches up. Hence, smart sorting concepts upfront will contribute to the qualification of recycled resins for specific applications. Sorting devices using camera systems for object recognition and identification combined with machine learning concepts are ready to be implemented (Berry 2022). However, the market for technical purification concepts is also facing innovation. A promising approach is to further expand vacuum degassing during extrusion. The idea is to increase the surface of the molten phase and the residence time under vacuum. Different mechanical/physical concepts already offer decontamination efficiencies significantly beyond the values shown in today's conventional solid-state processes (Starlinger 2023).

7.2 PP Post-Consumer Recyclates for Plant Pots and Automotive Parts

Benjamin Kampmann and Frank Schockemöhle

7.2.1 About Pöppelmann

Pöppelmann, headquartered in Lohne, Germany, was initially founded in 1949 as a manufacturer of corks entered the plastics processing sector in 1955 with the production of a handle cork. Today, around 2500 employees in Germany, France, and the US develop and produce plastic products for customers in over 90 countries on appr. 650 injection molding machines, thermoforming systems, and extruders. The product portfolio includes plastic protective elements (approx. 3,000 standard items and approx. 2,000 custom-made products), customized technical injection-molded parts for the automotive and electrical industries, and for mechanical and equipment engineering, plant pots and plant pot trays for commercial horticulture, and technical functional parts and packaging for the food and pharmaceutical industries and for medical technology.

7.2.2 Recyclate Use over Time

The use of recycled materials has been common practice at Pöppelmann for decades. The materials used, the areas of application, and the reasons for using recyclates have continued to develop over the years. The first recycled plastics were used back in the early 1980s, mainly regrind from the punching grids used in the production of yogurt containers.

The first application for recyclates at Pöppelmann was plant pots, for which the main goal was to procure a cost-effective material. For many years, post-industrial recyclates (PIR) were the preferred material source for plant pots. New material sources were constantly developed, e.g. waste from production processes that used films and tiles as semi-finished products, such as in the label industry.

From the mid-2010s, the focus shifted to the circular economy. A circular economy is only created through the recyclable design of plastic products, the recovery of plastic from used products in the form of post-consumer recyclates (PCR), and the recycling of PCR for the manufacture of new plastic products. Pöppelmann's plant pots were pioneering products in the use of PCR in recyclable products.

The use of recyclates in the technical sector grew from the 2010s onwards. The focus here was on the use of PIR materials in the production of technical plastic parts for the automotive industry, particularly in combination with physical foaming (MuCell). The objectives were reducing product weight, cutting costs, and conserving resources. For this reason, the use of glass fiber-reinforced polypropylene recyclates from post-industrial sources was also used directly in the development of the technology. PCR materials are now also being used for the first products.

7.2.3 Availability of Post-consumer Recyclates

Purchasing virgin plastics, post-industrial recyclates, and post-consumer recyclates faces the procurement department with different challenges. While virgin material is often sold directly by large plastics manufacturers and trading companies, the procurement market for recyclates is much more fragmented. Many small and medium-sized recyclers are active here, offering a large variety of recyclates from a wide range of sources. The prerequisite for the continuous availability of a particular recyclate is permanent access to the waste stream the recycled material is stemming from. For example, a PP recyclate from packaging waste, e.g. from fraction 324–0 (Green Dot 2024), is available in relatively constant and consistent quantities. However, the availability depends on the waste disposal of consumers and not on the needs of processors. In times of high demand, for example, due to short supply and/or high prices for virgin material, this leads to a high demand for recyclates and rising prices.

Fraction 324–0 (ibid) is one of the plastic fractions from the household collection of packaging wastes by the Dual Systems in Germany, carried out with the Yellow Bag or Yellow Bin. In 2020, a total of 6,547,743 t of lightweight packaging (packaging made of plastic, metals, and composites) was collected by Dual Systems (Öko-Institut 2022). A total of 790,217 t of plastics of various fractions were sorted out of this stream in the sorting plants of the Dual Systems (cf. Fig. 7.11).

For Pöppelmann, PP fraction 324–0 is one of the most important sources of PCR material, with a total of 134,873 t (sorting plant output) available from the Dual System in 2020. The output of a sorting plant usually consists of bales containing a mixture of various thermoformed and injection-molded, transparent, or colored, and in some cases printed or labeled PP packaging. These bales also contain filling residues and other adhesions and impurities that need to be removed before processing.

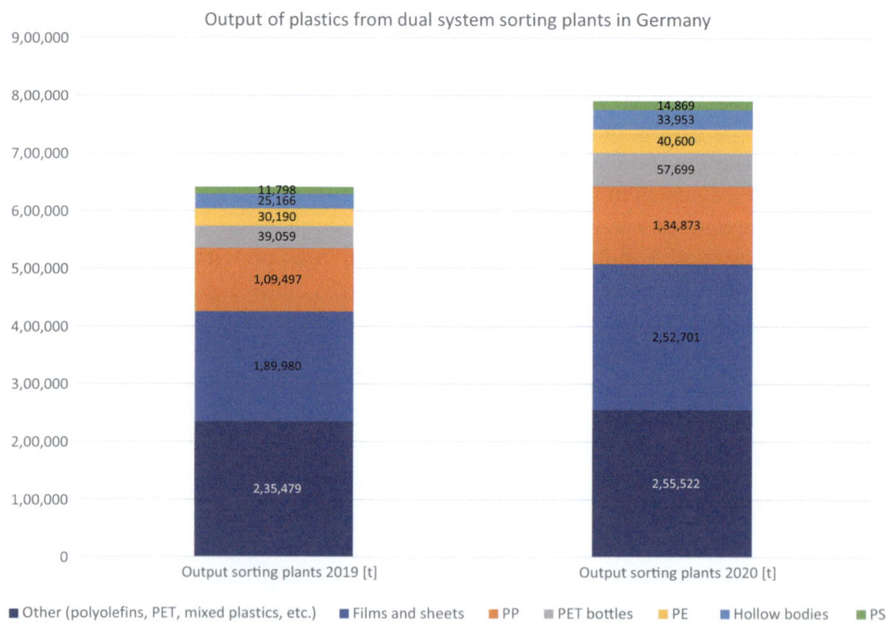

Fig. 7.11 Output of plastic fractions from the sorting facilities of the dual systems. (*Source* Own presentation based on Öko-Institut 2022 data)

The baled goods are usually processed at separate recycling sites into a processable PP post-consumer recyclate in granulate form. For this purpose, the bales are opened, shredded, washed, sorted (e.g. using air separation, float-sink separation, color sorting, NIR sorting), dried, and then agglomerated, granulated, or compounded (cf. Sect. 10.1). The properties of the material are determined by the mixture of the input. Various properties can be changed on this basis through different steps in processing and compounding. The removal of impurities and odors from the material is directly visible on the material. Different washing techniques (cold washing, hot washing, several washing stages) and the process technology in extrusion (degassing, filtration) can contribute to increasing purity and yield and decreasing odor. However, adapting the rheological and mechanical properties of the recyclates is crucial for processing and subsequent product use.

Table 7.2 shows examples of the properties of various PP post-consumer recyclates to be produced from fraction 324–0. The injection molding compound produced from this source without further modification has an MFR of 13 g/10 min and thus a medium to low flowability. One way of adapting the PP recyclate to processing needs is to reduce viscosity by adding additives. Thus, an MFR of 55 g/10 min can be achieved, e.g. making the material suitable for thin-wall injection molding. However, this process leading to shortened polymer chains has a negative effect on the stiffness and toughness of the material, as shown by the change in the tensile modulus of elasticity and the notched impact strength. Another modification option is the addition of glass fibers to obtain a stiffer material for technical applications.

7.2.4 Processing of Post-consumer Recyclates in Volume Production

Pöppelmann now produces many products partially or completely from PP post-consumer recyclate from packaging waste. Overall, Pöppelmann currently (2022) covers over 40% of its material requirements across all products with post-consumer recyclates. The plant pots and the "Holder Sound Generator" are presented here as examples.

Plant pots are produced using the inline thermoforming process (cf. Fig. 7.12). In this process, a flat film is extruded, which then is fed directly to the thermoforming machine, where the plant pot is formed and cut out. The base holes are then punched and finally, the plant pots are stacked and packaged. The grid created in this process is ground directly and fed back into the extrusion line. The plant pots made from post-consumer recyclate are produced on the same systems and with the same tools being used to produce plant pots from post-industrial recyclate. The challenge is to enable reliable production through adapted processes. Starting with extrusion, the relatively low viscosity requires adjustments to the process parameters to produce a film suitable for thermoforming. In addition, degassing must be adapted to the content of volatile substances and the respective odor of the material; the use of

Table 7.2 Overview of the properties of various polypropylene post-consumer recyclates based on fraction 324–0). *Source* Own presentation

Feature	New PP example	Injection molding compound from fraction 324–0 (Systalen 2024)	Injection molding compound from fraction 324–0 with modified flowability (Systalen 2024)	Injection molding compound from fraction 324–0 for technical applications with 30% glass fiber
MFR (230/2.16) (g/cm^3)	16	13	55	20
Density (g/cm^3)	0.90	0.92	0.92	1.10
Tensile strength (MPa)	26	25	25	65
Modulus of elasticity (MPa)	1450	1200	1150	6000
Charpy notched impact strength (23°C) (kJ/mm^2)	10	5.5	4.5	>7
Charpy notched impact strength (-20°C) (kJ/mm^2)	6	2.2	2.0	–

entraining agents can support processing here. Filtration must also be adapted to the materials. In thermoforming, the challenge is to react to the changed forming behavior of the materials and, despite the higher toughness compared to the materials previously used, to carry out cutting and punching cleanly. The overarching focus is on maintaining the cycle time. A particular challenge in the extrusion and thermoforming of post-consumer recyclate is the odor of the material, intensively spread over the production area due to large throughputs and surfaces. In addition to reducing odors by optimizing material preparation and processing procedures, it may also be necessary to adapt ventilation and exhausting systems in the production halls. One solution is to install supply and exhaust air devices directly in the employees' work area to protect them against the most significant odors.

The "Holder Sound Generator" is an example of a product for which the use of recyclates was considered from the outset during development (cf. Fig. 7.12). The product is used in electric vehicles and is responsible for holding the sound generator,

Fig. 7.12 Products made from PP post-consumer recyclate from packaging waste. (*Source* Pöppelmann)

which produces an artificial driving noise at slow driving speeds. Together with a compounder, a suitable recyclate was developed for this product containing 30% glass fiber and whose plastic content is entirely PP post-consumer recyclate from packaging waste. Simulation is a crucial step during the development of technical components. If the necessary characteristic values cannot be provided by the material supplier, Pöppelmann determines them using our own testing tools. In such a way, the article and tool can be optimally adapted to the rheological and mechanical properties specified by the material.

7.2.5 Quality Assurance in Post-consumer Recyclate Processing

The use of post-consumer recycled material also requires an adaptation of quality assurance, particularly the incoming goods inspection. Here, too, there are differences between the production of plant pots and the manufacture of technical components, such as the sound generator holder.

In plant pot production, decades of using post-industrial recyclates have led to the development of a testing routine continuously being refined for the use of post-consumer recyclates. Many material deliveries from a wide variety of suppliers must be checked every day. Mixtures are produced from these deliveries according to internal specifications to meet production requirements. The basic test in this area always includes the evaluation of flowability, stiffness, and toughness. Specifically, a flow spiral and one test specimen each for determining flexural modulus and Charpy impact strength are produced in a test specimen tool. The PP post-consumer recyclate is also subject to further tests. The residual contamination of the material is tested, and the density of the material is determined to ensure recyclability. When testing recycled materials, the focus is on fast and reliable determination of reproducible and meaningful characteristic values. Testing according to ISO-standards is often neither possible nor necessary to carry out an effective material test in the incoming goods department or during the process. In particular, the conditioning of test specimens is often not possible due to time constraints. However, standardized processes and procedures and adapted test methods can be used in practice to determine sufficiently reproducible characteristic values in quite a short time.

A compound developed jointly with the material supplier is used for the production of the "Holder Sound Generator". The supplier also determines the characteristic values documented in a manufacturer's test certificate. In addition, at least one test measurement is always carried out at Pöppelmann's. In the case of PP post-consumer recyclates, e.g. this involves determining the portion of PE in the material using a DSC analysis.

7.2.6 Circular Economy with Post-consumer Recyclates is Possible

The use of post-consumer recycled material is possible in many applications if additional challenges are mastered. To this end, suppliers, processors, and users must develop a common understanding of quality and further develop processes and materials on this basis. The question "Does recycled material have the same quality as virgin material?" should not be the benchmark. It is crucial to think of quality in terms of the function of the component and to develop material, design, and process in such a way that the function of the product is guaranteed.

However, the circular economy can only be successful if there is a constant and long-term availability of the recyclates required. To achieve this, close cooperation and coordination between all players in the circular economy is essential, from processors to distributors to retailers to end users to recyclers and back to processors.

The plastics industry can only make an effective contribution to tackling the climate crisis and decoupling from the use of (mostly fossil) primary raw materials if the circular economy is as efficient as possible. To this end, energy-efficient material recycling must be used, expanded, and further developed.

7.3 Challenges for Flexible Packaging Manufacturers

Johannes Wedi

7.3.1 Introduction

This section describes the implementation of sustainable packaging development from the perspective of a flexible packaging manufacturer, Bischof + Klein, one of Europe's leading full-range suppliers of flexible packaging and technical films made of plastic and plastic composites. Bischof + Klein employs a total of around 2,700 people at five locations in Germany, the UK, France, and Poland.

Particular attention is paid to the question of how and where the post-consumer recyclate (PCR) granules produced from recyclable film packaging for FMCG products can be directly reused in film packaging.

The classic packaging used by film packaging manufacturers are multi-layer polyethylene or polypropylene films and combinations of different film materials such as polyester (PET), oriented polypropylene (OPP), polyethylene (LDPE) in mono and coextruded multi-layers. Such films may be combined with each other depending on the requirement profile of the packaging, or according to the requirements of the product packaged. Common are also combinations with paper and aluminum foils or other barrier materials such as PA, EVOH, and PVDC. Barrier coatings are possible, too. Flexible packaging developed in recent years should be recycled, i.e. PCR granules produced thereof may be used again in respective packaging applications.

The following examples are discussed to illustrate the development of recyclable packaging and the use of PCR recyclates:

1. Ground coffee: vacuum packed or whole beans
2. Liquid household cleaners: stand-up pouches (doy packs)
3. Toilet paper rolls (wicket bags)
4. Industrial products: protective covers for pallets (SmartFlex)

7.3.2 Switch to Recyclable Flexible Packaging

Market analysis of recyclability

Most classic composite materials established on the market are not mechanically recyclable, as they are often combinations of polyester films with polyethylene or polypropylene. This does not include combinations of LDPE/LDPE or OPP/PP or OPP/PE. PET in combination with polyolefins is not considered recyclable as it cannot be mixed. The same applies to combinations of all materials that are not freely miscible. Composite packaging with aluminum foil is often considered recyclable since the aluminum component is recycled. The plastic component is only recovered energetically and hence lost for mechanical recycling. Exact data about this is not available, as composite materials were not assessed in terms of recyclability in the 2010s.

The overview shown in Fig. 7.13 is the composite film portfolio of a major packaging manufacturer in 2016, implying that the proportion of non-recyclable composites is above 90% (all composite materials containing aluminum and/or PET). However, aluminum composites were considered recyclable in many studies (Consultic 2015).

The following composites in different material thicknesses are established standards on the market: PET/PE; PET-met/PE; PET/Alu/PE; PET/OPA/PE; OPP/PP; OPP/PE; OPP-met/PP(PE); PE-PA-PE; PET/PE-EVOH-PE. All these laminates have special functions intended to fulfil respective product requirements. For decades laminate development aimed at optimizing product protection and minimizing material use; recyclability didn't play a role. It was not until 2015, with the amendment of the German Packaging Ordinance and the passing of the Packaging Act in 2017, that the demand for recyclability of flexible packaging came into focus.

Corresponding tools and guidelines, such as the Eco-Design Round Table (IK 2019) or the requirements of the ZSVR minimum standard for the design of recyclable composite materials and packaging design (ZSVR 2024) for paper/plastic and plastic film/aluminum combinations, were an important aid for further development of recyclable packaging.

After almost 10 years of development activities towards recyclable flexible composite packaging, approximately three-quarters of all consumer-related flexible packaging is now recyclable and some of it is already being recycled. In the medium term, this figure should increase to 90%. However, this requires a design for recycling

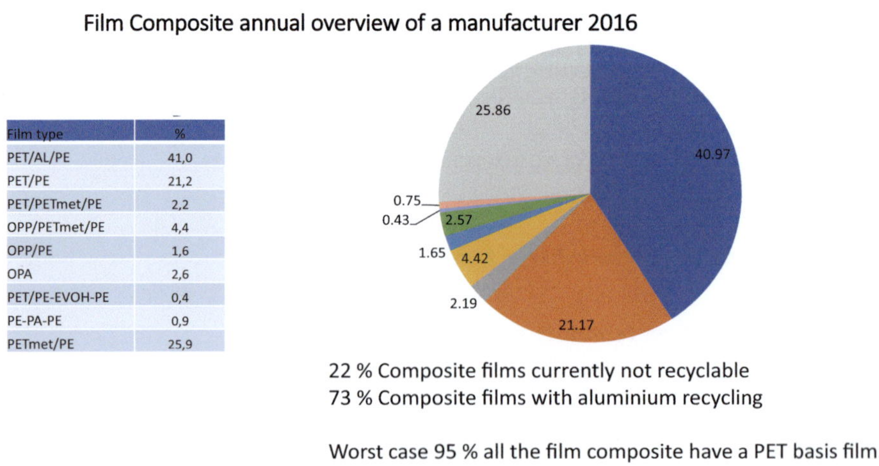

Film type	%
PET/AL/PE	41,0
PET/PE	21,2
PET/PETmet/PE	2,2
OPP/PETmet/PE	4,4
OPP/PE	1,6
OPA	2,6
PET/PE-EVOH-PE	0,4
PE-PA-PE	0,9
PETmet/PE	25,9

Fig. 7.13 Composite film annual overview of a manufacturer in 2016. (*Source* Own presentation)

taking into account higher cost and higher material input, as the following examples show (Consultic 2016).

The switch to recyclable composites has not yet taken place for all applications, as some distributors are still hesitating. In fact, recyclable solutions are possible for almost all applications. In the food sector, there are still many composite materials with aluminum and metallized film in use, of which only the aluminum content is recycled, but solutions are to come.

Development of Recyclable Composites with Barriers

Today, many recyclable mono-plastic alternatives are significantly more material-intensive, as was the case with polyester films over more than 50 years (the first oriented polyester films were produced in 1957). PET has essential functions, namely stiffness, strength, and temperature resistance, which polyolefin alternatives such as Machine Direction Oriented PE film (MDO) or Biaxially Oriented PE film (BOPE) do not easily provide. Consequently, this may lead to 15–25% more material used than with current non-recyclable composite materials (GVM 2018).

The challenge hence is to replace the large quantities of non-recyclable PET composites with mono-polyethylene composites with both sufficient functionality for the product packaged and sufficient process capability on the existing packaging machines.

The following trends towards recyclable composites can be observed on the market:

- PET/PE replacement by MDO/PE and MDO/PE-EVOH-PE/PE;
- PET-met/PE replacement by MDO/Barrier lacquer/PE;
- PET/ALU/PE replacement by MDO/Barrier lacquer + metallization/PE;
- EVOH and EVOH barrier lacquers became established as barrier materials.

If these coating systems are combined with metallization, oxygen barriers can be achieved comparable with non-recyclable films. Other barrier coatings based on polyvinyl alcohol (PVOH) are also emerging. Through skillful use of oriented polyethylene three-layer films with MDO or BOPE, it was possible to bring properties close to those of polyester composites. These alternative materials can also be used for stand-up pouches and customized packaging with side gussets. There is already a large proportion of stand-up pouches made from mono-polyethylene composites on the market today, especially for detergents and cleaning agents.

The changeover in market sectors such as coffee packaging, pet food packaging, stand-up pouches, liquid detergents, and household cleaners, food packaging such as confectionery, meat, sausage, and cheese packaging is in full swing. These areas represent the vast majority of the market share of film packaging for FMCG.

In some cases, the running properties of the filling lines still need to be optimized in order to achieve a performance comparable to that of PET composites. Suitability for the product and fulfilment of the minimum shelf life are generally fulfilled.

Application Requirements for Recyclable Composites with BARRIERS

In general, product manufacturers and retailers require recyclable composites to have. identical or comparable properties as the (usually non-recyclable) composites used to date. Frequent conflicts of interest are related to.

- Barrier properties: oxygen barrier, water vapour barrier, and CO_2 barrier are sometimes lower;
- Material thickness: recyclable composites are often up to 10–15% thicker in material structure;
- Cost parity with the previous composite material is often not given;
- running properties and filling speeds of the filling systems are sometimes not identical to the previous material.

Due to their mono-structure, the new recyclable composites are weaker in terms of rigidity and have lower barrier properties compared to the previous composites. Compensation is then achieved by using more material and by skillfully combining barrier layers leading to increased cost. The following examples may illustrate this.

- In the pet food packaging sector, PET/LDPE 12.130 μm (wall thickness distribution: a laminate of 12 μm PET and 130 μm LDPE) is commonly used today for packaging sizes of 2–3 kg. As PET already has a grease barrier due to its physical properties, no further barrier is needed. The alternatives now being considered are the recyclable composites MDO/barrier varnish/LDPE 25.130 μm or a multilayer film of MDO/PE-EVOH-PE/PE 25.50.90 μm with EVOH < 5% by weight.
- In the food sector, PET/ALU/PE 12.7.85 μm was previously used for 500 and 1000 g coffee packaging units. New barrier films were developed here to fulfil the high requirements for the oxygen barrier: MDO/barrier-varnish-metallization/PE or solutions based on OPP/barrier/PP with barrier coatings plus metallization.

In principle, it was possible to develop recyclable composites for the following applications:

- Pet food;
- Food packaging such as confectionery, coffee, meat/sausage/cheese;
- Stand-up pouches for household cleaners, detergents;
- Films as roll goods with applications for a wide range of products.

Some of this packaging has already been converted or is still awaiting conversion due to higher costs for the composite material, since there is no legal obligation to do so. In addition, there is also no legally binding bonus-malus regulation favoring the use of recyclable materials.

7.3.3 *PCR Polyolefins for Flexible Packaging*

Examples of PCR Use

Post-consumer recyclates have long been used in rigid packaging such as canisters, buckets, bottles, or IBCs, having thicker walls than films. As the plastic packaging market—depending on the national situation—is made up of 40–50% flexible packaging, opportunities must be considered to use PCR recyclates in large quantities here, too (PPWR 2023).

However, there are limits as to where PCR recyclates can and should be used. The following examples for three market sectors demonstrate where this has already been accomplished and where the availability of PCR grades is limiting:

1. Wicket bags pencil bags for tissue: example toilet paper
2. Stretch hood pallet covers
3. Stand-up pouches for laundry and cleaning products.

Example 1: Wicket bags for toilet tissue paper

Wicket bags (cf. Fig. 7.14) are extruded in a material thickness of 25 μm in 3 layers of coextruded film as a blown film using flexographic printing and made up into wicket bags with handles. This type of construction requires the relatively thin film to have a high degree of strength, as otherwise the side seals or the handle could tear. When using PCR recyclates, a gel point is often found to be the cause of tearing handles or sealing seams.

In the case of these films, a complex qualification process was used to ensure the use of PCR in the centre layer of the three-layer film. Through the supplier qualification process for certain properties of the PCR recyclates, strengths comparable to

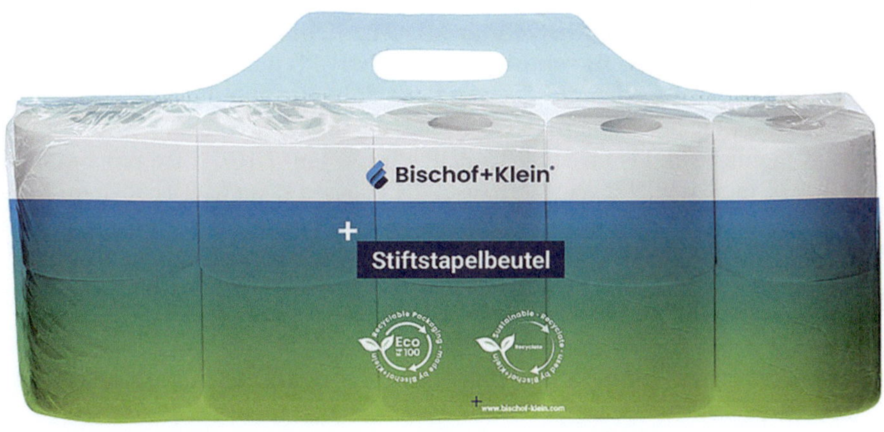

Fig. 7.14 Wicket bags for toilet tissue paper. (*Source* Bischof + Klein)

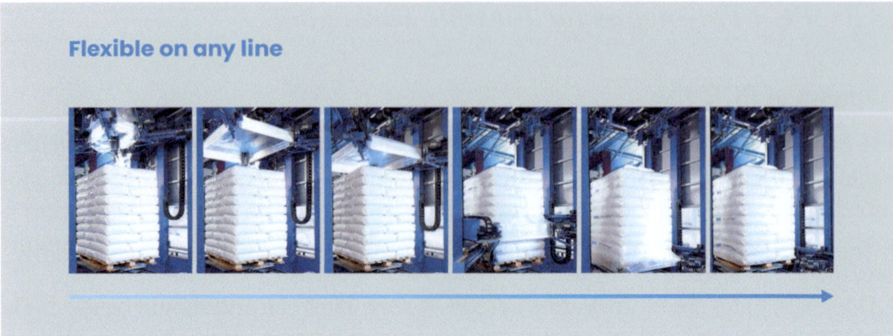

Fig. 7.15 Stretch hood films (*Source* Bischof + Klein)

those of the virgin version were achieved. The use of PCR is now at 30% and trials are underway for a PCR use of up to 50%.

Example 2: Stretch hood pallet covers SmartFlex®

In case of industrial packaging for pallets, transparent stretch bonnets are used being approx. 90 μm thick and having high tear resistance and elongation or stretchability in the longitudinal and transverse directions (cf. Fig. 7.15). In the case of 90 μm films, already used with several hundred tons of films with PCR per year, it has been possible to qualify them on a large scale with very good performance at the same material thickness. A qualified market standard has been achieved with 30% PCR content. Further development is underway to increase the PCR content up to 50% and 70%.

Example 3: Stand-up pouch for laundry and cleaning products

PET/PE composites in various designs have been used for these products to date. For mono-composite materials, MDO/PE or MDO/barrier/PE are predominantly used. The barrier can be an EVOH varnish or a PE-EVOH-PE film.

For products that are not close to food and do not require food approval, PCR can be used for the inner film so that a bag has a total PCR content of 30%. For this purpose, the inner film is made of PE-PCR-PE (cf. Fig. 7.16). If the proportion of PCR is to be increased, it is possible to use a PCR-based spout with a cup or to create the lamination layer of the inner PE film with a PCR proportion. This allows a total PCR content of the stand-up pouch of 40% to be realised.

Assuring PCR quality

Quality assurance for PCR recyclates is a basic prerequisite for the safe use of recyclates. A distinction must be made between external and internal influencing factors (cf. Fig. 7.17).

The following items in the supplier specification must be checked to achieve a high-quality PCR granulate:

Fig. 7.16 Stand-up pouch for laundry and cleaning products. (*Source* Bischof + Klein)

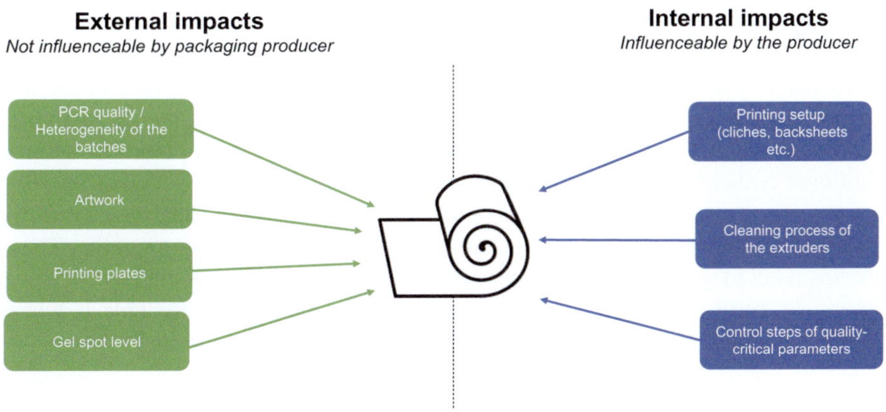

Fig. 7.17 Impacts on the quality of PCR-containing films. (*Source* Bischof + Klein)

- MFR
- Density
- Residual moisture content
- Recycled content
- Ash content
- HDPE content
- Heavy metal content.

To maintain or increase the quality with different suppliers and increasing quantities, it makes sense to introduce additional safety screenings for these applications. Granulates are regularly checked for the presence and concentration of the following substances:

- Phthalates
- Organic tin compounds
- Heavy metals
- Alkylphenols
- Polybrominated biphenyls/diphenyl ethers.

Production problems especially in extrusion may occur when using PCR. Film extrusion has to struggle with two challenges, strong formation of gels and migration of low molecular weight components out of the PCR. The main problems are:

- Variation of MFI melt flow index
- High gel spot level
- Intrinsic color of the films (yellowish, brownish, greyish)
- Off-odor both during production and in the final product

Possible solutions include the use of stabilizers during production with PCR (cf. Chap. 10.2), preventing polymer degradation, thus limiting the formation of additional gels, and the use of optimized filter technologies. In general, PCR extrusion requires more time and money to maintain quality. For instance, extruders must be additionally cleaned and the schedule for when the systems have to be cleaned and the tools such as dies and screws have to be re-chrome-plated is significantly shorter than for extruders that only produce films based on virgin material. One can assume additional costs of 10 to 15% due to this effort.

Subsequent printing of PCR films also causes production problems previously unknown. In film printing, video systems are used to check the print quality. These systems react to gels as errors and can only be used with PCR-optimized processes. Otherwise, fully automatic print monitoring would be impossible. This is where solutions are used to eliminate gels as a source of error from the printed image.

7.3.4 Conclusions and Outlook

Many composite films are now designed to be recyclable but with subsequent higher efforts and higher material use. This is the reason why product manufacturers have not yet implemented PCR use in general. It can be assumed that European legislation will exert strong pressure here, especially when the Packaging and Packaging Waste Regulation (PPWR 2023) will come into force. It will then be mandatory in some cases to use PCR in packaging films and to design flexible packaging films to be recyclable. Nine out of ten applications in the area of flexible composite packaging may be designed to be recyclable, but the final legal impetus to implement this everywhere is lacking.

The use of PCR will possibly be regulated under the political pressure of PCR content quotas, provided that enough material of sufficient quality is available. For many applications, only PCR recyclates with high transparency can be used stemming from flexible transport packaging. However, the majority of available PCR granulate qualities cannot be used for all film applications, which is also not practical. The use of certain grades of film PCR in thicker rigid packaging such as buckets, plates, etc. makes a lot of sense and can also be considered as closing the circle.

PCR from mechanical recycling of polyolefins in current quality cannot be used for food applications. Fulfilment of corresponding recycled content quotas for food applications is only possible by using products from chemical recycling. From today's perspective, it is difficult to predict whether the necessary quantities will be available as of 2030 in accordance with the forthcoming PPWR (2023).

7.4 PVC in Building and Construction

Thomas Hülsmann

7.4.1 Introduction—Plastics in B&C

Curved facades or customised insulation for different construction elements: plastics are the ideal choice to realise architects' requirements. It is not without reason that the building and construction industry with a market share of 21.3% is one of the most important applications of plastics in Europe after the packaging industry with a market share of 39.1% (cf. Fig. 7.18) (PE 2022). Traditionally, the material is used from the cellar to the roof and for infrastructure applications: for example, for floorings, windows, pipes, cable conduits or for thermal insulation and sealing. This makes it also an important supporter of the energy transition in the B&C sector. Plastics are also much appreciated for giving buildings a special radiance—for example, the Allianz Arena in Munich with its uniquely playful, luminous shell.

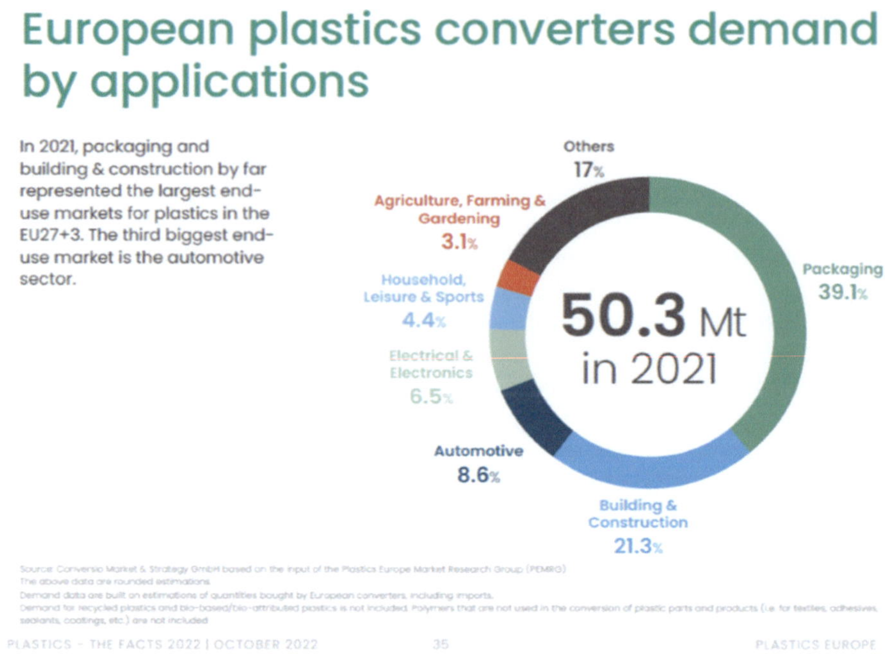

Fig. 7.18 Share of plastics demand in B&C applications. (*Source* Plastics Europe 2022)

7.4.2 PVC Products in B&C—Main Applications

In the building sector, PVC is one of the most important plastics (cf. Fig. 7.19) (PE 2022): The material is used in durable building products such as windows, pipes, floorings, cable insulations and sheathings, coated fabrics, and roofing membranes. The very wide range of use of PVC is possible because of its excellent combination of properties which can be enhanced or modified by additives to fulfil specific requirements for products.

The great majority of products are found in long-life applications with a service life of well over 20 years. These characteristics offer architects and planners a wide range of design options when realising living spaces and meeting high sustainability standards at the same time. According to a Conversio Study (Conversio 2022) of the life cycle of PVC in Germany 2021 the main application sector with about 75% of the total converted volume is building and construction. Plastics and PVC applications, especially, in the B&C sector enable tailor-made solutions: sensual design aspects combined with functionality, resource efficiency, and established recycling make the material an economic and eco-efficient choice in this sector.

Window Profiles

Energy-saving windows made of PVC (cf. Fig. 7.20) are a success story, PVC is the leading window profile material in Europe, in Germany the market share is about

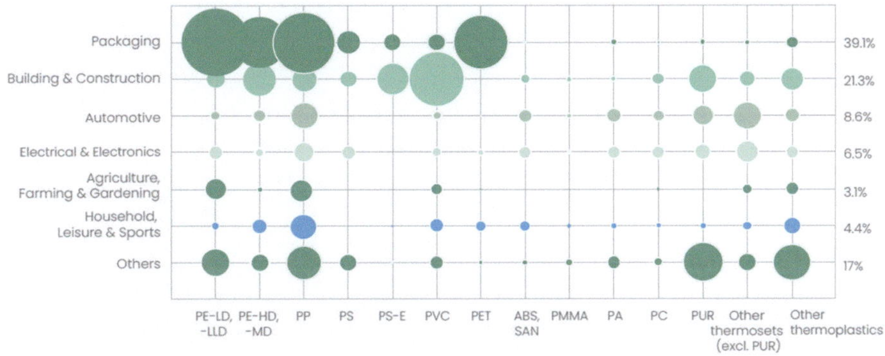

Fig. 7.19 Share of PVC demand in B&C applications. (*Source* Plastics Europe 2022)

60% (Prowindo 2018). The high molecular weight of PVC and the selected additives (e.g. heat stabilisers, impact modifiers, pigments) provide the optimum combination of processing as well as mechanical and weathering performance. Window profiles are light, tough and stiff with excellent weathering resistance, they do not support combustion and have a low thermal conductivity. The windows have outstanding durability, and a service life of 40 years is estimated in the Assessment System for Sustainable Building (BNB 2015), for example. Established quality standards (RAL 2023; DIN 2020) and Environmental Product Declarations (EPD 2022) consider already recyclate contents. Due to its chemical and physical properties, PVC profiles can be coextruded with PMMA or easily laminated with foils offering an enormous variety of colours and surface structures and can thus be adapted to almost every requirement for the facade design.

Floor Coverings

Resilient PVC floor coverings are considered particularly easy to clean, hygienic, pleasant to walk on and durable—especially important for high-traffic areas such as in hospitals, schools, or apartment buildings. The selected PVC grade (high molecular weight) in combination with the additives (e.g. heat stabilisers, plasticisers, fillers) is the basis for this broad performance spectrum. PVC floorings can be printed and embossed particularly well, so that an almost inexhaustible variety of different designs is possible. Homogeneous, heterogeneous, CV or LVT floorings are produced by different processes and may contain various shares of recyclates (EPD 2024a, b). The service life can be between ten and 30 years or more and depends not only on durability.

Pipes

Robust plastic pipes with smooth inner surfaces transport drinking water, drain roofs, and dispose of waste water. They are an integral part of our water and gas networks, and they require little maintenance. The service life of PVC water pipes can reach

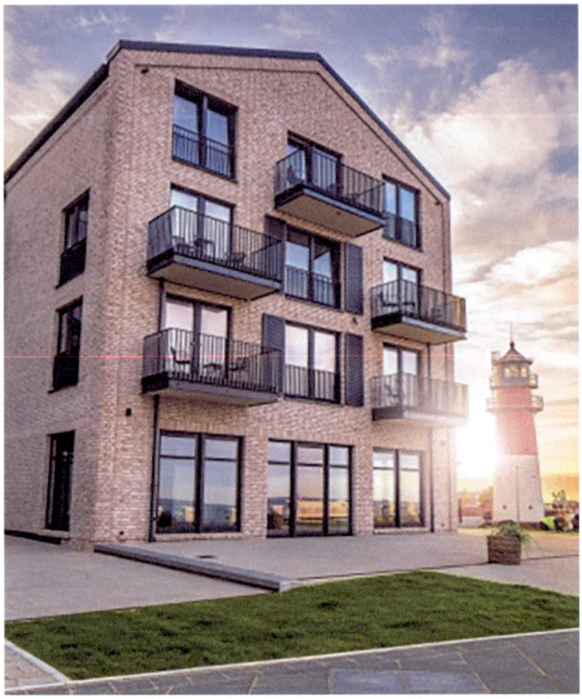

Fig. 7.20 Modern energy-saving PVC windows. (*Source* VEKA AG)

hundred years (TEPPFA 2019) or more. Light weight and ease of installation, toughness, good deformation resistance, excellent durability, and good corrosion resistance to a range of chemicals and oxidising agents as well as aggressive soil conditions are benefits of PVC pipes. Thus, they enable the efficient transport of gas and water over long distances—safely and without contamination. Taking existing quality standards into account there are a variety of possible applications to incorporate recyclate. These are pipes for wastewater disposal, rainwater management and cable protection (KRV 2019).

Coated Fabrics

In textile architecture, PVC-coated fabrics open new possibilities for facade and roof design. The most impressive examples include stadium roofs, which protect both spectators and athletes from the weather and often become landmarks for entire regions due to their unmistakable shape. In certain cases, the reuse or recycling of construction products such as coated fabrics is already a requirement when renovating or building new stadiums.

Cables and Wires

The insulation and sheathing of cables and wires have long been made from PVC, which is favoured in Europe for its versatility, electrical insulation properties, inherent flame retardancy and cost-effectiveness. The market share of PVC is currently around

40% (PVC4Cables 2022). The European Construction Products Regulation (CPR) (EU 2011) sets harmonised technical conditions for the free circulation of products within the European Union, including the reaction-to-fire performance of products in the event of a fire. PVC cables can adapt to the individual specifications of intended-use/fire-risk. They represent the best choice in terms of cost-efficiency and fire performance, in full compliance with the CPR and related standards (PVC4Cables 2019). The service life of 25 years is estimated for cables and electrical equipment in B&C in the BNB criteria (BNB 2015).

Roofing and Waterproofing Membranes

PVC roofing and waterproofing membranes have been used for many decades to waterproof buildings of various types. The single or multi-layer PVC roofing membranes, for example with glass or polyester fibres, protect the building structure against both weathering and mechanical damage. They are essential to ensure the durability and safety of buildings (VinylPlus 2024). The BNB criteria (BNB 2015) assume a service life of 20–30 years, while the manufacturers' EPD (EPD 2024c), for example, state up to 35 years.

7.4.3 Circularity of PVC

Through VinylPlus®, the European PVC industry´s commitment to sustainable development uniting about 200 leading companies in Europe—PVC manufacturers, additive producers, and PVC converters together with a network of about 150 recyclers—more than 810 kt of PVC waste were recycled in 2022 (cf. Fig. 7.21). This amount represents about 27% of the total PVC waste generated in 2022 in Europe (EU27 + 3). A total of 8.1 Mt of PVC waste has been recycled since 2000, reducing CO_2 emissions by 16.2 Mt (VinylPlus 2023).

The main sources of recycled PVC wastes are window and related profiles (408 kt), cables (101 kt), and flexible applications including films and floor coverings (213 kt) (VinylPlus 2023). In a 2020 traceability survey recycled PVC was mainly used in windows and profiles (35%), in traffic management products (15%), in pipes (13%), and floorings (10%). The remaining 27% are used in other applications such as rigid sheets, roof coverings, coils, and mandrels (VinylPlus 2021). The applied PVC processing of extrusion/co-extrusion or the lamination of different layers enables the use of recycled PVC in specific areas of the product where virgin polymer compound is substituted, for example, in the core of window profiles, in mono- or multilayer pipes, or in certain layers of floor coverings. Recyclates from cables are substituting, e.g. metal or concrete in traffic management products because of improved performance and handling. The mechanical recycling of PVC is well established, and it can be recycled several times without a substantial loss of key mechanical characteristics (Leadbitter and Bradley 1997). VinylPlus is committed to recycle at least 900 kt per year by 2025 supporting the European Commission's Circular Plastics Alliance

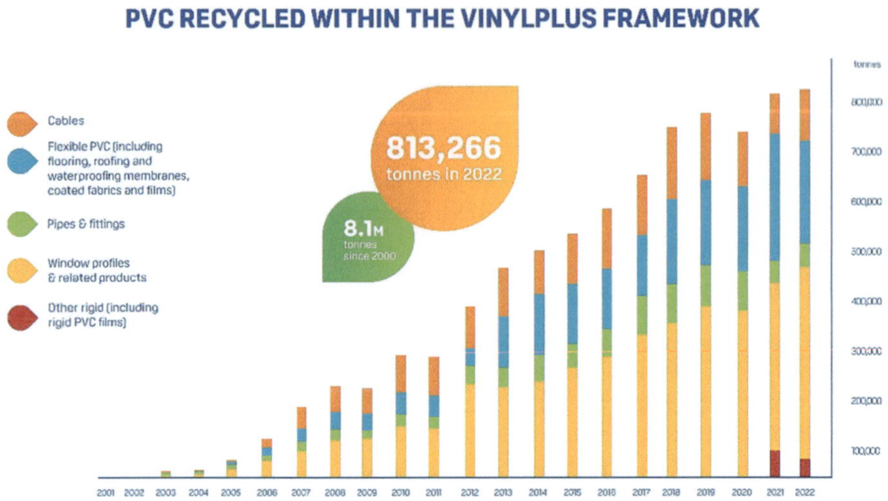

Fig. 7.21 PVC recycled within the VinylPlus® framework (*Source* VinylPlus 2023)

(CPA) initiative (VinylPlus 2018). Recovinyl® with its data collection system Recotrace™ has been recognised by the European Commission as one of the approved data collectors for the CPA to monitor, verify, and report European PVC recycling (CPA 2018).

According to a Conversio report (Conversio 2021), the total generation of PVC waste in Europe (EU27 + 3) for 2020 was 2.9 Mt. About 1.0 Mt of the PVC waste was recycled, 1.3 Mt was incinerated with energy recovery, and 0.5 Mt was landfilled in Europe. Due to the long and very long service life of PVC products (cf. Sect. 7.5.2) the total amount of waste is significantly lower compared to the amount of PVC currently processed of around 5.4 Mt. R&D projects and studies are attempting to model the development of waste generation from B&C products based on this service life effect (JRC 2024). In combination with data on previously used additives (legacy additives), more accurate estimates can be made about their future phase-out (Laner 2022). Such modelling also improves the planning of investments in the circular economy.

7.4.4 Challenges, Perspectives, and Developments

To accelerate towards circularity innovative recycling technologies have been explored by the industry—from waste sorting and separation to mechanical, physical, and chemical recycling and incineration with energy recovery. Transferring best practices among European countries and between different application sectors also contributes to achieving the committed CPA recycling targets.

The long service life of PVC products is a significant advantage in terms of material efficiency. However, previously used additives can pose a challenge when it comes to recycling. The industry developed the 'Additive Sustainability Footprint' (ASF) (VinylPlus 2022) tool in collaboration with The Natural Step (2024) allowing users to assess and promote the sustainable production and use of PVC additives across entire product life cycles and supporting forward-looking decisions about the use of additives (Everard 2022). Over the last 25 years the European PVC industry has deliberately moved away from additives classified as hazardous and considered a risk to health through its voluntary commitments. Major transitions took place from SVHC Low Molecular Weight (LMW) phthalates to High Molecular Weight (HMW) phthalates and other plasticisers; from cadmium and lead-based stabilisers to calcium-based stabilisers. REACH restrictions ensuring a level playing field for products manufactured in Europe with imported products allow, for example, the continued use of recyclates containing cadmium or lead in very specific applications such as building profiles, windows, or sewage pipes under defined conditions through corresponding derogations (REACH 2011, 2023) Different R&D projects—industry and VinylPlus projects (VinylPlus 2023) as well as EU funded consortia projects (e.g. Circular flooring (CF 2024; Remadyl 2020)—aimed to remove classified legacy additives from end-of-life PVC.

The development of solvent-based physical recycling technologies capable of separating PVC composite waste and removing contaminants is another important element to close the loop. Vinyloop in Ferrara, Italy has demonstrated for several years on a production scale (10 kt/year) the recycling of cable waste and the separation of fibres from PVC composites, but the classified plasticiser DEHP could not be extracted and remained in the rPVC, which is why the plant was closed in 2018 (VinyLoop 2018). New developments indicate that legacy additives can be extracted by solvent-based processes, e.g. in the Circular Flooring project (CF 2024).

Chemical recycling is complementary to mechanical recycling for the treatment of difficult-to-recycle mixed plastic wastes being highly contaminated or currently incinerated with energy recovery and not suitable for mechanical recycling. In the case of PVC, consisting of 43% hydrocarbon and 57% of chlorine, both parts of the molecule can be chemically recycled. The hydrocarbon part can be recycled by e.g. pyrolysis or gasification, processes which are suitable for hydrocarbon polymers (JRC 2023). The chlorine part can be recycled and used as feedstock for PVC production or other substances. The European PVC industry is investigating processes as part of the VinylPlus RecoChlor project (VinylPlus 2023), by which the chlorine is recovered either in the form of sodium chloride or as hydrochloric acid. In case of flue gas treatment, the recovery of sodium chloride is recognised as a recycling process in the BAT Reference Document for waste treatment (BAT 2010). The RecoAcid (VinylPlus 2023) process is based on the FLUWA technology, which uses the generated acid in municipal waste-to-energy incineration plants for the recovery and recycling of metals contained in filter ashes. Currently, the waste-to-energy plants do not generate enough raw acid by treatment of household wastes. This acid gap can be closed in the plants by in-situ generated hydrochloric acid from mechanically non-recyclable PVC wastes. The trial at the MSWI plant of AVAG Umwelt AG in

Thun, Switzerland, has confirmed that the increased chlorine concentration in the flue gas showed no corrosion or other negative effects in the plant. In 2026 the FLUWA technology will be obligatory in Swiss municipal waste-to-energy plants (VinylPlus 2023).

To promote the recyclability of plastic products in the framework of the Circular Plastics Alliance (CPA) the European Commission issued a Standardisation Request M/584 in 2022 and mandated the European Standardisation Committees CEN and CENELEC to develop a series of design-for-recycling (DfR) standards (EU 2022). The aim is to create conditions for defining and harmonising quality requirements for plastic waste, recyclates, and product families to achieve the target of 10 Mt of recyclates. PVC product sectors (e.g. EPPA profiles, ERFMI floorings) are directly involved and EPPA has been chairing the CEN/CENELEC coordination group "Circular Plastics" since 2023 in charge of the development of the DfR guideline principles in major plastic product sectors (Feigenbutz 2024).

7.5 The Electrolux Experience

Karl Edsjö

7.5.1 Introduction

Electrolux has successfully launched several products made with recycled plastic and is on the way to shape the white goods sectors' use of the circular material. Today Electrolux is one of the leading manufacturers when it comes to replacing virgin plastic with recycled in appliances (Recycling Today 2021).

The story behind this success is long and has many components, stakeholders, and contributors. Let's begin by having a closer look at where the company is today in terms of using recycled plastics, and let's take some help from Harvard economist Rebecca Henderson. In 2015 Henderson published an article titled "Making the Business Case for Environmental Sustainability" where she presented the case for two main routes towards more sustainable business: either demand-driven, e.g. consumers ready to pay extra for the more sustainable products, or supply driven, by which the greener supply options are cheaper for the producer or in other ways make economic sense (Henderson 2015).

Electrolux has pursued both routes that Henderson envisaged for the use of recycled plastics, the demand driven, and the supply driven. As it happened, the first steps were taken along the demand-driven path, and it began with the vacuum cleaners. While there had been attempts earlier, the journey started in earnest with the Vac from the Sea campaign (cf. Fig. 7.22) and the first Green Range vac.

The Green Range vac was at its launch a package of several "green" features: energy efficient motor, low noise, sustainable packaging, and was made with a

Fig. 7.22 Two of the six vacs from the sea, here the North Sea and Indian Ocean editions. (*Source* Electrolux)

considerable amount of recycled ABS and PP. Electrolux is still today marketing the successors of the very first green range (cf. Fig. 7.23).

The Vac from the Sea campaign that accompanied the launch of the Green Range vacs was focused on raising awareness about plastics in the oceans (Electrolux 2010). In the campaign, Electrolux joined with beach clean-up organizations around the globe to run dedicated beach plastic collection efforts and then used the collected plastic debris to decorate a set of six vacuum cleaners as an art and communication project. The campaign and the resulting arty vacs created considerable attention around the globe and was in fact the most successful PR campaign of Electrolux ever.

In Henderson's model, marketing and sales of the Green Range vacs used the demand-driven approach where the market is ready to pay something extra for the green features offered. For large appliances, mainly in the wet category—dishwashers and washing machines—up until 2022, the supply-driven approach became the main route, led by the company's internal ambitions and supply opportunities for recycled PP. Finally, in 2023 a fridge with an inner liner mainly made with HIPS from recycled fridges made it to the market, combining the routes of demand and supply opportunity

Fig. 7.23 The 2023 high-end version of the Electrolux Green Range vacuum cleaners, with 70% recycled plastic. (*Source* Electrolux)

(though the supply opportunity this time not necessarily being cheaper but finally technically possible).

In the following sub-sections, some of the key drivers pushing the Electrolux teams forward, as well as some of the challenges that faced those working on recycled plastics and the solutions that emerged, will be reviewed.

7.5.2 Drivers of Replacing Virgin Plastic with Recycled Materials

When the Green Vacs were launched Electrolux had already since many years made its first trials with recycled plastics, first on fridge inner liners, only prototyping, then on vacs, for small-scale commercialization. For the Green Range a few different parts, made from either PP or ABS, were replaced with recycled material. Most of the exterior, the bottom plate in PP and the cover in ABS were made from recycled plus parts of the interior. The development work was done in the R&D office in Stockholm and in the vacuum cleaner factory in Hungary.

One important element of the successful finalization of the project was to have the right manager with the right powers in place, able to drive the project from many angles, from supplier discussions to R&D to marketing. Another was to find the right suppliers, and having an agile product organization in place, ready to work intensively with a short time frame to do the R&D on the different components, and with marketing at the same time preparing the campaign.

While sustainability was established as a strong part of the Electrolux brand already since the 90's and played a key role in the development of the Green Range products, another key to success was the marketing and branding approach of the whole project. Additionally, equally transformative as the recycled plastic was the fact that the product featured a very energy-efficient motor. At the time this was contrary to any advertising around vacs, which usually contained references to the high-powered, and thus energy-hungry motors, implying better cleaning.

The Green Range instead came with a promise of high cleaning efficiency, low noise, and low energy consumption. While the energy efficiency indeed came to be at least as transformative as the recycled plastic (today the energy consumption of vacs is highly regulated through eco-design rules), the visually striking use of black recycled plastic, with green details, came to have a bigger impact on the legacy of the product, still today referred to as the "green" vac, though it's actually black (cf. Fig. 7.23).

Due to Electrolux' organization, the small appliance division to which the vacs belonged, and the white goods division, with washing machines and fridges, operated isolated from each other, with separate divisions for R&D, product development, marketing, and sales.

Therefore the achievements of the vac colleagues, while obviously well noted, did not have a direct impact on the large appliances departments, other than possibly inspirational. Partly due to this division, but also due to the different characteristics of the products, with larger appliances having relatively less visible surface, being more technical, having less plastic in some categories, being more long-lived, and more difficult to replace, the route for introducing recycled plastic in large appliances looked quite different, and took longer time to reach as big an impact externally, if even achieved today.

The drivers for recycled plastic in large appliances came from a few different directions. Still rooted in the company's dedication to sustainability, one particular source of these ambitions was Electrolux' intense lobbying in pushing an ambitious EU producer responsibility legislation through the WEEE Directive (EU 2012). Together with several mainly non-white goods brands such as Sony, HP, and Gilette, Electrolux formed a coalition asking for so-called individual producer responsibility, rather than a shared responsibility, in order to create a real driver for designing products that were easy to recycle at the end of life, thus enabling more recycled materials (Ökopol 2007).

While the final legislation did not really result in individual producer responsibility in practice, one legacy of this lobbying effort was to establish consciousness in parts of the company about the growth of problematic e-waste, and the challenges to make something useful out of this waste. While legislators clearly pointed to the concerns about e-waste they did not, and still don't require producers to use recycled materials. Instead, the drive to tackle the challenge of re-using materials from e-waste came from Electrolux ambition to be at the forefront of sustainability, to be and stay a sustainability leader.

Of the two main fractions in e-waste, metals thanks to their intrinsic value, are already being recovered and taken care of. The second biggest fraction, the plastic on the other hand has usually not been taken care of at all, and tackling that fraction indicated an opportunity where the company could make a difference. Now started a new journey, about the same time as the success with the vac, in the large appliance division to identify useful streams of waste plastic as well as to decide on what components to work on.

7.5.3 How to Make It Happen

With the drivers in place to move towards using recycled plastic it's time to look at the actions required to actually make it happen. A central element is the supply base and how to set the right requirements to get the material needed.

The first step is about identifying suitable supplier candidates, scouting for companies with the right type of material, from the right source at right cost level. With a candidate supplier identified, the next step is to make the initial tests of chemical compliance and material performance (mechanical properties). The chemical compliance is a critical component of the supplier approval process.

Chemical Compliance

In parallel with the WEEE Directive, the EU published the RoHS Directive (Restriction of certain Hazardous Substances), banning initially four substances (EU 2011). To secure compliance with the RoHS requirements Electrolux spent large resources on establishing a system to screen all types of suppliers and incoming materials to make sure there were no banned substances in the final products, including in recycled plastics (Electrolux 2024).

To secure the recycled plastics the company put a system in place to minimize any risk of unwanted chemical contamination in the recycled materials, and this system came to stand on three legs: safe sources, traceability, and regular testing. The safe source requirement had the impact that in the beginning recycled material from the company's own type of products was excluded. Too many different products in one stream (a mix of different sorts of e-waste) and not yet well-developed sorting practices meant that material from this stream posed too big a risk. Instead, plastics from e-waste was either treated as waste or used in applications without RoHS requirements, as RoHS is only applicable to electrical and electronic equipment. Over time, as handling and sorting had improved, and levels of chemicals diminished, plastics from e-waste has today become a safe material for EEE, too.

Food packaging waste was identified as a safer, and better accessible source. Especially the PP was interesting for some of the larger components. The traceability requirement meant that Electrolux wanted to be sure that the supplier of recycled plastic always used the same sort of waste materials for its production of recycled material. Electrolux also established relations with suppliers of food packaging waste material, securing an insight and overview of the whole supply chain.

Selection of Components

When selecting what components to work on there are a few key factors (cf. Table 7.3):

- mechanical properties required
- visibility of component
- food contact.

When having a choice, one would rather start with a non-visible component with low requirements on mechanical performance, and certainly without food contact as this is difficult to manage with recycled materials. Subsequently, the first component to be made of recycled plastic for large appliances was a PP cover for the service-hole in the back of washing machines, which is a large piece of plastic with low requirements on aesthetics and mechanical performance. When work was going on to tune the recycled material for this first application, some deemed it all but impossible to replace virgin plastics in the more demanding components. Fortunately, they turned out to be wrong.

Like most other makers of wet appliances such as washing machines, dishwashers, and tumble dryers, Electrolux relies on filled PP for many demanding and some less demanding components. However, Electrolux is unusual in using calcium carbonate as the filler rather than talcum, which is the preferred filler elsewhere. The purpose is anyhow the same, to achieve certain mechanical properties while at the same time keeping down cost.

What might be even more unique is that the company is using its own, patented blend of PP and calcium carbonate, called Carboran®. This material is used for several different applications within the wet category, for example the base of tumble

Table 7.3 Selecting components suitable for recycled material. (*Source* Own presentation)

Considerations when selecting components for recycled plastic	
Mechanical performance	Working temperature, exposure to strong chemicals, movement—in a washing machine or dishwasher there are many moving parts, in hot or cold water, and under at least part of the cycle with detergents
Aesthetic requirements	Front and top are often visible; side, back, and internal parts hidden or almost hidden. Most visible parts are typically white, placing very high demands on the plastic. Internal parts have much lower aesthetic requirements (but usually not entirely without requirements)
Food contact	Food contact regulations are strict and hard to meet for recycled. Mainly a concern for fridges and small kitchen appliances
Type of material	Something that can be found in sufficient volumes from traceable and acceptable sources at the right cost

dryers and the outer drum or the tub in washing machines, both large components weighing several kilos. This was the next level of challenges with introducing recycled material. To get it right, the material R&D team experimented with a few different blends of virgin and recycled, but always with the same amount of filler, around 40% of calcium carbonate.

In the end, they landed at two versions of the new material Carborec®, one high-performance, low-level recycled blend, and a high-level recycled blend for components with less requirements on mechanical performance. The filler is still "virgin", though there are now different projects, outside Electrolux, for $CaCO_3$ recovery from industrial waste which could become a source of recycled $CaCO_3$ (Electrolux 2014).

7.5.4 Reflections and a Look Ahead

Electrolux has a long history of driving sustainability topics within the white goods industry, beginning already in the early nineties. Since then a number of different initiatives and ideas have been launched and many of these have sooner or later reached the market. One early study showed the need for, or potential of recycling scrapped appliances, painting a circular future more than fifteen years before the term circular economy became the buzzword in Brussels and elsewhere. Also already in the late nineties were the very first manufacturing tests done using recycled plastic materials—inner liner to inner liner—with positive results.

Anyhow it would take more than ten years after that initial test before Electrolux successfully launched products with a significant amount of recycled plastic, the Green Range vacuum cleaners. And it would take more than a further ten years before the successful launch of a commercially available fridge with inner-liner to inner-liner recycling. Why did it take so long time, and what happened between the very first trials and the final launch to the market?

Many different factors played a role in affecting the process along the way, enabling factors, and those adding a bit more complexity to an already challenging task. Arguably a key factor, that led to finally enabling a commercially available fridge with a recycled inner liner, was the WEEE Directive mentioned earlier with the introduction of producer responsibility for e-waste across the EU member states.

In addition to Electrolux putting the focus on using recycled material, the key outcome of producer responsibility legislation such as WEEE was that producers had to make sure that there was both collection infrastructure, enabling separate collection of dedicated waste streams, and treatment in place for scrapped EEE products. This lead to considerably increased collection volumes and improved recycling processes, which in turn made recycled materials available on the market. With similar legislation in place for packaging, also materials from this stream became available.

While Electrolux' sustainability targets and an ambition to be at the forefront in this area were crucial factors in driving the implementation and overcoming all the challenges to reach the goal, another important enabler was the economic opportunities that came with recycled. It was an important factor in motivating staff and convincing management and is also a key driver to expand the use of recycled—the higher the virgin volume replaced, the bigger the economic benefit. Cost benefit for recycled material was not always the case but recycled plastic from costly suppliers in practice phased out itself and was replaced with more economical alternatives (Electrolux 2016). In addition, recycled plastic in appliances has been met with a lasting positive market response, as seen with the Green Range vacs launched in 2010, continuously updated, and still on offer 14 years later.

The future will look good if joint efforts by suppliers and users of recycled plastics continue to develop at the same pace. As seen from the Electrolux journey, recycled plastics have, step by step, entered into more and more demanding applications, beginning with dark colours and low requirements, over more mechanical components to white food-grade plastic in the fridge inner-liners. Each of these steps has required extensive work, and each step achieved has built confidence enabling the next level.

References

BAT 2010: Best available techniques (BAT) reference document for waste treatment (5.5.1.4.1). https://op.europa.eu/en/publication-detail/-/publication/782f0042-d66f-11e8-9424-01aa75ed71a1/language-en (Last access Aug. 20, 2024)

Berry 2022: Berry Global press release Sept 1, 2022: Berry Recycling Technology Receives FDA Letter of No Objection. Berry Global Achieves FDA Food Packaging Approval (Last access: Aug. 20, 2024)

BNB 2015: Bundesministerium für Wohnen, Stadtentwicklung und Bauwesen Assessment System for Sustainable Buildings (BNB-BN2015.Zip, Anlage 1 Nutzungsdauern). https://www.bnb-nachhaltigesbauen.de/en/assessment-system/office-buildings/ (Last access: Aug. 20, 2024)

CF 2024: Cirucular Flooring. https://www.circular-flooring.eu/ (Last access: Aug. 20, 2024)

Consultic 2015: Analyse/ Beschreibung der derzeitigen Situation der stofflichen und energetischen Verwertung von Kunststoffabfällen in Deutschland. (Analysis/description of the current situation of material and energy recovery from plastic waste in Germany. Dates from 2013)

Consultic 2016: Produktion, Verarbeitung und Verwertung von Kunststoffen in Deutschland 2015. (Production, processing and utilisation of plastics in Germany 2015)

Conversio 2021: PVC waste in EU 27+3 countries 2020, Conversio, September 2021; VinylPlus

Conversio 2022: PVC Material Flow in Germany in 2021. Facts and Figures on the Life Cycle of PVC, Conversio, December 2022

COSPATOX 2023: Cospatox consortium website. https://cospatox.com/the-project/ (Last access: Aug. 20, 2024)

CPA 2018: Circular Plastics Alliance (CPA) – a European Monitoring System. https://single-market-economy.ec.europa.eu/industry/strategy/industrial-alliances/circular-plastics-alliance/commitments-and-deliverables-circular-plastics-alliance_en (Last access: Aug. 20, 2024)

DIN 2020: DIN EN 12608–1:2020–11 Unplasticised poly vinyl chloride (PVC-U) for the fabrication of windows and doors. https://www.beuth.de/de/norm/din-en-12608-1/320181378. (Last access: Feb. 24, 2024)

Electrolux 2010: Electrolux Group: Electrolux unveils five Vacs from the Sea, Nov 1, 2010. https://www.electroluxgroup.com/en/electrolux-unveils-five-vacs-from-the-sea-8687/ (Last access: Feb. 24, 2024)

Electrolux 2014: Electrolux Sutainability 2014 on Carborec. https://www.electroluxgroup.com/annualreports/2014/en/sustainability/snapshots/carborec/ (Last access: Feb. 24, 2024)

Electrolux 2016: Electrolux Group Annual Report 2016: Recycled material is making it big. https://www.electroluxgroup.com/annualreports/2016/en/sustainability/case-studies/recycled-material-is-making-it-big/ (Last access: Feb. 24, 2014)

Electrolux 2024: Electrolux Group: Electrolux Group Restricted Materials List v20. https://www.electroluxgroup.com/en/restricted-materials-590/ (Last access: Feb. 9, 2024)

EFSA 2022: EFSA Panel on Food Contact Materials, Enzymes, and Processing Aid (CEP): Safety assessment of the process Starlinger recoStar HDPE FC 1 – PET2PET used to recycle post-consumer HDPE closures into food contact closures. EFSA Journal 2022; 20(1):7001

EFSA PANEL 2015: EFSA Panel on Food Contact Materials, Enzymes, Flavorings and Processing Aids (CEF): Scientific Opinion on the safety assessment of the processes 'Biffa Polymers' and 'CLRrHDPE' used to recycle high-density polyethylene bottles for use as food contact material. EFSA Journal 2015; 13(2):4016

EPD 2022: EPD EPPA Kunststoff-Fenster mit 3-Scheiben-Isolierglas. https://epd-online.com/EmbeddedEpdList/Detail?id=15560 https://ibu-epd.com/veroeffentlichte-epds/ (Last access: Mar. 11, 2024)

EPD 2024a: EPD ERFMI Homogeneous polyvinyl chloride floor coverings. https://ibu-epd.com/veroeffentlichte-epds/ (Last access: Feb.26, 2024)

EPD 2024b: EPD ERFMI Heterogeneous polyvinyl chloride floor coverings. https://ibu-epd.com/veroeffentlichte-epds/ (Last access: Feb. 2024)

EPD 2024c: EPD Sika Sikaplan® G – multi-layer synthetic waterproofing membrane. https://ibu-epd.com/veroeffentlichte-epds/ (Last access: Mar. 11, 2024)

EU 2011: Directive 2011/65/EU of the European Parliament and of the Council of 8 June 2011 on the restriction of the use of certain hazardous substances in electrical and electronic equipment. https://eur-lex.europa.eu/legal-content/EN/TXT/?uri=celex%3A32011L0065 (Last access: Aug. 20, 2024)

EU 2011: Construction Products regulation – EPR. https://single-market-economy.ec.europa.eu/sectors/construction/construction-products-regulation-cpr_en (Last access: Aug. 20, 2024)

EU 2012: Directive 2012/19/EU of the European Parliament and of the Council of 4 July 2012 on waste electrical and electronic equipment (WEEE). https://environment.ec.europa.eu/topics/waste-and-recycling/waste-electrical-and-electronic-equipment-weee_en (Last access: Aug. 20, 2024)

EU 2022: Commission implementing decision C (2022)5372 – Standardisation request M/584 Plastics recycling and recycled plastics in support of the European Strategy for Plastics in a Circular Economy. https://ec.europa.eu/growth/tools-databases/enorm/mandate/584_en (Last access: Aug. 20, 2024)

EU 2023: Commission Delegated Regulation (EU) 2023/1616 of 3 May 2023 supplementing Regulation (EU) 2021/23. https://eur-lex.europa.eu/legal-content/EN/TXT/PDF/?uri=OJ:JOL_2023_199_R_0004 (Last access: Aug. 20, 2024)

Everard 2022: Everard, M: Assessment of the sustainable use of chemicals on a level playing field. https://setac.onlinelibrary.wiley.com/doi/epdf/ https://doi.org/10.1002/ieam.4723 (Last access: Aug. 20, 2024)

Feigenbutz 2024: Feigenbutz, G: "Umsetzung des EC Normungsmandat M/584"; VinylPlus Deutschland und PlasticsEurope Deutschland; PVC-Workshop 11. Januar 2024, Bonn

Franz et al 2022: Franz, R, Welle, F: Recycling of post-consumer Packaging Materials into New Food Packaging Applications – Critical Review of the European Approach and Future Perspectives. Sustainability 2022,14, 848.

Green Dot 2024: Specifications. https://www.gruener-punkt.de/en/downloads. (Last access: Jan. 11, 2024)

Grüner Punkt 2023: Document on enterprise webpage: Rohstoffspezifikation 329 Polyethylen (Dokument 4.7.1.2.4.14 - Rohstofffraktionsspezifikation 329 Polyethylen v0.01.0005 (gruener-punkt.de)

GVM 2016: Recyclingbilanz für Verpackungen, Berichtsjahr 2015. (Recycling balance for packaging, reporting year 2015)

GVM 2018: Potenziale der Substitution von Kunststoff/Kunststoff-Verbundfolien durch recyclingfähige Mono- oder Verbundfolien. (Potential for substituting plastic/plastic composite films with recyclable mono or composite films. (Study commissioned by BKV)

Hallaji 2017: Hallaji, K.: Process of producing PCR pellets; EP 2 507 022 B1

Henderson 2015: Henderson, Rebecca, Ranjay Gulati, and Michael Tushman (eds), Leading Sustainable Change: An Organizational Perspective. Oxford, 2015; online edn, Oxford Academic, 23 Apr. 2015), https://doi.org/10.1093/acprof:oso/9780198704072.001.0001 (last access: Feb. 26, 2024)

Heyde 2016: Heyde, M.: Polyolefin-Rezyklate aus Post-Consumer Abfällen in endverbrauchernahen Anwendungen – Anforderungen und technische Lösungen 1. Recycling -Tagung Süddeutsches Kunststoffzentrum Würzburg 2016

IK 2019: Round Table Eco-Design für Plastics Packaging. https://ecodesign-packaging.org (Last access: Jan. 1, 2024)

JRC 2023: Environmental and economic assessment of plastic waste recycling, JRC Technical Report 2023. https://doi.org/10.2760/0472

JRC 2024: JRC 135901 Modelling plastic products flows and recycling in the EU, JRC Report 2024. https://publications.jrc.ec.europa.eu/repository/handle/JRC135901(Last access: Jan.1, 2024)

Kitzberger 2017: Kitzberger, C.: Odor out – Treating Post-consumer Recyclates to remove odors. Kunststoffe International 6–7/2017

KRV 2019: Kunststoffrohrverband: Impulse Mai 2019 Kreislaufwirtschaft. https://www.krv.de/sites/www.krv.de/files/pdfflip/krv_impulse_mai_2019/krv_impulse_mai_2019.html (Last access: Aug. 20, 2024)

Laner 2022: Laner, D: DFG-Projekt 490612391.Integriertes Modell der Materialeffizienz und Umweltauswirkungen von Materialkreisläufen in Gebäuden und Infrastruktur. https://gepris.dfg.de/gepris/projekt/490612391 (Last access: Aug. 20, 2024)

Leadbitter and Bradley 1997: From 'Closed loop recycling opportunities for PVC' published by J. Leadbitter and J. Bradley at the IPTME Symposium, Loughborough University, 3–4 November 1997

Natural Step 2024: The Natural Step. https://thenaturalstep.org/ (Last access: Aug. 20, 2024)

Öko-Institut 2022: Life cycle assessment of the performance of the dual systems in the area of packaging recycling; Öko-Institut Berlin/Darmstadt; 24.04.2022. https://www.oeko.de//fileadmin/oekodoc/Duale_Systeme_Oekobilanz_Endbericht.pdf (Last access: Jan. 17, 2024)

Ökopol et al. 2007: The Producer Responsibility Principle of the WEEE Directive. https://ec.europa.eu/environment/pdf/waste/weee/final_rep_okopol.pdf (cf. p. 43) (Last access: Aug. 20, 2024)

PE 2022: PlasticsEurope. Plastics—the facts 2022. https://plasticseurope.org/knowledge-hub/plastics-the-facts-2022/ (Last access: Feb. 27, 2024)

PPWR 2023: Council of the European Union. General approach on the proposal for a PPWR. Adopted on 18 December 2023. https://ST-16946-2023-INIT_en.pdf (Last access: Jan. 1, 2024)

Prowindo 2018: Initiativen für Markt, Nachwuchs und Umwelt, März 2018, Prowindo PK: Das Fenster zur Zukunft: Leistungsbilanz und Herausforderungen". https://www.vinylplus.de/prowindo-pressekonferenz-das-fenster-zur-zukunft-leistungsbilanz-und-herausforderungen/2295 (Last access: Aug. 20, 2024)

PVC4Cables 2022: PVC4Cables: European Cables Market 2021, Plastic Consult. https://pvc4cables.org/wp-content/uploads/2023/08/PVC4Cables_Leaflet_Sept2022.pdf (Last access: Aug. 20, 2024)

PVC4Cables 2019: PVC4Calbes: Fire-Safety—choosing PVC cables under the CPR. https://pvc4cables.org/wp-content/uploads/2023/08/PVC4Cables-Fire-Safety-Brochure_Oct2019_EN.pdf (Last access: Aug. 20, 2024)

RAL 2023: RAL-GZ 716 Kunststoff-Fensterprofilsysteme—Gütezeichen (April 2019). https://www.beuth.de/de/technische-regel/ral-gz-716/313778272 Technischer Anhang, Jan. 2023 https://www.bnb-nachhaltigesbauen.de/en/assessment-system/office-buildings/ (Last access: Aug. 20, 2024)

REACH 2011: EU Commission Regulation – REACH Annex XVII (Cadmium). http://data.europa.eu/eli/reg/2011/494/oj (Last access: Aug. 20, 2024)

REACH 2023: EU Commission Regulation – REACH Annex XVII (Lead in PVC). https://eur-lex.europa.eu/eli/reg/2023/923/oj (Last access: Aug. 20, 2024)

Recycling Today 2021: Electrolux eyes increased recycled-content plastic. https://www.recyclingtoday.com/news/electrolux-sweden-appliances-plastic-recycling-target/ (Last access: Feb. 23, 2024)

Remadyl 2020: Remadyl - Summary of Key achievements of Remadyl H2020 project. https://www.remadyl.eu/2024/02/26/summary-of-key-achievements-of-remadyl-h2020-project/ (Last access: Aug. 20, 2024)

Starlinger 2023: Starlinger viscotec company communication. https://www.viscotec.at/fileadmin/user_upload/products/viscoZERO/viscotec-productsheet_viscoZERO_2022.pdf (Last access: Aug. 20, 2024)

Strangl et al 2019: Strangl, M, Ortner, E, Buettner, A: Evaluation of the efficiency of odor removal from recycled HDPE using a modified recycling process. Ressources, Conservation and Recycling 146 (2019) 89–97

Systalen 2024: Technical Data Sheets Systalen®. https://www.gruener-punkt.de/en/systalenr-recyclate/systalenr-application-product-areas/systalenr-technical-data-sheets (Last access: Jan. 17, 2024)

TEPPFA 2019: TEPPFA and PVC4Pipes: 100-years lifetime for PVC pipes (June 2019). https://www.teppfa.eu/wp-content/uploads/TEPPFA-PVC4Pipes-Position-on-100years-lifetime-of-PVC-Pipes.pdf (Last access: Mar. 11, 2024)

VinylLoop 2018: VinyLoop closed PVC recycling project in Italy. https://www.plastech.biz/en/news/VinyLoop-closed-PVC-recycling-project-in-Italy-12649 (Last access: Mar. 11, 2024)

VinylPlus 2018: VinylPlus press statement 2018. https://www.vinylplus.eu/news/recycling-commitment-900000-tonnes-of-pvc-a-year-by-2025/ (Last access: Mar. 11, 2024)

VinylPlus 2021: VinylPlus Progress Report 2021. https://www.vinylplus.eu/resources/?sftresourcescategories=progress-reports (Last access: Mar. 11, 2024)

VinylPlus 2022: VinylPlus. ASF Introduction Webinar, 2022. https://www.vinylplus.eu/resources/webinar-asf-introduction/ (Last access: Mar. 11, 2024)

VinylPlus 2023: VinylPlus Progress Report 2023. https://www.vinylplus.eu/resources/?sftresourcescategories=progress-reports (Last access: Mar. 11, 2024)

VinylPlus 2024: Wissenswert—PVC-Dach- und Dichtungsbahnen, VinylPlus Deutschland. https://www.vinylplus.de/wp-content/uploads/2021/08/WissensWERT-PVC-_Dach-und-Dichtungsbahnen.pdf (Last access: Mar. 11, 2024)

Welle, F.: Recycling of post-consumer Packaging Materials into New Food Packaging Applications – Critical Review of the European Approach and Future Perspectives. Sustainability 2022, 14, 848.

ZSVR 2024: https://www.verpackungsregister.org/en/foundation-authority/minimum-standard-pursuant-to-section-21/2023-edition (Last access: Jan. 1, 2024)

Dr.-Ing. Michael Heyde was Head of the Technical Department of Der Grüne Punkt – Duales System Deutschland GmbH. He started the operative recycling business at Der Grüne Punkt as managing director of Systec Plastics GmbH, developing recycling technologies and closed-loop solutions for plastics packaging wastes. In 2018, he joined ALPLA Werke Lehner GmbH & Co KG as Head of Recycling Technology. He studied Chemical Engineering at the Technical University of Munich and worked as a scientist at Fraunhofer's IVV Institute in Freising/Germany.

Benjamin Kampmann graduated from the University of Applied Sciences Osnabrück as Dipl.-Ing. (FH) in Plastics Technology in 2009. He joined plastics converter PÖPPELMANN Lohne in 2010. His core tasks are the central management of the *initiative PÖPPELMANN blue*, promoting the circular economy and climate transformation, and the management of other sustainability projects.

Frank Schockemöhle has been an industrial foreman for plastics and rubber since his training at Pöppelmann Kunststoff-Technik GmbH & Co. KG. He has built up and managed various departments (e.g. material coordination department, development division) and has been Head of Technology Management since 2023 at Pöppelmann's.

Johannes Wedi studied chemical engineering in Münster and started his professional career in the food industry where he was responsible for product development of fats and oils. Since 1990 he has been responsible for product development at Bischof + Klein, initially in all product areas, such as industrial films, technical films, and consumer goods. From 1997, he has been responsible for product development and application technology for consumer goods.

Thomas Hülsmann is Managing Director of VinylPlus Deutschland e.V., the PVC value chain association for sustainable development in Germany, since 2014. He is also Managing Director of associated project and communication company AGPU MEDIA GmbH. On a European level, he is involved in various committees of VinylPlus®, the sustainability initiative of the European PVC industry, where he is a member of the Leadership Team.

Karl Edsjö is a biologist by training and a circular economy expert by experience. He is working as an independent consultant within sustainability management and communication, focusing on resource efficiency strategy and circular economy. He worked with Electrolux where he developed the company's circular economy strategy and coordinated the efforts to replace virgin plastic with recycled making Electrolux the European white-goods brand with the highest share and biggest volume of recycled plastic in its products.

Chapter 8
Market—Retailers and Consumers

Dagmar Glatz, David Stakemeier, Sarah Grede, Anna Ziessow, and Dietmar Böhm

Contents

D. Glatz (✉)
dm-Drogeriemarkt GmbH & Co. KG, Karlsruhe, Germany
e-mail: Dagmar.Glatz@dm.de

D. Stakemeier
dm-Drogeriemarkt GmbH & Co. KG, Edewecht, Germany
e-mail: david.stakemeier@dm.de

S. Grede
GS1 Germany GmbH, Cologne, Germany
e-mail: sarah.grede@gs1.de

A. Ziessow
Reyda GmbH, Freiburg, Germany
e-mail: anna@recyda.com

D. Böhm (✉)
PreZero Stiftung & Co. KG, Neckarsulm, Germany
e-mail: dietmar.boehm@prezero.com

P. Orth et al. (eds.), *Circular Economy of Plastics*,
https://doi.org/10.1007/978-3-031-78829-1_8

Summary

Chapter 8 highlights the influential role of retailers and consumers—as presented in the market segment of the circuit diagram (cf. Fig. 8.1)—in the development of a circular economy for FMCG packaging. The function of the retail sector as an important dialog partner for end consumers regarding environmental issues relating to products and their packaging is also shown.

The experience report of a drugstore retailer on the communicative implementation of its packaging strategy vis-à-vis the end consumer in Sect. 8.1 allows many conclusions to be drawn that go beyond the specific packaging area and shows how consumer perceptions of sustainability can be influenced in a dialog at eye level.

Despite the many obstacles and challenges, how can a retailer ensure that all of its competing suppliers are using recycled content in their packaging and steadily increasing recycled content? The reader will find answers to this question in Sect. 8.2. through examples of coopetition initiatives that lead to the systematic processing of challenges by all participants in the drugstore packaging value chain.

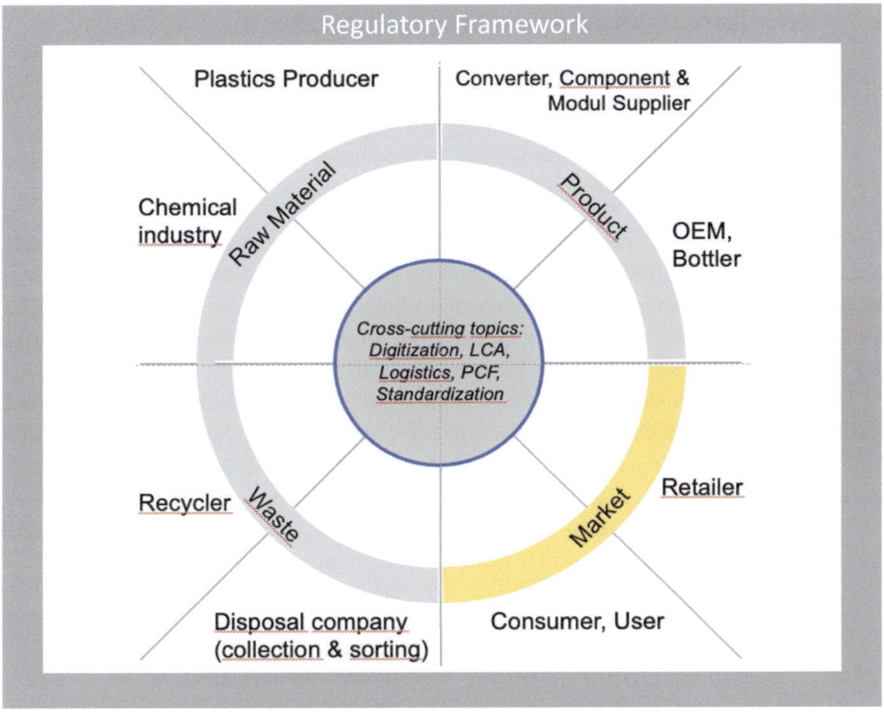

Fig. 8.1 Schematic diagram of the plastics cycle (*Source* Own presentation)

A key prerequisite for establishing a circular economy for plastic packaging is the monitoring of important attributes such as recyclability and recycled content across various stages of the value circle through communication with the end consumer. This requires standardized data models for automated data flows. Section 8.3 shows an example of the development and implementation of such a data flow system, an illustrative example of the crucial role of digitization for a functioning cycle management of individual packaging types.

Section 8.4 shows how the retail sector assumes responsibility for the circular economy of packaging using the example of one of the world's largest retailers, which has set up its own environmental service provider. This company is acting as a Dual System in Germany with licensing activities for sales packaging to meet the Extended Producer Responsibility (EPR), as an internal and external service provider for customers, and as a waste management and recycling company producing secondary raw materials. To meet tightening legal requirements and respective customer needs intelligent service propositions, infrastructure, and equipment are pushed on by means of digitization, automation, and artificial intelligence in close cooperation with upstream and downstream partners in the value circle.

8.1 Retailer/Consumer Dialogue on Sustainable Packaging

Dagmar Glatz and David Stakemeier

8.1.1 *Introduction*

dm-drogerie markt, as a retailer with strong brands that enjoy the trust of customers, can actively contribute to the optimisation of sustainable products. The relationship between retail companies and consumers plays a crucial role in shaping sustainable consumption. The optimization of packaging solutions and the establishment of a functioning circular economy are key aspects of this.

This section discusses the topic of sustainable packaging in the context of retail, particularly regarding the communication between retail companies and consumers. Communication with customers is crucial to convey sustainable packaging solutions, create trust, and raise awareness. The challenges and opportunities arising from different consumer perceptions of sustainability are presented. A successful dialogue makes it possible for the consumer to become a change agent.

8.1.2 *Customer Understanding of Sustainability in Retail*

On closer inspection, the issues surrounding sustainability in the retail sector are more individualised than might be assumed at first glance. Above all, they are a question of perspective. There is no such thing as one customer who pays attention to sustainability. Several influencing factors lead to very different assessments of more sustainable offers. These include demographic characteristics, consumption patterns, purchasing options and barriers and individual attitudes towards doing without (Wellbrock and Ludin 2021).

The German Federal Environment Agency emphasises the German government's goal of increasing the market share of environmentally friendly products to 34% by 2030. Accordingly, products that are sold with state eco-labels are classified as environmentally friendly products. In 2020, this share was 13.4%. There is a particular need to catch up in the food and car sectors (UBA 2022).

Customers are willing to act more sustainably. However, willingness to pay remains the biggest barrier when it comes to adapting behavior. Across various product groups, two-thirds of customers do not inform themselves about sustainability aspects before making a consumption decision. 34% of consumers consider environmentally friendly or recyclable packaging to be an important aspect of sustainability. In the same survey, 24% stated that sustainability seals were an important sustainability aspect for them (Lehmann et al. 2022).

In order to better understand the sustainability preferences of its own customers in relation to drugstore products, dm-drogerie markt conducts its own annual survey with the help of its customers (dm 2023). This study from 2023 confirms that customers factor four sustainability aspects into their purchasing decisions:

- A product should generally be good for the environment (39%);
- A product should be manufactured in an environmentally friendly way (34%);
- A product should be managed in a cycle (30%);
- A product should have sustainable packaging (26%).

When explicitly asked in this study about the use of sustainable packaging material, 29% of customers stated that this was their most important sustainability criterion when making a purchase decision. Only 16% of customers stated that sustainable packaging was the least important sustainability criterion for them. At 35%, only the criterion of sustainable ingredients is mentioned more often as the most important criterion.

However, customers consuming drugstore products are as diverse as our society. There is no such thing as the drugstore customer, just as there is no such thing as the sustainability-orientated customer for everyday drugstore products. Animal welfare, regional ties, efficiency, innovation, performance: these are all needs that different customers may prioritise when deciding for or against a product. And as different as the need is, the consumer decision can be just as different.

However, the products for all these customers have one thing in common: they are almost always packaged. And so, it is a worthwhile task to take a closer look at the customer's requirements in terms of sustainable packaging design.

8.1.3 Packaging Strategy and Its Communication to the Customer

dm-drogerie markt consistently pursues a packaging strategy that minimises its ecological footprint and strives for a functioning circular economy. This packaging strategy comprises four guidelines:

- Choose the right packaging material;
- Reduce packaging material;
- Promote packaging made from recycled materials;
- Promote the recyclability of packaging.

The customer can also take a holistic view of these factors and their consistent implementation and, by purchasing optimised packaging, commit to a functioning circular economy and thus to the conservation of resources.

So, a lot depends on the customer's perception. To ensure that sustainability factors can be incorporated into the purchase decision at the shelf, dm-drogerie markt communicates the sustainability aspects of dm brand products via several levels of information (cf. Fig. 8.2). Both the packaging strategy (choose the right

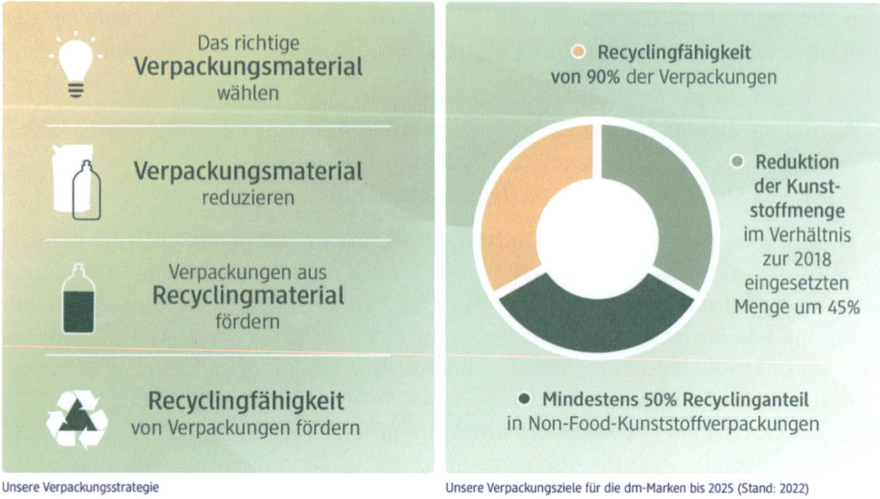

Fig. 8.2 dm infographic on packaging strategy (*Source* dm)

packaging material, reduce packaging material, promote packaging made from recycled material; promote the recyclability of packaging) and the packaging targets for 2025 (recyclability of 90% of packaging; reduction in the amount of plastic used by 45% compared to 2018; at least 50% recycled content in non-food plastic packaging) are presented to the customer. A particular challenge for the company is that the customer's individual assessment of what makes packaging a sustainable packaging solution is not always congruent with the actual, fact-based optimal packaging solution. However, the solution cannot be to adapt the packaging to the customer's understanding, but rather a communicative task to close this gap, educate and contribute to a long-term rethink. This will be briefly illustrated using examples of packaging strategy.

Choosing the Right Packaging Material

dm always selects materials according to the latest scientific data and endeavours and communicates this to the customer. Plastic packaging is heavily scrutinised by customers, who often ask for paper or glass packaging. dm does not follow trends but assesses packaging according to its life cycle from the raw material through production to the end of the packaging's life. So, there are many reasons why dm-drogerie markt uses plastic for packaging. Figure 8.3 shows the advantages of plastic packaging under the heading "Plastic packaging: Not always the worst choice: It requires less energy for manufacturing than aluminum or glass; it is recyclable if designed properly; it has very good product protection; it is very light and saves CO_2 during transport". This makes good communication even more important: make our decisions on the use of plastic packaging transparent, keep plastic packaging in the cycle and thus avoid littering and reduce CO_2 emissions.

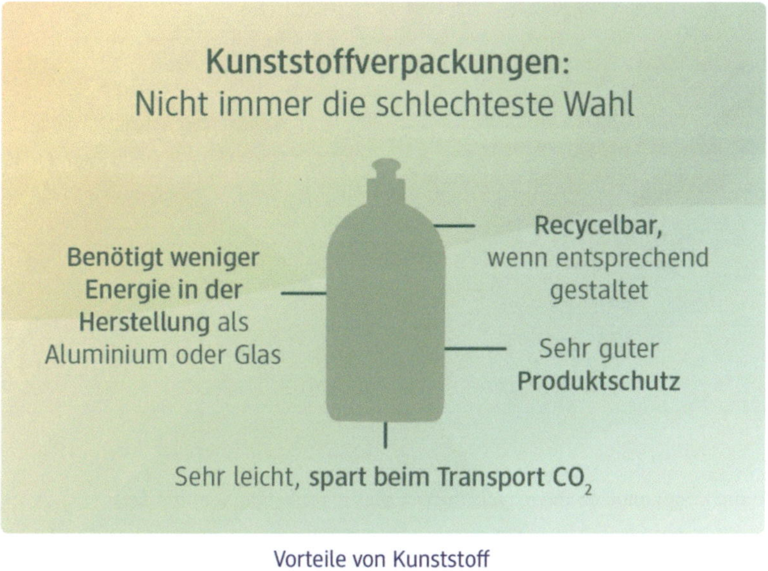

Fig. 8.3 dm infographic on the advantages of plastic as a packaging material (*Source* dm)

Reduction of Packaging Material

dm repeatedly offers customers products with reduced packaging (e.g. shampoo bottles with a 20% weight saving or refill bags), and sometimes also concepts where the consumer is involved. For example, a plastic-reduced shower gel pouch for direct use was tested in 2021. Until then, customers were only familiar with this type of packaging as a refill-at-home solution. Sales of this new concept did not show any increased customer interest. Despite various communication measures, the customer decided against the concept. The pouch was subsequently discontinued.

Use of Recycled Material

The claim for recycled content in the packaging is printed on the front of many products. The percentage of recycled material is often also shown and labelled with a standardised symbol. In addition, information relating to sustainability has been bundled on the back of the packaging for several years already. Recycled content, recyclability, and savings are communicated here and labelled with a cloverleaf.

Improving Recyclability

For packaging to be truly recyclable, it must be harmonised with the entire cycle. As early as the design phase, dm selects recyclable materials and material compositions for which collection and recycling systems are available or their development is imminent. The infographic in Fig. 8.4 explains to the customer that dm uses recyclable materials and, where ever possible, mono-material packaging (e.g. tube, cap, lid made of PE).

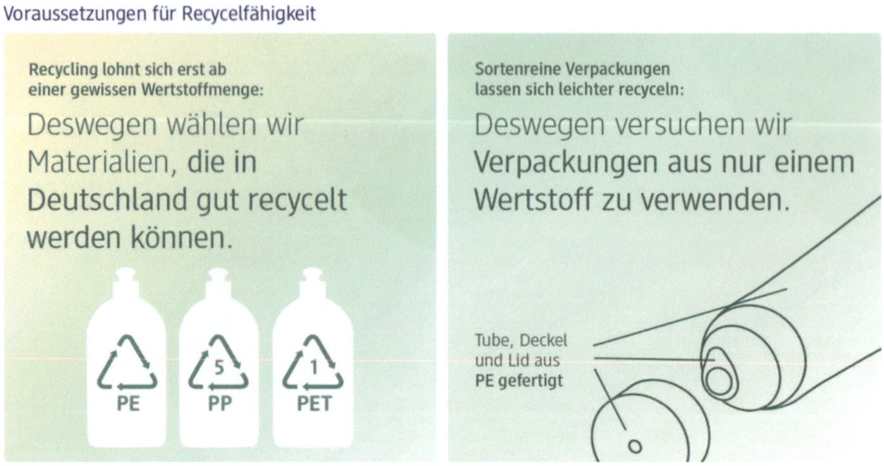

Fig. 8.4 dm infographic on the recyclability of plastic packaging (*Source* dm)

If the packaging is made of just one material the customer can clearly assign it to the correct collection stream based on its look and feel. Scientific studies on consumer behavior have shown that the separation rate for manually separable packaging components made of paper and plastic (paper sleeve around plastic cups) is below 50%, even among sustainability-oriented customers, despite clear, haptically supported separation instructions (CSCP 2022). This means that although customers want separation and sorting instructions, they do not act accordingly (here). For this reason, dm will continue to pursue its strategy of enabling consumers to sort the mono-material packaging according to its look and feel.

Communication at Eye Level

Due to the change in customer needs in relation to more sustainable consumption described above, retail companies are faced with the challenge of providing truthful information, explaining what they offer and communicating the meaningfulness of some developments, while at the same time not being accused of greenwashing.

dm-drogerie markt has set itself the premise of only communicating what has been realised and is available to the customer as an offer. Although transparency and dialogue at eye level also include explaining and communicating goals for the coming years, these are always incorporated into the communication of concrete solutions that signal that dm-drogerie markt has already set out on its path.

dm drogerie markt pursues the following communication activities with customers:

- At least once a year, dm drogerie markt enters dialogue with its customers with a large-scale communication campaign on sustainability. These broad-based, integrated multi-channel marketing campaigns utilise paid, owned, and earned media and take place both digitally and in dm stores throughout Germany. Each campaign

has a different marketing focus and attracts customers' attention with its unique design and headline mechanics. Thus, the dialogue on sustainability issues is ongoing. A sustainable background noise should have an internal and external effect, keep the topic present, and enable developments. Social media posts and newsletters focussing on sustainability as well as selective communication at the point of sale are used for this purpose. dm reaches out to experts and partners through technical discussions, interviews, and presentations;
- As the "customers of tomorrow", children represent a particularly important target group within sustainability communication. To reach this target group, dm offers special workshops in its shops called "Polli and the plastic" designed to familiarise children with the concept of sustainability and the importance of environmentally friendly packaging solutions in a fun and accessible way. In the "Polli and the plastic" workshops, the little polar bear Polli becomes the living protagonist of an interactive story;
- To support smart decisions by individuals, videos with extensive further information have been established providing information on various aspects of sustainable consumption in the recognisable "dm-erklärt" (dm-explains) format and are published on the dm YouTube channel.

The marketing campaign "Because external values also count (2021)" is used as an example to illustrate the principles and guidelines applied. The aim was to develop a message and emphasise dm's pioneering role in the field of packaging. Enthusiasm for co-design was to be created and combined with the communication of the goals of the packaging strategy. An instructive appearance patronizing or emphasising crisis scenarios should be avoided. A wide range of target groups was addressed for packaging topics. Both product-, efficiency- and innovation-oriented customers were reached with packaging optimisation topics.

The campaign was based on the following communication principles:

- Transparency and honesty
- Relief and orientation
- Reliability and concreteness

The observance of these principles should ensure that communication with the customer is at eye level and creates trust. The campaign motifs tell stories of success. Real packaging solutions are explained, and the goals set are communicated in a way that everyone can understand. This creates trust and an understanding that can simplify the decision on the shelf in favor of a more sustainable packaging solution. The work on the packaging itself and the issues behind it are also presented and illustrated with examples of products that have just been given optimised packaging.

Under the heading "We scrutinise everything we know about packaging to find the most sustainable solution", questions are asked about the selected packaging material, material savings, and the specific use of recyclate and answered with concrete packaging examples. The result of packaging optimisation is not always immediately apparent to the customer. For this reason, the campaign emphasised the importance of demonstrating the mechanisms using concrete product examples to be found in every dm shop. Once seen and understood, these optimisations can be taken into account when making future purchases. They can also be transferred to other products, as the mechanisms are the same (cf. Fig. 8.5).

It is not least the perception of sustainability that leads to the purchase decision in favour of or against a packaging option. Perception can be optimised with transparent information and actual sustainability facts that go beyond a blanket "glass is better than plastic". The implementation of the campaign has shown how educating customers can be harmonised with the company's own strategic considerations. Only through communication at eye level and the communication of real solutions, to be

Fig. 8.5 dm infographic on typical questions and answers based on optimised packaging (*Source* dm)

understood as such, a joint development towards a more pure collection of recyclable materials and ultimately a higher recycling rate is possible. The knowledge conveyed in the campaigns and in the digital background noise is repeated on the packaging through appropriate labelling and can be applied here by the customer: Labelling and information on the product enable an informed purchasing decision. Important components of this communication on the product are:

- The sustainability cloverleaf with a summary of the most important sustainability-related information (e.g. recycled content or recyclability) on the reverse side of the packaging;
- Recyclate content claim, often prominently displayed on the front of the packaging;
- Packaging reduction with reference to the original packaging and the corresponding material savings (e.g. refill bags) on the front of the packaging.

dm-drogerie markt attaches particular importance to this type of communication, as it is proof of the implementation of the packaging strategy on the product itself.

8.1.4 Outlook

In the future, communication will continue to make an important contribution to improving consumers' sorting behaviour, as there is still untapped potential here on the way to higher recycling rates. In addition to communicating the packaging strategy, a retailer such as dm-drogerie markt can also help to increase confidence in the recycling collection systems by using more and more recyclate in packaging. Thus, the image of plastic packaging we need for climate protection can also change. However, this is a joint task and all stakeholders along the value chain are required to ensure greater transparency.

The product and its packaging are closely linked when it comes to analysing the potential for conserving resources. In future, the ability of packaging to be emptied completely will become even more important. This is because consumers will learn more about the ecological footprint of a product and thus relate emissions not only to the packaging but also to the product itself. Transparent and honest communication with customers is necessary to convey the principles underlying these endeavours. Only by building trust and raising awareness of the levers we can all move as consumers through our purchasing decisions will more sustainable packaging solutions become mainstream.

8.2 Retailer Commitment: Case Study on the Use of Recyclates in Drugstore Packaging

Dagmar Glatz

8.2.1 Introduction

This article aims to explore two key aspects:

- The possibilities within the drugstore sector to contribute to a functioning circular economy through the use of recycled materials.
- The ways in which the application of recycled materials can make a positive impact and promote sustainability in this industry.

The sustainability of a product depends not only on its ingredients, but also on its packaging. Particularly in the realm of drugstore products, the focus on sustainability has long surpassed being just a trend. Customers now expect brands to design both their products and packaging in a sustainable and environmentally friendly manner. While most drugstore items like shower gel or shampoo enter the biological cycle after use, being ideally biodegraded in wastewater treatment plants, their packaging, on the other hand, often comprises plastic and should ideally be recycled through technical cycles.

The example of dm-drogerie markt is used to show how the use of recyclates has been strategically anchored and what has been done to achieve the set goal. The three types of plastic commonly used in the packaging sector—polyethylene terephthalate (PET), polyethylene (PE) and polypropylene (PP)—are examined.

8.2.2 Recyclates from Closed Loops

8.2.2.1 rPET from a Closed Loop

When the closed-loop collection of disposable drinks bottles via their deposit was launched in Germany in 2003, dm recognized the opportunity to replace virgin material in its PET bottles with recycled plastic. The deposit collection made a large, homogeneous material stream accessible. PET is widely used in the drugstore industry, mainly due to its low gas permeability, high transparency, comparatively easy processing and low cost. It is an excellent packaging material for protecting the product from environmental influences, i.e. preventing the migration of foreign substances from the environment into the product. However, before using recycled plastic, the risk of contamination from the recycled material itself must be ensured that it is excluded.

The advantage of a closed-loop collection of disposable beverage bottles is that the collection system reduces the risk of contamination by foreign substances. In addition, the material is processed through a high-quality recycling process. In PET bottle recycling, the bottles are shredded into flakes, and the caps and labels are separated. This cleans the PET flakes of foreign plastic or paper. The likelihood of direct printing is minimal, so there is no risk of contamination from inks or adhesives. The shredded flakes are washed at a higher temperature prior to extrusion (super-clean process) to remove any ink or adhesive residue. The FDA confirmed as early as 1992 that PET can be recycled into high-quality food-grade material using the Super Clean recycling process (FDA 1992). In collaboration with partners, dm tested rPET in bottles and was able to integrate it into production without any problems. After extensive quality tests with a product from the laundry and homecare range, dm began to gradually replace virgin PET with rPET in this range in 2009. In 2008, the EFSA (European Food Safety Authority) initiated the development of a framework for the recycling of recyclates for food packaging (EFSA 2008). In 2011, it summarized existing EU regulations for plastics in food contact materials in a single regulation. In 2012, the EFSA published the first expert opinions on the safety of PET recycling processes for use in food contact materials and thus also confirmed the high quality of rPET (EFSA 2012). Based on these opinions, rPET could subsequently also be used in packaging for cosmetic products without any concerns.

8.2.2.2 rHDPE from a Closed Loop

Rigid packaging in the drugstore industry, such as bottles, is also often made from polyolefins such as HDPE and PP. HDPE is commonly used for bottles or tube bodies, while closures or jars are almost always made of PP. It was therefore necessary to explore additional sources of recyclates.

A solution was found in the United Kingdom, where milk is packaged in natural-colored, meaning uncolored, HDPE bottles. This resulted in a source of sorted, unlabeled HDPE returns from food packaging. Since milk packaging is widely used in the UK, a continuous supply of such bottles in substantial quantities became available. Due to the short shelf life of milk, the packaging quickly re-enters the recycling loop. By 2008, HDPE milk bottles in the UK already contained approximately 10% recycled material, with plans to increase this proportion to 15% by October 2010. The high purity of the recycled material was ensured through the Super-Clean process. By 2020, the goal was to reach 50% recycled material content in milk bottles (Welle 2010). Based on these predictions for 2010 and the experience with rPET, dm decided to support the use of rHDPE, while recognizing the need for more intensive quality control efforts. In the case of cosmetic products, the EU Regulation No. 1223/2009 only required that the products be safe and did not impose specific requirements on packaging materials (EU 2009). Therefore, there were no legal obstacles to the use of recycled materials. Experts agreed that the quality of recyclates for cosmetics and detergents could be lower than food standards. Encouraged by consultation with various experts, dm initiated the first processing tests with rHDPE in 2011. During the production of bottles on the extrusion blow molding line, the high purity of the recyclate was quickly verified, as the weld line remained stable on these thin-walled bottles, indicating a high degree of homogeneity of the material. The sampled bottles made from this recycled material were successfully processed on the filling line without any issues. Stable bottle strength is crucial for filling operations, and any inconsistencies in the mechanical properties would have been readily apparent. In an initial series of tests, the 250ml alverde NATURKOSMETIK bottle was produced with varying levels of recycled material and made available to different manufacturing partners for filling and long-term stability testing. As even bottles made from 100% recycled material yielded excellent results, alverde introduced natural cosmetic products in the first bottle made from 100% rHDPE in the market in 2012. EFSA also carried out further safety assessments of the material and the recycling process and in 2015 approved the use of 50% rHDPE in packaging for certain food products (EFSA 2015). dm's goal was to enable the use of recycled materials not only in bottles but also in other packaging formats. Plastic tubes, due to their significant quantity, presented a large potential for utilization. At that time, tube production took place at various locations. Promoting the use of recycled materials at all these locations was deemed economically unfeasible. The transition from virgin material to recycled material production can be resource-intensive, involving investment in additional granulate conveying systems and logistics spaces. It wasn't until 2017 that dm was in a position to proceed with the next step, after consolidating certain orders at a single production site. Unlike blow-molded bottles, spots do not migrate to the inner wall but are visible and noticeable on the outer surface. As tubes are directly printed, spots can quickly compromise the print quality, accelerate plate wear, and increase production waste. Only after clearly defining the limits between maximum possible recycled material usage and sustainable production, could tests for long-term stability commence to verify the product's safety. Successful completion of these tests occurred in 2018, and the first alverde NATURKOSMETIK products in tubes with recycled material content were introduced to the market. This was followed by the rollout of other dm own-brand products in early 2019.

8.2.3 Use of Recyclates from Open Loops Through Coopetition

In 2018, the media increasingly confronted the public with images of packaging waste in the ocean and on land. These images had a strong impact on the employees at dm and made them think about further measures that could prevent packaging from ending up in the environment. The obvious next step was to ask the manufacturing partners of dm's own brands and industrial brands about the use of recycled material in plastic packaging.

At that time, packaging containing recyclate was still relatively rare in the drugstore sector. dm wanted to better understand the reasons behind this and sought insights from its partners regarding the obstacles to using recycled material. Limited availability of closed-loop recyclates, such as rHDPE from milk bottle collection in the UK, and the unknown quality of open-loop recyclates that were already readily available, were mentioned as factors.

However, some manufacturers had tested the use of open loop rHDPE for laundry and home care products and were able to use it proportionally, as the risk of human health effects from product use is very low for this product group. In the case of cosmetics, on the other hand, most companies saw little opportunity to use open-loop recyclate. Due to the direct contact with human skin, the use of recyclates and the potential impact on human health must be assessed much more critically. Regulation (EC) No 1223/2009 on cosmetic products states that "manufacturers shall place only safe products on the market" (EU 2009). However, the conditions under which recyclates may be used are not defined. Due to the highly diverse material composition of packaging from household recycling and the wide range of product residues found in these packages, the risk of contamination was initially considered uncontrollable. Studies have shown the higher diffusivity of polyolefins, which allows for easier absorption, adsorption, and release of contaminants, see also (Rung et al. 2023). It was clear that the toxicological quality of the recyclates needed to be more thoroughly investigated. This, in turn, meant that significant investments and extensive analyses were required, both in terms of cost and time.

Few market players were willing to make these investments alone, especially considering the unpredictable duration of the necessary analyses, which had to be conducted per recyclate batch. Recycling companies regarded their products as compliant with REACH regulations and did not understand at first why they should participate in this investment as demanded by brand manufacturers.

These challenges associated with the use of recycled material in the drugstore sector led to the initiation of two initiatives advanced by dm along with other stakeholders in the value chain:

- The establishment of the ForumRezyklat initiative.
- The establishment of the CosPaTox initiative.

8.2.3.1 The ForumRezyklat Initiative

In 2018, the ForumRezyklat was established to develop a deeper understanding of the entire value chain and address upcoming issues in an agile manner. It brings together 70 stakeholders from the retail, production, disposal, recycling, science, and politics sectors (Wellbrock and Ludin 2021).

To enable customers to make sustainable choices, the idea of highlighting packaging with recycled content of over 70% at the point of sale (POS) with a special icon was born. The term "recycled content" was defined uniformly, only considering post-consumer recycled (PCR) content in accordance with ISO 14021:2016–07 (ForumRezyklat 2022). This standardization was essential for fair competition among companies (cf. Fig. 8.6).

Through the Global Data Synchronization Network (GDSN), a global database, this information can now be transmitted by suppliers to dm monthly and made

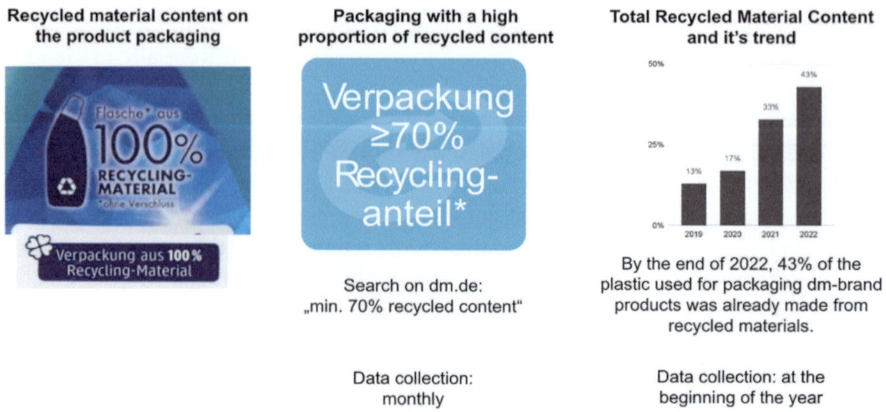

Fig. 8.6 Examples for highlighting the recycled content on the product packaging, at the point of sale, and on the dm website (*Source* dm)

accessible to customers at the POS, on the dm website, and in the sustainability report.

With the introduction of this attribute, the recycled content in many packaging formats, particularly PET bottles, saw a significant increase.

8.2.3.2 The Cospatox Initiative

The use of larger quantities of recycled polyolefins in cosmetic packaging was still not feasible in 2019/2020. The outlined testing scenarios seemed difficult to implement due to their lengthy and extensive nature, especially regarding the chemical analysis of the recycled material, packaging, and products. It was unclear what acceptable limits for contamination in recycled materials could apply to different areas of application in cosmetics. The safety of products and users should never be compromised, as it could have potentially halted the recycling of packaging materials for this product segment. Some stakeholders called for additional standards for recycled materials in areas such as detergents, cleaning agents, and cosmetics to ensure legal certainty. The discussion about the legal certainty of recycled materials lasted for months until the idea was proposed to set up a separate consortium with industry experts called CosPaTox.

Within 24 months, experts from the industry and scientific community in the field of packaging recycling were to develop toxicological safety standards in a focused work program and ultimately consolidation them into a voluntary industry standard. In addition to establishing toxicologically safe threshold values, the goal was also to develop testing and measurement methods that recycling companies could quickly implement on-site. Efficient analysis was seen as essential to the future success of higher levels of recyclate use, as it took weeks to assess the quality of recyclate, e.g. through migration testing. It was believed that this would also benefit companies that had previously refrained from using non-food PCR materials due to the lack of toxicological assessment and standardization. The CosPaTox consortium, also composed of stakeholders from the value chain, started its work in early 2021 and was able to publish its results in the form of a guideline in April 2024. (CosPatox 2024). Companies that have so far refrained from using recycled material in their packaging due to the lack of safety assessment guidelines and recyclers that have refrained from offering recyclates for this purpose due to the lack of quality definition should benefit from the guideline.

8.2.4 Recyclates from Open Loops Through In-Depth Dialog in the Supply Chain

In addition to the establishment of CosPaTox, dm, together with its manufacturing partners, made efforts to expand the use of recycled materials. Despite exchanging information with various recycling and packaging companies, the question of the toxicological assessment of the recyclates always remained.

In 2019, dm initiated an examination of five HDPE recyclates from household recycling available on the market, with the results being positive. This prompted dm to foster a more intensive dialogue between manufacturers and recyclers.

A key moment occurred in 2020 when dm and one manufacturing partner engaged in discussions with a recycling company and were impressed by their expertise, openness, and high-quality standards. All partners sought to understand each other's needs, and initial toxicological tests were commissioned. The manufacturing partner developed their own test strategies for products in recycled material bottles and established a comprehensive testing plan. The importance of having an in-house analytical department at the manufacturing partner became evident, as external laboratories were unable to meet the timetables for lengthy tests satisfactorily.

In the cosmetics industry, a distinction is made between leave-on products, which remain on the skin, and rinse-off products, which are rinsed off after application. Since rinse-off products only have brief skin contact, they were the ideal starting point for testing. Furthermore, the large contact surface between product and packaging minimized the risk of migration.

Therefore, dm's manufacturing partners initiated processes for ensuring the quality of the recyclates, conducting risk analysis and implementing assessments in their own control systems. In fact, despite initial concerns, these new processes were quickly implemented within medium-sized manufacturing companies.

By mid-2021, the first shower gel bottles from Balea made from 100% rHDPE recyclate, collected from the German household "yellow bag" collection, were launched on the market. This provided proof of the safe handling of rHDPE from the open-loop process, and more and more manufacturer partners of the dm brands began to test recyclates and use them in their packaging. However, no progress was made with PP closures for two reasons: the available rPP was often medium to dark grey and, initial guidelines for "design for recycling" recommended replacing PP with HDPE for mono-packaging.

On the other hand, there was progress in the recycling of PET; rPET was recycled from the "yellow bag" thanks to the commitment of a market companion from a German medium sized enterprise. dm was able to participate in 2022 with its own dm brand Denkmit.

In 2020, progress was already made in using recyclate in LDPE film packaging for goods such as toilet paper or diapers. It should be noted that this recyclate comes from commercial collections, such as films used in logistic processes. Due to their undecorated state, these films are easier to recycle and are of good visual quality. However, films pose a particular challenge due to their thinness, making the emphasis of quality testing primarily on mechanical properties.

8.2.5 *Further Challenges When Using Recyclate*

8.2.5.1 Grey Recyclate Color

Packaging made of HDPE is often printed or through-colored. As a result, rHDPE has a very strong gray coloration. On the other hand, PET packaging is predominantly transparent or only slightly colored when recycled. Consequently, rPET has only a subtle gray tint.

To better understand customer acceptance of gray-colored packaging, dm conducted numerous tests. The slight gray tint of rPET was well-received by customers for products such as detergents and cleaning agents, particularly as these products are typically not prominently displayed at home (cf. Fig. 8.7).

Fig. 8.7 Grey tint with rPET versus grey tint with transparent label or overlap with full body sleeve (*Source* dm)

However, for cosmetic products, the grey color became a critical consideration, especially in 2021 when customers focused on the coordination between product packaging and their bathroom decor, as the bathroom became a personal wellness oasis during the COVID-19 pandemic. The gray color of the recyclates sparked extensive discussions.

Alverde, under the motto "Grey is the new green," took a bold step by deliberately using transparent labels to emphasize the packaging's sustainability and demonstrate the brand's commitment to customers. On the other hand, Balea customers are accustomed to vibrant color designs. The gray color of the bottle is completely covered with a full body sleeve, minimizing its impact on the overall color concept.

Different solutions such as these offer opportunities to integrate the grey color into the design and create an attractive appearance that enhances the aesthetics of the packaging.

8.2.5.2 Polyolefin Recyclates for Food Packaging

Approximately 70% of packaging from household collection is food packaging (ForumRezyklat 2023), with the majority consisting of film packaging. The circularity of these films currently poses the greatest challenge. The gray color of the recyclates plays a minor role since these packaging types are often fully printed. However, it is precisely these printing inks that significantly impair the quality of the recyclate in terms of odor and toxicity.

In bottle recycling, the printing inks are removed along with the labels or sleeves made of the target material. In the case of film packaging, they remain in the process. The packaging is simply shredded, cleaned, and then melted and regranulated in the extrusion process. As the printing inks and binders are usually not heat-stable above 110 °C, they decompose during recycling, which requires at least 140 °C. The decomposition products are responsible for the reduction in recyclate quality. Currently, these recyclates can only be used for products with lower requirements, such as wastewater pipes or road underlay materials (Conversio 2022).

Potential solutions include the development of heat-stable printing inks and deinking technology (PrintCyc 2021). The thermostability of printing inks has now become part of the German minimum standard (ZSVR 2023).

Many food film packaging is still composed of multilayer structures. However, the DfR guidelines clearly point towards monomaterial packaging. Companies are in the process of making this transition (ForumRezyklat 2023 b), cf. Sect. 7.3). Once DfR becomes mandatory, these challenges should be overcome.

8.2.6 Conclusion and Outlook

In 2022, the German dual systems achieved a recycling rate of 67.5% for plastic packaging, an increase of more than 17.5% compared to the previous four years (UBA

2023). Despite this progress, a complete circular economy will only be achieved if recyclates are also used in new products and packaging. In phases when virgin material is available on the market at a lower price than recyclate can be produced, some (potential) recyclate customers opt for virgin material. This means that recycling companies, especially during periods of low virgin material prices, face economic challenges. Consequently, little investment is made in technology to improve the quality of recyclates (ForumRezyklat 2023a).

As a retailer and producer of private labels, dm supports the development of the circular economy through its efforts in using recyclates in drugstore packaging. dm recognizes the need to establish a level playing field in the market between virgin plastics and recyclates as quickly as possible.

Furthermore, dm promotes the overcoming of communication barriers between different stakeholders and aims to increase the level of standardization and digitalization in the circular economy while addressing specific technical challenges. By collaborating on projects such as the ForumRezyklat, Cospatox, PRINTCYC, or PACK2theLoop, efforts are made to develop solutions for concrete issues in partnership with all stakeholders.

To enable the use of recyclates in food packaging, further development work is required. Currently, the quality limits of recyclates, regardless of whether they are produced through mechanical, physical, or chemical recycling processes, considering ecological and economic aspects, are not yet clearly defined.

The commitment of a few market participants is not sufficient for achieving a higher use rate of recyclates; only through joint efforts and the exchange of knowledge can this be achieved.

8.3 Digitization of Circular Economy Data for Packaging along the Value Chain

Dagmar Glatz, Sarah Grede and Anna Ziessow

8.3.1 Introduction

This article explores the increasing importance of digitalized data flow in the packaging industry along the whole packaging network, focusing on attributes like recycled content and recyclability. It underscores the need for a comprehensive transformation within the packaging industry towards an interoperable data ecosystem. Additionally, the article introduces an open-source data model developed by ForumRezyklat (cf. Sect. 8.2.3.1), which has a leading role in facilitating this transformation and continues to do so. In 2019, images of ocean-littered plastic packaging made a profound impact on many, leaving a lasting impression (GFK 2019). These visuals have motivated brands to establish objectives regarding packaging reduction, the use of recycled materials in their packaging, and the recyclability of their packaging itself. As a result, numerous countries across the globe, initiated processes to assess the existing laws and guidelines regarding packaging, recognizing the necessity for a transformative shift in its management. In Germany, the revision of the Packaging Act was accelerated in 2019, and other countries in Europe are also setting increasingly higher recycling goals and incentivizing the use of recycled materials or investments in recyclable packaging through plastic taxes or eco-modulation. In addition to many individual regulations within each country's legislation, the new packaging regulation by the EU is underway. Indeed, to ensure that companies can effectively monitor their packaging development and meet future requirements, it is crucial to have additional data from various network points. This necessitates ensuring their semantic consistency and interoperability. Digitization plays a vital role in achieving the goal of transparent packaging characteristics and advancing the circular economy by processing and sharing substantial data throughout the value chain, extending beyond individual company boundaries.

8.3.2 Status of Digitization in the Circular Economy of Packaging

One notable example of how digitization drives the circular economy is by promoting the use of recycled content in packaging. The ForumRezyklat alliance recognized this potential and devised a strategy at the end of the year 2018 to emphasize this aspect at the point of sale. By highlighting the recycled material content, customers should be empowered to make informed and sustainable purchasing choices. The ForumRezyklat is an agile alliance, launched in 2018 by dm-drogerie markt. Meanwhile, more than 70 members—consisting of retailers, brand owners and packaging manufacturers, dual systems, waste disposal and recycling companies, and representatives from the science community and the world of politics—covering the entire circular value chain, work together to foster circular economy for packaging (ForumRezyklat 2024). To ensure the consistent and automated provision of this sustainable packaging attribute to customers, two steps were essential. First, the required data set for "recycled content" had to be developed in compliance with ISO 14021:2016–07 standard. Second, this data had to be implemented within the Global Data Synchronization Network (GDSN), a system that facilitates the automatic flow of product information between brand owners and retailers. Through this interface, brands can seamlessly communicate information regarding recycled content of packaging to retailers. A big advantage of the GDSN is that attributes can be shared in an efficient and simultaneous way by the distribution of information to all involved partners using the one-to-many approach. As this standardization and implementation process typically takes up to two years, data transmission was initially introduced through another data management tool to expedite the process and avoid any delays. At the end of 2020, the data set became available in the GDSN, providing access to all market participants in Germany. Following this, the ForumRezyklat strongly promoted its rollout in the European market (GDSN 2014). Since then, retailers and brand owners in Europe have the possibility to provide customers with this information. In the case of dm-drogerie markt, customers found it at this time at the point of sale on the packaging itself, through dedicated stickers on price tags, and daily on the dm.de website (cf. Sect. 8.2.3.1). Parallel the acceleration of the transformation to a circular economy as one of the key priorities in the EU was initiated. The European Commission summarized this ambitious roadmap in 2019 within the framework of the European Green Deal (EGD). In line with this ambition, the Circular Economy Action Plan (CEAP) was published in March 2020. This action plan structures the measures and paves the way towards a transition from a linear to a circular economy. This includes a call to action in terms of mobilizing data infrastructures to:

- track products and their material use
- track the journey of products, components, and materials
- make recycled content in products measurable.

All experts agreed that the implementation of these ambitious goals could only succeed based on a fundamental change in value chain networks. In the FMCG industry digital data flow between brand owners and retailers was state of the art for years via the GDSN. The GDSN was developed to optimize logistics processes, aiming to achieve efficiency and cost savings. For these processes, the provision of the packaging attributes weight, size, and material have long been implemented in GDSN, but the now addressed attributes, like the recycled material content of a packaging or its recyclability, were missing. Therefore, the participants of the ForumRezyklat started to foster the implementation in the GDSN of further packaging attributes, which seemed to be essential for the circular economy. In 2021, the same group of stakeholders initiated the standardization of the attribute "recyclability". That means, the experts defined the necessary approach and design of the requested attribute, like the recycling process, recycling design standard, the level of the calculated recyclability and the date of the data provision, through the GS1 standardization process (GDSN 2022). Additionally, the option should be given to load up the proof of recyclability, for example, a certificate. The basis for assessing recyclability is the German minimum standard for assessing the recyclability of packaging subject to system participation obligations, according to § 21 VerpackG of the Central Agency Packaging Register (ZSVR). This standard was chosen as there were no existing legal European standards available. However, once a legal European standard is published, it can easily be incorporated or even replace the German minimum standard. During that time, it was recognized that participants in the circular economy did not speak the same language. The data was often entered by a key account manager, collected by a packaging engineer, and the data manager within the company might struggle to make sense of the new data attributes due to a lack of technical understanding. Therefore, the experts provided additional concise guidance in a free handbook on how companies can best provide packaging data for the GDSN (ForumRezyklat 2023). Starting from 2023, the data attribute "recyclability" was also available and could be exchanged across companies through the GDSN.

8.3.3 *A Standardized Data Model for Automated Data Flows Across the Circular Value Chain*

When calculating attributes such as the percentage of recycled content or recyclability of a packaging, a significant amount of primary packaging data from the upstream supply chain is required. This data collection process starts with raw material suppliers and continues through converters and processors until the packaging components are ultimately assembled by the product fillers to create the final packaging. At this point, the recycled material content and the recyclability of packaging can be calculated. Currently, those required data are mainly manually maintained in different systems or even stored in archives on paper within the companies. This low level of digitalization in the industry across the entire value chain at that time is also underscored in the Roland Berger study (Büchele et al. 2021). The study reveals that as of 2022, the level of digitalization in the circular economy was less than 10%. It became apparent that manual management of large quantities of data for accurate calculations and assessments of sustainable data attributes, as well as their traceability, would not be feasible in the long term. While closing the interface between brand owners and retailers marked an important initial step, it became evident that additional actions were necessary to establish a holistic interoperable solution. It was also crucial to enable the upstream supply chain to transfer their data in a standardized and automated way, thereby ensuring seamless integration and comprehensive sustainability practices.

The participants of ForumRezyklat developed a vision where all the necessary data for calculating the two crucial attributes of packaging would be provided in a timely manner and automatically delivered to relevant stakeholders within the Circular Packaging Value Chain. This vision emphasizes the importance of seamless and automated data flow to enable effective collaboration and decision-making in promoting circular packaging practices.

All experts agreed that an open-source data model ensuring interoperability in the circular packaging value network should be developed. This "Open Packaging Standard", which digitally represents packaging specifications, should form the basis to pass all data along the value chain in a compliant manner (cf. Fig. 8.8). This will simplify data exchange along the value chain and enable applicability in various existing software systems and data pools, but also enable the development of data entry systems for small market participants. The ForumRezyklat also strives to stay informed about the work of other platforms and align their vision, such as those of relevant stakeholders like the DIN—Circular Economy Network. This collaboration aims to promote a unified understanding of the circular economy and to save resources by avoiding duplication of work.

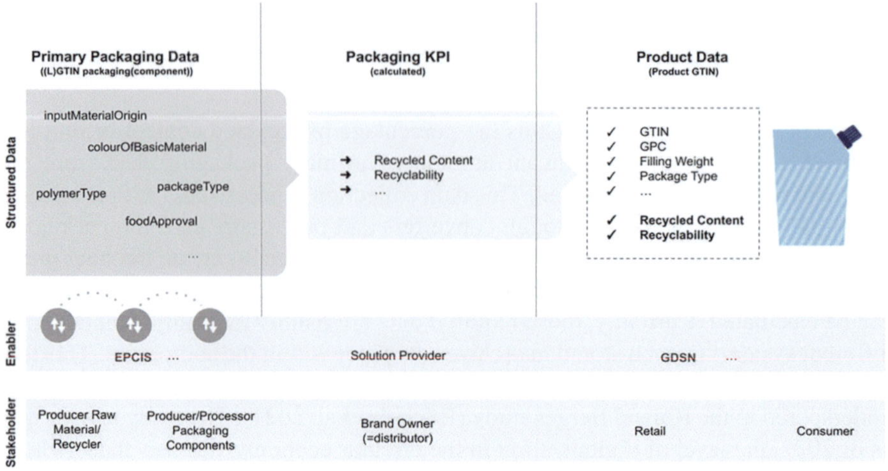

Fig. 8.8 Vision of a data flow along the value chain from primary packaging data to essential packaging attributes: (*Source* GS1 Germany; design modified)

8.3.3.1 Designing a Standardized Data Model

The use of existing standards saves effort and investment for all users and facilitates interoperability. The initial focus was therefore on reviewing and evaluating the existing data attributes, code lists, and code values which are used already today in GDSN against the new circularity requirements regarding the upstream supply chain, e.g. the assessment of recyclability. This examination aimed to ensure alignment and compatibility between the existing data structure and the evolving needs of the industry towards a circular economy. The existing GDSN structure was initially assessed, and the definitions were found to be clear and suitable for the purpose (cf. Fig. 8.9):

- **Attribute**: an attribute refers to a specific characteristic or piece of information associated with an item. Attributes provide details about the properties, features, or specifications of a product. Each attribute is typically associated with a specific data type, such as text, numerical value, dates, or code list.
- **Code list**: A code list is a predefined set of values that can be used to populate certain attributes through data pools. Code lists help standardize and ensure consistency in the data exchanged between trading partners. Instead of allowing free-form text for certain attributes, a code list restricts the possible values to a predefined list. This helps prevent discrepancies and improves the accuracy of data sharing.
- **Code value**: A code value is a specific entry or option within a code list. Each code value represents a particular attribute value that conforms to the predefined set established by the code list.

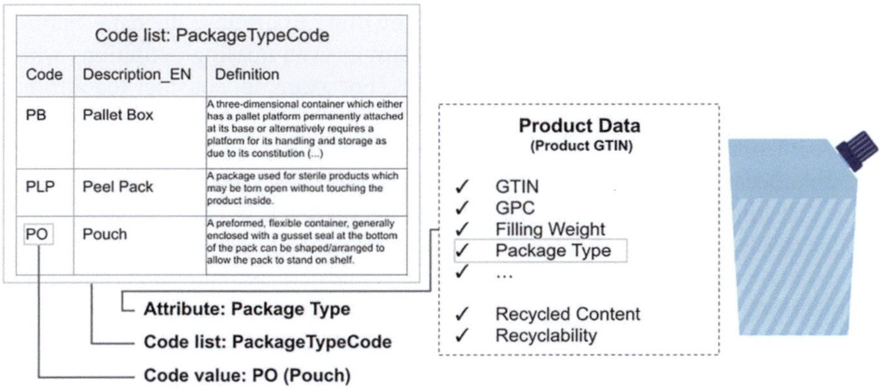

Fig. 8.9 GDSN structure (*Source* GS1 Germany; design modified)

Already these definitions are very important for a common understanding regarding what data attributes, code lists, and code values mean.

The experts of ForumRezyklat encountered obstacles in the following areas:

- Existing data structure with a strong focus on the product
- No completeness of data due to e.g. voluntary data input
- Different, outdated, or missing definitions

In existing data structures, it can be observed that for packaging they exhibit varying hierarchical levels, and the depth of information varies significantly. As already mentioned earlier (master) data has historically been tailored to applications such as logistics processes and compliance requirements, with a focus on product and not packaging information. As a result, existing structures are not easily transferable to the now required packaging use cases and are depicted on a different level. An example is the availability of packaging weights, where in the past, weights were only needed at the component level (e.g. weight of bottle body, weights of bottle caps). However, for recycling feasibility calculations, the weights of individual material layers in packaging are required (e.g. weight of the main material in the bottle body, weight of adhesives, printing inks, etc.) to calculate the recycled content. There are not only gaps in terms of missing detail or data information not previously queried but data in existing structures is often incomplete or erroneously filled. Some existing data fields are mandatory to maintain, meaning they must be filled in as they are legally required or essential for logistical processes. However, there are also optional data fields, referred to as “should” fields, which may vary in terms of how they are maintained and populated. Consequently, the data in these fields can be more individually managed by different companies and key account managers. This poses a significant problem for automated calculations and evaluations, especially as sustainability compliance requirements become more stringent. High data quality is essential as incorrect or missing data can lead to substantial fines. Additionally, many data are not available in the required level of detail because either the necessity was not recognized until recently, or companies are unwilling to disclose the data. Furthermore, a common problem is the lack of uniform or even missing definitions which may lead to a different interpretation of terms. For example, consider the attribute "separability". Does an adhered label already count as separable if a consumer could remove it from the packaging with significant effort? Or does a bottle cap count as separable because this component is not firmly glued and can be easily separated by the consumer? Or does a lid foil, as in toothpaste tubes, count as separable, as it must be irreversibly separated from the rest of the packaging for the product's use? This example clearly illustrates how quickly a single data attribute can become very complex, emphasizing the importance of a unified understanding. Such consistency is often lacking even within a company. For instance, Data Expert A might classify the adhered label mentioned above as separable, while Data Expert B might indicate it as non-separable. Such inconsistencies pose a significant problem for data reliability. To address this issue, at least in the calculation of recycled material usage, the ForumRezyklat published the Guideline Recycled Material in 2022, demonstrating with numerous examples how to accurately specify recycled content (ForumRezyklat 2022).

The ForumRezyklat experts unanimously agreed that for the success of a new data model, intensive consideration and solution development for the mentioned problem areas are imperative. Conversely, the following success criteria can be derived.

The data model must

- be flexible and expandable in the data structure to accommodate changing requirements in a dynamic environment.
- represent various levels of detail and hierarchy.
- provide clear, well-defined definitions for data attributes and values based on circularity requirements to prevent interpretation ambiguities.
- should be built upon existing models and standard structures to facilitate quick implementation.

The following "surface coverage" example illustrates this in detail.

To assess the recyclability of packaging according to the German minimum standard, information regarding the surface coverage of a sleeve or label on a packaging type is required. Since the data attribute "surface coverage" does not currently exist in the Global Data Synchronization Network (GDSN), the expert group of the ForumRezyklat has defined the requirements of "surface coverage" to establish a new attribute set with a common understanding within the group. The data type in this case is numeric, as it pertains to a factor calculated from the projected surface area of the packaging and the projected surface area of the sleeve and label—so no code list or code values were necessary. According to the business need the information of "surface coverage" is a primary packaging attribute as it plays a role in the assessment of the recyclability. Once this process step is completed, GS1 will evaluate the appropriate path for standardization. It should be noted that GDSN will only be affected if the packaging data in question is published between industry and retail in the future. Otherwise, additional standardization options will be explored within the global context. As the specifications for calculating recyclability change annually, the data attributes, code lists, and code values must continue to develop flexibly (adaptation, addition, deletion). This approach should enable efficient data exchange of primary packaging data and seamless integration with existing systems and processes for the next approach, the automatization of the data flow.

8.3.3.2 Implementing Automated Data Flows in a Circular Packaging Value Chain

Building upon the groundwork laid out, the next step of the journey involves implementing automated data flows grounded in the developed model. The focus then shifts towards seamlessly integrating the standardized data model into practical, efficient processes. This entails establishing automated mechanisms that facilitate the smooth and accurate exchange of information throughout the circular packaging value chain for the calculation of the packaging's key performance indicators (KPIs) from the primary packaging data. Building automated data flows also requires a close network and communication among stakeholders. It is essential that not only large corporations implement the model but also that pathways to digitization and improved data management are presented to small and medium-sized enterprises. Therefore, effective communication strategies will be employed to share insights, address concerns, and provide guidance, promoting a collective and inclusive approach to digital transformation within the entire spectrum of industry stakeholders. It is also worth emphasizing that the solutions will vary in scope, and small businesses will have different requirements for technical implementation compared to large enterprises. The diversity of participating companies calls for flexible and scalable solutions that cater to individual needs. While larger companies may have resources for comprehensive technologies and training, solutions for smaller businesses must be cost-effective, adaptable, and easily understandable. Therefore, a differentiated approach to technical implementation and training will be necessary to ensure that all businesses, regardless of size, can benefit from the advancing digitization and improved data management (cf. Fig. 8.10).

An approach to driving progress in the field has been introduced through the publication of the "Circular Plastics Traceability Guideline" by GS1 Germany on the GitHub platform in 2023. This guideline, developed in collaboration with machine builders, focuses on enabling the parallel transfer of data throughout the complete supply chain of plastic packaging, aligning with the actual flow of data. The purpose of this guideline is to facilitate a consistent flow of data throughout the entire process of plastic packaging production, enabling the calculation and tracking of key performance indicators (KPIs) such as the recycled material content of a packaging. To achieve this and to exchange relevant data attributes at each step of the supply chain, it is recommended to utilize the EPCIS (Electronic Product Code Information Services) system, along with the Core Business Vocabulary (CBV) data standard. EPCIS and CBV are data mechanisms and languages widely used by GS1 members and various stakeholders, enabling the exchange of packaging data that includes comprehensive information of relevant process steps with the associated primary packaging data. It

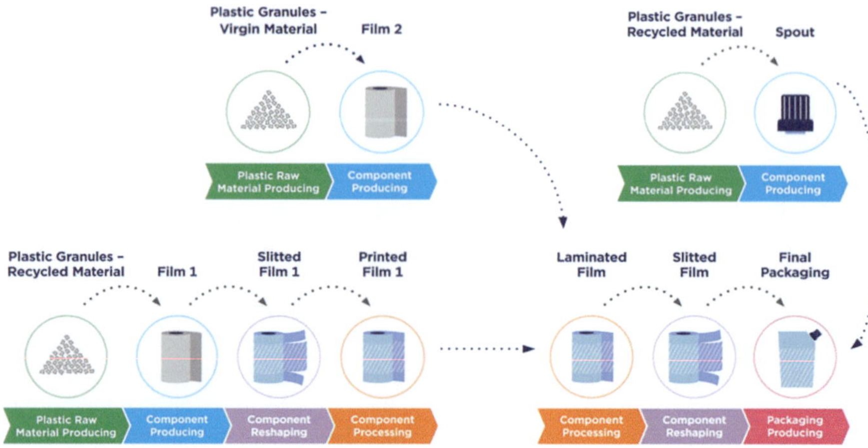

Fig. 8.10 Data flow for various production process steps for a pouch. (*Source* GS1 Germany, Guideline Circular Plastics Traceability: design modified)

offers a common language providing organisations with the what, when, where why, and how of objects traversing through business processes. The guideline introduces six crucial data points, which are known as EPCIS events, from the start of plastic packaging manufacturing to the final packaging stage. Furthermore, an example of how data collection for film production, specifically for a pouch, or a bottle could be implemented, is provided within the guideline. The approach described is open to everyone and is already being put into practice by the initial software as a service provider.

Another use case for this data model is employed by Recyda (2024), which has developed its own methodology for analyzing data to calculate recyclability KPIs tailored to the needs of the packaging value chain it serves. Recyda's methodology encompasses the evaluation of diverse recyclability standards within the packaging landscape, yet it ensures seamless compatibility with GS1 standards. This ensures that the calculated packaging attributes are communicated to the subsequent stakeholders in a standardized GS1 format. Thus, Recyda's software aligns with the open-source model and supports the launch of standardized primary data attributes in line with GS1 standards.

8.3.4 Conclusion

Reporting obligations regarding recyclability and circular product design are already extensive and will continue to increase. This results in a tremendous effort in every company along the entire value chain, which can only be reduced through standardized and automated digital solutions. Digitalization has the potential to reshape the circular economy, particularly in the packaging industry. The integration of Artificial Intelligence (AI) into the Packaging Value Chain can speed up the process, particularly in standardizing how data are shared across the chain. AI has the potential to change the way we handle and distribute packaging information, making data processes more efficient and practices more consistent. This progress is crucial for enhancing the flow of information and improving overall efficiency within the Packaging Value Chain. Careful consideration is necessary to balance the costs and benefits of AI implementation, ensuring that AI's strengths are leveraged while minimizing its environmental impact. As the packaging industry integrates AI, companies must remain adaptable and continuously learn to keep up with advancements. The industry is in the early stages of incorporating AI, and the field is constantly evolving. The fluid nature of AI brings uncertainty, emphasizing the need for industry players to stay informed about the latest progress. Adopting a strategic and flexible approach, companies should be prepared to embrace new technologies that align with their goals. Constant learning and adaptability are essential as the industry explores the potential of AI and aims to reap its full range of benefits. At present, maintaining an overview of various individual initiatives proves challenging due to the high degree of agility. Despite this, the aspiration is to work transparently and extensively, which necessitates constant coordination with internal and external actors. Moreover, there is a high need for coordination along the packaging value chain itself due to the community's diversity. Achieving mutual understanding and establishing it widely even among experts in master data and packaging engineering, who speak different languages, will take time. Standardization at the international level is of great importance here. GS1 also assumes this role and coordinates these engagements at a global level. Several studies or initiatives (JRC 2022; Friedrich Ebert Stiftung 2022, Röchling Stiftung 2023), underscore the need to further advance digitalization in the circular economy. This progression will enhance the traceability, transparency, and efficiency of packaging systems, marking a significant milestone in sustainable practices.

8.4 PreZero—A Retailer's Environmental Service Provider

Dietmar Böhm

8.4.1 PreZero—A Company of Schwarz Group

Schwarz Group

In 2022, the companies of Schwarz Group including the retail divisions Lidl and Kaufland generated over 575,000 employees[1] and a revenue of 154.1 billion Euros

[1] All descriptions of functions and persons always refer to all genders. For the sake of legibility, the following text does not distinguish between genders.

across all business areas, an increase of 15.4% compared to the previous year (Schwarz Unternehmenskommunikation 2023a). The family-owned company from Neckarsulm, Germany, is one of the largest retailers in the world but has long been more than just a trading company: The companies of the Group have set themselves the goal of covering the entire material cycle within the Group—from trading, waste management, and recycling to the production of new products and packaging. This makes it possible to develop and implement material cycles and thus minimize the consumption of natural resources in the long term.

To implement this holistic approach, the Schwarz Group has initiated efforts in several areas. Products made from PreZero recycled materials can already be found in many places, not just on the retail shelves. They extend to the design of store fixtures such as checkout dividers and protective wall guards made from recycled plastics, illustrating a commitment to embedding circular practices in every aspect of operations. An early example is the conversion of plastic film recyclables from the stores into new garbage bags, which are reused by the Schwarz Group again.

Another key area of focus is the redesign of plastic packaging. PreZero can not only supply recycled materials but also contribute its expertise as a circular economy consultant, advising on the design of sustainable packaging. This includes evaluating alternatives to reduce plastic use without compromising safety and innovating mono-material packages from non-recyclable multilayer, demonstrating how PreZero's recycling and sorting know-how can lead to sustainable packaging solutions. The results of this work are already visible: for example, the production of detergent, shower gel, and fertilizer bottles for Lidl's private labels has already been switched to using 80 to 100 percent recycled material (rHDPE) from the PreZero recycling plant in Grünstadt, Germany.

From a Recyclable Materials Manager to an International Environmental Service Provider

As part of Schwarz Group, PreZero is an international environmental service provider in Europe and North America. With the founding of the company, Schwarz Group bundled its waste management expertise to manage recyclables from retailer Lidl, later also from retailer Kaufland, and Schwarz Produktion and handle them within closed loops. From the very beginning, the intention was to assign an appropriate value to waste. Waste that generated costs was minimized through various strategies and recoverable waste—from now on referred to as "recyclables"—was accordingly maximized. In 2023, PreZero is operating in 11 countries with almost 30,0000 employees providing waste management services to around 40 million citizens in Europe as well as numerous industrial customers. This makes the company one of the ten biggest waste management companies in the world and, as a plastics recycler, it plays a leading role in the recycling of post-consumer plastics waste in Europe.

At 480 locations, the recycling expert takes over waste collection, sorting, treatment, and recycling of recyclable materials as well as the licensing of sales packaging and consultation on the recyclability of such packaging. The goal is to conserve resources and reduce the amount of waste that cannot be recovered to zero (Schwarz Unternehmenskommunikation 2023b).

8.4.2 *Create Closed Loops*

The business units

With a structural division into Packaging, Waste disposal, and Recycling business units, the company already covers key stages along the value chain (cf. Fig. 8.11). The Packaging unit includes the Dual System with its licensing activities for sales packaging being the strategic link between manufacturers and waste management companies and thus offering services to meet the Extended Producer Responsibility (EPR) in Germany. PreZero Sustainable Packaging on the other hand focuses on recyclability and sustainable design of packaging for its customers. In the Waste disposal unit, the environmental service provider ensures the separate collection, sorting, and correct treatment of waste in various fractions, which are then reprocessed to create new raw materials in the Recycling unit, which can be used for new products or packaging.

In total, around 15 Mt of recyclable materials are collected and processed for recovery by PreZero every year. Including appr. 150 kt of plastics waste, that are fed into the recycling market in Europe. With these business units, PreZero sees itself not just as a classic waste disposal or waste management company, but as a modern and innovative recyclable materials manager.

Fig. 8.11 Circular economy and business units (*Source* PreZero)

Sustainable Packaging

Throughout Europe, product manufacturers are responsible for their sales packaging in terms of avoidance, reuse, and recovery. With PreZero Dual, PreZero has successfully established its own dual system (EPR system) on the German market and is a reliable partner for customers for the licensing of sales packaging. By making effective use of the existing waste management structures and expanding them with innovative approaches, the aim is to actively help shape and drive forward the EU's requirements for the further development of the circular economy.

An important step is the design of recyclable packaging. PreZero therefore also offers services dedicated to creating more recyclable packaging designs for customers. For example, this means that packaging is ideally made from a single material for optimal recyclability. Wherever possible, the industry also needs to reduce the use of composite materials and increase the use of mono-material packaging. Where such materials are required for packaging, they must be designed to be as technically recyclable as possible. For example, the detectability of materials can be improved so that they can be correctly identified and assigned in sorting systems. Inks and adhesives used need to be easier to dissolve and remove. The path to a functioning circular economy starts with product design.

Waste management—from separate collection through to sorting and treatment

In the area of waste management, PreZero exploits the potential of waste as far as possible. The separate collection, sorting, and processing of waste form the basis for its subsequent recycling. A modern vehicle fleet, a wide range of different container types, and digital technologies support the employees in this process. Once the collected waste has arrived at a sorting plant (cf. Fig. 8.12), the aim is to separate the material streams as best as possible for further recycling. PreZero currently processes lightweight packaging at seven different plants in Europe. Here, various fractions such as plastics, tinplate, cardboard composites, and aluminium are sorted using innovative, highly efficient separation processes.

However, recyclable materials from industry and commerce, such as wastepaper, wood, glass, scrap metal, and mixed commercial waste, are also professionally sorted, processed, and returned to the market as high-quality raw materials. Wherever possible, waste that can no longer be recovered is used as a substitute fuel. The company operates several thermal waste treatment plants in Europe. The waste-based fuel is incinerated, and the resulting heat is used to generate process steam and energy. This reduces the use of fossil fuels and the production of CO_2. The resulting energy production is climate-friendly, conserves resources, and contributes to the security of supply.

Recycling—using recovered recyclables to create new raw materials

The Recycling unit processes recycled materials into secondary raw materials for the industry. In this way, the company contributes to the development of closed loops. The subsidiary PreZero Polymers is a leader in the recycling of post-consumer plastics in Europe. Those recycled materials (cf. Fig. 8.13) are the raw materials for new

Fig. 8.12 Modern sorting plant for lightweight packaging in Eitting, Germany (*Source* PreZero)

applications in industrial production and are used in the gardening sector, furniture manufacture, household appliances, construction, home and office furniture, and the automotive industry, among others. Biological waste is processed into biogas and compost, while other material streams such as aluminium, glass, and wood are also recycled in various countries.

Other Activities

The subsidiary GreenCycle runs the recyclable materials management for the other companies of Schwarz Group: for the retail divisions Kaufland and Lidl as well as for Schwarz Produktion. They support all countries and group companies with strategic consulting through to operational responsibility and the bundling and marketing of waste and recycling management. Another focus is on the management and monitoring of the quality, environmental, and energy management system for the PreZero Group companies at the Heilbronn site.

Intelligent reusable pallets and pooling services are being developed under the PreTurn brand. These are reshaping the entire supply chain efficiently and transparently to provide a clear optimization.

Finally, OutNature is also part of PreZero. OutNature develops and markets innovative and sustainable fiber and paper products for packaging solutions in retail and industry. In doing so, OutNature is using a new material: the fibers of the silphium plant, a raw material that offers an economically and ecologically viable alternative to virgin wood fibers used in paper production.

Fig. 8.13 Recycled post-consumer polypropylene-based polymer with added fillers (calcium carbonate/alkali), additives, and masters for use in injection moulding and extrusion moulding solutions (*Source* PreZero 2023)

A Design for Circularity forms part of the development of all products.

8.4.3 Best Practice—Some Implementation Examples

TSG Hoffenheim solution—the first certified "Zero Waste Stadium" in Europe

It started with a vision of a stadium where no recyclables are wasted. To achieve this ambitious ideal, TSG Hoffenheim, a 1st league football club in Germany, partnered with PreZero in 2019 to create Germany's first "Zero Waste Stadium". This means that at least 85% of the waste generated is reused, recycled, composted, or converted into biogas. It is now the first stadium in Europe to be certified for its operational waste and recyclables systems. The Zero Waste vision is based on DIN SPEC 91436-05 (DIN 2021), a holistic reference model for measuring and improving the level of maturity of waste and recyclable materials management (DIN = Deutsches Institut für Normung e.V.). It describes a process that takes a critical look at the status quo in terms of consumer habits and waste production of different stakeholders and takes a step-by-step approach to optimizing sustainability. An important part of the Zero Waste strategy is creating awareness. This must be done both internally, with all employees, and externally with partner organizations and customers, to create an understanding of the changes to be made and to generate a multiplier effect. In addition, continuous improvement is an essential feature of a successful Zero Waste strategy.

Silphium-based packaging

With the OutNature brand, PreZero develops and markets new and sustainable fiber and paper products for packaging solutions in retail and industry made from the Silphium plant (cf. Fig. 8.14). Regional cultivation reduces transport distances and associated carbon emissions. In addition, the processing of the plant requires little water and energy, and the fibers are obtained without the use of chemicals. In 2021 OutNature received the prestigious WorldStar Award from the World Packaging Organization (WPO) in the "Packaging Materials & Components" category for their Silphium-based paper.

Further research is expected to provide silphium paper with additional properties as a barrier, for example against water or grease. In this way, the novel paper can become a possible substitute product for plastic packaging in the future and thus contribute to the vision of the international Schwarz-Group-wide plastics strategy REset Plastic "Less Plastic—Closed Loops".

Recycled Plastics resins in products and packaging

Since 2021, recycled products with resins from PreZero have been offered on a regular basis at Kaufland and Lidl (Kaufland Dienstleistung 2023; Schwarz Unternehmenskommunikation 2023c). Starting with garden and household products such as storage boxes, laundry baskets, and clothes hangers as well as flowerpots and garden

Fig. 8.14 Sustainable silphium packaging by OutNature (*Source* OutNature 2023)

furniture, projects have also recently been implemented in the children's toy sector. With its recycled sand toys, retailer Kaufland offers its customers high-quality products in collaboration with PreZero. They are made of plastics from the recycling loop and thus generate a much lower carbon footprint in production compared to standard manufactured products. PreZero takes over waste management as well as the subsequent production of plastic granulate from packaging waste. While the toy vehicles are made from up to 80% recycled and 20% new resins, all the other products are made entirely* from recycled resins (*with the additional use of additives and colour (max. 5%)) (cf. Fig. 8.15).

To produce the resins, plastic waste is collected from private households, which is then cleaned and reprocessed to produce plastic regranulate. This is used to create new, high-quality, and environmentally friendly products. The entire recycling and manufacturing process takes place in Europe and is particularly sustainable thanks to the short transport routes involved. The product range is constantly being reviewed and expanded.

Another focus and important lever in the retail sector is sales packaging. Here, too, the packaging of Lidl and Kaufland private labels is currently being reviewed, and, wherever possible, virgin material is being replaced by recycled material from PreZero. In particular, rHDPE packaging in the cosmetics and drugstore sector, such as detergent bottles, is already being delivered to stores.

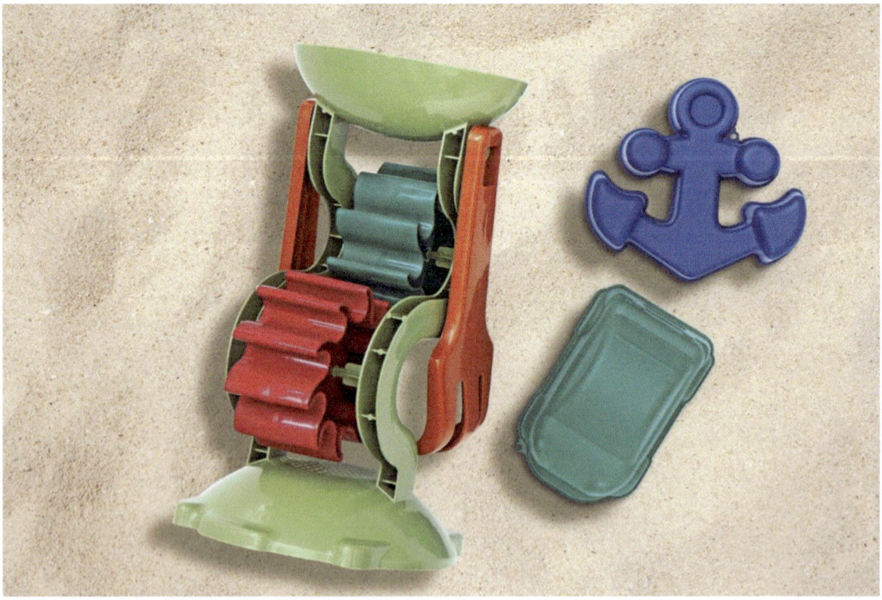

Fig. 8.15 Sand toys made from 80% rPP (recycled Polypropylene) from the post-consumer stream (*Source* Kaufland Dienstleistung 2023)

8.4.4 Future Challenges

In the future, both customers and companies will face several challenges. Customers, whether individuals, corporations, or municipalities, are becoming increasingly environmentally aware. This is forcing all parties in the developing circular economy to align their products and services more closely with the environmental needs of their customers. Companies that demonstrate genuine environmental awareness will be preferred by consumers and partners. On the other hand, companies that only pretend to be environmentally friendly (so-called greenwashing) will face considerable reputational damage. Even the seemingly best product at the best price will not achieve long-term market success if it lacks environmental sensitivity.

Significantly higher customer environmental awareness

Waste prevention, resource preservation, and climate protection are terms that have been receiving much greater attention from both consumers and industry in recent years. Today, consumers are much more interested in how products are manufactured, the underlying production conditions, and what happens to the products after use.

Waste disposal is becoming more environmentally friendly

The amount of recyclable and reusable products is constantly increasing. However in many countries, deposit systems or dual systems are not yet established. However, the evidence shows that a higher recycling rate is linked to a healthier economy. Waste intensity is defined as total waste generation as a percentage of gross domestic product over time. It is used as a KPI to determine sustainability. If the waste intensity is significantly below economic growth, this is referred to as a sustainable waste economy (BMUV 2023). In other words, the better an economy is performing, the more likely it is to place a much higher value on the environment and closed recycling loops.

Tightening legal conditions (incl. ESG reporting)

In the future, the EU Green Deal, EU Taxonomy, CSRD (Corporate Sustainability Reporting Directive), SFDR (Sustainable Finance Disclosure Regulation), etc. will regulate, among others, the obligation to report negative ESG (environmental, social, and governance) effects. This will examine areas of responsibility in companies that are required to operate sustainably according to certain criteria. From 2024 onwards, these provisions apply for all large, listed companies and will be followed by small and medium-sized enterprises (SMEs) from 2025 onwards, if they have at least 250 employees and generate revenues of more than 40 million Euro. Rating agencies will have to analyse and assess a company's strengths and weaknesses based on its social and environmental commitment.

Intelligent regional distribution centers, vehicles, and containers

While digital route planning may have seemed a novel concept for waste management companies in the past, it has since become a standard feature offered by a wide range of companies in the sector. These days, the future challenges for regional distribution centers are fleet tracking plans with live mode options as well as geocoded telematic and performance data. Event displays with emptying information and emptying weights require smart containers with RFID detection as well as fill-level sensors that are connected to fleet management. At the same time, access controls and the usage status of containers as well as the online status monitoring of all containers located in the regional distribution center area are necessary to ensure professional and advanced recyclable materials management.

Digitization, automation, and artificial intelligence

Just like in other industries, the elementary opportunities in the waste management industry come primarily from the areas of digitalization, automation, and artificial intelligence. These will lead to significant improvements in business processes. Traditional mindsets, which are predominantly focused on optimizing internal value chains, will be replaced by platform-oriented business models. This will necessarily lead to a further increase in consolidation amongst market participants in the waste management and recycling industry, while new and innovative solution providers will have the opportunity to disrupt the market and establish themselves. Disruptive business models are often based on exponential innovation that leads to scalable business models. These business models in turn offer potential revenue streams—not just the original sale of products and services. They also provide the opportunity to generate additional income through pay services, transaction fees, and advertising. To successfully establish these kinds of business models, it is necessary to collect much more information about consumers than has been the case in the past.

8.4.5 Outlook

In general, the waste management industry, as a singular player in the complete value circle, can influence neither the amount of waste nor the composition of wasted products and waste that is transferred at the end of the product lifecycle. This is making it increasingly important for producers to tackle the topic of recyclability and circularity at a strategic and operational level early on. The companies of the Schwarz Group were quick to grasp that for a retail group, environmental awareness does not end with the purchase of recyclable packaging or products, but that the suppliers upstream and the waste management and recycling industry downstream will be essential partners for a more environmentally friendly future.

References

BMUV 2023: Waste legislation - German Circular Economy Act with regulations concerning waste shipment law. Retrieved from https://www.bmuv.de/en/law/circular-economy-and-safeguard-the-environmentally-compatible-management-of-waste (Last access: Aug. 20, 2024)

Büchele et al 2021: R. Büchele et al: Greentech made in Germany 2021. Umwelttechnik-Atlas für Deutschland. https://www.rolandberger.com/en/Insights/Publications/Green-tech-industry-remains-on-course-for-growth.html (Last access: Aug. 20, 2024)

Conversio 2022: Stoffstrombild Kunststoffe in Deutschland 2021. https://www.bkv-gmbh.de/1150-studie-stoffstrombild-kunststoffe-in-deutschland-2021.html (Last access: Feb. 07, 2024)

CosPaTox 2024: CosPaTox-Guideline-April-2024.pdf

CSCP 2022: Collaborating Centre on Sustainable Consumption and Production (CSCP) & Die VERBRAUCHER INITIATIVE eV. (VI, Bundesverband): Praxistest Verpackung: Wahrnehmung von Trenn- und Sortierhinweisen aus Sicht von Verbraucherinnen. Studie des Clubs für nachhaltige Verpackungslösungen. (Practical packaging test: Perception of separation and sorting instructions from the consumer's perspective. Study by the Club for Sustainable Packaging Solutions) https://www.cscp.org/wp-content/uploads/2022/03/Studie_trennen_sortieren_de.pdf (Last access: Feb. 07, 2024)

DIN 2021: Retrieved from https://www.din.de/de/din-und-seine partner/presse/mitteilungen/die-vision-lautet-zero-waste—797972 (Last access: Feb. 07, 2024)

dm 2023: Nachhaltigkeit 2023. Representative baseline study. (internal study dm drogerie markt)

EFSA 2008: Guidelines on submission of a dossier for safety evaluation by the EFSA of a recycling process to produce recycled plastics intended to be used for manufacture of materials and articles in contact with food - Opinion of the Scientific Panel on food additives, flavourings, processing aids and materials in contact with food (AFC). https://efsa.onlinelibrary.wiley.com/doi/10.2903/j.efsa.2008.717 (Last access: Feb. 07, 2024)

EFSA 2012: PET-Recyclingverfahren für Lebensmittelkontaktmaterialien: EFSA verabschiedet erste Gutachten. https://www.efsa.europa.eu/de/press/news/120802 (Last access: Feb. 07, 2024)

EFSA 2015: Scientific Opinion on the safety assessment of the processes 'Biffa Polymers' and 'CLRrHDPE' used to recycle high-density polyethylene bottles for use as food contact material. https://efsa.onlinelibrary.wiley.com/doi/https://doi.org/10.2903/j.efsa.2015.4016 (Last access: Feb. 07, 2024)

EU 2009: Regulation (EC) No 1223/2009 of the European Parliament and of the Council of 30 November 2009 on cosmetic products. https://eur-lex.europa.eu/legal-content/EN/ALL/?uri=CELEX%3A32009R1223 (Last access: Feb. 07, 2024)

FDA 1992: Points to consider for the use of recycled plastics in food packaging: chemistry considerations. https://corporate.findlaw.com/law-library/points-to-consider-for-the-use-of-recycled-plastics-in-food.html (Last access: Feb. 07, 2024)

ForumRezyklat 2022: Guideline Recycled Material. https://www.forum-rezyklat.de/en/publications/ (Last access: Feb. 07, 2024)

ForumRezyklat 2023: Guideline Recyclability. https://www.forum-rezyklat.de/en/publications/ (Last access: Feb. 07, 2024)

ForumRezyklat 2023a: Vorstudie Vergleich innovativer Identifikationsverfahren für PostConsumer Leichtverpackungen; Hamburg Institute for Innovation, Climate Protection and Circular Economy GmbH (HiiCCE). https://www.forum-rezyklat.de/en/publications/ (Last access: Feb. 07, 2024)

ForumRezyklat 2023b: Umfrage vom ForumRezyklat zum Stand Monomaterialverpackungen - 20.11.2023

ForumRezyklat 2024: https://www.forum-rezyklat.de/en/about-us/ (Last access: Feb. 07, 2024)

Friedrich Ebert Stiftung 2022: Daten für die Circular Economy. https://www.fes.de/themenportal-bildung-arbeit-digitalisierung/artikelseite/daten-fuer-die-circular-economy (Last access: Feb. 07, 2024)

GDSN 2014: GDSN attributes. Release 3.1.14

GDSN 2022: GDSN attributes. Release 3.1.22

GFK 2019: Who cares? Who does? Wahrnehmung und Relevanz von Umweltaspekten aus Sicht der Käufer und welche Potentiale sich daraus für FMCG ableiten lassen (Perception and relevance of environmental aspects from the buyers' point of view and which potentials can be derived from this for FMCG)

JRC 2022: Towards a green & digital future. https://publications.jrc.ec.europa.eu/repository/handle/JRC129319 (Last access: Feb. 07, 2024)

Kaufland Dienstleistung 2023: *Für eine Welt mit weniger Plastik*. Retrieved from www.kaufland.de/plastik (Last access: Feb. 07, 2024)

Lehmann et al 2022: Lehmann, K., Renz, D., & Huber, F. Nun sag', wie hast du's mit der Nachhaltigkeit? (So, tell me, what's your attitude to sustainability?) EY-Parthenon GmbH. https://www.ey.com/de_de/consumer-products-retail/studie-nachhaltigkeit-deutscher-konsument-innen. (Last access: July 30, 2023)

PrintCyc 2021: https://www.flexotiefdruck.de/gravure-global/printcyc-provides-major-input-for-design-for-recycling-guidelines/ (Last access: Feb. 07, 2024)

Recyda 2024: https://www.recyda.com/about-us (Last access: Feb. 07, 2024)

Röchling Stiftung 2023: Polyproblem. Der Circularity Code. https://www.roechling-stiftung.de/polyproblem-report-the-circularity-code/ (Last access: Feb.07, 2024)

Runge et al 2015: https://www.researchgate.net/publication/368413466_Identification_and_Evaluation_of_Non-Intentionally_Added_Substances_in_Post-Consumer_Recyclates_and_Their_Toxicological_Classification (Last access: Feb. 07, 2024)

Rung et al 2023: https://www.researchgate.net/publication/368413466_Identification_and_Evaluation_of_Non-Intentionally_Added_Substances_in_Post-Consumer_Recyclates_and_Their_Toxicological_Classification (Last call: Feb. 07, 2024)

Schwarz Unternehmenskommunikation (2023a): Retrieved from https://gruppe.schwarz/presse/archiv/2023/die-unternehmen-der-schwarz-gruppe-wachsen-auf-stabilem-niveau (Last access: Aug. 20, 2024)

Schwarz Unternehmenskommunikation 2023b: Retrieved from https://gruppe.schwarz/geschaeftsfelder (Last access: Aug. 20, 2024)

Schwarz Unternehmenskommunikation 2023c: Kreislaufsysteme. Retrieved from https://gruppe.schwarz/nachhaltigkeit/kreislaufsysteme#verpackung (Last access: Aug. 20, 2024)

UBA 2022: Indikator: Umweltfreundlicher Konsum (Environmentally friendly consumption) https://www.umweltbundesamt.de/daten/umweltindikatoren/indikator-umweltfreundlicher-konsum#die-wichtigsten-fakten (Last access: July 30, 2023)

UBA 2023: Umweltbundesamt: Aktuelle Recyclingbilanz: Erfolge, Verfehlungen und Herausforderungen. https://www.umweltbundesamt.de/presse/pressemitteilungen/aktuelle-recyclingbilanz-erfolge-verfehlungen (Last access: Feb. 07, 2024)

Wellbrock and Ludin 2021: Wellbrock, W, and Ludin, D (Ed.). Nachhaltiger Konsum: Best Practices aus Wissenschaft, Unternehmenspraxis, Gesellschaft, Verwaltung und Politik. (Sustainable consumption: best practices from science, business practice, society, administration and politics), p.6. 1. Edition. Springer Gabler Wiesbaden. https://doi.org/10.1007/978-3-658-33353-9

Welle 2010: Welle, Frank: Recycling von Kunststoffverpackungen - PET Getränkeflaschen und andere Anwendungen. Präsentation vom 28. Oktober 2010. https://www.bfr.bund.de/de/praesentationen_zu_den_vortraegen_vom_28__oktober_2010-52899.html (Last access: Feb. 07, 2024)

ZSVR 2023: Minimum standard for determining the recyclability of packaging subject to system participation pursuant to section 21 (3) VerpackG 31.08.2023. https://www.verpackungsregister.org/en/foundation-authority/minimum-standard-pursuant-to-section-21/2023-edition (Last access: Feb. 07, 2024)

Dagmar Glatz has expertise in materials and packaging production and in recyclable packaging design, too. She advises dm's brand teams and manufacturing partners on packaging development. She graduated in plastics engineering from Montanuniversität Leoben/Austria and has more than 15 years of experience in the plastics value chain. She held management roles in the pharmaceutical and automotive industries before joining the drugstore chain dm in 2019.

David Stakemeier is part of the Sustainability Department at dm-drogerie markt and is responsible for sustainability communication. He works in interdisciplinary teams developing omnichannel communication campaigns and, in close cooperation with the specialist departments, on the company's sustainability report.

Sarah Grede is Senior Manager Sustainability with GS1 Germany. Since 2018, she has been working on solutions for a circular economy and supply chain transparency, based on the identification, collection, and communication of data along the entire (circular) value network. Standardization, digitalization, and cooperation are essentials of her activities.

Anna Ziessow is a Co-Founder of the start-up company Recyda GmbH, an innovative software platform for real-time analysis of packaging portfolios. She is a driving force for sustainable solutions and is helping to transform the packaging industry towards a circular economy through innovative technologies and intensively acquired expertise in the field of sustainability.

Dietmar Böhm is a member of the Management Board of PreZero Stiftung & Co. KG. He is responsible for PreZero Dual, Intragroup Services & New Business. Prior to PreZero, he held various management positions at Lidl in Germany. He is also Vice President of the Federal Association of the German Waste, Water and Raw Materials Management Industry (BDE).

Chapter 9
Wastes—Collection, Sorting, and Preparation for Recycling

Thomas Krupp, Kathrin Hesse, Anni Luu, Andreas Hanus, Michael Perl, Michael J. Schneider, and Herwart Wilms

Contents

T. Krupp · K. Hesse (✉) · A. Luu
Technische Hochschule Köln, Cologne, Germany
e-mail: kathrin.hesse@th-koeln.de

T. Krupp
e-mail: thomas.krupp@th-koeln.de

A. Luu
e-mail: jieying.luu@th-koeln.de

A. Hanus · M. Perl (✉)
Sesotec GmbH, Schönberg, Germany
e-mail: Michael.Perl@sesotec.com

A. Hanus
e-mail: Andreas.Hanus@sesotec.com

M. J. Schneider (✉)
Remondis Sustainable Services GmbH, Lünen, Germany

H. Wilms
Remondis Service International GmbH, Lünen, Germany
e-mail: herwart.wilms@remondis.de

P. Orth et al. (eds.), *Circular Economy of Plastics*,
https://doi.org/10.1007/978-3-031-78829-1_9

Summary

This chapter focuses on the contribution of collection and logistics, sorting and waste management to the development of a circular plastics economy as presented in the waste segment of the circuit diagram (cf. Fig. 9.1).

The basic prerequisite for utilising plastic waste as a raw material is the separate collection of as much waste as possible in the private and commercial sectors with the highest possible purity, so that the largest possible homogeneous mass flows are

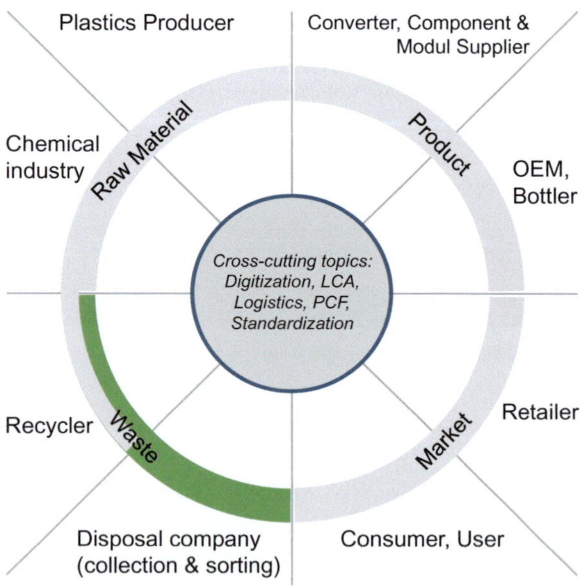

Fig. 9.1 Schematic diagram of the plastics cycle. *Source* Own presentation

created. This poses numerous challenges for waste logistics. As mixed waste predominates, particularly in the private sector, sorting the collected waste plays a decisive role. To this end, sophisticated optical sorting technologies are being developed at full speed and corresponding investments are being made in the recycling industry. Further progress is being made through the increased use of artificial intelligence methods. The aim in all areas of waste and resource management is to increase yields and qualities in order to provide the market with the required quantities and ensure fulfilment of the recyclate usage quotas required by law.

These aspects are analysed in this chapter 9 from the perspective of science, method development and a large waste management company.

9.1 Logistics—General Aspects

Thomas Krupp, Kathrin Hesse, and Anni Luu

9.1.1 Importance of Logistics for a Circular Economy

For enabling a worth living future for today and future generations worldwide a transformation of our current way of living and economy towards more resilience, sustainability and social responsibility is essential. Central aspects of this transformation include achieving a post-fossil energy supply and implementing continuous material cycles (cf. EUC 2021). In this context logistics plays an essential role, representing the planning and executing function of material transfer processes within companies as well in between different players in Supply Chains (cf. Hesse und Clausen 2019; Goepfert 2019).

Generally speaking, logistics is defined as the spatial, temporal and structural transformation of goods to ensure the supply and disposal of goods in line with requirements. The origin of business logistics is optimizing the logistics operations themselves, i.e. the professionalization of operative logistics functions, such as placing (transport), pacing (storage, warehousing), picking/packing (changing the order/arrangement of things) (cf. Klaus and Mueller 2012) or—put in a nutshell—**t**ransportation, **h**andling and **s**torage (cf. Pfohl 2023), also called "THS process").

The current supply chains in the plastics industry are predominantly linear, starting with the extraction of raw materials passing through the various production stages in the chemical and plastics industries to industrial customers, wholesalers, retailers and finally reaching the end customer. However, the disposal and recycling of plastic wastes has up to now mostly been considered separately from this "traditional" value chain. To achieve circular supply chains and consequently closed material cycles aspects of recirculation and recycling will become increasingly more important. The predominantly linear "one-way logistics" will develop into "multiple-way" reusable systems and cycles. This transformation requires efficient disposal logistics.

This backwards-oriented logistics or disposal logistics means "[...] the application of the logistics concept to residues in order to design an economically and ecologically efficient flow of residues with all activities of spatio-temporal transformation, including changes in quantity and type." (cf. Pfohl 2018, p. 246 ff.) Therefore, the disposal logistics processes include the **c**ollection, **t**ransport, **h**andling, **s**torage and **t**reatment of wastes, residual materials or end-of-life materials (CTHST-Process) (cf. Hesse and Clausen 2019, p. 239 f.) Along with supply-oriented logistics a closed-loop logistics is established on the base of disposal logistics. In this case, the flow of goods is further developed into a circular flow of goods. By doing so the recycling of products is placed at the center of logistics management (cf. Goepfert 2019, p. 167 f.).

Based on the "traditional" meaning of logistics this article focuses on the functions transport, handling and storage of goods ("THS logistics") or when including the backwards-oriented logistics the collection, transport, handling, storage and treatment of wastes (CTHST logistics).

In the modern context of logistics, the perspective is broadened from the pure consideration of waste disposal—or the question "how do I get rid of my waste without dumping it?"—towards a broader understanding of logistics in a circular economy closing the gap to the demand side of recycling based feedstock—or the question "how can the availability of recycling based feedstock be ensured to establish a resilient and sustainable recycling supply chain?"

The challenge and opportunities for logistics service providers (LSP), recycling companies, and chemical companies are then to effectively and efficiently use "logistics" to improve the value of plastics in a circular economy for all involved actors.

9.1.2 General Aspects of Logistics

Traditionally logistics was limited to the areas of procurement, production and distribution. While procurement logistics serves as a link between the suppliers 'distribution logistics and the customers' production logistics in a market-linked logistics system (cf. Pfohl 2018, p. 189), the task of production logistics is to secure an efficient flow of materials from the supply depots (procurement logistics) through production processes to the finished goods warehouses (distribution logistics) (cf. Pfohl 2018, S. 189). Distribution logistics represents the connection between production logistics and customers by providing goods, which can take place in two different ways (cf. Pfohl 2018, p. 222).

In the circular economy context, the procurement department in companies has a direct impact on the products to be purchased and their packaging materials along with their characteristics, e.g. recyclability, sortability, or waste avoidance under the premise of favorable purchase prices, supply safety, high quality as well as the lowest possible storage costs and inventories (cf. Pfohl 2018, p. 189; cf. BVL 2023a).

Since then waste disposal has not only become a cross-sectional-function but has also developed into the fourth branch of logistics in the overall process (cf. Wannenwetsch 2021, p. 745). The typical functions of transport, handling, and storage as well as the link in terms of information technology system are transferred to waste disposal, to the CTHST-process.

Disposal logistics differs from other logistics systems such as procurement, production, and distribution logistics especially relating to the objects being handled (cf. Pfohl 2018, p. 247). While those refer to production goods, disposal logistics focuses on residual and waste materials generated in production, distribution, and consumption processes (cf. Pfohl 2018, p. 247) along with end-of-life products.

The main task of disposal logistics is to collect, sort, pack, if necessary, treat, store and transport for further handling all types of accruing wastes from various sources such as service providing industry, retail, manufacturing industry as well as end consumers (cf. Pfohl 2018, p. 251). Accruing residual and waste materials include scrap, excess, overage stock of finished goods, recyclable materials, unwanted by-products, or defect-loading equipment such as pallets or packaging. Suitable redistribution strategies enable a structured and economic capturing and consolidation of used end-of-life products. In the end these (treatment) processes are applied depending on the used material or end-of-life product.

The transformation from linear to circular supply chains and hence the complete integration of disposal logistics into the previously end-customer-directed logistics also means changes in the current supply chains: the increasing recycling of waste directly causes additional recirculation and recycling processes and a growth of transport and sorting processes. At the same time, changes in the characteristics of existing flows of (raw) materials are expected due to the greater heterogeneity and smaller volumes of products. The resulting increase in logistics activities makes a resource-efficient way of working essential.

Finally, the savings and benefits of a circular economy should not be counteracted by the efforts and energy consumption of transport and sorting processes (cf. Fig. 9.2).

Considering efforts and energy the "THS" logistics or "CTHST" logistics therefore plays an essential role in the executing function of transport and sorting processes for returning plastic wastes and their reprocessing into recyclates. In the long term, the objective is to continuously reduce the efforts, energy consumption and emission associated with the recirculation and to improve resource efficiency. At the same time, the demand for and the utilization of recyclates will be increased when the efforts for recirculation are below the production of new materials from fossil resources. Therefore, a holistic review of the system in terms of effort, energy, and emission is necessary.

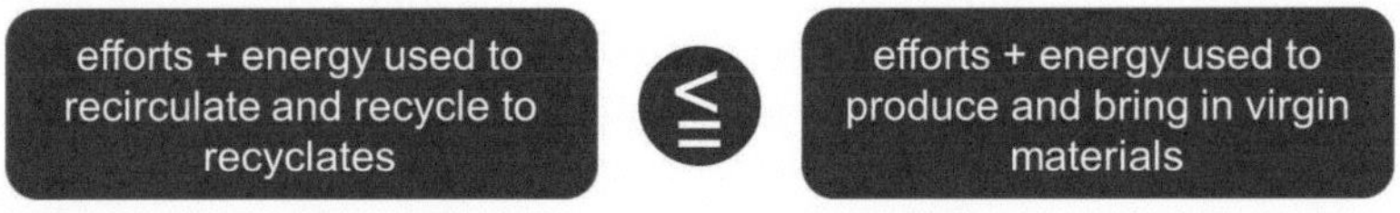

Fig. 9.2 Central goal—minimize efforts and energy in recirculation and recycling processes for recyclates. *Source* Own presentation

9.1.3 Plastic Regulations—Legal Framework and Requirements for Logistics

The transboundary movement of waste is the transport of waste between different countries. The Basel Convention of 22 March 1989 on the Control of Transboundary Movements of Hazardous Wastes and their Disposal came into force in 1992 and has since been ratified by around 170 countries. The Basel Convention generally stipulates that waste may only be shipped across borders for recovery and disposal with the consent of all countries involved and that the fate of the waste must be monitored. In addition, the OECD member states have introduced a control system for the transboundary movement of waste destined for recovery, which classifies waste into red, amber, and green lists depending on its hazard potential, and regulates the behaviour of states towards each other when authorising waste shipments (cf. Basel Convention 1989). Regulation (EEC) No. 259/93 of 1 February 1993 on the supervision and control of shipments of waste within, into, and out of the European Community transposes above all the essential contents of the Basel Convention and OECD Council Decision C(92)39 (e.g. the list classification of waste) into directly applicable Community law. This regulation was replaced on 12 July 2007 by Regulation (EC) No. 1013/2006 of 12 June 2006 on shipments of waste (cf. Regulation (EC) No 1013/2006). At the beginning of 2023, the European Parliament voted on its position on the rules for shipments of waste, which are intended to promote the reuse and recycling of plastics and minimise pollution. MEPs called for the export of plastic waste to non-OECD countries to be banned and for shipments to OECD countries to be stopped within four years (cf. EUP 2023a).

9.1.4 Operations and Logistics Activities—Elements of Logistics Systems for Collection of Recyclable Fractions

Based on the scope of the system logistics has different meanings and tasks. As explained above, the first meaning of logistics refers to the fundamental activities transport, handling and storage (also called THS-logistics). In the second meaning, as a cross-functional and cross-organizational coordination function, logistics focuses on systemic thinking, coordination and integration of materials management, production, distribution and marketing between interfaces. The third meaning refers to Flow Management. This means designing, planning along with implementing harmonious and synchronized flows, and processes in terms of material, goods, information and people (Pfohl 2018, p. 23; Delfmann et al. 2010). This scope covers the entire supply chain starting from the origin of raw material (chemical industry) to the utilization of a product at different value stages (plastics industry and consumers), the disposal (waste management companies) and finally the recirculation (recycling and plastics producer) into the product life cycle.

Considering the circular economy, the third meaning of logistics corresponds directly to its core idea of creating end-to-end plastics cycles and supports the implementation of the right logistics strategy.[1] From a more practical point of view logistics, in its first meaning, enables the movements of goods across the value chain by transport, handling and storage activities. The disposal logistics serves as a bridge between (end) consumers and plastics manufacturers/processors while closing the physical gap between these players. For implementing the physical waste stream and the flows of secondary material disposal logistics executes the activities collection, transport, handling along with storage and treatment (CTHST-logistics). Following this process, this paragraph is structured according to the CTHST-activities.

Waste Collection

Disposal logistics starts with the collection of wastes from all stages of the value chain originating from different sources such as service providers, retailers, manufacturers, and end consumers. As a rule, waste is roughly categorized according to where it is generated, e.g. production waste and municipal waste. Production waste, also known as post-industrial waste, is often generated in quantities that do not require collection. Instead, it can be transported directly from the company to the selected treatment or disposal facility, provided the production waste is not fed directly back into the production process. This process is referred to as point disposal (cf. Hesse and Clausen 2019). Municipal waste, also referred to as post-consumer waste, on the other hand, is mainly generated in small quantities, meaning that economical removal can only be achieved through collection and joint transport. This process is called area disposal (cf. Hesse and Clausen 2019).

Waste reaches the waste management industry via the collection process. The main purpose of collection is to collect the waste at defined transfer points. These are households (e.g. for the collection of household packaging waste), locations of depot containers (e.g. used glass collection) but also recycling centers where private individuals can hand in the recyclable waste generated in households, also known as recyclable materials (including used glass, used paper, used metals, used plastics, etc.) (cf. Hesse and Clausen 2019, p. 271). Ultimately, the separate collection of waste for material recycling is essential in terms of compliance with waste stream-specific quality requirements and is now enshrined in law in the German Circular Economy Act (cf. KrWG 2020) or the German Commercial Waste Ordinance (GewAbfV 2017). According to the Circular Economy Act, waste streams collected from households, such as waste paper, used glass, biowaste and plastic waste, must be collected separately since 2015 (cf. UBA 2022). In addition, the amended Commercial Waste Ordinance requires that wood, textiles, and biowaste must also be collected separately in commercial municipal waste in addition to the waste fractions paper, cardboard, cardboard packaging, plastics, glass and metals.

The collection methods are also differentiated according to the type of container used. Empty containers are mainly used for the collection of household waste and

[1] R0 Refuse, R1 Rethink, R2 Reduce, R3 Reuse, R4 Repair, R5 Refurbish, R6 Remanufacture, R7 Repurpose, R8 Recycle, and R9 Recover.

household-type commercial waste. They are taken from the collection site to the collection vehicle, emptied there and then returned. The transport of the containers depends on the system, user or personnel of the waste disposal service provider. There are now many emptying bins on the market (cf. Hesse and Clausen 2019). Disposable containers, such as the yellow bag for collecting used plastic packaging, are used for the disposal of packaging waste in the collection system, for example. Abroad, however, disposable containers are used much more frequently. The disposable containers are disposed of together with the waste. Plastic bin liners with a capacity of 35 to 1,100 L are used as collection containers (cf. Hesse and Clausen 2019).

Transport of Wastes

Different vehicle variants are available for the collection of packaging waste. The selection of the appropriate one depends on the waste fractions and quantities to be disposed of, the container systems used, and the structure of the disposal area. In addition to the container systems, the building structure is a decisive criterion for the selection of vehicle types. While front and rear loaders are particularly suitable for use in city centers, side loaders and combined front and side loaders are ideal for use in less densely populated urban peripheral areas and rural regions (cf. Bilitewski et al. 2018; cf. Hesse and Clausen 2019). The rear loader is the traditional design for emptying technology. The waste containers are emptied at the rear of the vehicle, which means that, in addition to the driver, additional personnel are required to bridge the distances between the parked waste containers and the loading point (cf. Bilitewski et al. 2018; cf. Hesse and Clausen 2019; cf. Kirchhoff 2022). A distinction is made between rotary drum and compactor refuse collection vehicles for waste compaction in the vehicle (cf. Kirchhoff 2022).

Handling and Storage of Waste

The term "handling" describes all conveying and storage processes during the transfer of goods to a work equipment, during the outflow of goods from a work equipment, and during the transfer of goods between work equipment (DIN 30,781–1:1989–05). In the waste management industry, handling refers to the transfer of wastes for utilization, recycling and disposal between means of transportation, conveyance, and storage. Depending on the respective handling area, a combination of different equipment is required to carry out the transfer processes.

Storage refers to the planned positioning of work items in the material flow (cf. VDI 3629:2005-03). Thus, warehouses are rooms or areas for storing work items which are recorded in terms of quantity and/or value. In the waste management industry, they take on the task of stockpiling, buffering, and distributing waste for disposal, utilization, recycling, or deposition and incineration. Depending on the type, warehouses are primarily used to bridge a period of time or to change the composition structure between incoming and outgoing waste. Storages of wastes enables improving the capacity utilization of treatment and disposal plants and achieving economically and ecologically more favorable disposal by grouping similar types of waste (cf. Hesse and Clausen 2019).

9.1.5 Logistics as the Linking Element in the Plastic Cycle

The following illustration (cf. Fig. 9.3) visualizes the material and information flows in an end-to-end plastics cycle from the logistics' point of view. As explained in the previous paragraphs the task of logistics is, based on the flow principle, to plan, coordinate, control, and implement these flows between all players at each stage of the value chain.

For realizing a closed-loop plastics cycle the integration of the players from the waste management industry and disposal logistics (companies for collecting, sorting and treatment of wastes) poses an essential prerequisite. Only by including them the previously linear organized supply chain will transform into a circular supply chain. The returning processes focus on the waste and secondary raw material stream originating from the consumers and directed towards production. From this perspective, the consumers and waste management enterprises take on the role of raw material suppliers. They provide new input materials for the production phase by returning the waste streams at the end of the utilization phase and turning wastes into recyclates during the disposal phase. There are partly substantial information deficits between the actors within the supply chain, e.g. the waste collection companies do

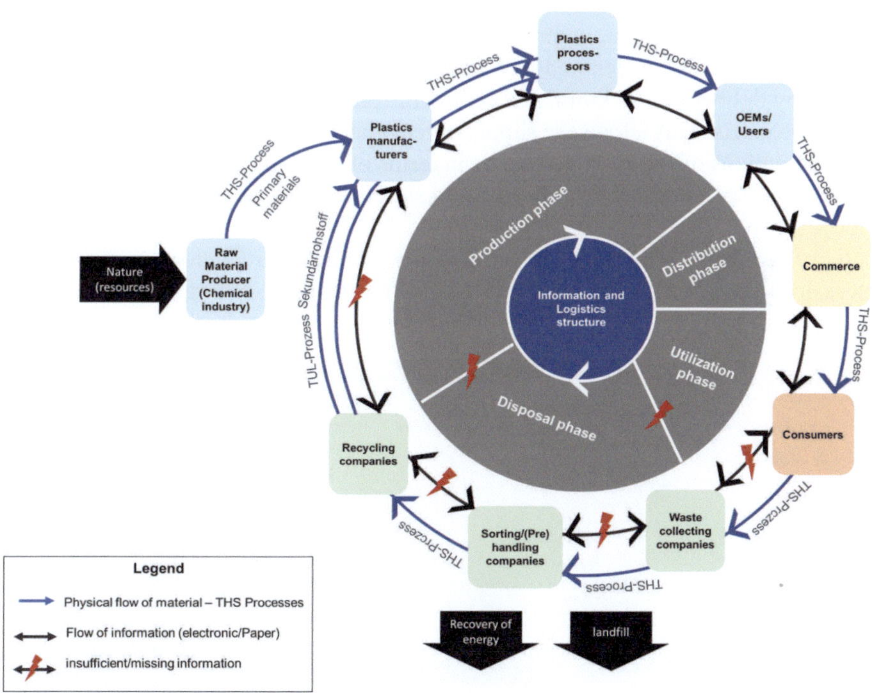

Fig. 9.3 End-to-end plastics cycle from the logistics perspective. *Source* Own illustration, based on VDI (2021, p. 7) and Fennemann et al. (2018, pp. 18)

not receive any information about the volume of waste or the degree of pre-sorting, making the (planning) subsequent processes "sorting" and "recycling" more difficult (cf. Fennemann et al. 2018).

9.1.6 Transformation Towards an Order-Oriented Stream of Secondary Materials

The plastics manufacturers and processors demand a secured volume of recyclates in the required quality level and at competitive prices (cf. VDI 2021). Thus, for waste management enterprises and logistics service providers, there is a need to initiate concepts during the disposal phase (cf. Fig. 9.4) aiming at transforming post-consumer waste streams into a recyclate stream, which fulfills the plastics manufacturers' and processors' requirements and demands resource-efficiently.

An important logistical lever is transforming the previous push-oriented supply of recyclates into an order-oriented flow of secondary raw material. Up to now, the companies involved in waste recovery/recycling process the waste from consumers to secondary material and "push" them into the market without always being able to reliably meet the quality requirements, delivery quantities, and delivery times defined by the processors (cf. Fennemann et al. 2018).

The realization of a demand-and order-oriented supply of recyclates is made possible by deliberate stockpiling/storaging of (pre-)sorted plastic wastes and intermediate products. This allows the supply of recyclates for production to be decoupled from fluctuations of the collected waste in terms of type and quality. Hence, the raw material for recyclates can be collected and stored in different quality levels until an order is received.

Recyclate prices and thus the profitability of the recirculation (returning and processing) are among others influenced by the required sorting or separation of

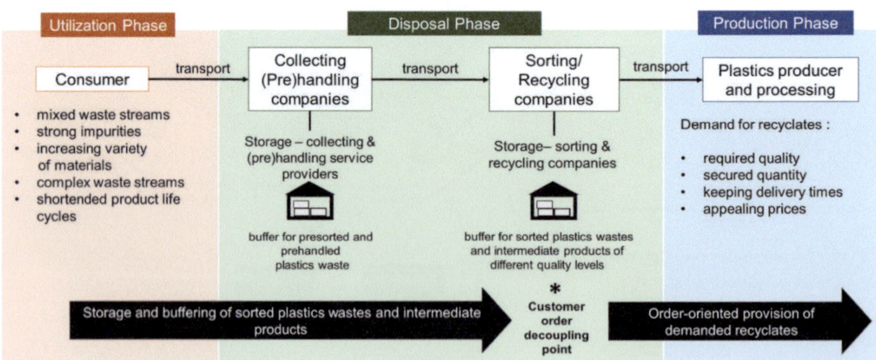

Fig. 9.4 Change from push-orientated to order-orientated provision of recyclates. *Source* Own illustration with information from VDI (2021, p. 14) and Remondis (2018) and EU Recycling (n.d.)

the waste input along with the plant utilization in the recycling process. A continuous exchange of information between plastics manufacturers/processors and recycling enterprises enables coordinated sorting and production processes. The recycling company receives information about at what time and what quantity of sorted intermediate products must be processed in order to ensure the customer's demanded output. These information support production planning as well as optimal plant utilization. Consequently, there is a need to optimize the information flow between the plastics manufacturers/processors and recycling companies.

Generally speaking, the potential lies in the efficient handling of existing logistics demands and the coordination of procurement, distribution, and returning processes across all supply chain players. Other potentials include the development of "logistics and recycling-compliance" of packaging and products.

9.1.7 Chances and Challenges for Logistics Service Providers

For the logistics industry as a provider of transport and warehouse infrastructure and at the same time as the link between the players in the end-to-end plastics cycle new challenges and chances will arise as part of the transformation towards circular supply chains.

Logistics service providers benefit from the potentials to establish new business areas or to expand existing business areas in disposal logistics. Reinforced by the current legal endeavors for circular economy the demand for logistical services will increase along the recirculation process starting from the waste collection to building feedstocks for customers. According to the "CTHST-functions" these logistics services include collecting, sorting, packing, handling, stocking, if required pretreatment and transport for further processing of wastes of all types originating from different sources such as service providers, retailers, industry and end consumers. Additionally, the current legal endeavor of politics to establish a circular economy offers chances for a secured stream of waste representing the input objects for disposal logistics services.

Today, the waste management industry already represents a significant market: in 2020 the global market volume of the circular economy is quantified at €148 billion. The majority, around €97 billion is accounted by the area's waste collection, waste transport and waste separation. With an annual growth of 5.9% the global circular economy, meaning the waste management industry in this context, is expecting a market volume of around €263 billion by 2030 (cf. BMU 2020). In 2019 Germany counted about 1,668 companies in the waste management industry of which 39.5% were active in waste collection, around 29.6% in recycling and ca. 27.2% in waste treatment and disposal (cf. Destatis 2022). In addition to the public companies in the waste management industry, Remondis, Suez, and Veolia are the largest players in the private sector (cf. Statista 2024).

Specifically for packaging waste, the German dual systems BellandVision, Duales System Deutschland, Eko-Punkt, Interseroh + along with PreZero (a subsidiary of the Schwarz Group) represent important players in this dynamic market (cf. ZSVR 2022). The global logistics volume for the end-to-end plastics cycle can be indirectly derived from the production volume of plastics. Taking this into account around 360 Mt new plastics and 30 Mt recyclates were produced globally in 2018 (cf. Chap. 5). The cumulated logistics volume is significantly higher as the accounted 390 Mt are transported, sorted, and stored several times across the different stages of the value chain and companies. These figures provide an indication to estimate the market potential.

The growing volume of plastics waste to be processed is already leading to an increase in collection and sorting processes during the disposal phase. As a result, there is a general need to establish or expand the existing recirculation systems for these increasing volumes and to efficiently operate the systems in terms of costs and energy consumption. Logistics is faced with the challenge of providing an efficient, cost-effective and reliable transport and warehouse infrastructure for recycling processes.

In the further development, the supply gap needs to be closed between the quantity, quality and prices of recyclates between consumers/waste management enterprises representing raw material suppliers and plastics manufacturers showing the demand side for recyclates. A fundamental rethinking of supply chains is necessary: the logistics service providers and recycling companies are faced with the challenge to transform the previously push-oriented material stream towards a customer and order-oriented flow of secondary raw material. An order-driven continuous supply to plastics manufacturers/processors is realized by decoupling the processes and stocking (buffering and storing) presorted plastics waste and intermediate products.

Along with this transformation of supply chains, new and in some cases significantly altered streams of material and goods will develop starting from the extraction of raw materials which also changes the processing and requirements in transport and storage. Overall, the material stream will become increasingly more heterogeneous with the stockpiling of presorted types of plastics. At the same time mass and bulk transports will be reduced in supply processes—simply by replacing crude oil with recyclates, this flow decreases considerably or is completely avoided. The different flows of materials require specific transport and stocking processes and thus an accordingly adapted logistics infrastructure to handle bulk streams of unsorted waste and compartmentalized (pre) sorted flows of waste and intermediate products must be established.

Established disposal specialists will benefit in the short-term from the increase in processing volume, allowing them to expand their business. Logistics service providers (LSPs) not involved in waste management industry yet such as generalists (still) have the chance to enter the market of the waste management industry timely and to broaden their service portfolio with disposal services. Apart from their existing role as the executing operator of transports, they can also take on the planning, coordination, and implementation of entire supply chain sections as a contract logistics service provider in returning materials to the plastics cycle in the next step. This results in different synergies such as the potential to optimize the vehicle utilization for return journeys in existing logistics systems, the possibility to implement a vendor-managed inventory, and along with that a continuous supply of recyclates on the plastics processor's side. In addition, service providers new to this industry branch bring expertise from other industries into the plastics cycle reinforcing the development of new concepts and ideas. These potentials for holistically integrating service providers in the supply chains within contract logistics projects can be compared to the developments in the chemical industry, where this has provided new impetus to further develop logistical systems (cf. Krupp et al. 2013, 2017). A loss of order volumes in primary raw material streams is expected for established providers serving the procurement processes in the chemical and plastics industry. An early strategic realignment is therefore necessary.

In the long term, the existing logistics infrastructure and systems must be rebuilt. A differentiated collection and sorting of wastes representing future resources can only be achieved by involving all participants in the end-to-end plastics cycle. Recycling and disposal logistics must be integrated and kept in mind by all players starting from the product design to establishing procurement, production, and distribution processes. Considering the goal of providing recyclates based on actual demand to fulfill the customers' required quality and quantity the material streams need to be developed in differing valences as the raw material for the various stages in the value chain.

Logistics service providers, industrial players, and waste management companies are offered chances for long-term cooperation for developing the required logistics infrastructure and systems.

Nevertheless, it should be kept in mind that logistics contribute substantially to the overall global greenhouse gas (GHG) emissions of companies who are now all heading to net zero and are focusing more and more on the reduction of "Scope 3" emissions. These emissions are not directly produced by their own operations but embedded in their supply chains accounting for most of their emissions (cf. Chap. 18).

A significant portion of these scope 3 emissions is generated from supply chain and logistics activities and regarding waste handling of any kind from road freight and warehousing operations, particularly. In total, logistics emissions from freight and warehousing account for at least 7% of global GHG emissions (cf. Fig. 9.5). Logistics and supply chain activities in the circular economy will thus need to address them as part of a company's holistic environmental, social, and governance (ESG) strategy (McKinsey 2024).

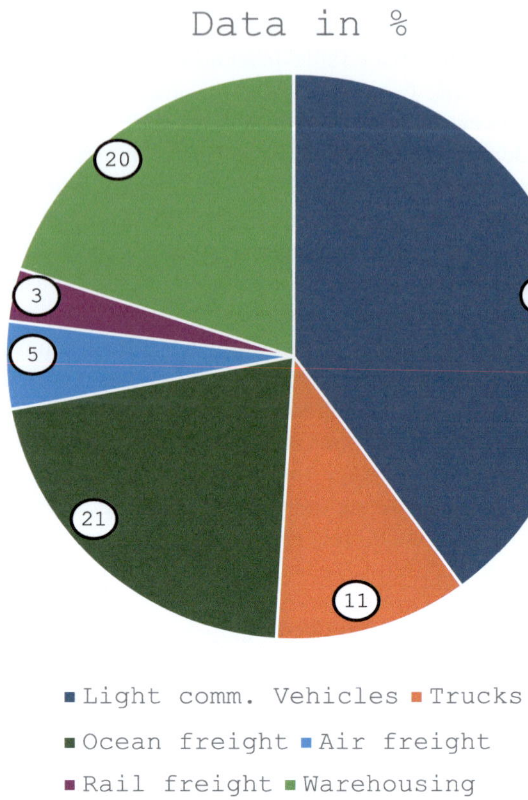

Fig. 9.5 Share of global annual logistics CO_2 eq. emissions. *Source* Own presentation, based on EMIT database from McKinsey (2024)

9.2 Sensor-Based Sorting of Plastic Waste Streams

Michael Perl and Andreas Hanus

9.2.1 *Introduction*

Sensor-based sorting is a technology used in many industries such as mining, food, recycling and waste management. It is by far the dominant method of sorting waste, particularly plastic waste, in the European waste management and recycling industry.

Other sorting methods that often complement each other in recycling processes are screening techniques, separating tables, air separation, density separation in liquids, electrostatic sorting and separation methods using magnets. Depending on the size of the parts to be sorted or separated, there are tailored system solutions or machines for each sorting technology. In optimized recycling plants, the various sorting methods are matched to the material streams to be sorted. Thus, they ensure the highest possible material purity and the lowest possible material losses. Ideally, this results in the maximum potential added value, in a solution for an important contribution to achieving recycling quotas, as well as in an optimal solution for reducing actual waste fractions that cannot be recycled.

In the recycling industry, material flows are often divided into "post-consumer waste" and "post-industrial waste" fractions. The industry segmentation itself is divided into the areas of household waste sorting, plastics recycling, waste of electrical and electronic equipment recycling (WEEE), scrap metal recycling, automotive recycling, glass recycling, textile recycling, wood recycling and construction waste recycling.

This chapter is intended to provide an overview of the economic, legal and technological framework conditions for sensor-based sorting of plastic waste streams in the plastic recycling industry. The aim is to convey all the important basics of different sensor technologies and their applications to understand possible future technological innovations resulting from the digital transformation and the new strategies that arise as a result.

9.2.2 Framework for Sorting in Europe

9.2.2.1 European Recycling Market

The development of the European plastics recycling market is closely linked to the use of sensor-based sorting machines. In 2021, 236.8 Mt of municipal solid waste (MSW) was generated in Europe, of which a total of 49.6% was recycled. In addition to plastic waste, this also includes paper products, biodegradable materials, glass and other waste materials (Statista 2024).

In 2022, the plastic waste stream contained 32.3 Mt of material (post-consumer plastic waste) within the EU (27 + 3). 27% of this was recycled, 50% recovered as energy and 23% landfilled (PlasticsEurope 2024).

Looking at the recycling rates for plastic packaging of the individual EU countries in 2022, Belgium (56%), the Netherlands (55%), Germany (51%), Spain (50%) and the Czech Republic (45%) are at the top for packaging plastics, while Hungary (26%), Finland (27%) and France (23%) are at the bottom of the table (ibid). This is partly due to the different approaches to waste separation and collection in the respective countries. While established waste separation and collection mechanisms have already been implemented in Belgium, the Netherlands, Germany, e.g. yellow bag, and recycling centers (El Green Mall 2023; Radio Prague International 2009), countries such as Hungary and France have only just initiated corresponding measures with regard to new EU regulations (BVSE 2022; GTAI 2023). This becomes clear when looking at the different collection options. In the case of mixed collection (e.g. residual waste), on average only 1% of the material is recycled (62% thermal recovery, 37% landfill). The low residual waste recycling rate is due to the heterogeneity of the contents (mixed collection). This only contains 6.7% plastics. Organic waste accounts for the largest share at 39.3% and hygiene products at 13.5%. Other recyclable materials such as waste paper (5.2%), used glass (4.6%) and metals (2.0%) also make up a smaller proportion (BMUV 2020). In contrast, with separate collections (e.g. recycling centers), an average of 80% of plastic packaging is recycled and only 18% is incinerated or 2% landfilled. In comparison, the pre-sorted fraction of the Yellow Bag requires less effort for further re-sorting. The guideline on what should be thrown into the yellow bag also leads to less contamination by obstructive substances. However, the proportion of incorrect throws in the Yellow Bag is still around 30%. This also makes it more difficult to return waste to the recycling loop. Around 40% of the contents of the Yellow Bag are still sent for energy recovery (Mülltrennung wirkt 2021). Nevertheless, the implementation of collection systems is necessary for successful recycling and a building block for further increasing recycling rates. In addition, however, appropriate incentives must be created to ensure the availability and acceptance of high-quality recycled material.

9.2.2.2 Legal Requirements as a Driver

Legal requirements are a strong driver for the use of recycled material instead of virgin material of petrochemical origin and consequently also for the demand for sorting machines. This applies in particular to recycled content requirements in certain packaging segments. From 2025 onwards, 25% and from 2030 onwards 30% of all PET beverage bottles must be made of rPET (EU 2019). This will gradually increase the demand for high-quality PET from recycled sources. When using recycled plastics for food packaging in particular, it must be ensured that the recycled material is also suitable for use with food and has undergone appropriate processes to ensure that the material is decontaminated, among other things (EU 2022). A distinction is also made here, for example, between "food-grade" and "non-food-grade" materials. This can be illustrated with PET beverage bottles or PET trays. New food-grade articles can be produced from PET beverage bottles. However, the proportion of non-food PET articles, such as trays for electronics packaging "that have been used in contact with materials or substances other than food", must be below 5% (ibid). Limits such as these are often checked by analyzing compressed bales of plastic packaging and beverage bottles. If necessary, bales or batches that contain more than the permitted amount of non-food materials must be rejected and re-sorted. In many countries PET waste streams contain more and more non-food PET so there is a need to differentiate and separate these specific pieces of PET. Thinking about the separation of pieces of other types of plastics like HDPE this could become even more important in the future. The good thing here is that there are already technologies like AI-based material detection or different kinds of marker solutions. More information about this will follow in the outlook of this paragraph.

The specifications in the draft PPWR (e.g. 30% recycled content for non-contact sensitive packaging in 2030) will also increase the demand for recycled materials, especially for polyolefins (PPWR 2024).

A further aim of the EU regulations (EU 2019) is also to ensure that less recycled material is exported. This is already having a visible impact on the European recycling market.

9.2.2.3 Investment Trends of the European Sorting Machinery Industry

Increasing investment in recycling facilities in countries such as Greece, Croatia, Romania, Bulgaria, and Hungary suggests that the destination for European waste exports is shifting from Asia via Turkey to Southeastern Europe and the Balkan states. The advantages over Central Europe are low labour costs and EU membership. The latter makes it possible to comply with EU regulations and directives while keeping costs low.

The European industry for sorting and recycling equipment also has a strong presence in non-European markets. The entire global recycling market will have a total volume of 29.6 billion $ in 2021. Assuming a compound annual growth rate (CAGR) of approx. 7%, the market will reach a volume of approx. $78 billion by 2031 (Allied Market Research 2023).

The Asian market represents the largest market at around $17 billion. In addition to the enormous volume of waste due to the high population figures and the increasing initiatives against environmental pollution, this is still partly due to waste imports from the US, Japan, and Australia (ChemAnalyst 2023). Low wages and energy costs also increase the attractiveness of collecting and processing plastics. China and, increasingly, India are the largest markets for recycling plants and plastic waste. The first recycling plants are also being designed according to European standards that meet the requirements of the European Food Safety Authority (EFSA) (Brunn 2023). This means that the recycled material or regranulate can also be used for bottle-to-bottle applications, also with the aim of exporting this regranulate to the EU.

In addition to the Asian market, recycling is also becoming increasingly interesting in the Middle East. Countries such as the United Arab Emirates, which previously had little to do with recycling, are increasingly recognizing the potential for the environment and the economy (Sustainable Plastics 2023).

In Africa, too, the African Circular Economy Alliance (ACEA) is striving to achieve economic benefits through recycling and a functioning circular economy (ACEA Africa 2023). However, only around 4% of African waste was recycled in 2022, for example. 90% of the waste ended up uncontrolled in landfills or in the countryside. The African Union's goal is to increase the recycling rate to 50%. The highest priority is to minimize health risks by providing a functioning waste management system and to reuse, collect and recycle valuable raw material (Labels and Labeling 2022). Accordingly, more and more recycling plants are being put into operation, particularly in Nigeria, Ghana, and South Africa. European companies are also driving recycling forward with corresponding investments (Packaging Europe 2023).

9.2.3 Technological Principles of Sensor-Based Sorting

The aim of paragraph 9.2.3 is to present the recycling process as a whole and to understand the background to sensor-based sorting and to place it in the context of the recycling industry. The focus is on the main components of a sorting machine and the introduction of the various sensor systems. Only sorting machines for the recycling industry and in particular for the recycling of plastics are considered. Sorting machines for food or in mining applications are based on the same principles, but differ in terms of applications, robustness, and the sensors used. The information on the recycling process, sorting machines, and sensors represent the current state of the art in the recycling industry. The following paragraph includes a description of the current market situation and is based on information obtained at trade fairs such as IFAT 2022, K-Messe 2022, ChinaPlas 2023, PRSE 2023, and Fakuma 2023 and experience gained from relevant activities in the recycling industry. Information on plastics recycling that goes beyond the basics in this section can be found in the literature (Niessner 2022). Further information on sensor systems can be found, for example, with Meskers et al. (2023). The latest developments in the field of sensor-based sorting are also presented, for example, every two years at the "Sensor Based Sorting and Control" conference (RWTH Aachen 2024).

9.2.3.1 Sensor-Based Sorting in the Recycling Process

The design of a recycling line can differ depending on the application. As the material to be sorted in the field of plastics recycling has different origins and also comes from different collection sources, this leads to a wide variety of applications. Large quantities of plastics packaging such as bottles, canisters, cups, buckets, films and other containers are produced, which can be in whole or shredded form. Plastics also accumulate in electrical and electronic scrap, from the automotive sector, the construction industry, or the agricultural sector. Plastics from these sectors also come in various forms, sizes and shapes. Basically, it can be said that several and often different sorting steps and techniques are used to achieve the required purity for recycling.

The following section takes a closer look at a PET recycling plant to illustrate the application of sensor-based sorting in a recycling process (cf. Fig. 9.6) The PET fraction, which is obtained from collection stations, recycling centers or through pre-sorting, comes in the form of compressed bales or briquettes, but also in loose form. In Germany and other countries, this material can also come from deposit return machines. The material is initially fed into a bale breaker for separation. The PET beverage bottles are then separated from smaller impurities such as broken glass or stones in a drum. The drum serves as a kind of sieve and the bottles then move on in the process chain. A so-called ballistic separator is often used as an alternative to the drum. In the next step, various units are used to separate larger metal contaminants (magnet system) and the labels as far as possible. A non-ferromagnetic separator

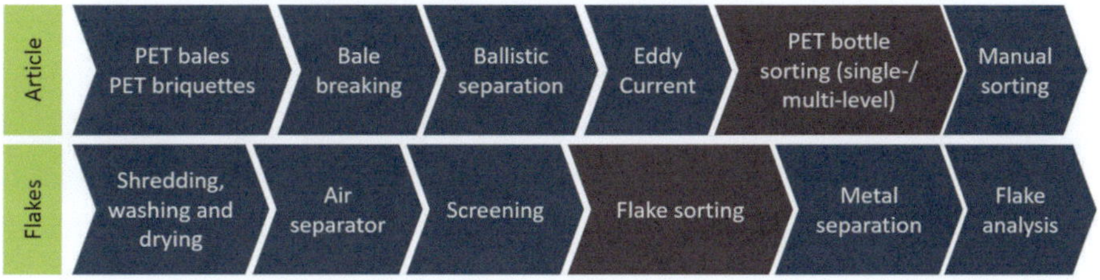

Fig. 9.6 Schematic diagram of the PET plastic recycling process steps. *Source* Sesotec

(for non-ferromagnetic metals) is then used to separate aluminum beverage cans, for example. The bottles are then sent to the first sorting stage. Here, sensor-based sorting machines with various sensors, for example by colour or material, are used to differentiate. Depending on the composition of the input material, further sorting stages follow. The aim at this point is to create fractions that are as pure as possible in terms of type and colour for further recycling. The loss of plastic should also be kept as low as possible. For this reason, recovery sorting stages are installed in many recycling plants, which recover good material from the actual bad material.

The bottle fraction is then shredded by a shredder. The resulting flake material is sieved to separate out small particles. Also, the light fraction (e.g. labels) is removed by an air separator. After optional intermediate storage in silo tanks, the flake material is cleaned in a hot wash or washing system. This is followed by the separation of foreign matter by density using a float-sink separator. At this point, plastics with a density $< 1\ g/cm^3$ (e.g. HDPE/PP caps) are separated from plastics with a density $> 1\ g/cm^3$ (e.g. PET flakes).

In a subsequent drying process, the wet flakes are dried to a residual moisture content of $> 3\%$. Optionally, the material is then subjected to further sieving or air separation before the flakes are sorted again by colour and material using a flake sorter. The different fractions are usually filled into big bag stations or temporarily stored directly in silos for extrusion. These sorted PET flakes, with a correspondingly high level of purity, are then processed into new pellets in extruders. Before this, however, metal detectors are again used to remove any metallic foreign bodies that could damage the extruder. The extrusion process itself includes decontamination, which ensures that the PET regranulate produced meets the relevant food-grade limits and can be reused for food packaging or drinks bottles.

9.2.3.2 Design of a Sorting Machine

In the recycling sector, sorting devices are used both in the article sector (e.g. PET beverage bottles, HDPE/PP containers) and in the flake sector. The integrated sensors are similar, but the design and size of the machines are different.

Conveyor belt devices such as the SESOTEC VARISORT + (cf. Fig. 9.7) are used for article sorting. The items are conveyed here via 3- to 8-m-long conveyor

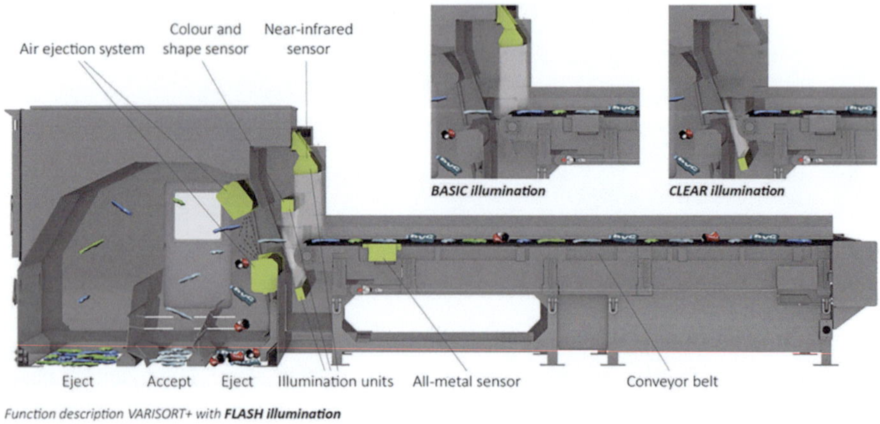

Fig. 9.7 Cross-section of SESOTEC VARISORT + sorting machine. *Source* Sesotec

belts to a separation housing. Integrated into this are the sensors for colour and material classification as well as corresponding lighting, valve bars for sorting and image processing computers and user interfaces. Machines are available in working widths from one meter up to three meters. In addition to the integration of sensor technology and evaluation in the separator housing, there is an increasing trend towards the use of sensor-measuring bridges. Compared to complete devices, these are easier to retrofit or integrate into systems and are also easier to maintain due to less soiling. Together with the valve bars, these sensor-measuring bridges are combined with existing conveyor belts and separator housings—or integrated into appropriately planned sorting systems. The advantages of the complete sorting device, on the other hand, are fewer space requirements and more options for special lighting and sensor systems. A detailed list of the various sensor technologies, their functional principle, and benefits as well as their cost estimations is shown in Table 9.1.

In flake and regrind sorting (cf. Fig. 9.8), chute devices are usually used, whereby the material to be sorted is detected and ejected in free fall. These sorting systems are characterized by a compact design, a very high sensor resolution and a small valve and discharge nozzle grid of the discharge unit. With this combination, even very small material and colour impurities can be detected and separated.

9.2.3.3 Sensors and Signal Acquisition

In the field of plastics recycling, line scan or area scan cameras for colour differentiation are among the most established sensors. In the flake area, only line scan cameras are used, while for article sorting sometimes also area scan cameras are used. While in the flake area, the focus is on colour differentiation and high resolution for finding the smallest impurities, in the article area line scan or area scan cameras are also used to differentiate between shapes and other characteristics.

Table 9.1 Sensor technologies in the recycling industry

Sensor technology	Functionality	Benefit	Costs
EM sensor	Measurement of disturbances of an electromagnetic field by conductive objects	Detection and identification of iron, stainless steel and non-ferromagnetic materials	$-$$
Height profile measurement	Laser triangulation measurement	Method for identifying object boundaries, finding black objects on the conveyor belt and determining shapes	$-$$
Laser fluorescence	Laser excitation and subsequent measurement of the fluorescence spectrum of the material under investigation	Best option for detecting and distinguishing foreign plastics in small concentrations	$$$
LIBS	Laser-induced plasma spectroscopy. "Vaporization" of material and subsequent analysis of the spectrum	Advantages in metal recycling and especially in the differentiation of metal alloys	$$$
MWIR spectroscopy	Hyperspectral image processing in the mid-wave infrared range (2.7–5.3 μm)	Sorting of certain black plastics	$$$
NIR spectroscopy	Hyperspectral image processing in the near-infrared range (900–1900 nm)	Sorting of non-black plastics	$$
SWIR line scan camera	Multispectral point-based classification in the range of 900–1700 nm in combination with corresponding wavelength filters	Application-optimized sorting of non-black plastics, restriction to relevant support points	$-$$
Terahertz/radar	Measurement of terahertz radiation	Research area for the classification of foreign substances and black plastics in particular	n.a
VIS	Colour line scan camera or colour area scan camera with neutral white illumination in the VIS range (450–700 nm)	Differentiation by colour and shape	$
VIS-UV-IR	Gray scale line scan camera in combination with filter arrangements or pulsed LED illumination (320–1000 nm)	More precise colour differentiation and detection of special glasses and additives	$-$$
VIS fluorescence measurement	Measurement of fluorescence with colour line scan camera and UV illumination for excitation	Detection of lead glass	$-$$

(continued)

Table 9.1 (continued)

Sensor technology	Functionality	Benefit	Costs
XRT	Measurement of X-ray transmission usually with two reference points (dual energy method)	Detection and classification of metals, glass, plastics or other materials	$$-$$$
XRF	Measurement of the X-ray fluorescence spectrum	Analysis of the chemical composition of materials. Advantages for sorting metals and metal alloys	$$$

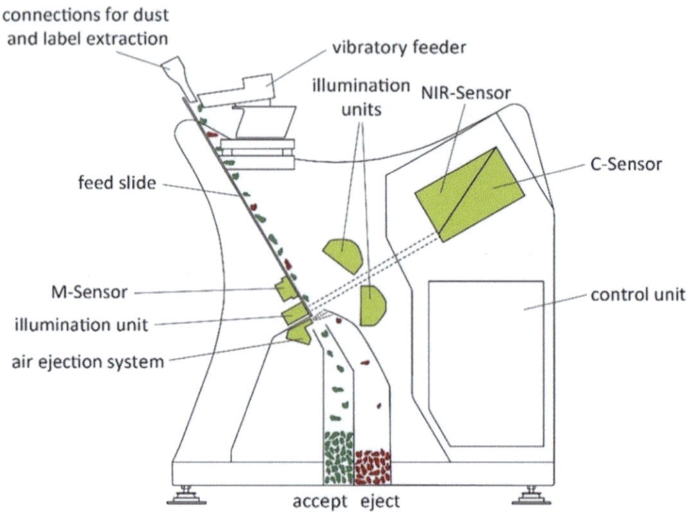

Fig. 9.8 Cross-section of a SESOTEC FLAKE PURIFIER + sorting machine. *Source* Sesotec

To sort non-black plastics, near-infrared (NIR)/shortwave infrared (SWIR) spectrometers or line cameras are used to differentiate between polymers. Line scan cameras in combination with filters or pulsed illumination represent a more cost-effective approach compared to spectrometers, with lower detection accuracy. The sensors cover a subsection of the wavelength range from 780 to 3000 nm. The NIR range extends from 780 to 1400 nm, the SWIR range from 1400 to 3000 nm. Colloquially, however, the terms SWIR and NIR are often mixed up and equated (Hamamatsu 2023). The wavelength range from 900 to 1700 nm or 1900 nm is generally used to differentiate between plastics and other materials such as paper, wood or other cellulose-based substances. Characteristic bands of plastics can be found in this range. In the flake area, this near-infrared sensor is responsible for detecting even the smallest impurities. In the article area, the larger objects can also be used to differentiate between very similar plastics.

At the same time, laser fluorescence technology also offers an approach for maximum purity of the final plastic fraction, although this is also associated with considerable additional costs. This technology is generally only used for the highest food-grade requirements of recycled material.

Metal detectors (all-metal sensor/EM sensor) are used for metal separation (e.g. of beverage cans or parts thereof). Depending on the frequency at which they are operated, these can also be used for metal separation. In metal scrap recycling itself, LIBS (Laser-Induced Plasma Spectroscopy), XRT (X-ray Transmission) or XRF (X-ray Fluorescence) are used to distinguish metals (cf. Table 9.1).

To detect black parts (plastics) on a black conveyor belt, height profile information is often also collected using laser triangulation. This is particularly important for measuring bridge solutions to relativize the disadvantages compared to the arrangement of the sensors in the separator housing. However, there are also possibilities for shape differentiation and object detection.

It is not possible to sort black plastics (carbon black) by type using NIR or SWIR technology, as the carbon black contained in the material absorbs all information about the type of plastic. However, this is made possible by mid-wave infrared (MWIR) spectroscopy, which can differentiate between different black plastics based on characteristic bands in the mid-wave infrared range. However this technology is expensive and requires a high level of maintenance. Yet, the sorting of black plastics would be of great interest, particularly in the WEEE and automotive recycling sectors. In this respect, there is an opportunity for further innovation if a cost-effective solution can be found to separate black plastics by type.

European suppliers of sorting machines can set themselves apart from the ever-increasing competition from Asia primarily by using technologically sophisticated sensors. When it comes to pure colour sorting with visual inertial system (VIS) line scan cameras, European manufacturers are increasingly finding themselves competing with their Asian counterparts. However, there are still competitive advantages in NIR and MWIR as well as metal recycling with LIBS, XRT, and XRF. There is also expertise in the corresponding analysis and software development.

9.2.3.4 Evaluation Unit and Blow-out System

In a sorting machine, these sensors collect the necessary information and send it to a central evaluation unit. A decision on sorting is made there using logical links. The result is sent to one or two valve bars. Depending on the position of the conveyor belt, a valve is then activated, and the respective object is sorted out. The valves must be controlled with correspondingly fast-acting valve drivers in the millisecond range. At the same time, these extremely short reaction and valve opening times also place demanding requirements on the hardware side. Depending on the type of material (flake, metal, electronic scrap, film or article sorting), different valves and valve grids are used, and the pressure of the air supply is adjusted. In conveyor belt devices, up to three fractions can be generated with two blow bars. In the flake sector, only two fractions are usually possible. A high signal frame rate is required to achieve the high sorting rates and throughputs. This requires high real-time computing power, using CPU-, DSP- and FGPA-based image processing computers.[2] At the same time, however, sufficient signal strength is also required, which is why, in addition to sensitive sensors, appropriately optimized excitation sources and illuminations must also be used.

The introduction of increasingly complex evaluation algorithms in the field of plastics sorting is also increasing the demands on real-time evaluation units. For example, GPU[3]-based evaluation systems are required to be able to execute very computing-time-intensive algorithms for AI-based object recognition and instance segmentation in real time in sorting machines. Enormous progress has been made in this area in recent years. The availability of cost-effective, industrial-grade evaluation systems with an integrated GPU has increased significantly, which is another driver of progress in article sorting.

9.2.4 Outlook

HolyGrail

Another possibility that has opened up as a result of increased computing power is the use of digital watermarks for sorting plastic packaging waste. With the HolyGrail 2.0 initiative of the AIM (European Brands Association), there is an initiative of over 160 companies with the aim of acting as a European driver for the establishment of digital watermarks. The initiative includes packaging manufacturers, consumer goods manufacturers, and sorting machine manufacturers. With the help of a digital watermark, which is integrated into labels as a 2D code or incorporated into trays as a 3D code, for example, a connection is to be established to a database in which information on plastic composition, manufacturers and other properties is stored. In addition to increased recyclability, this should also enable data to be collected along the entire life cycle of a product. The code itself is not visible to the human eye, especially as a 2D code, but it inevitably requires the integrity of the packaging labels. The aim is to enable the differentiation of food and non-food materials, for example, in addition to the existing material flows, thereby closing the cycle for food plastic packaging, for example. A correspondingly high resolution of the camera systems is required for detection in the sorting machines (Holy Grail 2.0 2023).

---

[2] CPU: Central Processing Unit; DSP: Digital Signal Processing; FGPA: Field Programmable Gate Array.

[3] GPU: Graphics Processing Unit.

Digital passport

At the same time, data collection and the use of codes offer significant advantages along the value chain and life cycle. As a result, digital watermarks could play their part in achieving the goals of a circular economy. The R-Cycle Initiative is therefore advocating a digital product passport for plastic packaging—which can be implemented using digital watermarks on the products, among other things. This should make it possible, for example, to determine data such as the recyclability rating or the CO_2 footprint of packaging or to provide further product information. At the same time, it should be possible to track plastic packaging over its entire life cycle and check its quality characteristics. By indicating the composition of the contents, it should be possible to check political requirements and ensure more precise waste sorting (R-Cycle 2024).

It is critical to consider whether the sorting results of the prototype tests of the digital watermark sorters can be confirmed, as the integrity of the labels and as little deformation of the products as possible are required here. However, this is difficult to implement due to the mechanical pre-processes. It also remains questionable whether the high number of material flows envisaged in sorting centers can be successfully generated from an economic point of view.

Fluorescent markers

An alternative solution that is not based on digital watermarks is fluorescent markers such as the TagTec technology from Sensor Instruments (2023) and Gabriel Chemie (2023). Here, a corresponding marker masterbatch is integrated into the plastic which is characteristic of a certain material or product property. Excitation at certain wavelengths of light then results in a fluorescence response, which can be evaluated. The decay after excitation is measured (tau measurement). This marking can be used, for example, to label food items with marker A and non-food items with marker B, making it possible to separate them accordingly. A process has also been developed to ensure that no negative accumulation of markers in the plastics is possible. For example, only as much new marker masterbatch is added to recyclates as is necessary to achieve the target values (Kunststoffe 2022). At the same time, the insertion of such a marker masterbatch into the plastic is also a not insignificant disadvantage of this technology. Due to enrichment and impurities as well as toxic requirements, plastics manufacturers have to overcome hurdles here. At the same time, there are again additional costs in the production of plastics. In addition, compared to digital watermarks, (almost) no data collection along the life cycle is possible. In contrast, however, there are advantages, as the integrity of the markers is guaranteed and therefore no labels are necessary and deformed or shredded items can also be sorted. In theory, this also makes it possible to examine flakes and granules (Gabriel Chemie 2023).

Artificial intelligence

In addition to the use of markers and watermarks on plastic packaging and plastic items, sorting solutions based on artificial intelligence open up possibilities for differentiating between food contact articles or manufacturer attributes. Initially, start-ups in particular focus on tapping into artificial intelligence in the recycling sector. In the meantime, almost all major sorting machine manufacturers have followed with their own solutions or have built joint ventures. AI evaluation methods have their advantages in the area of object detection or segmentation, particularly for sorting tasks in which humans can detect differences relatively easily. In addition to optimized sorting by colour, shape, texture or other properties that are recognizable to humans, new opportunities for understanding waste streams arise here. The decisive factor here is the corresponding database. The input material of sorting plants often fluctuates seasonally or due to the respective region of origin. The more precise these different flows can be mapped, the greater the value of the target fractions. Analyses and records of the various items in the input composition are crucial here, as is the analysis of the output stream and any problem articles it contains. Optimized sorting can also make manual re-sorting obsolete and at the same time increase the quality of the recycling fractions. This could make a significant contribution to the circular economy of plastics. There is potential here in PET recycling, for example, where new opportunities arise in tray recycling or future challenges can be addressed. As with HDPE/PP sorting, it will also be necessary to differentiate between food and non-food articles in the future, as more and more PET-based cleaning articles are being put into circulation, for example, which have to be cleaned below a certain limit. In the HDPE sector, in addition to these articles from household cleaning liquids, oil canisters or non-empty silicone cartridges, as well as a range of cosmetic articles, also pose problems that can be solved with AI object recognition. However, new applications are not only emerging in the field of plastic packaging. Purer fractions can also be generated in the WEEE sector, for example from circuit boards or other valuable material streams.

Challenges arise from the database required to train the corresponding AI models. The enormous number of different articles and the constant changes in packaging design pose challenges. Accordingly, the challenges here will lie less in the algorithms themselves but more in the area of optimized data collection and data processing of waste streams. At the same time, AI-based article sorting will only be able to reliably predict plastic types to a limited extent. A combination with material classification using NIR will therefore also be necessary in the future.

In addition to new developments in the field of sensor technology and evaluation methods, the importance of process analysis, process control, and process certification in the sorting process will continue to grow. Accordingly, new requirements will also arise here to ensure better, automated, more stable and transparent processes. Material flow analyses at the relevant points will be just as necessary as a central platform for collecting data and information.

9.3 Collection, Sorting, and Preparation for Recycling

Michael J. Schneider and Herwart Wilms

9.3.1 About Remondis

Remondis is a European environmental service provider/resource manager that generated sales of €12.1 billion in 2023 by collecting, sorting and processing waste from minerals, metals, glass, paper/cardboard and plastics. The company operates its own dual system, Eko-Punkt. The family-owned company, based in Luenen, Germany, has set itself the goal of making these materials increasingly available as secondary raw materials for further use, thus contributing to closing material cycles and reducing the consumption of natural resources. The company provides its services to local authorities, industry, trade and commerce, and has a customer base of 30 million people. The company collects and processes over 30 Mt of different materials, making it one of the largest companies of its kind in the world.

9.3.2 Plastic Waste and Its Remaining

The global production and marketing of plastic (cf. Fig. 9.9) has grown exponentially in seven decades since 1950—from 1.5 Mt in 1950 to around 400 Mt in 2022, a development that has been followed and documented for many years by the plastics producers' association PlasticsEurope (PlasticsEurope 2023). The OECD forecasts a continuous increase in production and consumption to around 1.2 Gt by 2060, due to the further growth of the world population and its increasing prosperity (cf. Sect. 5. 8).

Of the 400 Mt of plastics produced worldwide in 2022, an estimated 240 Mt will become waste after a more or less short period of use, of which only around 10% is recycled today, while around 90% is incinerated, landfilled or released into the environment in an unorganised manner (Geyer et al. 2017).

In Europe (EU 27 + 3),[4] on the other hand, we observe a stagnating market of around 60 Mt per year since the early 2000s, which corresponds to a per capita consumption of around 110 kg for a population of around 540 million people in these countries. It would be reasonable to expect a steady state between marketed plastic and collected and recycled waste after around 20 years, but in fact, only 32.3 Mt of plastic waste was collected in 2022 (PlasticsEurope 2024), which corresponds to 54%. There is therefore considerable potential here.

[4] EU 27 + 3: This definition by PlasticsEurope refers to the 27 member states of the European Union as well as Switzerland, Norway and the United Kingdom.

Fig. 9.9 Global and European plastics production 1950–2022. *Source* PlasticsEurope (2023)

In the EU 27 + 3, recycling rates vary between approx. 20% (LU, FR, PL, HU, CY, FI, LV, HR, GR, AT and MT) and > 30% (BE, ES, NL, DE, SE, DK, NO and IT) (cf. Chap. 5). A number of countries incinerate > 50% of the plastic waste collected (BE, NL, DE, SE, DK, NO, UK, IE, CH, LU, FR, FI and AT), while others landfill > 30% (MT, GR, HR, LV, CY, HU, PL, RO, BU, PT, SL, LT, CZ, ES) (cf. Fig. 9.10).

The following requirements can be derived from this:

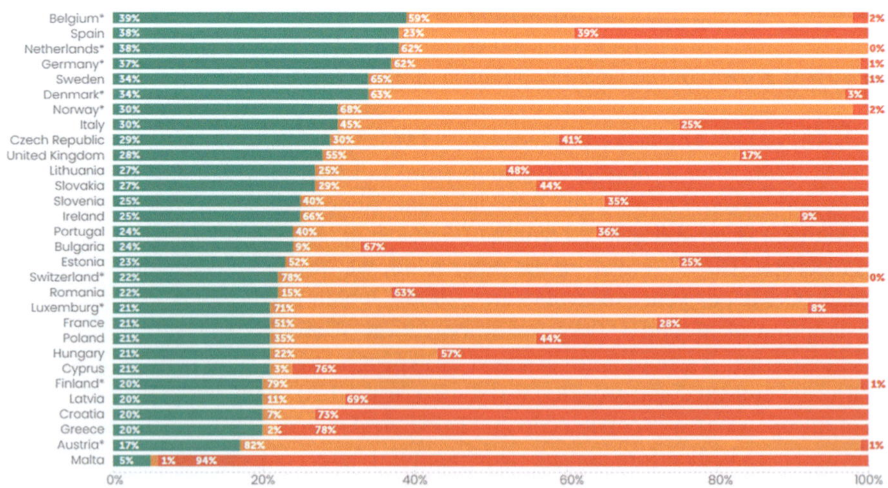

Fig. 9.10 Plastics waste management per country in Europe. *Source* PlasticsEurope (2024)

- Landfill ban for organic waste (including plastics) (as in BE, NL, DE, DK, NO, CH, LU, FI and AT);
- Minimising incineration, e.g. by pricing CO_2 emissions from incineration plants as with the Fuel Emissions Trading Act in Germany (BEHG 2019);
- Increase recycling to secure resources and mitigate climate change by consistently implementing the European Waste Framework Directive (WFD 2008) for plastic waste.
- Expansion of plastics recycling capacity. According to estimates by analysts at Pareto Securities, this would have to increase by 5% annually by 2030, which would require considerable investment and innovation in the entire value chain from collection to sorting and preparation for recycling (Pareto Securities 2023).
- In addition, the creation of incineration capacity in a number of countries or the authorisation of shipments of plastic waste within the EU 27 + 3 for incineration in existing plants.

9.3.3 Separate Collection with Sorting Requirements—History and Status Quo

The separate collection of recyclable waste is an essential prerequisite for its further processing. The aim is to increase the efficiency of the recycling process by reducing material contamination and ensuring a higher quality of the recovered material. Of the 32.3 Mt of plastic waste collected in the EU 27 + 3 in 2022, 15.9 Mt were collected via mixed waste collection and 16.4 Mt via separate waste collection. Figure 9.11 shows the dramatic impact this has.

In Germany, the separate collection of recyclable materials is of great importance and is a central component of waste management. In addition to beverage bottles from the deposit system as the highest possible separate collection level, various recyclable materials are collected in designated containers and collection points. Examples include lightweight packaging, paper and cardboard, glass, but also organic waste, textiles, shoes, batteries and electronic waste.

From the 1990s to the 2000s, the country played a key role in the development and establishment of a circular economy for plastics and other materials. The disposal and recycling of plastic products, whether with or without food contact, packaging, toys, household appliances and much more, is regulated in Germany by laws and directives, two of which are mentioned here.

The Packaging Ordinance (now the Packaging Act (2019)), first issued in 1991, regulates the handling of packaging waste of all kinds. The focus is on producer responsibility for packaging and packaging waste, symbolised by the Green Dot of the Dual System. Parallel to public waste disposal systems, the 'Green Dot' system was set up to take back and recycle packaging. This take-back system can be found in households, e.g. as yellow bags or yellow bins, paper bins and household collection containers for used glass. Manufacturers fulfil their responsibility by paying a licence fee to label their products with the Green Dot; this fee guarantees collection

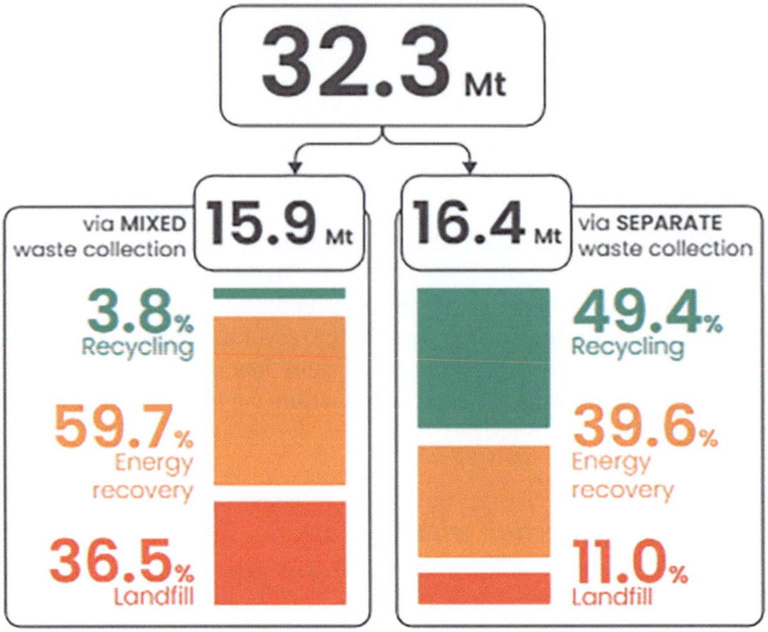

Fig. 9.11 Post-consumer plastics waste[5] and treatment. [6] *Source* PlasticsEurope (2024)

and recycling. Another important law is the Circular Economy and Safeguard the Environmentally Compatible Management of Waste of 1994, last amended in 2023, (Circular Economy Act 2024) which fundamentally regulates the handling of waste and promotes product responsibility and improved separate collection of waste. To this end, the German legislator introduced a now established system for this material flow, which serves as a model throughout Europe.

Lightweight packaging made of plastic, metal and composite materials is disposed of via systems such as the Yellow Bag or the Yellow Bin under extended producer responsibility. After collection, the packaging waste is sorted by material and composition in sorting plants. The waste fractions produced are specified and labelled and then sent to a recycling plant. The yield of a sorting plant and the accuracy of the sorting technology depend heavily on the composition of the collected waste. The fewer different components are collected in a collection structure and the lower the proportion of incorrectly disposed materials, such as organic waste in the yellow bag,

[5] Post-consumer plastics waste: waste generated by households or by commercial, industrial, and institutional facilities in their role as end-users of the product which can no longer be used for its intended purpose. This includes returns of material from the distribution chain or the installation of plastic products.

[6] Recycled quantities were previously measured upon leaving the sorting centres. The measurement is now done when the recycling actually takes place, according to the Packaging and Packaging Waste Directive (PPWD) (EU) 2018/852. The calculation point therefore lies after impurities and unsuitable substances have been removed from the sorted materials.

the more valuable the sorted fraction is. The number of containers to be used and the composition of the recyclables to be placed in them depend on several factors. A customised approach can be particularly useful in industry. Optimising the distances, the space required by the container, the likelihood of contamination or incorrect disposal and analysing the value of the individual waste components can lead to a reorganisation of processes, but also offer great financial opportunities in addition to reducing the volume of waste itself.

The common goal is to continuously increase material recycling rates. In the long term, almost 100% of all raw materials from packaging should be returned to the production cycle. However, from today's perspective, in 2024, this goal is still a long way off.

9.3.4 Transition to a Circular Economy—Prerequisites and Challenges

In Europe, only 19.5% of all plastics currently come from circular raw materials. The European plastics producers represented by PlasticsEurope have set themselves a target of 65% by 2050. By then, 65% of fossil resources are to be replaced by recyclable raw materials from recycled materials, biomass and CO_2 from Carbon Capture and Utilisation (CCU) (PlasticsEurope 2023).

However, energy recovery is still the most common way of utilising plastic waste throughout the European Union. Recycling is a distant second and around 25% of all plastic waste is still disposed of in landfill. Some of the plastic waste collected for recycling is still exported to countries outside the EU for treatment. The main reasons for this are the lower costs of treating and landfilling the sorted waste and the lack of domestic recycling facilities and technologies. Total waste exports from the EU to non-EU countries amounted to 32.7 Mt in 2020. The majority of this waste consisted of ferrous and non-ferrous metal scrap as well as paper, textiles and glass. The amount of plastic waste at that time was around 1.6 Mt and had fallen to around 1 Mt by 2022 (cf. Chap. 5.5).

Meanwhile, the EU is stepping up its efforts to find circular and climate-friendly ways to manage its plastic waste. The common goal is to ensure that all plastics already in the technosphere are fully and permanently recycled in the future (EU Monitor 2023; European Parliament 2024). According to the current draft of the EU Packaging Regulation, plastic packaging should contain up to 35% post-consumer recyclate by 2030 and 50–65% by 2040, depending on the type of packaging. In the future, companies will have to indicate the proportion of recycled material (cf. Chap. 22) in the form of a declaration of conformity in the technical information on the packaging. Companies whose packaging falls below such a mandatory threshold will have to pay penalties into a fund, the so-called eco-modulation levy.

In addition to the measures mentioned under 9.3.2, binding recycling quotas and an extension of extended producer responsibility are now being introduced. The private recycling industry must then invest in modern facilities for collecting, sorting and recycling plastic waste, while producers, manufacturers and recyclers must work hand in hand on the further development of products: designed for circularity and recycling especially and containing successively increasing proportions of recyclate. This is the only way to develop and produce climate-neutral and globally competitive products in the future.

9.3.5 Plastics Recycling—Processes and Challenges

The main problems that still need to be solved for plastics recycling are the quality and price of recyclates compared to the fossil-based virgin materials used today. To promote the use of recycled material, plastics processors need a continuous supply of large quantities of recycled plastic produced to strictly controlled specifications and at a competitive price. Plastics' great advantage of being easily customisable to the functional or aesthetic needs of a product or its respective manufacturer leads to heterogeneous waste streams, which complicates the recycling process, makes it more expensive and can affect the quality of the final product. In 2018, recycled plastics only accounted for ar-ound 6% of the demand for plastics in Europe. 6 years later, demand is still volatile and unsatisfactory.

A fundamental approach to dealing with the heterogeneity of waste streams is to use different recycling technologies to be able to recycle different input materials with different qualities and yields. For this reason, we will briefly discuss various recycling processes for plastics that are currently in use or under development.

The term 'recycling' refers to the process of reprocessing materials into a new product. There are two fundamentally different processes for recycling plastics:

Mechanical Recycling

Here, plastic waste is processed into recycled secondary raw materials while preserving the polymer chain. Post-industrial waste (PIR) and post-consumer waste (PCR) are separated by type, mechanically shredded, processed and fed back into the production cycle. The quality of the recycled raw materials depends heavily on the precise sorting and cleanliness of the waste stream. Optimised and new sorting technologies are being developed to improve the sorting process and the quality of the source material. Mechanical recycling has been the established recycling process for plastics for many years and is used in most of all plastics recycling plants (Mandelatz et al. 2024).

Chemical Recycling (Depolymerisation, Pyrolysis or Gasification)

Various chemical recycling processes are used to break down the macromolecules of plastics into basic building blocks from which, ideally, new plastics, chemicals or fuels can be produced. There is an ongoing debate between industry experts and EU policymakers about the merits of one process or another. At the centre of this debate is the question of how the proportion of recyclate in products manufactured using pyrolysis oil from chemical recycling, for example, can be measured at all. Various approaches to mass balancing—as described in Chap. 16—are being discussed.

To date, however, chemical recycling is more of a promise than a real prospect, which is why the recycling industry is continuing to focus intensively on mechanical recycling, while remaining committed to technological openness. This is in the knowledge that chemical recycling can be an important addition to the recycling of plastic waste such as PUR-based foams (mattresses, seat cushions, etc.), which cannot be mechanically recycled (cf. Chap. 11.6).

9.3.6 Artificial Intelligence in Waste Management

The use of artificial intelligence (AI) in waste management has the potential to revolutionise collection and recycling processes. AI-supported systems and facilities automate and optimise the sorting of materials without manual effort, thus increasing actual recycling rates.

Waste sorting, one of the biggest challenges in the recycling industry, is being catapulted to the next technological level using AI, in particular machine vision. These intelligent systems recognise and classify different materials such as plastic, paper or metal with a previously unattainable level of sorting depth. Artificial intelligence and deep learning algorithms significantly increase recycling rates across almost all material streams.

A recent study carried out by Eunomia (2024), which focused on the perspective of advanced sorting technologies (object recognition by AI (OR), use of chemical markers and digital watermarks) for sorting different plastic packaging waste, showed that OR has many advantages in terms of efficiency, cost-effectiveness, and speed of implementation.

In Germany, the innovative use of AI technologies in waste management is about to fundamentally change the way plastic packaging is sorted. For example, REMONDIS Recycling GmbH and RE Plano GmbH have launched a plant in Bochum, North Rhine-Westphalia, that sorts and recycles plastic packaging using the latest AI technology (Remondis 2023). The technology enables a more precise separation of materials, which has so far been challenging due to mixing and contamination.

AI systems also offer enormous potential for optimising logistics and energy consumption in the entire waste management system. Intelligent route planning for collection vehicles and service providers can reduce not only fuel consumption but also CO_2 emissions. This more efficient design of waste logistics also leads to a higher collection rate and supports the goal of an energy-efficient circular economy. Furthermore, intelligent control of the facilities enables a more precise adjustment of the energy demand, which further reduces environmental pollution and increases the economic efficiency of the recycling processes. The integration of AI into waste management is thus a crucial step in increasing resource efficiency and promoting a more sustainable use of our planetary resources.

9.3.7 *Outlook—Motivation and Further Development of Recycling Technology*

The further development of existing recycling technologies as well as the search for new processes involves a complex field of tension between several interdependent factors that need to be optimised together. The four most important are yield, quality, availability and sustainability.

Yield

- Collection, sorting and recycling should result in the highest possible yield so that most of the plastic waste can be converted into valuable secondary raw materials. The suitability of plastics and the products made from them for recycling (see other chapters in Part II) depends on the ability and willingness of product owners, who determine the end of life of their products and some waste properties, to provide the waste as cleanly and separately as possible, as well as optimised collection or return logistics to the sorting or recycling facilities;
- Sorting technologies and facilities that enable safe, high-quality sorting with high throughput and minimised rejects;
- Recycling plants and operations that produce high-quality secondary raw materials or compounds with low rejects from the sorted materials (regrind, flakes, etc.—cf. Chap. 10.1) for reuse in the plastics processing industry.

Only if it is possible to achieve a yield of 80% in each of the three sub-processes mentioned will this result in an overall yield of appr. 50% and thus enable the proportion of recyclate intended by the legislator for several markets.

Quality

The quality of the recyclate is crucial to ensure that the recycled materials fulfil the requirements for use in downstream processes and new products. Lower quality will affect the performance, durability or safety of the end products. However, striving for higher recyclate quality may require additional or more complex processing steps, leading to higher costs, energy and resource consumption. Product design and waste collection have a significant influence on the resulting recyclate quality, so strict eco-design regulations will also be of great importance from this perspective in the future.

Availability

Certified recyclates—similar to today's virgin material—must be reliably available on the market with consistent quality for the applications for which they are listed, possibly over a longer period of time, in order to persuade a producer to purchase them. The generation of waste mass flows that are as uniform and large as possible is an indispensable prerequisite for this.

Sustainability

Recycling technologies must be sustainable to minimise their ecological footprint. This includes the use of energy, water and auxiliary materials as well as labour during collection, sorting and recycling. If an advanced recycling technology is particularly energy or water intensive, this can reduce the positive impact of recycling on the environment. It is therefore important to find an optimum solution to minimise the environmental impact.c

The selection of suitable processes and technologies requires careful consideration of these four factors. Overcoming the resulting challenges requires a new type of co-operation between the parties involved in the cycle at all stages, as shown in the cycle diagram at the beginning of Part II and in Fig. 9.1. Without mutual knowledge and understanding of the needs of all those involved in the cycle, there can be no qualitatively new, climate-friendly and world-leading products from Europe.

References

ACEA Africa 2023: Circular Africa. Acea Africa. https://www.aceaafrica.org/circular-africa-1 (Last access: Feb. 11, 2023).

Allied Market Research 2023: Plastic Recycling Market Size, Share | Industry Forecast 2031. Allied Market Research. https://www.alliedmarketresearch.com/plastic-recycling-market-A08207 (Last access: Feb. 11, 2023).

Basel 1989: https://www.basel.int/theconvention/overview/tabid/1271/default.aspx

BEHG 2019: Fuel Emissions Trading Act (in German: Brennstoffemissionshandelsgesetz). https://www.gesetze-im-internet.de/behg/BJNR272280019.html

Bilitewski et al. 2018: Bilitewski, B.; Wagner, J.; Reichenbach, J. (2018): Bewaehrte Verfahren zur kommunalen Abfallbewirtschaftung Informationssammlung ueber Ansaetze zur nachhaltigen Gestaltung der kommunalen Abfallbewirtschaftung und dafuer geeignete Technologien und Ausruestungen. https://www.umweltbundesamt.de/publikationen/bewaehrte-verfahren-zur-kommunalen-0 (Last access: Dec. 04, 2023).

Brunn, M. 2023: Recycled PET for India's first 100% rPET beverage bottle. RECYCLING magazine. URL https://www.recycling-magazine.com/2023/08/01/recycled-pet-for-indias-first-100-rpet-beverage-bottle/ (Last access: Feb. 11, 2023).

BMUV 2020: Zusammensetzung des Hausmülls in Deutschland - BMU-Infografik. bmu.de. URL https://www.bmu.de/MD1349 (Last access: Oct. 11, 2021).

BVSE 2022: Frankreich will Rückstände im Recycling rasch aufholen. https://www.bvse.de. URL https://www.bvse.de/recycling/recycling-nachrichten/8875-frankreich-will-rueckstaende-im-recycling-rasch-aufholen.html (Last access: Jan. 11, 2023).

BVL 2023a: Bundesvereinigung Logistik e.V.: Beschaffungslogistik: Ein wichtiger Baustein zum Erfolg. https://www.bvl.de/service/zahlen-daten-fakten/logistikbereiche/beschaffungslogistik (Last access: Nov. 21, 2023).

ChemAnalyst 2023: Asia Pacific Plastic Recycling Market Analysis. ChemAnalyst. https://www.chemanalyst.com/industry-report/apac-plastic-recycling-market-545#:~:text=Asia%20Pacific's%20Plastic%20Recycling%20market,(PS)%2C%20and%20Others (Last access: Feb. 11, 2023).

Circular Economy Act 2024: in German: Kreislaufwirtschaftsgesetz. https://www.bmuv.de/en/law/circular-economy-and-safeguard-the-environmentally-compatible-management-of-waste.

Delfmann et. al. 2010: Delfmann, W., Dangelmaier, W., Gunthner, W., Klaus, P., Overmeyer, L., Rothengatter, W., Weber, J., Zentes, J. and the Working Group of the Scientific Advisory Board of German Logistics Association (BVL): Towards a science of logistics: cornerstones of a framework of understanding of logistics as an academic discipline. Logist. Res. 2, (2010), pp. 57–63, https://doi.org/10.1007/s12159-010-0034-5.

Destatis 2022: Statistisches Bundesamt (Hrsg.): 2021 wurde 25 % weniger Plastikmuell exportiert als im Vorjahr. Pressemitteilung Nr. N 035 vom 7. Juni 2022. https://www.destatis.de/DE/Presse/Pressemitteilungen/2022/06/PD22_N035_51.html (Last access: Nov. 21, 2023).

Destatis 2022: Beschaeftigte, Umsatz, Produktionswert und Wertschoepfung der Unternehmen in der Energie- und Wasserversorgung. Deutschland. Jahre. Wirtschaftszweige (Code 43221–0001), gefiltert fuer die Wirtschaftszweige WZ 08–381, WZ 08–382, WZ 08–383, WZ 08–390. https://www-genesis.destatis.de/genesis//online?operation=table&code=43221-0001&bypass=true&levelindex=0&levelid=1643042853619#abreadcrumb.

DIN 30781–1:1989–05: Deutsches Institut fuer Normung (DIN): DIN 30781 Teil 1: Transportkette – Grundbegriffe. Ausgabe Mai 1989.

EL Green Mall 2023: Wie man in den Niederlanden Müll sortiert und recycelt. EL Green Mall. https://elgreenmall.com/de/blogs/sustainable-and-ethical-products-in-europe/how-to-sort-trash-and-recycle-in-the-netherlands (Last access: Jan. 11, 2023).

EU 2019: Richtlinie (EU) 2019/904 des Europäischen Parlaments und des Rates vom 5. Juni 2019 über die Verringerung der Auswirkungen bestimmter Kunststoffprodukte auf die Umwelt (Text von Bedeutung für den EWR) OJ L.

EU 2021: Proposal for a Regulation of the European Parliament and of the Council on shipments of waste and amending Regulations (EU) No 1257/2013 and (EU) No 2020/1056.

EU 2022: Verordnung (EU) 2022/1616 der Kommission vom 15. September 2022 über Materialien und Gegenstände aus recyceltem Kunststoff, die dazu bestimmt sind, mit Lebensmitteln in Berührung zu kommen, und zur Aufhebung der Verordnung (EG) Nr. 282/2008 (Text von Bedeutung für den EWR), OJ L.

EU Monitor 2023: Plastic waste and recycling in the EU: facts and figures. https://www.eumonitor.eu/9353000/1/j9vvik7m1c3gyxp/vkufaw0m3zy4?ctx=vjxzjv7ta8z1.

Eunomia 2024: Advanced Sorting for Circularity Report 2024. https://eunomia.eco/reports/advanced-sorting-for-circularity-report/.

EUP 2023a: Europaeisches Parlament. Plastikmuell und Recycling in der EU: Zahlen und Fakten. https://www.europarl.europa.eu/news/de/headlines/society/20181212STO21610/plastikmull-und-recycling-in-der-eu-zahlen-und-fakten (Last access: Dec. 04, 2023).

European Parliament 2024: How the EU wants to achieve a circular economy by 2050. https://www.europarl.europa.eu/news/en/headlines/society/20210128STO96607/how-the-eu-wants-to-achieve-a-circular-economy-by-2050.

Fennemann et al. 2018: Fennemann, V., Hohaus, C., Kopka, J., Fraunhofer IML (Hrsg.) 2018: White Paper. Circular Economy Logistics. Fuer eine Kreislaufwirtschaft 4.0. https://www.iml.fraunhofer.de/content/dam/iml/de/documents/101/07_Whitepaper_CE_REPRO_Auflage_02_WEB.pdf (Last access: Jan. 31, 2022).

Gabriel Chemie 2023: Gabriel-Chemie Masterbatch. https://old.gabriel-chemie.com/TagTec (Last access: Feb. 10, 2023).

German Packaging Act 2019: in German: Verpackungsgesetz. https://www.bmuv.de/gesetz/gesetz-ueber-das-inverkehrbringen-die-ruecknahme-und-die-hochwertige-verwertung-von-verpackungen.

Geyer et al. 2017: Geyer, R., Jambeck, J. R. , Law, K. L., Production, use, and fate of all plastics ever made. Sci. Adv. 3, e1700782 (2017) (Last access: Nov. 18, 2023).

GewAbfV 2017: Gewerbeabfallverordnung (GewAbfV), Verordnung ueber die Bewirtschaftung von gewerblichen Siedlungsabfaellen und von bestimmten Bau- und Abbruchabfaellen in der Fassung vom 18. April 2017, Bundesgesetzblatt Teil I Nr. 21 vom 21.04.2017, pp. 896 ff.

Goepfert 2019: Goepfert, I., Logistik der Zukunft – Logistics for the future. Springer Gabler Wiesbaden. 2019, pp. 167 ff. https://doi.org/10.1007/978-3658-23805-6.

GTAI 2023: Ungarns Abfallbranche vor großem Umbruch | Branche kompakt | Ungarn | Abfallwirtschaft. https://www.gtai.de/de/trade/ungarn/branchen/ungarns-abfallbranche-vor-grossem-umbruch-808496 (Last access: Jan. 11, 2023).

Hamamatsu 2023: NIR and SWIR Questions and Answers | Hamamatsu Photonics [WWW Document]. URL https://hub.hamamatsu.com/us/en/ask-an-engineer/imaging/nir-and-swir-questions-and-answers.html (Last access: Sep. 12, 2023).

Hesse and Clausen 2019: Hesse, K., Clausen, U. Entsorgung und Kreislaufwirtschaft. In: Furmans, K., Kilger, C. (Ed.): Betrieb von Logistiksystemen. Berlin, Heidelberg: Springer (Fachwissen Logistik), pp. 239–303. https://doi.org/10.1007/978-3-662-57943-5_8.

Holy Grail 2.0 2023: Pioneering Digital Watermarks | Holy Grail 2.0. https://www.digitalwatermarks.eu (Last access: Feb. 10, 2023).

Kirchhoff 2022: Kirchhoff, J. F., Intelligente Konzepte für Sammelfahrzeuge. In: Kurth P, Oexle A, Faulstich M, Hrsg. Praxishandbuch der Kreislauf- und Rohstoffwirtschaft. Wiesbaden: Springer Fachmedien Wiesbaden, pp. 717–738.

Klaus and Mueller 2012: Klaus, P., Mueller, S., Towards a Science of Logistics: Milestones along Converging Paths in the Roots of Logistics (pp. 3–26).

Krupp et al. 2013: Krupp, T., Suntrop, C., Kille, C., Veres-Homm, U., Heeg, L., Chemielogistik – Bedeutung, Struktur, Dynamik. Studie der Bundesvereinigung Logistik, Hamburg: Deutscher Verkehrsverlag 2013.

Krupp et al. 2017: Krupp, T., Suntrop, C., Veres-Homm, U., Differentiated Logistics Services – Two Powerful Tools to Increase Competitive Advantages in the Chemical Industry, Research Paper DHL, available online: https://www.dhl.com/content/dam/dhl/global/dhl-global-forwarding/documents/pdf/glo-dgf-dhl-chemical-whitepaper-differentiated-logistics-22-june-2017.pdf.

KrWG 2020: Kreislaufwirtschaftsgesetz - Gesetz zur Foerderung der Kreislaufwirtschaft und Sicherung der umweltvertraeglichen Bewirtschaftung von Abfaellen (KrWG) vom 24. Februar 2012, BGBl. Nr. 10 vom 29.02.2012, S. 212 ff. unter Einbeziehung von dem Gesetz zur Umsetzung der Abfallrahmenrichtlinie der Europaeischen Union vom 23. Oktober 2020, BGBl. I Nr. 48 vom 28.10.2020, S. 2232 ff.

Kunststoffe 2022: Pigmente statt digitaler Wasserzeichen für die Abfallsortierung. Kunststoffe: Nachrichten, Stellenmarkt, Produkte für die Kunststoffindustrie | Kunststoffe.de. https://www.kunststoffe.de/a/fachartikel/leuchtende-marker-statt-heiligem-gral-2856290 (Last access: Apr. 01, 2024).

Labels and Labeling 2022: Recycling in Africa | Labels & Labeling. http://www.labelsandlabeling.com/features/sustainability/recycling-africa (Last access: Feb. 11, 2023).

Mandelatz et al. 2024: Mandelatz, R., Rabanizada, N., Harting-Pfeifer., M., Correct collection and sorting can turn waste into a recyclable material (in German: Korrekte Sammlung und Sortierung können aus Abfall einen Wertstoff machen), in Granzow, M. (ed.) Circular Excellence, Berlin 2024.

McKinsey 2024: McKinsey & Company, June 2024: Decarbonizing logistics. Charting the path ahead.https://www.mckinsey.com/capabilities/operations/our-insights/decarbonizing-logistics-charting-the-path-ahead.

Meskers et al. 2023: Meskers, C., Worrell, E., Reuter, M.A.: Handbook of Recycling - 2nd Edition. Elsevier 2023.

Mülltrennung wirkt 2021: Mülltrennung wirkt! https://www.muelltrennung-wirkt.de/fragen-und-antworten/was-passiert-mit-dem-gelben-sack/ (Last access: Oct. 23, 2021).

Niessner 2022: Niessner, N., Recycling of Plastics. Hanser Publications 2022.

Packaging Europe 2023: ALPLA to open PET bottle recycling facility in South Africa. Packaging Europe. https://packagingeurope.com/news/alpla-to-open-pet-bottle-recycling-facility-in-south-africa/9928.article (Last access: Feb. 11, 2023).

Pareto Securities 2023: Future of plastics recycling in Europe – New study by Pareto (fee-paying). Securities highlights opportunities and challenges. https://www.eqs-news.com/de/news/corporate/pareto-securities-as-future-of-plastics-recycling-in-europe-new-study-by-pareto-securities-highlights-opportunities-and-challenges/1899411 (Last access: Nov. 1, 2024).

Pfohl 2018: Pfohl, H. C., Logistiksysteme. Betriebswirtschaftliche Grundlagen. 9. Auflage. Darmstadt. Springer Vieweg 2018.

Pfohl 2023: Pfohl, H. C., Logistics Management. Conception and Functions. 1. Auflage. Darmstadt. Springer Vieweg 2023.

PlasticsEurope 2023: Changing Plastics for Good—The Plastics Transition Roadmap. https://plasticseurope.org/changingplasticsforgood/the-plastics-transition/.

PlasticsEurope 2024: The Circular Economy for Plastics—A European Analysis 2024. https://plasticseurope.org/knowledge-hub/the-circular-economy-for-plastics-a-european-analysis-2024/.

PPWR 2024: Compromise Paper between Council and EU-Parliament. https://www.europarl.europa.eu/meetdocs/2014_2019/plmrep/COMMITTEES/ITRE/DV/2023/07-19/12-CA-PPWR_EN.pdf (Last access: Nov. 03, 2024).

Radio Prague International 2009: Gelb, blau, grün – die Mülltrennung. Radio Prague International. https://deutsch.radio.cz/gelb-blau-gruen-die-muelltrennung-8586853 (Last access: Jan. 11, 2023).

R-Cycle 2024: Unsere Lösungen für Abfallsortierung & Recycler. R-Cycle. https://www.r-cycle.org/ (Last access: May 01, 2024).

Remondis 2018: Neue Wege bei Recycling und Logistik. https://www.remondis-aktuell.de/012018/recycling/neue-wege-bei-recycling-und-logistik/ (Last access: Jan. 24, 2022).

Remondis 2023: Remondis press release, New standards in the sorting of plastics packaging. https://en.remondis-aktuell.de/recycling/new-standards-in-the-sorting-of-plastic-packaging (Last access: Nov. 01, 2024).

RWTH Aachen 2024: RWTH Aachen University. Sensor-Based Sorting & Control. https://www.sbsc.rwth-aachen.de/go/id/sfzp/ (Last access: Mar. 1, 2024).

Sensorinstruments 2023: https://www.sensorinstruments.de/pdf/TAGTEC%20-%20Die%20ID%20fu%CC%88r%20die%20Kunststoffindustrie%20v01-2023.pdf.

Statista 2024: Topic: Recycling in Europe [WWW Document]. Statista. URL https://www.statista.com/topics/9617/recycling-in-europe/ (Last access: Jan. 11, 2023).

Sustainable Plastics 2023: Plastics Recycling Show Middle East and Africa kicks off in Dubai. Sustainable Plastics. https://www.sustainableplastics.com/news/plastics-recycling-show-middle-east-and-africa-kicks-dubai (Last access: Jan. 11, 2023).

UBA 2022f: Umweltbundesamt. Abfallrecht. https://www.umweltbundesamt.de/themen/abfall-ressourcen/abfallwirtschaft/abfallrecht.(Last (Last access: Dec. 4, 2023).

VDI 2021: Verein Deutscher Ingenieure: Circular Economy fuer Kunststoffe neu denken. Wie die Transformation zur zirkulaeren Wertschoepfung gelingen kann. https://www.vdi.de/fileadmin/pages/mein_vdi/redakteure/publikationen/VDI-Green-Paper-Circular-Economy-fuer-Kunststoffe-neu-denken.pdf (Last access: Jan. 24, 2022).

Wannenwetsch 2021: Wannenwetsch, H., Integrierte Materialwirtschaft, Logistik, Beschaffung und Produktion, Supply Chain im Zeitalter der Digitalisierung. 6. ueberarbeitete Auflage. Springer 2021. https://doi.org/10.1007/978-3-662-61095-4.

WFD 2008: Directive 2008/98/EC of the European Parliament and of the Council of 19 November 2008 on waste and repealing certain Directives (Waste Framework Directive). https://eur-lex.europa.eu/legal-content/EN/TXT/?uri=celex%3A32008L0098.

ZSVR 2022: Zentra.le Stelle Verpackungsregister (ed.) Vorlaeufig zuzuordnende Marktanteile der Systeme fuer das erste Quartal 2022. https://www.verpackungsregister.org/fileadmin/files/Marktanteile/vorlaeufige_Marktanteile_Q1_2022.pdf (Last access: Jan. 24, 2022).

Prof. Dr. Thomas Krupp is Professor of Transport and Traffic Logistics at the Cologne University of Applied Sciences. Since 2016, he has been the head of the master's programme (M.Sc.) Supply Chain and Operations Management. He has many years of experience in teaching, consulting and applied research from his positions as Professor of Logistics Management at the European University of Applied Sciences in Brühl, as a management consultant in the Competence Center Transportation at Horváth and Partners and as a research assistant at Fraunhofer SCS in Nuremberg.

Prof. Dr. rer. nat. Kathrin Hesse is Professor of Waste Management, Logistics and Environmental Management Systems at the Faculty of Vehicle Systems and Production at Cologne University of Applied Sciences. In research and teaching, she deals with topics from the fields of waste management logistics, waste disposal technology, environmental management systems, in particular green logistics, as well as sustainability in logistics, production and chemical logistics.

Jieying Luu is research assistant for logistics and supply chain management at Cologne University of Technology. She studied Business Administration focussing on logistics at the Bonn-Rhein-Sieg University of Applied Sciences and at Hunan University in Changsha (China) and completed M.Sc. in Supply Chain and Operations Management in Cologne. At Thyssen Krupp Rasselstein, she wrote her bachelor's and master's theses which dealt with issues of logistics in the process industry.

Andreas Hanus holds a degree in electrical engineering and information technology from the Technical University of Munich and a master's degree in digital business engineering. He started his professional career in 2016 as a development engineer for sensor-based sorting machines at Sesotec GmbH. Since 2024, he has been working as a Team Leader for the development of artificial intelligence for the company's various business units.

Michael Perl After studying business administration, Michael Perl has been working for Sesotec GmbH in various positions since 2007. Initially working as Regional Sales Manager in various regions, he has been responsible for the global sales activities of the Sorting Recycling Division since 2013. The focus of the product and application portfolio is on plastics, glass, electrical and metal scrap recycling.

Michael J. Schneider did an apprenticeship as typesetter before studying Communications, English literature, film and TV-science at the Ruhr University Bochum. He then joined EASY Software AG and later on, RWE Solutions AG (later SAG) where he acted as Manager Press and Public Relations. In 2007, he joined REMONDIS where he now is active as Manager of Press and Public Relations, REMONDIS Sustainable Services GmbH and Co. KG.

Herwart Wilms has a degree in economics and social sciences, and German studies from Ruhr University Bochum. He joined RWE Umwelt AG where he became Head of Sales and Dual Systems. After the takeover by Remondis, he later became Managing Director, REMONDIS Sustainable Services GmbH & Co. KG, and then Managing Director of Remondis International Service GmbH. He acts as VP of BDE - Federation of the German Waste, Water and Raw Materials Management Industry, as VP of FEAD – The European Federation of Waste Management and Recycling Industries, and as Chairman of the Raw Materials Committee of BDI—Federation of German Industries.

Chapter 10
Raw Materials—Recycler, Chemical Industry, and Plastic Manufacturer

Herbert Snell, Rudolf Pfaendner, Michael Carus, Lars Krause, Peter Orth, and Ulrich Liman

Contents

H. Snell (✉)
Snell GmbH, Tittling, Germany
e-mail: herbert@snellweb.de

R. Pfaendner (✉)
Fraunhofer Institute LBF, Darmstadt, Germany
e-mail: rudolf.pfaendner@lbf.fraunhofer.de

M. Carus (✉) · L. Krause
nova-Institut GmbH, Hürth, Germany
e-mail: michael.carus@nova-institut.de

L. Krause
e-mail: lars.krause@nova-institut.de

P. Orth
OPC - Orth Plastics Consulting, Cologne, Germany
e-mail: opc.cologne@t-online.de

U. Liman (✉)
ULC - Ulrich Liman Consulting, Langenfeld, Germany
e-mail: druliman2@gmail.com

P. Orth et al. (eds.), *Circular Economy of Plastics*,
https://doi.org/10.1007/978-3-031-78829-1_10

Summary

This Chapter deals with the recyclate production based on technologies currently in use and under development—as presented in the raw materials segment of the circuit diagram (cf. Fig. 10.1). Depending on the technology used, recyclates are produced as a secondary raw material for further processing by various players such as mechanical recyclers, the chemical and plastics manufacturing industries.

The fundamental technological principles of mechanical recycling of plastic waste are shown in Sect. 10.1. The various pre-treatment processes are covered

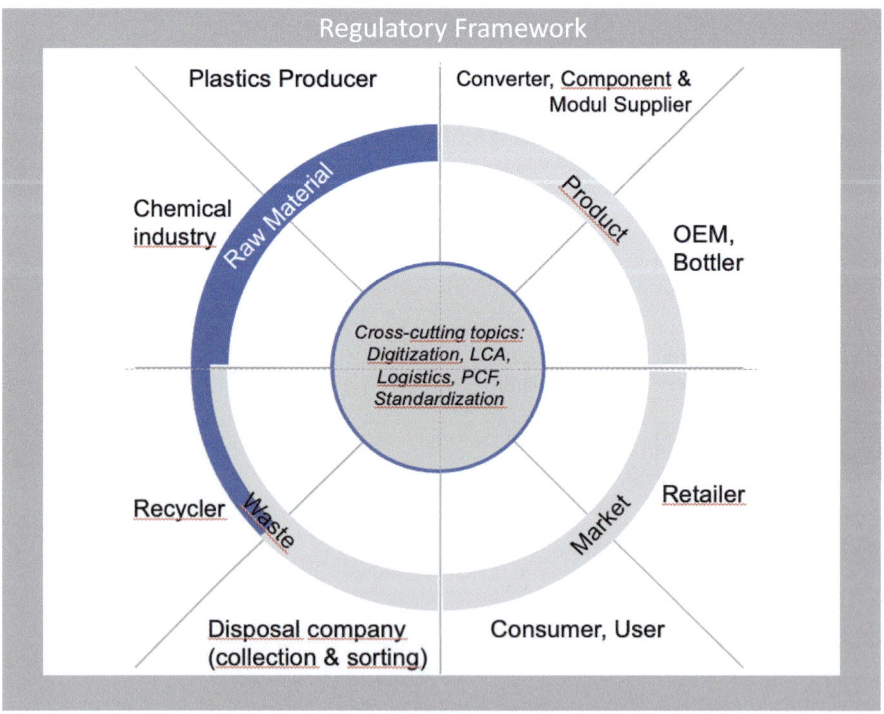

Fig. 10.1 Schematic diagram of the plastics cycle (*Source* Own presentation)

in detail, as well as extrusion, melt filtration, odor removal and process water treatment. Beyond the technological description, the outlook draws conclusions on the further development of mechanical recycling in a changing economic and regulatory environment.

Section 10.2 gives an overview of the quality improvement methods of recycled plastics to obtain recyclates for closed-loop and high-value applications. Key steps for high-grade materials in the thermal transformation process from regrind to recyclate compound are smooth processing, degassing steps to remove volatiles, and the addition of stabilizers to avoid polymer degradation and maintain the properties.

Additives such as tailor-made products for restabilization, compatibilizers, chain-extenders or odor-reducing agents are described in terms of their functionality and significance for quality improvement of recyclates.

Section 10.3 provides an up-to-date overview of the development status of various chemical recycling processes, including the overlap with physical recycling processes. Based on the current input and output capacities of chemical and physical recycling plants in Europe, estimates of future capacity development in Europe and the world are made based on investment announcements.

In Sect. 10.4, the reader will find a case study of how a global chemical company is pursuing its own strategy to actively shape and participate in the circular economy to become a future supplier of circular feedstocks. Time, capital investments, and organizational changes are outlined based on published communication. This demonstrates that ambitious timelines are key to establishing a circular economy based on business models.

10.1 Mechanical Recycling of Plastics Waste

Herbert Snell

10.1.1 Introduction

Mechanical recycling is the recovery of plastic waste while preserving the polymer chain of the plastic. This value creation process is important for thermoplastics especially because they can be melted and processed again.

The process has been developing since the beginning of the 1980s (Michaeli et al. 1993) and since then new processes and methods have been constantly developed and new capacities established across Europe. Mechanical recycling gained particular momentum through the establishment and introduction of Expanded Producer Responsibility systems (EPR) (cf. Chap. 11) for plastic packaging waste. The need to recycle these and use them to produce new plastic items promoted global innovation.

Recyclates produced in this way compete economically with primary plastics (virgin materials) produced by the chemical industry. Their price is driven by the price of oil as the raw material, the global demand for plastics, existing capacities, logistics costs and local energy costs. The costs of mechanical recycling, on the other hand, essentially result from the collection of plastic waste, its sorting, and the various stages of the recycling process. Hence, recycled materials have a different price structure; they cannot follow the often- volatile prices of virgin materials and come under pressure when prices for virgin materials are low (cf. Sect. 13.2) (Langen 2023).

The use of recyclates instead of virgin materials contributes considerably to the reduction of CO_2 emissions. Depending on the process, polymer, and location, up to 87% less CO_2 is emitted during recyclate production compared to virgin material (IK 2022; DIW 2021; ALPLA 2023).

The objective of mechanical recycling of plastic waste is to obtain recyclates that are as homogeneous as possible. This requires various process steps that aim at removing foreign substances such as other polymers or other materials from the target material.

Depending on the origin of the plastic waste, a distinction is made between post-industrial and post-consumer waste. Recyclates made from this are referred to as PIR (Post Industrial Recyclate) or PCR (Post Consumer Recyclate) (DIN EN ISO 14021:2016). Instead of post- industrial, the synonymous term pre-consumer is occasionally used.

PIR is mostly production waste that is available without mixing and is largely pure so shredding and a few treatment steps are sufficient to produce a usable recyclate.

Post-consumer waste occurs after a period of use. Due to its function, it is usually dirty and mixed with other polymers and/or substances. It comes from collections of household or commercial and industrial waste. Complex separation and cleaning processes are used to recover them.

Only thermoplastics are suitable for mechanical recycling. Thermosets can be used as fillers after being shredded accordingly. There are similar uses for elastomers, e.g. rubber from old tires. In Chap. 5 the quantity distributions of individual plastic waste streams are discussed in more detail.

The product of the recycling process is a recyclate, a term under which various manifestations are subsumed:

i. Ground material (flake): Grinding in a mill produces different and irregular particles, which are classified by sieving to sizes of 5–20 mm.
ii. Agglomerate: The shredding of films, fibers, or foams results in large-volume materials. Flowable goods are created from them through agglomeration.
iii. Regranulate: A melt process is used to produce granulates of homogeneous particle size that can be used directly in the plastics processing industry.
iv. Regenerate: The regrind is also produced using a melt process, with additives being added to improve properties.

Mechanical recycling is currently the most important process for recycling plastic waste and reintroducing it into the cycle.

The following sections describe various processes that are used in different ways depending on the starting material (plastic waste) (cf. Fig. 10.2).

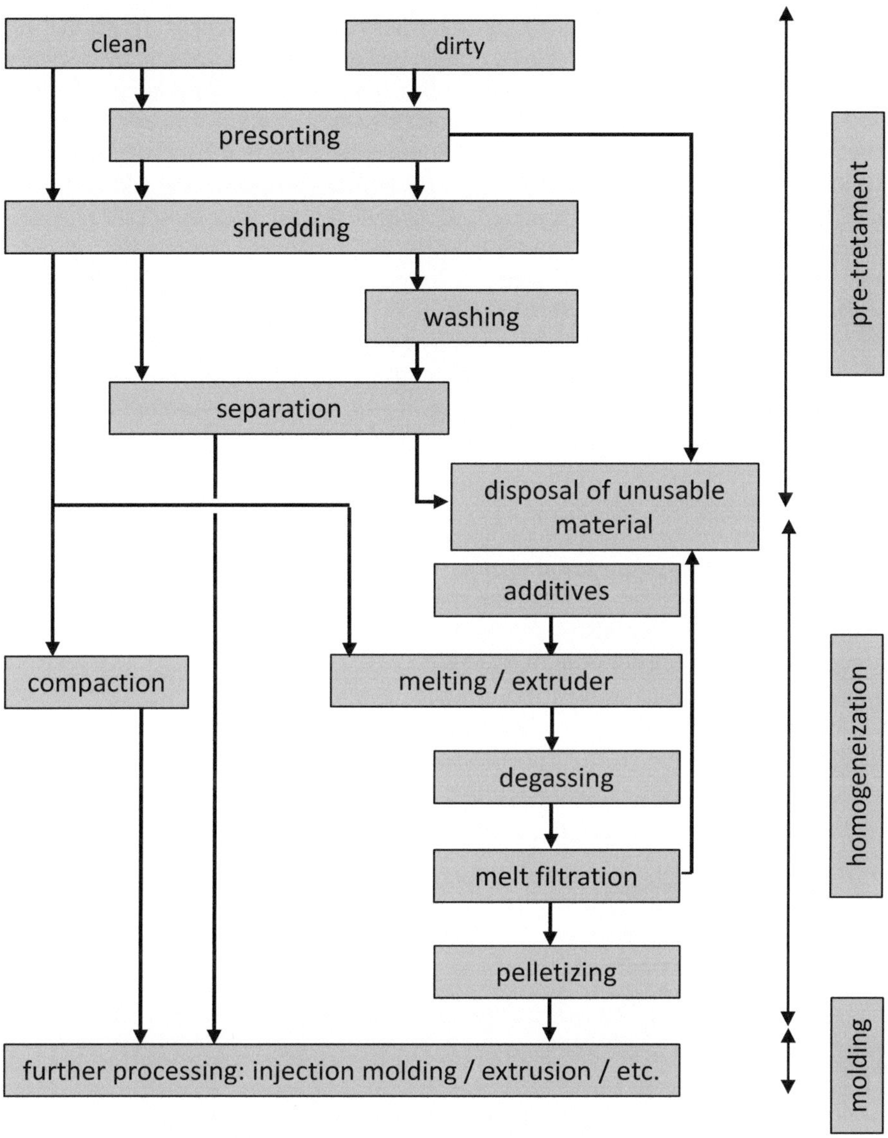

Fig. 10.2 Processing of plastic waste: process diagram (*Source* Own illustration)

10.1.2 Pretreatment

Pre-sorting

On the one hand, pre-sorting takes place as external sorting when separately collected waste such as light packaging (yellow bag/yellow bin) is sorted by the consumer or user. Usually, sorting is carried out by article or item, respectively, so that the material no longer must be laboriously processed at the article level by the recycler. Item sorting is when sorting is carried out based on unshredded packaging or other products. On the other hand, there is a complex sorting of items for certain material flows, such as PET from German deposit bottle collections at the recycler. Different deposit containers such as cans (Fe/NF), PET bottles (mixed colors) and collection bags (LDPE mixture) are usually mixed. To separate this mixture, most PET recyclers in Germany use in-line sorting systems, while in Scandinavia the recycler provides largely homogeneous material flows. To improve the purity of an HDPE bottle/hollow body fraction from household collections, items are sorted in line. In addition to classification units (sieves, flat/round separation, etc.), NIR sensor technology is increasingly being used (cf. Sect. 9.2). As a result of such sorting processes, entire items are available sorted by polymer, color, or size, such as transparent PET bottles, white HDPE bottles or transparent films.

Shredding

A shredding operation to defined particle sizes is used to further process pre-sorted plastic waste. For this purpose, either mills that work according to the cutting principle and are equipped with sieves or cutting roller shredders are used (cf. Fig. 10.3).

Impact shredders are also occasionally used, in which the shredding occurs through the mutual impact of individual particles.

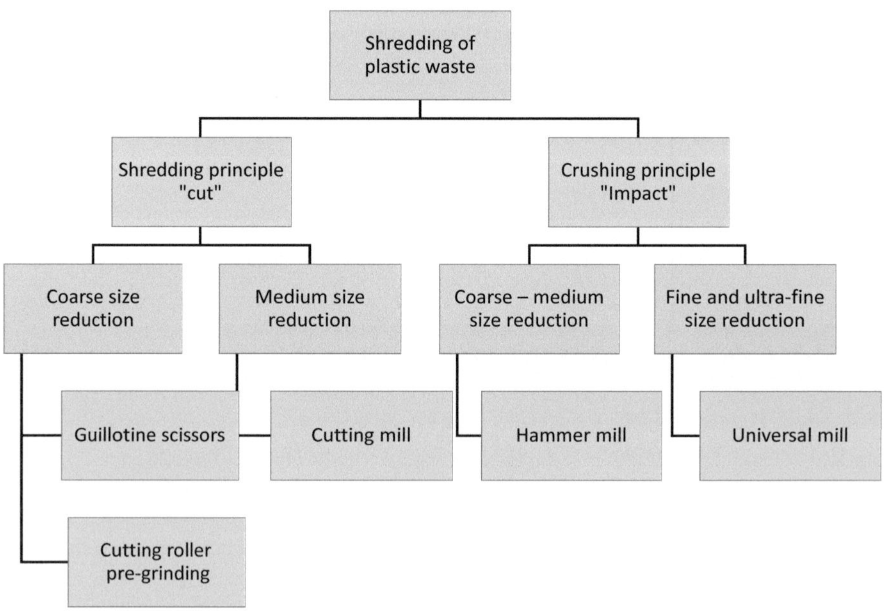

Fig. 10.3 Shredding of plastic waste: process diagram (*Source* Own illustration)

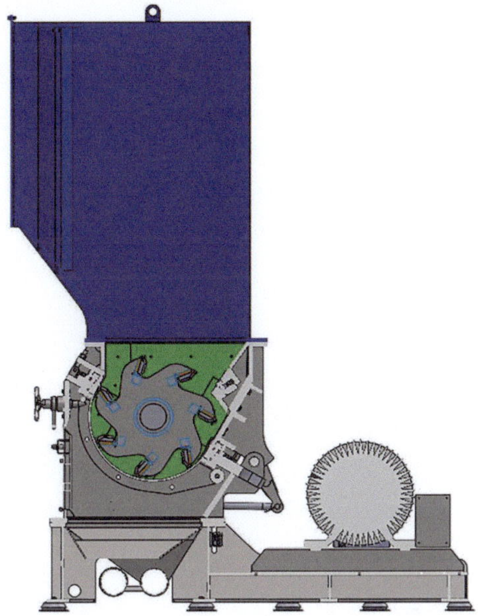

Fig. 10.4 Shredder (*Source* AMIS 2023)

Shredding in a mill (cf. Fig. 10.4) often takes place with the addition of water; it is then a "wet mill". The use of water during shredding has positive effects on cleaning the material, but also serves to cool the process.

Shredders are often used for pre-shredding. Either because the items to be shredded are too large for a mill or to obtain a particle size range that enables subsequent efficient automatic sorting. This is often followed by further comminution in a cutting mill. The common particle size for plastic recycling ranges from 8 to 20 mm.

Washing

The shredded material is freed of adhering contamination in scrubbers, primarily using the friction principle. Trough screws or containers with one or more chambers with stirrers are usually used.

Hot washers (cf. Fig. 10.5) are traditionally used for PET recycling. Derived from the washing of reusable beverage bottles, the hot washing process uses caustic soda and other additives to remove adhesives and labels from the PET bottles. The process temperatures are around 90 °C.

The experience in PET recycling is used for the recycling of HDPE hollow objects such as bottles, canisters, etc. or for PP packaging. This also involves removing adhesives and labels as well as removing grease. Positive effects on the reduction of contamination have been proven, but a significant reduction in odors could not be achieved using hot washers. Hot washing systems for polyolefinic waste are still in intensive development.

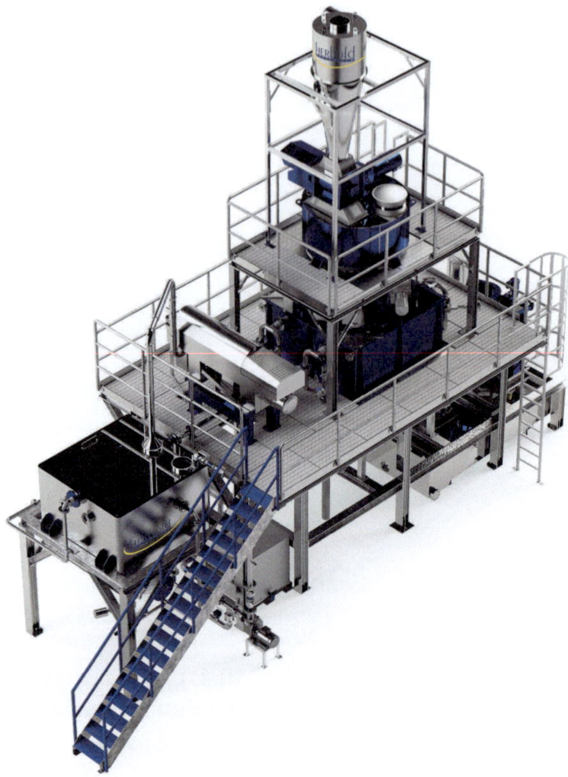

Fig. 10.5 Stand-alone hot washer (*Source* Herbold 2023)

Separation

When separating substances, a distinction is made between different starting mixtures (cf. Fig. 10.6). When it comes to plastic recycling, all separation tasks are relevant:

- Solid/gaseous separation for dedusting the system
- Solid/liquid separation for drying and in the area of water treatment
- Solid/solid separation in different process steps

Following sorting according to material properties is discussed. Classification methods come from classic plastics processing.

As explained at the beginning, when it comes to plastic recycling, it is important to achieve polymer flows that are as homogeneous as possible. Although there is a certain compatibility of polymers so that they can be mixed, such as PE with PP or PS/HIPS with ABS/SAN, this always results in a loss of quality. Another criterion is the molecular weight distribution of the polymer, on which the melt flow depends. Parts manufactured using extrusion or injection molding processes should therefore be separated at the item level.

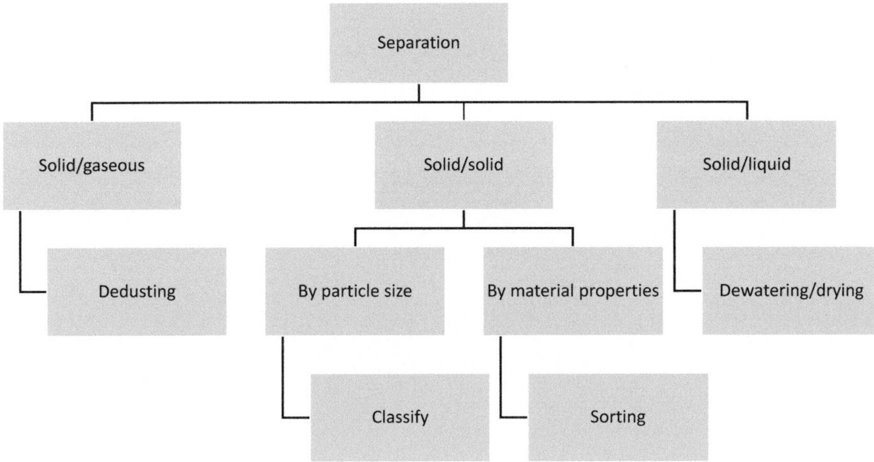

Fig. 10.6 Separation of plastic waste: process diagram (*Source* Own illustration)

Density separation

The commonly used separation process is density separation with water. Here, polymers with a density <1 g/cm^3 float, while all other polymers sink. The use of separation media other than water to achieve other separation cuts has not yet become widespread because the effort involved in adjusting the density of the separation medium is high and can hardly be covered by the sales revenues of the recyclate.

The state of the art are various processes, all of which make use of the difference in density and, in individual cases, multiply it, for example by using centrifuges. It should be noted that fillers often cause a change in the density of the plastic and thus influence the separation result, which is often the case with PP.

Below are some examples of aggregates for density separation: The classic unit is a floating/sinking tank (cf. Fig. 10.7), a basin on the surface of which screws or paddles ensure the propulsion of the floating polymer and a screw or "clearing chain" ensure the discharge of the sinking fraction. The design of the basin depends heavily on the polymers to be separated. When treating PE or PP ($d < 1$ g/cm^3) the focus is on the floating fraction, while with PET ($d > 1$ g/cm^3) the focus is on the sinking fraction. In this case, approximately 90% of the input material is transported away as a sinking fraction.

Hydrocyclones, which originally stem from the processing of mineral materials, have been further developed for plastic recycling since the 1990s and used for sophisticated separation operations (cf. Fig. 10.8). Depending on the design of a hydrocyclone, the separation cut may vary from 1 g/cm^3 when using water.

The so-called flow difference method was developed towards the end of the 1990s. In a container with a cylindrical middle part and conical extensions at the top and bottom, the separation with water also takes place at $d = 1$ g/cm^3. The materials are separated in a quiet zone in the cylindrical part of the container.

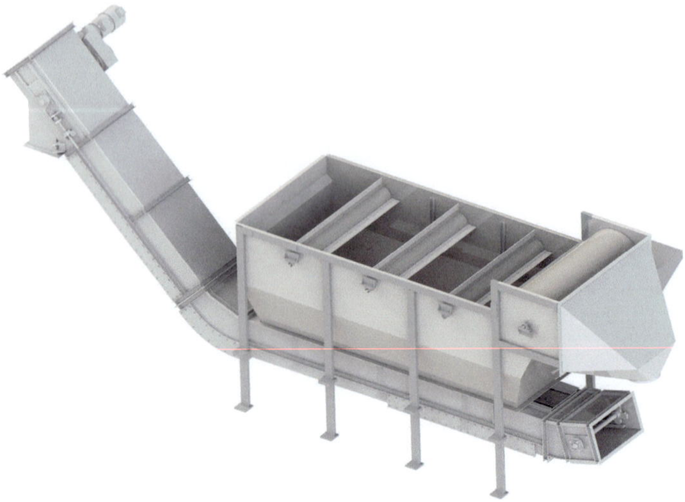

Fig. 10.7 Floating/sinking tank for density separation (*Source* Herbold 2023)

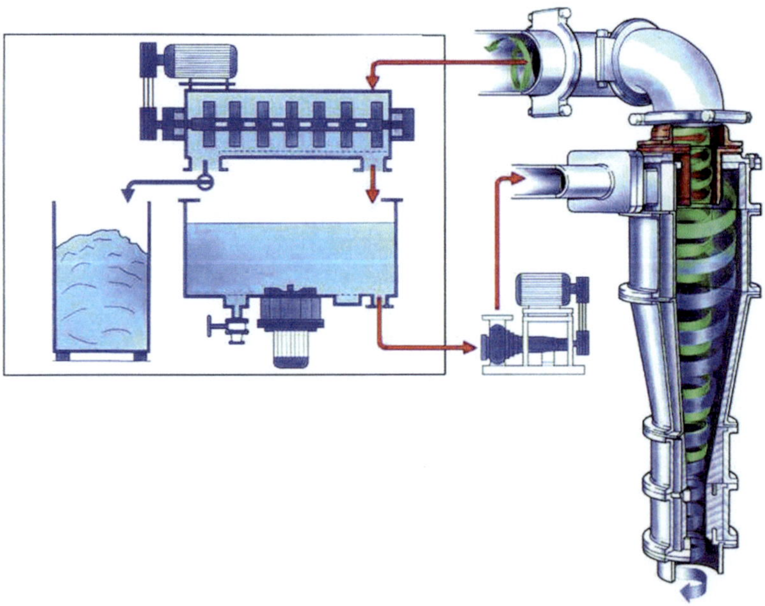

Fig. 10.8 Hydrocyclone process. Green: inner vortex—light fraction; blue: external vortex—heavy fraction (*Source* Herbold 2023)

Fig. 10.9 Separation centrifuge. Red: heavy fraction; yellow: light fraction (*Source* Andritz 2023)

At the same time, dewatering centrifuges (cf. Fig. 10.9) were also further developed into separation units. The resulting centrifugal forces increase the density differences and thus enable the separation of polymer mixtures. The material is also cleaned by friction.

In the first decade of the 2000s, a technology for the density separation of plastics was developed in the Netherlands, which makes it possible to use magnets to adjust the density of a magnetic separation medium so that different polymers can be separated.

Several material streams can be separated at the same time in a separation unit. This means that PP, PE, PET can be separated from foams or PS, PMMA, PC in one step (TU Delft 2023).

Multilayer materials, which often consist of different polymers to achieve certain properties (oxygen barrier, etc.), represent a particular challenge for mechanical recycling.

Other separation processes

In recent years, a delamination technology has been developed to separate certain multilayer systems (cf. Fig. 10.10). Multi-layer composite films are separated using a physical–chemical separation process. The process is suitable, for example, for PE/paper/PE/aluminum/PE composite board (beverage cartons) or for PE/aluminum/PET composite films (coffee packaging). Special aqueous-formulated separating fluids serve to overcome intermolecular interactions and counteract the adhesion forces between the individual film layers (Saperatec 2023).

In addition to solid/liquid separation processes based on differences in density, processes are also used that work with clean ground material with a uniform particle spectrum and a high degree of dryness. The NIR technology commonly used in item sorting is used here, too, while electrostatic processes are also used. This makes use of the different electrostatic charges of individual polymers. The polymers are

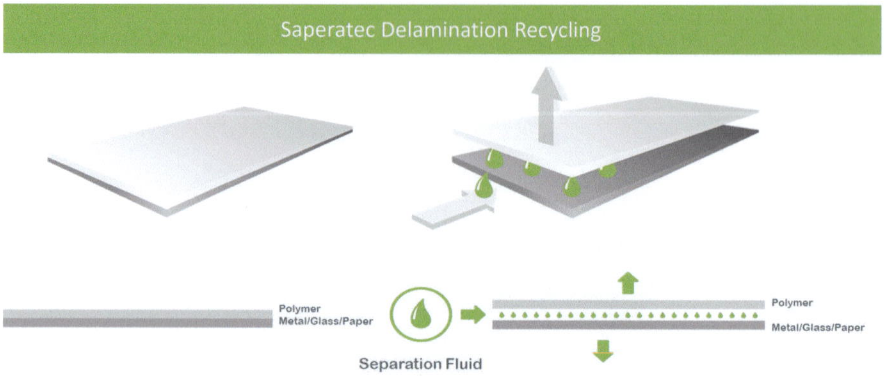

Fig. 10.10 Delamination process (*Source* Saperatec 2023)

charged between differently charged plates, so that the polymers are deflected and sorted depending on their charge (Hamos 2023).

In addition to the classic units for metal separation in plastics processing, overband magnets and eddy current separators are also used. Such units are used both in (pre-)sorting and for refining the regrind. To separate light mixture components such as dust or film, units for wind sifting or gravity sifting are used. These are derived from similar aggregates in agriculture or adopted from classic plastics processing. In a countercurrent, light particles are separated from heavier ones. There are different versions depending on the throughput and separation problem. In addition to countercurrent sifters, crossflow sifters are also on the market. Aerocyclones are also suitable for removing particle dust from the air flow.

Drying

Both mechanical (cf. Fig. 10.11) and thermal dryers are used to dry the ground material after separation and washing processes. Mechanical dryers work on the principle of a centrifuge and are particularly suitable for materials with larger wall thicknesses. Thermal dryers are suitable for materials with thin walls. It is used instead of mechanical drying or in a two-stage process following this. The material is dried in a warm air stream, with warm and possibly hot air being supplied in a variety of ways.

Post-separation and upgrading

Depending on the application, in addition to density separation, further separation steps are required to deplete impurities or other polymers. When recycling PET, the dried ground material is usually further cleaned using a so-called flake sorter (cf. Fig. 9.6). Metals, off-colors or foreign polymers can be separated in this way. These devices identify the particle to be sorted out in a stream of regrind, blow it out at a nozzle bar and thus remove it from the stream. Camera systems in the NIR range (cf. Sect. 9.2) or in the visible range (color recognition) as well as inductive systems for metal detection are used for identification. Some devices are equipped with all the

Fig. 10.11 Mechanical dryer (*Source* Herbold 2023)

above-mentioned detection options to remove various types of contaminants in one step.

Such flake sorters work in defined regrind size ranges. If the particle size distribution is too large, a pre-separation to the optimal distribution must first be carried out using sieves to then be able to sort more efficiently.

To evaluate the efficiency of these processes, a distinction is made between purity and output. Purity describes the proportion of contaminants still remaining in the material flow, while output describes the proportion of contaminants that have been discharged.

Deinking

There are further requirements for the use of recycled plastic in packaging. In addition to a material that is as odorless as possible (cf. Sect. 10.1.4), a light-colored or uncolored recyclate is often required. Recent developments are therefore aimed at using colors that can be removed from the polymer matrix during the recycling process, or printing inks that can be removed from the plastic surface. In both cases, the colors used must be dissolved.

Water-based or solvent-based paints can be removed with a chemical process using water-based reactants and friction. The surface tension is changed so that the paint comes off and collects in the process water (Keycycle 2023). After the deinking process, the material is rinsed with clean water and dried. The deinking process includes the following steps: comminution, decolorization (deinking), rinsing and drying; it precedes the extrusion operation. It is important to treat the water in a closed system to remove the paint components from the washing water and to ensure that paint residues do not remain in the process water. In principle, this process is suitable for all polymers. However, it cannot be used for laminated films, as the color layer is applied between the material layers and cannot be removed.

10.1.3 Moulding

Extrusion

Part of the recycling of plastics from post-consumer collections is extrusion, which is often coupled with melt filtration, which serves to homogenize the material. Since the 1980s, existing extruder systems have been adapted and optimized for use in the recycling industry to overcome the following challenges, among others:

- Processing of non-contaminant-free plastics
- Insensitivity to residual moisture
- Degassing of the melt
- Melt filtration

Some companies have specialized in building extruders for recycling, using their experience from processing virgin material to adapt existing systems. New developments have been made specifically for the recycling of PET. It is important to maintain the reduction in viscosity (degradation, cf. Sect. 10.2) as low as possible during the extrusion process. Such extrusion systems are often equipped with pre-drying and/or post-crystallization units. Depending on the challenge, single-screw extruders are used combined with agglomeration units, twin-screw extruders, or cascade systems.

The wear on the screw and barrel is significantly higher compared to processing virgin material, so regular replacement is advisable to maintain the performance of the system. The better the cleaning in the preceding process steps, the longer the change intervals and the shorter the downtime.

Melt filtration

The use of melt filters in extrusion is now standard in plastics recycling (cf. Fig. 10.12). Only when processing PIR can it sometimes be omitted. Usually, however, one or even two melt filters are used to remove impurities (metals, wood, mineral impurities, or polymers that do not melt at the processing temperature) from the plastic melt. These filters are unsuitable for thermoplastic contaminants that have already melted or for cross-linked contaminants, such as foamed and partially cross-linked PP, PE, or elastomers.

Depending on the application, different filtration systems such as cassette screen changers, single-piston screen changers, multi-piston screen changers, swivel and screen disc filters, belt filters, drum filters or laser filters are used. The systems

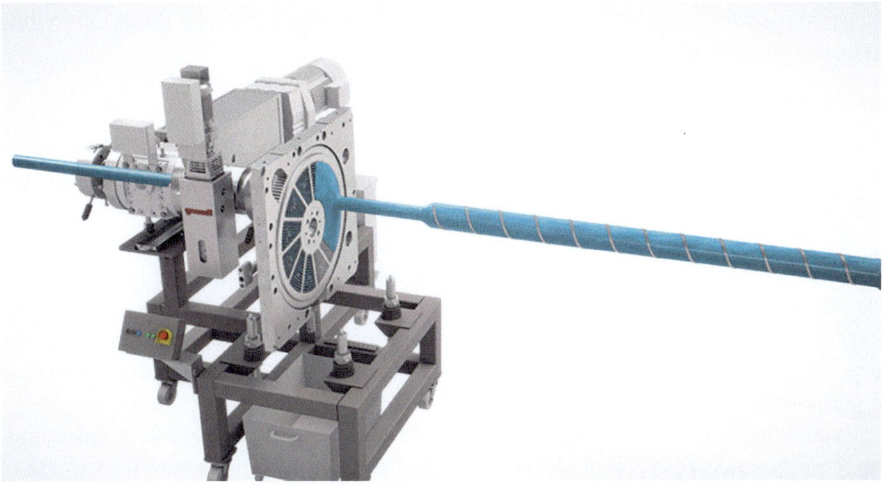

Fig. 10.12 Rotary filtration system (*Source* Gneuss 2023)

mentioned differ in terms of costs, continuity of production, operation, influence on the pressure conditions and shear in the extruder, etc. Depending on the system (e.g. in twin-screw extruders), a melt pump is also required to build up pressure.

10.1.4 Other Aspects

Odor removal

Due to the growing use of recyclate in packaging, the recycler must deal with the odors of plastic recyclate and install processes to reduce them. Removing odors from plastics is not new, hence, established processes are available that are used for virgin plastics.

The smell of recyclate is usually caused by residual components from previous use. The topic is particularly relevant for polyethylene (PE) and polypropylene (PP), as migration processes into these polymers are often the cause of odors due to the polymer structure. Migrated substances from packaging contents or residues from degradation processes lead to unpleasant odors during and after the recycling process. Printing inks that reach the extrusion stage can also cause a "burnt" smell, for example.

Odors are volatile organic compounds (VOCs) that must be reduced to address the problem. In recent years, processes have been developed to eliminate VOCs in a thermal cleaning process. Hot air is passed through the regranulates absorbing and removing volatile organic compounds. This process follows extrusion, during which some of the VOCs are separated with sufficient degassing. In contrast to the occasional addition of additives, degassing removes the causes of the odors, while additives simply cover up existing odors.

Food contact

The use of recyclate in direct food contact with plastic packaging is regulated in Regulation (EU) 2022/1616 (EU 2022). In comparison to the regulation that was valid until then, some innovations were made.

While the use of PET in direct contact with food is state of the art, solutions for other plastics such as PE, PP and PS are missing. The challenge is to prove, in accordance with EFSA requirements (European Food Safety Authority), that migration has taken place during the previous usage phase after decontamination in the recycling process does not have any negative effects on the packaging contents (food in particular) in the new usage phase.

Even if these rules do not explicitly apply to other sensitive uses such as cosmetics, the requirements from Regulation (EU) 2022/1616 are also used here for the use of recyclates.

Water treatment

Many process steps in plastic recycling require water, so washing, separating, and cooling are impossible without the use of water. The water quality is impaired by the entry of organic and inorganic contaminants.

The quantities of water used require that its usage phase be extended as long as possible for economic reasons, as the contamination of the process water with impurities prevents it from being discharged directly into the public wastewater system. A plastic recycling plant therefore mandatory includes process water treatment.

The requirements for such water treatment differ from system to system, as the entry of contaminants depends on the type and history of the plastic waste used. HDPE packaging for cleaning products has a significantly different level of contamination than PET bottles from the beverage industry, and films from agriculture differ significantly in the degree of contamination from packaging films for commercial use.

In addition, the local requirements for the quality of the wastewater to be discharged and for the reuse of the water in the process must be considered. As a rule, the local authorities specify parameters such as temperature, COD[1] and BSB[2] contents, pH value, dry matter content, for the discharge conditions. Based on this, a water treatment system must be individually planned and built for each location. The process water treatment at a location must therefore be given special consideration during site planning.

---

[1] COD: Chemical Oxygen Demand.

[2] BSB: Biochemical Oxygen Demand.

10.1.5 Quality Assurance and Standardization

The subject of quality assurance in a recycling company is input materials, products, and processes.

The input materials of a plastic recycling company are usually pre-sorted waste. The quality (composition and proportion as well as the quality of the permitted contaminants) of the materials to be supplied must be agreed upon between the supplier of the plastic waste and the recycler, whereby reference can be made to various definitions. For this purpose, the Plastics Recycling Association in the German Bundesverband Sekundärrohstoffe und Entsorgung e.V. (bvse 2000) created a grade list in 2000 in which different fractions from different areas of origin were defined. Publications refer to this list of grades when reporting on the market.

With the introduction of a separate collection of packaging plastics, detailed specifications were created for the resulting sorted fractions, which were continuously updated and adapted over the years. Many of these specifications developed in Germany in the 1990s were adopted in other European countries or material descriptions adapted from this model were created.

Specifications for sorted plastic waste are also available for collection systems from other areas of origin, such as agriculture (cf. Sect. 11.3).

In addition, there are also bilateral agreements between the supplier and the recycler regarding the plastic waste to be supplied. Various measures are useful to check compliance with these specifications. In addition to the visual assessment of a truck load during incoming inspection, samples are also taken for analysis. As a rule, plastic waste is delivered in bales, some of which are removed from the load to separate sample sizes of 40–80 kg. These samples are sorted and tested according to the corresponding specification. Such analysis can be carried out by third parties (GBP 2024), which ensures an independent assessment of the results.

This analysis is particularly relevant in the production of recyclates for food contact. According to European regulations, a certain amount of the input material must have already been used as food packaging on the market. If there is no separate recording of such packaging, e.g. in a deposit system, it must be checked regularly whether the required conditions are being met, for which Plastics Recycler Europe (PRE 2021) has developed a testing scheme. During the recycling process, the quality of the intermediate products and the efficiency of the individual process steps should be continuously checked, for which sampling points on various units are helpful. The quality control of the recyclates is carried out using analytical methods known from plastics processing to determine the material properties. In addition, analyses are carried out to determine any contaminants or foreign substances that are still present. Recyclates that are designed for food contact require more testing effort to identify undesirable substances.

Standardization of recyclate is only possible in relation to the determination of defined properties, but not defined values, for example, the melt viscosity, since the input material would also have to be standardized in this regard. It is the mixing of plastic items with different properties during collection and during the recycling process that ultimately determines the properties of the recyclates. The recycler's know-how consists of knowing the quality and composition of the input streams with sufficient precision to produce defined recyclate properties of consistent quality from the mixture.

Various CEN standards specify how common recyclates should be characterized and describe the test methods to be used:

- DIN EN 15342: Characterization of polystyrene (PS)
- DIN EN 15344: Characterization of polyethylene (PE)
- DIN EN 15345: Characterization of polypropylene (PP)
- DIN EN 15346: Characterization of polyvinyl chloride (PVC)
- DIN EN 15348: Characterization of polyethylene terephthalate (PET).

Further relevant descriptions are set out in the following standards:

- DIN EN 15343: Traceability in plastic recycling and assessment of conformity and recyclate content
- DIN EN 15347: Characterization of plastic waste.

10.1.6 Conclusions and Outlook

- Current legal regulation activities for the collection of plastic waste in Europe are leading to a higher volume of plastic waste. On the one hand, there is a final ban on landfilling plastic packaging waste, and on the other hand, deposit systems are being introduced for beverage packaging, among other things. Both bundles of measures increase the quantities recorded and are made available for plastic recycling. There is also the obligation to increase the proportion of recyclate in other applications or markets. This increases the requirements for collection and sorting as well as the entire recycling process.
- Since the use of recyclate in the original application is to be increased (closed loop recycling), the requirements for recyclates are increasing, especially for the high-volume plastics PE, PP and PS, especially for the packaging sector.
- Experiences with PET recycling are transferred to the recycling of other packaging materials: these include color sorting, hot washing, and migration analyses, which require investments in appropriate machines and units. The packaging design will also draw on experiences from PET recycling: for example, the use of adhesives or the use of labels.
- Printing inks are coming into focus, whether for printing on packaging or labels. Pigments and color components lead to restrictions in the target color or, as with nitrocellulose-based colors, to undesirable effects during extrusion. Further developments in this area are foreseeable.
- There is a close connection between the design of plastic products and the quality of the recyclates. Recyclate properties depend largely on the "Design for Recycling (D4R)" of the primary products. Quantitative and qualitative parameters

for recyclates can only be defined depending on the "original product". Standardization of recycled materials is only possible if the original products are also standardized. Just as new material is not standardized, this will not be possible for recyclables either.
- The technologies of mechanical plastic recycling have developed dynamically in recent years in close dialogue with machine and plant manufacturers. The focus remains on the adaptation and further development of processes and units from application areas such as agriculture or food processing. A significant development is sorting using NIR sensor technology and artificial intelligence.
- With increasing demands from the plastics processing industry, recycling processes are becoming more demanding and cost intensive. To be able to operate recycling plants for commodity plastics such as PET, PE, or PP economically, plant capacities of significantly >30 kt/a (input quantity) are required. Constant additional investments are necessary to keep pace with technological developments on the one hand and changes in the waste stream on the other. The resulting investment volumes can hardly be sustained by medium-sized businesses.
- A consolidation process has been taking place in Europe for several years. Medium-sized recyclers are either taken over by companies in the plastics manufacturing industry or large waste disposal companies. Both are doing this with the intention of extending their value chains. This development is ongoing.
- The prerequisites for market development of recyclates of defined quality and availability based on supply and demand are not yet sufficiently met but are necessary for planning security and long-term conversion of materials and processes and building trust in the market.

10.2 Polymer Degradation and Quality Improvement of Recycled Plastics

Rudolf Pfaendner

10.2.1 Introduction

Although sorting, cleaning, and purification processes of post-consumer recyclates have been improved within the last decade recycled plastics do often not fulfill the requirements to allow closed-loop applications. Except for well-defined post-industrial scrap recycled and virgin plastics differ fundamentally because of aging, blends, impurities and variations of the input stream. Methods to enhance and maintain recyclate properties (mechanical properties, visual appearance, stability over time) comprise technical processes such as compounding to support material purification and the addition of suitable additives to compensate for structural deficits of the recyclate.

10.2.2 Scope and Limits of Quality Improvement

Even under the best purification processes the recycled plastic is not identical to the virgin analogue. During use and aging by oxidation and photooxidation the polymer chain is chemically modified, degraded, or even cross-linked. New functional groups such as carbonyl or acid groups are formed on the polymer backbone or as end groups functioning as initiator sites for further oxidation or degradation during processing and over the lifetime of the new application. Even a well-separated single recyclate stream is a mixture of different grades, manufacturers, and formulations with individual ingredients. Despite an optimal separation, the recyclate may contain impurities of other polymers, glue residues, paper, inks, contact media, and degradation products (Vilaplana et al. 2008; Pfeisinger 2017; Vogt 2021; Pfaendner 2022).

Aging of Polymers

Many polymers are sensitive to oxidation taking place during storage, processing, and use. The autoxidation starts with an initiation step forming radicals during thermal processing or through UV-light. Defect structures are ab initio part of industrially manufactured polymers. Polyethylenes, e.g. contain unsaturated groups like vinyl, vinylidene, and allyl groups. As polymer processing and use takes place in the presence of oxygen the initially generated radicals react with oxygen to peroxy radicals followed by hydroperoxide formation through H-abstraction. Hydroperoxides are thermally unstable and decompose into alkoxy and hydroxyl radicals. Although autoxidation reactions are more complicated in detail the polymer chain is finally modified through oxygen-containing species, e.g. carbonyl-, alcohol- and carboxyl-groups (Smith et al. 2017). Through elimination and disproportionation reactions unsaturated groups and low molecular weight products are formed, recombination of radicals may result in branching or in cross-linked polymer chains. Additives such as antioxidants will reduce or delay the oxidation process, however, it cannot be avoided. Autoxidation is crucial for polyolefins and polymers containing double bonds such as elastomers. Other formulation ingredients such as pigments or impurities, e.g. transition metals may catalyze and accelerate the autoxidation process.

Polyesters and polyamides and all high-performance plastics (polyetherketones, polyethersulfones, polysiloxanes) are less sensitive to oxidation. However, polyesters, polyamides, and all other polycondensates are susceptible to hydrolysis causing molecular weight degradation and finally loss of mechanical properties. Any changes in molecular weight (MW) and molecular weight distribution may influence the mechanical properties until complete loss of strength. Moreover, rheological properties are modified and, therefore, processing, and potential application fields are compromised. Degradation of polyvinylchloride (PVC) takes place during processing where HCl is split off, starting from defect structures, causing strong discoloration (Starnes et al. 2002). During aging PVC is stable and, therefore, an example for long-term use, however, plasticized PVC loses plasticizers over time resulting in embrittlement (Gumargalieva et al. 1996).

Differences Between Recyclates and Virgin Plastics

Because of aging recyclates differ substantially from virgin plastics depending on chemical structure and composition, processing conditions (temperature, shear), formulation ingredients, use time and the environment (outdoor/UV light, elevated temperatures, contact media) (cf. Table 10.1). Structural changes in the polymer backbone, e.g. carbonyl, hydroxy, carboxylic groups in polyolefins are irreversible, therefore the overall range of oxidized structures in aged polymers or recyclates can be several orders of magnitude higher compared to virgin polymers. The decisive factor is not the influence of comparatively minor structural changes on the overall polymer properties, but the oxidized structures function as initiator sites or prodegradants for further oxidation. Obviously, a recyclate is more sensitive to oxidation than a virgin polymer as experimentally proven for polyolefins (Luzuriaga et al. 2006; Maxwell et al. 2008; Craig et al. 2006; La Mantia et al. 2004; La Mantia et al. 2023) and for styrenics (Signoret et al. 2020). As such structural changes cannot be reversed in a simple way the only strategy is to "neutralize" the oxidized structures of the recyclate through suitable additives.

Recyclates are a mixture of different manufacturers, grades, production processes, and technologies with different catalysts and different additives. The latter may result in synergistic or antagonistic effects, difficult to predict in a mix of materials. Moreover, the composition of such a "single" stream may vary over time, source, and season. For example, HDPE shows different degradation mechanisms in multiple extrusion cycles simulating recycling depending on the catalyst used for its synthesis. HDPE produced via Ziegler–Natta catalysts (Moss et al. 1989) or metallocene catalysts (Allen et al. 1999) and LLDPE tend to molecular weight degradation during processing, whereas HDPE via Phillips catalysts (Moss et al. 1989; Pinheiro et al. 2004) and LDPE tend to molecular weight increase through cross-linking as the concentration of initial double bonds is higher. A recyclate mix of both types may show no effect at all although both processes—degradation, and cross-linking—take place in parallel (Pfaendner et al. 1995). A potential strategy needs additives to cope with the different individual behavior of "single" mixtures and should be capable to balance any fluctuation.

Impurities in the recyclate stream are another area to be considered to explain different behaviors of recycled and virgin plastics. Whereas virgin polymers may contain only small concentrations of catalyst residues, monomers, oligomers, and incorporated application-specific additives, recycled plastics will often exhibit more impurities not removed through previous purification processes. In any case, the recyclate contains residual additives from its first use, as well as their degradation products. Incomplete separation results in impurities from other polymers in a single stream, e.g. PP in PE or PE in PP. Residues from packed content (e.g. oil, fats, flavoring agents) or from printing inks, glues, paper fibers may be present, influence the overall properties and contribute to odor formation. Transition metal impurities, e.g. iron from rust can cause a dramatic loss of long-term thermal properties since they act as oxidation catalysts (Day et al. 1995). Depending on the impurities different strategies to improve quality can be chosen; volatile components

Table 10.1 Differences between virgin and recycled plastics (*Source* Own presentation)

Cause	Effects	Consequences	Remediation
Oxidation, Photo-oxidation	• Molecular changes • New chemical groups: carbonyl, acid, alcohol, hydroperoxide groups • MW changes (degradation, branching, cross-linking)	• Accelerated oxidation and degradation • Discoloration • Loss of mech. properties • Modified rheological and processing behavior	• Restabilization
Hydrolysis	• Molecular weight degradation in polycondensates (polyesters, polyamides)	• Loss of mechanical properties	• Chain extension • Functionalized compatibilization
Insufficient or limited polymer separation	• Polymeric impurities, polymer blends, mixed plastics	• Inferior mech. properties • Processing issues	• Compatibilization
Insufficient or limited cleaning	• Presence of residues/contaminants (printing inks, paints, glues, paper fibers, oil, fats) and their degradation products	• Increase of volatiles and odor • Inferior mech. properties • Discoloration	• Restabilization • Compatibilization • Addition of anti-odorants • Degassing during processing
Others	• Macroscopic impurities (elastomers, metal residues, inorganics)	• Inferior mech. properties • Processing issues	• Melt filtration

can be removed during extrusion in a degassing stage optionally in combination with additives. Macroscopic impurities can be removed through melt filtration, other incompatible impurities can be incorporated in the matrix polymer through reactive additives or through compatibilizers.

Analytical Assessment of Recyclate Quality

To assess the chemical status of a recyclate, to select the best cost/performance upgrading route, and to identify the potential application scope, information about the starting material should be generated via analytical tools. Typical plastic characterization values such as melt viscosity (MFI/MVR) (ISO 1133), density (ISO 1183), and mechanical values (tensile strength and elongation (ISO 527), flexural strength and modulus (ISO 178), Izod impact strength (ISO 180)) are often specified by recyclers. However, more sophisticated information about impurities and aging status of the recyclate may be needed, therefore thermal analysis (TGA, DSC), spectroscopic and chromatographic techniques are applied (Klingenberg et al. 2024). Elemental analysis gives an indication of foreign material and deviations from the ideal composition compared to analogous virgin plastics (Petrovic et al. 2022).

To get an indication of the aging status and residual presence of stabilizers Oxygen Induction Time (OIT, ISO 11357) is applied, where via DSC heat flow (or weight increase via TGA) under inert temperature and subsequently in the presence of oxygen at elevated temperatures is determined.

Spectroscopic methods such as Infrared (FTIR), Raman and NMR reveal information on contaminants, composition of blends, and the aging status. For example, through increased autoxidation carbonyl bands occur in the FTIR spectrum (not existing in virgin polyolefins) as a function of aging time providing the so-called Carbonyl Index (CI (Almond et al. 2020)).

Chromatographic methods are helpful to get detailed information about the polymer itself and again about contaminants. Size exclusion chromatography (SEC) describes molecular weight and molecular weight distribution of a polymer. Changes in the molecular weight distribution can be correlated to rheology properties and processing.

Gas chromatography in combination with mass spectrometry (GC–MS) is used to analyze and to quantify volatile organic compounds (VOC) (Cabanes et al. 2020). Furthermore, extraction methods are used to analyze additives, contaminants, and degradation products via subsequent chromatographic und spectroscopic methods.

10.2.3 Improvements Through Technical Processes

The regrind from a purification process is usually recompounded via extrusion into granules generating a homogeneous recyclate processable on state-of-the-art equipment (e.g. injection molding). Besides incorporating additives to adjust properties during compounding the processing equipment itself is the first option to improve the quality of the recycled material.

Melt Filtration and Degassing

Melt filtration is an essential standard method to remove unmeltable particles, e.g. cross-linked polymers, metal residues, and other impurities whereas continuous melt filtration with automatic filter change and filter cleaning is established today. Vacuum degassing during extrusion is standard to remove volatiles, reduce emissions and undesired migration of plastic ingredients and to avoid processing issues and defects influencing aesthetics. The obtained level of volatiles reduction is around 50% (Höftberger et al. 2023). Multiple degassing may improve the values, however elevated temperatures increase the concentration of volatiles through polymer degradation. Stripping through water injection during extrusion enhances degassing and can also reduce more difficult contaminants (Monti et al. 2023). Even water-carrying additives together with surface-active substances adsorbed on polymers in solid form can be used instead of water injection (Lok et al. 2022). Also, solvents may be used in extrusion to extract contaminants, whereas supercritical CO_2 as a solvent is likely the most powerful and environmentally benign solvent (Dufresne et al. 2002). In general, the addition of scavenging or adsorbing additives and their combination with a degassing step can minimize volatiles, emissions, and odor (Bledzki et al. 1999, Garofalo et al. 2023).

Solid-State Polycondensation (SSP)

Solid-state polycondensation is the process of choice to increase the molecular weight of selected polycondensation polymers (polyesters, polyamides) and to remove volatiles at the same time. As the process takes place in the solid phase at lower temperatures than the melting point of the particular polymer and under inert gas (or vacuum), the polymer as such is less affected. Degradation, oxidation, and discoloration are substantially reduced. Within the recycling area, SSP is an essential part of PET recycling for food contact applications such as bottles (Pinter et al. 2021). Acceleration of the SSP process and modification of the polymer structure were demonstrated through the addition of catalysts (Vouyiouka et al. 2006) and of chain-extenders (Rastin 2016).

10.2.4 Quality Improvement through Additives

Additives are the most valuable tool to improve the quality of recycled plastics. Like in virgin plastics, additives are essential to maintain polymer properties in the processing steps and during use and to adjust the properties according to the application. The entire bunch of additives known for virgin polymers can be used in recyclates to adjust properties and to fulfill aesthetic aspects, e.g. pigments, fillers, reinforcement agents, flame retardants and others, mainly if the recyclate is used in a different application as the first one. However, there is a substantial group of additives which is designed for recycled plastics resulting in restabilization, repair, compatibilization, and antiodorization.

Restabilization

Nearly all polymers and in particular polyolefins are sensitive to oxidation and photooxidation. Antioxidants and in the case of outdoor applications UV/light stabilizers are added to protect the polymer from degradation. As these stabilizers (depending on the processing and use conditions and time) are consumed fulfilling their protective role all recycled plastics are "under-protected". Therefore, restabilization is the first step for quality improvement of recycled plastics to approach the properties level of virgin plastics (Vishnu Sankar et al. 2017, Pfaendner 2022). Minimum is the replacement of the consumed quantity of stabilizers from the previous application. Also, a sufficient basic stabilization for the first application ("design for recycling") taking a priori into account potential material recycling is an important prerequisite to avoid severe degradation. It is obvious that the replacement alone will be not sufficient if a recyclate from short-term applications like packaging is foreseen for long-term use in the second life. Because of the newly formed chemical structures from the autoxidation process recycled plastics often require additional protection to avoid accelerated degradation. This can be fulfilled through increased stabilizer loadings or preferably specific stabilizer combinations addressing the weak structures of the recycled plastic.

Standard stabilizers for processing and long-term thermal stability comprise synergistic blends of phosphites (or phosphonites) and phenolic antioxidants often in combination with antiacids such as fatty acid salts, metal oxides, or hydrotalcites. Sterically hindered amines (HALS) usually piperidine derivatives and/or UV absorbers based on benzophenones or benzotriazoles protect the recyclate from the negative influence of UV radiation. Additive manufacturers offer tailored restabilization systems for recycled POs and PET to compensate the increased oxidation sensitivity and/or impurities and/or the larger product fluctuations (Pfaendner 2023). These restabilization systems outperform virgin stabilizers and can provide cost-performance advantages.

PVC is an exception from typical stabilization systems with antioxidants as elimination of hydrochloric acid must be avoided, whereas so-called heat stabilizers, e.g. metal soaps or organo tin compounds with inorganic and organic costabilizers are applied for processing. During its lifetime at ambient temperature degradation of PVC is only marginal and predestines PVC for long-term applications. Even without additional heat stabilizer PVC recycling is feasible (Ditta et al. 2004, Kelly et al. 2005), however recycling of PVC containing phased-out stabilizers, e.g. based on lead can be a matter of discussion.

Restabilization can provide several benefits for processing and use of recyclates, e.g. a consistent and stable production process, increased throughput, fewer defects such as avoiding gel formation in film production, maintenance of material properties during processing, extended long-term thermal stability, less discoloration, compensation of the negative influence of impurities such as acidic residues and potential for multiple recycling. Depending on processing parameters and the intended application, predamage and previous use conditions the range of restabilization agents is from 0.1 to 1.0%, if needed for outdoor applications UV/light stabilizers up to 1.0% can be added.

To prove the efficiency of restabilization the methods of choice are multiple extrusion, accelerated heat aging, and for outdoor applications artificial weathering (cf. Fig. 10.13). Multiple extrusion simulates several processing steps such as compounding and injection molding and judges the molecular weight changes of a polymer at each processing step. Accelerated aging is the method of choice to predict the long-term properties of a plastic formulation. Specimens are stored at elevated temperatures for a certain time in a forced air oven and, e.g. mechanical properties are screened versus the storage time. Artificial weathering offers reliable information about the photostability of plastic formulations. In general, specimens are irradiated and stored at a certain temperature in defined wet/dry circles (e.g. according to ISO 4892). Mechanical properties or surface changes (chalking, crack formation, yellowing) are recorded over irradiation time.

<table>
<tr><th>Objective</th><th>Evaluation method</th><th>Example</th></tr>
<tr><td>Processing stability</td><td>Multiple extrusion at typical processing temperatures: Recording of molecular weight related changes e.g. melt flow, mechanical properties, color</td><td>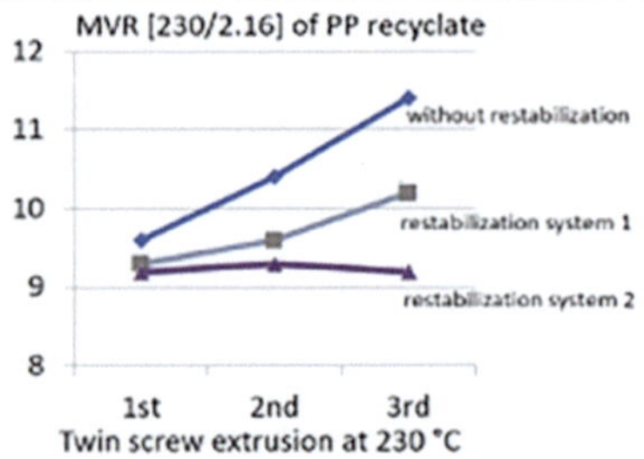
</td></tr>
<tr><td>Long-term thermal stability</td><td>Accelerated aging at elevated temperatures: Recording of molecular weight related changes e.g. melt flow, changes of mechanical properties, surface changes, discoloration</td><td>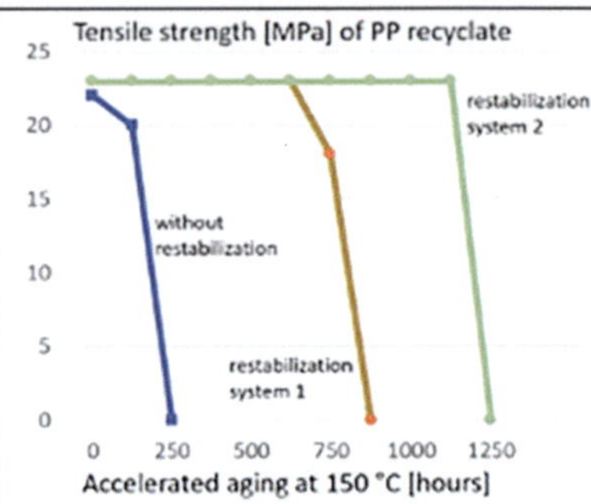
</td></tr>
<tr><td>Outdoor/light/UV stability</td><td>Artificial weathering: Recording of molecular weight related changes e.g. melt flow, changes of mechanical properties, surface changes, discoloration</td><td>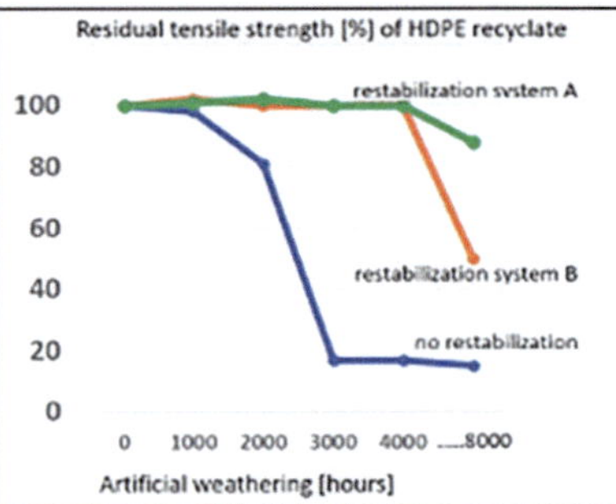
</td></tr>
</table>

Fig. 10.13 Evaluation of stability of recycled plastics (*Source* Own presentation)

Chain Extension and Repair

Aging of polycondensation polymers (polyesters, polyamides) induces molecular weight reduction mainly through hydrolysis and to a lesser extent through oxidation reactions. As the mechanical properties of the polymer are directly linked to the molecular weight loss of mechanical values is the consequence. Fortunately, hydrolysis results in functional end groups (–OH and –COOH in polyesters, $-NH_2$ and –COOH in polyamides) useful to build-up again the molecular weight again through a condensation reaction with suitable partners. Thus, the hydrolytic degradation is "repaired", and the molecular weight of the virgin polymer is regained. As reactive molecules to increase the molecular weight diepoxides, bis-oxazolines, bis-oxazolones, bis-oxazines, diisocyanates, tetracarboxylic acid dianhydrides, bis-acyllactames, bismaleimides, dicyanates or carbodiimides have been reported. PET (Awaja et al. 2005) and Polyamide-6 (Ozmen et al. 2019) or Polyamide-66 (Pfaendner et al. 1998) are good examples of a linear chain extension with bifunctional additives.

Reactive additives with higher functionalities or polymers with reactive side groups generate branching and, depending on the concentration, cross-linking, although mechanical properties are increased. To facilitate and support control of the chain-extending reactions masterbatches are available and combinations of chain extenders with moderators and catalysts were described (Kiliaris 2007). Concentrations of chain extenders must be adjusted to the molecular weight of the recyclate, end group concentrations, and targeted molecular weight and are often found between 0.05 and 1.0%.

Defined molecular weight build-up of polymerization polymers such as PP, PE, or PS is hardly possible as functional end groups do not exist. Nevertheless, it is in some cases possible to use the formed chemical groups from aging processes such as alcohol, acid, ketone, or double bond functions as reaction partners to link polymer chains. Thus, PE greenhouse films could be repaired through the addition of an ethylene-glycidyl methacrylate-copolymer or a hydroxylamine radical generator (Scaffaro et al. 2006).

Compatibilization

Polymers of different structures are as a rule not compatible with each other. Therefore, blending polymers or recyclates often results in inferior mechanical properties far away from the parent polymer properties because of the formation of inhomogeneous mixtures where large particles of the polymer in lower concentration are dispersed in the matrix of the polymer with higher concentration. This is already the case for polyolefin mixtures of PP and PE but is even more significant for blends of unpolar (PP, PE, PS) and polar (PA, PET) polymers. To enhance the mechanical properties of mixed plastics and to compensate for the negative influence of polymeric impurities, the use of compatibilizers is a potential solution (Dorigato 2021, Maris et al. 2018, Nechifor et al. 2018, Self et al. 2022). Compatibilizers are capable of modifying polymer interphases by reducing the interfacial tension in the melt and increasing the adhesion at phase boundaries, thus preventing delamination in the solid phase. During processing agglomeration of the dispersed phase is avoided and the morphology is stabilized. As a result, e.g. the mechanical properties like tensile strength and elongation at break are considerably improved.

To fulfill the function of a compatibilizer the additive should contain chemical structures of the polymers to be blended, e.g. a PP-PE block copolymer for blending PP/PE blends or PS/PE blends through styrene-ethylene/butylene-styrene block copolymers (SEBS). Mixtures of non-polar polymers with polar polymers are preferably compatibilized through reactive compatibilizers, where a non-polar chain contains functional reactive groups, e.g. maleic anhydride, acrylic acid or glycidyl(meth)acrylate to react with the end groups of the polar polymer, e.g. polyamide or polyester. Examples of the latter type of compatibilizer are polyethylene-graft-maleic anhydride, polypropylene-graft-maleic

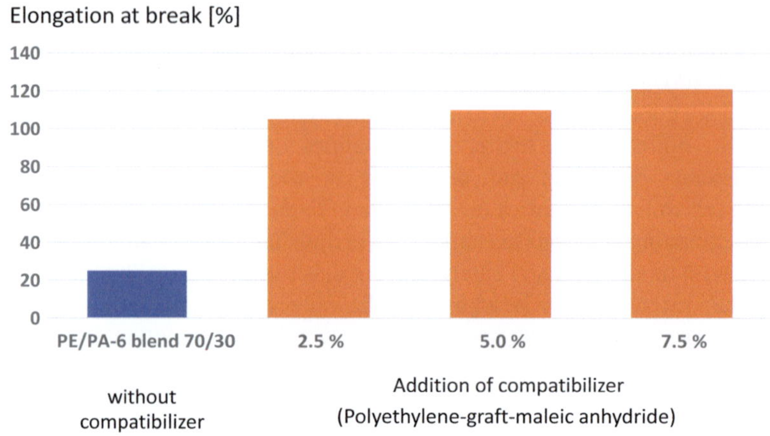

Fig. 10.14 Effect of compatibilizers in polymer blends

anhydride or glycidyl(meth)acrylate-(meth)acrylate-polyolefin-copolymers. For example, mixtures of 30% polyamide-6 and 70% polyethylene, simulating coextruded packaging films, were compatibilized through the addition of a maleic-acid-modified-polyethylene resin. The elongation at break can be substantially increased by adding 2.5%, with higher concentrations additional improvements are accessible (cf. Fig. 10.14).

In general concentrations of compatibilizers are above 2% up to 15%. As high concentrations of compatibilizers increase the overall compound cost improved separation to manufacture single-stream materials may be an advantageous alternative in some cases.

Anti-odorization

Contrary to virgin material odor can be a severe issue with recyclates. Reasons for strong and/or unpleasant odor are low molecular weight molecules from polymer and/or additive degradation, from substances migrated into the polymer, from contact media (food, detergents, perfumes), as well as their thermal decomposition products in the compounding step (Paiva et al. 2021, Cabanes et al. 2020). An intensive washing and purification process is certainly a prerequisite and as already mentioned degassing is an established technique to remove volatiles.

Selected additives can contribute significantly to odor reduction in recycled plastics. The challenge of odor reduction lies in the many possibilities of odor origin, the diverse chemical components, and the often low concentrations causing the odor. Therefore, a universal additive to cope with all potential odor sources is hardly to be found. Antiodorization additives follow two potential strategies: Functional molecules react with the odorous components forming chemical bonds or strong complexes. The volatile substance is transferred into a non-volatile and non-odorous product remaining in the polymer. Examples of this concept are the zinc salt of ricineolic acid, metal ionomers (Shoaib et al. 2002, Spigaroli et al. 2017) and epoxides

(Pfaendner 2006). Furthermore, the addition of antioxidants can avoid the formation of low molecular weight volatile degradation products from polymers.

Adsorbents are the second possibility to reduce odors in polymers. Examples are based on silicates such as zeolites (Gustafsson 1992, Höhne et al. 2022) or cyclodextrins (Zanichelli 2018). Adsorbers do not form a bond with the odorous substances but act via complex or cage formation. The resulting interaction is weaker than a chemical bond and desorption may take place again under heat or through extraction.

Other Additives

Any additive used in virgin polymers can be applied in the same manner for recycled plastics in addition to the more specific additive classes mentioned above. Further additives used frequently in recycled plastics comprise pigments, lubricants, fillers, reinforcement agents, metal deactivators, impact modifiers and markers.

Pigments are incorporated to provide an attractive color, to correct previous color fading or to overdye previous colors or color mixtures which are often grey or green in recyclates.

As fillers and reinforcement agents modify the mechanical characteristics of recyclates providing increased tensile strength, modulus and enhanced heat distortion temperature and thus open new applications.

Metal deactivators form complexes with transition metals such as copper or iron and compensate in this way for the influence of ions catalyzing oxidation reactions. These metal deactivators are manly used in wire and cable applications but can be helpful to compensate for the negative influence of metallic impurities in recycled plastics.

Lubricants and other processing aids will reduce the melt viscosity, increase throughput, and save energy during processing to some extent. Typical lubricants are based on waxes, e.g. PE waxes and long-chain fatty acids such as calcium stearate. Titanates and zirconates are more sophisticated forms of processing aids (Monte 2002).

Impact modifiers based on elastomeric components, e.g. styrene-butadiene block copolymer (SB) or styrene–butadiene–styrene block copolymer (SBS), ethylene-propylene-diene copolymer (EPDM) or methacrylate-butadiene-styrene copolymer (MBS) are incorporated in recycled plastics to improve impact strength and elongation at break. E-modulus and heat distortion temperature are lowered.

Markers on the molecular basis of fluorescent dyes or rare earth elements can be used to identify type and sources of polymers as well as to distinguish recycled plastics from virgin material (Woidasky et al. 2020).

10.2.5 Conclusions and Outlook

Quality improvement of recycled plastics in mechanical recycling processes is crucial to obtain recyclates for closed-loop and high-value applications. Key steps for high-grade materials in the thermal transformation process from regrind to recyclate compound are smooth processing, degassing steps to remove volatiles, and the addition of stabilizers to avoid polymer degradation and maintain the properties. To guarantee recyclate performance in the intended application tailor-made products for restabilization are available in the market. To enhance the properties of recyclate blends compatibilizers are suitable additives. Polycondensation products such as polyesters or polyamides can be repaired through the range of chain-extenders. As post-consumer recyclates suffer often from odor issues, additives to reduce odor are gaining importance.

Demand for processes and additives to increase the quality of recycled plastics will further grow in parallel with novel applications and recyclate consumption. In the field of additives, environmentally friendliness and product safety are important criteria. With increasing recycling cycles and recyclate quantities in the market, the additive load (and their degradation products) will accumulate. To cope with migration and incompatibilities a need for additives already efficient in low concentrations will arise, showing permanent activity but being degradable to non-toxic substances in the environment. Potential additives of the future might be bio-derived ("bioadditives") reducing at the same time the carbon footprint of plastic applications and of recycled plastics.

10.3 Chemical Recycling

Lars Krause and Michael Carus

10.3.1 Introduction

Chemical recycling is providing various technologies that are essential for expanding and improving the circular economy of plastics as well as supporting the green transition of Europe and beyond. In contrast to established recycling definitions from the plastic sector which mainly include closed loops (e.g. plastics-to-plastics or polymer-to-polymer), the term "recycling" is understood differently in the context of chemical recycling and the renewable carbon economy. To maximize waste elimination and value generation both closed and open recycling loops are intrinsically linked and can be realized via chemical recycling. The purpose of open-loop recycling is derived from the renewable carbon concept. This concept serves to counteract anthropogenic climate change, driven by the emission of CO_2 from fossil carbon. To mitigate climate change via greenhouse gas emissions, the inflow of further fossil carbon from the ground must be reduced and eventually eliminated. The chemical and plastics sector cannot be decarbonized, because its products are mainly based on carbon. So, another approach is required, in which the carbon is kept (or cycled) as renewable carbon preventing its loss from the system—creating sustainable carbon cycles. In doing so, less virgin fossil carbon is used. There are three pillars of renewable carbon which are bio-based, CO_2-based, and recycling-based. Chemical recycling supports this concept by realizing closed and open recycling loops in which the carbon from different wastes is recycled into plastics and polymers as well as into valuable feedstocks for the production of any carbon-containing products with functional use (e.g. for the chemical industry). At the same time the production of energy and fuels would not fall under the concept since the carbon would be lost from the system into the atmosphere.

10.3.2 Definition and Categorization

ISO 15270:2008 defines chemical recycling as "[...] conversion to monomer or production of new raw materials by changing the chemical structure of plastics waste through cracking, gasification or depolymerization, excluding energy recovery and incineration [...]". The depolymerization of polymers into its monomers can be realized via biochemical (e.g. enzymolysis), chemical (e.g. solvolysis and solid-state hydrolysis), and thermochemical (e.g. thermal depolymerization) processes. The production of new raw materials is realized via thermochemical processes using pyrolysis and gasification.

Aside the ISO definition other non-standardized terms and synonyms co-exist including advanced recycling and molecular recycling. Especially in North America the term "advanced recycling" is often used as a synonym for chemical recycling. However, since a common definition is lacking, the advanced recycling term may also be used as a broader definition for other new and innovative recycling technologies including chemical and physical recycling. The dissolution is a physical recycling technology and currently lacking an official definition. Basically, physical recycling differs from chemical recycling in that the chemical structure of the polymer is not significantly changed.

10.3.3 Chemical and Physical Recycling Technologies

The current recycling landscape (cf. Fig. 10.15) offers a versatile set of different technologies that serve as an interface between a wide range of different waste streams and products to further improve existing or establish new value chains. It offers material recycling to feed plastics and polymers back into the cycle via mechanical recycling and dissolution. Furthermore, a wide range of chemical recycling technologies is available to feed monomers and other renewable feedstocks back into the cycle of plastics and other carbon-based materials and chemicals by utilizing biochemical,

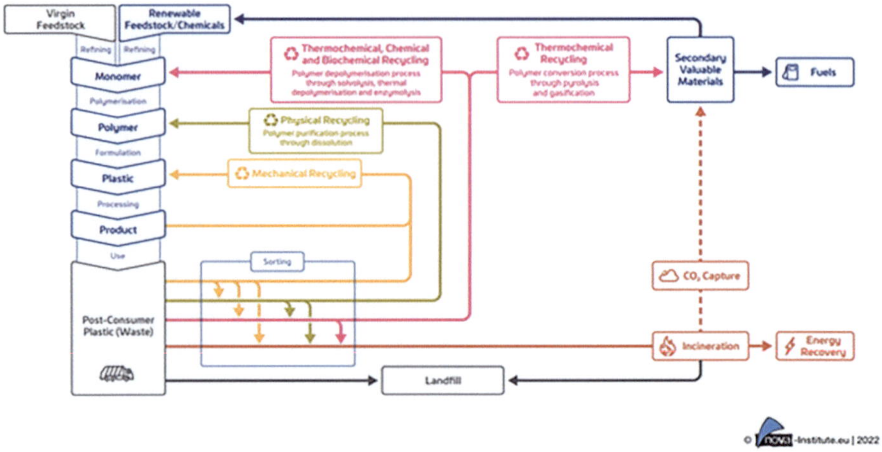

Fig. 10.15 Life of a polymer from production to its disposal (e.g. landfill) including various recycling and recovery routes indicated in different colored arrows. The different recycling technologies are divided by their basic working principles and their products. Adapted from Krause et al. (2022) with kind permission from nova-Institute

chemical, and thermochemical processes whereby the borders of some processes may be floating. In an integrated approach, the variety of technologies is capable to address qualitatively different waste streams transforming them into a variety of products that can be fed back at different positions along the value chain of plastics, chemicals, and other materials.

While mechanical recycling is already well established and characterized by its simplicity it still has its limitations that can be complemented by chemical and physical recycling. Only thermoplastics can be addressed via mechanical recycling, contaminations cannot be removed, and the waste needs to fulfill certain requirements before it can be processed. One of the critical requirements of mechanical recycling is to sort the waste by its polymer types. For pre-consumer waste characterized by its homogeneity and known composition, this might be not critical, but post-consumer wastes remain a challenge since these are often characterized by their heterogeneity as well as their unknown composition and level of contamination. Post-consumer wastes therefore need to be characterized, sorted, and washed before they can enter mechanical recycling. Furthermore, with each mechanical recycling loop, the quality of the recyclate decreases (cf. Sect. 10.2). The mechanical recycling of PET, for instance, is well established, nevertheless it shows after three repeated recycling loops a reduction of the molecular weight, impact strength, and tensile strength by 56%, 13% and 8%, respectively (Frounchi 1999). In contrast to mechanical recycling, chemical and physical recycling is not yet established and is currently in development under which the first plants are realized and commercially operating. In contrast to mechanical recycling, these technologies are more robust against the heterogeneity of a waste stream (e.g. processing of multilayer material can be realized) although it also has its requirements to ensure higher product yields and (virgin) qualities. Furthermore, with chemical recycling not only thermoplastics can be addressed but also thermosets, elastomers and composites. Another advantage of chemical and physical recycling is its capability to remove contaminations and hazardous substances from the waste stream preventing their accumulation in recycled products. With that chemical and physical recycling are enabling the production of recycled plastics with approval for food contact applications for which a high demand is expected in Europe. Primarily, the demand is driven by the political framework conditions that are currently being developed. For instance, the revision of the Packaging and Packaging Waste Regulation (PPWR) considers mandatory recycled contents for contact-sensitive packaging.

Pyrolysis

Pyrolysis (cf. Sect. 13.3) is capable of processing nearly all kinds of organic material through thermal decomposition in presence of heat (depending on specific process ranging between 350–650 °C) and absence of oxygen into a broad range of products such as liquids, gases, or solids that can be utilized as feedstocks for the production of new polymers and chemicals. From the liquid fraction, products such as naphtha and monomers can be used to produce new polymers. The performance of the pyrolysis and yield of products strongly depends on the adjusted reaction conditions and the composition of the processed polymers. A wide range of polymer wastes can be processed via pyrolysis including mixed or multilayer thermoplastics and thermosets, bio-based plastics, and rubber. Besides the operating conditions, the specific polymers are critical factors for the initiated reactions and obtained products. Usually, mainly polyolefins are addressed via pyrolysis. While some polymers, such as PE and PP, can be processed into a wide range of products via random radical chain reactions, the polymers PS and polymethylmethacrylate (PMMA) can be processed into their monomers via end-scissions. The presence of PET is usually avoided or kept at lower amounts since it is characterized by the formation of coke as a by-product (Ramesohl et al. 2020) and it negatively impacts the production of condensable products (Genuino et al. 2023). For pyrolysis, the majority of 81 worldwide known providers is from Europe, followed by North America (Krause et al. 2024). Overall, the majority of technology providers on the market is represented by small companies (10–50 employees) (Krause et al. 2024). Exemplary representatives for larger companies are Blue Alp (Eindhoven, Netherlands), Demont (Millesimo, Italy), INEOS Styrolution (Frankfurt, Germany), Neste (Espoo, Finland), Österreichische Mineralölverwaltung (OMV) (Vienna, Austria), Repsol (Madrid, Spain), Unipetrol (Prague, Czech Republic), VTT (Espoo, Finland) and Chevron Phillips (The Woodlands, TX, USA). At 40 kt per year, pyrolysis currently achieves the second-largest capacity for a single plant.

Solvolysis and Solid-State Hydrolysis

The solvolysis is based on the depolymerisation of specific polymers with solvents which is partially conducted under near-critical or supercritical temperature conditions. Depending on the utilized solvent the technology can be further categorized into alcoholysis including glycolysis (e.g. ethylene glycol as solvent) and methanolysis (methanol as solvent), hydrolysis (aqueous solvents), and ammono- and aminolysis (ammonia and primary amines as solvents). Solid-state hydrolysis utilizes a solid chemical (such as NaOH) instead of a solvent to initiate the depolymerization. In these processes, polymers (mainly PET and PUR) are broken down into their building blocks (e.g. monomers, dimers, oligomers). The depolymerization by solvolysis does not affect C–C bonds, but can cleave bonds between carbon and heteroatoms, therefore suitable polymers are those with heteroatoms in their backbone (Vollmer et al. 2020). After the breakdown, the building blocks are separated from the other plastic components (e.g. additives, pigments, fillers, non-targeted polymers) and purified. After purification, the building blocks are polymerized to synthesize new polymers. Compared to pyrolysis, there are fewer suppliers active on the market, which also offer smaller capacities of up to 10,8 kt per year for a single plant. Of 24 identified solvolysis technology providers, the majority is based in Europe, followed by North America (Krause et al. 2024). With nine companies, the majority of suppliers is among the small companies (10–50 employees) (Krause et al. 2024). Exemplary representatives for larger companies include Aquafil (Arco, Trentino, Italy), Eastman Chemical Company (Kingsport, TN, USA), IFP Energies Nouvelles (IFPEN) (Rueil-Malmaison, France), International Business Machines Corporation (IBM) (Armonk, NY, USA), DuPont Teijin Films (Tokyo, Japan) and Dow (Midland, MI, USA).

Gasification

Gasification is another thermochemical process that can convert mixed plastic waste and biomass into syngas ("synthetic gas") and CO_2 in the presence of heat (depending on specific process > 700 °C) and oxygen. The produced syngas is a mixture composed of H_2, CO, CO_2, CH_4 and N_2 in different concentrations depending on the processed feedstock. Based on the intended use of the syngas, different purification and processing steps are required to eliminate specific components such as N_2 or, e.g. to adjust the concentration and ratio of CO and H_2. The advantage of pyrolysis is a higher flexibility in processing mixtures of different plastics and/or plastic mixed with other wastes (Lopez et al. 2018). Currently, the largest capacities achieved are up to 100 kt per year for a single plant. Most of the 13 plants on the market are based in North America. With four companies each, most of the suppliers are small (10–50 employees) and medium-sized (51–250 employees) enterprises. The only representative of a larger company is Eastman (Kingsport, TN, USA).

Enzymolysis

An alternative route is offered by enzymolysis, a technology based on biochemical processes that use various types of biocatalysts to depolymerize a polymer into its building blocks. The process can be fine-tuned via enzyme- and metabolic engineering and the depolymerization mechanism can be developed to achieve a high (chemo-, regio-, enantio- and stereo-) specificity. The technology is currently in an early stage of development and at a laboratory scale mainly addressing PET. Currently, only one commercial supplier of enzymolysis technology is on the market, which is Carbios (Clermont-Ferrand, France).

Dissolution

Dissolution describes a physical recycling process that utilises solvents. In dependence of the utilized solvent the technology is capable of addressing specific polymers such as PA, PE, PP, PS or PVC but also bio-based polymers such as PLA. The process allows the targeted dissolution of polymers from mixed plastic waste in a suitable solvent while retaining the chemical structure of the polymer. Other plastic components (e.g. additives, pigments, fillers or non-targeted polymers) remain undissolved and can be separated from the dissolved target polymer. An antisolvent is then added to change the physicochemical properties of the solvent and initiate precipitation of the target polymer. The polymer can thus be obtained directly; unlike solvolysis, this process does not require a polymerization step. Currently, the process reaches a maximum capacity of 8 kt per year for a single plant, with most of the nine technology suppliers on the market coming from Europe, followed by North America (Krause et al. 2024). With four companies, most of the suppliers are small companies (10–50 employees) (Krause et al. 2024). A representative large company for dissolution is Shuye Environmental Technology (Shantou, China).

10.3.4 Available Capacities in Europe

In Europe,[3] there is already a considerable potential of know-how and providers for chemical and physical recycling technologies which is reflected in the comparison with the globally installed plants and capacities. Of all installed chemical and physical recycling plants worldwide more than 60 and therewith the majority is operating in Europe (cf. Fig. 10.16), followed by the Rest of the World, CIS,[4] North America, Middle East & Africa, China, Latin America, and Japan. Europe is capable to address 358 kt plastics waste[5] per annum via chemical and physical recycling. This corresponds to nearly one-quarter of the worldwide input capacity of advanced recycling which ranks Europe at the top of the global comparison. With 275 kt per annum, pyrolysis has currently the largest share of the installed input capacity in Europe, followed by solvolysis, dissolution, and enzymolysis (cf. Fig. 10.17). In contrast to regions such as Japan and North America, gasification in Europe is in early development and therefore currently no relevant capacities are installed.

Overall, 271 kt of products can be obtained from chemical and physical recycling under consideration of yields and losses of the installed plants. A detailed evaluation (cf. Fig. 10.18) reveals that 41% of the products can be utilized for the production of

[3] EU 27 + 3 including Norway, Switzerland, and United Kingdom.

[4] Commonwealth of Independent States.

[5] All kind of synthetic plastics and polymer wastes such as plastic packaging, tyres, polyester textiles.

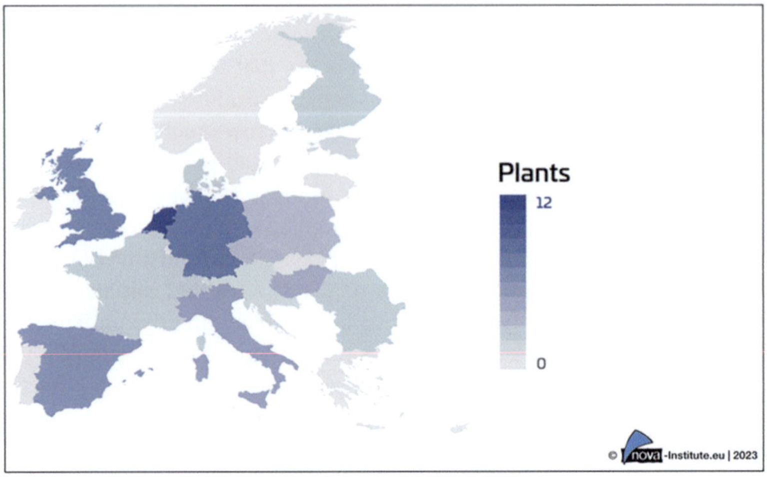

Fig. 10.16 Number of installed chemical and physical recycling plants in EU 27 + 3. Adapted from Krause et al. (2024) with kind permission from nova-Institute

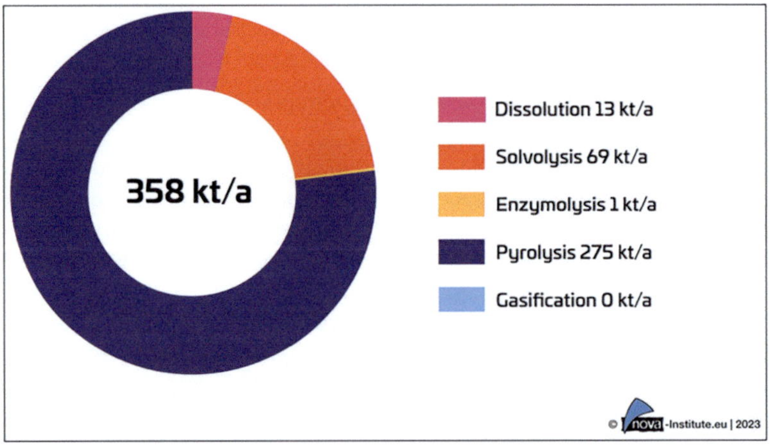

Fig. 10.17 Chemical and physical recycling input capacities by technology. Adapted from Krause et al. (2024) with kind permission from nova-Institute

new plastics and polymers in the form of purified polymers, as well as monomers/oligomers and naphtha. Additionally, 35% of the products are representing Secondary Valuable Chemicals (SVC, such as pyrolysis oil, carbon black, wax and others) that may be utilized for a wide range of applications. No specific end applications can be directly assigned to SVC due to the complexity of possible product composition and quality as well as downstream process paths. The major share of SVC is pyrolysis oil of unknown fate since it can probably be further refined to naphtha, fuel or other chemicals. For a conservative estimate, it has therefore to be assumed that the oil

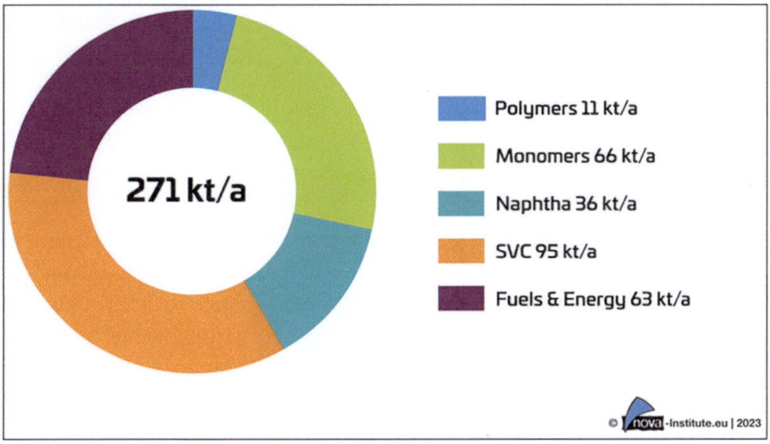

Fig. 10.18 Chemical and physical recycling output capacities by product group. Adapted from Krause et al. (2024) with kind permission from nova-Institute

from the SVC fraction is not primarily used for the production of new plastics or fuels. With that, only 24% of the obtained products fall in the fuels & energy category where the end-applications are known.

Europe's circular strategy becomes evident by setting the product shares of polymers, monomers, naphtha, and SVC from chemical and physical recycling into the global context. Here, Europe is capable of covering 36% of the installed global capacity.

10.3.5 Outlook

In the coming five years, strong market growth is expected in which the amount of installed chemical and physical recycling plants will steadily grow. The first indicator is the announcements of the technology providers for the construction of new plants. An analysis of these announcements shows that the input capacity in Europe in 2027 will more than triple while globally the capacity will double. Considering announcements of plants without any specific date the capacity will more than quadruple while the global capacity will triple. However, the projection for Europe might change in dependence on additional political measures, such as the revision of relevant directives or the establishment of incentives and investment programs.

10.4 Plastic Waste as Circular Feedstock—A Case Study

Peter Orth and Ulrich Liman

10.4.1 LyondellBasell as an Example

Today's climate change mitigation is a challenge for the chemical and plastics manufacturing industry compelling a radical change from a linear business model of converting fossil raw materials into energy and plastics to a circular economy based on renewable ("green") carbon and energy. This transformation needs to be executed within a short time by the standards of the chemical industry, operating energy-intensive assets running continuously to supply the materials to manufacture innumerable products indispensable for our daily life.

The following case study for one of the large chemical companies so far processing fossil raw materials to base chemicals to produce plastics illustrates the first steps to prepare that transformation from their current fossil raw materials-based business model (Ullmann 2003) to a circular one where plastic waste becomes a feedstock for the future success of the company. The company we chose—LyondellBasell (LYB)—in its current shape mainly stems from the polyolefin businesses of Shell, BASF, and Hoechst in Europe and from Arco demerger Lyondell Chemicals in the US. LYB carries the history of the above-named petrochemical and chemical companies that changed their business models over time according to the needs of the markets. Other large chemical companies and plastics manufacturers are also preparing for the transformation from a linear to a circular economy, e.g. BASF, Borealis, Covestro, Dow, Ineos, Repsol, Sabic in Europe, hence, the "LYB Case Study" we compiled is prototypical for the change in the plastic industry in Europe and elsewhere in the world.

LYB was chosen in our context as the company several years in its business communication and especially in its sustainability reports clearly communicates concrete goals and first steps on its way to substitute fossil raw materials.

With a turnover of some $ 50 bn in 2022 LYB belongs to the largest chemical companies in the world and is a leader in polyethylene and polypropylene manufacturing. The company is legally headquartered in Rotterdam the Netherlands with administration centres in Houston, Hongkong, and London.

10.4.2 Elaborating a Strategy for Circularity

In their sustainability reports 2022 and 2023 (LYB 2023a, 2024) clear targets are stated on how to prepare the company portfolio for the upcoming challenges within the next 10 years and beyond. The focus is on three critical sustainability challenges:

(1) Taking climate action
(2) Ending plastic waste
(3) Supporting a thriving society

The first two items will be dealt with in the paragraphs following and the strategic changes are explained that are visible in the market.

(1) Taking Climate Action

With regard to this goal, LYB has defined the following measurable targets:

- Achieve net zero greenhouse gas (GHG) emissions from global operations by 2050
- Reduce absolute scope 1 and 2 GHG emissions to 42% by 2030
- Reduce absolute scope 3 GHG emissions to 30% by 2030
- Procure a minimum of 50% of electricity from renewable sources by 2030.

This is of utmost importance for the company's petrochemical activities where naphtha from the refinery is broken down into olefins (ethylene and propylene) and aromatics (benzene, toluene, xylenes). The basic process therefore is steam cracking taking place in large furnaces ("steam crackers") of which LYB in Europe runs 4 facilities with an overall capacity of some 1,900 kt ethylene per year at three sites (Petrochemicals Europe 2024). The cracking process requires a temperature of more than 850 °C attained by combustion of fossil fuels (gas especially), emitting of course large amounts of CO_2, and is by far the biggest contributor to scope 1 emissions of the company.

In June 2023, LyondellBasell signed a memorandum of understanding with Chevron Phillips Chemical and Technip Energies to potentially design, construct and operate a demonstration unit using Technip Energies' electric steam cracking furnace technology to produce olefins. Technip's technology could enable LyondellBasell to use renewable electricity as a heat source for olefin cracking in the future while reducing the GHG footprint of this process. The demonstration unit will be sited at LyondellBasell's petrochemicals complex in Channelview, Texas, and is designed to prove the technology at an industrial scale. Construction of the demonstration unit is a pre-condition for the potential future construction of a full-scale unit (OGJ 2023). The deployment of an industrial-scale electric cracking furnace is an important development option as it can reduce furnace GHG emissions by up to 90% compared to a conventional furnace.

(2) Ending Plastic Waste

With regard to this goal, the following measurable targets were set:

- Produce and market 2 Mt of recycled and renewable-based polymers annually by 2030
- For every dollar invested in venture funds addressing the plastic waste challenge, another five dollars from co-investors are catalyzed,
- Zero loss of plastic pellets to the environment from our operations.

To accelerate efforts to innovate, scale, and deliver solutions to turn post-use plastics wastes into everyday products and enable a circular economy several measures were announced starting in October 2022.

Circular & Low Carbon Solutions (CLCS) is a new business unit being built up within the group's Olefins & Polyolefins (O&P) segment to "... provide clear accountability and enable LyondellBasell to have a differentiated operating model." The unit will focus on growing the olefins and polyolefins group's Circulen® renewable and circular solutions and accelerating the development of scalable sustainable and circular technologies. To align business strategy with execution, the CLCS business unit has global responsibility for manufacturing operations. LyondellBasell expects CLCS to grow to at least $1 billion in incremental EBITDA by 2030 with attractive returns by leveraging the company's existing strengths of scale, market access, and technology leadership (LYB 2023i). CLCS is active in making investments to secure feedstock supply, expand the recycling footprint, and develop scalable technologies to grow the Circulen® family of polymers.

The Circulen® family of polymers is produced using raw materials derived from mechanical recycling, advanced recycling, or renewable materials. The goal is to produce and market at least 2 Mt of recycled and renewable-based polymers annually by 2030. In pursuit of this goal, the CLCS business unit is investing upstream in plastic-waste sorting and recycling operations in the United States, Europe and Asia to help build a complete offering of sustainable solutions across multiple recycling processes.

To differentiate the new circular products from classical products a new brand name Circulen® for a full range of circular polymers was introduced for differentiated marketing. Thus, LYB is supporting the reduction of plastic waste through the use of recycled content, and a lower carbon footprint through the use of renewable-based content as compared to feedstock from fossil-based sources.

The Circulen® portfolio marketed globally is structured into three primary categories:

- CirculenRecover polymers are made from both pre- and post-consumer wastes using mechanical recycling processes to produce recycled resins. By processing the plastic waste through the shortest recycling loop, a lower carbon footprint can be achieved.
- Using advanced (chemical) recycling technologies, CirculenRevive polymers are intended to become a solution to address at scale the challenge of hard-to-recycle plastic waste, thus being complementary to mechanical recycling. Wastes are converted into feedstock to produce new polymers, which allows for larger volumes of plastic waste to return back into the value chain as high-quality polymers while accomplishing a lower CO_2 footprint. LyondellBasell is developing its own advanced recycling technology under the trade name MoReTec (see below).
- With renewable-based raw materials, such as used cooking oil, as a feedstock, CirculenRenew polymers offer a variety of polypropylene (PP) and polyethylene (HDPE and LDPE) grades equivalent to virgin resin quality while reducing

fossil feedstock use and helping to reduce CO_2 over the product life cycle. These renewable-based polymers offer the same properties in terms of product performance as well as regulatory approvals. This allows drop-in solutions for products like food packaging or surface protection films.

MoReTec is LYB's proprietary pyrolysis technology for which a final investment decision was taken to build a first of its kind, commercial-scale advanced recycling plant, MoReTec–1, located at the Wesseling, Germany site. Once operational, this 50 kt unit will convert hard-to-recycle, post-consumer plastic waste into feedstock for new polymers, marketed under the CirculenRevive brand. Investments alike in additional projects are considered for the near future. The MoReTec technology is highly energy efficient thanks to its innovative process design and use of proprietary catalysts. These characteristics create favorable economics along with a low environmental footprint and high plastics-to-plastics yield (cf. Sect. 13.3).

10.4.3 Strategy Implementation—Published Milestones

To have access to carbon from circular resources different pathways can be chosen. Purchasing specified material is one way, another is to build alliances or ventures. As LYB assets must run continuously this effort is demonstrated in the publications listed below.

Equipped with a new strategy, a new organization, and a new vision the following milestones have been published within fiscal year 2023 after announcement in 2022 (PlastEurope 2022).

February 08, 2023 (LYB 2023b)

LYB and KIRKBI A/S, the family-owned holding and investment company of the LEGO® brand, signed an agreement to invest in APK AG located in Merseburg, Germany, which specializes in a unique solvent-based recycling technology for low-density polyethylene (LDPE). APK aims to increase the recycling of multi-layer flexible packaging materials—which today make up most of the mixed plastic waste from the private consumer sector. To this end, APK has developed the solvent-based Newcycling® process, which separates the different polymers of multi-layer packaging materials and produces recycled materials with a high degree of purity suitable for new packaging materials. LyondellBasell and KIRKBI will become minority shareholders in APK and together with other co-investors will invest approximately € 130 million in APK.

Further Newcyling® Plants are planned to be built to increase the production capacity. Products made using the Newcycling® technology are intended to be added to LYB's Circulen® product portfolio. At its site in Merseburg, APK is operating a pilot plant, where the industrialization of this technology is taking place. In June 2022, APK took over the previous Clariant technical center at the Frankfurt Hoechst location, Germany, via its subsidiary APK Newcycling Competence Center GmbH

(APK NCC), which is now run as APK's research and development centre. APK AG, founded in 2008, currently employs almost 220 people. Around 200 of these are employed at the Merseburg site. The two plastic recyclate products created in Merseburg are marketed as Mersalen® and Mersamid®

April 28, 2023 (LYB 2023c)

LyondellBasell and Veolia Belgium have agreed to restructure the ownership of Quality Circular Polymers (QCP) BV located in Geelen, The Netherlands, with locations in Belgium and the Netherlands. Veolia has sold its 50% share to LyondellBasell which became 100% owner of QCP. The full ownership of QCP supports LYB's goal to produce and market at least 2 Mt of recycled and renewable-based polymers. Thus, LYB is progressing its strategy to build a profitable circular and low-carbon solutions business to meet customer demand for more sustainable products and solutions. The QCP mechanical recycling plants with capacities of about 55 kt of recycled materials make blends using household plastic waste as a feedstock. Those recycled materials are used to make items such as bottles, buckets, caps and closures, as well as strollers and suitcases. LyondellBasell will continue to offer QCP polymers under its CirculenRecover brand. QCP is considered a growth platform to enable circular solutions.

June 16, 2023 (LYB 2023d)

LyondellBasell and AFA Nord, a subsidiary of Mensing Group, a leading recycler of agricultural films, located in Hohenwestedt, Germany, have agreed to create a 50:50 joint venture that intends to recycle post-commercial flexible packaging waste. The joint venture company LMF Nord GmbH plans to build a mechanical recycling plant in Northern Germany to turn Linear Low-Density Polyethylene (LLDPE) and Low-Density Polyethylene (LDPE) waste into recycled plastic materials for use in flexible packaging. Secondary plastic packaging, such as stretch or shrink film, is mainly used to hold packed consumer goods together and protect them during transportation and storage from contamination and damage. Until today, this material is difficult to recycle into high-quality raw materials that can meet the requirements of the packaging industry regarding the strength and transparency of their final product. The start of production is expected in early 2025. The new recycling unit is expected to produce 26 kt per year of recycled LDPE and LLDPE grades that LyondellBasell will market through its CirculenRecover family of products for use in applications like stretch films or collation shrink films.

July 05, 2023 (LYB 2023f)

LyondellBasell has completed the acquisition of Mepol Group, a manufacturer of recyclate-based technical compounds located near Treviso, Italy, and in Poland. Mepol and its subsidiaries will become subsidiaries in LyondellBasell's Advanced Polymer Solutions (APS) segment. Mepol has a compounding capacity of some 40 kt in Italy and some 27 kt in Poland. The APS segment produces and markets compounding solutions, such as polypropylene compounds, engineered plastics, masterbatches, engineered composites, colors, and powders and is a global leader in compounding solutions. With Mepol's expertise in recycled compounds and LyondellBasell's strategy and scale, significant opportunities are instituted to accelerate the growth of the APS business by leveraging macro trends in circularity and creating solutions for everyday sustainable living. The acquisition forms part of LyondellBasell's strategy to build a leadership position to address the rapidly growing demand for circular and renewable solutions with innovative products and technologies through its integrated value chain approach.

July 19, 2023 (LYB 2023e)

LYB announced the acquisition of a 50% stake in Stiphout Industries B.V. (Stiphout) located in Montfort, The Netherlands. Stiphout is involved in sourcing and processing of post-consumer plastic packaging waste. The company operates a facility located with an annual processing capacity equivalent to the amount of plastic packaging waste generated by over 500,000 Dutch citizens per year equivalent to some 15 kt.

Investment in Stiphout aligns with LYB's strategy to invest in recycling and plastic waste processing companies that support existing assets in the Netherlands and Germany and fits with the integrated hub model. Through this collaboration, local synergies with Quality Circular Polymers (QCP) in terms of logistics and operations can be leveraged. It also unlocks possibilities to further expand the CirculenRecover portfolio.

Oct.16, 2023 (LYB 2023g)

LyondellBasell has acquired a 50% stake in Rodepa Vastgoed B.V., the holding company of De Paauw Sustainable Resources (DPSR). The Dutch plastic waste recycling company is involved in sourcing, processing, and trading of post-consumer and post-industrial plastic packaging waste. DPSR operates production facilities located in Hengelo and Enschede, the Netherlands, with an annual processing capacity equivalent to the amount of plastic packaging waste generated by over 1.7 million Dutch citizens per year. The capacity at the two locations in Hengelo and Enschede are at 55 kt for regranulates, as well as 15 kt of regrind. The group also trades in plastic waste with a throughput of around 60 kt annually. In addition, De Paauw Recycling specializes in the processing of PET and PVC as well as the technical polymers ABS and polycarbonate.

Investing in DPSR aligns with LYB's strategy to increase access to plastic waste feedstock, which supports the integrated hub strategy to build scale and expand the production and offering of CirculenRecover products. Demand for recycled plastics continues to grow as consumers and brand owners ask for sustainable materials. Through this collaboration, the CirculenRecover portfolio of mechanically recycled polymers will be further expanded, creating solutions for customers.

Dec. 4, 2023 (LYB 2023h)

LYB has acquired a minority share in Source One GmbH, located near Braunschweig, Germany, a plastic waste sourcing and engineering company, that specialized in developing technical solutions for hard-to-recycle post-consumer plastic waste. Both

LYB and Landbell will be shareholders in Source One. Landbell is a global environmental service provider operating a comprehensive network of Extended Producer Responsibility (EPR) and take-back organizations worldwide.

This strategic investment provides LYB access to Source One's engineering and plastic waste-sourcing services. It also establishes a relationship with Landbell to provide plastic packaging waste volumes for LYB's advanced recycling operations. Processing plastic waste volumes from EPR Service Providers such as Landbell reinforces the commitment by LYB to support brand owners to close the loop for plastic packaging.

50% of the shares in Source One will be held by 23 Oaks Investments, Leiferde, Germany, which together with LYB has formed the joint venture Source One Plastics in 2022. Source One Plastics currently builds an energy efficient, advanced plastic waste sorting and recycling facility in Germany.

10.4.4 Conclusions and Outlook

Strategy and operative European activities of LYB described here show that the company has set ambitious goals and is taking concrete investment steps for its implementation. A dedicated organizational unit is busy to take over and integrate knowledge and experiences which have developed in the market for decades in the form of acquisitions and joint ventures. This unit reports economic success and has ambitious future targets. Own development is carried out making use of the company's vast chemical and process know-how and its scale. Plastic waste is becoming a growing resource that is procured and processed by almost all technologies available to serve as secondary raw material to feed a growing comprehensive portfolio of recovered products even for demanding applications oriented at changing market needs. Simultaneously, research and development are taking place to significantly lower the Product Footprint of the company with a focus on scope 1 and 2 for the time being.

A classical chemical company is driving ecological and economical value and thereby the circular economy. The speed and the size of actions and changes within one fiscal year after the announcement demonstrate that LYB is positioning itself to continuing their current business while preparing to use and grow a circular economy business model.

References

Allen et al. 1999: Allen NS, Edge M, Holdsworth D, Rahman A, Catalina F, Fontan E, Escalona AM, Sibon FF, Ageing and spectroscopic properties of polyethylenes: comparison with metallocene polymer, Polymer Degradation and Stability 1999, 67, 57–67.

Almond 2020: Almond J, Sugumaar P, Wenzel MN, Hill G, Wallis C, Determination of the carbonyl index of polyethylene and polypropylene using specific area under band methodology with ATR-FTIR spectroscopy, e-Polymers 2020, 20, 369–381

ALPA 2023: https://blog.alpla.com/de/pressemitteilung/newsroom-recycling/recycelter-kunststoff-senkt-co2-verbrauch-um-bis-zu-87-prozent (Last access: March 22, 2024)

AMIS 2023: AMIS Maschinen-Vertriebs GmbH. https://www.amis.de/en/welcome/ (Last access: Nov. 7, 2023)

Andritz 2023: ANDRITZ Separation GmbH. https://www.andritz.com/separation-en/locations/andritz-separation-gmbh-germany (Last access: Nov. 16, 2023)

Awaja et al. 2005: Awaja F, Pavel D, Recycling of PET, Eur. Pol. J. 2005, 41, 1453-1477

Bledzki et al. 1999: Bledzki AK, Kessler A, Lieser J, Odour reduction on plastics and its measurement, Polymer Testing 1999, 18, 63–71

Britas 2023: BritAS Recycling-Anlagen GmbH. https://www.britas-solutions.com (Last access: Sep. 12, 2023)

Bvse 2000: Grade list for used plastics. bvse August, 2000. https://www.bvse.de/fachverband-kunststoffrecycling.html (Last access, March 22, 2024)

Cabanes et al. 2020: Cabanes A, Valdes FJ, Fullana A, A review on VOCs from recycled plastics, Sustainable Materials and Technologies 2020, 25, e00179.

Craig et al. 2006: Craig IH, White JR, Mechanical properties of photo-degraded recycled photo-degraded polyolefins, J. Mater. Sci. 2006, 41, 993–1006

Day et al. 1995: Day M, Cooney JD, MacKinnon M, Degradation of contaminated plastics: a kinetic study, Pol. Degr. Stab. 1995, 48, 341–349

DIN EN ISO 14021:2016: Environmental labels and declarations - Self-declared environmental claims (Type II environmental labelling). https://www.beuth.de/de/norm/din-en-iso-14021/253836567 (Last access: Aug. 20, 2024)

Ditta et al. 2004: Ditta AS, Wilkinson AJ, McNally GM, Murphy WR, A study of the processing characteristics and mechanical properties of multiple recycled rigid PVC, J. Vinyl. Addit. Technol. 2004, 10, 174–178

Dorigato 2021: Dorigato A, Recycling of polymer blends, Advanced Industrial and Engineering Polymer Research, 2021, 4, 53–69

Dufresne et al. 2002: Dufresne L. A., Agrawal R. D. (inv.), Methods for removing contaminants from polymers, US 6369192 (2002) assigned to Praxair Technology Inc.

EU 2022: Commission Regulation (EU) 2022/1616 of 15 September 2022 on recycled plastic materials and articles intended to come into contact with foods, and repealing Regulation (EC) No 282/2008.

https://eur-lex.europa.eu/legal-content/EN/TXT/?uri=CELEX%3A32022R1616 (Last access: Aug. 20, 2024)

Frounchi 1999: Frounchi, M. Studies on degradation of PET in mechanical recycling. Macromolecular Symposia, Vol. 144 (1), 465–469. https://doi.org/10.1002/masy.19991440142

Garofalo et al. 2023: Garofalo E, Taurino L, Di Maio L, Neitzert HC, Incarnato L, Assessment of melt compounding with zeolites as an effective deodorization strategy for mixed plastic wastes and comparison with degassing, Polymers 2023, 15, 1858

GBP 2024: GBP Quality GmbH, https://www.gbp-quality.eu (Last access: Nov. 6, 2023)

Genuino et al. 2023: Genuino, H. C., Pilar Ruiz, M., Heeres, H. J. and Kersten, S. R. A. Pyrolysis of mixed plastic waste: Predicting the product yields. Waste Management, Vol. 156 208–215. https://doi.org/10.1016/j.wasman.2022.11.040

Gneuss 2023: Gneuss GmbH. https://www.gneuss.com/en/ (Last access Nov. 9, 2023)

Gumargalieva et al. 1996: Gumargalieva KZ, Ivanov VB, Zaikov GE, Moiseev JV, Pokholoc TV, Problems of ageing and stabilization of poly(vinyl chloride), Pol. Degr. Stab. 1996, 52, 73–79

Gustafsson et al. 1992: WO 9213029, Elimination of odor- and taste-producing substances in polyolefin packaging materials using hydrophobic aluminum silicates, inv.: Gustafsson B, Olsson S, Friman B, assigned to Neste Oy, 1992

Hamos 2023: hamos Recycling and separation technology GmbH. https://en.hamos.com.

Herbold 2023: Herbold Meckesheim GmbH. https://www.herbold.com/ (Last access: Nov. 13, 2023)

Höftberger et al. 2023: Höftberger T, Burgstaller C, The influence of processing on the mechanical properties and emissions of recycled polyolefin, AIP Conf. Proc. 2023, 2607, 140001

Höhne et al. 2022: Höhne CC, Menrath A, Pfaendner R, Metzsch-Zilligen E, Lok B, Keeping the smell in, Kunststoffe International 2022 (6), 28–31

IK 2023: Industrievereinigung Kunststoffverpackungen: Initiative ERDE fördert Kreislaufwirtschaft für Erntekunststoffe.

https://kunststoffverpackungen.de/wp-content/uploads/2023/02/IK-Positionspapier-ERDE-220727.pdf (Last access: Dec. 12, 2023)

ISO 15270:2008: Plastics - Guidelines for the recovery and recycling of plastics waste. ISO copyright office (Ed.), Switzerland, 2008–06. Download at https://www.iso.org/standard/45089.html

Kelly et al. 2005: Kelly AL, Rose RM, Spares R, Coates PD, Recycling of uPVC window profile waste, J. Vinyl Addit. Technol. 11, 2005, 119–126

Keycycle 2023: Keycycle GmbH. Decolorizing films on a gentle cycle. Z. Kunststoffe 7/23, pp. 56–58.

https://www.kunststoffe.de/a/fachartikel/folien-im-schonwaschgang-entfaerben-4877709 (Last access: Aug. 20, 2024)

Kiliaris et al. 2007: Kiliaris P, Papaspyrides CD, Pfaendner R, Reactive-Extrusion Route for the Closed-Loop Recycling of Poly(ethylene terephthalate), J. Appl. Pol. Sci. 2007, 104, 1671–1678

Klingenberg et al. 2024: Klingenberg PCA, Fell T, Ortner E, Soll M, Bruell R, Material Characterization of Recyclates in Circular Economy Handbook, Springer 2024, in print

Krause et al. 2022: Krause, L., Carus, M., Raschka, A. and Plum, N. Mapping of advanced recycling – Providers, technologies, and partnerships. nova-Institut GmbH (Ed.), Hürth, Germany, 2022–06. https://doi.org/10.52548/ITZE5668

Krause et al. 2024: Krause, L., Carus, M., Raschka, A., Özgen, A., Kern, J. and Das, S. Mapping of advanced recycling – Providers, technologies, partnerships, and worldwide capacities. nova-Institut GmbH (Ed.), Hürth, Germany, 2024–01. TBA at https://renewable-carbon.eu/publications/ (Last access: Aug. 20, 2024)

La Mantia et al. 2004: La Mantia FP, Tzankova Dintcheva N, Time-carbonyl groups equivalence in photo-oxidative aging of virgin/recycled polymer blends, Plastics, Rubbers and Composites 2004, 33, 184–186

La Mantia et al. 2023: La Mantia FP, Scaffaro R, Baiamonte M, Ceraulo M, Mistretta MC, Comparison of the recycling behavior of a polypropylene sample aged in air and in marine water, Polymers 2023, 15, 2173

Langen 2023: Langen, M : Considerations for Economic Success in Plastic Recycling Structure. HTP GmbH & Co. KG. https://www.htp.eu/publikationen/ (Last access: March 28, 2024)

Lok et al. 2022: Lok B, Mueller G, Buettner A, Bartel M, Erdmann J, Odor-reduced HDPE-Lignin blends by use of processing additives, Polymers 2022, 14, 2660

Lopez et al. 2018: Lopez, G., Artetxe, M., Amutio, M., Alvarez, J., Bilbao, J. and Olazar, M. Recent advances in the gasification of waste plastics. A critical overview. Renewable and Sustainable Energy Reviews, Vol. 82 576–596. https://doi.org/10.1016/j.rser.2017.09.032

Luzuriaga et al. 2006: Luzuriaga S, Kovarova J, Fortelny I, Degradation of pre-aged polymers exposed to simulated recycling: Properties and thermal stability, Pol. Degr. Stab. 2006, 91, 1226–1232

LYB 2023a: Unlocking possibilities. 2022 Sustainablity Report. https://www.lyondellbasell.com/498cad/globalassets/sustainability/2022-lyb-sustainability-report.pdf (last access: Apr. 14, 2024)

LYB 2023b: https://www.lyondellbasell.com/en/news-events/corporate--financial-news/lyondellbasell-and-kirkbi-invest-in-apk-to-develop-recycling-technology/ (last access: Apr. 14, 2024)

LYB 2023c: https://www.lyondlellbasell.com/en/news-events/corporate--financial-news/lyondellbasell-and-veolia-restructure-their-plastics-recycling-joint-venture-qcp/ (last access: Apr. 14, 2024)

LYB 2023d: https://www.lyondellbasell.com/en/news-events/corporate--financial-news/new-joint-venture-to-boost-circularity-of-secondary-plastic-packaging/ (last access: Apr. 14, 2024)

LYB 2023e: https://www.lyondellbasell.com/en/news-events/corporate--financial-news/lyondellbasell-acquires-50-share-in-dutch-recycling-company/ (last access: Apr. 14, 2024)

LYB 2023f: https://www.lyondellbasell.com/en/news-events/corporate--financial-news/lyondellbasell-completes-acquisition-of-mepol-group/ (last access: Apr. 14, 2024)

LYB 2023g: https://www.lyondellbasell.com/en/news-events/corporate--financial-news/lyondellbasell-acquires-stake-in-de-paauw-sustainable-resources/ (last access: Apr. 14, 2024)

LYB 2023h: https://www.lyondellbasell.com/en/news-events/corporate--financial-news/lyb-acquires-minority-stake-in-source-one/ (last access: Apr. 14, 2024)

LYB 2023i: https://www.lyondellbasell.com/en/news-events/corporate--financial-news/lyondellbasell-launches-new-company-strategy/ (last access: Apr. 14, 2024)

LYB 2024: Everyday sustainability. 2023 Sustainability Report. https://www.lyondellbasell.com/49688c/globalassets/sustainability/2023_lyb_sustainability_report.pdf. (last access: Apr. 14, 2024)

Maris et al. 2018: Maris J, Bourdon S, Brossard JM, Cauret L, Fontaine L, Montembault V, Mechanical recycling: Compatibilization of mixed thermoplastic wastes, Pol. Degr. Stab. 2018, 147, 245–266

Marrone et al. 1996: Marrone M, La Mantia FP, Re-stabilization of recycled polypropylenes, Polymer Recycling 1996, 2, 17–26

Maxwell et al. 2008: Maxwell AS, Weathering of recycled photo-degraded polyethylene, Pol. Eng. Sci. 2008, 48, 381–385

Michaeli et al. 1993: Michaeli, W, Bittner, M, Wolters, L: Stoffliches Kunststoff-Recycling. Hanser, München, 1993

Moss et al. 1989: Moss S, Zweifel H, Degradation and stabilization of high-density polyethylene during multiple extrusions, Polymer Degradation and Stability 1989, 25, 217–45.

Nechifor et al. 2018: Nechifor M, Tanasa F, Teaca CA, Zanoaga M, Compatibilization strategies toward new polymer materials from re-/up-cycled plastics, International Journal of Polymer Analysis and Characterization 2018, 23, 740–757

Neuhoff 2021: Neuhoff, K. Hochwertiges Recycling hat hohes Einsparpotenzial an Treibhausgasen: Interview. DIW 2021. diw_01.c.820724.de

OGJ 2023: https://www.ogj.com/refining-processing/petrochemicals/article/14294627/lyondellbasell-to-test-electric-furnace-technology-for-olefins-production (last access: Apr. 14, 2024)

Ozmen et al. 2019: Ozmen SC, Ozkoc G, Serhatli E, Thermal, mechanical, and physical properties of chain extended recycled polyamide 6 via reactive extrusion: Effect of chain extender types, Pol. Degr. Stab. 2019, 162, 76–84

Paiva et al. 2021: Paiva R, Wrona M, Nerin C, Bertochi VI, Gavril GL, Cruz SA, Importance of profile of volatile and off-odors compounds from different recycled polypropylene used for food applications, Food Chemistry 2021, 350, 129250

Petrochemicals Europe 2024: https://www.petrochemistry.eu/about-petrochemistry/petrochemicals-facts-and-figures/cracker-capacity/ (last access: Apr. 14, 2024)

Petrovic et al. 2022: Petrovic A, Colnik M, Prsa A, Van Fan Y Skerget M, Knez Z, Klemes JJ, Cucek L, Comparative analysis of virgin and recycled thermoplastic polymer based on thermochemical characteristics, Chemical Engineering Transactions 2022, 94, 94, 3121–1326]

Pfaendner et al. 1995: Pfaendner R, Herbst H, Hoffmann K, Sitek F, Recycling and restabilization of polymers for high quality applications, An Overview, Angew. Makromol. Chemie 1995, 232, 193–227

Pfaendner et al. 1998: Pfaendner R, Herbst H, Hoffmann K, Innovative concept for the upgrading of recyclates by restabilization and repair molecules, Macromol. Symp. 1998, 135, 97–111

Pfaendner 2006: Pfaendner R, How will additives shape the future of plastics? Pol. Degr. Stab. 2006, 91, 2249–2256

Pfaendner 2022: Pfaendner R, Restabilization – 30 years of research for quality improvement of recycled plastics review, Pol. Degr. Stab. 2022, 203, 110082

Pfaendner 2023: Pfaendner R, Vitamins for the circular economy, Kunststoffe International 2023 (3), 56–61

Pfeisinger 2017: Pfeisinger C, Material recycling of post-consumer polyolefin bulk plastics: Influences on waste sorting and treatment processes in consideration of product qualities achievable, Waste Management and Research 2017, 35, 141–146

Pinheiro et al. 2004: Pinheiro LA, Chinelatto MA, Canevarolo SV, The role of chain scission and chain branching in high density polyethylene during thermo-mechanical degradation, Polymer Degradation and Stability 2004, 86, 445–453.

Pinter et al. 2021: Pinter E, Welle F, Mayrhofer E, Pechhacker A, Motloch L, Lahme V, Grant A, Tacker M, Circularity study of PET bottle-to bottle recycling, Sustainability 2021, 13, 7370

Plasteurope 2022: https://www.plasteurope.com/news/LYONDELLBASELL_t251266/

PRE 2021: Plastics Recyclers Europe. Methodology guidelines to check the quality of baled PET waste. Brussels, Januar, 2021. https://www.plasticsrecyclers.eu/wp-content/uploads/2022/10/Methodology-to-check-the-quality-of-baled-PET-waste-1.1-January-2021.pdf (Last access: Aug. 20, 2024)

Ramesohl et al. 2020: Ramesohl, S., Vetter, L., Meys, R. and Steger, S. r. Chemisches Kunststoffrecycling - Potenziale und Entwicklungsperspektiven. IN4climate.NRW GmbH (Ed.), Gelsenkirchen, Germany, 2020–07. Download at https://www.in4climate.nrw/fileadmin/Downloads/Ergebnisse/IN4climate.NRW/G-Papiere/2020/in4climate.nrw-diskussionspapier-chemisches_kunststoffrecycling-web.pdf (Last access: Aug. 20, 2024)

Rastin 2016: Rastin H, Ahmadi Z, Pakdel AS, Saeb MR, Abbasian Y, Liravi M, Eslahi A, A physicochemical route for compensation of molecular weight loss during recycling of poly(ethylene terephthalate), Journal of Vinyl and Additives Technology 2016, 22, 387–395

Saperatec 2023: Saperatec GmbH. https://www.saperatec.de/de/ (Last access: Oct. 18, 2023)

Scaffaro et al. 2006: Scaffaro R, Dintcheva Tzankova N, La Mantia FP, On the effectiveness of different additives and concentrations on the re-building of the molecular structure of degraded polyethylene, Pol. Degr. Stab. 2006, 91, 3110–3116

Self et al. 2022: Self JL, Zervoudakis AJ, Peng X, Lenart WR, Macosko CW, Ellison CJ, Linear, graft, and beyond: multiblock copolymers as next-generation compatibilizers, JACS Au 2022, 2, 310–321

Sesotec 2023: Sesotec GmbH. https://www.sesotec.com/emea/en/index (Last access: Nov. 13, 2023)

Shoaib et al. 2002: EP 1250938, Odor absorption and deodorization, inv.: Shoaib A, Raushanah R, assigned to Goldschmidt Chemical Corporation, 2002

Signoret et al. 2020: Signoret C, Edo M, Lfon D, Caro-Bretelle CS, Lopez-Cuesta JM, Jenny P, Perrin D, Degradation of styrenic plastics during recycling: Impact of reprocessing photodegraded material on aspect and mechanical properties, Journal of Polymers and the Environment 2020, 28, 2055–2077

Smith et al. 2018: Smith LM, Aitken HM, Coote ML, The fate of the peroxyl radical in autoxidation: how does polymer degradation really occur, Acc. Chem. Res. 2018, 51, 2006–2013

Spigaroli et al. 2017: Anti-odour multilayer packaging films and packages made therefrom, inv.: Spigaroli R, Stafyla E, WO 201709354 (2017), assigned to Bell Flavors and Fragrances Inc., 2017

Starnes 2002: Starnes WH, Structural and mechanistic aspects of the thermal degradation of poly(vinyl chloride), Prog. Pol. Sci. 2002, 27, 2133–2170

Steinert 2023: Steinert GmbH. https://steinertglobal.com/de/ (Last access: Oct. 13, 2023)

TUD 2023: TU Delft. Plastics refined. Feb. 2023. https://www.tudelft.nl/citg/onderzoek/stories-of-science/magnetisch-plastic-scheiden/. (Last access: Oct. 28, 2023)

Ullmann 2003: https://onlinelibrary.wiley.com/doi/book/https://doi.org/10.1002/14356007

Umincorp 2023: Umincorp B.V. https://umincorp.com/ (Last access: Sep. 22, 2023)

Vilaplana et al. 2008: Vilaplana F, Karlsson S, Quality concepts for the improved use of recycled polymeric materials: a review, Macromol. Mater. Eng. 2008, 293, 274–297

Vishnu Sankar et al. 2017: Vishnu Sankar S, Anil Kumar S, Common Additives used in recycling of polymers in Recycling of polymers, 11-53, edited by Francis R, Wiley 2017, 11–53.

Vogt et al. 2021: Vogt BD, Stokes KK, Kumar SK, Why is recycling of postconsumer plastics so challenging, ACS Appl. Polym. Mater. 2021, 3, 4325–4346

Vollmer et al. 2020: Vollmer, I., Jenks, M. J. F., Roelands, M. C. P., White, R. J., van Harmelen, T., de Wild, P., van der Laan, G. P., Meirer, F., Keurentjes, J. T. F. and Weckhuysen, B. M. Beyond Mechanical Recycling: Giving New Life to Plastic Waste. Angewandte Chemie International Edition, Vol. 59 2–24. https://doi.org/10.1002/anie.201915651

Vouyiouka 2006: Vouyiouka SN, Papaspyrides CD, Pfaendner R, Catalyzed solid-state polyamidation, Macromol. Mater. Eng. 2006, 291, 1503–1512

Weissermel 1997: https://www.wiley-vch.de/de/fachgebiete/naturwissenschaften/industrial-organic-chemistry-978-3-527-28838-0 (Last access: Aug. 20, 2024)

Woidasky et al. 2020: Woidasky J, Sander I, Schau A, Moesslein J, Wendle P, Wacker D, Gao G, Kirchenbauer D, Kumar V, Busko D, Howard IA, Richards BS, Turshatov A, Wiethoff S, Lang-Koetz C, Inorganic fluorescent marker materials for identification of post-consumer plastic packaging, Resources, Conservation & Recycling, 2020, 161, 104976

Zanichelli et al. 2018: A process to remove bad smell and odours from plastic materials using cyclic oligomeric carbohydrates, inv.: Zanichelli A, Tamburello D, Jicsinszky L, Rosetti L, Cravotto G, Martina K, Camporaso MN, WO 2018047205 (2018), assigned to Resilia S.r.l.

Dr.-Ing. Herbert Snell is active in plastics recycling for over 30 years and has held various positions in the waste management and plastics recycling industry. He is involved in the collection, sorting, recycling and recyclate use. He was founding shareholder of MultiPet GmbH/Bernburg, Germany, and managing director of MultiPet and Multiport GmbH until 2023. He holds honorary management positions in the German and European Association of Plastics Recyclers.

Prof. Dr. Rudolf Pfaendner studied chemistry at the University of Bayreuth, Germany, and received his Ph.D. in 1985 with a thesis on electrically conductive polymers. For 25 years he held positions in R&D at the specialty chemicals company "Ciba" i.a. as Global Head of Research for "Plastic Additives" and as Senior Research Fellow. He was Director Division Plastics at Fraunhofer's LBF Institute until his retirement in 2024 working now as Senior Consultant for the LBF Institute. Furthermore he is Honorary Professor at the Technical University of Darmstadt.

Michael Carus is one of the leading European experts, market researchers and policy advisers in the renewable carbon economy – including bio-based, CO2-based materials, and recycling. In 1994, he together with other scientists founded the private and independent nova-Institute for Ecology and Innovation. Ever since the beginning, Carus has been involved in the company as owner and one of the Managing Directors.

Dr. Lars Krause passed his Ph.D. studies in the field of colloid- and environmental chemistry at RWTH Aachen University. Besides his academic education, he collected several years of industrial experience in the field of project management, science, and technology with companies active in aerospace, crop science, and ecotoxicology. As a senior expert at nova-Institute's technology and markets department, he is responsible for monitoring technologies and innovations in the sector of chemical recycling.

Dr. Peter Orth is a chemist by education; with OPC - Orth Plastics Consulting he leads a group of senior consultants with decades of professional experience in leading positions in the German and European plastics industry and its associations. His main focus is on innovation and sustainability in cooperation with science, industry and politics. He is currently working in the fields of CO2 as a feedstock, resource efficiency, and recycling technologies.

Dr. Ulrich Liman is a chemist by education, he joined Bayer AG in 1989 in the marketing of Polyurethane and Polycarbonates and held various managerial positions in Japan, China and Germany. Until 2022 he was responsible for polyurethane R&D and held a seat in the Supervisory Board of Covestro AG.

Chapter 11
Extended Producer Responsibility

Jan Bauer, Günter Krieglstein, Michael Löwe, Lorena Fricke, Michael Vetter, Peter Orth, Ulrich Liman, and Jürgen Bruder

Contents

J. Bauer (✉)
RIGK GmbH, Wiesbaden, Germany
e-mail: Bauer@rigk.de

G. Krieglstein
Tomra Systems ASA c/o Tomra Leergutsysteme Ges.m.b.H, Brunn am Gebirge, Austria
e-mail: guenter.krieglstein@tomra.com

M. Löwe (✉)
Tomra Systems ASA c/o Tomra Systems GmbH, Langenfeld, Germany
e-mail: michael.loewe@tomra.com

L. Fricke
Industrievereinigung Kunststoffverpackungen e.V, Bad Homburg vor der Höhe, Germany
e-mail: frickelorena@gmail.com

M. Vetter (✉)
Rewindo GmbH, Bonn, Germany
e-mail: michael.vetter@rewindo.de

P. Orth
OPC - Orth Plastics Consulting, Cologne, Germany
e-mail: opc.cologne@t-online.de

U. Liman (✉)
ULC - Ulrich Liman Consulting, Langenfeld, Germany
e-mail: druliman2@gmail.com

J. Bruder (✉)
BPC- Bruder Plastics Consulting, Potsdam, Germany
e-mail: plastics@jbruder.de

P. Orth et al. (eds.), *Circular Economy of Plastics*,
https://doi.org/10.1007/978-3-031-78829-1_11

Summary

According to Thomas Lindhqvist "Extended Producer Responsibility (EPR) ... is a policy principle to promote total life cycle environmental improvements of product systems by extending the responsibility of the manufacturers of the product to various parts of the entire life cycle of the product, and especially to the take-back, recycling and final disposal of the product" (Lindhqvist 2000). The perception of Extended Producer Responsibility (EPR) is illustrated using selected examples of plastic products from the packaging, agriculture, building & construction, and furniture markets. These examples may be suggestions for application in other markets considering local conditions, or for seeking similar solutions for other products.

Section 11.1 on industrial packaging gives an impression of how EPR systems initiated by manufacturers function. The complex monitoring of material, information and financial flows is demanding for all parties involved, in particular the system operator. A valuable overview of EPR systems operating in Europe and a quantitative comparison with private consumer packaging in terms of waste treatment is given.

Section 11.2 on PET beverage bottles provides a worldwide overview of the development of existing and planned Deposit Return Systems (DRS) with selected return rates. The cooperation between bottle producers and bottlers (mostly one company), retailers, logistics companies, recyclers, managed by a Central System Operator in a social environment of political frameworks and consumer behavior varies from country to country. Interestingly the deposit fee can also be used to stimulate design for recycling.

Section 11.3 on agricultural plastics (silage films, harvesting twine and nets, greenhouse, asparagus and mulch films, etc.) compiles successfully operating EPR systems and systems under development in Europe and Canada. With one exception only, they are voluntary, i.e. initiated and continuously financed by agricultural plastics manufacturers. Collection and recycling rates accomplished demonstrate the environmental successes being achieved with professional system management, a high level of participation by market players in each country, and effective collaboration with farmers and related stakeholders.

Section 11.4 on PVC products in the B&C markets describes the successful development of the EPR system Rewindo in Germany and its integration into the European sustainability initiative Vinylplus. The results of the take-back of PVC window profiles, shutters, and doors using a network of stakeholders in the entire B&C industry are shown based on recycling quantities and quotas, as well as the reuse of recyclates in various PVC products, namely window profiles again.

Section 11.5 on EPS insulation deals with the very special requirements for the recycling of that material. Due to the flame retardant HBCD, now banned in Europe, the recycling of waste containing HBCD poses very specific challenges. Despite setbacks, an initiative by manufacturers and a scientific institute has developed a recycling technology that reduces the proportion of HBCD-containing EPS waste that is incinerated.

Section 11.6 on PUR flexible foam focuses on products in the home and furniture sector in terms of product responsibility. PUR flex foams are used in furniture upholstery and mattresses, but as an addition, polymer PUR cannot be mechanically recycled. Driven by legally required EPR systems for furniture and mattresses in France and Belgium, intensive work is being done on chemical recycling processes through various corporate collaborations described.

The final Outlook in Sect. 11.7 discusses the need for further EPR systems and possibilities and frame conditions of rolling out voluntary and statutory Extended Producer Responsibility systems on a collective or individual basis for other products or product groups.

11.1 Collection and Recycling of Industrial Packaging

Jan Bauer

11.1.1 Introduction

Plastic packaging in the industrial and commercial sectors plays a significant role in protecting and preserving products (cf. Fig. 11.1). From hazardous substances like chemicals to raw materials for the food industry, various products require tailored packaging concepts. Whether it's solid and powdery products packed in LDPE film bags and PP big bags, or liquid substances filled into HDPE drums or IBCs, individual packaging solutions are essential.

Besides sales packaging, various transport packaging materials such as shrink and stretch films for securing, as well as strapping bands and plastic pallets, play an indispensable role in optimizing industrial logistics processes. Therefore, it is crucial to systematically take back these packaging materials after their use and keep them within the material cycle.

Fig. 11.1 Industrial plastic packaging (*Sources* 1. Ari Wid/shutterstock.com; 2. akturer/shutterstock.com; 3. Srinuan hiranwat/shutterstock.com; 4. Salamahin/shutterstock.com; 5. & 6. RIGK)

In the (EU + 3) 37% of all plastic packaging waste is industrial packaging. 41% of the packaging waste collected is treated in mechanical recycling plants, almost the same amount is going to energy recovery and 17% of this waste stream is still going to landfills (cf. Table 11.1).

Table 11.1 Waste management of packaging waste in the EU + 3 (*Source* PasticsEurope, Conversio 2023)

Eu27 + 3 2022 applications	Waste collected in kt	Waste treatment in kt				Disposal in kt		
	Mt	Total	Mechanical rycling (NMP)	Chemical recycling	Energy recovery	Total	Landfill	Incineration without EfW
Packaging	18.5	**15.3**	6.8	0.2	8.3	**3.2**	3.2	0.0
Household packaging	**11.4**	**9.49**	3.9	0.1	5.5	**1.9**	1.9	0.0
C and I packaging	**7.0**	**5.80**	2.9	0.0	2.8	**1.2**	1.2	0.0

11.1.2 Collection, Sorting, and Recycling

Industrial packaging collection versus household packaging collection

The process of collecting and recycling packaging materials for industry and commerce differs significantly from household packaging recycling. Industrial and commercial waste sources vary widely, depending on the sector and company size, in terms of packaging waste composition, quality, and quantities, necessitating specialized collection systems. In the field of industrial packaging disposal, customized solutions are needed, which can vary significantly depending on the sector and product, while household packaging collection processes and logistics are more standardized. Another peculiarity is the generation of relatively high quantities of hazardous substance packaging requiring collection and recycling processes tailored to their needs.

In industry, unlike in households, the complex sorting process is not required. When industrial packaging materials are systematically collected by specialized return systems, sorted mono-material streams can be generated. The role of the industrial waste generator is paramount since they typically decide the disposal path for their waste and are economically involved in the disposal process. Manufacturers of industrial packaging materials are often involved in the return of their packaging. Successful examples include the reconditioning systems for IBCs and barrels offered by renowned manufacturers (Mauser 2023; Schütz 2023). Similarly, waste management companies catering to industrial clients, often medium-sized, are highly specialized and tailored to specific industries and packaging types.

The material flow of industrial packaging

Following customized, sorted collection by waste management companies, industrial packaging materials are typically roughly sorted and then compacted into stackable bales using channel baling presses. This compaction increases transport efficiency, allowing for more effective use of storage and transport capacity. The plastic waste fractions produced by this process are then transported to specialized recycling plants, where the actual recycling process begins. Typically, the process starts with shredding or milling to reduce the plastic materials into smaller pieces and to increase the surface for the washing process. Subsequently, the shredded plastic parts go through a multi-stage washing process until they are mostly free of contaminants and are then subjected to drying. In a final step, the cleaned and dried plastic flakes or granules are filtered and remolten (extruded) in an extruder. To adjust the quality and properties of the recyclates according to customer requirements, additives are usually added. The produced recyclates can now serve as raw material for various plastic applications.

For some time now, recyclates produced from industrial packaging have increasingly been used in new industrial packaging materials like IBC´s, drums, Big Bags, films, and heavy-duty sacks.

Exceptions to the process described above include all industrial packaging materials that cannot be mechanically recycled due to their filling material or material composition. An example of this is empty packaging of hazardous substances usually collected separately at the industrial end consumer in designated collection systems. The recycling of such materials is carried out in facilities authorized for these substances, often involving energy recovery or highly specialized processes like the cryogenic facility operated by Nehlsen in Bremen, Germany (Nehlsen 2023).

11.1.3 Extended Producer Responsibility

To organize and finance the process of collecting and recycling industrial packaging materials, the tool of "Extended Producer Responsibility" (EPR) is used, as with other waste streams. The concept, now applied worldwide, was initially defined by Thomas Lindhqvist in a report to the Swedish Ministry of the Environment in 1991 (Lindhqvist 2000). Regarding industrial packaging, this extended producer responsibility means that the manufacturer, usually the initial distributor of industrial packaging, is responsible for organizing the return and recycling of these materials.

Germany was the first country to introduce the concept of extended producer responsibility for packaging in 1991. It was implemented within the framework of the Packaging Ordinance, which led to the German industry establishing a system for the collection of household packaging and voluntary manufacturer-run return systems for industrial packaging. Up to the present day, the German Packaging Ordinance has been amended seven times and transformed into a Packaging Law in 2019.

On the European level, a packaging and packaging waste directive was introduced in 1994. An important element of this directive is extended producer responsibility. Since packaging usually has a short lifespan and quickly becomes waste, packaging manufacturers are responsible for its collection and disposal, including the associated costs.

The implementation of the European Packaging and Packaging Waste Directive in the respective member states differs, particularly with regard to the collection of industrial packaging, cf. Table 11.2 (DIHK 2023).

By the end of 2022, the European Commission published a proposal for a regulation on packaging and packaging waste (EU 2022). This proposal was adopted on

Table 11.2 Classification of industrial packaging in national packaging legislation

Packaging legislation		
National legislation on packaging waste **does not** differentiate between household and C and I packaging	National legislation on packaging waste **does** differentiates between household and C&I packaging	The recovery of C and I packaging is not regulated by national leg islation
BG, CY, EE, FI, GR, IE, IT, HR, PL, RO, SK, SI, UK	AT, BE, CZ, DE, LU, NL, NO, FT, SE, ES	DK, FR, HU, CH

(*Source* Own presentation)

December 16th 2024, it will need to be implemented by each member state which leads to a more harmonized internal European market for industrial packaging.

11.1.4 Collection Systems

Role of manufacturer-run return systems

In the field of industrial packaging, manufacturer-run return systems, are very successful. These systems represent the most direct form of practised producer responsibility. In 1992, the German RIGK system was established for the collection and recovery of plastic packaging. RIGK GmbH, now operating a total of eight return systems for packaging and plastics, was founded by plastic manufacturers and producers of industrial packaging (RIGK 2023). Table 11.3 gives an overview of collection systems for industrial plastic packaging in Germany.

Successful collection systems set up by industry and trade have also been established in other European countries. An example is the Belgian system Val-I-Pac established in 1997. It operates successfully as a monitoring system in parallel to the Belgian household system Fost Plus (Val-I-Pac 2023). In many other European countries, industrial packaging materials are also collected through the household system, such as in Austria and Finland. Not all European countries already have collection systems for industrial packaging in place. In Spain and France, industrial collection systems are still in the early stages of development. In these countries, manufacturer-run return systems like the Spanish system ENVALORA and its French counterpart twice are currently evolving (Envalora 2023).

The concept of extended producer responsibility for industrial packaging is also gaining ground internationally. In Chile, there was a system establishment in 2023, PROREP, a manufacturer-run return system developed by the Chilean industry (PROREP 2023). However, not only the developmental status but also the models of these systems differ, depending on legal conditions and industry structures in respective countries.

System models

In essence, there are two common system models: a monitoring and an operational one. In a monitoring system, such as the one implemented by Val-I-Pac, the focus is on achieving recycling quotas through documentation and incentivization. Monitoring systems do not actively interfere with waste disposal and recycling markets, the system does not become the owner of the collected materials, and companies are free to choose their disposal partners. The system pays an incentive fee per ton to waste management companies and, in return, receives information and documentation on the collection and recycling of the processed packaging material flows by the waste manager (cf. Fig. 11.2). Additionally, incentives are provided for initiating separate collection of various materials for industrial end-users. Monitoring systems operate usually at a lower cost than operational ones. They are the preferred option in well-functioning existing recycling markets.

Table 11.3 Manufacturer-run collection systems in Germany for industrial packaging (*Source* Own representation)

Name of the system	Products	Producers that take responisbility	Plastic packaging types	Website
RIGK-system	Chemicals, plastics, construction	Plastic producers, producers of chemicals, producers of construction chemicals, producers of plastic packaging	Containers, drums, sheet sacks, big bags, transport packaging	www.rigk.de
RIGK-G-system	Chemicals, clastics, construction	Producers of chemicals, producers of construction chemicals, producers of plastic packaging	All plastic packaging for hazardous products	www.rigk.de
GVO	Mineral oil, lubricants, anti-freezing agent	Producers of mineral oil, lubricants, operating supplies	Oil carts, drums, container	www.gvoe.de
PAMIRA-system	Pesticides, liquid fertilizers	Producers of pesticides liquid fertilizers	Containers, sheet sacks	www.pamira.de
Beize system	Seed treatment	Producers of seed treament chemicals	Drums	www.pamira.de/verpackungen-abgeben/pamira-beize/
Verena system	Seeds, fertilizers, animal feed products	Producers of seeds, fertilizers, animal feed products	Big bags, film, transport packaging	www.verena-recycling.de
PVS	Automotive parts, oil, lubricants	Producers of automotive parts, oil, lubricants, operating supplies	All plastic packaging	www.partslife.com

In the operational system model, the system becomes the material owner. The system operator typically manages the complete reverse value chain of the packaging materials. This makes it possible to customize the collection system to the needs of different sectors and industries. Especially for packaging requiring specific take-back management operational models are more flexible and customizable (cf. Fig. 11.3). They offer various possibilities. As the system operator becomes the material owner, the system provides full transparency over material flow and can control it. This supports the circulation of materials back into the industry and ensures that the recycling routes of the collected materials remain transparent and monitorable. Additionally, investments in recycling infrastructure can be promoted as the system can provide constant material supply. Drawbacks of operational systems include higher financial responsibility and capital commitment, which can lead to financial risks, especially in volatile markets. Operational systems are typically operated in

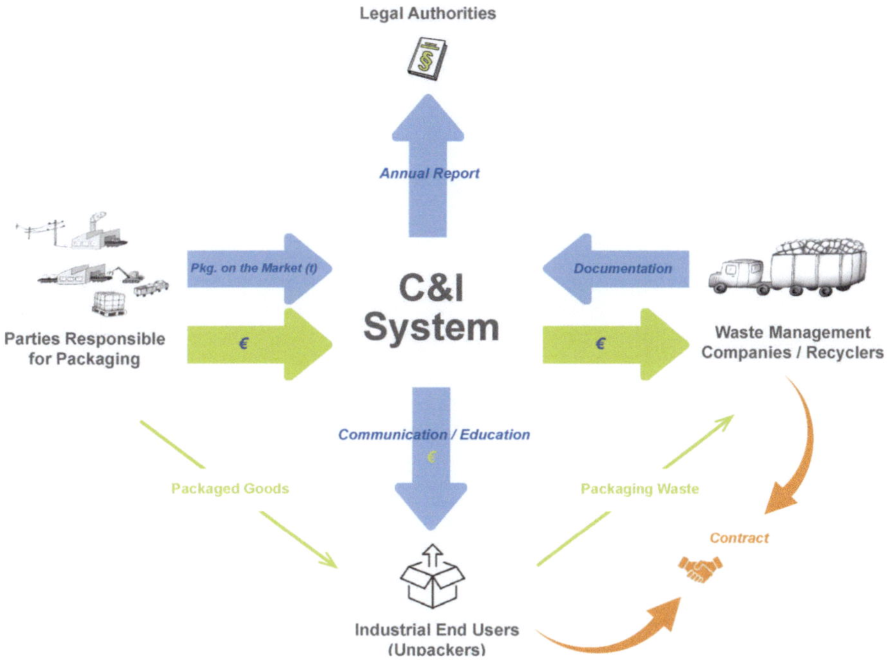

Fig. 11.2 Flowchart monitoring system (*Source* Own presentation)

a full-cost mode, meaning the system covers 80–100% of the operational costs. As a result, contributions from participating manufacturers in operational systems are often higher than in monitoring systems. Good examples of operational systems are the take-back and recycling of oil containers, or the globally operating systems for the collection of pesticide packaging under the management of the Crop Life Manufacturers' Association (Crop Life 2024).

There are also systems that operate a hybrid model combining both approaches. For instance, there are numerous systems in the industrial packaging sector that are operated as so-called "bring systems." These systems only take over the packaging from the waste management site onwards. This means that the system handles the operational step of recycling. Other hybrid systems limit their operational activities to specific regions within a country where collection and recycling do not function adequately or at all without operational system support.

A general statement cannot be made about which of the two system types is preferable. Different system architectures are needed depending on the type of packaging, market, country, and infrastructure. As for industrial packaging collection systems, the heterogeneity of system configurations is also related to the varying industry structures of countries.

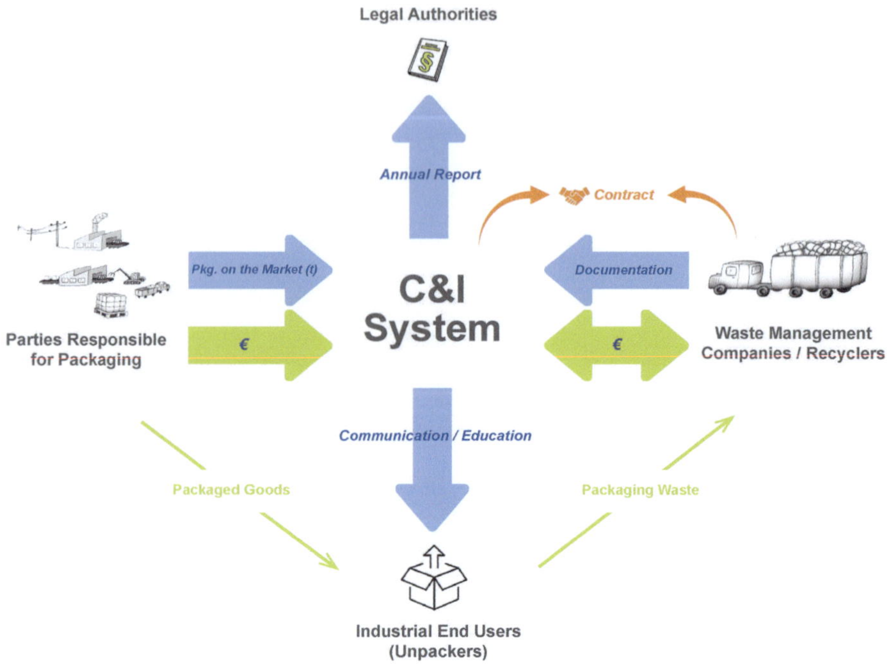

Fig. 11.3 Flowchart operative system (*Source* Own presentation)

11.1.5 Shared Responsibility

Upon closer examination of operational and industry-specific system models, one recognizes a pattern constituting the key to the success of these models: the shared responsibility within the value chain. While manufacturers bear the financial burden of the system and assume responsibility for management and achieving objectives, such as recycling quotas, other segments of the respective value chain also play crucial roles. In the German take-back system for pesticide packaging, PAMIRA (cf. Fig. 11.4), it is the farmer who brings the rinsed packaging to collection points. Without this active involvement of the end-users in agriculture, the system could neither achieve a high collection rate nor the attained collection quality.

The collection points of the system, in turn, are organized by private and cooperative agricultural trade. Collections occur on specific dates once or twice a year. Without the support of the trade in the collection process and its role in actively communicating acceptance conditions and collection dates to the farmers, the system processes would not be feasible with the same efficiency and cost-effectiveness.

The used packaging is collected and compressed by waste disposal companies. Subsequently, transport companies convey them to recycling facilities. These specific and tailored systems are typically managed by a system manager with extensive know-how. Like the example of RIGK GmbH, system operator of PAMIRA, they

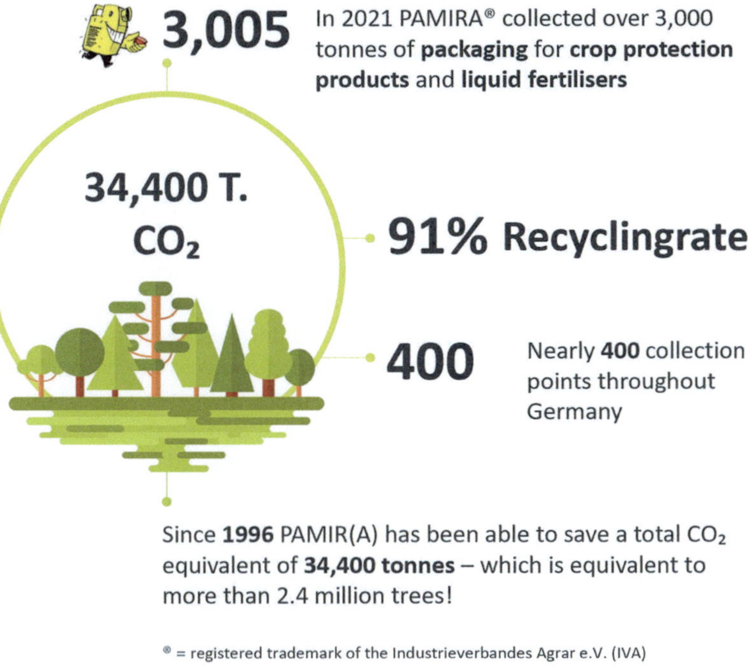

Fig. 11.4 Results of the PAMIRA-SYSTEM (*Source* Own presentation)

often possess both long-term industry experience in the target sector and the necessary waste management expertise in organizing take-back systems.

Without the allocation of various responsibilities to participants in the value chain involving product manufacturers, customers, distributors, and waste management industry, an efficient industry system would be inconceivable.

11.1.6 Future Challenges

Modern take-back systems not only have the central task of retrieving packaging from the market and meeting recycling quotas; they are also responsible for closing material cycles. To achieve this goal, take-back systems assume the role of an intermediary between the various actors in the value chain. This process begins with the design of system contributions within the concept of eco modulation, which means fees not only include operational and overhead cost components but also design for recycling and the use of circular raw materials. The system can support this transformation with advice on eco-friendly design and information on obtaining packaging with circular raw materials.

To increase the percentage of recycled materials in packaging, systems can play the role of a catalyst. This can be achieved by bringing recyclers and processors together or by conducting pilot projects in terms of recycled material testing. The holistic approach aims to adapt the collection and sorting system of a material stream, allowing the production of a regranulate that meets the requirements of a specific target application.

In summary, it can be stated that specialized manufacturer-driven Extended Producer Responsibility (EPR) systems are the best solution for managing used industrial packaging. This applies to the provision of the necessary funds to support collection, sorting, and recycling, as well as the coordination an EPR system can provide. This coordination can encompass the collection, sorting, and recycling of industrial packaging, as well as facilitating the use of recycled content and promoting better product design within an eco-friendly design framework (e.g. through fee modulation). EPR systems can also stabilize the supply chain through long-term contractual relationships and provide flexibility to respond to fluctuating market conditions. EPR systems also play a crucial role in conveying best practices and educating industrial end-users and other stakeholders.

For the efficient take-back of industrial packaging, it is of utmost importance that all parties in the reverse value chain cooperate, and the central task of the system manager is to organize this cooperation.

11.2 Deposit Systems for PET Beverage Bottles

Günter Krieglstein and Michael Löwe

11.2.1 The Importance and the Functioning of Deposit Systems

At the heart of the vision of a circular economy, shared by politicians, industry, and technology providers, is the rediscovery of plastic as a valuable resource. Recycling is a key component for the reuse of plastics in a closed loop. To meet these challenges, many regions considered and are considering the implementation of Deposit Return Schemes (DRS) for single-use beverage containers as Extended Producer Responsibility (Lindhqvist 2000), which can see up to 98% of all drink containers returning for recycling. Deposits on beverage containers are by no means new, they have been around for more than a century. Today there are more formalized programs known as Deposit Return Schemes (also known as deposit return systems, container deposit schemes, or bottle bills) that play a significant role in encouraging recycling around the world. They work by adding a small refundable deposit to the price of a drink, which consumers get back when they return the empty drink container for recycling (cf. Fig. 11.5).

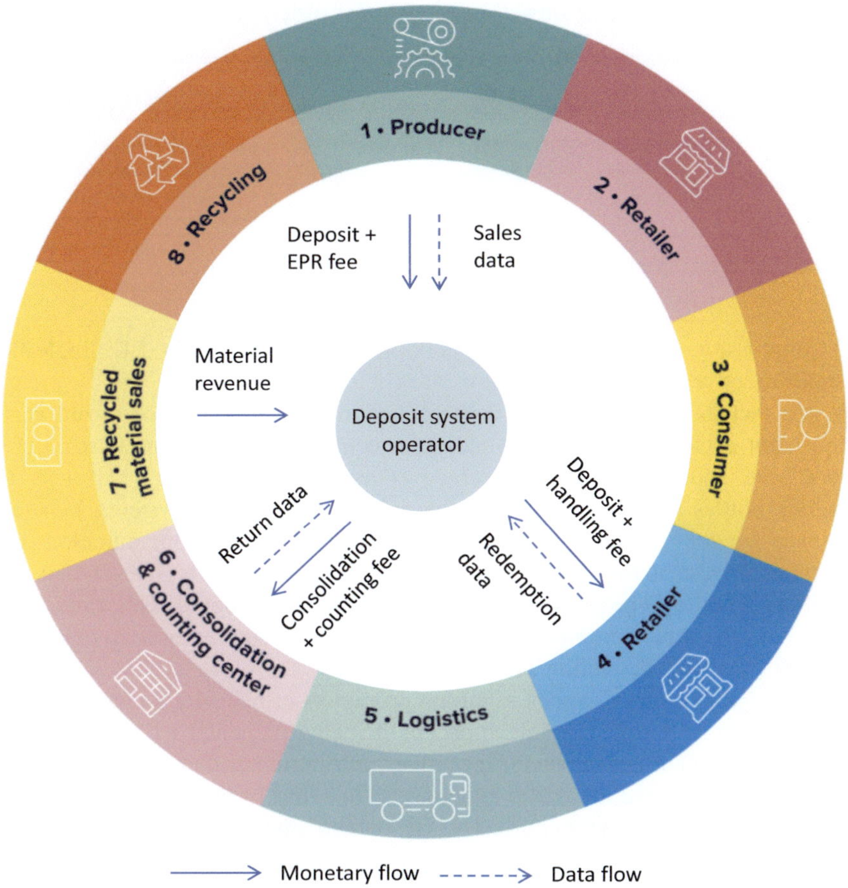

Fig. 11.5 How a deposit system works (*Source* TOMRA 2021)

> Producer pays deposit and EPR fee, and sends data for each container sold to DSO Producer charges price + deposit for each container sold Retailer charges deposit upon sale (regaining their deposit)—Consumer returns beverage container, redeeming their deposit DSO arranges and pays for transport of containers and receives data—Consolidation, counting and validation of all manually-returned containers 050 sells material as feedstock for recyelers Material recycled and converted into new packaging

The deposit value must be added to the cost of the beverage in a transparent manner and be fully refundable (i.e. no tax). Deposit systems guarantee high collection and recycling rates of high-quality materials that can be used as valuable secondary raw materials in the production process of new containers. International experiences show that this is the only realistic way to achieve collection rates of at least 90%

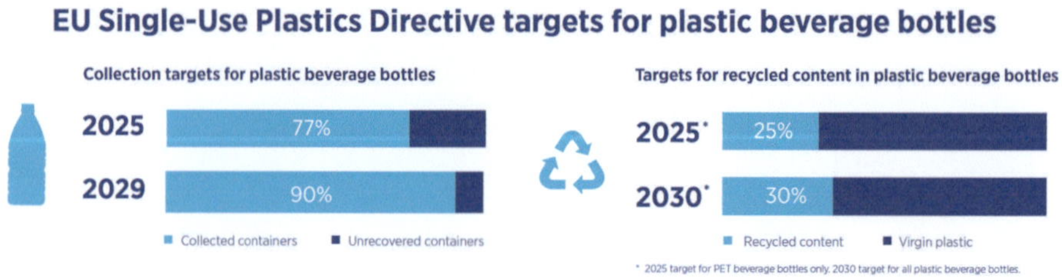

Fig. 11.6 EU targets for plastic beverage bottles (*Source* TOMRA 2021)

(EU Commission 2018) of plastic drinks bottles, as required by the EU Single-Use Plastics Directive (cf. Fig. 11.6).

It is a fact that DRS, separate waste collection and recycling of mixed waste complement each other. Reverse vending systems provide an automated way to collect, sort, and process used beverage containers for recycling or reuse. By collecting bottles and cans separately through reverse vending systems, beverage containers are collected in a safe and clean way. This prevents contamination from other types of waste in the recycling garbage bin and maintains a high level of quality that allows a bottle to be returned through recycling into a bottle. This clean cycle reduces both the dependence on raw materials and the waste that ends up in landfills or as litter in nature (Recycling Markets 2020).

Deposit systems are also seen as strong promoters of behavioral change as they help consumers realize the tangible impact of their environmental actions and reward them with emotional and financial incentives. Furthermore, deposit systems are a tangible way to achieve a circular economy by moving from a linear 'take-make-waste' approach to a 'reuse' culture. The fact that no other method of collection can effectively recover 90% (Reloop 2022) and above of drink containers for closed-loop recycling explains why governments around the world have shown a renewed interest in DRS in recent years; introducing new programs or amending existing legislation to improve scope, participation, and/or increase redemption rates.

11.2.2 Evolution of Deposit Systems Worldwide

Over the last 50 years, deposit systems have been used consistently and have proven to be an unrivaled success for consumers, businesses, and the environment.

The combination of DRS and the recycling of municipal waste helps to prevent littering (WEF 2016, CSIRO 2014, Longwood 2020, Law et al. 2020, RDC 2011), promotes reuse, and guarantees the highest recycling quantity and quality (Eunomia 2020, CRI 2017, Recycling Markets 2020). The positive results in markets where both systems are in place support the fact that several governments are looking to introduce new deposit regulations worldwide (Collins 2020). Today deposit schemes for single-use beverage containers exist in 54 countries (Reloop 2022) around the world. The following table gives an overview of the countries with DRS in place.

Besides the 15 European countries with implemented DRS in 2024 (cf. Table 11.4), several DRS programs are still expected in Europe to roll out over the next couple of years, including Turkey (2024), Scotland (2025), Austria (2025), Poland (2025) and Kosovo (2025). The following figures show the countries and regions with DRS schemes in Europe (cf. Fig. 11.7), Canada (cf. Fig. 11.8), USA (cf. Fig. 11.9) and Australia (cf. Fig. 11.10).

In the USA lawmakers are looking into a draft bill for a nationwide DRS. The legislative language and makeup of supportive stakeholders are still evolving. At the same time, lawmakers in several states including California, Iowa, Massachusetts (MA 2018) and Vermont are working to debate and potentially update their existing legislation to increase container deposits and/or to add more types of containers or beverage categories to the program.

Latin American countries are exploring DRS for single-use beverage containers, with Uruguay leading the way. By the end of 2024, Uruguay plans to implement a comprehensive deposit scheme covering plastic, metal, glass, and even Liquid Packaging Board (LBP) containers. Once in operation, Uruguay's DRS has the potential to serve as a role model for other countries in Latin America interested in adopting a similar system.

Parts of Asia have jumped on the DRS bandwagon, too: Singapore has announced to implement a deposit for drinks containers by April 1, 2025. Similarly, the government of Hong Kong has decided to move forward with a new producer responsibility scheme focused on plastic drink bottles. This scheme incorporates a rebate arrangement where individuals receive a small refund for each plastic bottle returned for recycling at designated collection points.

Momentum around DRS is also growing in Oceania, where several small island countries are pursuing implementation of DRS as a sustainable financing mechanism for waste management (cf. Table 11.4) Samoa is currently preparing for the development of regulations that aim to introduce deposits on a broad scope of materials including PET and aluminum drinks containers. These initiatives are being supported by the PacWastePlus Program (SPREP 2021, PacWastePlus Steering Committee 2022), a 72-month project (2018–2023) funded by the European Union and implemented by the Secretariat of the Pacific Regional Environment Program (SPREP). By the end of 2025, Australia will be the first continent fully covered by DRS.

11.2.3 Benefits and Success Factors of Deposit Systems

The positive impacts of deposit systems cover the political, economic, environmental, and social landscape with the benefits presented in Table 11.5.

So far, all existing deposit refund schemes, with one exception, are based on legislation. The one exception, the Norwegian deposit scheme, is a voluntary deposit refund system based on an environmental tax scheme.

Table 11.4 Deposit schemes worldwide (*Source* TOMRA 2023)

Country/DRS informations					
Region	Country/state	DRS implemented	Region	Country/state	DRS implemented
Canada	British Columbia	1970	Europe	Estonia	2005
Canada	Alberta	1972	Europe	Netherlands	2005 + 2021 + 2023
USA	Oregon	1972	Canada	Northwest Territories	2005
USA	Vermont	1973	USA	Hawaii	2005
Australia	South Australia	1977	Oceania	Kiribati	2005
USA	Michigan	1978	Europe	Croatia	2006
USA	Maine	1978	Canada	Ontario	2007
USA	Iowa	1979	Canada	Prince Edwardian Island	2008
USA	Connecticut	1980 + 2009	Canada	Manitoba	2010
USA	Massachusetts	1983	Oceania	Palau	2011
USA	New York	1983 + 2009	Oceania	Pohnpei	2012
Canada	Quebec	1984 + 2023	Australia	Northern Territory	2013
Europe	Sweden	1984 + 1994	Europe	Lithuania	2016
USA	California	1987	Australia	New South Wales	2017
Canada	Saskatchewan	1988	Australia	Queensland	2018
Europe	Iceland	1989	Australia	Australian Capital Terri	2018
Oceania	Kosrae	1991 + 2027	Oceania	Rep. of the Marshall Islan	2018
Canada	Yukon	1992	USA	Barbados	2019
Canada	New Brunswick	1992	Oceania	Tuvalu	2019
Oceania	Yap	1995 + 2009	Australia	Western Australia	2020
Europe	Finnland	1996 + 2008 + 2012	Europe	Slovakia	2022
Canada	Nova Scotia	1996	Europe	Latvia	2022
Canada	Neufundland and Labrador	1997	Europe	Malta	2022
Europe	Norway	1999	Australia	Victoria	2023
Asia	Israel	2001	Europe	Romania	2023
Europe	Denmark	2002	Europe	Hungary	2024
Europe	Germany	2003 + 2022	Europe	Ireland	2024

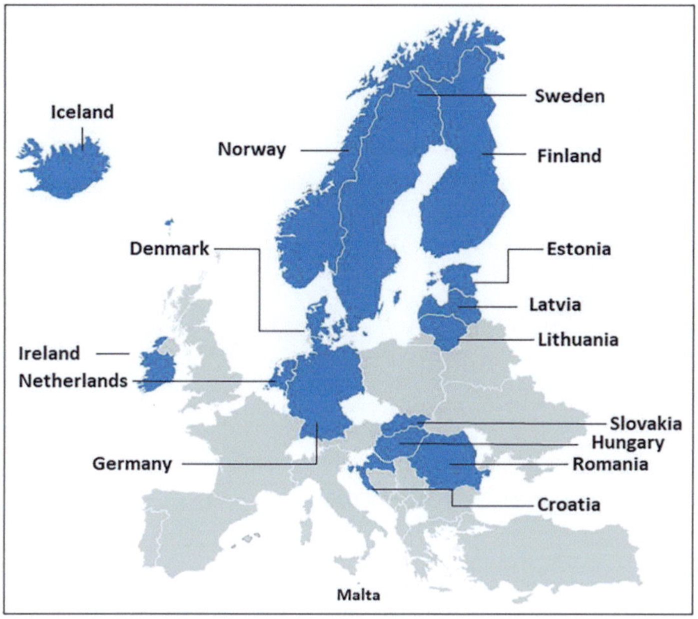

Fig. 11.7 DRS countries in Europe (*Source* TOMRA 2023)

Fig. 11.8 DRS counties in Canada (*Source* TOMRA 2023)

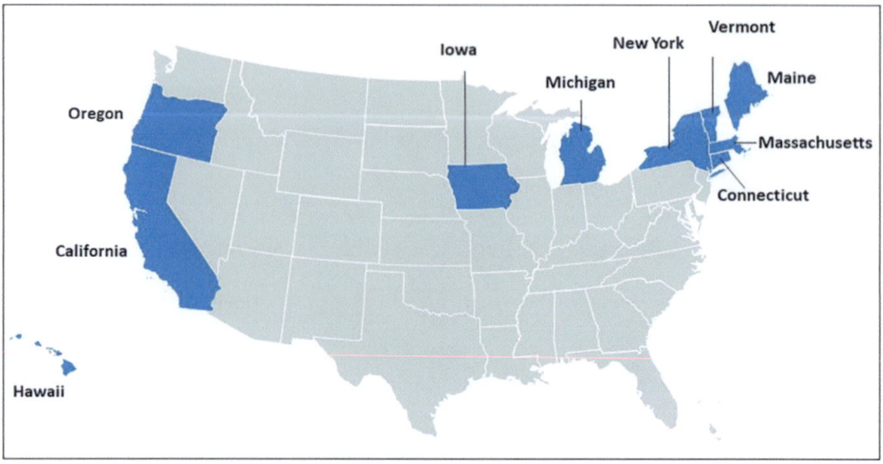

Fig. 11.9 DRS in US states (*Source* TOMRA 2023)

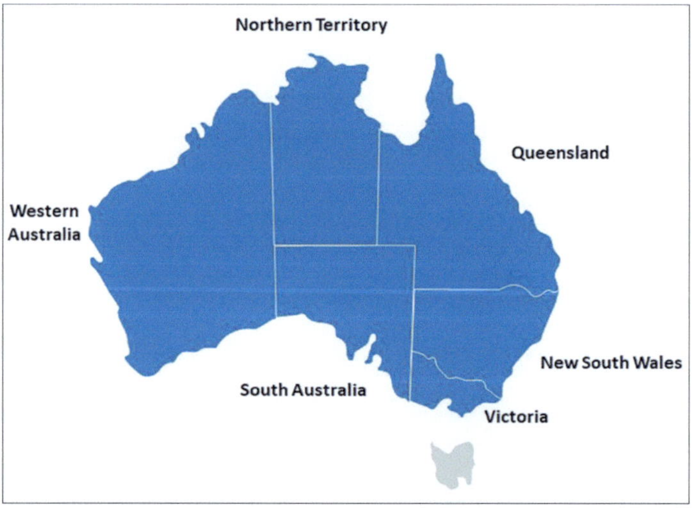

Fig. 11.10 DRS counties in Australia (*Source* TOMRA 2023)

There are no two of the aforementioned deposit refund schemes that are 100% identical – there are always some slight variations between the schemes. Deposit scheme setup is flexible and can differ in one or more key parameters, which define a system. These parameters can be set according to local needs, legislation, recycling targets, demographics, industry structure, existing agreements, etc. So why are some deposit systems succeeding while others are failing? The **success factors** for high-performing deposit return systems prioritize four design principles.

Table 11.5 Benefits of deposit systems (*Source* Own representation)

Benefits	Comments
1. Reduction of litter and ocean plastic leakage	Beverage container litter as a proportion of all litter is 66% less in regions with a DRS than without
2. More material captured for recycling and "saved" from disposal	European collection rate for PET beverage bottles via curbside collection is 47%; via DRS 94% (Eunomia 2020)
3. Guaranteed recycling	Deposit stream ensures the material's value throughout the recycling process
4. Climate benefits	1.1–3.0 t of CO_{2e} emissions reduction by recycling compared to producing 1 t of plastics from virgin fossil feedstock (EMAF 2019)
5. More material recycled in a closed loop	High material quality results in more demand from manufacturers
6. Waste disposal cost savings	Saving taxes or private waste services
7. Litter clean-up cost savings	Saving costs on municipalities, regional governments, and private property owners for dealing with littered material
8. Jobs	DRS creates job demand for collection, sorting, counting, processing and recycling services (CRI 2011)
9. Enjoyment of local environment	Studies have shown that people are willing to pay to live in areas without litter (RDC 2011)
10. Facilitate the transition to refillables	Example Germany: One of the world's most successful refillable container programs, boasting a 41% refill quota and collecting 98% of refillable containers (25.4 billion annually) alongside 98% of one-way containers (20.5 billion)
11. Creation of a local circular economy	DRS is also likely to spur the creation of local material processors
12. Access to recycling	High-performing deposit systems allow all households - regardless of demographic or income - equal access to recycling services. Increasing convenient access is key to increasing recycling

- Performance: A collection target for all beverages plus a meaningful deposit deliver strong results.
- Convenience: A redemption system that is easy, accessible, and fair for all users.
- Producer Responsibility: Producers finance and invest in the system using unredeemed deposits, commodity revenues, and an eco-modulated EPR fee.
- System Integrity: Trust is built into the system's processes through transparent management, a data-driven clearinghouse, and reliable redemption technology.

In practice, these four design principles are brought to life through 12 key policy or program elements (TOMRA 2021) Prioritizing one but not the other will disrupt a deposit system's performance and cost-effectiveness. As with all policies, local culture, infrastructure and politics need to be factored in to shape the system that works best for each market. The 12 key elements of high-performing deposit return systems include:

1. **Broad scope of beverages and containers:** Legislation precisely outlines the covered beverages, material types, and sizes for the program. Leaving out a beverage category could mean billions of recyclable cans or bottles are wasted and potentially littered.[1] Expanding the range of beverage types minimizes consumer confusion at redemption points and enhances economies of scale for the system.
2. **Minimum deposit value:** DRS stands out by providing a financial incentive for recycling, proven to boost container returns significantly.[2] Successful systems set a meaningful minimum deposit, allowing producers to raise for optimal performance. Deposit value is crucial for success - higher deposits correlate with increased return rates.
3. **Return-rate target:** Setting a return-rate target defines a policy's goal, aligns producers with incentives, and triggers convenient redemption options. In this context, "producer" refers to the initial seller of the deposit container (e.g. producer, importer, or distributor). Regulators enforce performance measures.[3]
4. **Convenient redemption system for consumers:** High-performing DRS prioritizes consumer convenience in redemption. Consumers deserve easy access to their deposit refunds, and producers and retailers have an obligation to make that possible. The design should also consider high-volume redeemers and the informal economy. The widely adopted "return to retail" model, where retailers selling beverages have to take back empty containers, proves effective. About 9 out of 10 top-performing global deposit return systems employ this model, boasting an average return rate of 91% (Reloop 2022).
5. **Separately charged and fully refundable deposits**: Effective deposit systems clearly indicate the deposit value on receipts and store shelves while ensuring full refundability. A true deposit is intended to be returned in full upon completing the required action, maintaining a robust financial incentive (cf. Fig. 11.11). The world's top five performing deposit systems (DE, NL, FI, DK, and LT) all feature fully refundable deposits, averaging an impressive 94%

[1] New York's 2009 expansion of the DRS, to include water, resulted in a 100% increase in captured PET plastic containers - water containers now constitute approximately 25% of all containers collected.

[2] Michigan's 10 cent deposit yields a remarkable 90% return rate, while Connecticut's stagnant 5 cent deposit since 1978 has seen rates drop from 88% in 2002 to 50% in 2018. To address this, Connecticut raised the deposit to 10 cents in January 2024 (State of Connecticut 2021).

[3] In Oregon, a 2011 legislative update included a performance target. When the return rate fell below 80% for two consecutive years by 2016, the deposit value automatically increased from 5 to 10 cents, resulting in a rise from 64% in 2016 to 86% in 2019 (Oregon Beverage Recycling Cooperative 2020).

return rate (Reloop 2022). Separating the deposit value from the sales price on shelves and receipts serves to educate consumers and prevent confusion.

6. **Container deposit markings for consumers and manual returns, barcodes for accurate accounting**: High-performing deposit systems enhance consumer awareness and facilitate manual returns by incorporating standardized markings, such as text or logos, on eligible beverage containers (cf. Fig. 11.12). These markings also aid manual redemption locations in easily identifying deposit-worthy containers. Barcodes play a parallel role, enabling automated technology to accurately recognize, count, and account for each deposit container. This ensures precise payments, security, and transparent financial tracking. Unique deposit marks and market-specific barcodes prevent fraudulent redemption and reduce costs.
7. **Extended producer responsibility financing**: EPR is an environmental strategy holding producers accountable for a product's entire lifecycle, emphasizing take-back, recycling, and disposal. DRS involves beverage producers in financing system operations. Ineffective models, unredeemed deposits and material sales revenues are reinvested in the system. If costs exceed revenues, producers must pay the net costs through an "EPR fee". These fees, based on eco-modulated principles, cover remaining system costs, promote circular packaging design, and prevent cross-subsidization among producers.
8. **Reinvestment of unredeemed deposits and material revenue within the system**: Producers gain a "license to operate" through participation in the DRS

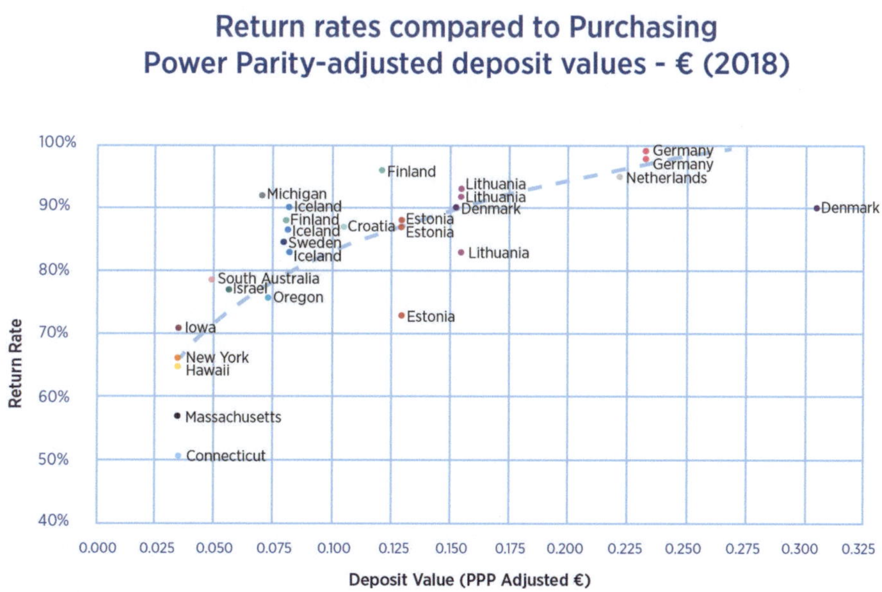

Fig. 11.11 Return rates compared to deposit values (*Source* Own presentation)

Fig. 11.12 Examples for visual container deposit markings (*Source* TOMRA 2021)

and coverage of the net costs—unredeemed deposits and material revenue will be reinvested.[4]

9. **Recycled content requirements**: High-performing deposit systems empower operators to retain revenue from material sales, reducing overall costs with stable commodity values. Recycled materials face market price volatility, posing uncertainties for the net costs.[5] Mandates, such as the EU's Single Use Plastic Directive to include 30% recycled content by 2030, will stabilize commodity values and encourage high-quality recycling (Reloop 2018a, b).
10. **Centralized, non-profit administration and operations**: To prevent increased overall system costs from individual compliance services, high-performing systems advocate centralization under a non-profit entity, known as the "Central System Administrator" (CSA) (cf. Table 11.6). Legislation typically designates system design, such as "return-to-retail" redemption or assigns it to an independent "network operator" responsible for prerequisites like a specified number of inhabitants per collection point.
11. **Government reporting and consumer communication:** Cf. Table 11.7, Right Column
12. **Government enforcement**: In high-performing DRS, government enforcement is clearly outlined in statutes and regulations, specifying penalties and the relevant enforcing agency. An active government agency oversees program success, ensuring compliance, maintaining a fair competitive landscape, and communicating performance. Substantial penalties higher than the savings from non-compliance incentivize stakeholders' adherence and encourage system efficiency investments. Legislation also establishes auditing protocols and designates the enforcing agency.

[4] Norway exemplifies this approach with a 92.3% return rate in 2021. Unredeemed deposits and material sales covered 95% of system costs, with 42% from unredeemed deposits, 45% from material sales and 7% from other revenues, just 5% (net costs) required an EPR fee (Infinitum 2020).

[5] For instance, in the US, food-grade recycled PET prices were 7% cheaper than virgin PET in January 2018 but rose to around 35% more expensive by mid-2020 (Alasdair Carmichael 2020). While brand owners commit to increased recycled content, past targets often yield minimal progress.

Table 11.6 Norway's CSA EPR fee structure (*Source* TOMRA 2021)

Eco-modulated EPR fee structure for Norway's Central System Administrator, Infinitum				
	Aluminium can	Steel can	PET bottle	HDPE bottle
Basic fee	−0.08 NOK	0.21 NOK	0.10 NOK	0.10 NOK
Surcharge for standard barcode (also sold outside Norway)	0.06 NOK	0.06 NOK	0.06 NOK	0.06 NOK
Surcharge for light blue container			0.08 NOK	0.08 NOK
Surcharge for colored container or a sleeve that covers 75% or more of the packaging			0.15 NOK	0.15 NOK
Surcharge for label or sleeve that covers 75% or more of the packaging	0.03 NOK	0.03 NOK		

1 NOK = € 0.096/US $0.12

11.2.4 Conclusion and Outlook

Encouraging stakeholders to view the 12 elements as an ecosystem, not just as options, is crucial. Prioritizing one over another risks compromising system performance and cost-effectiveness. While there's no one-size-fits-all solution, these concepts aim to guide system design based on evidence.

In the face of challenges such as plastic pollution and climate change, stakeholders can find reassurance in a recognized approach to beverage container waste. Thoughtful deposit system design can combat plastic pollution, address climate change, and deliver on the promise of the circular economy. As world leaders grapple with waste challenges, it's clear that public policy ambitions must expand to meet the moment.

As we enter the sixth decade of the DRS, there's room for improvement. Public policies need to be regularly adjusted to reflect changes in products, packaging, inflation, and innovation. Modernizing traditional deposit return models with a thoughtful approach based on performance, convenience, producer responsibility, and system integrity can foster higher performing systems.

Table 11.7 CSA responsibilities (*Source* TOMRA 2021)

CSA responsibilities			
System operation	Data management, deposit clearing and reporting	Sale of collected material	Government reporting and Public communications
• Fulfilling collection performance targets • Managing the system's finances Including setting any EPR (administration) and handling fees • Designing and funding the redemption infrastructure. Including return to retail and where applicable branded redemption centers, to enhance the number of convenient redemption points • Registering sellers and new containers Into the system • Assessing fraud risk and developing countermeasures (e.g. container markings) • Developing and signing contracts with • Approving and conducting quality assurance of manual and automated collection procedures and technology	• Maintaining a central database for all participating products' barcodes (provided by producers) and provision to parties in need • Aggregating data from automated and manual collection points • Oeailng of deposits across the different trade levels • Administering handling fees/ compensations • Reporting program performance to government	• Negotiating conditions/prices and sale of materials • Conducting quality assurance and product development	• Establishing branding and communication guidelines • Conducting public awareness campaigns regarding 1st a minimum) deposit value, covered containers and how to participate • Providing standardised marketing packages to every collection point

Table 11.8 Overview of operational National Collection Schemes for non-packaging plastics 2023

National collection scheme	Country	Status	Collection Portfolio
Svepretur	Sweden	Voluntary	Silage and stretch film, bale nets, baler twine, horticultural films
IFFPG	Ireland	Mandatory	Silage and stretch film, bale nets, baler twine
APE	France	Voluntary	All agri-films, bale nets, baler twine, non-wovens
ERDE	Germany	Voluntary	Silage and stretch film, horticulture film, bale nets, baler twine, non-wovens
CleanFarms	Canada	Voluntary	Silage and stretch film, bale nets, baler twine
ERDE Schweiz	Switzerland	Voluntary	Silage and stretch film, bale nets, baler twine

(*Source* Own representation)

11.3 Collection and Recycling of Agricultural Films and Plastics

Lorena Fricke

Non-packaging agricultural plastics, such as agricultural films, nets and twines, irrigation systems, and non-wovens are indispensable in modern farming techniques. They provide many advantages such as boosting yields, improving efficiency, controlling pests and weeds, managing soil moisture, and curbing food and feed loss and waste. These materials contribute significantly to enhancing food and feed production and sustaining agricultural practices.

Despite their benefits, the improper disposal of these plastics poses severe environmental threats. From causing pollution to emitting microplastics and endangering wildlife, their haphazard management can lead to ecological damage. To counter these risks, it is crucial to adopt robust collection and recycling strategies specifically tailored to handle agricultural plastics worldwide.

11.3.1 Scope and Characteristics of Agri-Plastics

According to FAO (2021) every year 12.5 Mt of plastic products are used in plant and animal production globally, about 3.5% of total plastic production. FAO anticipates a significant increase in consumption over the next 10 years. Asia is the largest market for agricultural plastics today, accounting for roughly 70% of the market. In Europe, an estimated 722 kt of agricultural plastics are put on the market, with nearly half in crop production and the remainder in feed production for livestock (Eunomia 2021). The most common types of agricultural plastics used in Europe are mulch film, greenhouse film, irrigation systems, and protective netting in crop production, as well as silage and stretch film, bale nets and baler twine in feed production.

For films, materials like LDPE and LLDPE are primarily used, with LDPE silage films occasionally incorporating small amounts of EVOH as a barrier layer. Bale nets are predominantly made of HDPE, drip lines of LDPE and MDPE, while PP is commonly employed in yarns and non-wovens. Occasionally, PE/PA films are found in the market. Post-consumer recycled material constitutes a significant share of materials, reaching up to 30% for stretch and up to 80% for silage film in several cases. However, closed-loop recycling systems are still rare.

After use agricultural plastics accrue as waste on farms. The material itself is usually mono-material—mostly polyolefins with added colour and UV protection—but can be heavily soiled with sand, stones, soil, and plant residues. These residues can significantly exceed the weight of films, especially in the case of mulch films, posing a primary challenge in disposal. Eunomia (2021) estimates that on average, contamination accounts for roughly 40% of all agricultural plastics.[6]

Documentation on disposal routes for agricultural plastics worldwide remains inadequate. Available data suggests that only a small share of these plastics is collected and recycled, mainly in high-income countries. In Europe, approximately two-thirds of the generated agricultural plastic waste is reported as "collected," while the fate of the remaining third remains unclear (cf. Fig. 11.13). The recycling rate for agricultural plastics across Europe is estimated at 24% (Eunomia 2021). Improper disposal practices or the absence of convenient recycling options might lead to the incineration or (illegal) landfilling of agri-plastics, posing environmental risks such as soil and water pollution and microplastic emissions.

The solution to inadequate waste management of used agricultural plastics and many environmental concerns are so-called *National Collection Schemes* (NCS)—organizations focussed on providing a nationwide disposal and recycling structure accessible for growers and farmers. These specialised recovery schemes are prepared to deal with the specificity of agricultural plastic waste streams, that may not be compatible with collection schemes for consumer or industry packaging.

11.3.2 National Collection Schemes

Existing collection schemes for agri-plastics are typically based on EPR principles and may or may not be combined with a legislative underpinning.[7] Both, existing

---

[6] Soilage rate for agricultural film and other plastics ranges from 5% to more than 80% depending on type and application (Eunomia 2021; FAO 2021).

[7] In Europe, Ireland stands out as one of the few countries that have implemented dedicated laws aimed at fostering the recycling of agricultural plastics. In France, Sweden, Germany, and Switzerland, the collection schemes are set up voluntarily.

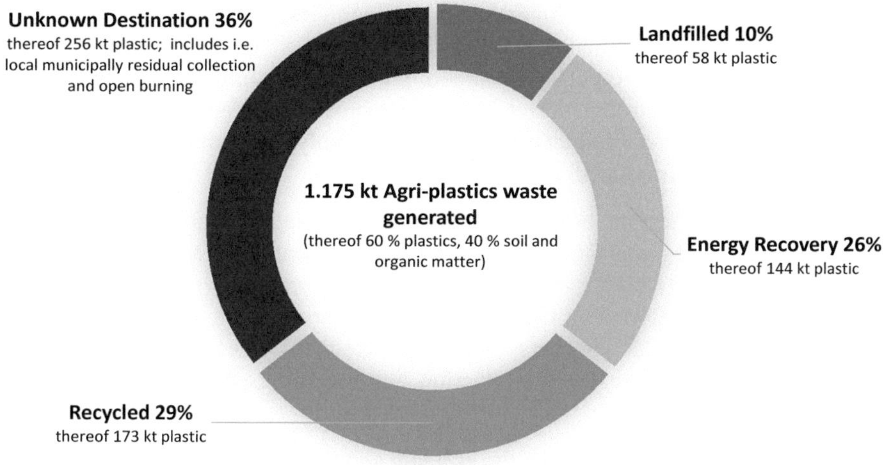

Fig. 11.13 Agri-plastic waste destinations in the EU 2021 (*Source* Eunomia 2021)

voluntary and mandatory schemes, are successful due to their coordinated national approach. They vary in terms of collection portfolio and internal setup but generally work on the same principles.

First collection schemes for agri-plastics date back to the early 2000s when in Europe, the Swedish non-profit industry association *Svensk Ensilageplast Retur* (Svepretur), and the *Irish Farm Film Producers Group* (IFFPG) were founded, followed by the *French Agriculture Plastique Environment* (APE) as part of the *Committee for Plastics in Agriculture* (CPA) in 2009, the *IK-Initiative Erntekunststoffe Recycling Deutschland* (ERDE) as well as *Cleanfarms* in Canada in 2013. The Swiss ERDE collection scheme was founded in 2021 out of the German ERDE system.[8] At the point of writing this chapter, systems in Spain, Finland, Denmark, and the UK were in the process of being set up amongst others (cf. Table 11.8).

According to APE (2023), the fundamental principles that underpin NCS encompass:

- Individual and collective commitment: The success of collection systems hinges on the individual commitment of stakeholders, encompassing manufacturers, distributors, and farmers. Managing used agricultural plastics aligns with Corporate Social Responsibility (CSR), necessitating a collective commitment to optimize methods and costs.
- Shared responsibility: Each participant actively engages in the scheme's operational functions, fostering enhanced technical and economic efficiency.

[8] Svepretur (2023); IFFPG (2023); APE (2023); ERDE (2023); CleanFarms (2023); ERDE Schweiz (2023).

  - Farmers acquire and utilize contributing and recyclable products, prepare waste in compliance with minimum technical requirements post-use, and make it available to the responsible recovery entity.
  - Distributors play a crucial role in communication, training, and operations.
  - Manufacturers, through designated service providers, oversee the recovery and processing of used products. They ensure that only recyclable or biodegradable products are introduced to the market.

- Monitoring and Certification: A monitoring system and external audits scrutinize the collection of used plastics. Recyclers offer certifications for recycled material, ensuring its quality before it is integrated into new products.
- Research and Development: Continuous research projects are pivotal to ensure the sustainability of collection schemes, aiming to reduce waste at its source, enhance waste quality, and streamline collection and recycling processes for seamless integration.

These principles strive to ensure the efficient and sustainable management of agricultural plastics, fostering their recycling and incorporation of recycled material into new products, while minimizing environmental impact and optimizing the efficient use of resources.

How these principles are incorporated into the set-up of NCS differs.

The German ERDE for example is mainly an organization for manufacturers who decide on the strategy, budget, and collection goals of the scheme. Shared responsibility and collective commitment are ensured through bodies such as the ERDE advisory board, where representatives of farmers' associations, distributors organizations, environmental NGOs, research, and governmental agencies engage in regular exchanges and formally advise the ERDE board. Another ERDE-specific instrument to engage actors across the value chain is the "Voluntary Self-Commitments" dating 2019 and 2024. These include concrete commitments on recycling quotas and the extension of the NCS to other product groups. They are formally submitted to the Federal Ministry of the Environment (BMUV) by ERDE jointly with several agricultural associations. On the one hand, these commitments tie the national collection scheme to the agricultural value chain and on the other hand, keep engagement on a high level through the setting of concrete goals.

The Swedish *Svepretur* is an association focussed on manufacturers, importers, and retailers. In the French APE and the Swiss ERDE, the principle of shared responsibility is reflected in a very broad membership group, including farmer's associations, distributors, recyclers, and other representative form the agricultural value chain who jointly make strategic decisions on the collection scheme.

Also, research and development as an integral part of NCS is incorporated in different ways: In France, APE's umbrella association CPA and its system operator ADIVALOR directly invest in the development of new technology for the preparation and the recycling of used agri-plastics such as the RAFU project for mulching film.[9] In Germany, a more indirect approach is chosen, in which expert advice and support (e.g. in the form of recycling tests) is given by ERDE and its system operator RIGK GmbH.[10]

11.3.3 Challenges in Recycling Agri-Plastic Waste

The main challenges in collecting and recycling agri-plastics stem from the processing complexities due to high soilage rates, geographic dispersion of agricultural activities, transportation difficulties, limited specialized recycling facilities, market demand for recycled material, and the economic viability of recycling these materials. Addressing these challenges requires innovative solutions, improved infrastructure, and increased collaboration among stakeholders involved in the agricultural and recycling sectors.

The extensive geographic expanse of agricultural territories poses a challenge in providing adequate and easily accessible collection points in remote or sparsely populated areas. Collection schemes usually set up collection points, strategically positioned in rural regions for accessibility. In 2024, ERDE set up around 700 collecting points across Germany. These collection sites are commonly found at agricultural supply stores, cooperatives, recycling centres, or even specific drop-off locations. The primary objective is to facilitate the disposal of used plastics for small-scale farmers and agricultural producers. For larger scale producers systems may offer on-site container collections—ERDE organized more than 3,000 mobile collections in 2024 alone.

Upon collection, the transportation logistics of used agri-plastics to recycling facilities pose notable challenges. The expense of conveying relatively small amounts of plastics from remote locations to recycling facilities can be economically challenging. This transportation obstacle may restrict recycling options, impacting the practicality and cost efficiency of handling smaller material quantities. NCS addresses this challenge by setting up hubs where collected material from several smaller scale collections is bundled and prepared for transport to specialized recycling facilities.

Post-use agricultural plastics often carry soil, organic remnants, or other residues, posing challenges to their recycling. The requirement for thorough cleaning and sorting prior to processing further complicates and increases the cost of recycling. The recycling process for agri-plastics often demands specialized facilities tailored to manage these distinct characteristics and relatively high soilage rates. In Europe,

[9] RAFU is a project of the Committee for Plastics in Agriculture (CPA) and APE's system operator ADIVALOR to reduce the level of contamination of used agricultural films and improve the conditions for their recycling.

[10] For example in the project SpaFo (2022).

these facilities are scarce and unevenly distributed. Consequently, collected plastics will be transported over extended distances to reach these specialized recycling facilities, intensifying the logistical challenges and expenses.

Today's oftentimes limited market demand for recycled agricultural plastics can also hinder the development of extensive recycling facilities, although demand for recycled material, in general, is estimated to increase in the coming years and decades. Specific recycled plastic markets might not be well-established or lucrative enough yet to justify the investment required for establishing widespread recycling facilities, further limiting recycling options.

National collections schemes, however, are well prepared to face these challenges. In recent years, existing collecting schemes have reported recycling rates of 65% and beyond especially for films used in feed production.[11]

11.3.4 The Future of Agricultural Film and Plastic Recycling

The management of agricultural films and plastic has been on the international political agenda for years.

The UNEP Plastics Treaty, currently being negotiated, aims at confronting the worldwide issue of plastic pollution by establishing a framework for global collaboration, aiming to minimize plastic waste and its environmental consequences. The Council of the Food and Agricultural Organisation of the United Nations (FAO) recommended in 2022 to explicitly consider agricultural plastics in this legally binding treaty, along with developing a complementary *Voluntary Code of Conduct on the sustainable use of plastics in agriculture* (FAO 2025).

Even before that, the *European Circular Plastics Alliance,* committed to boosting the EU market for recycled plastics, included an agricultural working group that resulted in a mandate to not only revise existing product standards for agri-plastics regarding design-for recycling principles, but also to develop a new standard for agri-plastics lifecycle management.[12]

The future of agricultural film and plastic recycling is advancing towards greater sustainability and circularity. Efforts across the industry are aiming to close the loop of agricultural films to create new products with recycled content. The industry's focus remains on developing sustainable and functional solutions, with targets such as utilizing a minimum of 30–50% recycled content in products.

National collection schemes are underlining the importance of collecting and recycling agricultural plastic waste, offering more and more sophisticated solutions for agricultural film and closing the recycling loop. This shift toward sustainability

[11] IFPPG reported a recycling rate for silage and stretch film of 88% in 2022. In the same year ERDE Germany achieved a 68.7% recycling rate for this product group. The French APE reports a collection rate of 74% over all licensed agri-plastic including the horticulture sector in 2022.

[12] CEN/TC 249 WG26: Agricultural plastic products — Installation, use, removal, sorting, collection, preparation for recycling and design-for-recycling guidelines.

and circularity is motivated by the increasing demand for recycled materials, the impact of climate change, and addressing waste challenges within the agricultural sector.

11.4 Collection and Recycling of PVC Windows, Shutters, and Doors

Michael Vetter

Establishing an industry solution

In 2002, Bonn-based Rewindo GmbH Fenster-Recycling-Service became the first company in Germany to set up a nationwide collection and recycling system for used PVC windows, doors, and shutters. In this way, the plastics industry has pioneered the recycling of construction products in the spirit of recycling management and product stewardship. Since its inception, Rewindo's mission, supported by its recycling and logistics partners, has been to promote the material recycling of old windows by providing service, information, and communication to waste producers, in particular window construction and demolition companies, but also to waste management companies, housing construction companies, and the public sector.

Recycling quantities—and quotas

Since its inception, the recycling system has seen an annual increase in recycling volumes from 3 kt of PVC recyclate from old windows to almost 45 kt today, which is returned to the material cycle in the form of PVC recycling profiles, equivalent to more than 2.5 million old windows. In addition, there are over 99 kt of profile waste from window production. In total, almost 145 kt of secondary raw materials are recycled, saving resources and CO_2. The recycling rate for old windows—based on the measurable quantity available—is now around 88% (Conversio 2023).

Logistics and processing

The recycling process for old PVC windows starts with the collection of dismantled and end-of-life plastic windows (cf. Fig. 11.14).

For larger construction projects, Rewindo's recycling and logistics partners collect the old windows directly from the construction site or commission local waste management companies (cf. Fig. 11.15). In addition, Rewindo has set up more than 70 regional collection points in recent years. Small craftsmen can drop off their old windows there. Other PVC windows, doors, and shutters in the area are collected and recycled (Rewindo 2024a).

The recycling and reuse of materials have been technically optimized to such an extent that window profile manufacturers have long been extruding recycled PVC

Fig. 11.14 Removal of old windows (*Source* Rewindo 2023)

profiles with a surface coating of virgin PVC from the recyclers' regranulates (co-extrusion). Once the end-of-life components, some of which have been pre-sorted by logistics partners, have been received, the glass, metal, and other materials are separated and the PVC is recovered in the modern facilities of the recycling partners. The recycling process consists of several steps: First, the old PVC windows are shredded (cf. Fig. 11.16), washed, and milled.

The material is then separated into metal, rubber, glass residue, and plastic using various processes. The latter is processed in an extruder and pressed through a filter to remove any remaining foreign particles. In the end, an almost pure PVC pellet is left as a raw material for profile manufacturers (Veka 2024).

Reuse and conservation of resources

About 70% of these recyclates go back into the production of new window and door profiles (cf. Fig. 11.17). 20% goes into the production of PVC pipes and around 10% is processed into other building profiles. The material cycle is therefore almost 100% closed.

Material flow proof

Further, increasing recycling rates is a common goal of the European PVC industry and its sustainability initiative VinylPlus®. The voluntary commitment is to recycle 1 million tonnes of PVC per year for all applications across Europe by 2030. As a partner of VinylPlus, Rewindo is making an important contribution to achieving

Fig. 11.15 Loading old windows into containers (*Source* Rewindo 2023)

these ambitious targets at the national level. From the beginning, Rewindo and independent auditors have been responsible for monitoring and proving the material flow of the annual recycling volumes in Germany. A few years ago, data collection for old windows and related products was harmonized across Europe with the European Profile Association EPPA (2024a), Brussels, Recovinyl® (2024), Brussels, and Rewindo. This harmonized data collection and independent third-party verification is based on the web-based platform "PolyRec" (2024) and the data collection system "RecoTrace™" (2024).

Product stewardship across the value chain

The increased recycling targets and activities at the European level have led to the creation of a Rewindo network of premium partners since 2016. Over the past few years, the initial beginnings in the field of window construction have grown into a much-noticed alliance of active participants and supporters of the recycling of old PVC windows. Rewindo currently has a total of 26 Premium Partners along the PVC window value chain, including partners from the window manufacturing

Fig. 11.16 Ground material from old windows (*Source* Rewindo 2023)

Fig. 11.17 PVC window profiles with recycled core (*Source* Rewindo 2023)

industry, extrusion technology, film manufacturers, the glass industry, fittings, and steel reinforcement, as well as recycling machinery companies.

The network is currently being expanded to include additional sponsor partnerships. The potential is great. Some 300 window manufacturers have signed an eco-agreement with Rewindo for many years, obliging them to recycle their window waste. In total, there are long-term contacts with more than 1,600 window construction companies, 650 demolition companies, 1,300 municipal and private waste disposal companies, and 700 housing companies (Rewindo 2024b).

Similar schemes in Europe

In addition to the well-established window recycling in Germany, similar organizations for used plastic windows are active in Poland, France, and the UK (EPPA 2024b). There are also a number of recycling schemes for PVC flooring, pipes, roofing membranes, tarpaulins, and cables as part of the "PVC Recycling Campaign" (APR 2024) or via the "PVC Recycling Finder" (PRF 2024).

11.5 Collection and Recycling of EPS Waste from Building Insulation

Peter Orth

Expanded polystyrene (EPS), commonly known under the brand name Styropor®, is a coarse-pored rigid foam made from polystyrene that is used as a lightweight, usually white packaging and insulation material. EPS is used in densities from approx. 15 kg/m^3 (insulating material) to approx. 90 kg/m^3 (helmets). The material is shock-absorbing, cold and heat insulating, waterproof and water-repellent, it is also food-safe, germ-free, odorless, and tasteless.

In 2022, a total of around 7.2 Mt of EPS were marketed worldwide, of which 53% went into building and construction and 41% into packaging applications (Ceresana 2024).

EPS is used in the construction industry for.

- External insulation of facades,
- Core insulation in double-shell masonry,
- Perimeter insulation (thermal insulation of parts of buildings in contact with the ground on the outside).

Because of its flammability, EPS must be adjusted flame-retardant for fire protection reasons in buildings, which until the mid-2010s was predominantly done with hexabromocyclododecane (HBCD) (UBA 2017). However, HBCD was classified as a Persistent Organic Pollutant (POP) in 2013 based on the Stockholm Convention (Stockholm 2004), resulting in a worldwide ban on trade and use. For this purpose, the EU has issued Regulation (EU) 2019/1021 on persistent organic pollutants, which regulates the ban and restriction of the production, placing on the market, and use of POPs and makes provisions for the disposal of corresponding waste (EU 2019a, b). EPS manufacturers have since replaced HBCD with a harmless "polymer FR" (a brominated styrene-butadiene copolymer).

It is expected that HBCD-containing waste from building renovation and demolition will be generated in the order of several 100 kt/a in Europe over the next few decades. This requires an economically and ecologically sound recycling process. With the ban on the use of EPS flame-retardant with HBCD and requirements for the disposal of such construction and demolition waste, the only option initially left was to incinerate this waste. To establish an alternative solution the PolyStyreneLoop initiative was founded in 2017 to develop a safe recycling process for EPS waste from building insulation (PSLoop 2022). The then 75 partners from the entire value chain in 15 European countries worked on the development of such a process until 2021 as part of a project funded by the EU's LIFE program (LIFE 2017), for which funds were provided by the partners, the EU, and the Dutch province of Zeeland, supplemented by loans were used. During this time, a total of over € 17 million was invested in the development and construction of a technical facility at the Terneuzen site, Netherlands, which opened on April 1, 2021, and is operated by PolyStyreneLoop B.V.

Based on the CreaSolv® process developed at the Institute for Process Engineering and Packaging (IVV) of the Fraunhofer Society in Freising, Germany, the EPS waste is first ground and dissolved with a suitable solvent. By adding a precipitant, the polystyrene precipitates while HBCD remains in solution. The precipitated polystyrene is dried and extruded and can then be used again in the same way as the new material. The HBCD solution is distilled and the remaining HBCD is chemically recycled in the adjacent Bromine Recovery Unit of flame-retardant manufacturer ICL-IP.

Parallel to the technical development, a return process is being set up in cooperation with waste disposal companies for the comprehensive collection of building demolition EPS waste, in which the waste is decontaminated (separation of bituminous and asbestos-containing residues as well as plaster and mortar and the like), dried and compacted.

The process is included as a Best Available Technique (BAT) in the categories of pretreatment and hazardous waste incineration in accordance with the technical guidelines of the Basel Convention of the United Nations (UNEP 2017), so that cross-border transport is possible and can therefore be collected across Europe and transported to Terneuzen.

In a life cycle assessment study by Ifeu (2022) titled "The building stock is facing a wave of renovation—insulation materials must open up the material cycle" the outlined solution process is rated as good as regranulation (mechanical recycling). However, since HBCD-contaminated EPS may not be recycled mechanically the solution process is the only way to recover the polystyrene, so that "...across all impact categories/indicators considered, there is a clear advantage for material recycling via PolyStyreneLoop (CreaSolv® Process)".

For liquidity reasons, PolyStyreneLoop B.V. went bankrupt in March 2022 in the wake of the pandemic. As a result, it was taken over by the German consortium German EPS Converters (GEC) in June 2022 and converted to the new PS Loop B.V. and has since continued as a 100% subsidiary of GEC. The GEC group consists of the German EPS processors Bachl (2024), Brohlburg (2024) and Rygol (2024). GEC is supported by the EPS raw material suppliers BASF, Bewi, Sunpor, Synthos, and Unipol, the Austrian EPS processor Hirsch, the flame-retardant manufacturer ICL-IP, and the German rigid foam industry association (IVH) as part of a cooperation on the platform of the European Manufacturers of Expanded Polystyrene Association (EUMEPS 2024). The plant in Terneuzen should now be able to recycle 3 kt of HBCD-containing material annually. For Europe, industry experts expect an annual dismantling of 200 kt of EPS insulation materials (KI 2022).

The driver of success here is that innovation from science is used and that the companies and institutions involved in a value chain cooperate in order to jointly solve the problems associated with the use and exploitation of their product. These include plastics producers and processors as well as construction companies and specialized craft companies, waste disposal and logistics companies, and recyclers, who are often organized in an association and assume their product responsibility cooperatively.

11.6 Recycling of PUR Flex Foams

Ulrich Liman and Peter Orth

Soft PUR foams (flex foams) are mainly used as upholstery material for furniture or car seats, as mattress foam, or as carpet backing material. 1 Mt appr. of raw materials (the isocyanates MDI and TDI as well as polyols) enter the European market every year. Appr. 50% of such foams are used in furniture production, and a further 35% in mattress production (Europur 2021).

The linear mattress value chain comprises (in terms of PUR):

- Raw material producers (isocyanates, polyols, etc.);
- Sales partners and/or system houses;
- Approx. 200 medium-sized foam manufacturers;
- Mattress manufacturer from the components listed below;
- Mattress retailers—established or online;
- Furniture stores.

According to Recyc-Matelas (2021), mattresses are made of the following materials on average:

– PUR flex foam	23%
– Metal (springs)	18%
– Latex	11%
– Textiles	10%
– Polyester fibres	8%
– Others	30%

EUROPUR, the association of European PUR foam manufacturers (Europur 2021) states a theoretical annual waste volume of furniture foams of 350 to 400 kt, that of mattresses of 250 to 300 kt. However, there is hardly any reliable information on the quantities and fate of corresponding waste streams. Currently, when mattresses or upholstered furniture become waste, they mostly end up in bulky waste, to which they contribute 6–8% in France (Recyc-Matelas 2021). According to the European Bedding Industry Association (EBIA), in the EU in 2020, 49% of mattresses were landfilled, 33% were recovered for energy, and 17% were recycled (EBIA 2020). In unit terms, appr. 40 million mattresses containing an average of 15–20 kg of foam became waste in 2019 (Maze 2019); however, no distinction is made here between PUR flex foam and latex foams.

As a cross-linked polymer like rubbers, polyurethanes cannot be reprocessed like thermoplastics once the polyaddition between isocyanate and multifunctional polyols has happened. Instead, PUR flexible foams from collected mattresses (RetourMatras 2021) are shredded, and the resulting foam flakes are reused as composite foam for carpet backing, gymnastic mats or noise insulation (Recticel 2021)—a variant of materials recycling that has been established for many years. However, the market only absorbs limited quantities of such recycled materials, which is why more than 80% of the mattresses discarded by users are incinerated or landfilled (see above).

Initiated by Grenelle Environnement (Légifrance 2008), which began in France in 2007, Extended Producer Responsibility (EPR) systems for furniture and mattresses have since been set up in some EU member states—such as eco-mobilier in France (2021) or Valumat in Belgium (2022), contributing to the European Green Deal (EU 2019a, b). Consequently, concepts for the chemical recycling of PUR foams from mattresses have now also been worked on for several years.

Several major plastics producers are currently reporting on projects for chemical recycling of flexible PUR foams from mattresses, most of which they launched together with major foam manufacturers and mattress collectors: BASF and Greiner (2022), Covestro and Recticel (2021a), Dow and Vita (2021), Evonik and Vita (2021), Repsol and Rampf (2021). The flexible foam derived from old mattresses is degraded by solvolysis processes (alcoholysis, aminolysis, glycolysis (cf. Sect. 10.3)) to polyols and amines (base product for isocyanates). Pilot plants are under way, and the first technical plants are being planned. Covestro describes under Evocycle a process that recycles the long-chain polyether (polyol) as well as the isocyanate TDI (Covestro 2022) to form the desired raw materials for the PUR reaction.

After reprocessing, this opens the way to the reuse of PUR foam production together with corresponding newly synthesized PUR prepolymers. The proportion of recyclates in the resulting virgin material is determined by mass balancing (Covestro 2021b) (cf. Chap. 16).

The partnership of the large raw material and foam manufacturers with specialists for processing and sorting plant construction such as H&S Anlagentechnik (H&S 2022) or BT-Wolfgang Binder Anlagenbau (Binder 2022), with specialized waste disposal companies such as eco-mobilier (2021), RetourMatras (2021) or ReMatress (BASF and Greiner 2022) is remarkable, so that the above-mentioned linear value chain closes into a cycle. Starting points for such collaborations are sometimes EU research projects such as "PUReSmart" (Cordis 2019) or "Polynspire" (Cordis 2021).

In parallel, work is being carried out by stakeholders to improve the suitability of new mattresses for recycling, for example, to reduce the variety of materials or to develop alternatives for bonding mattress layers (Niaga 2022). It remains to be seen whether raw materials used will be standardized.

11.7 Outlook

Jürgen Bruder

The examples of extended producer responsibility in plastics value chains presented in this chapter show, among other things, that

- Collection is the key for any EPR-System in the post-use phase of a product or packaging.
- the holders of producer responsibility can be very different depending on the product group and life cycle phase; in plastic value chains, they are plastic producers, manufacturers of plastic products and packaging, distributors of goods packaged with plastic (e.g. brand owners), OEMs or retailers.
- The assumption of product responsibility through voluntary initiatives by industry can lead to outstanding collection and recycling rates, which are unlikely to be higher even if producer responsibility is prescribed by law.
- The voluntary assumption of financial and operational responsibility for the collection and recovery of used products is the most demanding and riskiest form of product responsibility for the companies involved. For this reason, system operators are often commissioned by the producers to carry out the operational work of the EPR system.

As a rule, the collective fulfillment of a voluntary assumption of extended producer responsibility can only be accomplished as a joint task by companies at the same stage of the value chain for a product/product group, cf. Sects. 11.1 or 11.3, or by companies operating on different stages of the same value chain (cf. 11.2 or 11.6). This places high demands on the compliance work of the companies involved and the platform (e.g. an association) on which such an initiative comes together. A major problem for voluntary initiatives of several companies in one product area is the insufficient market representation of the participants, especially at the beginning of an initiative. In particular, the higher costs per product unit for the members of an initiative due to their "voluntary fee" represent a competitive disadvantage compared to the so-called free riders. Targeted communication work is required in both horizontal and vertical directions, i.e. towards non-participating companies and into the market, to clarify the advantages of such an initiative for all participants, manufacturers, customers through to the private or commercial end consumers.

In terms of the circular plastics economy, there is undoubtedly a need to initiate and install further EPR systems by producers in the sense of those responsible for the products. This is especially relevant for other mass-produced plastic products, to enable these products to be recycled. It is also clear that, for reasons of enforceability alone, the legislator cannot enforce a large number of product-related end-of-life regulations to tackle the shortcomings. Enforcement is already reaching its limits with the laws that have been passed or are to be passed.

The draft PPWR reflects this situation with Articles 41 and 42 (PPWR 2024), which are intended to facilitate the formation of EPR systems and place control in the hands of independent, non-governmental bodies. Here, the PPWR gives the Member States a lot of leeway, whereby EPR systems (collective or individual by a company) should always cover the entire territory of a country. With the obligation addressed to producers of a certain product group to participate in a certified collective EPR system or to be certified with an individual system, both the free-rider problem can be solved, and bogus solutions can be prevented.

Another possible approach to promote EPR systems would be to include an obligation to participate in an accredited EPR system in selected CEN product standards. The new Ecodesign for Sustainable Products Regulation (cf. Chap. 21) also opens possibilities to initiate separate collection of products as a precaution for collective or individual EPR systems.

References

Alasdair Carmichael 2020: Communication with Alasdair Carmichael, Program Director, National Association of PET Container Resources (NAPCOR). November 2020

APE 2023: https://www.plastiques-agricoles.com/agriculture-plastique-et-environnement-ape-commission-ape/ (Last access: Nov. 12, 2023)

APR 2024 : https://www.aktion-pvc-recycling.de/ (Last access: Sep. 6, 2023)

Bachl 2024: https://www.bachl.de/de/ (Last acces: Feb. 27, 2024)

BASF and Greiner/neveon 2022: https://www.neveon.com/de/news-downloads/detail/basf-und-IKA 2022; https://www.ika-wolfen.de/de/pvc-stabilisatorenneveon-arbeiten-zusammen-beim-recycling-von-matratzen/ (Last access: Feb. 27, 2022)

Binder 2022: www.bt-anlagenbau.at (Last access: Feb. 27, 2022)

Brohlburg 2024: https://www.brohlburg.com/

Ceresana 2024: https://ceresana.com/produkt/marktstudie-expandbaren-polystyrol-eps

Clean Farms 2023: https://cleanfarms.ca/ (Last access: Nov. 21, 2023)
Collins 2020: Collins, Susan: Container Recycling Institute NAPCOR Webinar July 2020
Container Recycling Institute 2017: "Beverage Market Data Analysis 2015," Container Recycling Institute. 2017. https://www.container-recycling.org/index.php/publications (Last access: Aug. 20, 2024)
Conversio 2023: Determination of the waste volume and recycling scenario in Germany for PVC windows and related products 2022, Conversio, Mainaschaff, August 2023
Cordis 2019: https://cordis.europa.eu/project/id/814543/de (Last access: Aug. 20, 2024)
Cordis 2021: https://cordis.europa.eu/project/id/820665/de (Last access: Aug. 20, 2024)
Covestro 2022: https://www.covestro.com/press/neues-konzept-fuer-Kreislaufloesungen-mit-alternativen-rohstoffen (Last access: Jan. 14, 2024)
Covestro and Recticel 2021a: https://www.covestro.com/investors/news/de/den-kreislauf-fuer-polyurethan-matratzen-schliessen/ (Last access: Jan. 14, 2024)
Covestro 2021b: https://www.covestro.com/press/de/covestro-unterzeichnet-erste-globale-vereinbarung-zur-lieferung-von-massenbilanziertem-tdi-an-sinomax/ (Last access: Feb. 27, 2022)
CRI 2011: Returning to Work: Understanding the Domestic Jobs Impacts from Different Methods of Recycling Beverage Containers, Container Recycling Institute, 2011. https://productstewardship.net/sites/default/files/PDFs/libraryContainers-Jobs-CRI-Morawski-Morris-Dec2011.pdf (Last access: Aug. 20, 2024)
Croplife 2024: https://croplife.org/our-work/promoting-stewardship/ (Last call: Feb. 09, 2024)
CSIRO 2014: Understanding the effects of marine debris on wildlife, CSIRO. 2014
DIHK 2023: https://www.dihk.de/resource/blob/33922/72bdaa7b9bbd8438b3fe617ef3d332ae/dihk-broschuere-verpackungen-in-europa-data.pdf Last call 03.12.2023
Dow and Vita 2021: Closing the Loop for Mattresses: A New Beginning with RENUVA-Qumu Player (dow.com) Last accessed Feb. 27, 2022.
EMAF 2019: "The New Plastics Economy: Rethinking the future of plastics and catalyzing action," Ellen MacArthur Foundation 2019: https://emf.thirdlight.com/file/24/_ABkCs_skP18I_Am1g_JWxFrX/The%20New%20Plastics%20Economy%3A%20Rethinking%20the%20future%20of%20plastics.pdf (last call: Nov. 12, 2023)
Envalora 2023: https://envalora.es/ Last call 03.12.2023
EPPA 2024a: https://www.eppa-profiles.eu/ (Last access: Sep. 6, 2023)
EPPA 2024b: https://www.eppa-profiles.eu/sustainabiltity-energysaving (Last access: Sep. 6, 2023)
ERDE Schweiz 2023: www.erde-schweiz.de (Last access: Nov. 12, 2023)
Erntekunststoffe Recycling Deutschland 2023: www.erde-recycling.de (Last access: Nov. 12, 2023)
EU 2019a: https://ec.europa.eu/info/strategy/priorities-2019-2024/european-green-deal_de (Last access: Aug. 20, 2024)
EU 2019b: https://eur-lex.europa.eu/legal-content/DE/ALL/?uri=CELEX%3A32019R1021 (Last access: Aug. 20, 2024)
EU 2022: https://environment.ec.europa.eu/publications/proposal-packaging-and-packaging-waste_en' Last call 03.12. 2023
EU Commission 2018: Single Use Plastic Directive. https://environment.ec.europa.eu/topics/plastics/single-use-plastics_en (Last access: Aug. 20, 2024)
EUMEPS 2024: https://eumeps.org/psloop-project-eps-recycling-breakthrough/ (Last access: Aug. 20, 2024)
Eunomia 2020: DRS: Derived from Global Data sales and redemption data from European deposit system operators. 2019. Available upon request. Curbside: "PETCORE Europe Presentation 2020," Eunomia. 2020. https://743c8380-22c6-4457-9895-11872f2a708a.filesusr.com/ugd/dda42a_0f5a9363bfd44caa8b890d6f43529aa9.pdf
Eunomia 2021: Conventional and Biodegradable Plastics in Agriculture, For the European Commission DG Environment. Project conducted under Framework Contract No ENV.B1/FRA/2018/0002 Lot 1; Luxembourg: Publications Office of the European Union
Europur 2021: file:///C:/Users/orth_/AppData/Local/Temp/EoL_Brochure_2021_EUROPUR.pdf

Evonik and Vita 2021: https://corporate.evonik.com/de/presse/pressemitteilungen/speciality-additives/evonik-entwickelt-gemeinsam-mit-the-vita-group-ein-effizientes-verfahren-zum-recycling-von-matratzen-165404.html (Last retrieval 27.02.2022)

FAO 2021: Assessment of Agricultural Plastics and their Sustainability, Rome: Food and Agricultural Organization of the United Nations

FAO 2025: A Provisional Voluntary Code of Conduct on the sustainable use of plastics in agriculture; https://openknowledge.fao.org/handle/20.500.14283/cd3004en (Last access: May 2, 2025)

Farm Film Producers Group 2023: https://farmplastics.ie/ (Last access: Nov. 12, 2023)

H&S 2022: www.hs-anlagentechnik.de. (Last access: Feb. 27, 2022)

Ifeu 2022: https://www.ifeu.de/fileadmin/uploads/pdf/2022_04_07_D%C3%A4mmstoffe_II_JR_CV_FK.pdf (Last access: Aug. 20, 2024)

Infinitum 2020: "Annual Report 2019," Infinitum. 2020. https://infinitum.no/media/t2nfd30f/infinitum_annual_report_2019_01_low.pdf (Last access: Aug. 20, 2024)

KI 2022: https://www.kiweb.de/Default.aspx?pageid=199&docid=250559 (last access: Aug. 20, 2024)

Law et al. 2020: "The United States' contribution of plastic waste to land and ocean," Law, et al. Science Advance. 2020. https://oceanconservancy.org/wp-content/uploads/2022/03/Law-et-al-2020-Fact-Sheet.pdf (Last access: Aug. 20, 2024)

Légifrance 2008: https://www.legifrance.gouv.fr/jorf/id/JORFTEXT000020243534 (Last access: Feb. 24, 2022)

LIFE 2017: (https://webgate.ec.europa.eu/life/publicWebsite/project/LIFE16-ENV-NL-000271/polystyrene-loop

Lindhqvist 2000: Thomas Lindhqvist: Extended Producer Responsibility in cleaner production. https://lup.lub.lu.se/search/files/4433708/1002025.pdf (Last call: Febr. 09, .2024)

Longwood U. 2020: Littered Bottles and Cans: Higher in Virginia than in States with Bottle Bills. Longwood University. 2020

MA 2018: Massachusetts Department of Environmental Protection 2018: "Municipal Benefits of an Expanded Bottle Bill". Commissioned 2009. https://www.amherstma.gov/DocumentCenter/View/3352/635-REQUEST-FOR-SUPPORT-OF-BOTTLE-BILL?bidId=

Mauser 2023: https://mauserpackaging.com/about-us/ (Last call: Nov. 22, 2023)

Nehlsen 2023: https://www.nehlsen.com/recycling-entsorgung/anlagen/cryogenanlage/ (Last access: Oct. 01, 2023)

Niaga 2022 : https://www.niaga.world/what-we-do/product-design-principles.html. (Last retrieval 27.02.2022)

PacWastePlus Steering Committee 2022: 12 May 2022. "Meeting Report." Zoom Teleconference. https://pacwasteplus.org/wp-content/uploads/2022/05/PacWastePlus-Steering-Committee-Meeting-Report-2022.pdf

Polyrec 2024: https://www.polyrec.eu/ (Last access: Sep. 6, 2023)

PPWR 2024: Comporomise Paper between Council and EU-Parliament, 15 March 2024. https://data.consilium.europa.eu/doc/document/ST-7859-2024-INIT/en/pdf . Last access: March 25, 2024

PRF 2024: https://www.pvcrecyclingfinder.de/ (Last access: Sep. 6, 2023)

ProRep2023: https://prorep.cl/ (Last access: Nov. 10, 2023)

PSLoop 2022: https://www.psloop.eu/downloads/

RDC 2011: RDC Environment (2011) "Évaluation contingente du coût des désagréments visuels causés par les canettes dans les déchets sauvages en Wallonie, Raport Final, Etude pour l'Office Wallon des Dechets, Décembre 2011." Can litter in 2011 assumed to also be beverage-related litter in today's environment

Recotrace 2024: https://recotrace.com/auth/login (Last access: Sep. 6, 2023)

Recovinyl 2024: https://www.recovinyl.com/ (Last access: Sep. 6, 2023)

Recticel 2021: https://www.recticel.com/sites/default/files/newsroom/News_articles/2021_flip-book/Recticel_report_2021_0504.pdf. (Last access: Feb. 27, 2)2

RecyclingMarkets2020: https://www.recyclingmarkets.net/ (Last access: Aug. 20, 2024)

Recyc Matelas 2021: http://www.recyc-matelas.fr/comprendre-le-recyclage-chiffres-cle.html. (Last access: Feb. 27, 2022)

Reloop 2018a: "Closing the Loop on the Circular Economy," Reloop. 2018. https://www.reloopplatform.org/wp-content/uploads/2018/06/RELOOP_POSITION-ON-RECYCLED-CONTENT_June-2018.pdf (Last access: Aug. 20, 2024)

Reloop 2018b: Update on Europe's New Waste Legislation: Single Use Plastic Directive https://www.reloopplatform.org/wp-content/uploads/2018/12/SUPD-Backgrounder.pdf (Last access: Aug. 20, 2024)

Reloop 2022: Global Deposit Book 2022, Reloop. 2022, Access via https://www.reloopplatform.org/wp-content/uploads/2023/05/RELOOP_Global_Deposit_Book_11I202.pdf

Repsol and Rampf 2021 : https://www.rampf-group.com/de/aktuelles/newsroom/2021/repsol-und-rampf-eco-solutions-partnerschaft-fuer-das-polyurethan-recycling/ (Last access: Feb. 27, 2022)

RetourMatras 2021: https://www.retourmatras.nl/ (Last access: Feb. 27, 2022)

Rewindo 2024a: https://rewindo.de/recycler-und-annahmestellen/ (Last access: Sep. 6, 2023)

Rewindo 2024b: https://rewindo.de/netzwerk/ (Last access: Sep. 6, 2023)

RIGK 2023: https://rigk.de (Last access: Dec. 04, 2023)

Rygol 2024: https://www.rygol.de/ (Last access: Aug. 20, 2024)

Schütz 2023: https://www.schuetz-packaging.net/schuetz-germany-hq/de/ticket-service/ (Last access: Oct. 05, 2023)

SpaFo 2022: Umweltgerechte Folienentsorgung und Vermeiden von Mikroplastik im Spargelanbau; https://www.atb-potsdam.de/de/forschung/forschungsprojekte/projektsuche/projekt/projekt/spafo (Last access: Nov. 12, 2023)

SPREP 2021: Secretariat of the Pacific Regional Environment Program. 19 November 2021. "Clean Pacific Roundtable Enables Pacific Islands to Enhance Understanding of Waste Management Sustainable Financing Systems." Accessed from https://www.sprep.org/news/clean-pacific-roundtable-enables-pacific-islands-to-enhance-understanding-of-waste-management-sustainable-financing-systems (Last access: Jan. 12, 2024)

Stockholm 2004: https://chm.pops.int/default.aspx

Svensk Ensilageplast Retur (2023): https://svepretur.se/ (Last access: November 2023)

TOMRA 2021: TOMRA White Paper Rewarding Recycling, 2021 Rewarding Recycling: Learnings from the world's highest-performance deposit return systems (tomra.com) (Last access: Aug. 20, 2024)

UBA 2017: https://www.umweltbundesamt.de/sites/default/files/medien/421/publikationen/faq_hbcd_de_17.pdf (Last access: Aug. 20, 2024)

UNEP 2017: UNEP-CHW.13-6-Add.1-Rev.1.English-2.pdf

Val-I-Pac 2023: https://www.valipac.be/en/ (Last access: Feb. 09, 2024)

Veka 2024: https://www.veka-umwelttechnik.de/recyclingprocess/ (Last access: Sep. 6, 2023)

WEF 2016: "New Plastics Economy: Rethinking the Future of Plastics," World Economic Forum. January 2016. https://www.weforum.org/publications/the-new-plastics-economy-rethinking-the-future-of-plastics/ (Last access: Aug. 20, 2024)

Jan Bauer serves as the Managing Director of RIGK GmbH, a company established by the plastics industry in 1992. RIGK GmbH is headquartered in Wiesbaden and has been dedicated to the development and operation of Extended Producer Responsibility (EPR) systems for plastic packaging and plastic products for over 30 years. He holds a degree in International Business Studies from the University of Paderborn.

Günter Krieglstein is Vice President of Public Affairs of System Design Europe with TOMRA SYSTEMS ASA, Norway. He has a degree in Business and Economics and brings eight years of experience in TOMRA as a consulting and business professional for various legislators and stakeholders. He has wide experience in deposit schemes worldwide. All in all, he has 25 years of professional experience on an international level.

Michael Löwe is Vice President Public Affairs, Head of System Design in TOMRA Systems ASA, Norway. He started his career with TOMRA in 1994, right after graduating in industrial engineering studies. With experiences in deposit schemes worldwide he consults and supports governments, legislators, and involved stakeholders. He was involved in TOMRA's preparation for the Germany deposit implementation in 2003/2006 and was also part of the deposit preparation team in New South Wales/Australia in 2016/2017.

Dr. Lorena Fricke was Managing Director of IK's "ERDE" initiative from 2021 until 2024. ERDE is the collection and recycling scheme for agricultural plastics in Germany. She completed her doctorate in economics on "Regime Shifts in Marine Ecosystems" at Kiel University. In 2019 she joined IK, the Association for Plastic Packaging and Films in Germany, where she oversaw the economics unit and the ERDE initiative. She is distinguishing herself as an expert on sustainability and circular economy in agriculture.

Michael Vetter studied mechanical engineering and energy and environmental technology in Aachen and Julich. As of 1993, he worked in the plastics recycling department of a leading German waste disposal company. In 2002, on behalf of the German plastic window profile industry, he founded the nationwide industry solution for old window recycling Rewindo GmbH Fenster-Recycling-Service, whose managing director he is today.

Dr. Peter Orth is a chemist by education; with OPC—Orth Plastics Consulting he leads a group of senior consultants with decades of professional experience in leading positions in the German and European plastics industry and its associations. His main focus is on innovation and sustainability in cooperation with science, industry, and politics. He is currently working in the fields of CO_2 as a feedstock, resource efficiency, and recycling technologies.

Dr. Ulrich Liman is a chemist by education, he joined Bayer AG in 1989 in marketing of polyurethane and polycarbonates and held various managerial positions in Japan, China, and Germany. Until 2022 he was responsible for polyurethane R&D and held a seat in the Supervisory Board of Covestro AG.

Dr. Jürgen Bruder is a chemist with an additional degree in business. He worked in higher education and the chemical industry, before becoming Director General of IK, the German plastic packaging association. He held positions in national, European, and global committees including the Global Plastics Alliance. Today he deals with the transformation of the plastics industry to a circular economy focusing on plastic packaging and plastic applications in agriculture.

Chapter 12
Plastics Leakage into the Environment

Jürgen Bruder and Stephanie Cieplik

Contents

Summary

The vast majority of the world's plastic wastes are currently not (yet) recovered; they are either landfilled or released into the environment in an uncontrolled manner. This chapter focuses on the latter kind of plastic waste representing a global challenge for all actors in the plastics cycle and policymakers due to the huge volumes entering the environment and their negative environmental impact. Where unavoidable, these plastic wastes must first be collected and then gradually introduced as a resource into a circular plastics economy. Special attention is given to the plastic pollution of the oceans—the marine litter problem. Based on the extensive literature on this topic, the causes and the ecological and further consequences of this global environmental

J. Bruder (✉)
BPC- Bruder Plastics Consulting, Potsdam, Germany
e-mail: plastics@jbruder.de

S. Cieplik
BKV GmbH, Frankfurt am Main, Germany
e-mail: stephanie.cieplik@bkv-gmbh.de

P. Orth et al. (eds.), *Circular Economy of Plastics*,
https://doi.org/10.1007/978-3-031-78829-1_12

problem are summarized, together with the current status of sources, input pathways and quantities, and existing and planned measures to combat the input of plastics into the environment. The need to differentiate between microplastics and larger plastics and the need for in-depth research on the possible effects of microplastics is emphasized. Finally, an attempt is made to give an overview of plastic pollution in Europe. Due to lacking data across Europe, available data for Germany are used exemplary.

12.1 Global Pollution and Pathways

The introduction of plastics into the environment is a huge environmental problem, which in terms of its relevance is one of the biggest global environmental problems after the climate and biodiversity crises. While much of the focus has been on marine litter, input to the terrestrial environment, e.g. in soils, especially through microplastics, is also increasingly being studied.

Studies on the amount of input often refer to the marine or aquatic environment. Analyses on the general fate of plastics that are not recovered and not properly landfilled are rather rare. The Conversio study with the reference year 2018 (Conversio 2020) examined the global material flows including unmanaged plastic waste. In 2018, of the approximately 250 Mt of plastic waste worldwide, around 200 Mt was not recycled and only around 50 Mt was recycled. Around 75 Mt was disposed of properly in landfills and around 50 Mt was recovered as energy. About the same amount as the properly landfilled plastics (75 Mt) ended up in the environment as unmanaged waste. This quantity was further subdivided into improper disposal and leakage.

Improper disposal of plastic waste includes not only emissions into the sea (marine litter), but also the untreated plastic waste that remains on land. In 2018, this amounted to 62 Mt globally, i.e. five to six times the amount that ended up in the sea (Conversio 2020). This waste stream includes unmanaged landfills (as opposed to managed landfills), the disposal of waste in nature or (open) incineration of plastic waste without energy recovery.

Waste leakage is essentially caused by the uncontrolled or unintentional release of plastic waste into the environment. The sources of such leakage are manifold. Examples include leaks from landfills or industrial facilities and other various human activities (ibid.). However, the material flow referred to as leakage in Fig. 12.1 mainly ends up as marine litter in the sea.

The commonly used term Littering can be seen as a specific form of waste leakage, as it is a source of uncontrolled waste disposal. Littering is generally understood as the improper disposal of waste into the environment. This also includes the disposal of waste through domestic wastewater, forgetting or losing waste. All materials are affected by littering—metals, glass, paper or plastics. However, there is no legal definition of the term littering; an overview of the various definition approaches with their subtle nuances can be found, for example, in a report by the German Federal

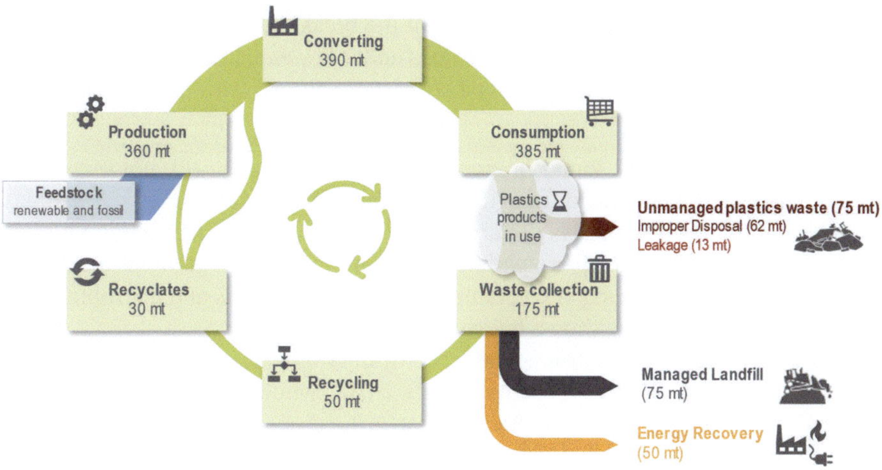

Fig. 12.1 Global plastics flow 2018 (*Source* Conversio 2020)

Environment Agency (UBA 2020). However, the lack of such a legal definition is irrelevant to the public and political debate on cleanliness in public spaces. However, it is also clear that the terms improper disposal and littering overlap each other.

The term Plastic litter has become established for plastic waste that is not disposed of properly; if this waste ends up in the sea, it is referred to as marine plastic litter (Bertling et al. 2022).

Another category to describe parts of the plastic emissions are so-called intentionally released plastic applications, e.g. tyre abrasion, paints or laundry fibres. These applications contribute significantly to the emission of microplastics.

12.2 Microplastics and Larger Plastics

With plastic litter, a distinction is made between microplastics and larger plastics. Microplastics are plastic particles between 1 μm and 5 mm in size, whereby all polymers are counted as "plastic", i.e. rubber, textile fibres, paints, varnishes, etc. in addition to plastic as material. A distinction is made between two types of microplastic:

– Primary microplastics enter the environment directly through any of various channels—product use (e.g., personal care products being washed into wastewater systems from households), unintentional loss from spills during manufacturing or transport, or abrasion during washing (e.g., laundering of clothing made with synthetic textiles). Examples of primary microplastics include microbeads found in personal care products, plastic pellets (or nurdles) used in industrial manufacturing, and plastic fibres used in synthetic textiles (e.g. nylon);

- Secondary microplastics are formed from the breakdown of larger plastic parts; this typically happens when larger plastics undergo weathering, through exposure to wave action, wind abrasion, and ultraviolet radiation from sunlight (Thompson 2015).

Microplastics are now found in many places around the world, not only in the oceans, but also in Antarctic ice or in the food chain. The largest source of microplastics entering the environment is tire abrasion, e.g. 1.3 Mt in the EU or 1.1 Mt in the US (Wagner et al. 2018). Only a small fraction of these tire wear particles reaches the oceans. Previous estimates of the amount of both larger plastics and microplastics in the oceans vary widely, as shown below.

The highly complex environmental issue of microplastics came to public attention only in the mid-to-late 2010s. In the early 2010s, while the public and the media focused on the highly visible quantities of plastics in the oceans and on beaches (packaging, disposable products, fishing gear, etc.) as an environmental problem, scientists were already drawing attention to the then unexplored field of microplastics with its unknown consequences for the environment and human health. Thompson is considered the initiator and "godfather" (Uni Plymouth 2024) of microplastic research with the famous publication "Lost at sea: where is all the plastic?" (Thompson et al 2004). The first comprehensive scientific assessment of microplastics was published in 2015 (GESAMP 2015). Since then, research on microplastics has grown by leaps and bounds. While around 100 scientific publications were published worldwide on the topic of microplastics in 2015, around 3400 publications were registered in 2022 (Nyadjro et al. 2024).

12.3 Marine Litter

Marine littering (the uncontrolled discharge of waste into the marine environment) has not only been a topic of marine and environmental science since the mid/end of the 2000s, it also became a global environmental problem gaining increased attention and importance. Looking back, this can also be seen in certain events whose impact has reached far beyond the field of science.

The pollution of the oceans received the necessary public attention for the first time in 2009 with the report of the UN environmental organization UNEP identifying plastic waste as the main component of marine debris and a main threat to marine flora and fauna (UNEP 2009). Data on ocean and beach pollution had already been collected in some regions since the 1980s, but the 2009 UNEP report became even more powerful of the fact that plastic was the dominant component of litter. The proportion of plastic waste in total marine debris was estimated at 80% in 2009 and has since risen to around 85% (UNEP 2021, p. 14).

In 2011, as a result of the alarming UNEP report (2009), the 5th International Maritime Debris Conference addressed the issue of marine litter caused by plastic waste in detail and adopted the so-called Honolulu Strategy for Marine Protection for about 10 years (IMDC 2011). In June 2012, the United Nations Conference on Sustainable Development (Rio+20) launched the Global Partnership on Marine Litter (GPML) in response to the global problem of marine litter (GPML 2012).

However, it was only at the beginning of 2016 that this global environmental problem gained media and political relevance with the warning issued at the World Economic Forum in Davos: without serious changes to the global plastic cycle, there will be more plastic than fish in the world's oceans by 2050 (EMAF 2016). Images from around the world of floating plastic debris in bays, estuaries and even open ocean areas, beaches covered with plastic debris, or marine animals dying from plastic parts have made this global problem visible, to the point that it has sometimes been ranked as the world's number one or number two environmental problem (before or after climate change) in public opinion polls.

Since then at the latest, the entire plastics sector, but in particular the plastics industry and producers of consumer goods packaged in plastic (FMCG industry) have been under public and political pressure to tackle a fundamental transformation towards a circular plastics economy together with all stakeholders and, as a first step, to prevent the uncontrolled release of waste into the environment. This pressure was accompanied by a "plastic bashing" initiated by the media in many European and Asian countries, partly supported by representatives of politics, retail, and civil society. It was not until the COVID-19 pandemic and the subsequent (re)recognition of the obvious benefits of plastic products for health, hygiene, and nutrition that a largely objective debate about the causes of and solutions to the plastic waste problem emerged (FAZ 2020).

The mandate adopted by the United Nations Environment Assembly in March 2022, and prepared over several years, to negotiate a binding agreement under international law to reduce and prevent the discharge of plastics into the ocean points to a politically achievable, long-term solution to the global marine litter problem (UNEP 2022). This agreement, which is planned to be concluded by the end of 2025 and is comparable to the Paris Agreement on climate change, will undoubtedly have a strong impact on the circular plastics economy (cf. Sect. 12.3.5).

The problem of marine litter continues to have far-reaching consequences for the image of plastics in general and plastic packaging in particular.

12.3.1 Causes of Marine Litter

The causes of the global marine litter problem have been studied extensively and are manifold due to the complexity of the issue and the very different regional conditions (McKinsey et al. 2015). The main problem in countries with a very high input of plastic waste into the sea is the lack of adequate waste infrastructure, i.e. insufficient or no waste collection, organized landfilling or material and energy recovery due to economic, political, or social structures in the countries concerned. A key factor is the lack of incentives for private or commercial end-users to behave in a sustainable manner. A rapidly growing middle class and the associated increase in consumption of plastic products and packaging without adequate waste infrastructure is a major problem in countries with high growth in a linear economy, such as China or Vietnam. An example of the different conditions in the densely populated Southeast Asian countries is the leakage points. According to McKinsey (ibid.), 84% of the waste released into the environment in China was not collected, while in the Philippines 74% of the waste released into the environment comes from waste that has already been collected but ends up in the sea due to inadequate, often illegal dumping and the ubiquitous proximity of water.

Socio-cultural aspects—such as certain behaviors and traditions—are also significant, as the following example shows: An internal investigation by a shipping company with 75 vessels revealed that 500,000 PET bottles were thrown overboard within one year (IMarEST 2019).

12.3.2 Adverse Effects of Marine Litter

The UNEP report of 2021 (UNEP 2021) summarizes in a compact form and with the evaluation of about 1000 literature sources the extensive scientific results on the causes and effects of the input of plastic waste, which have been developed in particular after the last UNEP report on marine litter (UNEP 2016). The impacts are presented in terms of threats to the entire marine ecosystem, risks to human health and economic damage.

On the environmental effects

The plastic waste causes "lethal and sublethal effects in whales, seals, turtles, birds, and fish as well as invertebrates such as bivalves, plankton, worms and corals. Their effects include entanglement, starvation, drowning, laceration of internal tissues, smothering and deprivation of oxygen and light, physiological stress, and toxicological harm" (UNEP 2021, p. 14).

Plastics in the ocean can also alter the global carbon cycle through their impact on plankton production, the growth of mangroves, seagrasses, corals, and salt marshes, which play an important role in sequestering carbon (ibid.).

Regarding the effects of plastic debris on marine wildlife, despite intensive research worldwide, (only) three cause-and-effect relationships have been scientifically established to date:

- Inhibited food intake and/or reduced nutritional value of the species' diet;
- internal physical damage to the species;
- external physical damage (ibid., p. 24).

An assessment of over 100 studies by de Ruijter et al. (2020) found that suspected further negative biological effects of virgin microplastics, including toxic substances and pathogens that can be ingested by animals via microplastics as a substrate, have not (yet) been scientifically substantiated. Further research is needed to investigate the effects of microplastics and larger plastics (both flexible and rigid) on habitats, species communities and ecosystem functions in greater depth.

There is a consensus that plastic marine litter acts as a multiplier of environmental damage in conjunction with other factors that have a negative impact on the marine environment, such as climate change and overfishing. This makes the damage significantly greater than it would theoretically be if the environmental impacts of marine litter were considered in isolation (UNEP 2021, p. 15).

On the effects on health and well-being

There is a large amount of literature on this topic, most of which does not allow any clear conclusions to be drawn. Similar to the suspected effects of plastic debris on animals, the relationship between exposure to chemicals that can enter the human body via microplastics, e.g. through the consumption of seafood, and the effects on human health is still unclear. (UNEP 2021, p. 14).

Several studies are investigating hypotheses according to which microplastics—not only from marine debris—could lead to health damage in the human body through a variety of mechanisms—such as possible cell damage or the introduction of pathogenic or resistant bacteria via microplastics as a vector (UNEP 2021, p. 36).

The negative impact of floating and washed-up plastic debris on the well-being and quality of life of the people living in coastal regions with littered beaches and bays or on the banks of heavily polluted rivers is beyond doubt. This also includes injuries caused by litter on beaches (Campbell et al. 2016). Isolated shipwrecks, sometimes with fatalities, are due to lost fishing gear. One known incident of a vessel sinking as a result of debris entanglement resulted in 292 human fatalities. The sinking of the Korean 110 GT ferry M/V Soe-Hae in 1993 was caused partly by fishing ropes around the propellers (GESAMP 2021, p. 25).

On the economic effects

It is extremely difficult to assess the economic impact of marine litter on a global level. Various approaches can be found in the literature, e.g. based on the reduction of marine ecosystem services or based on additional costs in economic sectors affected by marine litter.

Beaumont et al. (2019) have derived a reduction in the ecosystem services of marine natural capital due to plastic in the sea of 1–5% (based on 2011) using a conservative approach. With an estimated benefit of marine ecosystem services for society in 2011 of around $49.7 trillion per year, equal to an annual loss of $ 500–2,500 billion. Considering the estimated quantities of plastic in the oceans of 75–150 Mt in 2011, this results in an economic loss of ecosystem services of $ 3,300 to 33,000 per metric ton of plastic in the sea.

Estimates of the additional global costs of dealing with marine litter range from $6 billion to $19 billion in 2018 (Deloitte 2019). This includes impacts on fishery and aquaculture, tourism, shipping and port activities, beach cleaning costs, and more, but does not include impacts on human health and marine ecosystems.

While global assessments clearly show a wide range of economic impacts, regional studies need to be more precise. In the Asia–Pacific region, annual damages to tourism, fishing and shipping amounted to about $ 10.8 billion in 2015 (UNEP 2021, p. 42). In Europe, the cost of cleaning coasts and beaches alone is around € 630 million per year (UNEP 2018, p. 9). In the Mediterranean region, the annual losses to fishery and aquaculture, shipping and tourism due to marine litter are estimated at $ 696 million (Dalberg Advisors, WWF Mediterranean Marine Initiative 2019).

Need for research on the effects of marine litter

Based on the evaluation of numerous (meta-) studies, the UNEP report (2021) identifies several critical research areas that need to be prioritized. These include "quantification of the volumes of different plastic fractions from key sources and their fate in different marine habitats; quantification of the damage and economic costs of marine litter and plastics to maritime industries and in terms of ecosystems and human health; improvements to recycling technologies and standards for plastics recyclates; the development of circularity and eco-design for plastic products, including the use of alternative materials and biodegradability outside industrial conditions; improved risk assessment frameworks; a deeper understanding of the impact of marine litter and plastics on societal norms, attitudes and behavior" (ibid, p. 20).

12.3.3 Sources, Input Pathways, and Quantities

Regarding the origin of plastic debris in the sea, a fundamental distinction must be made between land-based and sea-based litter. The ratio of the two sources of origin is predominantly stated in the literature as approx. 80% to 20% (McKinsey et al. 2015), although there is no scientific evidence to support this estimate (GESAMP 2021). Research on quantity inputs into the environment initially focused on land-based litter as the (supposedly) largest and easier-to-estimate proportion. Since the late 2010s, research on sea-based litter has intensified (e.g. Richardson et al. 2019[1]) with the result that previous estimates of inputs from fisheries, aquaculture, and shipping were probably far too low. As with land-based litter, the figures for sea-based litter also vary considerably depending on the specific regional conditions. In regions with high maritime traffic density, the percentage of sea-based litter is always higher, for example, in the North Sea.[2]

The main land-based inputs are rivers, wastewater, and so-called stormwater from the mostly densely populated coastal regions, and drift from coastal landfills. Until the publication of Kaandorp et al. (2023) to be discussed below, data on inputs via

[1] Richardson et al. determined the average proportion of lost fishing nets (5,8%) and traps (8,6%) globally, applicable to vessels and fleets.

[2] Schäfer et al. (2019) report that marine litter items collected at the German North Sea coastline between 2011 and 2017 with estimations that 60% of the litter was from local and regional sea-based sources.

the world's rivers ranged from 0.41 to 5 Mt/a (Schmidt et al. 2017; Strokal et al. 2023). Schmidt et al. (2017) identify 10 rivers accounting for 88–95% of river inputs (not total inputs):

- Asia: Yangtze, Indus, Huang He, Hai He, Ganges, Zhujiang, Amur, Mekong
- Africa: Nile, Niger.

In contrast, Meijer et al. (2021) use a different model to estimate that about 80% of global riverine plastic emissions into the oceans occur via more than 1000 rivers.

The article published in 2015 by Jambeck et al. (2015) is one of the most cited literature sources for the extent of marine litter from plastic waste. For 2010, the authors cite an amount of 4.8 to 12.7 Mt of plastic waste per year entering the oceans from land (land-based litter). Without further action, this amount would increase by an order of magnitude by 2025, a 2015 estimate that is highly unlikely to be met. The authors list 20 countries with the highest absolute amounts of plastic waste. Globally, the largest share of about 80% comes from Asian countries, and about 2% from Europe and North America. The five Southeast Asian countries of China, Indonesia, Vietnam, the Philippines, and Thailand alone are estimated to account for more than 50% of global emissions (UNEP 2021, p. 63).

The publication by Kaandorp et al. (2023) provides a new impetus for research into the quantification of marine pollution and its transport behavior in the sea. Using a "state-of-the- art three-dimensional global transport model for marine plastics" (ibid.), which includes more than 22,000 specific measurements from various marine areas and coasts, the sea surface, beaches, and the deep sea, an annual global input of approximately 500 kt of initially buoyant plastics such as PE, PP and PS is determined. If non-buoyant plastics such as PVC, PET and PA are added, based on their share of the amount of plastic emitted, the input would be a maximum of 850 kt, which is at least a magnitude less than the previous figures used in science, media, and politics. The proportions of the sources of buoyant marine litter also deviate significantly from previous studies, with 45–48% coming from fisheries, 39–42% from the coastline and 12–13% from riverine sources. However, at 3.2 Mt, the total amount of initially buoyant marine plastic is significantly higher than previous estimates. Only about 10% of the plastic sinks within two years, particularly due to the formation of biofilms. The annual increase in marine plastic input is estimated to be 4%. "Without further mitigation strategies and clean-up efforts, (the) estimated growth rate of 4% per year has the potential to double the plastic standing stock within two decades" (ibid.). If the results of Kandoorp et al. come closest to the plastic pollution that is so difficult to determine, the good news is that less plastic enters the oceans than previously assumed. But this is overshadowed by the realization that both the amount of buoyant marine plastic and the residence time are much higher than previously assumed, with the associated negative impacts on the ecosystem.

The sometimes widely varying amounts of global marine plastics and their sources in scientific publications show the need for further intensive research, taking advantage of advances in modeling that now incorporate available large data sets. Sea-based litter, especially from fishery, should receive much more attention from the scientific community, policymakers, and the public worldwide than it had in the past, as its impact has been underestimated. In the EU, the SUPD is the first set of regulations to prevent the discharge of ghost nets into the sea as far as possible.

12.3.4 Selected Industry Measures to Combat Plastic Waste Input

Chapters 6, 7, 8, 9, 10 and 11 of this book present a series of measures and initiatives along the entire plastics value chain heading for a circular economy and therefore also contributing directly or indirectly to reducing plastic pollution in the sea. Various studies show scenarios of how a reduction in emissions into the oceans can be achieved globally by applying certain measures. According to a stochastic model by PEW and SYSTEMIQ (2020), a reduction of up to almost 80% of the expected input into the oceans in 2040 compared to the expected input under business-as-usual conditions is considered feasible. This "System Change Scenario" derives a mix of measures already possible today with available technologies and quantifies them to achieve the goal, such as:

- Reduce the increase in plastic production and consumption to avoid almost a third of the projected plastic waste generation through elimination, reuse, and new delivery models;
- Increase the proportion of economically recyclable products and packaging from an estimated 21 to 54% through improvements in design for recycling;
- Expand waste collection in middle- and low-income countries to 90% in urban areas and 50% in rural areas;
- Double mechanical recycling to 86 Mt/a;
- Establish a global capacity for chemical recycling of 13 Mt/a;
- Reduce exports of plastic waste to countries with low collection and high emission rates by 90% (ibid.) (cf. Sect. 5.5).

The following is a brief overview of selected industry initiatives specifically aiming at reducing the input of plastics.

A central initiative of the plastics industry and its associations across countries and continents is "Operation Clean Sweep" (OCS), which has set itself the goal of using normative measures to prevent the entry of plastic pellets into the environment along the entire supply chain (i.e. in plastics production, logistics, processing, and recycling companies). The initiative is also known in Europe as "Zero Pellet Loss" (Cleansweep 2024) and in Germany as "Null Granulatverlust" (IK 2017). OCS was founded by the Plastics Industry Association in the US in 1991 and is managed in partnership with the American Chemistry Council's Plastics Division. In Europe, the initiative was launched by PlasticsEurope in 2013. Globally, more than 3,300 companies in more than 60 countries participate in OCS and have committed to the OCS Pledge (OCS 2024). All members of OCS can be found on the OCS homepage, divided by country.

The European Plastics Association EuPC und PlasticsEurope support the European Commission's objective of identifying policy options aimed at reducing unintentional releases of microplastics in the environment. At the beginning of 2023, both associations issued the first ever European pellet loss prevention certification scheme strengthening Operation Clean Sweep by creating harmonized processes and procedures for controlling and documenting pellet losses across the entire plastics supply chain (PlasticsEurope 2023). The intention was to give the EU Commission a blueprint for the planned Regulation. The expected Proposal for a Regulation on preventing plastic pellet losses to reduce microplastic pollution of the EU-Commission as of 17 October 2023 shall apply to all participants in the supply chain who handle plastic pellets in quantities of more than 5 tons and to transporters of plastic pellets (EU 2023). The measures and obligations proposed by the EU include informing national authorities about facilities and transports of plastic pellets, risk assessment plans and declarations of compliance, documentation of annually estimated release quantities, certification by an accredited body and, above all, taking measures to prevent, contain, and clean up spills (ibid.).

When the UNEP report on marine litter was published in 2009 (UNEP 2009), this was a completely new challenge for the plastics industry and its associations. In response, many activities were initiated by national and regional plastics associations on all continents and documented as of 2011 under the umbrella of the "Global Declaration on Marine Litter Solutions" adopted on the occasion of the 5th UN Marine Conference in 2011. Over 75 associations and organizations have signed this declaration, and over 350 projects were implemented worldwide within this framework from 2011 until the COVID-19 pandemic.

Many of these projects were implemented in countries such as Malaysia, Thailand, India, or South Africa and are exemplary in terms of their public impact, environmental education, and broad community involvement, including beach clean-ups.

Due to the limited resources of the initiators, such activities can at best address the "soft factors" of the marine litter problem; they have little influence on the "hard factors" of the infrastructure for environmentally sound disposal of municipal waste, including plastic waste.

Only in 2019, ten years after the UNEP report, the Alliance to End Plastic Waste (AEPW) was founded, creating a much-needed industry commitment to fundamentally change the hard factors, especially in the priority countries for marine litter. The Alliance convenes more than 60 companies from across the plastics value chain with local communities, civil society groups, intergovernmental organizations, and governments. The collective knowledge, experience, and resources of this global network allow for the current portfolio of more than 50 projects worldwide (AEPW 2024). The projects of the plastics value chain initiative with global companies from the chemical, plastics, branded goods and waste management industries can be viewed on the Alliance's website. At its inception, AEPW pledged to invest $1.5 billion in specific local and regional projects by 2025. In early 2024 for example, AEPW announced Playbook Projects, which are guidelines to be transferred and scaled to many regions with underdeveloped waste infrastructure. These guidelines relate to "Engaging Households in Segregated Municipal Waste Collection" and "Unlocking Value through Basic Manual Sorting of Municipal Waste" (ibid.).

12.3.5 Regulatory Measures—Global Plastics Treaty

In many countries, policies to address the "flood of plastics" on beaches and in public spaces have often been implemented, at the local level initially and later on at the national level, including taxes or fees, restrictions or bans on the use of single-use plastics (SUPs) such as PET bottles, carrier bags, fast food packaging, and more. According to a UNEP survey (UNEP 2018, p. 10), more than 60 countries had such regulations in place by 2018.

At the global level, the 5th UN Environment Assembly in Nairobi in March 2022 achieved a seminal breakthrough after years of preparation. With the adoption of the resolution "End plastic pollution: Towards an international legally binding instrument" (UNEP 2022), it was decided that a legally binding international treaty should be negotiated by the end of 2024. The goal is to adopt the treaty at a diplomatic conference in 2025. Preparatory meetings of the Intergovernmental Negotiating Committee (INC) on Plastic Pollution with representatives from governments, industry, science, and NGOs focus on drafting the text of the treaty. The draft for the fourth session (INC-4) in Ottawa at the end of April 2024 provides for six parts of the treaty (INC-4 2024).

Part II addresses, among other things, primary plastics and a discussed limitation of global production; chemicals of concern; problematic and avoidable products; product design, composition, and performance; non-plastic substitutes; EPR; emissions and releases of plastics throughout their life cycle; waste management (including fishing gear); existing plastic pollution; transparency, tracking, monitoring and labeling. Part III addresses issues such as financing, technical assistance, and technology transfer. Part IV deals with national action plans, implementation, reporting on progress, monitoring, and awareness-raising (ibid.).

One of the most difficult items of negotiation is the proposal by some countries and NGOs to limit polymer production worldwide.

12.4 Plastic Pollution in Europe

12.4.1 Overview and Data Situation

Littering is also a European problem, but studies aimed at reducing littering must always refer to specific countries or regions, as socio-economic and socio-cultural conditions vary greatly (Grobler et al. 2022). The amount of waste that is improperly disposed of varies from country to country, even within Europe. It depends on several factors, including population density, the availability of waste management systems, and the relevance of environmental protection in the consciousness of society and consumers of the respective country. As a result, there is little comprehensive data on the total amount of improperly disposed waste in Europe, including its plastic content, as it is difficult to quantify the extent of the problem. The Global Plastics Flow study (Conversio 2020), with reference year 2018, gives an idea of the general extent of plastic leakage in Europe (Table 12.1).

Of the approximately 30 Mt of plastic waste generated in the EU 28 + Norway and Switzerland in 2018, appr. 29.1 Mt were collected and recycled. Nevertheless, around 0.3 Mt of plastic waste was not disposed of properly and escaped into the environment in an uncontrolled manner (leakage). In addition, appr. 0.7 Mt were disposed of improperly in unauthorized landfills. SYSTEMIQ (2022) comes to similar conclusions: In 2020, a total of around 37% of the 25.2 Mt of plastic waste was properly landfilled (6.9 Mt), exported (2 Mt) or improperly disposed of (0.5 Mt). The study generally points out that plastics leakage is still seen as a problem in Europe, even though the continuous expansion of the collection and recycling infrastructure and EU regulations with bans and restrictions on some plastic products (cf. Chap. 21) have led to a reduction in open dumping, littering, or unauthorized landfilling. However, especially in less economically developed regions of Europe, improper disposal of plastic waste is more common due to inadequate waste management systems and landfills that are not state of the art and often located near watercourses (ibid.).

Table 12.1 Improper disposal and leakage of plastic waste in the EU 28 +2 in 2018 (*Source* Conversio 2020)

EU 28 + Norway, Switzerland in 2018								
Plastic waste as part of total waste	Generated		Collected for treatment					
	Total		Total		Thereof formal		Thereof informal	
Applications	mt	Share (%)	mt	Share (%)	mt	Share (%)	Mt	Share (%)
Packaging	18.4	61	17.8	97	17.6	99	0.2	1
Non-Packaging	11.6	39	11.3	97	11.1	99	0.2	1
Σ	**30.0**	100	**29.1**	97	**28.6**	99	**0.4**	1

Plastic waste as part of total waste	Treatment						Unmanaged waste			
	Recycling		Energy recovery		Managed landfill		Improper disposal		Leakage	
Applications	mt	Share (%)	mt	Share (%)	mt	Share (%)	mt	Share (%)	mt	Share (%)
Packaging	7.5	41	7.0	38	3.3	18	0.4	1.9	0.2	1.2
Non-Packaging	2.0	17	5.4	46	3.9	34	0.3	2.6	0.1	0.6
Σ	**9.4**	31	**12.4**	41	**7.2**	24	0.7	2.2	0.3	1.0

The release of plastic waste into the environment, especially into the sea, continues to be a major challenge for Europe. There are numerous studies on marine plastics pollution that focus on the pollution of defined coastlines, beaches, and sea areas.

A European Commission report (EU 2018), referring to Jambeck et al. (2015), referred to an estimate of 0.15–0.5 Mt of plastic waste entering EU seas in 2015. In view of more recent studies (Kaandorp et al. 2023), however, this widespread figure in the literature may be questioned. Intergovernmental commissions for the protection of European seas, such as HELCOM for the Baltic Sea and OSPAR for the North-East Atlantic and North Sea, report regularly on the state of the seas and thus also on the pollution with marine litter, in particular plastic litter.

OSPAR had already begun identifying the top beach litter items for the North Sea and the North-East Atlantic in 2009 through counting studies. The Quality Status Report of 2023 (OSPAR 2023) highlights that the number of litter items at 252 is well above the EU threshold value of 20 items per 100 m beach and that microplastics can be detected everywhere in the aquatic environment of the OSPAR area. Tire abrasion and larger plastic, which then becomes secondary microplastic, were identified as the largest land-based sources of microplastics in trawl tests with nets. This is followed by paints, pellets, cosmetics, and laundry fibers, in that order. OSPAR has also summarized findings on the production and loss of plastic pellets. It is estimated that the loss of pellets in the OSPAR marine area in 2015 was between 3,100 and 31,000 tons (OSPAR 2018). Marine litter pollution is measured over time using certain indicators. The 2023 report states: "Overall, some indicators show downward trends in some regions; however, general levels of marine litter remain high" (ibid).

HELCOM (2023) provides geographically detailed information on the condition of the beaches and bays of the Baltic Sea, showing both areas above and below the threshold value for beach litter of 20 litter items per 100 m of beach. During 2016–2021, eleven out of the sixteen sub-basins that could be assessed were above this limit and did not reach good status. Plastic litter, including single-use items, was the most common litter category, accounting for between 32 and 93% of the total number of litter items (ibid).

One of the main routes through which litter enters the marine environment is via rivers and other tributaries. An initial estimate of the total macro-litter entering the Greater North Sea from the six rivers Seine, Rhine, Meuse, Ems, Weser, and Thames is 10.5 to 220.6 tons per year (OSPAR 2023). The largest amount of this is attributable to plastic waste.

Compared to marine pollution, the data situation for terrestrial pollution is still more difficult. However, initial research and studies have been carried out in some European countries to gain insights into the type and quantity of improperly disposed waste (Maier 2019). The focus of such studies is usually on localized areas, such as analyses in Austria (Stoifi and Oliva 2020). Sorting analyses often show a high proportion of plastic in the collected litter (e.g. beverage packaging in countries without a deposit system, to-go packaging, cigarette butts).

12.4.2 Case Study Germany

In Germany, on the initiative of the plastics industry, a model has been gradually developed since the early 2010s that systematically and continuously deals with the input and fate of plastics from Germany in soils and waters, some of which then also end up in the sea and provides a quantitative estimate of plastics in the environment. This provides a holistic picture of the inputs and fate of improperly disposed plastic waste in the aquatic and terrestrial environment. Only inputs that enter the environment outside of regulated and systematically recorded waste are taken into account (cf. Fig. 12.2) Thus, to a certain extent, inputs from waste leakage, littering and from plastic products and plastic containing products intended to be used in the environment are recorded (Conversio 2021).

The most important results on the fate of all polymers in the environment are shown in Fig. 12.3.

- Approx. 100 kt of polymers remain in the terrestrial environment in Germany every year, 81% of which are tire abrasion, 13% larger plastics from packaging, agriculture, and other applications, 3% microplastics (such as pellet losses, compost, emissions from construction sites and waste disposal, consumer products and fibers in household wastewater) and 1% or less each from paints and lacquers, road markings, sports fields and playgrounds.
- Approx. 14 kt of polymers remain in the aquatic environment in Germany every year, of which 72% is tire abrasion, 21% larger plastics and 4% microplastics.
- Related to plastic as a material (incl. fibers) in the narrower sense, the fate in the environment is about 20 kt, i.e. about 20% of the total fate of polymers in the environment.

In the Conversio study (2021), model calculations show that of the 130 kt of tire abrasion generated annually in Germany, 10.1 kt remain in the aquatic environment and 80.7 kt in the terrestrial environment. As in OSPAR (2018), a key finding is that pellets are released into the environment along the entire value chain. For 2020, the pellet input remaining in the environment is estimated to be 624 t for Germany. According to the study, inland and seaports contribute the most to the direct release of pellets to the environment. Quantities released during loading and unloading can freely enter water bodies and the terrestrial environment. Road transport can be expected to have a similarly high direct input of pellets into the environment, especially when pellets are transported by tarpaulin trucks.

All previous studies have in common that the database is not sufficient to create a valid quantity structure. In addition, it is often not possible to compare different studies due to different research frameworks and definitions. This is also pointed out by a UBA study (UBA 2020) which, in addition to littering, examines the plastic applications to be used in the environment and thus makes an estimate per year for all plastics remaining in the environment. 30 most mass-relevant emission sectors are summarized in a table with the estimated plastic input, from tire abrasion being

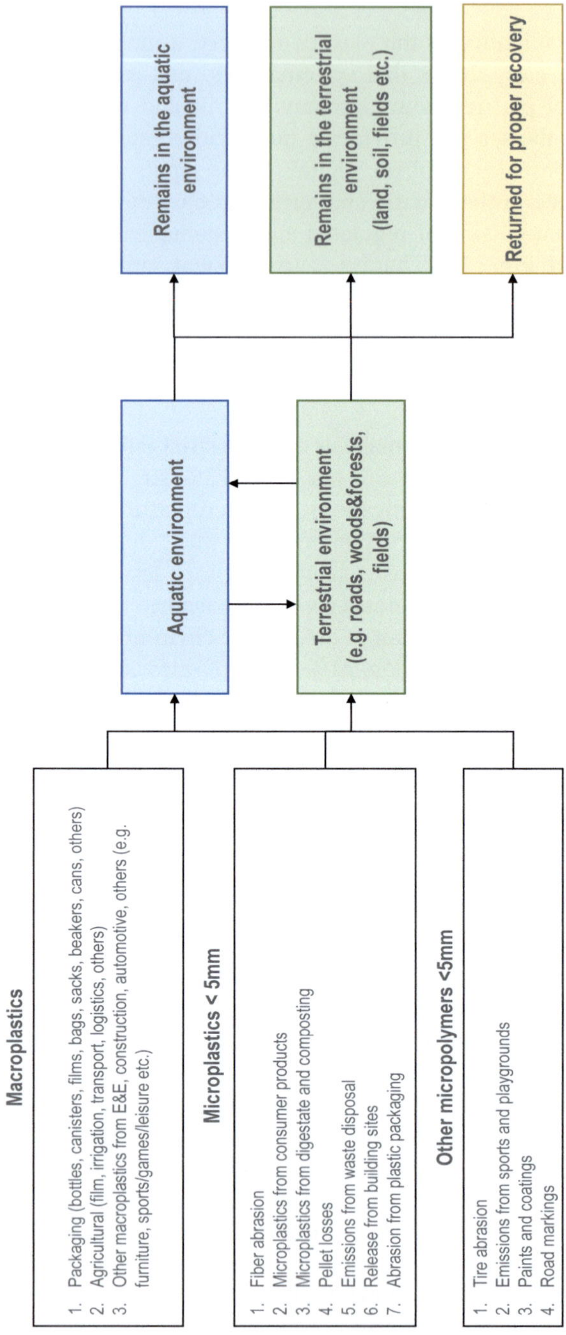

Fig. 12.2 Discharge and remains of land-sourced plastic litter—Structure of the study “Plastics in the Environment” (*Source* Conversio 2021)

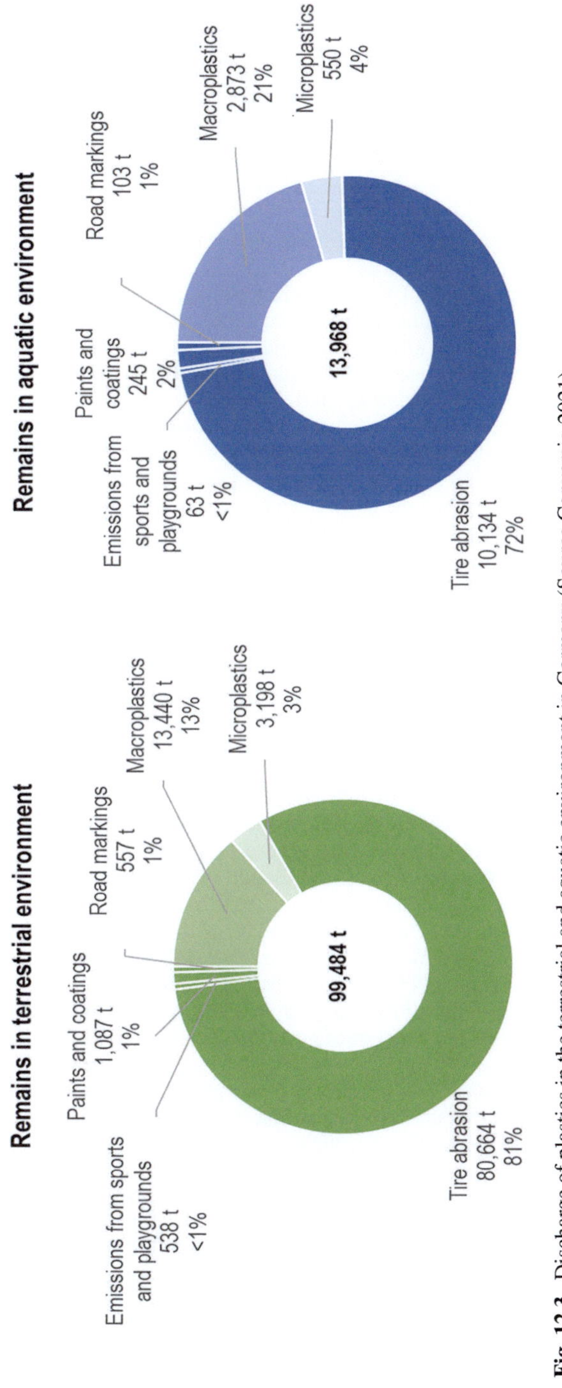

Fig. 12.3 Discharge of plastics in the terrestrial and aquatic environment in Germany (Source Conversio 2021)

No. 1 to granulates from artificial turf pitches (No. 9) and cigarette butts No. 15 to bite protection in forestry as No. 28 (ibid).

References

AEPW 2024: https://endplasticwaste.org/en/our-stories/solutions-for-change (Last access: Aug. 20, 2024)

Beaumont et al 2019: Global ecological, social and economic impacts of marine plastic. https://doi.org/10.1016/j.marpolbul.2019.03.022

Bertling et al 2022: Kunststoff in der Umwelt–ein Kompendium (Plastics in the environment—a compendium) 2nd edition 2022. https://doi.org/10.24406/umsicht-n-647638

Campbell et al 2016: Human health impacts from litter on beaches and associated perceptions: A case study of 'clean' Tasmanian beaches.https://doi.org/10.1016/j.ocecoaman.2016.04.002

Cleansweap 2024: https://www.opcleansweep.eu (Last access: Aug. 20, 2024)

Conversio 2020: Global Plastics Flow 2018. https://newsroom.kunststoffverpackungen.de/2019/10/15/transparente-weltweite-kunststoffstrome/?press=1 (Last access: Feb. 11,2022)

Conversio 2021: Kunststoffe in der Umwelt. Weiterentwicklung des Modells "Vom Land ins Meer", im Auftrag der BKV. (Plastics in the environment. Further development of the model "From Land to Sea", on behalf of the BKV) April 2021

Dalberg Advisors, WWF Mediterranian Marine Initiative 2019: Stop the Flood of Plastic: How Mediterranean Countries Can Save Their Sea. WWF-World Wide Fund for Nature. http://awsassets.panda.org/downloads/a4_plastics_reg_low.pdf (Last access: Feb. 12, 2024)

Deloitte 2019: The Price Tag of Plastic Pollution: An Economic Assessment of River Plastic. https://www2.deloitte.com/content/dam/Deloitte/nl/Documents/strategy-analytics-and-ma/deloitte-nl-strategy-analytics-and-ma-the-price-tag-of-plastic-pollution.pdf (Last access: Feb. 12,2024)

de Ruijter et al 2020: Quality criteria for microplastic effect studies in the context of risk assessment: A critical review. Environmental Scienc and Technology 54(19), 11692–11705. https://pubs.acs.org/doi/abs/https://doi.org/10.1021/acs.est.0c03057

EMAF 2016: The new plastics economy-Rethinking the future of plastics 2016. https://ellenmacarthurfoundation.org/topics/plastics/projects (Last access: Feb. 12, 2024)

EU 2018: Commission Staff Working Document SWD(2018) 16 final accompanying the document Communication from the Commission to the European Parliament, the Council, the European Economic and Social Committee and the Committee of the Regions—A European Strategy for Plastics in a Circular Economy https://eur-lex.europa.eu/legal-content/EN/TXT/PDF/?uri=CELEX:52018SC0016&from=EN (Last access: Feb. 12, 2024)

EU 2023: Proposal for a Regulation of the European Parliament and of the Council on preventing plastic pellet losses to reduce microplastics pollution. Brussels, 16.10.2023 SWD (2023) 333 final

FAZ 2020: https://www.faz.net/aktuell/technik-motor/plastik-und-corona-ist-kunststoff-doch-kein-teufel-16719484.html (Frankfurter Allgemeine Zeitung: Plastic and corona- Is plastic not the devil after all?) (Last access: Feb. 12, 2024)

GESAMP 2015: Microplastics in the ocean. A global assesment, United Nations Joint Group of Experts on the Scientific Aspects of Marine Pollution (GESAMP), Working Group 40, 2015. http://www.gesamp.org/publications/reports-and-studies-no-90 (Last access: Feb.12, 2024)

GESAMP 2021: Sea-based Sources of Marine Litter. GESAMP Working Group 43. http://www.gesamp.org/publications/sea-based-sources-of-marine-litter (Last access: Feb. 12, 2024)

GPML 2012: Global Partnership on Marine Litter–Framework document. https://wedocs.unep.org/handle/20.500.11822/41633 (Last access: Feb. 12, 2024)

Grobler et al 2022: https://www.researchgate.net/publication/361223487_Waste_Management_Littering_and_Illegal_Dumping_A_Literature_Review (Last access: Feb. 12, 2024)
HELCOM 2023: https://stateofthebalticsea.helcom.fi/findings/pressures/pollution/marine-litter (Last access: Feb. 12, 2024)
IK 2017: https://www.k-zeitung.de/erste-ergebnisse-der-ik-initiative-null-granulatverlust (First results of the IK initiative Zero Granule Loss) (Last access: Feb. 12, 2024)
IMar EST 2019: Steering towards an Industry Level Response to Marine Plastic Pollution: Roundtable Summary Report. London. https://www.imarest.org/policy-news/thought-leadership/1039-marine-plastics/file (Last access: Feb. 12 ,2024)
IMDC 2011: https://5imdc.wordpress.com/about/commitment/ (Last access: Feb. 12, 2024)
INC-4 2024: Revised draft text of the international legally binding instrument on plastic pollution, including in the marine environment. https://www.unep.org/inc-plastic-pollution/session-4/documents (Last access: Aug. 20, 2024)
Jambeck et al. 2015: Plastic waste inputs from land into the ocean, Science, Vol. 347, issue 6223, p.768–771, https://doi.org/10.1126/science.1260352
JEPSEN et al 2020: Kunststoffe in der Umwelt – Erarbeitung einer Systematik für erste Schätzungen zum Verbleib von Abfällen und anderen Produkten aus Kunststoffen in verschiedenen Umweltmedien. UBA Texte 198/2020. https://www.umweltbundesamt.de/publikationen/kunststoffe-in-der-umwelt-erarbeitung-einer (Last access: Nov. 11, 2021)
Kaandorp et al 2023: Kaandorp, M.L.A., Lobelle, D., Kehl, C. et al. Global mass of buoyant marine plastics dominated by large long-lived debris. *Nat. Geosci.* **16**, 689–694 (2023). https://doi.org/10.1038/s41561-023-01216-0
Maier 2019: Maier, N.: Littering. Discussion Paper from the interest groups plastics of the European network of the heads of Environment Protection Agencies (EPA-Network). November 2019. Littering-europa.eu
McKinsey et al. 2015: McKinsey & Company & Ocean Conservancy, Stemming the Tide: Land-based strategies for a plastic-free ocean, September 2015. https://www.mckinsey.com/business-functions/sustainability/our-insights/stemming-the-tide-land-based-strategies-for-a-plastic-free-ocean (Last access: Feb. 12, 2024)
Meijer et al 2021: More than 1000 rivers account for 80% of global riverine plastic emissions into the ocean. https://doi.org/10.1126/sciadv.aaz5803
Nyadjro et al 2024: The NOAA NCEI marine microplastics database. https://www.nature.com/articles/s41597-023-02632-y (Last access: Feb. 12, 2024)
OCS 2024: https://opcleansweep.org/ (Last access: Feb. 12, 2024)
OSPAR 2018: Background document on pre-production Plastic Pellets. https://www.ospar.org/documents?v=39764 (Last access; Feb. 12, 2024)
OSPAR 2023: https://oap.ospar.org/en/ospar-assessments/quality-status-reports/qsr-2023/thematic-assessments/marine-litter/ (Last access: Aug. 20, 2024)
PEW and Systemiq 2020: Breaking the Plastics Wave: A Comprehensive Assessment of Pathways towards Stopping Ocean Plastic Pollution. https://www.oneplanetnetwork.org/resource/breaking-plastic-wave-comprehensive-assessment-pathways-towards-stopping-ocean-plastic (Last access: Feb. 12, 2024)
PlasticsEurope 2023: https://plasticseurope.org/media/plastics-europe-and-eupc-launch-european-operation-clean-sweep-certification-scheme/ (Last access: Feb. 12, 2024)
Richardson et al 2019: Richardson, K., Hardesty, B. D., & Wilcox, C. Estimates of fishing gear loss rates at a global scale: A literature review and meta-analysis. Fish and Fisheries, 20(6), 1218–1231
Schäfer et al. 2019: Schäfer, E., Scheele, U. & Papenjohann, M. Identifying sources of marine litter: Application of the matrix scoring technique to the German North Sea region. Report on behalf of NLWKN and LKN-SH. https://doi.org/10.13140/RG.2.2.34106.85441
Schmidt et al. 2017: Export of Plastic Debris by Rivers into the Sea, Environ. Sci. Technol. 2017, 51, 12246–12253. https://doi.org/10.1021/acs.est.7b02368

Stoifi and Oliva 2020: Stoifi, B., Oliva, J.: Littering in Österreich. https://www.bmk.gv.at/themen/klima_umwelt/abfall/abfallvermeidung/publikationen/littering.html (Last access: Feb. 12, 2024)

Strokal et al 2023: River export of macro- and microplastics to seas by sources worldwide. https://www.nature.com/articles/s41467-023-40501-9 (Last access: Feb. 12, 2024)

SYSTEMIQ 2022: ReShaping Plastics: Pathways to a Circular, Climate Neutral Plastics System in Europe, 2022. https://plasticseurope.org/changingplasticsforgood/reshaping-plastics/ (Last access: Aug. 20, 2024)

Thompson et al. 2004: Lost at sea: where is all the plastic? Science 304 (5672), 838–838

Thompson 2015: Thompson, R.C.: Microplastics in the Marine Environment: Sources, Consequences and Solutions. https://doi.org/10.1007/978-3-319-16510-3_7 (Last access: Feb. 12, 2024)

UBA 2020: Jepsen et al: Kunststoffe in der Umwelt–Erarbeitung einer Systematik für erste Schätzungen zum Verbleib von Abfällen und anderen Produktenaus Kunststoffen in verschiedenen Umweltmedien.https://www.umweltbundesamt.de/sites/default/files/medien/5750/publikationen/2020_11_24_texte_198_2020_kunststoffe_in_der_umwelt.pdf (Last access: Feb. 12, 2024)

UNEP 2009: Marine Litter: A Global Challenge. Nairobi, April 2009. https://wedocs.unep.org/handle/20.500.11822/7787;jsessionid=6EEBBBE0E8F283CA6939C5D788A50479 (Last access: Feb. 12, 2024)

UNEP 2016: Marine Plastic Debris and Microplastics: Global Lessons and Research to Inspire Action and Guide Policy Change. https://wedocs.unep.org/20.500.11822/7720. (Last access: Feb. 12, 2024)

UNEP 2018: Single-use Plastics: A Roadmap for Sustainability (Rev. ed., pp. vi; 6). https://www.unenvironment.org/resources/report/single-use-plastics-roadmap-sustainability (Last access: Feb. 20, 2024)

UNEP 2021: From Pollution to Solution. A global assessment of marine litter and plastic pollution. https://www.unep.org/resources/pollution-solution-global-assessment-marine-litter-and-plastic-pollution (Last access: Feb. 12, 2024)

UNEP 2022: Draft resolution. End plastic pollution: Towards an international legally binding instrument. https://wedocs.unep.org/bitstream/handle/20.500.11822/38522/k2200647_-_unep-ea-5-l-23-rev-1_-_advance.pdf?sequence=1&isAllowed=y (Last access: Feb. 12, 2024)

Uni Plymouth 2024: https://www.plymouth.ac.uk/whats-on/university-of-plymouth-event-at-inc-4 (Last access 07.04.2024)

Wagner et al 2018: Tire wear particles in the aquatic environment—A review on generation, analysis, occurrence, fate and effects. https://doi.org/10.1016/j.watres.2018.03.051

Dr. Jürgen Bruder is a chemist with an additional degree in business. He worked in higher education and the chemical industry, before becoming Director General of IK, the German plastic packaging association. He held positions in national, European and global committees including the Global Plastics Alliance. Today he deals with the transformation of the plastics industry to a circular economy focusing on plastic packaging and plastic applications in agriculture.

Stephanie Cieplik is Project Manager with BKV GmbH, Frankfurt, a thinktank of the German plastics industry. As a qualified lawyer and project manager she supports project activities with her legal background since 2013. Her focus is on plastics in the environment and marine litter. In particular, she coordinates the "From Land to Sea—Model for the Documentation of Land-based Plastic Litter" project and the ongoing study on "Plastics in the Environment".

Part III
Circular Feedstock

Part III of the book addresses the other major challenge for the plastics industry: the transition from fossil-based to circular feedstock to avoid greenhouse gas emissions and thus contribute climate change mitigation. This transformation of the feedstock base, known as defossilization, all existing carbon sources are included as feedstock. These are, on the one hand, the plastic waste that is already in the technosphere, secondly, the carbon bound in biomass stemming from the biosphere and, thirdly, the carbon bound in gaseous carbon dioxide. The four chapters of Part III describe the basic raw materials, the technologies and methods used to use them and their future contribution to the ongoing and necessary growth of plastic as a material.

Building on the content of Part II, Chap. 13 deals with plastic waste from which secondary raw materials are obtained through mechanical or chemical recycling. Since they are only created after the end of life of the corresponding plastic products based on the past, they cannot cover the raw material requirements for plastic production today and tomorrow. Nevertheless, they and the technologies described for their use form the cornerstone of defossilization, i.e. the feedstock transformation in the future plastics industry.

Chapter 14 deals with the use of biomass as a feedstock for the chemical and plastics industries. In our planet's natural carbon cycles, CO_2 is reduced and fixed as carbon in the structural polymers of plants through photosynthesis and can be used as a material instead of being released into the environment through combustion. In addition to dealing with the basic properties of biomass and its possibilities for plastic production, this chapter also deals with biowaste made of carbohydrates and natural fatty acid derivatives as well as the role of forestry as a carbon supplier for plastic production.

Chapter 15 discusses options for the direct use of carbon dioxide, where carbon is present in its highest oxidation state. The use of unavoidable exhaust gases from waste incineration or from basic industries such as the cement industry, whose processes inevitably lead to the release of CO_2, also serves to reduce greenhouse gas emissions. The chapter concludes with a consideration of the resulting innovations in the plastics industry in the future. By using CO_2 as a feedstock, the plastics industry is becoming a driving force in mitigating climate change.

In order to use such circular feedstock, basic chemicals and building blocks from different sources with established processes and existing systems in the chemical and plastics industry, suitable models and methods are required. This is achieved through the mass balance approach described in Chap. 16, which originally stems from the bioeconomy. Mass balancing as a chain of custody ensures traceability and transparency for companies, markets and regulatory authorities.

Chapter 13
Plastics Wastes

Frauke Hoffmann, Peter Jetzer, Andreas Neumann, and Daniel F. White

Contents

F. Hoffmann (✉)
Hoffmann + Voss GmbH, Viersen, Germany
e-mail: frauke.hoffmann@hoffmann-voss.de

P. Jetzer (✉)
Recycario Data Science Institute for Economical Plastics Recycling, Erlensee, Germany
e-mail: peter.jetzer@recycario.com

A. Neumann (✉) · D. F. White
LyondellBasell Industries, Basell Polyolefine GmbH, Wesseling, Germany
e-mail: Andreas.Neumann@lyondellbasell.com

D. F. White
e-mail: daniel.white@lyondellbasell.com

P. Orth et al. (eds.), *Circular Economy of Plastics*,
https://doi.org/10.1007/978-3-031-78829-1_13

Summary

Plastic wastes come to everybody's mind as a basic feedstock source for plastics in a circular economy. We designed this chapter to describe the available technologies as well as the markets in subchapters.

Section 13.1 takes care of mechanical recycling technology and availability from different industry sources. Technologies to sort and process are discussed as well as the differentiation between waste and residues.

Section 13.2 illustrates the dynamic of the changing market for recyclates and its values for the circular economy.

Section 13.3 focusses on catalytic pyrolysis as an important strategic option for the de-fossilization of the chemical and plastics industry. Mixed plastics wastes will be pyrolyzed by the proprietary MoReTec chemical recycling technology to be used as feedstock for subsequent steam-cracking in existing electrified cracking furnaces. Products like ethylene and propylene will then be used for the polymerization of respective polymers thus enabling the use of recovered plastic wastes even in safeguarded applications like food or drinking water contact materials.

The chapter makes clear that plastic waste is an increasing and important industrial feedstock source.

13.1 Recyclates from Mechanical Recycling

Frauke Hoffmann

13.1.1 Introduction and Background

"Raw materials in the circle"—what seems so modern as the title of the chapter has a history in Germany and Europe of over 60 years. As early as the early 1960s, mechanical recycling became established to reuse production residues from plastic production or processing as raw materials and thus reduce industrial waste flows. Plastic recycling comes from a time when sustainability and circular economy were not yet omnipresent buzzwords, and political pushes for recycling and moving away from single-use were not even remotely considered because "waste" was viewed as potentially valuable. If you keep this in mind, it becomes clear why raw materials should continue to be recycled, and recycling materials should be a matter of course in the future.

History

However, the long successful history of plastic recycling is offset by emotional messages and images, in which plastic pollutes the environment, birds use plastic waste as nesting material and marine animals die from it. This brings up a second argument for the plastic cycle: preventing improper disposal. Good plastic recycling can do two things: save resources and protect the environment, for 60 years and in the future.

If the goal is to circulate raw materials and it still doesn't always work, you have to ask many questions. Where does mechanical recycling work and where does it not, what strengths, weaknesses, opportunities, or risks does the process have?

Mechanical plastic recycling has one of its strengths in its long history. Since the 1960s, numerous companies have emerged across Europe and around the world. Many of them have emerged from metal processing and now carry out this kind of plastic recycling. Companies and employees have been able to build up know-how, and mechanical engineers have been able to further develop the required technology.

Detailed information about waste quantities in Germany and Europe, and about recycled volumes, must be discussed carefully, as reference is usually made to the recycling of plastic waste. However, waste is only a substance that "its owner gets rid of, wants to get rid of or has to get rid of" (Circular Economy Act 2020). So, anything that is not defined as waste is very likely not included in the statistics, even though a part of this plastic fraction is recycled!

Present Situation

Material flows for recycling include both: industrial flows, called post-industrial or pre-consumer, as well as end-of-life products that the owner wants to get rid of which are called post-consumer. This definition is broader than one would imagine: "Material from households, commercial and industrial establishments, or other end users of the product that can no longer be used for its intended purpose. This includes returned material from the supply chain (Federal Environment Agency 2021).

'Post-consumer' therefore also includes waste streams from industrial applications, but only material that is legally considered waste can be processed into a recyclate that can be counted towards the current and future quotas. For political reasons, the focus is on the recycling of post-consumer waste streams, while the established, well-functioning recycling of post-industrial materials is sidelined. A categorization that should be reconsidered in the light of today's resource efficiency.

The recycling of post-consumer waste is associated with many technical and economic problems. These will be discussed further below. At the same time, increased energy costs on the one hand and lower raw material prices for imported goods on the other have led to plant closures among producers, processors and recyclers in Germany and Europe, especially those who recycle post-consumer fractions. This development contradicts social necessities and political will.

13.1.2 Processing

Mechanical recycling includes a variety of processing technologies that do not change or even dissolve the polymer structure. In contrast, chemical recycling has exactly this goal: the dissolution of the polymer chain down to the molecular level (cf. Sect. 13.3).

The processing steps of mechanical recycling include sorting, shredding, homogenizing, dedusting, metal separation, agglomeration and the final processing and determination of properties of the newly obtained material using single or double-shaft compounders with integrated melt filtration. During compounding, different batches can be blended, finished and enhanced with pigments, additives, fillers and reinforcing materials. In this way, you achieve a level of quality that is very suitable for numerous applications—although not for all, as will be discussed below.

The processes mentioned can be accompanied by prior washing, the separation of mixed plastics, e.g. through density or infrared separation, paint stripping or decoating, or other complex mechanical steps such as manual removal of films or stickers.

Shredding and compounding inevitably lead to material loss. Shredding on cutting or hammer mills produces dust; the more brittle the plastic is, the higher the percentage of loss. Thorough dedusting is even more important in connection with homogenization, because remaining dust leads to black specks in the subsequent compounding (or in the direct injection molding processing of a regrind). The reason is that the dust melts faster than the regrind during the material heating and thus cracks. This is not only a visual flaw, but also negatively affects the mechanical properties. However, dust can be used again: just like foils or milling chips, dust from the grinding process can—if it is clean enough—be agglomerated with little thermal stress and returned to the circuit in a free-flowing state.

The processing steps described above are generally suitable for all thermoplastics. Limits are set if the production residues or waste consist of mixed plastics, be it through unwanted cross-contamination, through multi-component injection molding, co-extrusion or through rinsing lumps when changing material on the machine. If incompatible plastics were used in the latter processes, the ground material obtained would be of inferior quality, if not even unusable.

A positive aspect that is becoming increasingly important is the lower energy required to produce recycled plastics compared to prima goods and therefore lower CO_2 emissions. Compared to chemical recycling, mechanical recycling performs better in terms of energy consumption and CO_2 emissions.

Whether post-industrial or post-consumer waste, the material flows must first be collected, whereby this collection of post-industrial material is subject to consideration of effort, benefits, and costs. Even if rejects in natural colors or crystal clear are a valuable fraction for recycling, separate collection of small quantities can be uneconomical for everyone involved. When it comes to collecting, what is crucial is a neat separation into varieties and a meaningful name. A mixture of filled and unfilled production waste of a polymer must always be described as "filled", a mixture of FR and non-FR production waste always as a flame retardant. These circumstances usually mean that the quantities that enter the recycling cycle are not the optimum of what is possible and consist of different colors, which has an impact on the colors that can be achieved after recycling (cf. Table 13.1).

There is a functioning recycling market for post-industrial material flows. Recycling companies pay their suppliers a price for the goods that depend on the type,

Table 13.1 Types of material flows in mechanical recycling (*Source* Own presentation)

Input stream	Post-industrial materials	Household packaging wastes	End-of-live vehicles wastes	Wastes from electric and electronics
Source	Converters and controllable suppliers	Households	Garage, shredder companies	Waste mgmt. companies
Quality	Good	Low, contaminated, mainly films, bottles, etc	Large parts as specified, mixed quality in SLF	Large parts as specified
Composition	Pure fractions of many polymers and types with modifications, additives, fillers, flame retardants, etc.	Limited number of polymers, limited modifications	Large variety of polymers, many modifications, many additives, fillers, fibers	High kinds of Polymers, many modifications, pigments, fibers, flame retardants, etc.
Life cycle—time to recovery	Immediate	Short (some months)	Long, life time of vehicle (~20 years)	Medium, life time of appliance (~ 10 years)
Preparation for recycling and processing	Grinding, quality testing, regranulation and compounding	Sorting, washing, shredding, homogenization, quality testing, chemical analysis, compounding	Sorting, delamination, shredding, homogenization, quality testing, chemical analysis, compounding	sorting, delamination, adhesive removal (labels), demetalisation, shredding, homogenization quality testing, chemical analysis, compounding

purity, color, quantity, and processing effort. You can also complain about the goods if the agreed quality is not correct.

The market for large areas of post-consumer recycling works completely differently. Even in the ideal case, the source of the material is anonymous, waste from different sources mixes during the collection process, and there is no possibility for the recycling company to agree on quality or make complaints. However, where there are no recycling facilities available or collection does not work (cf. 9 'Logistics'), the material is worthless and, in the worst case, causes environmental pollution. This applies to post-consumer material from the packaging sector, but also to waste from Electric & Electronics (E&E) or waste from end-of-life (ELV) vehicles. Post-consumer waste streams generated in the industry as a pre-consumer recyclate, on the other hand, are more comparable in terms of market and characteristics to the post-industrial waste streams described above.

Mechanical recycling is the method of choice for many waste material fractions, but it also has weaknesses. First, the quality that can be achieved through mechanical recycling is largely determined by the quality of the raw material, i.e. plastic waste or production waste. Most production residues or secondary output are very clean fractions. Fractions that require more effort are laminated punching waste, painted industrial waste or parts with metal inserts. With problems in the processing process, be it due to overheating or excessive residual moisture, the polymer was most likely damaged, leading to limited subsequent use.

Post-consumer waste from the packaging sector is usually a contaminated mixture, waste from old electrical devices must be separated from metals, stickers, etc., plastic waste from end-of-life vehicles must be further separated as they come from the shredder light fraction.

The associated challenge is a black box: the recycler receives production residue or post-consumer waste but knows little about the quality of the material. This can usually only be determined at the end of the recycling process through laboratory tests. Properties of the ground material or agglomerate that are not tested in the laboratory remain an unknown quantity—with implications for further usability. Untested properties are unknown properties, which require a lot of experience and expertise on the part of the recycler and the user to assess whether the quality of the material is sufficient for a functioning part.

Finally, there are fractions that cannot be processed or cannot be processed sufficiently well using mechanical recycling. These include, for example, starting lumps with polymer mixtures or waste from multi-component injection molding with incompatible polymers.

Another unknown quantity is waste from long-lasting products, which only gradually enter the recycling cycle. The processing of old E&E devices is already established, the dismantling and recycling of plastic parts from old vehicles is partially established, and the recycling of used clothing is only just beginning. There are large amounts of used plastics lying dormant here, which can be used as a source of raw materials. However, every post-consumer waste from applications that do not come from the rather short-lived packaging sector (cf. 2.4) often has a long service life and associated with this, possible material degradation, e.g. through UV exposure. In addition, with such sources there is the possibility that the original raw materials contained additives, flame retardants or pigments that are now banned. Such substances can be detected on the ground material through high testing effort—because it is batch-related. However, every test carried out is reflected in the price of the processing and every kilo of plastic that contains such substances incurs costs for mechanical processing and laboratory tests but may not be put back on the market. Such problems have occurred in the past when recycling bottle crates, window profiles or vehicle interior parts that contained cadmium pigments, or electrical appliance housings with flame retardant concepts that were banned during the use period between production and recycling. Future problems of this kind may be caused by PFAS (polyfluoroalkyl substances), which are integrated into the polymer matrix, but which cannot be put back on the market as recycled plastic.

For the closed-loop use of recyclates in automobile construction, which will soon be required, the question arises as to whether the plastic fractions from dismantled, possibly decades-old end-of-life vehicles are of such good quality that they can be used again, and whether the ingredients comply with the current regulations so that they may also be used.

Since post-consumer waste enters the processing process with less information compared to industrial waste and, unlike industrial waste, there is neither a quality agreement nor a complaint option, the risk in this case lies solely with the recycling company. The person responsible for the waste is unknown; complaints or returns to the person responsible are neither possible nor desired! At the same time, the focus, and expectations of the public and regulators are almost exclusively on post-consumer waste, its processing and reuse.

13.1.3 Products

Which products can be obtained from thermoplastics through the described mechanical recycling? For the large group of thermoset and other cross-linked plastics, mechanical recycling remains almost insignificant. It depends primarily on the quality of the available thermoplastic raw material, i.e. the production residue or waste. Unreinforced, undyed qualities such as a natural-colored polyamide 6, a natural-colored ABS or a crystal-clear polycarbonate can be returned to the injection molding or extrusion process as regrind or agglomerate. When it comes to compounding, there are numerous possibilities with such raw materials; coloring and modification with any choice of fillers and reinforcing materials are conceivable. However, such undyed, unfilled qualities remain the exception. In principle, such recycled plastics can therefore be used as an alternative to virgin materials, as the processability and typical properties of the polymer do not differ from virgin materials. However, there are limitations in the details.

The quality level that can be achieved with recycled plastics depends largely on the quality of the post-consumer or post-industrial waste stream and will never come close to the fully guaranteed property profile of a prime grade. Only rarely will it be possible to specify the variety of mechanical and rheological or optical and thermal properties for such materials, as is possible and required for new materials, and one will often have to live with optical limitations. There are also a growing number of regulatory barriers affecting, for example, materials for food contact or medical devices.

Another challenge is the demand of globally operating companies for the global availability of identical, locally produced, recycled qualities since the quality of a recyclate or recycled compound initially depends on the available plastic waste. Where the composition of this plastic waste differs or where different primary goods have been and are used due to legislation, regional availability or other considerations, the recycled plastic will also differ in its properties. This fact cannot be ignored and must be resolved in dialogue between the OEM, supplier or processor and recycling company.

Furthermore, no matter what polymer and color is considered, there will be quantitative limits. Assuming that waste-free production is not possible, and that plastic waste is not always recycled immediately, more plastic waste would have to be recycled and available for reuse for a period X than is needed net for packaging or molded parts.

In addition, it is not per se possible to produce a part in the same color from a colored plastic part because colors are collected mixed together. Filled plastics cannot easily be used again in the same application.

And finally, no one produces post-consumer waste or an industrial waste fraction for recycling, because tomorrow a recyclate will be needed for any application. The available amount of recyclates from post-consumer waste or recompounds from post-industrial waste is not only limited from the outset by the amount of waste available, it can also hardly be planned and can only be guaranteed to a limited extent. Processors and recycling companies deal with this constructively; however, it is still unclear how legal quotas should and can be met under these circumstances.

Under these circumstances, recycled materials from mechanical recycling have their place in many applications for black non-visible parts in the automobile, the electrical and electronics industry and the construction sector. Very clean fractions or materials with light inherent colors can even be used for colored recyclates or recompounds and visible parts.

13.1.4 Opportunities and Risks

The demand for global availability, fractions with high processing costs and the quantitative limits are major challenges for the recycling market, which is currently still largely medium-sized and regional. Today, this means that despite an established recycling industry, some material flows cannot be recycled or recycled materials cannot be considered for some projects because they are not available in color, quantity, or quality.

If you look into the future from here, it becomes clear that several decisions must be made; material developments must consider the complexity of recovery and recycling, material combinations are minimized and optimized, material information is provided in a suitable form, collection and logistics are geared towards generating sufficiently large material flows—in short, "Design for Circularity" must make it possible to facilitate, optimize and maximize future recycling. Ultimately, it must be accepted that recycling larger and therefore more complex material flows is becoming more complex and expensive.

Having considered strengths—economic efficiency, energy savings and CO_2 avoidance—and weaknesses of mechanical recycling (cf. Table 13.2)—available volumes and qualities—the question arises as to whether the potential of this recycling process has already been exhausted.

Today we are in a contradictory situation in which many waste fractions are not processed or are processed only inadequately, while interest in processors is increasing, customers and consumers are willing to reward recycled materials and politically and legislatively binding use quotas for recyclates from post-consumer

Table 13.2 Strengths and weaknesses—SWOT analysis of mechanical recycling (*Source* Own presentation)

Strengths Infrastructure—established in EU Market demand—increasing Waste disposal—reduction CO_2 emissions lower vs virgin material Circularity—integral part	Weaknesses Competitivity—driven by qualified specs and costs Market access—limited as compared to virgin material Material sources—unspecified (no claims) Volumes—too small homogenous waste flow Degradation of polymer—aging and thermos-mechanical treatment Legacies—critical additives in long-living plastics
Opportunities Plastic wastes—large and increasing volumes globally Sorting—technology development Design for recycling—new technologies Fossil-based capacities—reduction for plastics Regulations—on plastic waste Market—growing via regulations	Threads Green premium—little customer acceptance on pricing only Cost structure of recyclers—unsatisfactory Capacities—for sorting and logistics too little Waste management—insufficient Barriers—via restrictive regulation

waste are on the rise be taken away. What sounds so contradictory is caused by an information, knowledge and financial gap.

The amount of recycled plastics could be larger if the resulting waste could be better recycled. Better "Design for Circularity" is required for innovative technical parts that will or shall be recycled in the next decade to overcome the major challenges associated with them. The amount of recyclates or recompounds put into production could be bigger if more expensive recycling with more efficient technology were paid for. The entire industry would have to operate in a circular manner and not only think about which recycled material is suitable for the application, but also about how well or poorly the production waste or packaging can be recycled. Mixed post-consumer waste from packaging can already be separated very well using separation processes due to the manageable range of polymers. This has so far been much more difficult for mixed technical plastics, regardless of whether they are post-industrial or post-consumer fractions.

The technologies required are expensive and since this processing step is not necessary for all fractions like laboratory testing or compounding, not every company in the predominantly medium-sized plastics recycling industry will be able to invest in it due to a lack of capital resources. Just like the limited flow of raw materials on the one hand and the desire for guaranteed delivery quantities on the other, this is probably one of the reasons for the strong consolidation of the industry over the last 5 years.

Limits will remain for mechanical recycling. The qualities that can be achieved through purely mechanical recycling are ideal for numerous applications but will never achieve optical transparency or the completely consistent property profile of a prime grade. Therefore, there will be a variety of applications and application categories that will remain closed to plastics from mechanical recycling. The availability of a globally identical recycled material combined with regional processing and production will hardly be possible. And finally, every processing step and the associated mechanical and thermal stress, both during production and recycling, have damaging effects on the polymer chain. At some point, the polymer has reached its natural end of life and can no longer be used in a comparable high-quality application.

Until this end of life, however, mechanical recycling is the method of choice wherever possible, as it produces cost-effective alternative raw materials for a wide range of uses with comparatively low energy input. It avoids waste being incinerated, and it avoids littering. In addition, mechanical recycling also ensures the regional availability of raw materials for the plastics processing industry without being dependent on fragile logistics chains.

13.2 Market Situation for Recyclates

Peter Jetzer

13.2.1 Introduction

At the beginning of the new millennium, the market for secondary raw materials could have been described in just a few sentences. Statements about the location and volume of plastic waste as well as the sorting quality would have been sufficient. However, a few years ago the market for granulates from recyclates changed fundamentally.

While the capacity for mechanical recycling in Europe at the turn of the millennium was still around 2 Mt (PE 2006), in 2021 it had already grown to over 11 Mt, meaning the supply of regranulate has grown sixfold within around 20 years. The European production of "fossil-based plastics" reached a volume of around 50 Mt in 2022 and has therefore only grown by around 10% in the comparable period. The production capacities for new goods were reduced through plant closures and dismantling, consolidation or relocation of production outside of Europe (PE 2022).

In 2022, an unprecedented situation occurred in the previous price constellation of mass plastics such as LDPE film nature and its recycled sibling rLDPE film nature regranulate (PCR). While the annual average prices for regranulate were previously around € 1,000/t, they approached the high of € 1,800/t in mid-2022, almost reaching the € 1,900/t level due to special influences of the pandemic, logistical turbulence and war in Ukraine. Under the pressure of falling prices for primary LDPE film material, the recyclers then reacted by reducing prices in order not to lose too much volume to the primary market (cf. Fig. 13.1.).

Added to this was the general deterioration in the macroeconomic outlook and the resulting reluctance to purchase among customers. The Plastics Recycler Europe

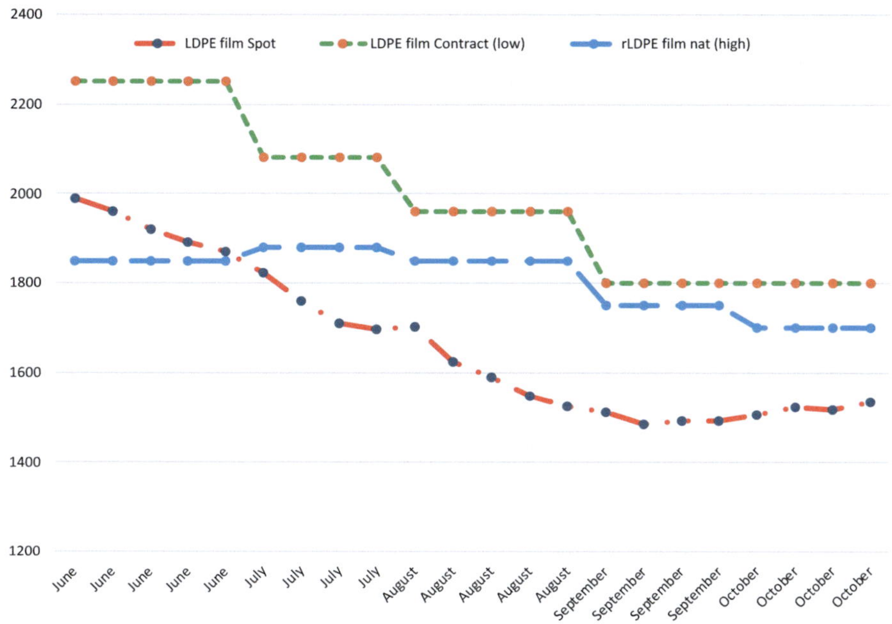

Fig. 13.1 Price dynamics in 2022 for different LDPE film grades (*Source* Own presentation)

Association (PRE 2023) stated that recyclate prices are generally decoupled from those of virgin materials.

Accordingly, ambitious sustainability goals for packaging materials were offset by structural bottlenecks in waste raw materials. Without changing the framework conditions, the differentiated demand for recyclates can only be ensured with strict adherence to operational profitability calculations as summarized by PRE (2023). Due to a lack of standardization, there were significant differences in quality. Consequently, top prices can be achieved for pure recycled material.

It is not yet clear whether this will break the "cheap ban" hanging over secondary materials (cf. Fig. 13.2). What is certain, however, is that the "formula" that has been used for decades: new material price −30% equals recyclate price has lost its validity.

13.2.2 Secondary Raw Materials Market

Waste market

While in 2010 there were 950 companies active in the waste disposal industry in Germany, the number had increased to more than 1,700 companies by 2020 (Statista 2022). During this period, the turnover of the waste disposal industry in Germany

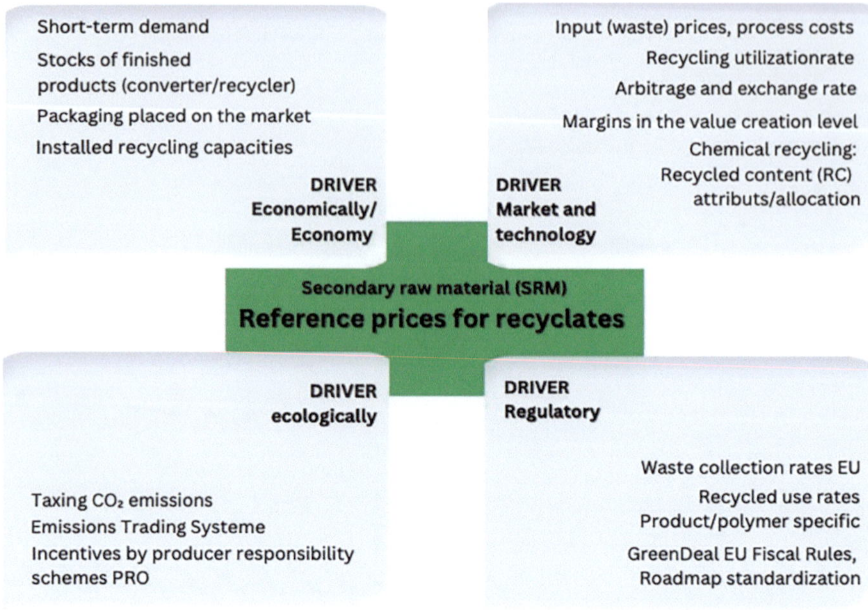

Fig. 13.2 Market price driver for recyclates (*Source* Own presentation, Copyright Recycario)

doubled from around 21 to around 40 billion €. Such an approximate doubling in just a decade demonstrates the interest in the ever-growing mountain of waste as well as the growth of waste disposal companies.

The approximately 700 European plastics recycling companies achieved sales of around € 9 billion € in 2021, after € 3 billion € in 2019. This exponential sales growth was made possible by investments of € 1.75 billion € in new plastics recycling capacities (PRE 2023). Nevertheless, the industry associations are dissatisfied because across Europe. There is still too little plastic waste being collected separately (cf. Fig. 13.3 in 2020: 29 Mt) and recycled (cf. Fig. 13.3 in 2020: 11 Mt) (KI 2023). Many of the 27 EU states therefore run the risk of missing the recycling targets for packaging plastics for 2025 (cf. Sect. 21). At the same time, the demand for recyclates from other sectors is also growing, which is being triggered by PCR quotas for automotive parts, among other things (Tecpart 2024).

The expanded recycling capacities (red columns) should contribute to the more extensive treatment of packaging waste.

Plastic waste prices

Economic aspects of recyclate production are just as important as ecological reasons. A competitive and reliable price for plastic waste is an essential basis for cost calculations in recyclate production. With mechanical recycling, in addition to the waste price, there are proportional process and material costs for shredding and sorting,

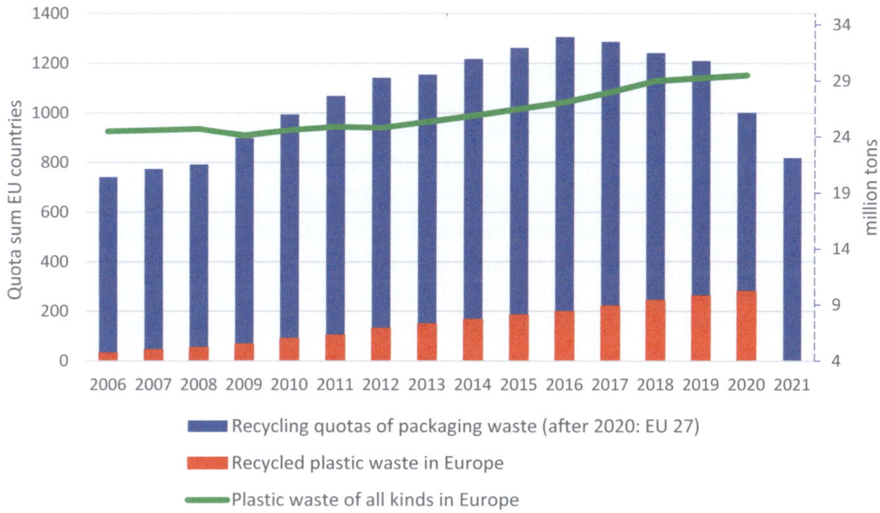

Fig. 13.3 Plastic waste in the EU (*Source* Own presentation[1])

washing and drying, for the disposal of raw material and process-related waste—i.e. for processing—as well as for extrusion, the additives and finally for quality control (cf. Sect. 13.1). The larger and more homogeneous the waste volume flows are and the more continuously they are available, the lower the specific costs for the process steps mentioned, the more likely a scaling-up will be possible and the more predictable the system utilization will be for the recycler. It is therefore in the recycler's well-understood interest to cooperate with waste suppliers, i.e. waste disposal companies, and to be familiar with their business models.

Waste categories as a basis for understanding

Plastic waste collections have been carried out in Europe for several years, divided into different categories. A quality directory for used plastics has long existed in northwestern Europe, which is maintained by the German associations Federal Association of Secondary Raw Materials and Disposal (BVSE) and Federal Association of the German Waste Management (BDE) as well as the Bureau of International Recycling (BIR) in Brussels and is available in the BIR's annual reports is (BIR 2023). The collected plastic waste reaches the market, for example, as film waste pressed into bales, as regrind from production waste in the manufacture of technical parts or after dismantling household appliances to produce recyclate, and as refuse-derived fuel (RDF) for energy producers or cement manufacturers. To make it easier for buyers of used plastics to differentiate and classify, type groups and a type code were introduced. In addition, reference should be made to the European Waste Catalog (EWC), which largely deals with the classification of plastic waste (UBA 2021).

[1] Data bases Eurostat and Statista.

Waste supply chain

The global trade in plastic waste is monitored in numerous publications, studies or digital services and published at regular intervals as price data overviews. A particularly valid source for this in the European area is the "European Economic Service" (EUWID). An editorial team records the actual over-the-counter (OTC) business in interviews with market players (EUWID 2023). Alternatively, monthly prices for bottle and film waste can be found on the British website "Let's recycle" (Letsrecycle 2024). In addition to price data for plastic waste that corresponds to the type names in the BIR list, EUWID also provides prices for waste that has already been treated, e.g. regrind. The waste nomenclature, which includes 36 types, is divided into the categories PE production waste, PE post user, PP production waste, PVC production waste, PVC window goods and PS production waste. Price histories going back decades are available online in a variety of forms. Current price information for waste plastic lots can also be found in various online portals.

13.2.3 Recyclate Supply

Different price developments

The installed recycling capacities (orange bars in Fig. 13.4) have been growing for 10 years, while the recycling rates for packaging waste (average rate for all countries) in the EU remain almost constant (blue bars). This is a clear indication that the packaging waste stream has been recycled to a greater extent for mechanical recycling and to a lesser extent for thermal recycling. Nevertheless, in view of increasing supply quantities, recyclers had to stimulate sales with price discounts, which caused prices for the secondary product to plummet. The move away from energy recovery to increase mechanical recycling must therefore remain the focus and is considered a relevant indicator. The higher demand for waste and specified regranulates led to higher prices in both cases.

Waste EU exports

The decline in waste exports from 2020 increased the feedstock supply in the EU27 +3 and led to higher recyclate production. However, the first months of 2023 show an irritating trend reversal. In March, exports of EU plastic waste to Turkey, Indonesia, Malaysia and Vietnam rose to their highest level since spring 2021. The recent increase counteracts the EP's ban on waste shipments to countries outside the OECD. Recycling received a significantly greater boost when exports of plastic waste decreased, which can be seen in Fig. 13.5 (green bars). This left more waste to produce regranulates on the EU continent with almost 12 Mt (brown bars) in 2022.

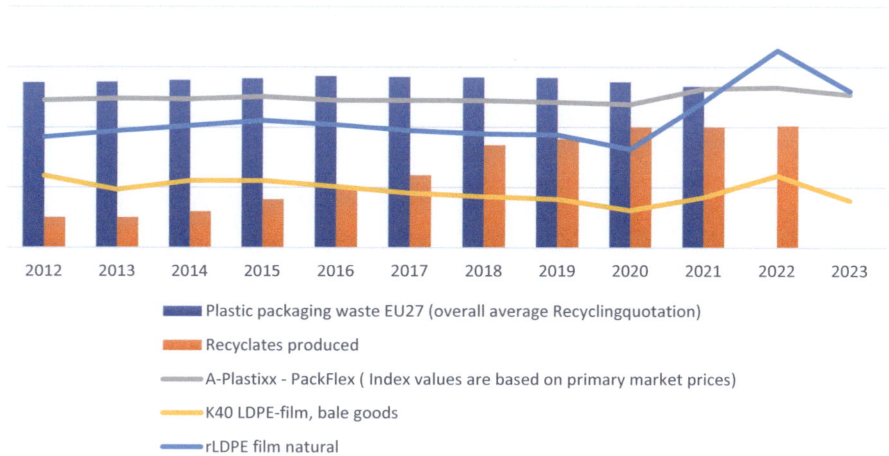

Fig. 13.4 EU 2012–2022 annual averages for prices, quotas and index points (*Source* Own presentation[2])

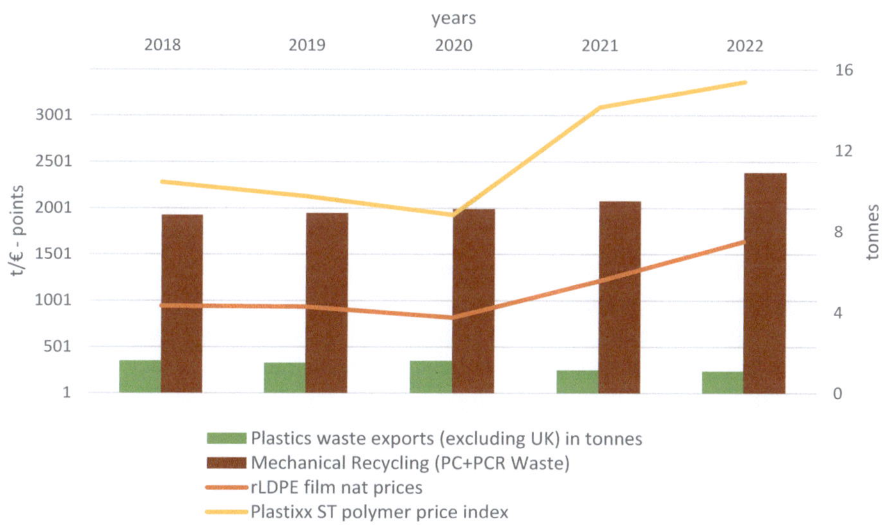

Fig. 13.5 Waste export to non-OECD countries by volume and price—prices and indices as annual averages (*Source* Own presentation[3])

Price acceptance from a competitive perspective

The granule price trends for primary and secondary goods as well as plastic waste resemble a rigid hierarchical order that has lasted for years. The "ranking" shown in

[2] Based on data from Eurostat, PIE-Web, EUWID, PRE, PlasticsEurope.

[3] Data base from Eurostat, PIE-Web, PRE.

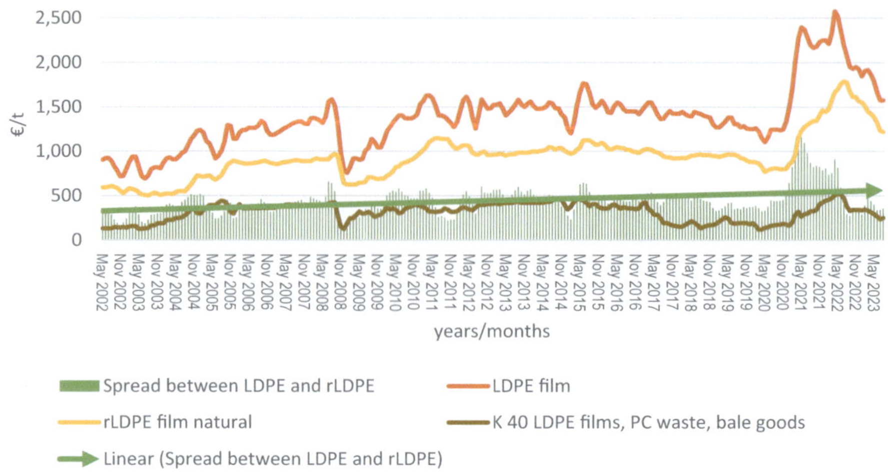

Fig. 13.6 Historical price gaps of virgin versus recyclate plastic (*Source* Own presentation presentation[4]) based on Database from PIE-Web, EUWID)

Fig. 13.6 can also be found for other types of plastic, only the spread varies minimally. For LDPE film (orange line) the "discount" to the secondary rLDPE film (yellow line) was on average 450 €/t (also shown as green bars and trend arrow in Fig. 13.6). It was impossible for recyclate manufacturers to "break" this unwritten "distance law" because the market did not accept a convergence of recyclate and virgin goods prices. Attempts to do this were blocked by cancellations and switching to alternative sources.

However, the first indication of an impending "collapse" of the old order is provided by the plastic waste price (brown line in Fig. 13.6), which shows significantly lower signs of decline compared to the price lines for granules above. If the situation becomes permanent, recyclers will pass on these higher feedstock purchasing costs in higher recyclate prices. The trend towards more secondary material could play into their hands, which would encourage the old distancing order to be broken! This would establish this market sustainably and more predictable.

Many indicators that reflect the pricing of recyclates have been highly volatile in recent years. The risk of being substituted by cheap offers remains despite attempts to shield the price decoupling. When there is favorable arbitrage, primary production surpluses from the Middle East, North America and Asia always come to €ope and undermine the €opean price structure for regranulates.

[4] Data base from PIE-Web, EUWID.

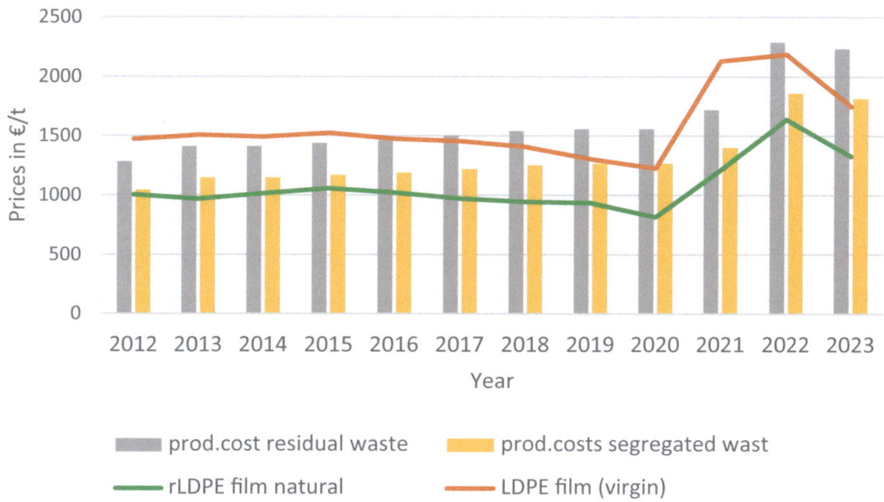

Fig. 13.7 LDPE price fluctuation in recycled and virgin plastics (*Source* Own presentation[5])

13.2.4 Value-Added Levels

Cost calculation of recyclate production

It is important to position the value of the secondary material among partners involved in the cycle, which helps to look at the upstream stages. Retailers as well as municipal and private waste disposal providers are responsible for collecting and sorting the feedstock. Other actors in the recycling structure are logisticians and/or sorters or plant operators for waste sorting. This is followed by the actual recycling process, with washing processes and the production of the granules by extrusion. The market or buyers then decide on the economic success of the secondary material.

The cost calculation in the following figure was based on the 2022 price (cf. Fig. 13.7), and the cost history was adjusted using the "Producer Price Index of Commercial Products" (Destatis 2023). Prices achieved on the market per ton of rLDPE (cf. Fig. 13.7) do not show much promise. The green line for the price trend of the secondary material would only approximately cover the production costs in 2021, but only when using pre-sorted plastic waste (orange columns). In previous years, the sales price of regranulate (green line) was usually slightly to very significantly below the production costs.

Compared to 2019, the annual volatility for the primary material has quadrupled with a standard deviation of 120 €/t. In the same period, the significantly cheaper rHDPE has had a six-fold increase in volatility to 60 €/t, but there is still not enough of an upswing to sustainably cover costs laid down. High volatility alone does not seem to be an effective means of price development.

[5] Data bases PIE-Web, http of Aachen, Germany.

Is the plastic cycle tumbling back into the old linear track?

Not every medium-sized recycler can survive a long-term industrial transformation process. Especially since in the case of the plastics industry, long-term safeguards for the purchase of recycled material are only hesitantly agreed upon by market players. Since 2022, European recyclers have been struggling with skyrocketing production costs, which they have not yet been able to pass on in their sales prices (cf. Fig. 13.7 LDPE price fluctuation).

On the contrary: More and more imports of inexpensive new goods from overseas contributed to "destabilizing" the plastic recycling market in Europe, PRE stated in November 2023 (www.kiweb.de, 11/2023). Here President Ton Emans at PRE pointed out that imported virgin material was offered at prices that were significantly lower than those for recyclate produced in the EU. Among many others, the German subsidiary of the waste management group Veolia, Paris, closed its plant in Rostock, Germany, by the end of 2023. The reason given was that economic operations in the volatile business environment were no longer possible (www.kiweb.de, 05/2023).

Prognostic possibilities[6]

Predicting price developments is crucial for companies in the recycling industry; profitability and sustainability are directly dependent on the volatility of waste prices. However, the market dynamics are poorly established and understood compared to the markets for primary materials. Predictive analytics—based on several indicators—can help to transfer experiences from known markets and predict future developments. Scenario forecasts can help provide additional insights. While the emerging market for recycled materials is more dynamic and fluctuating, the conventional market for primary plastics has more established processes and is therefore easier to assess. A comparable forecasting process is currently being refined (Prognostica 2023). However, it still needs to prove itself in practice in order to better understand the sensitivity of recyclate prices to key influencing factors—and ultimately to derive robust and plausible scenarios for future price developments. Encouragingly, initial model training conducted by the project partners Recycario and Prognostica GmbH has already shown that the gap between worst- and best-case scenarios is noticeably narrowing.

13.2.5 Summary

Recyclate market

The recyclate market still presents itself as a temporary solution that can be used as desired and avoided by the players depending on the general economic situation or geopolitical influences. The attempt to establish a recyclate business as a sustainable, economical business model in the competition between plastics suppliers has so

[6] Based on data from Prognostica, Würzburg, Germany.

far only been successful in certain areas. Pricing models for the established new goods market are unsuitable for the recyclate market or are still being developed. The attempt at an adequate calculation for recyclates fails due to numerous sudden market influences, so the required regular income can hardly be generated, and the profitability of medium-sized companies is constantly in question.

Dynamics of the waste market

The plastic waste market is a less than reliable player as a supplier of feedstock for the circular economy we are aiming for. The reasons for this are the collection of plastic waste and the exclusive disposal of it based on costs. This could change because of the German Fuel Emissions Trading Act (BEHG), which has now come into force, and which imposes high CO_2 taxes on the previously economical incineration of plastic waste. However, the continued export of significant quantities of the feedstock (despite an EU-wide ban) exacerbates the situation for European recyclers, as the export quantities are used directly for economic purposes without reappearing in the circular economy.

Investment needs

The investments of € 1.75 billion mentioned by PRE (2023a) at the beginning of the decade are just a beginning. To deal with inhomogeneous and volatile input and thereby increase quality, increase availability, and reduce costs, further investments and innovations are required that exceed the capital strength of the medium-sized industry. An Additional problem is the need for suitable, more highly qualified personnel.

Chemical recycling as an option

Chemical recycling is not yet available as a buyer for the poorly qualified and rejected feedstock due to a lack of valid technology and capacity. The PlasticsEurope Roadmap "The Plastics Transaction" only sees the "scaling of capacities for chemical recycling" in the medium term (2028 to 2030). The essential calculation and evaluation of the products from chemical recycling are at least among the immediate measures taken by plastics producers for: "Determination of the mass balance approach for chemical recycling" (cf. Chap. 16).

Profitability threatening the existence of recycling companies

Profitable recyclate production requires the establishment and protection of a market price that reflects values. In addition, there is the impact of many and rapidly changing impulses on the cost calculation, which can only be offset by further "price premiums". Protecting the high investments required is becoming a question of existence for the recycling industry. For many medium-sized recyclers, this is becoming an ever-increasing challenge.

13.3 Circular Feedstock from Chemical Recycling

Andreas Neumann and Daniel F. White

13.3.1 Strategic Importance of Chemical Recycling for Defossilization

Chemical recycling products as carbon source

As efficient mechanical recycling streams currently exist only for a limited number of highly sorted plastic waste streams, the remaining waste consisting mainly of mixed plastic or flexible materials is currently either downcycled to lower value applications, incinerated, or, in some countries, landfilled.

Chemical recycling as a technology is quickly emerging to close this gap, breaking down plastic waste materials into liquid feedstocks as raw materials to produce new plastic materials. These recycled products are intended to serve as a sustainable carbon source for various applications, reducing dependence on traditional fossil-based feedstocks and closing the loop, thus contributing to enable a circular economy of plastics.

However, there is no well-organized and efficient supply chain yet for chemical recycling operations and the European market for plastics waste is itself far from being fully integrated. Major feedstock volumes in Germany and other EU member states are being pulled from the incineration supply chain, which requires no treatment steps to sort out mixed plastic waste. Hence, separation and treatment assets will need to be built to support chemical recycling at scale. Building and expanding alliances with, and potentially investment in, mixed plastic waste suppliers is crucial to develop and secure supply.

An example of such a collaboration is Source One Plastics, a joint venture between 23 Oaks Investments and LyondellBasell. After its formation in 2022, this company has now successfully started up an advanced plastic waste sorting and recycling facility in Eicklingen, Germany (cf. Fig. 13.8). The new unit processes difficult-to-recycle post-consumer plastics waste such as mixed plastic packaging and flexible polyolefin materials that would otherwise be sent mostly to incineration. The annual expected processing capacity of 70 kt equals the amount of plastics waste produced by approximately 1.5 million German citizens a year.

The new plant uses an innovative dry processing method to reduce energy consumption by up to thirty percent compared to conventional plastic recycling technologies and is powered by locally generated renewable energy. The plastic waste processed by Source One Plastics will provide part of the feedstock for the first commercial-scale catalytic chemical recycling plant that LyondellBasell will build at its Wesseling, Germany, site. It is, however, important to evaluate the potential contribution of chemical recycling to the circular economy not only from an input perspective, emphasizing its potential to address the issue of plastic waste that is currently hard to recycle, incinerated or even landfilled. Output considerations are of

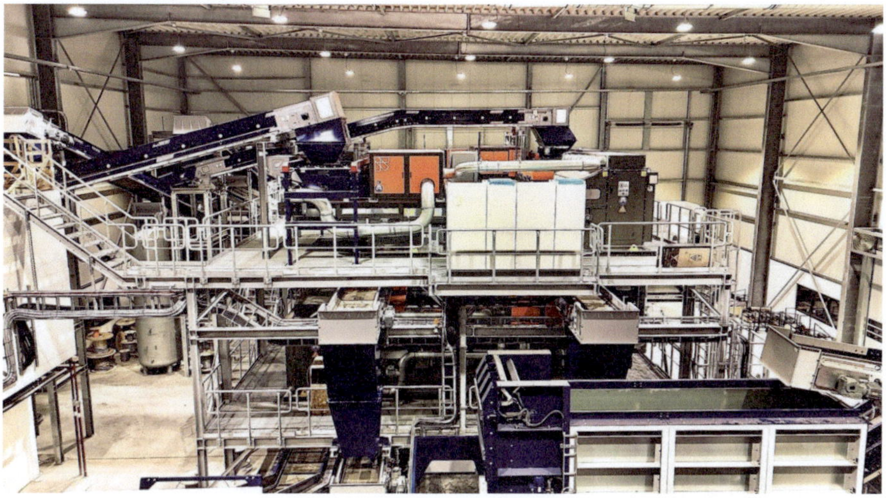

Fig. 13.8 Source One Plastics advanced sorting and recycling plant (*Source* Source One Plastics)

equal importance as a significant percentage of plastics is used in applications that are strictly regulated, which represents an important entry barrier for mechanically recycled materials.

Volume and mass aspects

A single naphtha cracking furnace requires 100 kt of feedstock per year (Oliveira and Van Dril 2021) and a typical steam cracker consists of multiple furnaces, feeding their products into a single, common fractionation section. Under these circumstances, it is most likely that in the future, steam crackers will process a blend of chemically recycled and fossil feedstock rather than exclusively operate with chemically recycled feedstock (Amghizar et al. 2017).

Currently, one of the main barriers to scale-up chemical recycling is the uncertainty about the method for calculating the recycled content in the final plastic product (Cefic 2022). In the case of chemical recycling, a mass balance approach is needed to leverage recycled feedstocks in existing infrastructure together with the virgin fossil-based feedstock. As the two different feedstocks cannot be physically separated once they are co-fed into the large-scale steam cracker installation, the mass balance methodology is required to accurately calculate and verify the amount of recycled content allocated to individual products (BASF 2020; Berggren 2024) (cf. Chap. 16).

13.3.2 Processing of Plastic Waste by Pyrolysis

Pyrolysis is defined as thermal degradation in the absence of oxygen. The process results in the breakdown of the polymeric hydrocarbons to short-chain molecules, the absence of oxygen prevents a partial or complete combustion to occur.

The process typically produces gases, liquids, and solids (char). The ratio of the product depends on the process conditions. As the process is endothermic (a process that requires the absorption of heat), some critical operational factors are temperature, residence time and heat transfer with higher temperatures and residence times typically favoring lighter products. Another characteristic of the products is that the polymeric feedstock is hydrogen deficient, therefore the products from this reaction are highly olefinic. However, the composition of plastic waste pyrolysis oil significantly differs from liquid fossil-based steam-cracking feedstocks. Work has been conducted to identify the composition of chemically recycled feedstock and the level of contaminants, particularly in the liquid phase (Kusenberg et al. 2022).

The gas produced is a mixture of hydrogen, methane and C2 + olefins and paraffins. It has a high heating value and is often used to generate the energy required for the pyrolysis process. However, this results in the combustion of the gas and related CO_2 emissions. More advanced recycling technologies recover the gas product from the reaction process and use this for chemical production, increasing the carbon efficiency of the process. This also facilitates the use of methane or electricity to heat the processes. As discussed earlier, pyrolysis is an endothermic process.

The liquid product typically has a wide boiling range, the lower range being used for chemical production and the higher range used for fuel or sent to a refinery for further operations. More advanced processes produce a product with a narrow distillation range and a lower upper boiling point, with the most advanced sending the full range product for chemical recovery further increasing the carbon recovery for the process.

The solid product from the process—or char—is a mixture of residues, typically consisting of materials introduced to the process with the plastic waste feed. This can include carbon but also metals and other fillers and additives. As a result, the overall yield of the process can vary widely as the composition of the feed as well as the process conditions offer a wide range of parameters.

Catalytic pyrolysis

Catalytic pyrolysis is based on the same pyrolysis process but adds a catalyst at some stage to promote the reaction. The catalytic effect allows the process to operate at lower temperatures, reduces the residence time of the plastic materials, or helps to depolymerize hard-to-recycle plastics. As with pyrolysis the process typically yields gases, liquids and solid (char). The ratio of the product depends on the process conditions as well as the concentration and type of catalyst. One challenge is the deactivation of the catalyst caused by contaminants in the feedstock, so catalyst consumption is a variable that must be monitored.

LyondellBasell has been conducting base research to develop its own chemical recycling technology since 2018. Initial work has shown that use of a catalyst in the pyrolysis of plastic waste can reduce the temperatures and residence time required to break down the polymer chains into small molecules, making the process faster and more energy efficient. Additionally, the use of a catalyst is improving the quality of the feedstock, and a higher oil content could be achieved, which makes the production of new polymers out of plastic waste more efficient.

LyondellBasell's *MoReTec* technology uses a proprietary catalyst-based approach to pyrolysis that converts mixed plastic waste into feedstock for the cracking process and eventually to produce new plastics for all applications including food contact and healthcare items. Since 2020, the company is operating a semi-industrial scale, catalytic chemical recycling plant in Ferrara, Italy and has made the decision to build an industrial-scale, catalytic *MoReTec* chemical recycling plant in Wesseling, Germany, based on the favorable results of the Ferrara operation.

This single-train unit with an annual capacity of 50 kt will operate continuously at relatively low process temperatures and high energy efficiency thus reducing energy costs and associated greenhouse gas emissions. In contrast to other pyrolysis technologies, which use the pyrolysis gas stream to heat the process, plastic-to-plastic yield is maximized by using both the liquid and the gas phase out of *MoReTec* process. This combination will lead to high energy efficiencies, high yields, increased scalability potential and optimized capacities, ultimately reducing the use of fossil-based feedstock in the production of new materials. The required heat inputs will also be supplied using 100% renewable electricity. The construction of this new plant is supported by a € 40 million grant from the European Union (EU) Innovation Fund.[7] The Innovation Fund is one of the world's largest funding programs for the demonstration and commercialization of innovative low-carbon technologies.

For LyondellBasell, the construction of this new chemical recycling plant represents the next step towards addressing the challenge of hard-to-recycle plastics at scale as illustrated in Fig. 13.9. This project is expected to provide the company with valuable operating experience and additional technological know-how that is needed to scale-up and fully commercialize the *MoReTec* technology.

13.3.3 Processing of Recyclates from Chemical Operations

Naphtha steam cracker operations

Steam cracking of hydrocarbons, a pyrolysis reaction, is currently the most important industrial process to produce olefins, representing the source of 98% of the production of ethylene and 60% of propylene (Zimmermann and Walzl 2012). A schematic

[7] Funded by the European Union. Views and opinions expressed are however those of the author(s) only and do not necessarily reflect those of the European Union or the European Climate, Infrastructure and Environment Executive Agency (CINEA). Neither the European Union nor the granting authority can be held responsible for them.

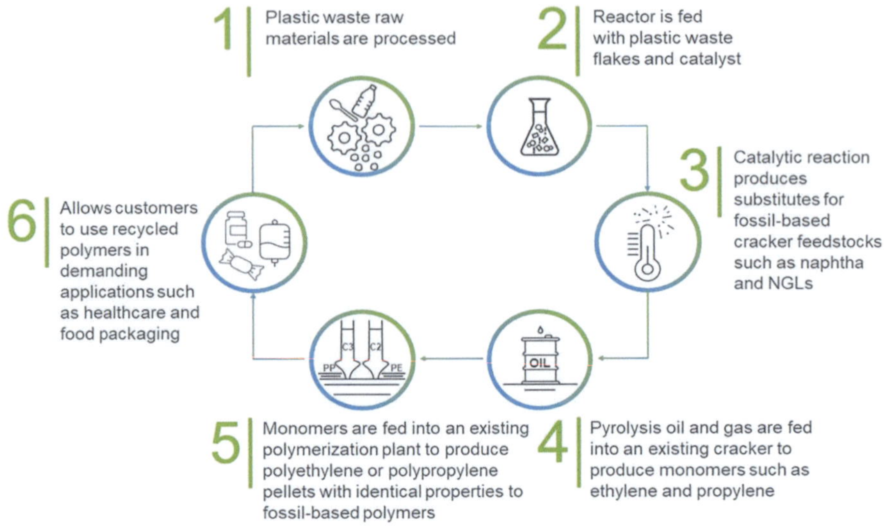

Fig. 13.9 Chemical recycling (*Source* LyondellBasell)

flowsheet of the steam cracking process is shown in Fig. 13.10, providing an overview of the main units of the process as well as the key products. The core elements of the steam cracking process are represented by the highlighted thicker lines, representing the hydrocarbon feed, the actual steam cracking furnace, which is the core of the steam cracker and the main products ethylene and propylene. The operations highlighted in blue are thereby specific only to plants running with liquid feedstocks such as naphtha.

The hydrocarbon feedstock enters the furnace of the steam cracker in the convection section, where it is pre-heated. The feedstock is then mixed with steam and heated further to temperatures of 500–680 °C. Thereafter the mixture of feedstock and steam enters the radiant section of the furnace, where radiant coils act as tubular reactors, submitting the hydrocarbons to cracking reactions for periods of 0.1–0.5 s.

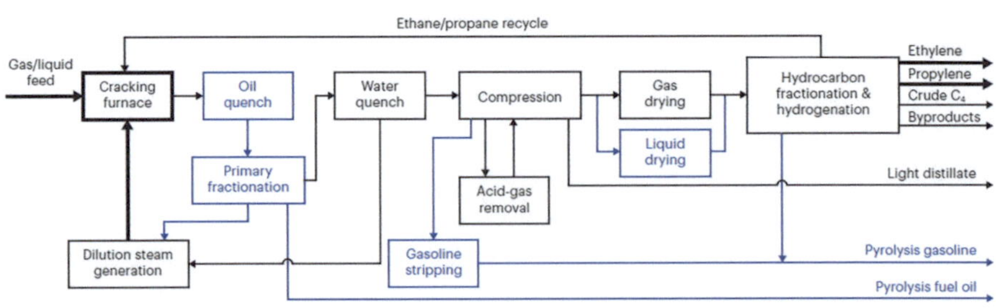

Fig. 13.10 Simplified flowchart of the steam cracking process (Reprinted from "Ethylene" in Ullman's Encyclopedia of Industrial Chemistry, Heinz Zimmermann, Roland Walzl, 2007, with permission from John Wiley & Sons)

The radiant section of the furnace operates at a temperature between 600–860 °C, which is maintained through heat transfer with the surrounding firebox, which is heated by fuel burners reaching temperatures around 1000–1200 °C. Due to the endothermicity of the occurring cracking reactions, a high heat flux is required. If the hydrocarbon feedstock is naphtha, the resulting product stream of the steam cracker is typically made up of (in *wt.* %) 25–35% of ethylene, 14–18% of propylene, 4–6% of butadiene as well as ~ 14% of methane and 5–10% of aromatics, namely BTX, along with hydrogen as an additional byproduct. This product composition can be modified to a certain extent by the operating conditions, through which the severity of the cracking can be influenced.

The cracked gas then exits the radiant coil at 800–860 °C and is rapidly quenched to terminate the reaction and prevent further cracking of valuable reaction products as well as the formation of coke, which tends to form at higher temperatures. This cooling process occurs in the transfer-line exchangers (TLE), by indirect quenching. An oil quench, which is only used for liquid feedstocks, due to higher temperatures at the exit of the radiant coil, is used to reduce the temperature down to around 230 °C (Zimmermann and Walzl 2012).

Next, a primary fractionator is used to separate the heavier hydrocarbon products from the main product stream. To be further processed, the remaining hydrocarbon product stream is then further cooled to near ambient temperature by means of a water quench tower. To compress the cracked gas, a series of subsequent compression stages with inter-stage coolers is used, allowing the cracked gas to reach pressures up to 35 bars, while maintaining temperatures below 100 °C (TVK 2010).

The condensates, as well as water and other heavier components, are removed during this cooling process, alongside Hydrogen Sulfide and Carbon Dioxide, which are removed through contact with an alkaline solution (Zimmermann and Walzl 2012).

The resulting gas is dried over molecular sieve beds before it proceeds to the fractionation towers. Here the cracked and now purified gas is chilled and separated into its individual main components by means of a fractionation train composed of the following distillation columns: de-methaniser, de-ethaniser, de-propaniser, de-butaniser, ethylene fractionator and propylene fractionator. To further increase the light olefin yield, catalytic hydrogenation beds are employed, which convert acetylene, methyl acetylene and propadiene into ethylene and propylene (Zimmermann and Walzl 2012; TVK 2010).

Cracking furnace

The cracking furnace is at the core of any steam cracking plant. The cracking of lower-value feedstocks into higher value products, such as olefins, is taking place here. Cracking furnaces consist of two main sections: the convection section and the radiant section (or firebox), as shown in Fig. 13.11 (Karimzadeh 2009).

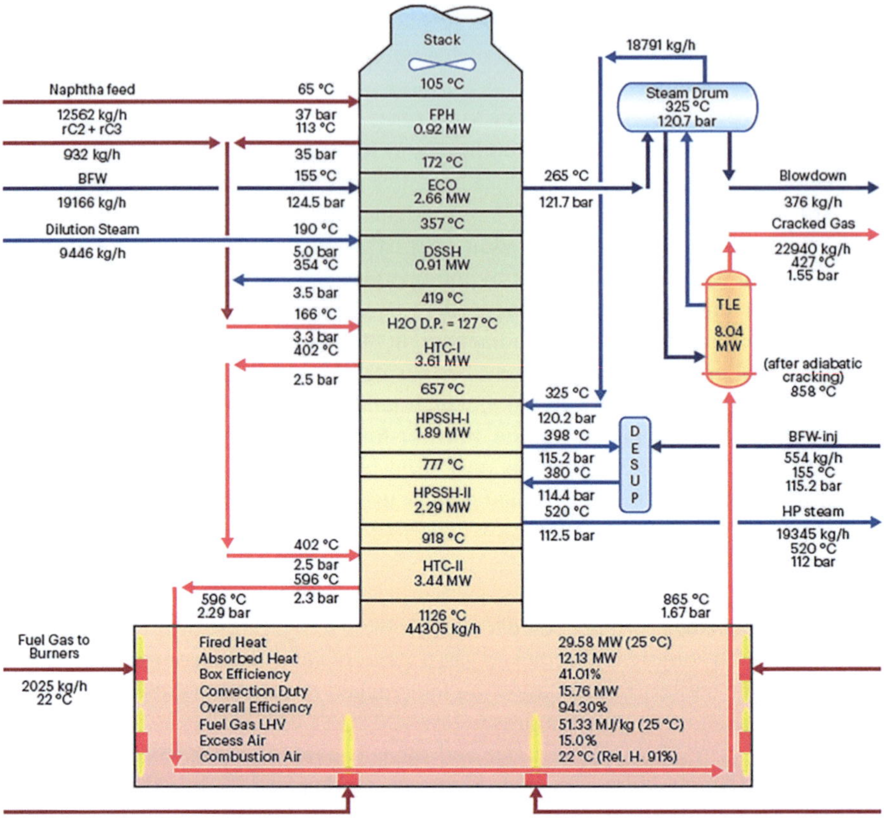

Fig. 13.11 Diagram of a cracking furnace (Reprinted from Chemical Engineering Research and Design, Vo. 87, No. 1, Ramin Karimzadeh, Hamid Reza Godini, Mohammad Ghashghaee, 2009, with permission from Elsevier)

Convection section

In this section of the furnace, the pre-heating of the hydrocarbon feed takes place by convective heat transfer with hot flue gases coming from the radiant section. This allows for energy integration, which is essential for the optimization of the plant.

Also, in this section, the mixing with steam occurs leading to a reduction of the hydrocarbon partial pressure inside the coils, which prevents high coke formation rates inside the coil. Typically Steam-to-Oil ratios (kg steam/kg hydrocarbon) of 0.25–0.40 for gaseous feed and 0.40–0.55 for naphtha are employed.

Radiant section

This section of the furnace is where the cracking of the hydrocarbon feed occurs inside the radiant coils at high temperatures (up to 860 °C in the case of naphtha cracking) and low pressure (1.50–2.75 bar) (Zimmermann and Walzl 2012). The preheated gases from the convection section (at temperatures of 500–600 °C) are fed into this radiant section, where they are heated to cracking-favoring temperatures, by heat transfer with the firebox burners, that reach temperatures as high as 1000–1200 °C. The burners can be mounted on the wall and/or floor, thus influencing the firing type and the heat transfer between the firebox and the radiant coils. Since the cracking reactions are highly endothermic, high heat fluxes (75–85 kW/m^2 of coil) are needed to keep the radiant coil at the high temperatures that favor pyrolysis. Nearly 90% of the heat transfer in this section is accomplished by radiation mechanisms, namely between hot flue gases/coil and between refractory walls/coil (TVK 2010). Cracking coils can have several different configurations, ranging from single tubes to branching parallel tubes of different diameters, depending on the furnace design and on the feedstock. Typical coil dimensions are usually 40 − 90 m in length, diameters from 3−20 cm.

The electrical steam cracker

While steam crackers play a central role in the production of basic chemicals, they require a significant amount of energy to break down hydrocarbons into olefins and aromatics. By using electricity from renewable sources (cf. Fig. 13.12) instead of fossil fuels, electrical steam crackers have the potential to reduce the CO_2 emissions of one of the most energy-intensive production processes in the chemical industry by at least 90% compared to technologies commonly used today (ChemistryViews 2024).

Two concepts for an electrical cracking furnace are conceivable: Direct heating would apply an electric current directly to the process tubes inside the reactor, while

Fig. 13.12 LyondellBasell Steam cracker at the Wesseling site (*Source* LyondellBasell)

indirect heating would imply radiative heat of heating elements placed around the tubes.

Several industry consortia have been formed in the recent past to develop the electrical steam cracking process (News 2024; ACS 2024).

13.3.4 Products

Ethylene and propylene are highly valued in the petrochemical industry as building blocks for several types of almost any chemical product. Ethylene is almost exclusively produced by the steam cracking process (TVK 2010), while 60% of propylene is produced by this process.

Ethylene is the major product of a steam cracking unit, with a world production of around 150 Mt/year in 2021 (Chemanalyst 2023) and is expected to continue to grow at an average rate of around 4% per year. Ethylene is mainly used to produce different plastics and pharmaceutical products, such as polyethylene, ethylene oxide, vinyl acetate, ethylbenzene, and ethylene dichloride (Siemens 2013). The primary driver of global ethylene demand is Polyethylene. In 2021, approximately 54% of global ethylene production was utilized by the polyethylene industry. Polyethylene has various industrial applications like food and beverage packaging, cable insulation, pipes, and housewares.

Propylene production reached 117 Mt in 2022 (Chemanalyst 2024). Most of the total propylene produced globally, about 90%, is consumed in the production of polypropylene via polymerization. Polypropylene is the most widely used thermoplastic polymer and is used in applications such as packaging, the automotive sector, consumer goods, and textiles.

The butadiene market was estimated to be 12 Mt in 2022 growing at an average rate of 4.8% per year. Butadiene is a key material to produce synthetic rubber for applications such as tires, footwear, and consumer goods and 55% of the global butadiene production was consumed in the styrene butadiene rubber (SBR) industry.

13.3.5 Safeguarded Consumer Applications

Many consumer-packaged goods (CPG) companies have made pledges to sell goods that have a lower impact on the environment. According to a recent survey by McKinsey, more than 80 global CPG, packaging, and retail companies have made public commitments to reach 15 to 50% recycled content in their packaging by 2025. These pledges affect a large portion of the plastic products people use or encounter in everyday life, including food packaging materials such as bottles, caps, meal trays, and flexible film wraps.

Like medical and drinking water applications, food contact materials are subject to stringent regulation in the European Union and other parts of the world to safeguard consumers.

For these applications, use of recycled content is safely achievable only with products produced using pyrolysis oil derived from waste plastic. CPG companies therefore see chemical recycling as the solution as these materials are highly suitable for regulated applications that require virgin quality raw materials.

In 2021 LyondellBasell launched *Circulen*Revive, a class of polymers made using materials from a chemical recycling process to produce new polymers (cf. Sect. 10.4). The chemically recycled feedstock is mixed with conventional feedstocks in the steam cracking process and allocated to *Circulen*Revive products using a mass balance approach certified according to the International Sustainability and Carbon Certification (ISCC) PLUS standard (cf. Chap. 16).

*Circulen*Revive polymers are already used in some marketed products, such as the packaging of the Almond Shower Scrub by L'OCCITANE en Provence, whose tube and cap are made of 93% recycled polyethylene (PE) content using a mass balance approach and the Greiner coffee capsules (cf Fig. 13.13), which meet the strict requirements of the food industry.

Fig. 13.13 Circular solution for coffee capsules (*Source* LyondellBasell)

13.3.6 Challenges and Perspectives

As an emerging technology chemical recycling is still facing many challenges:

- From an economic perspective investment into chemical recycling only makes sense if chemical recycling gets fully recognized as a recycling technology, its products counting towards recycled content quotas, especially for applications like food packaging that cannot be easily addressed with other recycling technologies.
- Furthermore, given the need to integrate chemical recycling into the existing infrastructure of the chemical industry with large-scale assets such as steam crackers, the recognition of mass balance to allocate recycled content is imperative.
- The chemical recycling plants currently under construction in Europe (LyondellBasell 2023; Dow 2023; Sabic 2023) are still rather small, especially when compared to the steam crackers they are intended to provide feedstock for. The technologies need to mature and be scaled up once demonstration units have proven to operate reliably.

Notwithstanding, chemical recycling is a circular technology that is complementary to mechanical recycling both in terms of feedstock requirements as well as final market application thus representing a key technology for the establishment of a truly circular economy.

References

ACS 2024: BASF, SABIC, and Linde to test electric cracker designs. C&EN, Vol. 100, Issue 32, Sep. 8, 2022

Amghizar et al. 2017: I. Amghizar, L.A. Vandewalle, K.M. Van Geem, G.B., Marin New Trends in Olefin Production Engineering, 3 (2) (2017), pp. 171–178, https://doi.org/10.1016/J.ENG.2017.02.006

BASF 2020: Mass-Balance-White-Paper-2020.pdf.assetinline.pdf (basf.com)

Berggren 2024: Berggren, A. Mass Balance Concept, Perstorp Holding AB, https://www.perstorp.com/en/products/pro_environment_solutions/mass_balance_concept

BIR 2023: https://www.bir.org/publications/annual-reports

BVSE 2000: https://www.bvse.de/fachverband-kunststoffrecycling.html

Cefic 2022: Policy Matters.Innovation.Chemical Recycling.Cefic 2022.

Chemanalyst 2023: Ethylene Market Size, Growth, Analysis & Forecast, 2035 (chemanalyst.com)

Chemanalyst 2024: Propylene Market Size, Share, Industry Analysis Report, 2030 (chemanalyst.com)

ChemistryViews 2024: First Demonstration Plant for Electrically Heated Large-Scale Steamcracker Furnaces - ChemistryViews

Circular Economy Act (2020): Section 3, Paragraph 1. Definitions. https://dejure.org/gesetze/KrWG/3.html

DeStatis 11/2023: https://www.destatis.de/DE/Themen/Wirtschaft/Preise/Erzeugerpreisindex-gewerbliche-Produkte/_inhalt.html#_vd3ox9fzk

Dow 2023: Dow and Mura Technology announce largest commitment of its kind to scale advanced recycling of plastics | Dow Investor Relations

EU 2002: https://ec.Europa.eu/environment/pdf/waste/studies/eucostwaste.pdf
EU 2018: https://environment.ec.€opa.eu/strategy/plastics-strategy_en#objectives
EU 2021: https://environment.ec.€opa.eu/topics/waste-and-recycling/waste-shipments/plastic-waste-shipments_en#:~:text=Exporting%20hazardous%20plastic%20waste%20%28A3210%29%20and%20plastic%20waste,countries%20will%20only%20be%20authorised%20under%20specific%20conditions
Eurostat 2021: https://ec.€opa.eu/€ostat/databrowser/view/env_waspacr__custom_8008024/default/table?lang=de
EUWID 2023: https://www.euwid-recycling.de/
Federal Environment Agency (2021): Handout on plastic recyclate procurement. https://www.umweltbundesamt.de/publikationen/beschaffung-von-kunststoffprodukten-aus-post
HTP 2023: https://www.htp.eu/publikationen/; https://cefic.org/policy-matters/innovation/chemical-recycling/
ICIS 2023: https://www.google.com/url?sa=t&rct=j&q=&esrc=s&source=web&cd=&cad=rja&uact=8&ved=2ahUKEwi9trnSt-yEAxVygP0HHV_UDj0QFnoECAkQAQ&url=https%3A%2F%2Ficis2023.aisconferences.org%2F&usg=AOvVaw1_byAR8SirHy6LKlCVDu6f&opi=89978449
IEA 2022: https://www.iea.org/reports/fossil-fuels-consumption-subsidies-2022
IK 2024: https://www.linkedin.com/pulse/ppwr-unpacked-part-14-parliament-decides-negotiating-martin-engelmann-dga3e%3FtrackingId=gEYZdCBCTkCgx3FW09ae6w%253D%253D/?trackingId=gEYZdCBCTkCgx3FW09ae6w%3D%3D.
IPPR 2023: Istituto per la Promozione delle Plastiche da Riciclohttps://www.ippr.it/pubblicazioni/materie-plastiche-riciclate-utilizzate-in-italia-analisi-quantitativa/
Karimzadeh 2009: Karimzadeh, R, Godini, HR, Ghashghaee, M. Flowsheeting of steam cracking furnaces, Chemical Engineering Research and Design, vol. 87, no. 1, pp. 36–46, 2009.
Kusenberg et al. 2022: Kusenberg, M, Eschenbacher, A, Djokic, M, Zayoud, A, Ragaert, K, De Meester, S, Van Geem, K. Opportunities and challenges for the application of post-consumer plastic waste pyrolysis oils as steam cracker feedstocks: To decontaminate or not to decontaminate? Waste Management 138 (2022) p. 83–115. Elsevier
Letsrecycle 2024: letsrecycle.com (last accessed 03 2024)
LyondellBasell 2023: LyondellBasell Makes Decision to Progress Advanced Recycling Plant in Wesseling, Germany | LyondellBasell. PR, Nov. 20, 2023
McKinsey 2022: Advanced Recycling: Opportunities for Growth. McKinsev report, May 16, 2022. Advanced Recycling: Opportunities for growth
News 2024: LyondellBasell, Technip Energies, and Chevron Philips Chemical to demo electric steam cracking furnace technology - News - The Chemical Engineer
OECD 2022: https://www.oecd.org/environment/monitoring-trade-in-plastic-waste-and-scrap-8f3e9c56-en.htm
Oliveira and Van Dril2021: Oliveira, C. & Van Dril, T. Decarbonisation options for large volume organic chemicals production, Sabic Geleen. Available: https://www.pbl.nl/sites/default/files/downloads/pbl-2021-decarbonisation-options-for-large-volume-organic-chemicals-production-sabic-geleen_3718.pdf
PE 2006: PlasticsEurope 2006 - Compelling-facts.pdf (plasticseurope.org)
PE 2022: Plastics Europe, Plastics – the Facts 2022, https://plasticseurope.org/knowledge-hub/plastics-the-facts-2022/
PIE 2023: https://pieweb.plast€ope.com/default.aspx?pageid=200
PIE 2023: https://pieweb.plasteurope.com/Default.aspx?pageid=399&docid=252143&query=(BAT:ton%20AND%20BAT:emans)
PRE 2021: Plastics Recycler Europe, Mapping of installed plastics Recycling Capacities 2021
PRE 2023a: Plastics Recyclers Europe, 2023 Study "State of Play"; https://www.plasticsrecyclers.eu/publications/
PRE 2023b: Plastics Recycling Industry in Europe 2023
Prognostica 2023: https://www.prognostica.de/de/unsere-angebote.html

Sabic 2023: SABIC - SABIC And Plastic Energy Set To Start Construction Of....PR, Jan. 21,2021
Siemens 2013: "Petrochemical industry ethylene plant." , Siemens, 2013, https://support.industry.siemens.com/cs/attachments/109770563/PIAAP-00002-0118-Ethylene.pdf
Stapf 2022: D. Stapf. Thermochemical Processes in: N. Niessner, Recycling of Plastics. ISBN: 978–1–56990–856–3, 2022, Carl Hanser GmbH & Co. KG
Statista 2021: https://de.statista.com/statistik/daten/studie/257673/umfrage/unternehmenszahl-in-der-entsorgungswirtschaft-in-deutschland/
TecPart 2023: https://www.tecpart.de/de/verband/aktionsfelder/tecpart-team
TVK 2010: Olefins production: Olefins by steam cracking, Tiszai Vegyi Kombinat (TVK), 2010, http://www.tvk.hu/repository/713326.pdf
UBA 2021: Österreichisches Umweltbundesamt: Sortierung und Recycling von Kunststoffabfällen in Österreich, Status 2019
Wolf Polymere Solutions 2023: https://www.wolf-polymere.de/
Zimmermann and Walzl 2012: Zimmermann, H., Walzl, R.: "Ethylene" in Ullmann's Encyclopedia of Industrial Chemistry, Wiley-VCH Verlag, 2012.

Frauke Hoffmann holds a master's degree from the University of Göttingen and is Head of Sales at Hoffmann + Voss GmbH, Viersen, Germany. She has a strong interest in finding practicable solutions both for the recycling of plastics and for the application of recyclates in the market.

Peter Jetzer is a freelance journalist on the transition to a circular economy. As head of the Recycario Data Science Institute for Economical Plastics Recycling, he develops price forecasts for regranulates, indices for advanced plastics recycling, and profitability models for recycling systems. Previously, he led the Polymer Prices division at Plastic Information Europe (PIE) and was Head of Communications at KUNSTSTOFF.swiss. He holds degrees in Finance Journalism from the Frankfurt School of Finance & Management and in Business Administration from BBS/BVS Business School Zürich Winterthur, Switzerland.

Andreas Neumann finished his education as a chemist at RWTH Aachen University. With almost 25 years of professional experience with LyondellBasell and their predecessor companies, he spent his career in senior roles for R&D, Technology/Licensing, and polyolefins business based in Germany, Italy, and Thailand. Being the R&D Strategic Program Lead at LyondellBasell currently, he is focusing on sustainability topics in advanced recycling and carbon reduction areas.

Daniel White holds a bachelor's degree from the University of Nottingham in England and a doctorate from St. Andrews University in Scotland. In his current role with LyondellBasell, he provides leadership for Global Catalyst Systems and Advanced Recycling technology development, maintaining LyondellBasell's leadership and intellectual property in the field of Z/N catalysts. He preserves and develops knowledge in polymer catalysis to develop process and catalyst technology for advanced recycling.

Chapter 14
Biomass

Christian Bonten, Lars Börger, Ace Fung, Miia Laamanen, Petri Lehmus, and Oona Koski

Contents

C. Bonten (✉)
Institut für Kunststofftechnik, Universität Stuttgart, Stuttgart, Germany
e-mail: christian.bonten@ikt.uni-stuttgart.de

L. Börger (✉)
nova-Institut GmbH, Hürth, Germany
e-mail: lars.boerger@nova-institut.de

A. Fung
Neste Oyj, Porvoo, Finland
e-mail: ace.fung@neste.com

M. Laamanen · P. Lehmus
Neste Corporation, Porvoo, Finland
e-mail: Miia.laamanen@neste.com

P. Lehmus
e-mail: petri.lehmus@neste.com

O. Koski (✉)
UPM-Kymmene Oyj, Tampere, Finland
e-mail: oonalotta.koski@gmail.com

P. Orth et al. (eds.), *Circular Economy of Plastics*,
https://doi.org/10.1007/978-3-031-78829-1_14

Summary

For the foreseeable growth of plastic consumption, the plastics industry needs feedstock sources replacing fossil raw materials, thus satisfying a sustainable existence and growth of the industry. Biomass provides carbon-based products that can be converted for the use in the plastics industry.

Section 14.1 presents selected biomass sources with proven records for usage and processing in the chemical and plastics industry. We look at bioplastics and their role in a circular economy. In Sect. 14.2 biowastes and bio residues from existing industries are analyzed and in Sect. 14.3 agricultural and forestry biomass are looked at. All types of biomass already play an industrial role in the feedstock supply of the chemical and plastic industry and depict respective efforts to go along with the growing demand for renewable raw materials while offering the chance to close the loop in a circular economy.

The authors demonstrate that the plastics industry does not essentially need fossil raw materials to produce modern plastics, but that the carbon needed can be derived from other sources. With appr. 10 Gt of biomass available there would be enough carbon on hand to feed the plastic industry. An established agricultural and forestry industry is preparing to invest in supplying the chemical industry with feedstock, thus delivering base chemicals, and needed building blocks for the plastics industry as well as an industry that so far was treating biowaste and bio residues from other industries as feedstock for plastics. Nevertheless, all addressed industries rely on an existing circular economy to be able to act economically and ecologically feasible.

14.1 Introduction to Bioplastics[1]

Christian Bonten

Bioplastics are on everyone's lips and generally achieve a positive response from consumers who want to behave sustainably. If one takes a closer look at the topic, it quickly becomes clear that the term "bioplastics" is an umbrella term, which does not differentiate the benefits of biodegradability and biological origin and conceals the fact that both properties are not always present in a bioplastic at the same time

[1] The text of this chapter is based on the chapter "bioplastics" in the author´s own book on Plastics and Plastics Technology (Bonten 2019). However, it has been greatly modified, actualized, and rearranged for the purpose in this book.

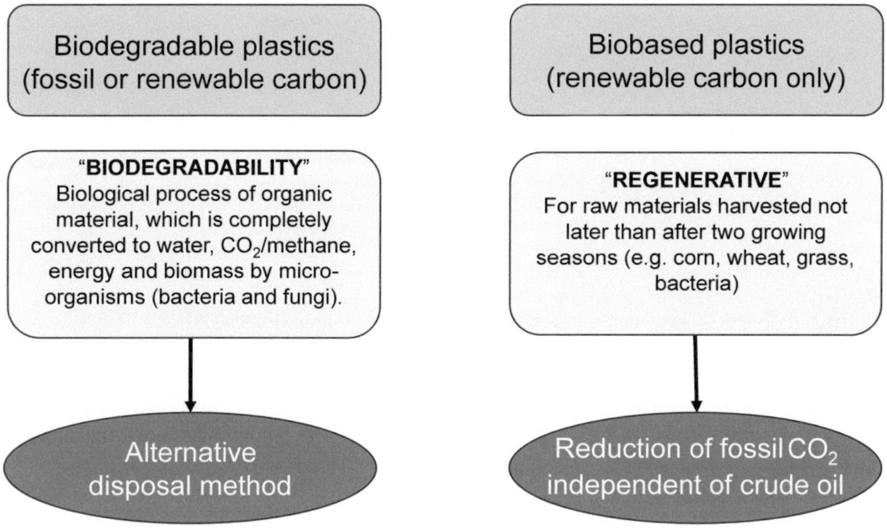

Fig. 14.1 Various characteristics of bioplastics. *Source* Bonten (2019)

(cf. Fig. 14.1). Despite the attractive prefix "bio" [Greek: bíos (βίος) 'Life'], it is important to understand what benefits it has for the consumer and thus for the plastics industry. To understand the benefits, the exact meaning of the prefix "bio" needs clear communication.

The author wants to point out that from his perspective the term "polymer" does not have the same meaning as "plastics". It needs to be clear that polymers do not necessarily have to be raw materials for plastics. Polymers exist in many forms in nature, for example. As explained later, proteins, spider silk, cotton, hair, but also the complex Deoxyribonucleic acid (DNA) are polymers, as are the complex Carbohydrates cellulose, starch, chitin, and wood (cellulose, hemicellulose, and lignin). Not all these natural polymers are used to produce bioplastics.

Polymers are not only raw materials for plastics, but also for other materials (cf. Fig. 14.2) like for chemical fibers (Polyamide, Polyester fibers) and for natural fibers (cotton or silk fibers made from natural polymers). The chemical bases of some polymers may be the same for plastics and chemical fibers, but for fibers, the molecular chains are specially aligned and arranged. Chemical fibers form part of textile technology, not of plastics technology.

Furthermore, polymers are used in large quantities for surface coatings. These are not always, but often, lacquers (also so-called powder lacquers). Adhesives are also a kind of very special coatings. Polymers are required as functional additives to improve the properties of coatings/lacquers, chemical fibers/textiles, plastics, cosmetics, and even food. However, these additives as well are not plastics. The term "functional polymer" may be used for this purpose; however, we do not speak of "functional plastics".

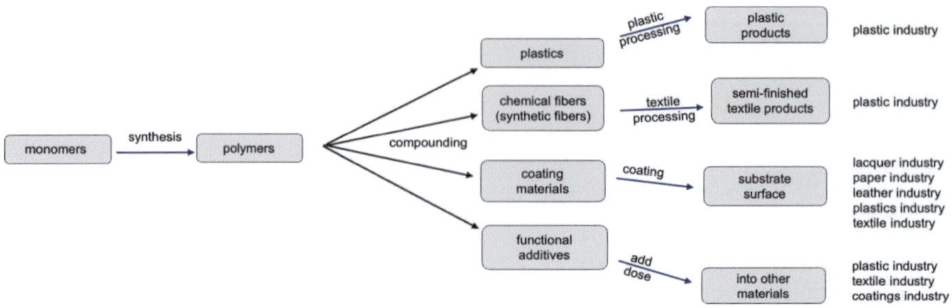

Fig. 14.2 Polymers—raw material not solely for plastics. *Source* Bonten (2019)

So, wherever the author means the chemical "backbone" of a plastic material and its characteristics, the term biopolymer is used. If the author means the applicable plastic material, the term bioplastics is used.

14.1.1 Bio-based Plastics

In the case of bio-based plastics, the prefix "bio" stands for renewable raw materials that are derived from nature. The origins of bio-based polymers, which are used for the manufacture of bio-based plastics can be divided into (cf. Fig. 14.3):

1. Biopolymers from plants (e. g. rubber trees, sugar cane).
2. Biopolymers from animals (e. g. spiders, crabs).
3. Biopolymers from microorganisms (e. g. the bacteria Aeromonas hydrophila).

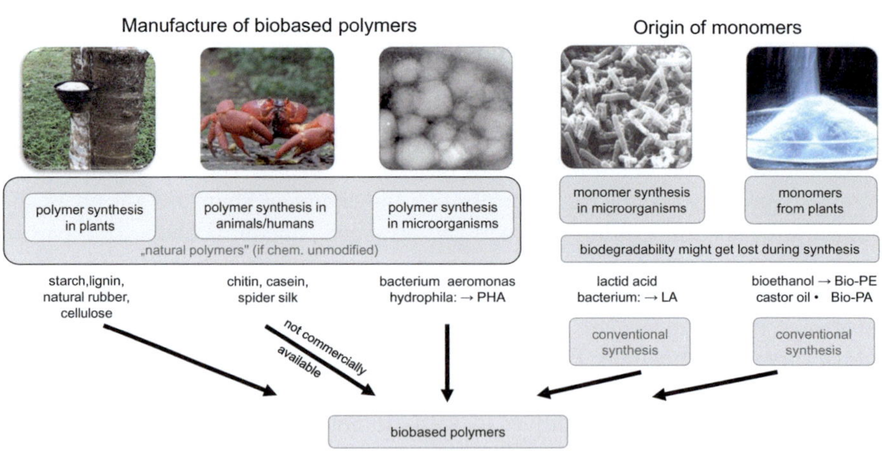

Fig. 14.3 Synthesis routes for bio-based polymers. *Source* Bonten (2019), modified

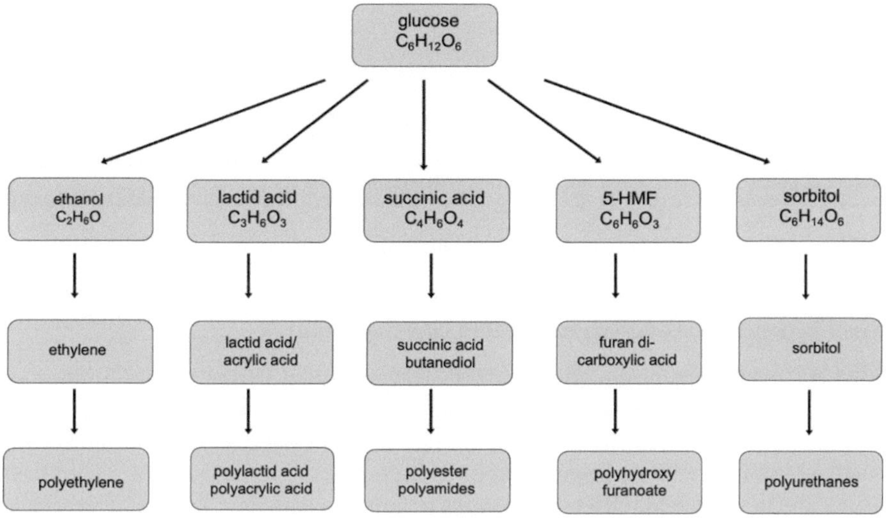

Fig. 14.4 Polymers based on glucose. *Source* IGVP,[2] University of Stuttgart

4. Bio-monomers from microorganisms (e. g. fermentation of starch to Lactic acid and subsequent synthetic polymerization to Polylactic acid (PLA)).
5. Bio-monomers from plants (e. g. bioethanol) from sugar cane becoming bio-polyethylene).

Polymerization either takes place in nature itself (Bonten 2019; Essel 2012a; BMELV 2013; examples of natural polymers are proteins, cellulose, spider silk and PHA-fats in bacteria) or bio-based monomers are polymerized in processes known from fossil-based plastics (Essel 2012b, Kliem 2020).

"Bio-based" means that the carbon atoms that make up the molecular chains stem from today's nature. The example of five possible routes (cf. Fig. 14.4) shows which bio-based chemicals and bio-based polymers can be produced from glucose, a simple form of sugar. Glucose is mainly produced by plants using photosynthesis from sunlight, water, and Carbon dioxide (CO_2) and can be used by all living organisms as a source of energy and carbon.

However, glucose is usually not freely available but occurs as a disaccharide (lactose, beet sugar) or in its long-chain polymer form (such as starch, cellulose, etc.), which are both reserve substances and components of the cell structure in plants. When humans, animals, fungi, and bacteria ingest food, enzymes first break down the long-chain saccharides into glucose monosaccharides before being metabolized.

[2] Institut für Grenzflächenverfahrenstechnik und Plasmatechnologie (Institute of Interfacial Process Engineering and Plasma Technology).

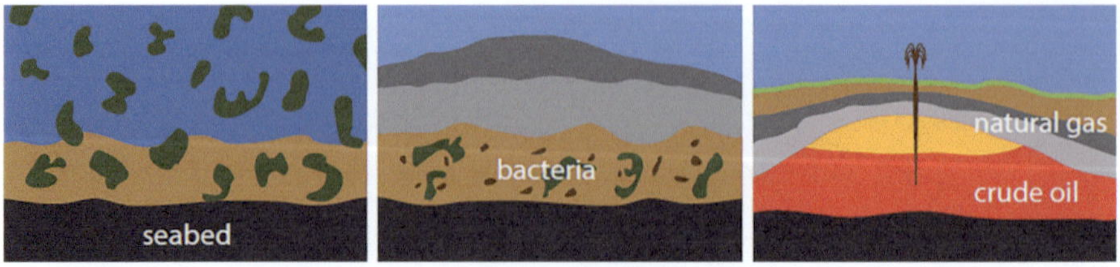

Fig. 14.5 Origin of fossil hydrocarbons (crude oil, natural gas, coal). *Source* Bonten (2019)

Fossil hydrocarbons versus renewable raw materials

However, why is the origin so important? Haven't fossil hydrocarbons from crude oil, natural gas, and coal ever been part of nature? As to our knowledge, carbon has existed since shortly after the beginning of the universe and within molecules, usually combined with hydrogen, carbon is the origin of all life on earth. There have been algae in the sea for 540 million years (Cambrian age), and fungi and lichens as well as the first animals for about 440 million years (Silurian age). After their death, algae and small marine animals sank on the seabed over millions of years in huge quantities (cf. Fig. 14.5). In the Triassic age, this seabed dried out and was covered by other strata as soil by geological changes. Bacteria, followed by the action of heat and pressure converted this biomass into "fossil" hydrocarbons in the absence of oxygen over many millions of years. These bacteria are different from those involved in biological degradation.

Humanity annually extracts the "fossil" biomass collected over a period of 1 million years in the form of fossil crude oil, natural gas, and coal, and brings it into the present. The combustion of such fossil hydrocarbons produces carbon dioxide that was taken hundreds of million years ago from the atmosphere of that time by living organisms during their growth. As a result, today's atmosphere has a strongly increasing concentration of CO_2, leading to manmade climate change. Hence, global society would do well to free itself from its dependency on crude oil, natural gas, and coal.

Although, there is "good" and "bad" CO_2—that from today and that from hundreds of million years ago, it is necessary to avoid any CO_2 emissions to reduce climate change. When using renewable raw materials, i.e. carbon atoms from biomass, exactly as much Carbon dioxide is removed from the atmosphere during plant growth as is later released into the atmosphere during biological degradation or thermal utilization. This material cycle is called "CO_2 neutral".

In Fig. 14.6 it is made clear that only 6% of Germany's crude oil demand is used to manufacture polymeric materials. With a polymer production in Germany of 20.7 Mt, 10.1 Mt (48.7%) is used for adhesives, resins, and coatings and only about 8.8 Mt (42.7%) to produce plastics. Transferred to oil demand, only 2.6% of crude oil is used to produce plastics (Essel 2012a).

Since it is accepted common sense that burning fossil fuels is the main reason for manmade climate change, it is inevitable to produce more and more polymers from

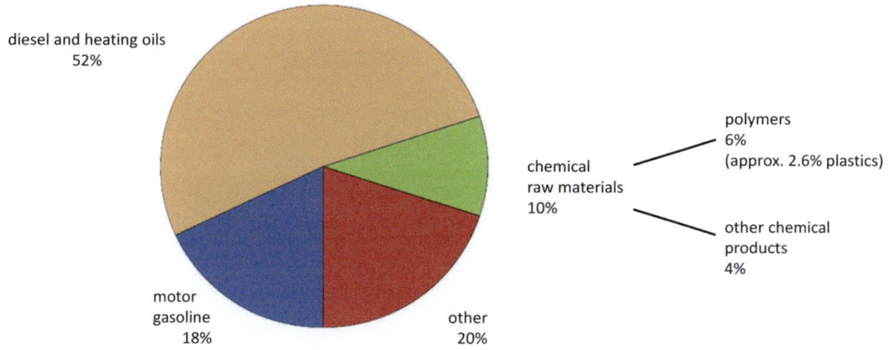

Fig. 14.6 Use of crude-oil in Germany. *Source* Bonten (2019)

renewable raw materials (cf. Fig. 14.1) as part of a de-fossilized Carbon industry. Therefore, the social benefit of bio-based plastics is to prepare society for the post-petroleum era because fewer fossil hydrocarbons are used.

However, it must be pointed out that despite an increasing use of renewable raw materials for biopolymers and bioplastics, the plastic industry will hardly contribute to leaving fossil resources underground, as only a few percent of crude oil is used to manufacture polymers for fossil-based plastics. Fossil resources are used to a far greater extent in the world for power generation and transportation (cf. Fig. 14.6). The plastics industry cannot "save the world" by using renewable raw materials. But it can save itself since fossil resources will be less and less available.

Food shortage

Fears that food shortages would result from the use of carbohydrates for bio-based plastics are unfounded. Here, the fact that the world's energy demand cannot be met by biomass carbon sources is confused with the much lower need for carbohydrates to produce biopolymers for bio-based plastics.

According to BMELV 2023, only 6% of the total usable agricultural area would be needed, which is less than the normal annual yield fluctuation to produce all plastics, i. e. about 300 Mt/a, based on biomass. In 2008, 0.7 Mt/a of bioplastics were produced, using only about 0.02% of the world's agricultural land (cf. Fig. 14.7) (Essel 2012b).

There are sources having calculated that even the fallow arable land in Europe would be sufficient for Europe's entire plastics demand. Then Europe would have a successful model of being able to operate its agriculture subsidy-free and at full capacity. In all these calculations it is also forgotten that the oceans are full of carbohydrates in the form of algae, the drying of which is very uneconomical at present, however.

Examples of Bio-based Polymers and Bio-based Plastics

Bio-based polymers from plants (cf. Fig. 14.3) are various carbohydrates, such as sugar, starch, cellulose, lignin, fats, and vegetable oils. Bio-based plastics, made from

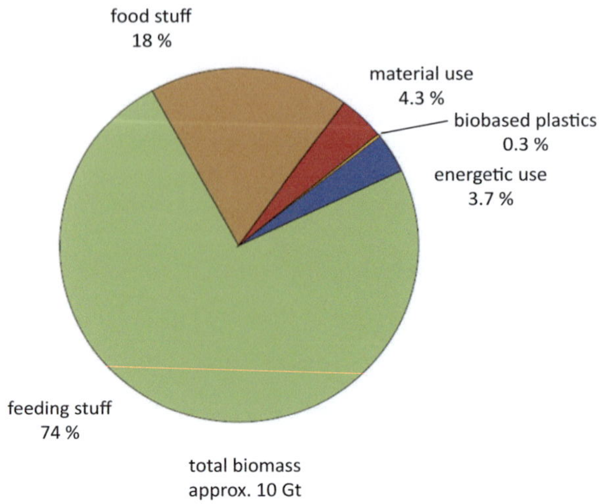

Fig. 14.7 Bioplastics—relative demands on the natural environment. *Source* Essel (2012b)

bio-based polymers are often chemically modified to, e.g. thermoplastic starch (TPS; denaturized starch), unsaturated natural rubber (cis-1,4-polyisoprene; often cured to elastic rubber), and cellulose acetate (CA; esterified with acetic anhydride) since natural polymers without chemical modification often do not match applications´ demands.

Examples of bio-based polymers from animals are chitin and spider silk, as well as proteins and animal fats. Although research is done on the use of some of these biopolymers for the manufacture of bioplastics, the commercial breakthrough did not happen, yet.

In addition, microorganisms like the bacteria Aeromonas hydrophila (and many more) hold natural polyhydroxyalkanoates (PHA) as intracellular energy reserves, especially when their source of nutrition becomes scarce. PHA is a family of polyesters, and the most common types are polyhydroxybutyrates like PHB, PHBV, and PHBH. PHBV is commercialized by, e.g. Tianan/China and PHBH by Kaneka/Japan.

Bio-based polymers that polymerize in nature (see above) are "natural polymers" as long as they are not chemically modified and pure. Recycling of bio-based plastics made from natural polymers is hardly possible, since on the one hand, the low volumes do not allow economic separation, and on the other hand, these polymers are thermally unstable meaning that they degrade strongly with every process step.

Polymers that do not polymerize in nature, but in standard polymerization processes with bio-based monomers, are much more stable due to the chemical process. Anyhow these polymers are not natural polymers. Often, such polymerized bio-based monomers lose their biodegradability (cf. 14.1.2).

An important example of a monomer derived from microorganisms, is lactic acid (LA), a natural acid, mainly produced by the fermentation of sugar or starch by means of microorganisms. LA can be polymerized to polylactic acid (PLA), one of the most important biopolymers. PLA has not fully lost its biodegradability due to the polymerization process, however, needs hydrolysis before it gets metabolized. There are many commercial grades of PLA on the market, often blended with other biopolymers to reduce brittleness. SoBiCo/DE is the first producer offering soft co-polymers based on PLA. Some PLA suppliers are working on getting back used PLA from the market, since chemical recycling is possible.

So-called drop-in-solutions describe bio-based polymers made from bio-chemicals like bioethanol obtained from sugar cane or 5-hydroxymethylfurfural (5-HMF) from glucose. The bio-polyethylene "I´m green™" from Braskem/BR is made entirely from renewable raw materials having properties identical to conventional polyethylene (but is by no means biodegradable anymore). Recycling bio-based polyethylene is very easy, since it is chemically identical to fossil-based PE and does not disturb already existing recycling schemes.

Some polyamide manufacturers offer bio-polyamides partly based on renewables like polyamide 4.10 or 6.10 or fully based on renewables like polyamide 5.10 and 10.10. There is a partly bio-based PET on the market (the glycol constituent being bio-based), and the newly introduced bio-based polyethylene furanoate (PEF), promoted by Avantium/NL as a PET substitute for beverage bottles, is based on 2,5-furandicarboxylic acid (FDCA). The author will monitor further developments and is curious to see what happens next in the market.

Measuring the Bio-based Content

To prove that there is bio-based carbon used for the manufacture of bio-based polymers and plastics, producers—especially producers of drop-in solutions—use the ^{14}C method (radiocarbon dating method), used by archaeologists for the determination of the age of finds. The measured biobased mass content includes other elements like hydrogen, oxygen, and nitrogen, while the measured biobased carbon content focusses solely on the carbon content. These values can be indicated with respective labels by certifiers, e.g. DIN CERTCO and TÜV Austria.

14.1.2 Biodegradable Plastics

There is another group of bioplastics, which has nothing to do with the origin of the carbon atoms but is often mixed up with plastics made from biomass. Since many people consider biodegradability as the ultimate solution for plastic litter and consider it an important part of circularity, we must deal with this group of bioplastics as well.

Biodegradability means that microorganisms (bacteria and/or fungi) can metabolize the polymers. Specific bacteria colonize the surface of a plastic component (cf. Fig. 14.8, top left) producing suitable enzymes with which they wet the plastic (cf. Fig. 14.8, top right). While common aging processes such as media and radiation attacks weaken the polymer chains, the enzymes can "cut out" small fragments that

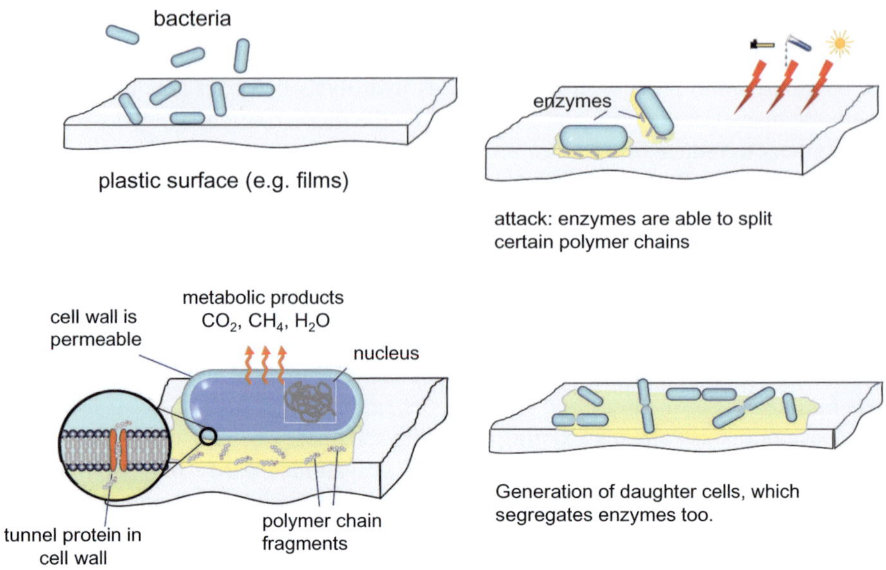

Fig. 14.8 Different stages of attack by bacteria on biodegradable polymers. *Source* Bonten (2019)

the bacteria can absorb by means of so-called tunnel proteins (cf. Fig. 14.8, bottom left).

Inside the bacteria, the polymer chain fragments are completely metabolized (to biomass, water, and CO_2 or methane) and thus are used as energy carriers. The bacterium grows and after sufficient cell growth, daughter cells are formed, forming the correct enzyme right from the beginning and then increasingly attacking the biodegradable plastic components (cf. Fig. 14.8, bottom right). The metabolism is conclusively verified by special experiments that correlate the CO_2 formation with the sample weight.

Some biodegradable polymers are metabolized by bacteria alone and some in combination with fungi only. There are also those requiring higher temperatures to be chemically degraded by hydrolysis beforehand, so that they then become vulnerable to the enzymes. This is why a group of biodegradable polymers biodegrade even under seawater conditions under which few active bacteria can be expected (cf. Fig. 14.9).

In fresh water there is a greater bacterial diversity, so the biological attack takes place more rapidly. If biodegradable polymers are used in composting at home, fungal attack will occur in addition to bacterial attack, making additional polymers vulnerable. The best conditions are in industrial composting; here the temperatures are higher, allowing not only faster metabolism, but also hydrolysis before the biological attack, which is important for the biodegradability of, e.g. polylactic acid (PLA; see above). Fears that biodegradation may occur during normal intended use, e.g. of a kitchen appliance, are unfounded.

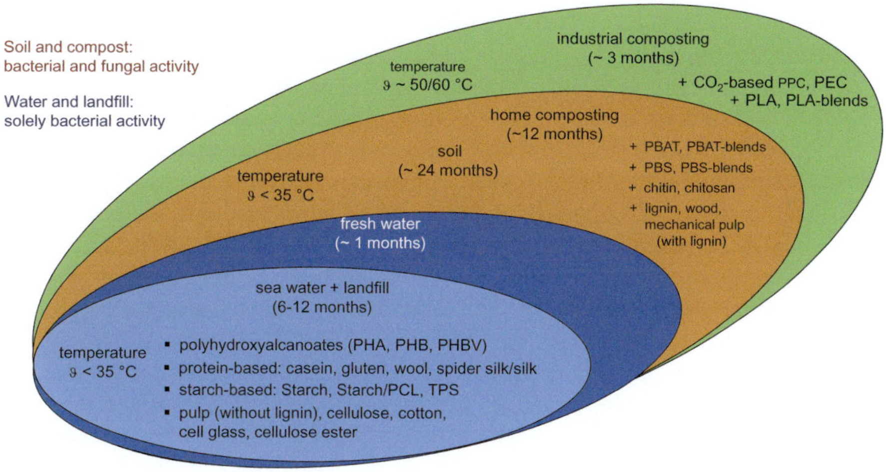

Fig. 14.9 Biodegradability of polymers under different environmental conditions. *Source* Kliem (2020)[3]

Unfortunately, the term "biodegradability" is used misleadingly and there are conventional plastics on the market that contain either small amounts of biodegradable or other rapidly decomposing substances. In the environment, these decompose under suitable conditions into smaller, hardly visible components for which, however, the complete metabolism by CO_2 formation was never proved. The plastic fragments—actually "plastic dust"—of these plastics accumulate over time in nature. It must be clearly pointed out that biodegradable polyethylenes (PE), polypropylenes (PP), polyethylene terephthalates (PET), and polyamides (PA) simply are not possible by their chemical structures!

The terms "composting" and "compostability" refer to a product (!) characteristic and are clearly defined in European (EN 13432), American (ASTM D6400), Japanese (JIS K 6950), and Chinese standards (GB/T 41010-2021). A biodegradable polymer is a necessary basis for compostability being defined as "the microbial transformation of a biodegradable material or product under controlled conditions to produce compost".

The conditions of EN 13432 are as follows (those of the other standards are quite similar):

- Chemical characterization: organic fraction > 50%; limit values for toxic components, e. g. heavy metals.
- Biodegradability: laboratory test: max. 5% not biodegradable or 1% per substance; rest degrades to CO_2, H_2O, salts, and biomass.
- Compostability: field test: 90% of fragments < 2 mm after 12 weeks.
- Ecotoxicity: germination rate of special plants > 90% of the untreated soil.

[3] Created in collaboration with Michael Carus, Nova Institute/Hürth, Germany, and Bruno de Wilde, OWS/Gent, Belgium.

To permit biodegradable plastics in organic waste they are not allowed to exceed certain concentrations of heavy metals and must not restrict the fertility of the soil. At least 90% of the entire plastic component must have disintegrated into fragments smaller than 2 mm within 12 weeks to allow for the economical and trouble-free operation of composting plants. In most cases, only films less than 0.1 mm thickness meet this criterion. Like the branch of a tree, thicker products made of bioplastics must be brought to such small dimensions before being introduced into the composting plant to be metabolized quickly enough (EN 13432; ASTM D6400). Only, if a product meets the requirements of EN 13432, the commercial use of the European seedling logo is permitted for an annual fee. This is intended to help ensure that the product is clearly labeled.

This clearly shows that the term "bio" in "biodegradable" stands for the fact that the plastic becomes part of nature or "life" again. The social benefit and therefore also the benefit for sustainable consumers is an additional disposal option via organic waste. In Germany, the decision as to whether bioplastics may be disposed of with bio-waste is made by local authorities. It is expected that by means of biodegradable bin liners, organic kitchen waste can be more consistently separated from household waste.

Examples of biodegradable polymers are polylactide (PLA), polyhydroxyalkanoate (PHA), cellulose derivatives such as cellulose acetate (CA), and some cellulose acetobutyrate (CAB), thermoplastic starch, but also petroleum-based polybutylene adipate (PBAT) and petroleum-based polybutylene succinate (PBS). All these biodegradable polymers form the basis for biodegradable plastics.

14.1.3 From Biopolymer to Bioplastic

Occasionally, plastics converters still encounter the conviction that bioplastics "do not work", but this certainly derives from the unfortunate results of experiments with pure biopolymers of the "first hour". Compounded bioplastics do indeed have very good and independent property profiles, as long as the biopolymers (be it either biodegradable or bio-based; this is up to the desired benefit) are blended with others and/or additives are added. While mixing/blending the "new polymers" by smooth compounding with suitable additives is key to market success (cf. Fig. 14.10), drop-in biopolymers only need conventional additives like their fossil-based pendants.

Bioplastics experienced their first successes in tear-proof, biodegradable waste bags, mulch films, and break-proof disposable cutlery. Now they are used on a large scale for biodegradable carrier bags in, e.g. France and Italy. More and more sophisticated packaging solutions made of biodegradable plastics followed and for some time now bio-based plastics have been used successfully even in automotive and consumer electronics applications.

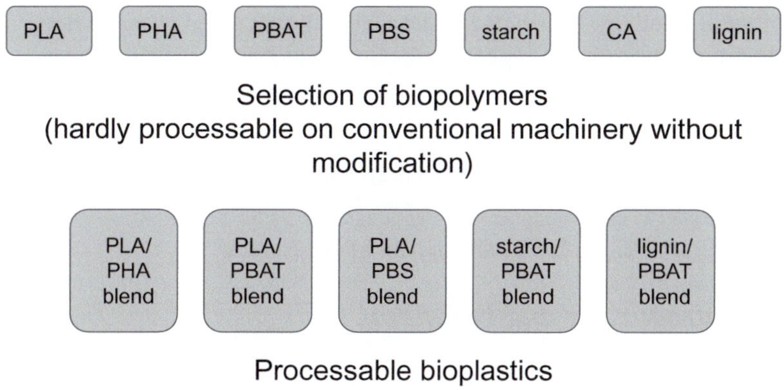

Fig. 14.10 Biopolymers are compounded into processable bioplastics. *Source* Bonten (2019)

14.1.4 Market Data

Market data from the European Bioplastics Association show an annual growth of 20% (cf. Fig. 14.11). However, this comes from a low initial value: bioplastics represent only 0.4% of the world's plastics production.

Despite the anticipated growth rate for bioplastics the market size exists as a niche market next to the fossil-based plastic market. Only by triggering the reduced

Fig. 14.11 World production of bioplastics—forecast until 2028. *Source* European Bioplastics (2024)

availability of fossil raw materials and the availability of green energy will the market for biobased plastics show a bigger competitiveness for the plastic industry.

14.2 Usage of Biowastes and Residues

Lars Börger, Ace Fung, Miia Laamanen and Petri Lehmus

14.2.1 Introduction and Definition

There is strong consensus when it comes to advancing circularity as the backbone for a more sustainable polymers and chemicals industry. A circular economy enables us to tackle the two most pressing sustainability-related issues polymers and chemicals are facing: reducing plastic waste pollution and reducing the dependence on fossil resources by keeping materials in the loop. To achieve true defossilization of the industry, we need to go one step further: while circularity can indeed provide the core of sustainable transformation, there will be a need to keep adding virgin material. The solution: creating a circular (bio-)economy.

The authors describe industrialized use of biogenic carbon. In a way, biogenic carbon is atmospheric carbon taking a detour through plants or biomass, which are now much more efficient in making CO_2 available as a chemical building block. However, biogenic carbon also doesn't come without challenges: the influence on land-use and the environmental impact of biomass production are essential parameters for a bio-based feedstock supply (cf. Chap. 14.3).

This chapter will therefore take a closer look at ways to overcome challenges related to biogenic carbon by describing a specific class of sources of feedstock: carbon from bio-based waste and residue sources. The molecules described in this chapter are derived from oils and fats and treated with hydrogen resulting in saturated aliphatic hydrocarbons, often referred to as "hydrogenated vegetable oils" (HVO) synonymously in the related industry.

The terms "waste" and "residue", by definition, both refer to secondary renewable, bio-based or biogenic materials. Many of these originate from various industrial and commercial processes. They are not the primary outcome the processes are designed to produce. The EU Renewable Energy Directive (EU, Renewable Energy Directive 2018) defines waste as something that is discarded or is required to be discarded, while residues are a substance that is not a primary target of a production process.

This chapter will describe the benefits of using bio-based wastes and residues as a source in general, followed by a description of the most important sources and the processing of those. It will be concluded by a view on availability and economics as well as an outlook into the future of these feedstocks.

14.2.2 *Rationale for Using Waste and Residue*

Bio-based wastes and residues are considered a valuable carbon source for various reasons, and the advantages relating to sustainability are particularly evident when compared to virgin biomass and recycling as a carbon source.

Renewable, sustainable and resource-efficiency: Bio-based wastes and residues are derived from organic materials such as agricultural residues, food scraps, and organic waste from various industries. Like with any biomass, atmospheric CO_2 has been captured during the growth phase of the materials. Additionally, leveraging wastes and residues as raw material maximizes the utilization of biomass as it is a discarded or byproduct of existing processes, making it a potentially more sustainable and renewable resource.

Reduction of process-related greenhouse gas emissions: Using bio-based wastes and residues as a carbon source helps mitigate the release of greenhouse gasses from the processes creating them. For some sources, when the wastes and residues decompose unutilized, they produce methane, a highly potent greenhouse gas. By utilizing them for materials, the release of methane is reduced, contributing to the overall effort to combat climate change.

Waste reduction and dependency: Incorporating these wastes and residues as a feedstock source can be considered a form of waste reduction since the biomass is further treated and utilized to increase carbon availability. Thereby the dependency on other materials, especially from fossil sources is reduced, which is particularly relevant for countries and regions that do not have their own fossil resources or more stringent climate protection targets. Wastes and residues thus can contribute to local feedstock security by providing a decentralized and distributed source of energy. Regions can harness the feedstock potential of their own bio-waste, reducing dependence on external energy sources.

Biodiversity and impact reduction: By optimizing the land-use of existing biomass production, the impact of biomass production per ton of utilized biomass is reduced and thus allows to increase yields from the already utilized land. This in return leads to the preservation of further land and respective biodiversity. That is also true for the related necessary infrastructure and logistics to be built for this purpose.

Producing high-quality products from waste and residue raw materials is an example of efficient use of resources. It also provides one of the means to reduce the pressure to clear new areas of land for the cultivation or production of virgin raw materials for various uses. Thus, bio-based wastes and residues stand out as a favorable carbon source reducing emissions while accelerating and utilizing the circular economy.

Molecular economics of biomasses (C:O ratio --> oils and fats preferred)

There are multiple classes of biomass and consequently also different classes of bio-based wastes and residues. There are principally three main classes of biomass:

Table 14.1 Carbon/oxygen ratio in sugar and fat molecules. *Source* Own representation

Molecule	Sugar	Fat
Molecular formula (exemplary)	$C_6H_{12}O_6$	$C_{55}H_{104}O_6$
Molecular weight	180.2	861.4
C:O ratio	1:1	9:1
% of carbon of total mass	40%	77%

M(C) = 12.0111u, M(O) = 15.999u, M(H) = 1.008u; *triglyceride consisting of one stearic, one oleic, and one palmitic acid

proteins, sugars (incl. lignocellulosics) and oils and fats. For simplification's sake, one can exclude proteins as there is a protein gap for serving human food and feed needs globally, thus this is the least favorable feedstock class with regard to sustainable availability. The remaining two classes can be represented by a sugar and fat molecule to indicate the desired logic as pointed out in Table 14.1.

The following figure (cf. Fig. 14.12) gives an overview of chemicals by chain length and carbon-to-oxygen ratio.

Depending on the target molecules it does make a difference from which biomass class the processes are started. It can be differentiated between target molecules, that offer new chemical functions, and drop-in molecules that are available in industry already before. Specifically for the path towards drop-in hydrocarbons that do not contain oxygen, it is more beneficial to start from oils and fats as carbon source than from sugars. The higher C:O ratio (cf. Table 14.1) corresponds to a lower demand for chemical removal of oxygen, respectively, reduction with, e.g. hydrogen. This

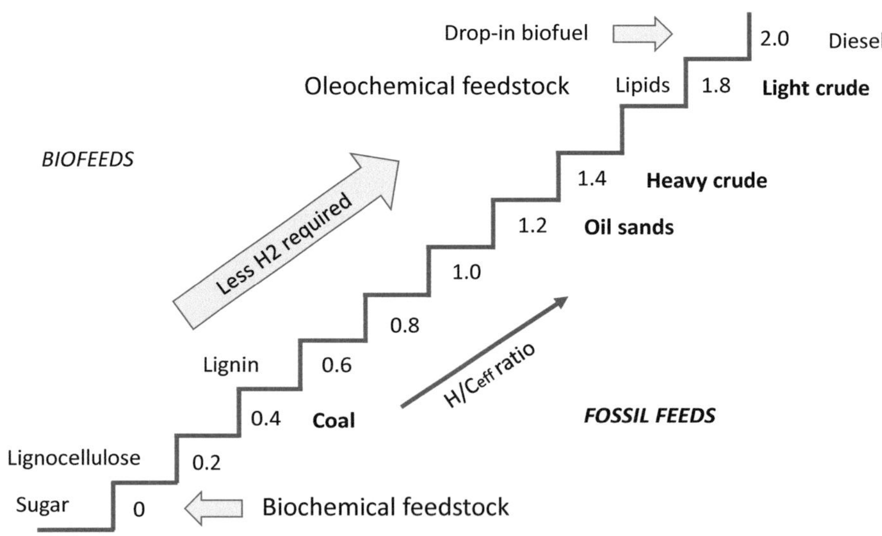

Fig. 14.12 Increase in effective hydrogen to carbon ratio represented as a "staircase" compared to fossil feedstocks in brackets and bold. *Source* IEA (2014)

will offer a better economic and ecological balance to achieve the targeted drop-in molecules for the chemical and plastic industries.

Sustainability Assessment of Bio-waste and Residues

The lifecycle emissions of products produced from waste and residues have shown to be lower than those produced from virgin vegetable oils, and significantly lower than lifecycle emissions of fossil oil-based alternatives. This latter comparison was highlighted as part of a scientific study that was conducted by Neste (2021) and was one of the case studies published by the Renewable Carbon Initiative (RCI 2023) and is summarized here as an exemplifying case standing pars pro toto:

A Life Cycle View

Part of the complex sustainability journey in finding new renewable sources and, where possible, lower carbon solutions to the current status quo with fossil fuels, is to understand a product's life cycle. A Life Cycle Assessment (LCA) serves as a systematic approach to evaluate the environmental impact of a product, process, or service throughout its entire life cycle. This assessment considers every stage, from raw material extraction to end-of-life disposal or recycling, offering a holistic perspective on alternatives and allowing stakeholders to make informed decisions.

A cradle-to-gate life cycle assessment to produce renewable hydrocarbons from waste and residues was conducted by Neste. The study followed ISO 14040/44 and ISO 14067 standards, with a critical peer review ensuring conformity with methodology, data, interpretation, and reporting requirements.

The study compared renewable liquid hydrocarbons with fossil naphtha, which is the most widely used fossil hydrocarbon feedstock for the chemical industry and is a refinery product derived from crude oil, as the reference for the assessment. The LCA covered all life cycle stages, incorporating a representative raw material mix of used cooking oil (UCO), animal fat (AF) and residues from vegetable oil processing. A small portion of fish fat (FF) was added to the study to widen the raw materials pool. These raw materials entered the process with minimal environmental impacts since they are treated as waste or residue streams.

While an LCA considers multiple impact categories, here the focus is put on the net Global Warming Potential (GWP) of these renewable liquid hydrocarbons, including biogenic carbon emissions at end-of-life. It was determined to be 0.45 kg CO_2 eq./kg of product. In comparison to fossil naphtha, these hydrocarbons exhibited an at least 85% lower GWP. The primary contributor to this reduction is the biogenic removals from CO_2 uptake during biomass growth and subsequent emissions during the life cycle.

This LCA and related findings represent the entirety of waste & residue oils and fats as feedstocks for hydrocarbon-based chemistry processes very well, however, the specific numbers differ from case to case.

14.2.3 Sources of Waste and Residue Oils and Fats

Animal and Fish Fat

Animal fat from food industry waste is derived from the food industry's meat processing waste as it is further processed into meat and bone meal, as well as rendered animal fat at rendering plants. Rendering is a heating process which also eliminates pathogens to protect the health of humans and animals.

Depending on the quality and safety of the fat there are different markets worldwide for animal fats. The three general qualities are food grade, feed grade and technical purposes and the best practice is to source mixed animal fat waste that is unsuitable for human consumption. Animal fat waste may be used as feedstock for renewable fuel production, in the chemical and energy industries, and partly as animal feeds. Animal fats feedstock production and market is overall well established, and the market is primarily determined by meat consumption.

Fish fat waste is separated from the gutting waste of, for example, freshwater fish pangasius after the parts suitable for human consumption have been removed for the food industry use at fish processing plants. Pangasius is grown particularly in Southeast Asia. Fish fat can also be used as feed.

Used Cooking Oil (UCO)

UCO consists of oils and fats of a vegetable or animal origin that have been used by the food industry, restaurants, households to cook food for human consumption. UCO is considered a waste as it is no longer fit for human consumption for food hygiene reasons. Qualities include in addition to UCO consisting of waste or spent cooking oil and frying oils, e.g. yellow grease (YG) comprising of UCO and rendered animal fat.

UCO is sourced globally from collectors and aggregators. It is predominantly collected from restaurants and food industry factories. There are hundreds of used cooking oil suppliers around the world, wherever oils and fats are used for cooking and the collection of such oil has been organized for recycling and reuse. Due to logistic reasons, UCO collection from the professional sector is much more developed and is practiced much more frequently than collection from households. There are still large untapped volume potentials that are either not yet collected or disposed of in low-value end uses.

One of the largest supply sources is China with a collection infrastructure of local collectors on the East coast of China, especially in the regions around Shanghai. UCO is also sourced from Europe and the US. An example of UCO as a source for the chemical industry in the US is the cooperation between McDonald's, LyondellBasell, Neste, and Berry which created a value chain that directs UCO from restaurants back to plastic cups that are used in the very same restaurants (McDonald's 2022; nova 2022).

Technical corn oil (TCO)

Technical corn oil (TCO) is a processing residue generated in the production of ethanol from corn, during which also animal feed is produced. The yields of TCO are small and this residue is not suitable for human consumption. TCO is cheaper than corn oil (or maze oil) produced for human consumption. TCO is used primarily as raw material in the oleochemical industry, as an animal feed additive, and as raw material in renewable diesel production. The need for ethanol as a bio-component in gasoline blends has simultaneously increased the availability of TCO as a raw material suitable for renewable diesel production.

Vegetable oil processing waste and residues

Many vegetable oil processing wastes and residues can be used as raw materials to produce renewable products. They are generated in different steps of the refining processes of various vegetable oils and examples include palm fatty acid distillate (PFAD), spent bleaching earth oil (SBEO), acid oils (AO) and palm oil mill effluent (POME).

PFAD is a processing residue derived from the final stages of refining food-grade palm oil. It consists of degraded fats, free fatty acids, that need to be removed before palm oil meets the food industry's quality standards in terms of taste, smell, color, and shelf life. When harvesting and handling oil palm fruit bunches, normal bruising occurs and fat in the fruit starts degrading. The longer it takes for the fruit to be transported, processed, and refined into palm oil, the larger part of the fats degrade. During the palm oil refining process, this degraded fat, about 3.5 to 5% of the raw material input, is removed as PFAD to produce edible, food-grade oil. PFAD is unsuitable for human consumption.

Several further wastes and residues of vegetable oil processing include SBEO which consists of oil waste recovered from spent bleaching earth that is used in the refining processes of various vegetable oils. Acid oils are free fatty acids that are also derived from refining processes of a variety of vegetable oils, for example, the refining of soybean oil. Wastes from palm processing include POME and the oil recovered from empty fruit bunches (EFBO). POME oil is waste oil derived from palm oil mill effluent. It consists of the remaining oily waste skimmed from the palm oil mill's wastewater. EFBO oil can be recovered from the empty fruit bunches that are the remains of the oil palm's fresh fruit bunches after the fruit has been stripped for oil pressing.

Others

There is an ongoing search for lower quality wastes and residues to be used in the production of materials, fuels, and chemicals. This is because companies being able to upgrade lower quality raw materials in an efficient way can potentially better withstand the margin pressure arising from growing demand and competition. Thus, the capability to capture the availability of emerging, lower-quality waste and residue raw materials, such as e.g. wastewater-derived grease (i.e. "brown grease") is of importance. Additionally, it becomes more relevant to continue to study other sources, e.g. lignocelluloses, such as agricultural and forest harvesting residues, municipal solid waste as well as novel vegetable oils, as future raw material options.

One example of an already well-developed alternative source is crude tall oil (CTO), which is a residue generated at pulp mills when pulp is produced from pine wood. Crude tall oil is further processed at tall oil refineries. The distillation process produces different fractions such as tall oil fatty acids, tall oil rosins and tall oil pitch. Crude tall oil and some of its derivatives can be used as raw material for renewable fuel production. Tall oil derivatives are also used for various chemical applications.

Representing an increasingly important source of raw material for further scaling up solutions, the production of novel vegetable oils (NVOs) with regenerative agricultural practices is explored to increase the availability of sustainable renewable raw material sources. Regenerative agriculture refers to methods that agricultural producers can use to protect or restore the viability of land and soil as well as the ecosystems they are a part of. Regenerative practices such as cultivating marginal lands or planting intermediate crops do not create additional demand for agricultural land and the novel oils are not in conflict with food production. With soil conservation at the center, regenerative practices can also reduce GHG emissions and unlock new opportunities in agricultural value chains.

14.2.4 Processing of Biowaste

Pre-treatment and Upgrading

The development of capabilities to pre-treat, i.e. purify, low-quality raw materials to remove impurities, is a decisive differentiation factor. Many of the low-quality raw materials inherently contain impurities such as phosphorus, metals, nitrogen, sulfur, or chlorine, which if not removed, would have an adverse effect on the catalytic upgrading process, for example, by reducing the catalyst lifetime and thus the overall economics of the process, or could even end up in the product and impact quality. Therefore, pretreatment technologies have been developed and optimized for better performance, high-end product quality and the capability to move towards lower quality raw materials.

The basic configuration of oils and fats pretreatment is known from the vegetable oil industry and comprises the steps of acid degumming and bleaching. The former is used for the removal of phospholipids and gums and the latter to trap impurities by adsorption. Advanced pretreatment processes contain additional steps and operational know-how, which is usually considered sensitive and confidential as it provides a competitive edge.

After the pretreatment the purified feed is upgraded in a catalytic process which comprises the steps of hydrodeoxygenation, i.e. hydrogenation to remove oxygen, followed by a step of further isomerization to adjust the molecular chain branching for optimizing properties.

Conversion into feedstock and polymers and chemicals

There are multiple ways to utilize bio-based wastes and residues to produce polymers and chemicals. Most commonly the wastes and residues are converted into drop-in hydrocarbons that can replace fossil feedstock at the very beginning of the value chain into the existing assets. Most commonly that is the replacement of fossil naphtha by equivalent renewable naphtha or diesel, to a lesser degree also gaseous feedstocks like LPG and propane can be replaced by their 100% bio-based alternatives. They are developed to be suitable for current production infrastructures and can be used to produce climate-friendly polymers either on their own or as a blend with crude oil-based feedstocks. The products resulting are of the same quality as the purely fossil alternatives. This includes the fact that they can be used in sensitive applications, such as food-packaging, for medical applications or as ingredients for cosmetics. This comes with the advantage of being a climate-friendly alternative to products based on virgin fossil oil. Reducing fossil oil dependency and consumption provides clear and measurable climate benefits. The replacement of fossil oil-based feedstock with alternatives made from bio-based wastes and residues can significantly reduce the carbon footprint of the end products. Life cycle assessments (LCA) studies (Neste 2021) show reductions of greenhouse gas (GHG) emissions by more than 85%.

14.2.5 Potential of Biowaste as Feedstock

Volume availabilities

Renewable waste and residue fats and oils will be indispensable in delivering emission reductions soon and over the longer term. It is expected that the volume potential of raw materials for use in the renewable products sector will continue to increase in the coming years and reach a global volume potential of around 40–50 million tons per year in the 2030s, as indicated in Fig. 14.13.

Roughly 11-13 Mt of used cooking oil are available each year, and more could become available as growing populations and rising wealth increase the demand for food. Collecting the oil presents challenges, but stricter regulation, for example, to avoid health hazards, can improve accessibility. Animal fats from industry rendering waste represent roughly a third of the global supply of waste and residue lipids—12-15 Mt each year, a figure that should grow as meat consumption rises, especially in Asia. The same holds true for fish oil, although only about 1 million metric tons are available each year. Other considered waste and residue lipids total about 11-14 Mt each year. The total practical availability of this feedstock category thus amounts to roughly 40 Mt per year. Numbers are conservative not including many smaller streams of waste and residue lipids in the analysis (WEF 2020, Clean Skies for Tomorrow). Regionally highest accessible waste and residue feedstock potentials are in the Asia Pacific region, North America, and Europe.

Biowaste and advanced renewable feedstocks hold significant long-term growth potential as a global feedstock pool via scale-up of new technologies. Novel vegetable

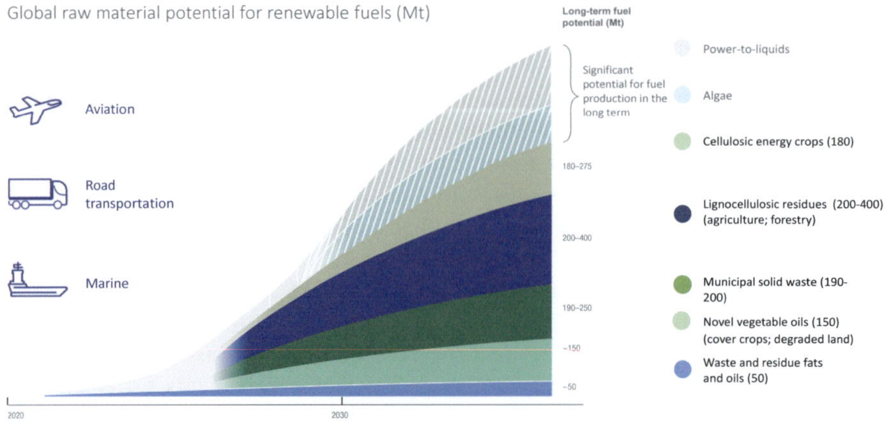

Fig. 14.13 Biomass potential. *Source* Neste analysis based on WEF (2020)[4]

oils such as cover crops cultivated on existing agricultural land during off-season represent an increasingly important source of raw material for further scaling up solutions. In addition, new conversion technologies will enable the use of currently untapped raw material pools such as municipal solid waste and lignocellulosic biomass. In the long term, feedstocks produced from electricity and waste CO_2, so-called power-to-liquids, will also increasingly play a role. The potential of these technologies could be substantial if innovation activities successfully enhance technology maturity and bring down costs.

Market overview and economic considerations

The key to successful market activities is the capability to get access to a broad range of feedstocks and the capability to process them into products that function as feedstock for fuels and chemicals depending on the attractiveness of the markets. From an economic standpoint, profitability is optimized by being able to purchase low-grade raw materials that can then be processed into products in the most efficient manner. Typically, the highest margins are achieved by the company with the best ability to turn low-quality raw materials into high-quality products.

The top five key players creating feedstock for materials (polymers and chemicals), as well as fuels (Diesel and Sustainable Aviation Fuel, SAF), are listed in 14.2. Currently, this landscape is in a hefty development with several new players entering the market. The role China is going to play is not clear yet but can be expected to increase in importance soon.

As shown in Table 14.2, which represents one of many possible future scenarios for the chemical industry's carbon sources, the need for bio-based solutions is going

[4] Biomass potential converted to fuel potential, using around 85% conversion efficiency (weight-based) for fats and oils and novel vegetable oils; around 25% efficiency for lignocellulosic biomass and municipal solid waste. (*80% organic waste, with 20% non-reusable, non-separable plastic waste).

Table 14.2 Top five key players for renewable feedstock. *Source* Estimations by authors

Company	Regions	Total capacity (2023) est.
Neste	EU, NA, AP	5.5 M t
Diamond green diesel	NA	4 M t
Marathon	NA	1.5 M t
ENI	EU	1.3 M t
HF	NA	1 M t

to be very high. The competition between solutions for transportation and material usage is dominating the market already today. The inhomogeneity of the legislative landscape is currently driving the demand allocation. The polymers and chemicals market are mostly a voluntary market, while the fuel sector is heavily regulated and carries incentives for the usage of bio-based solutions for fuels for road transportation and aviation. The key to the commercial success of companies involved in this market is thus believed to lay in the flexibility to serve different markets in different regions depending on economic conditions.

Strong upstream capabilities to ensure feedstock supply, ability to process broad ranges of feedstock and flexibility in the production of fit-for-purpose feedstocks for fuel and chemistry applications become the decisive differentiation factors.

Outlook

It is estimated that bio-based waste and residue oils and fats may reach some 50 Mt/a in the long-term (Neste 2022, The Change Book). Considering that in 2022, only some 6 million tons of plastics worldwide (PE 2022) were produced with bio-based materials. However, against a global plastics production of close to 400 Mt, it becomes obvious that waste and residue oils and fats can only carry part of the total transformation—especially since they are also a valuable feedstock for renewable fuels such as renewable diesel or sustainable aviation fuel.

In the mid to longer term, the use of other types of raw materials, such as algae and novel vegetable oils is expected to become more relevant. Lignocelluloses and municipal solid waste represent further potential sources for renewable feedstocks.

The most probable scenario is one in which various technologies and raw materials complement each other—In the same way, bio-based solutions will complement recycling and CCU to create a fully defossilized future for polymers and chemicals in the long term.

14.3 Agricultural and Forestry Biomass

Oona Koski

14.3.1 Introduction

This chapter introduces biomass from the agricultural and forestry industries for the use of raw materials to the plastic and chemical industries. The common interest is to have feedstock and supply chain without negatively impacting the climate and bringing together industries in a mutual effort for a sustainable future. The chapter describes the principles of biomass creation and examples of its use by mutual sustainability principles und industrial economics. A circular economy creates new business opportunities, linking together environmental targets and economic goals. In a sustainable supply chain, no raw material goes to waste and renewable sustainable biomass is used responsibly and efficiently.

14.3.2 Biomass-Based Feedstock

Biomass is a renewable carbon source readily available for use in the material sector. When using biomass, it is important to ensure that it is produced, processed, and used in a sustainable and efficient way, providing greenhouse gas savings, and maintaining carbon storage and biodiversity. Since biomass production is closely linked to different ecosystems possibly unique, there are different sustainability challenges to be taken into consideration.

Replacing fossil and mineral-based materials with biobased ingredients will typically reduce the carbon footprint of products and lock-in carbon sequestered from the atmosphere by trees and plants for the life of the product, including multiple recycling loops to come.

Material efficiency is key to the production of advanced biobased raw materials. Using waste and residue streams existing sustainably produced raw materials are put to even more efficient use. No new land is needed for production, and neither food production nor biodiversity is affected. However, additionally, new biomass is needed, too, wherefore best agricultural practices supporting, e.g. co-production of food and material as well as sustainable forest management practices are needed.

Improving efficiency cannot happen without changing also human behavior. Huge amounts of biowaste are generated worldwide. Every year nearly 30% (equal 1.3 Gt) of food produced is wasted (WFP 2020). By reducing this, biomass could be released for nourishing people and utilizing it in the material sector.

Social, economic, and environmental benefits of regional biomass

It is the beauty of biomass that its production can take place almost everywhere in the world, whereas fossil oil drilling can only be done in discrete places. Sustainable industrial use of wood supports reforestation and ecosystem-focused replanting and protection. Forest management and the environmental benefits of forests are not mutually exclusive.

The use of wood to produce ingredients for biobased chemicals represents a completely new market for the forest sector. Regional sourcing of raw materials and other services will enable local value creation and ensure compliance with high social and environmental standards. For the UPM biorefinery in Leuna, Saxony-Anhalt Germany for example, it offers significant market development and enables foresters and forest companies to participate in a future-proof use case for their biomass as depicted in Fig. 14.14. Using forest management thinning and industrial beechwood that is currently incinerated, along with side streams from sawmills, the facility contributes significantly to advancing the circular economy. A dedicated wood sourcing organization works with a broad range of forest owners, from major state-owned organizations in Saxony-Anhalt and Bavaria to private forest organizations and small owners.

Another example from the UPM ecosystem related to social, environmental, and economic benefit; UPM Biofuels has for several years been developing new concepts of sustainable land use and carbon farming practices in Uruguay. Carbon farming is based on the cultivation of brassica oil crops, outside the main growing season. Adding productive intermediate crops to the secondary farming season improves the quality and carbon sequestration capacity of the soil and prevents plant diseases. At the same time, valuable oil and proteins are produced as raw materials and, e.g. animal feed. Winter farming offers local farmers a novel opportunity for additional income.

Sustainability principles for biomass

When biomass sourcing is discussed, concerns may arise. What if it disrupts food production, reduces carbon sinks, or destroys biodiversity? To prevent the risk of

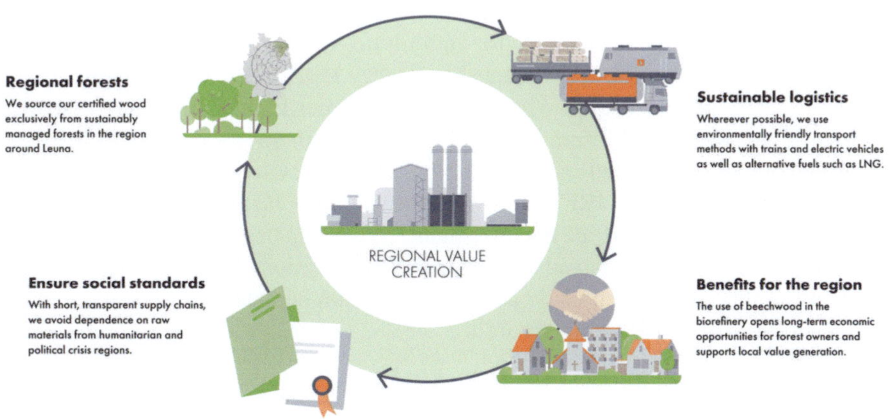

Fig. 14.14 Fostering regional value creation. *Source* UPM-Kymmene Oyj

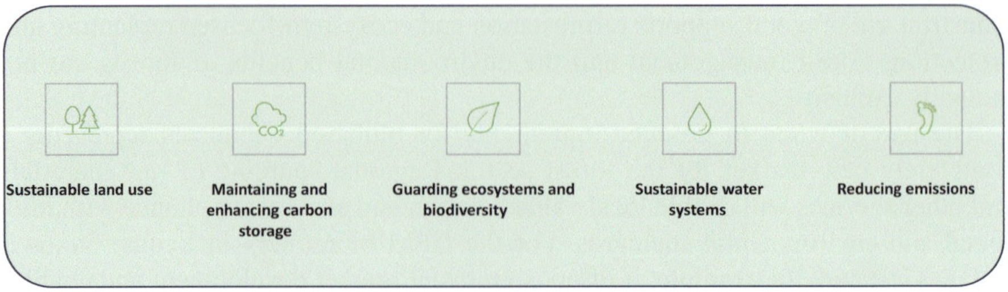

Fig. 14.15 Sustainability principles for renewable feedstock. *Source* UPM-Kymmene Oyi

unintentional consequences, sustainability principles for biomass cultivation and sourcing must ensure no over consuming of valuable renewable resources but using it in a responsible manner. The main principles specifically important for biomass production are listed below (cf. Fig. 14.15).

Sustainable land use

The United Nations defines sustainable land use as "the use of land resources, including soils, water, animals and plants, for the production of goods to meet changing human needs, while simultaneously ensuring the long-term productive potential of these resources and the maintenance of their environmental functions" (FAO 2020). Accordingly, sustainable land use management must maximize the resilience, towards climate change by minimizing land degradation, rehabilitating degraded land, and ensuring sustainable use of land resources. The responsible production of renewable feedstock through well-designed sustainable value chains can enhance soil health and promote green resilience practices in agriculture and forestry fostering the bioeconomy. To achieve sustainable land use, in both agriculture and forestry, a deep understanding of the impacts of different ecosystems and biodiversity is needed.

It is acknowledged that the demand for biobased feedstocks has the potential to indirectly incentivize land clearance for agriculture, and hence undermine their environmental credentials. The EU Renewable Energy Directive actively engages with this issue, stating: "Feedstock which has low Indirect Land-Use Change (ILUC) impacts when used for biofuels, should be promoted for its contribution to the decarbonisation of the economy" (COM 2018). Both the Global Biosphere Management Model (GLBIOM) (JRC 2023) and the Global Change Assessment Model (GCAM) assessments illustrate the importance of well-managed forests in mitigating ILUC emissions and increasing soil carbon sequestration. Furthermore, land use must ensure that unaccepted land occupation does not happen as well as indigenous people's rights to their land are secured.

No deforestation and limiting the risk of indirect land use change

According to the FAO's State of the World's Forests Report (FAO 2022), over 90% of global deforestation is caused by agricultural expansion, partly because there is a need for increased food production because the lands in use are not maintained sustainably and are therefore degraded and not feasible for food production. Deforestation has negative impacts on all aspects of sustainability: species and habitat loss; carbon storage and carbon sink loss; disappearance of local forest dependent livelihoods, etc. Companies therefore should strictly condemn deforestation. Their forest management and wood sourcing should not cause any deforestation thus ensuring that forests grow more than they are harvested.

To stop deforestation the EU Regulation on Deforestation-free Products (EUDR) (COM 2023) entered into force in 2023. It aims to ensure that a set of key products traded and consumed in the EU and globally no longer contributes to deforestation and forest degradation. The main driver of these processes is the expansion of agricultural land that is linked to the production of commodities like soy, palm oil, wood, cocoa, coffee, rubber, and derived products, such as leather, chocolate, tyres, or furniture. Under the Regulation, any operator or trader who places these commodities on the EU market, or exports from it, needs to prove that products do not originate from recently deforested land or have contributed to forest degradation. However, harvesting wood as part of sustainable forest management should not be confused with deforestation—since sustainable wood harvesting is always based on the regeneration of forests. It sees the forest itself as an invaluable asset to be managed and maintained in perpetuity and that forests are managed in a way that they are preserved for future generations respecting their environmental, social, and economic values.

National forest inventories allow foresters to access public information on forests including forest area, growth, and harvest, complementing environmental management and certification systems. Forest inventory data provide information verifying that deforestation does not take place in areas where wood is sourced. Certified forest management and chain of custody systems provide mechanisms for traceability and information linking the wood harvested and end products. The chain of custody verifies that the wood doesn't come from controversial sources, such as illegal logging and forest conversion.

Protecting soil quality and productivity

Intensive agronomy and monocropping focus on maximizing yields by using high inputs of fertilizers, pesticides, and water has been widely adapted over the last decades causing several challenges. Some of the challenges associated with intensive agronomy are environmental degradation, land acting as carbon source or health risks for workers. To address these challenges, there is a need to promote sustainable agriculture practices that are environmentally friendly, economically viable, and socially acceptable. Some of these practices include crop rotation, minimum tillage, integrated pest management, or limiting irrigation. These will help to protect soil health and ensure long-term productivity.

Maintaining and enhancing carbon storage

Forests absorb carbon from the atmosphere through photosynthesis, storing the carbon in a wide variety of biomass, including live trees, dead trees, soils, and even harvested trees as a product even short-lived products can retain carbon for the duration of their lifecycle, including through multiple rounds of recycling. Based on a recent study integrating ground and earth observation data, the world's forests sequestered almost twice as much carbon as they released into the atmosphere between 2009 and 2021 (WEF 2021). The total carbon stock in forests was estimated at an average of 163 t/ha (FAO 2020.) As the world looks to cut greenhouse gas emissions, forest experts are implementing measures to improve forest sustainability. Achieving a net positive carbon balance is a crucial part of sustainable forest management. When forests are at optimal density, they absorb more carbon. Therefore, the more fertile the forest, the higher the wood volume, the more carbon sequestered.

Guarding ecosystems and biodiversity

Biodiversity is part of a responsible business. A rich variety of species encourages healthy, resilient, and carbon-absorbing forests; in agriculture or forest plantations improved management practices can support the soil health and biodiversity as well as, e.g. biodiversity corridors can guard the biodiversity in the surrounding environment. Management of commercial forests is crucial to nurturing global biodiversity since the proportion of protected forests is small in comparison to commercial forests worldwide. This means that protected forests alone cannot secure biodiversity. Measures to enhance biodiversity in managed forests have an impact on much larger areas than protected areas.

Reducing emissions

Products made of biomass from forests or agricultural should all contribute to a reduced carbon footprint. However, many of the calculation methodologies are not currently considering correctly the biogenic carbon embedded into the products and therefore the calculation methodologies must develop for having reliable harmonized standard.

Verifying sustainability and traceability—certification

Sustainability certifications enable companies to prove their feedstock production and the supply chain are sustainable. Sustainability certification is a dynamic system with regularly updated requirements. The requirements of international regulations and frameworks, as well as a science-based understanding of the actions needed, are all considered in certification standards.

There are plenty of global voluntary schemes verifying raw materials and their value chain sustainability (cf. Fig. 14.16). Some of the schemes like International Sustainability and Carbon Certification (ISCC 2023), Roundtable on Sustainable Biomaterials (RSB), Better Biomass (Better Biomass 2023) or REDcert (2023), have implemented a multi-feedstock approach (cf. Chap. 16) meaning that different raw materials from agriculture, forest, or waste and residue streams can be certified utilizing the same standards. In addition, there are schemes focussing on single feedstock, like the Forest Stewardship Council (FSC 2023) and the Programme for the Endorsement of Forest Certification (PEFC) certifying forest-based products.

Fig. 14.16 Multi-feedstock and single-feedstock voluntary certification schemes. *Source* UPM-Kymmene Oyj

Bonsucro (2023) certifies sugar cane production and Roundtable on Sustainable Palm Oil (RSPO 2023).

14.3.2.1 Forest Biomass

Forests cover one-third of the world's land surface—more than 4.06 Gha but the area is still decreasing, especially in the tropics (FAO 2023) Within this vast area, there are all types of habitats, from the boreal forests of Canada to the jungles of Brazil. It is vital for companies utilizing forest products to have a holistic overview of their impacts and aim at a collaborative approach to sustainable forest management—working closely with NGOs, researchers, and government institutions to develop metrics for monitoring, e.g. biodiversity, to gain a deeper understanding of rare and endangered species, and to promote conservation and sustainable management practices around the world.

Sustainable forest management (SFM)

SFM is a dynamic and evolving concept aiming to maintain and enhance the economic, social, and environmental values of all types of forests and forest supply goods, and services for the benefit of present and future generations. Forest certifications are a great tool for promoting sustainable forestry.

SFM combines both conservation and sustainable use of forests. In forests, it is important to recognize areas and specific species that have a high biodiversity value and ensure that those are correctly protected, with a focus on quality. The most important thing is that the most valuable habitats are kept out of forestry use and, on the other hand, that efficient and effective nature management is carried out in commercial forests. Sometimes, improving living conditions in commercial forests means logging and nature management measures rather than conservation. Sustainable forest management methods are also an efficient tool to mitigate the risks of climate change and environmental disasters in forests. Contrary to many expectations, totally unmanaged forests can be also in very poor condition, for example with diseases and pests spreading in the absence of efficient forest management actions or significant forest fires. By investigating the major threats (e.g. climate change) that may impact forest ecosystems, better preventive management systems can be

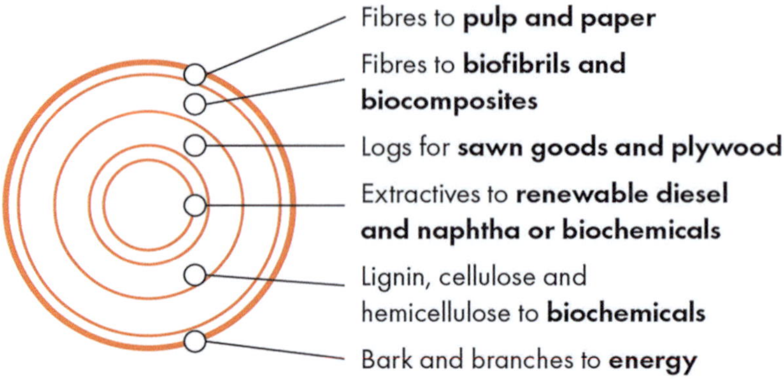

Fig. 14.17 How to utilize wood efficiently. *Source* UPM-Kymmene Oyj

developed. In other words, sustainable forest management accelerates CO_2 sequestration in the forest. In addition, in the sustainable forest management, it is required to ensure that overconsumption does not result. It can be avoided by reviewing the management plans for each forest area regularly and setting sustainable harvesting levels at a country or regional level.

Utilizing wood efficiently

For centuries wood has been used for different purposes from furniture, pulp and paper, to burning it for heat and cooking purposes. During the last decade, the innovations have also turned wood into renewable biofuel or biobased plastics. Figure 14.17 illustrates how the different parts of a tree can be used for different purposes. There are multiple by-products as well as waste and residue streams that can be utilized with high efficiency nowadays.

For example, a natural wood extract of mainly softwood origin, crude tall oil, is collected as a residue in pulp production. Crude tall oil is a completely renewable feedstock, so using it provides significant environmental benefits compared to fossil feedstock. The use of crude tall oil in a biorefinery does not increase wood usage and can significantly reduce emissions while committing to sustainable production methods.

14.3.2.2 Agricultural Biomass

There is quite some debate about whether biomass production causes disturbances for the food and feed markets, which may of course be the case sometimes, but typically there are more synergies and co-benefits producing food, feed, and feedstock altogether. For example, farmers can gain security for their production if they have more options for selling stock to different markets. At the same time market stability improves due to increased global availability of food and feed crops. Also, food security can improve due to better availability of edible crops that can be flexibly

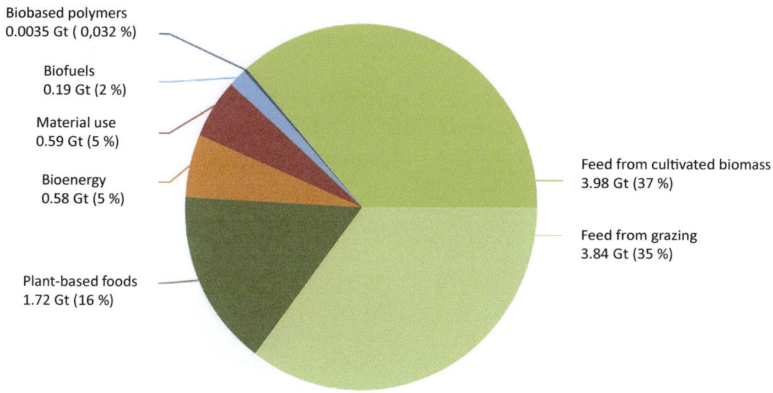

Fig. 14.18 Global harvested agricultural and grazed biomass demand by sectors. *Source* Nova-Institut (2023)

distributed between the sectors. The majority of agricultural biomass is being used for food and feed purposes, chemical and fuel uses are the minority, however. When utilizing the synergies of food, feedstock, and fuel production and ensuring the best agricultural management practices plenty of room is given to grow and scale up material production (cf. Fig. 14.18).

Sustainable agricultural practices

In agricultural negative land use change needs to be avoided, meaning to avoid converting any valuable area, e.g. high carbon stock or high biodiversity land into production. Instead, finding new ways to use existing agricultural land more productively is needed. Hence, there should be a focus on finding synergies and on regenerative agriculture, bringing back under-utilized, abandoned, or degraded areas back into usage. However, sustainable agricultural should not be practised in the industry sector specifically, but the same principles should rather be followed when producing food, feed, or feedstock as set in the EU deforestation regulation (COM 2023).

14.3.3 Innovative Examples to Transit Beyond Fossils

150-year-old UPM is part of the Finnish forest industry with pulp and paper production is on a journey from traditional industry to incorporating modern, innovative new businesses—such as biorefining—Into their portfolio. Innovation offering alternatives to fossil-based materials is at the core of their business strategy. At its biorefineries, UPM is pioneering sustainable chemistry—innovating in chemical processes, scaling biorefining, and unlocking the potential of biomass for transforming industries. Nowadays UPM is producing a variety of renewable fuels

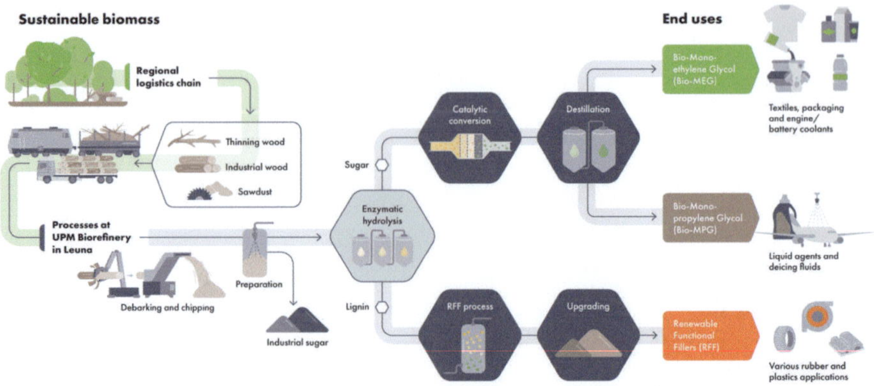

Fig. 14.19 Products from wood-based biomass UPM biorefinery in Leuna. *Source* UPM-Kymmene Oyj (2023)

and chemicals made from sustainable biomass. UPM's Lappeenranta Biorefinery produces 130 kt of advanced biofuels a year and is a commercial-scale biorefinery producing renewable wood-based diesel and naphtha (BioVerno™) using upcycled waste from UPM Kaukas pulp and paper mill. BioVerno™ is made from wood-based tall oil recovered from the pulp by a hydrotreatment process developed at UPM's Leuna Biorefinery to come onstream in 2024 (UPM 2023) and will turn wood and forest management by-products into next-generation biochemicals producing 220 kt of glycols and lignin-based renewable functional fillers as illustrated in Fig. 14.19. Currently, UPM is investigating a production expansion opportunity in advanced biofuels and biomaterials and is conducting basic engineering for a possible further biorefinery in Rotterdam/The Netherlands. This biorefinery with an annual max. capacity of 500 kt Is meant to produce advanced biofuels for transportation, including maritime and aviation, and for the petrochemical sector as feedstock for biobased plastics.

14.3.4 Summary and Outlook

Forestry and agriculturally based sustainable biomass are renewable feedstock to be used as fuels and raw materials for various applications for a circular bioeconomy and chemical industry. In our attempts to fight climate change, we will need sustainable biomass as basis to replace fossil-based carbon content.

The main advantages of producing forestry and agriculturally based sustainable biomass are:

- Reduction of greenhouse gas emissions and dependence on fossil fuels by providing alternative and cleaner sources of energy and materials.
- Regional availability offering already available alternative carbon streams.

- Enhancing rural development and creating new jobs and income opportunities for farmers, foresters, and other stakeholders in the biomass value chain.
- Supporting biodiversity and ecosystem services by promoting sustainable management of forests and agricultural lands.

To minimize any unexceptional and unintentional consequences, like indirect land use change or deforestation or biodiversity loss when utilizing biomass, harmonized sustainability principles are needed. In addition, solid external verification systems need to be in place to ensure sustainability principles are followed and all products can be traced back to their origins. Products from biorefineries are identical to products from fossil-based feedstock for the chemical industry and offer alternative feedstock for the plastics industry with a positive carbon footprint. UPM is committed to create a sustainable bioeconomy utilizing different waste and residue streams with the highest quality and developing new sustainable management practices for their forestry operations to maintain carbon stocks and improve biodiversity. Equally, UPM is committed to gaining new materials from new regenerative land use concepts. UPM delivers an infrastructure for a sustainable carbon-based raw material as a feedstock alternative to the presently used coal and crude oil-based industry.

Depending on market demand, regulations, and economic growth biomass will become a growing sustainable raw material satisfying the needs of tomorrow's world population.

References

Better Biomass 2023: Retrieved 2023, from https://betterbiomass.nl/.

Bonsucro 2023: Retrieved 2023 from https://bonsucro.com/.

BMELV 2013: Bundesministerium für Ernährung, Landwirtschaft und Verbraucherschutz, August 29, 2013. http://www.bmelv.de/SharedDocs/Downloads/Landwirtschaft/Markt-Statistik/ (Retrieved on Jan. 17, 2014).

Bonten 2019: Bonten, C, Plastics Technology—Introduction and Fundamentals, Munich: Hanser 2019, ISBN-10: 1569907676.

COM 2018: Renewable Energy Directive (2018/2001/EU). Retrieved 2023, from https://energy.ec.europa.eu/topics/renewable-energy/renewable-energy-directive-targets-and-rules/renewable-energy-directive_en (Last access: Aug. 20, 2024).

COM 2023: EU's Regulation on Deforestation-free Products (EUDR). Retrieved 2023, from https://eurlex.europa.eu/legalcontent/EN/TXT/uri=CELEX%3A32023R1115&qid=1687867231461 (Last access: Aug. 20, 2024).

EN 13432: Packaging—Requirements for packaging recoverable through composting and biodegradation—Test scheme and evaluation criteria for the final acceptance of packaging, 2000.

Essel 2012a: R. Essel, Consulting Marketing & Industrieberatung GmbH, "Produktion, Verarbeitung und Verwertung von Kunststoffen in Deutschland 2011", 2012.

Essel 2012b: Essel, R., Carus, M., Raschka, A. "Bio-Based Polymers—An Ecological Alternative?" Kunststoffe International, No. 8, pp. 54–57, 2012.

EU, Renewable Energy Directive, 2018: Directive (EU) 2018/2001 on the promotion of the use of energy from renewable sources https://eur-lex.europa.eu/legal- (Last access: Aug. 20, 2024).

European Bioplastics 2024: Market data. Retrieved 05 Feb 2024 from https://www.european-bioplastics.org/market/.

FAO 2022: Food and Agriculture Organization of the United Nations. State of the World's Forests. https://www.fao.org/documents/card/en?details=cb9360en (Last access: Aug. 20, 2024).
FAO 2020: Food and Agriculture Organization of the United Nations. Global Forest Resources Assessment 2020 – Main report. https://doi.org/10.4060/ca9825en.
FAO 2023: Food and Agriculture Organization of the United Nations. Sustainable land management. Retrieved 2023 from https://www.fao.org/land-water/land/sustainable-land-management/en/ (Last access: Aug. 20, 2024).
FSC 2023: Forest Stewardship Council. Retrieved 2023 from https://fsc.org/en. (Last access: Aug. 20, 2024).
GB/T 41010–2021: Degradability and identification requirements of biodegradable plastics and products. https://plasticseurope.org/knowledge-hub/plastics-the-facts-2022/.
IEA 2014: https://doi.org/10.1080/15567249.2017.1350770 (Last access: Aug. 20, 2024).
ISCC 2023: International Sustainability and Carbon Certification. Retrieved 2023 from https://www.iscc-system.org
JRC 2023: Global Biosphere Management Model. GLOBIOM. Retrieved 2023 from https://web.jrc.ec.europa.eu/policy-model-inventory/explore/models/model-globiom/ (Last access: Aug. 20, 2024).
Kliem 2020: Kliem, S, Kreutzbruck, M, Bonten, C: „Review on the Biological Degradation of Polymers in Various Environments", Materials (2020)13, https://doi.org/10.3390/ma13204586
McDonald's 2022: https://bioenergyinternational.com/neste-mcdonalds-netherlands-and-havi-launch-circular-economy-collaboration/.
Neste 2021: Neste, Life Cycle Assessment on Environmental Impacts of Neste Renewable Polymers and Chemicals (2021) https://www.neste.com (Last access: Aug. 20, 2024).
Neste 2022: Neste - The Change Book. How to create a renewable and circular future? Feb 2022. Retrieved from https://www.neste.com.
Nova-Institut 2021: Carbon Dioxide (CO_2) as Chemical Feedstock for Polymers.Technologies, Polymers, Developers and Producers. Renewable Carbon Publications. Retrieved 2023, from https://renewable-carbon.eu/publications/product/carbon-dioxide-utilisation-and-renewable-energy-preview/ (Last access: Aug. 20, 2024).
Nova 2022: McDonald's Tests Circular Clear Cups Sourced From Recycled and Biobased Materials, Including Used Cooking Oil. Retrieved 2024 from https://renewable-carbon.eu/news/mcdonalds-tests-circular-clear-cups-sourced-from-recycled-and-biobased-materials-including-used-cooking-oil/ (Last access: Aug. 20, 2024).
Nova-Institut 2023: The Use of Food and Feed Crops for Bio-based Materials and the Related Effects on Food Security. Retrieved 2023, from https://renewable-carbon.eu/publications/product/rci-paper-on-the-use-of-food-and-feed-crops-for-bio-based-materials-and-the-related-effects-on-food-security-recognising-potential-benefits-long-version.pdf.
PE 2022: https://plasticseurope.org Plastics Europe, The facts 2022, p. 16. (Last access: Aug. 20, 2024).
RCI 2023: https://renewable-carbon-initiative.com (Last access: Aug. 20, 2024).
RedCert 2023: https://www.redcert.org/ (Last access: Aug. 20, 2024).
RSPO 2023: https://rspo.org/de/ (Last access: Aug. 20, 2024).
UPM-Kymmene Oyi: https://www.upmpaper.com/about-us/paper-mills-offices/upm-kymi/ (Last access: Aug. 20, 2024).
WEF 2021: World economic forum 2021 Environment: How much carbon do forests absorb? Retrieved 2023 (Last access: Aug. 20, 2024).
WEF 2020: Clean Skies for Tomorrow; https://www.weforum.org/publications/clean-skies-for-tomorrow-sustainable-aviation-fuels-as-a-pathway-to-net-zero-aviation/ (Last access: Aug. 20, 2024).
WFP 2020: World Food Programme, 2020. 5 facts about food waste and hunger. https://www.wfp.org/stories/5-facts-about-food-waste-and-hunger (Last access: Aug. 20, 2024).

Prof. Dr.-Ing. Christian Bonten is Full Professor of Plastics Engineering at Stuttgart University in Germany. He studied Mechanical Engineering at the University of Duisburg and Plastics Technology at the University of Aachen. After working with BASF he joined bioplastics producer FKuR in the managing board. Since 2010 he chairs as Director of the 'Institut für Kunststofftechnik' (IKT) at the University of Stuttgart.

Dr. Lars Börger Ph.D. (Chem) had various managerial leadership positions at Neste, Germany, since joining the company in 2017. Previously he was with BASF as Head of Global Marketing Biopolymers. He was Board member and Chair at European Bioplastics and Advisory Board Member of Renewable Carbon Initiative. Since 2025 he is CEO of nova-Institute, the leading think tank for Circular Economy.

Ace Fung is Senior Manager, Sustainability and Regulatory Transformation at Neste. Prior to joining Neste, Ace played a role in leading the development and launch of the first bio-mass-balanced aseptic liquid carton packaging at SIG Combibloc. Hailing from Canada, Ace has cultivated a diverse professional background through his international experiences spanning over 15 years. Ace holds an MBA from Mannheim Business School.

Miia Laamanen is Head of R&D, Feedstock and Pretreatment at Neste leading a department that delivers solutions to expand renewable and circular feedstock bases. Since joining Neste in 2012, she has worked in a series of R&D positions related to renewable fuels and chemicals technologies development and scouting. Miia has an M.Sc. degree in chemical engineering from Helsinki University of Technology. Prior to Neste, she worked at Aalto University on industrial biotechnology development.

Petri Lehmus D.Sc. (Chem.Eng.) is Vice President of R&D at Neste since 2011. Previously spent more than ten years in the polymer industry, from 2007 to 2011 as Innovation Center Manager and 2004 to 2007 as Catalyst and Process Research Manager at Borouge in the United Arab Emirates. He is Board member of ClicInnovation and Chair of the Finnish Chemical Industry Federation´s innovation group.

Oona Koski has worked as sustainability expert at UPM for the past 10 years in very versatile teams, projects, and tasks. Currently, her key responsibilities at UPM Biorefining business area are to integrate feedstock sustainability as a strategical element to business and develop sustainable value chain processes and certifications. Moreover, she works with the newest innovations related for example sustainable forestry and new regenerative concepts like agroforestry and carbon farming.

Chapter 15
Carbon Dioxide (CO_2)—The Ultimate Feedstock for an All-Circular and Climate-Neutral Carbon and Plastics Economy

Reinhold W. Lang, Stefan P. Schleicher, Walter Leitner, and Guido Schroer

Contents

R. W. Lang (✉)
Johannes Kepler Universität Linz, Linz, Austria
e-mail: Reinhold.Lang@jku.at

S. P. Schleicher
Wegener Center for Climateund Global Change, Graz, Austria
e-mail: stefan.schleicher@wifo.at

W. Leitner (✉)
Max Planck Institute for Chemical Energy Conversion, Mühlheim, Germany
e-mail: walter.leitner@cec.mpg.de

G. Schroer
RWTH Aachen University—Power2Polymers GmbH, Aachen, Germany
e-mail: schroer@itmc.rwth-aachen.de

P. Orth et al. (eds.), *Circular Economy of Plastics*,
https://doi.org/10.1007/978-3-031-78829-1_15

Summary

This chapter sets the stage and explores the potential for a shift towards a greener future, focusing on using carbon dioxide (CO_2) to create plastics and other chemicals. It aligns with global efforts outlined in initiatives like the UN's Sustainable Development Goals, emphasizing the importance of safeguarding our planet.

Diving into the challenges and opportunities of CO_2 utilization, the chapter explores conditions and innovative solutions for an all-circular plastics and carbon economy. It also addresses the potential demand for CO_2 and methods for capturing it preferably from industrial point sources in Sect. 15.2. Furthermore, the chapter delves into the detailed processes of establishing a circular carbon cycle within various industries, highlighting the need for collaboration and policy support. It emphasizes the significance of regulatory frameworks outlined in Sect. 15.3 to support the transition towards sustainable development. Additionally, the chapter points out the transformation scale and ecological challenges associated with transforming the carbon source, as discussed in Sect. 15.4. Most importantly, it underscores the need for efficient strategies and adequate policies to overcome these obstacles.

Lastly, Sect. 15.5 highlights the opportunities for research and development in creating new polymers based on CO_2, offering additional avenues for innovation in sustainable materials production. Overall, it's about finding smarter, more sustainable ways to produce essential items while preserving the environment for future generations.

15.1 Setting the Stage—'*Searching a Way out of no Way*'[1]

Reinhold W. Lang and Stefan P. Schleicher

Sustainable development, energy and climate politics, and circular economy have become economic, environmental and resource political top priorities on a global and regional level (e.g., European Union (EU)), and in many countries also on a national level. Thus, meeting the desire of a still highly equity-divided[2] and growing world population for prosperity and justice has emerged as a guiding principle via several strings of arguments and policy developments:

- the path-breaking guidance of the United Nations (UN) Global Agenda 2030 with the 17 Sustainable Development Goals (SDGs) adopted in 2015 (UN 2015a), emphasizing global and inter-generational equity and justice as central value principles[3]
- the Paris Agreement, also negotiated and adopted in 2015 by the 21st Conference of Parties (COP 21) of the annual UN climate summits, aiming at reducing and limiting the risks and impacts of climate change (UN 2015a),
- visions of radical transformations integrating climate and circular economy policies, put forward with the European Green Deal by the European Union (EU), striving towards a climate-neutral and sustainable EU by 2050 (EU 2019),
- the EU Plastics Strategy adopted in 2018 [EU 2018a] and the EU Circular Economy Action Plan (CEAP) approved in 2020 [EU 2020], targeting issues of plastics and plastics waste management on an EU and a member state level, and
- on a global level, the UN Global Plastics Treaty (GPT) to combat plastic pollution, initiated and accepted by the UN Environmental Assembly (UNEP/EA) in 2022 (UN 2022b).

---

[1] The present chapters (1-4) represents a shortened version of a chapter in a report by the authors entitled *"Industrial Carbon Management: Navigating towards a climate-neutral, all-circular industrial Carbon and Plastics Economy"*; available online at: https://carboncyclecircle.at/ and https://www.rethinkconstruction.net/.

[2] Historically, and in broad terms, an estimated 20–30% of the world population in high(middle)-income countries are responsible for about 70–80% of the global environmental damages and future risks, with the 70–80% of the population in low(middle)-income countries bearing the brunt of current and future environmental damages and risks.

[3] Universal value principles and central, transformative promises of the UN 2030 Agenda expressing and phrasing the aim of diminishing inequalities as follows: *"Leaving no one behind"* and *"A world we want for all"* (United Nations, 2022a).

Considering these policies, we commence Sect. 15 on the use of carbon dioxide (CO_2) as a novel feedstock in a 'sustainable' circular plastics economy by posing the following question:

Is the establishment of a climate-neutral, 'all-circular' Carbon and Plastics Economy (c^2CPE) within the technosphere of the human enterprise both desirable and feasible?

To address this question, we will first explore the emerging facets of the radical transformations implied, before outlining the aims, the scope and the structure of this section. Upon examination, it becomes clear that the historical close linkage of plastics to fossil fuels from their beginning, both with fossil fuels being a feedstock and an energy resource, will remain a determining factor throughout this transformation, embodied and framed by the term "*climate-neutral* and 'all-circular' Carbon and Plastics Economy (c^2CPE)". Simultaneously, the transformative changes expected are so far-reaching and profound that a poignant parallel may be drawn to the African American expression, **'Making a Way Out of No Way'**,[4] posited and used by African Americans on their long march that ultimately led to the successes of the civil rights movement in the US in the 1960s. This mantra, which also has inspired the title of this introduction and to which we will return again at the end of the chapter, may indeed apply here as well, considering the complexity and scale of the challenges, which are fundamentally also about **equity and justice**.

15.1.1 Emerging Facets of Radical Transformations

Over the last decade, the call for and the urgency of a **Great Societal Transformation** has also gained increasing support from academics and the scientific community and society. There is now also a broad consensus that meeting the above political strategies and aims will require adequate technologies along with proper choices of materials and handling of matter/material streams. **Two principal technological pillars** for an innovation-driven, sustainable transformation have emerged (Lang 2017):

(1) the **transition of the energy system** from a fossil fuel-based energy system to an essentially all-renewable energy system, and
(2) the **transition of the matter/material resource system** from a currently linear resources-to-waste system to predominately circular material/matter streams in the future ("circular economy"), for which an **'all-circular' carbon & plastics management** is pivotal!

[4] https://nmaahc.si.edu/explore/exhibitions/making-way-out-no-way.

15.1.2 Challenges for the Hard-to-Abate Industries and the Role of the Plastics Industry

The aforementioned two technology pillars and their interaction, along with the anticipated significant role of the **hard-to-abate (HTA) materials industry** (steel/metals, concrete/cement, plastics, composites and hybrids) in these transitions, constitute the context and frame for Sect. 15. Special emphasis will be placed on the topic of cross-sectoral ‘industrial’ circular carbon management (cCM), as several such initiatives have recently gained momentum in Europe and worldwide. Specifically, the potential role of plastics and the plastics industry in these technology transitions will be explored by considering the question: can plastics assume a key role in the Great Societal Transformation, and will the plastics industry become a key driver of sustainable transformative change?

At first glance, and regarding both technology transformation pillars, plastics (i.e., synthetic polymer materials including thermoplastics, thermosets, elastomers, polymer-based composites and hybrid materials) will quite obviously have to bear and fulfill a crucial role, indeed. In brief, this is founded in and arises from the following facts and considerations (Lang 2017):

- Without doubt, plastics—with their tailor-made material property profiles and product performance—have contributed tremendously to the global prosperity rise since the mid-last century, now affecting and pervading all areas and aspects of modern life. It is reasonable to assume that this will remain so in the foreseeable future.
- Moreover, compared to alternative material solutions, specific features of plastics and plastics products are their competitive economic efficiency, typically also coupled with an advantageous ecological footprint performance over the entire life cycle.
- Also, as already mentioned, plastics historically were and still are closely linked to the dominant fossil primary raw materials of the current energy system (oil, gas, coal), not only energy-wise but also matter-wise as feedstock via their carbon/hydrocarbon raw material base.
- Finally, but nonetheless of great importance, plastics—as the youngest group among the major material classes—still exhibit exceptional innovation, growth and growth rate, and market penetration potential, both regionally and globally.

15.1.3 The Intermediate Role of Green Hydrogen and Derivatives Thereof

In the envisioned radical transformations, green hydrogen (H_2) based on renewable energies (wind and solar) will play a crucial role for several reasons. First, it will be vital for synthetic energies and feedstocks based on hydrogen, hydrocarbons and ammonia. Second, it will enable transport and storage to balance supply and demand fluctuations while building on and making the best use of existing infrastructure. Third, it will necessitate and ideally facilitate transnational cooperation with renewable energy-rich countries. These points align well with efforts towards a climate-neutral, circular economy. However, there are concerns about potential dependencies on partners not aligned with fair and resilient business practices. In any case, from an EU perspective, a sizeable amount of renewables will be needed via synthetic energies and feedstocks from countries outside the EU that are rich in renewable energies.

15.1.4 Making the **'Great'** *Transformation of the Plastics Industry Operational*

When striving towards the radical and bold vision of a fully circular carbon and plastics economy in the Technosphere, a rather specific future role for plastics in interconnecting the above two technology transformation pillars and their interaction may be derived. This interaction and interdependence of the energy system and a visionary, yet viable, all-circular plastics economy system of the future is depicted in Fig. 15.1.[5]

Given the current reliance on fossil raw materials for plastics, and apart from plastics waste causing numerous grave problems worldwide (including marine litter and risks associated with micro-plastics) that clearly need to be reduced and avoided, comprehensive recycling strategies are imperative for numerous reasons. This is particularly pressing amidst the shift away from fossil fuels in the energy system. With landfilling becoming untenable in the long run, mechanical and chemical-feedstock recycling, along with energetic-thermal processes, offer viable options for managing plastics waste. However, considering energy efficiency and environmental impacts, mechanical and conventional chemical-feedstock recycling may be limited to around 30–50% of plastics waste. This leaves some 50–70% (and globally long-term perhaps an even higher share) of plastics waste for incineration in cement or steel production plants or in waste incineration plants, thereby generating carbon dioxide, CO_2.

There are now intensive research efforts worldwide on the use of CO_2 as a building block for a wide variety of organic syntheses via so-called **carbon capture and utilization** [or use] **(CCU)** technologies.[6] This opens up a realistic future perspective for new forms of circular carbon management (cCM) via power-to-feedstock

[5] Apart from the quantitative range shares for circular carbon management vs. conventional mechanical and chemical recycling, essential features and aspects of this illustration are also incorporated in and aligned with the paper *"Circular EconomyPLUS"* by Plastics Europe, prepared in collaboration by an expert group with one of the authors of this section participating (PE 2022).

[6] CCU in this section is used equivalently to as pointed out elsewhere: *"The use of the term CCU generally refers to the utilization of carbon dioxide (CO_2), but can also include industrial carbon monoxide (CO) sources prior to flaring or other conversions to CO_2 before release to the atmosphere. In the US, CO_2 and CO are grouped together as 'carbon oxides' for purposes of Sect. 45Q CCUS*

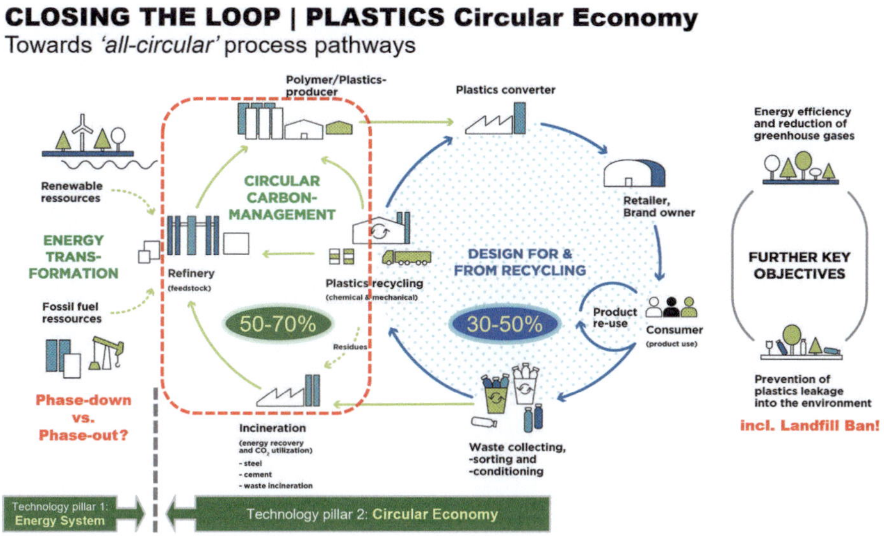

Fig. 15.1 Comprehensive illustration for an '*all-circular*' plastics economy indicating the interaction of the current fossil fuel-based energy and plastics feedstock system, also providing indications (estimates) for envisaged worldwide future shares of the mechanical and chemical recycling of plastics waste versus incineration coupled with circular carbon management technologies. (*Source* Adapted version of illustration by (Lang and Fischer 2021) also used in (PE 2022)

technologies (production of gaseous or liquid hydrocarbons), powered by renewable energy, and thus for a completely renewable/regenerative energy/material cycle for polymer materials (i.e., *all-inclusive* and *all-* or *full-circular* economy for plastics).

An excellent treatment of the important role of CCU in a future renewable carbon strategy and in replacing fossil carbon in the chemical and material industry is provided in a report by the **Renewable Carbon Initiative (RCI)** [Berg 2023a]. This remarkable platform initiative, founded in 2020 at the behest of the nova-Institut,[7] consists of industrial, scientific and other institutional members and partners and aims at effectively promoting the significance and proper industrial implementation of CCU [Berg 2023b]. In best agreement with our preceding argumentation above, it is convincingly posited that CCU not only aids in mitigating climate change but also aligns with and actively supports multiple SDGs of the UN Global Agenda [Berg 2023]. Furthermore, the report also emphasizes that the benefits of CCU can only be realized when powered by renewable energy. This is vividly portrayed by the series of illustrations of the nova-Institute collected in Fig. 15.2 (nova-Institut 2024), which again highlights the synergy between the renewable energy system and a CO_2-based renewable carbon system throughout the value creation chain, ultimately leading to the production of renewable polymers and other chemicals. Most notably, numerous

---

tax credits. In this report, 'CO_2 utilisation' is meant to also include other carbon oxides." (RCI 2023).

[7] Nova-Institut GmbH, 50,354 Hürth, Germany.

process routes from CO_2 to hydrocarbons and polymers and for Hydrogen H_2 exist that are assessed to be at technology readiness levels (TRL) of TRL 7 and above, with some already at TRL9 (cf. Fig. 15.2b).

Lastly, but significantly, the RCI report [Berg 2023a] also addresses and tackles a prevalent misunderstanding concerning carbon management policies and regulations, particularly the notion that CCU merely postpones emissions and thus fails to contribute to climate change mitigation or the attainment of net-zero targets. Of course, this latter argument is invalidated when considering the visionary objective of achieving an *all-circular* plastics economy, as depicted in Fig. 15.1, wherein CCU plays a pivotal role.

15.1.5 Preparing for an **All-Circular** *and* **Climate-Neutral** *Carbon and Plastics Economy*

A prime precondition to achieve an all-circular economy for plastics is the establishment of novel forms of **cross-sectoral collaborations** between point-source CO_2 emitters (waste incineration, cement, and steel sectors) and other providers (operating perhaps with direct air capture (DAC) technologies), on the one hand, and the polymer and petrochemical industry, on the other. As elaborated further in Sect. 15.3, establishing such collaborations requires adequate policies along with attractive incentive schemes. It should be emphasized again, that a particular advantage of these technologies is the ability to use and build upon the already existing, transnational, and national infrastructure for (chemical) feedstock transport and storage (e.g., pipelines and gas distribution/storage network).

A particular focus of the present section is on any interactions and interdependences of the above technology pillars 1 and 2 (i.e., left side in Fig. 15.1). Thus, an attempt is made to explore the mid/longer-term potential and feasibility of CCU technologies for plastics in relation to the energy systems transition, with the plastics industry not only striving towards establishing an *all-circular* plastics economy but also being pivotal in supporting the **defossilization of the energy system**, thereby enabling an ***all-circular*** **and** ***climate-neutral*** **carbon and plastics economy**. The following topics and related issues are addressed and covered:

- Gauging of the [principal] feasibility of an ***all-circular*** **plastics and carbon economy by around 2050 or sooner** along with comprehending and assessing the necessity and benefits of such strategies and policies on a global and regional level.
- Exploring and emphasizing the essential role of **CCU technologies** to achieve an all-circular plastics and carbon economy from the specific perspective of the polymer industry and, interconnected via cross-sectoral needs for a **circular Carbon Management** (i.e., cCM), the other **HTA industries** of which the polymer industry also is part of.

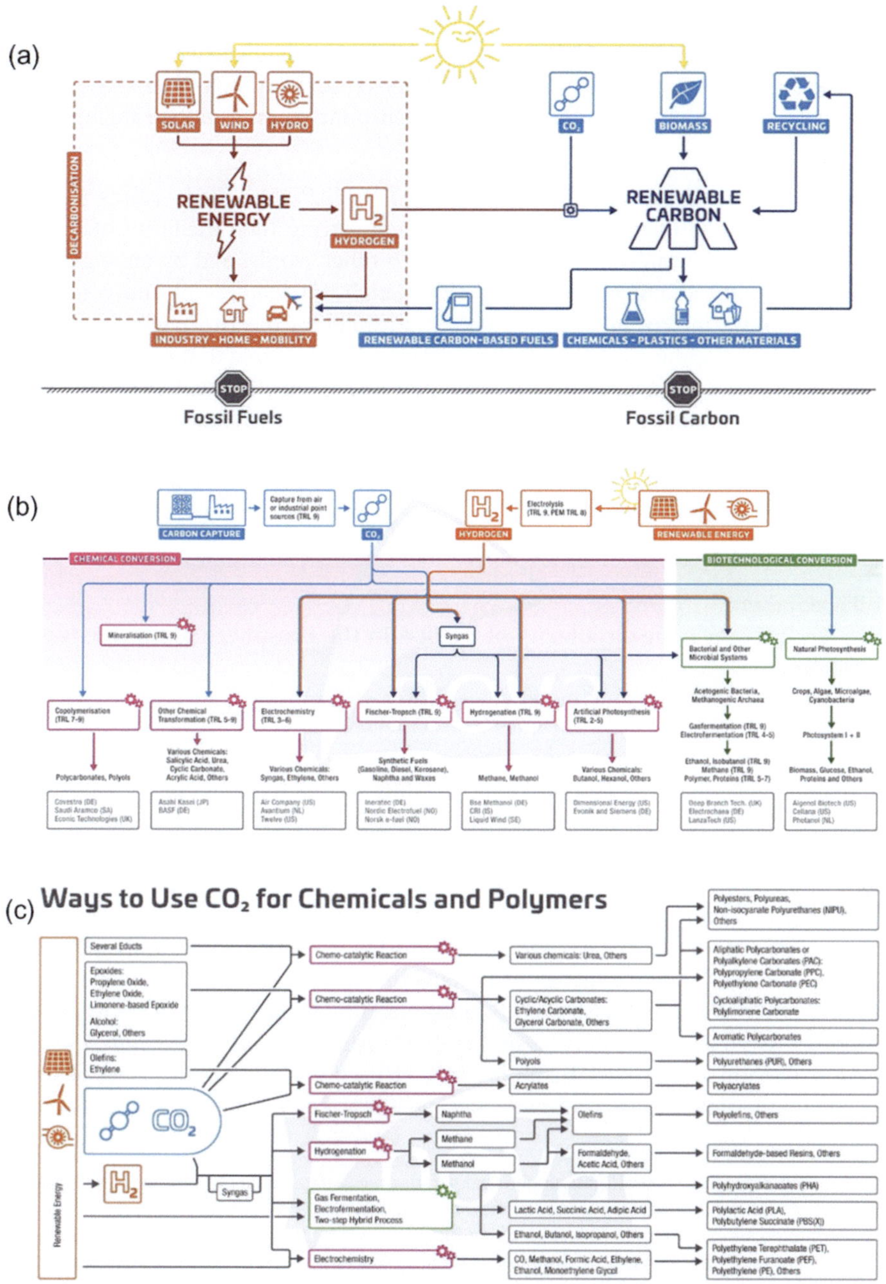

Fig. 15.2 **a-c** Series of illustrations depicting the synergy between the renewable energy system and a CO_2-based renewable carbon system throughout the value creation chain; **a** overview (2021), **b** process and product pathways with current technology readiness levels (TRL) and industrial stakeholders (2023), **c** specific pathways for CO_2-based renewable polymers and other chemicals (2023). (*Source* nova-Institute 2024; courtesy of M. Carus)

- Addressing, substantiating, and emphasizing the need for a **NEXT LEVEL energy and materials strategy** embedded in a comprehensive, consistent and integrated framework of CROSS-SECTIONAL & TRANSNATIONAL energy, climate and circular economy strategies and policies on a global, regional and national level.

Methodologically, a top-down (i.e., global) '*BIG Picture*' approach is applied, reflecting the underlying guiding vision of transferring the circularity of Carbon (C) in the Biosphere into the Technosphere. In other words, and by analogy, what is conceptionally attempted here is to also sketch out features of and conditions for a ***new*** **bioeconomy** in which essential attributes of the Biosphere (renewable energy base & circular use of matter/materials) are transferred to the Technosphere. Most importantly this applies to utilizing renewable energy to establish and drive the technospheric carbon cycle so that technologies such as CCU become a core element of a *new* bioeconomy understanding in which the HTA industries and cross-sectoral carbon streams play a crucial and central role.[8] Hence, Chapter 15 attempts to explore the following guiding vision:

The Guiding VISION:
To transpose the circularity of carbon in the Biosphere to the Technosphere,
with PLASTICS serving as key enabler and motor!

To conclude what was laid out in the heading of this introduction as '*setting the stage*', the section structure proceeds as follows. In Sect. 15.2, an evaluation of the '*fictitious*' CO_2-demand for worldwide future plastics production volumes in an *all-circular* plastics economy together with the CO_2-availability for emission pathways to climate neutrality by 2040 to 2060 is presented. These assessments are based on what are considered robust lower/[medium]/upper-bound scenario assumptions. The multi-faceted policy dimensions for enabling an all-circular plastics and carbon industry are then treated in Sect. 15.3. Subsequently, Sect. 15.4 attempts to provide an answer to the question of the feasibility of a 'sustainable', *climate-neutral* and *all-circular* Carbon and Plastics Economy (c^2CPE) within the Technosphere, posed at the beginning of this section.[9] Finally, the specifics of novel chemical process routes

[8] Concomitantly, and in best alignment with the paradigm of Sustainable Development, main future activities of the *traditional* bioeconomy sector in such a *new* bioeconomy should involve and concentrate on safeguarding and strengthening biosphere integrity (incl. protection of biodiversity etc.), combatting climate change by biomass and humus build-up (thus simultaneously establishing carbon sinks and supporting soil improvement), and - more generally - sustainable cultivation and use of land and forests.

[9] Supplementary remark: as is the case with this section, Sects. 15.2, 15.3 and 15.4 also represent shortened versions of chapters in a report by the authors entitled *"Industrial Carbon Management: Navigating towards a climate-neutral, all-circular industrial Carbon and Plastics Economy"*; available online at: https://carboncyclecircle.at/ and https://www.rethinkconstruction.net/.

for CO_2 as raw material for polymer innovations are covered in Sect. 15.5. Each section addresses relevant background, objectives, and, if applicable, methodological aspects.

15.2 Potential CO_2 Demand and Availability for Plastics: Technosphere Dimensions[10]

Reinhold W. Lang and Stefan P. Schleicher

15.2.1 Approach and Objectives

15.2.1.1 Setting a Proper Timeframe

As alluded to in Sect. 15.1, this section aims to assess and rationalize the fictitious CO_2 demand for a visionary future plastics production with *all-circular* carbon streams, and to juxtapose this assessment with the technospheric CO_2 availability for emission pathways to climate neutrality by 2040 to 2060. This time frame was chosen for three reasons.

First, and foremost, it corresponds to the time frame which is widely believed to be the window for achieving global climate neutrality as mandated by the **UN climate treaty** (Paris Agreement) of 2015 (UN 2015), with corresponding **policies for climate neutrality** derived thereof on a national and regional level (e.g., Austria 2040, Germany 2045, EU 2050, UAE 2050, and China 2060). Second, by assuming that cross-sectoral and circular Carbon Management (**cCM**) along with Carbon Capture and Utilization (**CCU**) technologies driven by the **hard-to-abate (HTA) materials industry** (steel/metals, concrete/cement, plastics, composites and hybrids) will need to play a central role in this transition, it becomes evident that policy makers and industry stakeholders have to act immediately in creating favorable policy and strategic conditions (including regulations) that allow for the broad and scaled implementation of these technologies in the time frame set by climate neutrality ambitions. Third, this time frame is not too far in the future so that still plausible projections and scenario assessments may be made with some confidence, validity, and merit.

[10] The present chapter represents a shortened version of a chapter in a report by the authors entitled *"Industrial Carbon Management: Navigating towards a climate-neutral, all-circular industrial Carbon and Plastics Economy"*; available online at: https://carboncyclecircle.at/ and https://www.rethinkconstruction.net/.

15.2.1.2 Data Generation and Juxtaposition

For the above timeframe setting, an attempt is made to gauge the principal feasibility of achieving an all-circular plastics and carbon economy on a global level by around 2050. This involves matching and converging data from future plastics growth scenarios with global climate goals and corresponding CO_2 emission pathways. Specifically, this covers:

- Assessment of the fictitious CO_2 demand by lower/upper-bound assumptions for plastics growth to 2060, also accounting for lower/upper-bound contributions of mechanical and (conventional) chemical recycling of plastics waste, and assuming an entirely defossilized plastics production.
- Assessment of the technospheric CO_2 availability by lower/upper-bound assumptions for (fossil) CO_2 emission reduction pathways to 2050/2060 corresponding to the Paris Agreement.
- Assessment of the technospheric CO_2 availability long-term by superimposing a 'pragmatic' phase-down versus phase-out scenario of unabated fossil fuels onto the lower/upper-bound (fossil) CO_2 emission pathways corresponding to the Paris Agreement.

Based on the findings, a comparative juxtaposition with data derived from literature on a global, regional (Europe), and national (Germany) level, is then provided in Sect. 15.2.4, including reflections on the top-down vs. bottom-up approaches.

15.2.1.3 Methodology and Quantitative Assumptions

Methodologically, a quantitative and yet comprehensible '*back-of-the-envelop*'-type[11] '*technology wedge*' approach for CCU for plastics is taken, based on clear and traceable future scenario assumptions defined below. This approach, when coupled with common sense and common expert knowledge (i.e., no special expertise is needed, and there are no hidden parameters), was chosen to make our case along with the conclusions and messages more comprehensible and acceptable, and thus more credible and convincing.

Also, while the focus here clearly is on assessing the CCU potential for plastics, which possibly may be one of the largest technology wedges, there may be other technology wedges for handling CO_2 emissions, the ramifications of which are considered but are outside the quantitative assessment scope of this section (e.g., carbon capture and storage (CCS), direct air capture (DAC), bioenergy with carbon capture and storage (BECCS)). This analogously applies to CCU being used in technology wedges beyond plastics such as other chemicals, textiles, pharmaceuticals, food stuff, synthetic fuels and alike.

In performing the scenario assessments for CO_2 demand for plastics and for technospheric CO_2 availability, robust global lower/[medium]/upper-bound scenario

[11] I.e., only a pocket/smartphone calculator is required to perform the underlying calculations.

assumptions for the time horizon 2040–2060 worldwide are made. These include specifically:

- lower/upper-bound plastics growth rates p.a.: 2/4%
- lower/medium/upper-bound non-CCU plastics recycling rates p.a.: 10/30/50%
- lower/upper-bound CO_2 emissions p.a. for pathways acc. to *Paris Agreement*: 1.5/ 2 °C

These lower/upper-bound assumptions are explained and substantiated in Sects. 15.2.2 and 15.2.3, thus providing additional information on the reasoning and justification for the various quantitative values, along with some implications on a global and regional level.

15.2.2 CO_2 Demand for Plastics Based on Plastics Growth Scenarios for 2040 to 2060

To assess the *fictitious* global CO_2 demand for plastics in a time horizon 2040 to 2060, assuming an entirely defossilized plastics production, in a first step, a quantification of the global plastics production volume over this time frame is needed. This is done by using recent data for plastics production worldwide, which is 400.3 Mt in 2022 (PE 2023). Taking this as a starting point, the above-defined plastics growth rates p.a. of 2% (lower-bound) and 4% (upper-bound), respectively, are then applied for the time horizon 2040 to 2060. These growth rate figures are rationalized as follows.

Since the beginning of global industrial production of plastics at around 1950, the annual average growth rates (AAGR) have evolved as follows[12]: 7.8% (1950 to 2020, 7 decades); 5.2% (1970 to 2020, 5 decades); 4.3% (1990 to 2020, 3 decades); 3.7% (2010 to 2020, 1 decade). The lower-bound assumption for AAGR from now to 2060 seems conservatively feasible based on this trend. The upper-bound assumption of 4% is justified when considering that global prosperity and equity considerations, as mandated by the UN Sustainable Development Agenda, require significantly higher growth rates over the next decades particularly in low- and middle-income countries with their still growing majority of the world population.

Based on these assumptions, current and projected annual plastics production data worldwide are listed in Table 15.1, from which the range for the projected annual plastics production volume for the time horizon 2050–2060 of 850 Mt p.a. (lower-bound) to 1,500 Mt p.a. (upper-bound) is deduced.

> **Projected annual plastics production volume by 2050/2060** (worldwide):
> **850 Mt p.a.** (lower-bound) **to 1,500 Mt p.a.** (upper-bound)

---

[12] Own calculations, based on worldwide plastics production data p.a. available from PlasticsEurope (PE 2023).

Table 15.1 Worldwide plastics production scenarios for 2040/2050/2060 based on the plastics production volume of 400.3 Mt in 2022 [PE 2023] and annual average growth rates (AAGR) of 2% (lower-bound) and 4% (upper bound), respectively (*Source* Own presentation)

Annual average growth rate (AAGR) in %	Plastics production in Mt p.a. (worldwide)			
	2022	2040	2050	2060
2 (lower-bound)	400.3	572	697	850
4 (upper-bound)	400.3	811	1,200	(1,777)

These numbers compare quite well with data of the more advanced approach in the Report *Global Plastics Outlook* by OECD, which for business-as-usual projects 1,072 Mt p.a. for plastics in 2060, plus another 159 Mt for polymer fibers (OECD 2022) that are not included in the PlasticsEurope data base (PE 2023). It is worth noting that according to this OECD Report, bio-based plastics, in the absence of new policies, are projected to represent only a negligible share, estimated at around 0.5% of total plastics usage by 2060 (OECD 2022).[13]

Based on the projected annual plastics production volume of 850 to 1,500 Mt p.a. for 2050/2060, in a next step, the projected *fictitious* global CO_2 demand for plastics for this time horizon can now be assessed assuming an entirely defossilized plastics production (i.e., all-circular plastics economy). Data of such calculations are provided in Table 15.2.

While the definitions of terms and assumptions made in the calculations of Table 15.2 are explained in the footnotes, the lower/medium/upper-bound assumptions of 10/30/50% for mechanical and (conventional) chemical recycling rates (footnotes 3–5 in Table 15.2) deserve some commenting. First, these recycling rates embrace the numbers provided in the already cited Report *Global Plastics Outlook* [OECD 2022], which assumes market-shares of recycled plastics for 2060 as 12% for the *Baseline* scenario (business-as-usual), 29% for what is denoted as the *Regional Action* policy scenario, and 41% for what is called the *Global Ambition* policy scenario.

Second, as indicated before in Sect. 15.1, any mechanical and conventional chemical recycling faces eco-efficiency limitations that arise from logistics of plastics waste management (i.e., collection, transport, and separation) as well as waste cleaning operations involving perhaps excessive water and detergent use. This, of course, puts limitations onto certain recycling routes, so that any eco-efficient (sustainable) recovery of post-use plastics should also meet the criteria of a product/process-specific life cycle (LC)/cost–benefit (CB) analysis.

Regarding the aforementioned considerations and assumptions, after 20 years of investigating the eco-efficiency of mechanical recycling of plastics packaging materials through representative case studies conducted by the denkstatt group, estimates

[13] In [OECD 2022] it is also argued, that a shift to biobased plastics only marginally reduces greenhouse gas (GHG) emissions, and only if land use impacts are mitigated. The impact on GHG emissions would remain ambiguous even if policy measures were to succeed to increase the share of biobased plastics to 5% by 2060.

Table 15.2 Worldwide longer-term CO_2 demand (2050/2060) based on the assessment of the projected, correspondent future plastics demand, adjusted and accounting for upper/medium/lower-bound mechanical and chemical recycling rates (*Source* Own presentation)

Projected plastics production in Mt p.a. as of 2050/2060 (worldwide)	Total net amount CO_2 required in Gt p.a	Total gross amount CO_2 required[2] in Gt p.a	Adjusted gross amount CO_2 required 'lower-bound'[3] in Gt p.a	Adjusted gross amount CO_2 required 'medium bound'[4] in Gt p.a	Adjusted gross amount CO_2 required 'upper bound'[5] in Gt p.a
850 (lower-bound)	2.5	3.1	1.6	2.3	2.9
1,500 (upper-bound)	4.4	5.5	2.9	4.0	5.2

[1]Total net amount of CO_2 required, i.e., embedded carbon not considering any CO_2/carbon losses; a plastics-to-carbon conversion factor of 0.72 is assumed (Kähler 2023)
[2]Total gross amount of CO_2 required, considering 28% [Markowitsch 2023] and 3% (Paulik 2024), respectively, of CO_2/carbon losses in the CCU transformation chain and in the subsequent polymerization
[3]Adjusted gross amount of CO_2 required, considering an upper-bound mech./chem. recycling rate of 50% in terms of the annual plastics production volume
[4]Adjusted gross amount of CO_2 required, considering a medium-bound mech./chem. recycling rate of 30% in terms of the annual plastics production volume
[5]Adjusted gross amount of CO_2 required, considering a lower-bound mech./chem. recycling rate of 10% in terms of the annual plastics production volume

for maximum eco-efficient mechanical recycling of plastics packaging are: input-based (waste collection) 35–53%; output-based (after sorting and pre-processing) 31–43% (Pilz 2014).

Providing further insights, and of particular relevance in the context of input/output analysis along the plastics waste management chain, is the following simple assessment. For instance, to meet the EU target of achieving a mechanical recycling rate for plastic packaging waste of at least 55% by 2030 [EU 2018b], the output efficiencies in the three essential process steps—(a) collection, (b) sorting and pre-processing, and (c) conversion & recovery—must fall within the range of 80–85% (!) for each of these steps.[14] Whether this is feasible while maintaining eco-efficiency levels remains to be determined.

Third, an issue of paramount importance, that significantly impacts the widespread application and market acceptance of plastics, relates to their tailored and well-defined, often multifunctional performance characteristics in terms of both processing and applications. This aspect, among others, has been a key success factor for plastics in the past and is anticipated to remain pivotal in the future. However, while next-generation 'virgin' plastics, derived from conventional (fossil) or chemically recycled feedstocks, which includes CO_2, are expected to enter future markets

[14] For example, as the overall output efficiency is obtained by multiplying the efficiencies of each process step, the EU target could be achieved by: $0.82 \times 0.82 \times 0.83 = 0.55 = 55\%$.

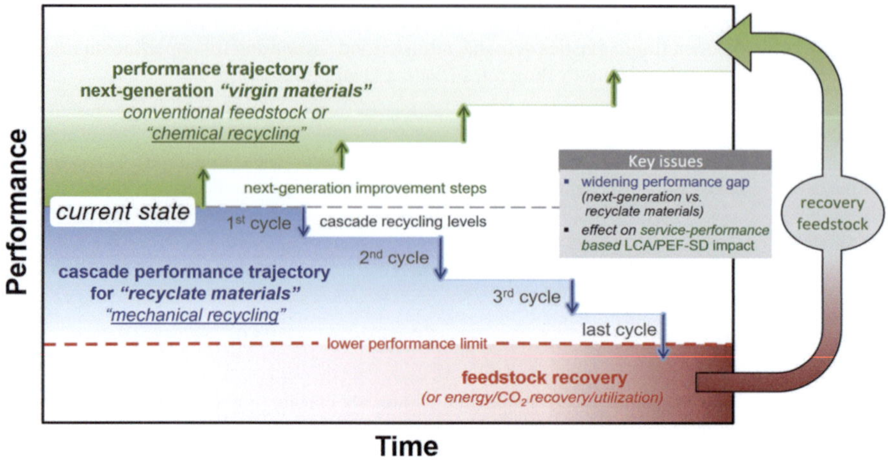

Fig. 15.3 Plastics recovery options and performance trajectories over time for next generation "virgin materials" versus cascade deteriorating mechanical recyclate materials. Considering the service relevant material performance evolution over time, there is a wideneing performance gap between mechanical recyclates versus other and novel forms of next generation "virgin" materials. (*Source* Lang, 2015)

with continuously enhanced performance profiles, a cascading degradation is probable for recyclate materials from pure mechanical recycling, particularly for repeated mechanical recycling cycles over time.[15] As depicted in Fig. 15.3, this further widens the performance gap between virgin-like plastics and mechanical recyclates, and along with it, possibly also exacerbates the life-cycle eco-efficiency gap.

Hence, referring again to the data in Table 15.2 and considering the above three fields of constraint for mechanical and/or conventional chemical recycling, the projected *fictitious* global CO_2 demand for plastics for a *defossilized*, *all-circular* plastics economy by 2050/2060 is thus estimated to be in the range 2 to 4 Gt CO_2 p.a., potentially even up to 5 Gt CO_2 p.a.

Projected fictitious global CO_2 demand for plastics by 2050/2060 (worldwide):

from 2 Gt p.a. (lower-bound) **to 4 (5) Gt p.a.** (upper-bound)

[15] A notable exception may be poly(ethylene terephthalate) (PET) which may be in-situ post-condensated.

15.2.3 CO_2 Availability for Plastics Based on CO_2 Emission Scenarios for 2040 to 2060

While there is currently a substantial surplus of CO_2 in the atmosphere due to historical emissions, which could potentially be accessed almost limitlessly through direct air capture (DAC) also for plastics production in the more distant future, the analysis here focuses on emissions generated within the Technosphere. These emissions are preferably accessed directly and more efficiently from so-called industrial point sources.

To begin, reference is made to Fig. 15.4 [UN COP28 2023a; includes own adaptations] which depicts various global CO_2 net-zero pathways for a global temperature increase to remain in the range of 1.5 to 2 °C, corresponding to the Paris Agreement (UN 2015), using remaining carbon budgets of the Sixth Assessment Report (AR6) of the Intergovernmental Panel on Climate Change (IPCC) (IPCC 2023).

While no global policy agreement on the issue of *phase-out* versus *phase-down* of (unabated) fossil fuels has yet been reached, the most recent UN Climate Summit in Dubai (COP28) has nevertheless brought a remarkable result and signal as to where global agreements may be headed. Of particular relevance to the evaluation here is §28 in the final document of COP28, which is paraphrased in Table 15.3. Most importantly, explicit reference is made under §28e to ***"accelerating zero- and low-emission technologies** … such as **carbon capture and utilization and storage, particularly in hard-to-abate sectors"*** [UN COP28 2023b].

Reflecting on the outcome of COP28 as presented in Table 15.3, and drawing from personal assessments and experience, an attempt is made to envision a possible, perhaps even plausible scenario for a transition to net-zero emissions, which could conceivably emerge in one of the forthcoming COPs. This scenario is illustrated on the right side in Fig. 15.4 and represents our *'pragmatic'* approach to *phase-out*

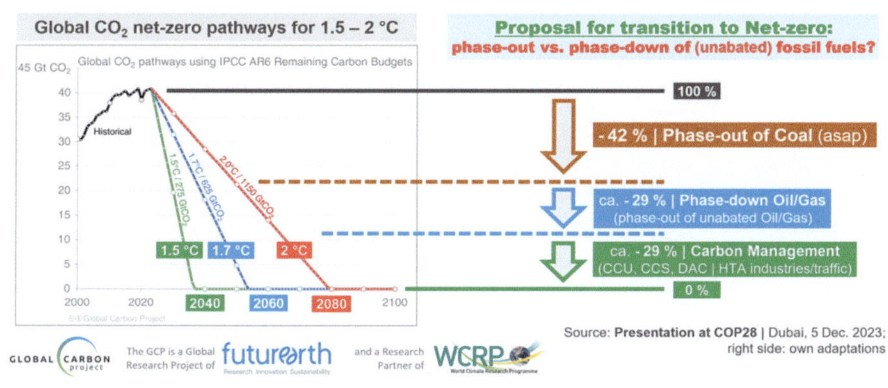

Fig. 15.4 Transforming the energy system. Keeping within a global temperature rise of 1.5–2 °C as mandated by the Paris Agreement (*Source* left side: Global Carbon Budget 2023; published at COP28 on 5 December 2023 (UN COP28 2023a); right side: own adaptations)

Table 15.3 Outcome of the first global stocktake at the COP28 climate summit in Dubai - §28 of the proposal by the President (UN COP28 2023b)

§28. *Further recognizes* the need for deep, rapid and sustained reductions in greenhouse gas emissions in line with 1.5 °C pathways and *calls on* Parties to contribute to the following global efforts, …:
(a) **Tripling renewable energy capacity globally** and doubling the global average annual rate of energy efficiency improvements **by 2030**;
(b) Accelerating efforts towards the **phase-down of unabated coal power**;
(c) Accelerating efforts globally **towards net zero emission energy systems, utilizing zero- and low-carbon fuels well before or by around mid-century**;
(d) **Transitioning away from fossil fuels in energy systems**, in a just, orderly and equitable manner, accelerating action in this critical decade, so as to **achieve net zero by 2050** in keeping with the science;
(e) **Accelerating zero- and low-emission technologies, including**, inter alia, renewables, nuclear, **abatement and removal technologies such as carbon capture and utilization and storage, particularly in hard-to-abate sectors**, and low-carbon hydrogen production;

Remark: especially relevant text passages for this section in bold letters and highlighted by an underline

versus phase-down of (unabated) fossil fuels, considering diverse policy conditions across different world regions.

Starting from the current global emissions of ca. 40 Gt CO_2 p.a., and assuming a near-term phase-out of coal (which contributes to around 40% of these emissions), while simultaneously intensifying efforts to halve the usage of oil and gas, would leave about 30% of the current total emissions for advanced carbon management technologies. This would amount to roughly 12 Gt CO_2 p.a., of which up to one-third, specifically emissions from industrial point-sources of the HTA industries and waste incineration plants, would be required by 2050/2060 for producing plastics. If more ambitious CO_2 reduction pathways were pursued, the share needed for plastics would correspondingly increase. In any case, this assessment demonstrates that an all-circular plastics economy scenario based on CCU may indeed be quite feasible from a CO_2 emissions perspective. To sum these findings up:

Estimated technospheric CO_2 potential for plastics based on emission scenarios

for 2050/2060: 20 to 30% share from up to 12 Gt p.a. worldwide (upper-bound)

15.2.4 Comparative Juxtaposition with Data from Other Sources and Literature

Inquiries into the necessity and potential of utilizing industrial CO_2 emissions as feedstock in the medium and long term have increasingly become a topic of both technological and academic research. To allow for a comparative juxtaposition with our data above, three studies are referenced here: one conducted on a global scale, another at a regional level (focused on Europe), and a third at a national level (i.e., in Germany).

Each of these studies provides valuable additional insights, not only into the future potential for utilizing industrial CO_2 emissions as feedstock at different geographic scales but also in comparing their findings with our own projected scenario data as well as with current overall CO_2 emissions in the respective geographical domains. This allows researchers and policymakers to gain a better understanding of the challenges and opportunities associated with future scenario assessments across different regions and contexts. Simultaneously, it underscores the merits and benefits of the upper/lower-bound approach chosen by us.

15.2.4.1 Global Scale Data by the *Renewable Carbon Initiative*

On a global scale, the *Renewable Carbon Initiative* (RCI) conducted a study to assess the current (2020) and projected future demand (2050) for carbon embedded in products (i.e., chemicals and derived materials including plastics) (Kähler 2023). The results are shown in Fig. 15.5. For projecting the total amount of embedded carbon needed in 2050, an AAGR (or CAGR) of 2.5% was chosen based on the 2020 data. Moreover, of the 550 Mt C reported for 2020, roughly 50% may be allocated to plastics according to our own estimates. As also depicted in Fig. 15.5, the projected total of 1,150 Mt carbon in 2050 is considered to be distributed to 55% from recycling, 25% being CO_2-based and 20% being bio-based. Assuming no losses for definition reasons (i.e., embedded), the 25% CO_2-based embedded carbon for 2050 in this RCI estimate would translate into a net-amount of about 1.1 Gt of CO_2.

Although these projected 1.1 Gt of CO_2 deduced for embedded products in 2050 are in the same order of magnitude as the estimated demand of 2 to 4 Gt of CO_2 p.a. in 2050/2060 by us above (Sect. 15.2.2), some disparities persist. These differences may primarily stem from three factors. First, our calculations account for a carbon loss of approximately 30% during the process of carbon capture to the actual carbon content embedded in CO_2-based plastics. Second, our upper-bound CO_2 demand figure was derived from a significantly higher AAGR value for plastics growth, that is 4%, compared to the average value of 2.5% in the RCI study. Third, and perhaps most significantly, we consider the assumptions regarding the proportion of embedded carbon in products sourced from conventional recycling (55%) and from bio-based materials (20%)—totaling 75%—rather high, perhaps too high. Nevertheless, despite these disparities, it is evident that there is a substantial level of agreement in the

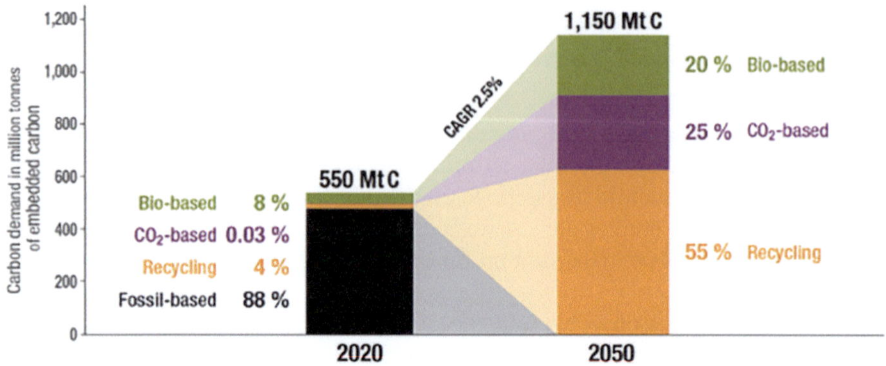

Fig. 15.5 Current (2020) and projected future (2050) global carbon (C) demand for chemicals and derived materials in Mt of embedded carbon ((Kähler 2023); *Source* nova-Institute 2023)

projected data, particularly regarding the overarching conclusion emphasizing the pivotal role of CO_2 for CCU-based plastics in the coming decades.

Considering the above, another way of interpreting the RCI data is as follows. Defining the 2050 embedded carbon data for CO_2-based carbon (25%) as lower-bound and the total carbon demand of 1.150 Mt carbon as fictitious upper-bound and scaling these data again with 30% carbon loss to the corresponding CO_2 emission demand, leads to a lower/upper-bound-range for CO_2 emissions of ca. 1.5–6 Gt CO_2. This range represents approximately 3.8 – 15% of the current annual global emissions of ca. 40 Gt CO_2. In other words, regardless of the actual distribution of sources contributing to the 2050 global carbon demand for chemicals and derived materials, as a good assumption, **some 10% of current global CO_2 emissions will be required**, with a margin of around plus or minus 5%.

15.2.4.2 European Scale Data by the European Cement Association

Turning now to the regional scale of Europe, inquiries into the future carbon demand for the European Union (EU27) were made by the research organization VITO under the auspices of the European Cement Association (CEMBUREAU) (CEMBUREAU 2023, vito 2023). From the perspective of the cement industry, together with the steel industry a major industrial emitter, the key findings of this study for the period spanning 2030 to 2050 are summarized as follows:

- circular use of carbon is needed not only to decarbonize the cement sector but also to aid decarbonization efforts of other sectors (fuels, chemicals, food & beverages, etc.);
- the annual CO_2 demand from these sectors ranges from 250—420 Mt by 2050;
- only 21—69 Mt of CO_2 (i.e., a fraction of the need) are estimated to be available from biogenic sources and from direct air capture (DAC) by 2050;

- CCU is vital for landlocked EU cement kilns, not located next to CO_2 storage sites.

Comparing the annual CO_2 demand range of 250 - 420 Mt by 2050 from the aforementioned sectors to the total CO_2 emissions of the EU27 in 2022, yields a **percentage range of approximately 9–15%**, comparable to what we have derived above for the worldwide case.

15.2.4.3 Data for the Chemical Industry Sector in Germany by VCI and VDI

Especially noteworthy and significant is the report titled *"Chemistry4Climate"*, prepared and compiled by the German Chemical Industry Association (Verband der Chemischen Industrie e.V., VCI) in collaboration with the Association of German Engineers (Verein Deutscher Ingenieure e.V., VDI). This report, funded by the German Federal Ministry for Economic Affairs and Climate Action, delves into specific issues and questions regarding the successful management and implementation of the net-zero transformation of the German chemical industry by 2045, aligning with Germany's national target date for net-zero emissions.

The report analyzes three net-zero scenarios for the German chemical sector: *Scenario 1* focuses on maximizing the use of direct electricity, *Scenario 2* prioritizes hydrogen and power-to-X (PtX) fuels and feedstock, and *Scenario 3* emphasizes secondary feedstock such as plastics waste and biomass. Here are key messages and findings from this report with some quantitative data for the German chemical industry:

- Quote of Robert Habeck ([former]Federal Minister for Economic Affairs and Climate Action, Germany), on the importance of CCU implementation in his opening address: *"Especially Circular Economy and emphasis on CCU vs. CCS offer great potential, which we want to exploit."*
- Huge amounts of renewable electricity and green H_2, but also of CO_2 are needed.
- Depending on the scenario, the CO_2 demand for the German chemical sector in 2045 amounts to 21–52 Mt CO_2 in 2045 (excludes synthetic fuels/e-fuels).
- The estimated point-source CO_2 availability of the HTA industries and of waste incineration in 2045 and subsequently, after all abatement measures are taken, is ca. 50–60 Mt CO_2 p.a.
- Due to (current) biased and skewed EU regulations, favoring bio-based CO_2 over non-bio-based CO_2, strong use conflicts for bio-based CO_2 are expected.
- Depending on the scenario, nominal investments needed range from 25–40 bill. Euro (for comparison, the German GDP in 2022 amounted to ca. 4 trill. Euro).

When comparing the scenario-dependent ranges of CO_2 demand projected for the German chemical sector in 2045 (21–52 Mt CO_2) and the estimated point-source CO_2 availability of the HTA industries and of waste incineration combined, also projected for 2045 (ca. 50–60 Mt CO_2), **percentage ranges of 3.2–8% and 7.6–9.2%**, respectively, are derived relative to the total German CO_2 emissions of 655 Mt in 2022 **just for the chemical industry's CO_2 demand** (excluding synthetic fuels/e-fuels) and the point-source CO_2 availability. This once more underscores the compelling case for a well-designed CCU strategy implemented early on.

15.2.5 Summary and Conclusions

Our assessment, presented in this section, focuses on estimating the CO_2 demand for an *all-circular* plastics production scenario through the implementation of well-dimensioned Carbon Capture and Utilization (CCU) technologies, while accounting for the availability of unabated, or otherwise unabatable, industrial CO_2 emissions to simultaneously achieve climate neutrality by 2040 to 2060. The chosen time-frame aligns with global climate targets and policies, necessitating urgent action from policymakers and industry stakeholders.

By analyzing data from future plastics growth scenarios and global climate goals, we assess the feasibility of an *all-circular* plastics economy based on techno-spheric CO_2. Methodologically, a quantitative upper/lower-bound scenario assessment approach was employed, considering factors such as plastics growth rates, recycling rates, and CO_2 emission reduction pathways. The projected global CO_2 demand for plastics in such a fictitious all-circular plastics economy ranges from 2 to 5 Gt CO_2 per year by 2050/2060, assuming an entirely defossilized plastics production. Additionally, we analyzed the CO_2 availability from industrial point sources, highlighting the need for advanced carbon management technologies in the future.

Comparative juxtaposition with other studies at the global level, and selected regional (EU) and national (Germany) levels, provides additional insights into remaining uncertainties with some of the data and, derived thereof, some of the challenges but also opportunities associated with CCU technologies. Most importantly, our findings emphasize the strong rationale and underscore the urgent need for **implementing a carefully calibrated and well-dimensioned CCU strategy** to match CO_2 availability and CO_2 demand to achieve an all-circular plastics production while transitioning to a climate-neutral future.

A **key takeaway** of our analysis thus is:

> Approximately **10% of current global or regional CO_2 emissions** will be **required by the carbon-dependent industry by around 2050,** with a margin of around plus or minus 5% depending on the region and remaining uncertainties.
>
> The above range **overlaps and aligns well with the remaining CO_2 emissions of sectors considered not abatable** after all decarbonization efforts are applied.

15.3 Enabling an All-Circular Plastics and Carbon Industry: Policy Dimensions[16]

Reinhold W. Lang and Stefan P. Schleicher

15.3.1 Background and Topics Covered

The ambitious goals of global treaties like the *Paris Agreement* and the *UN Sustainable Development Agenda* mandate the need for effective political strategies and harmonized policy frameworks at regional and national levels. Long-term resilience and flexibility are crucial for guiding corporate strategies and investments amidst evolving insights and technological innovations, ensuring a well-directed transformation process. The significance of this for a targeted and accelerated transformation process cannot be overstated.

While being of general merit for global future strategies and policies, these considerations are particularly pertinent for navigating towards the highly ambitious goal of a fully and thus *all-circular* plastics and carbon economy in the coming decades. In other words, such an aspiration clearly necessitates a **well-designed, effective, and coherent policy and regulatory framework at various levels and in multiple dimensions** so that such a transition is encouraged, facilitated, and fostered.

Below we further explore these issues, examining their current status and the need for reform, especially within existing European Union (EU) policies and from a perspective of the **hard-to-abate (HTA) sectors**, encompassing the production of steel, cement, plastics and chemicals, and their waste management. The focus on the HTA industry is justified as follows:

First, the HTA industries, collectively, account for about **25% of total CO_2 emissions globally** and **66% of those from the industrial sector** [UNECE 2022]. These are significant shares indeed, with CO_2 emissions primarily originating from so-called *'point-source'* locations in concentrated form, making them particularly accessible for carbon capture and utilization (CCU). Similar shares exist for Europe and the EU (2024).

Second, in a study on accelerating the low carbon transition [Victor 2019], the progress of ten sectors was analyzed. Among these ten sectors, the **HTA sectors ranked 8th (steel), 9th (cement), and 10th (plastics).**[17] So clearly, there is a great need to focus efforts for climate-compatible strategies and policies particularly in these sectors.

---

[16] The present chapter represents a shortened version of a chapter in a report by the authors entitled *"Industrial Carbon Management: Navigating towards a climate-neutral, all-circular industrial Carbon and Plastics Economy"*; available online at: https://carboncyclecircle.at/ and https://www.rethinkconstruction.net/.

[17] This ranking, after all, reflects why these industries are denoted as "hard-to-abate (HTA)".

Third, to propel the HTA sectors from trailing positions to leading roles in the global transformation endeavor, it requires profound and deep **managerial expertise and competencies**. Therefore, leveraging their exceptional technological innovation capabilities, coupled with their managerial competencies, the HTA sectors are well positioned, and thus even obligated, to assume a leading role in the development and large-scale (global) implementation of CCU technologies through cross-sectoral and, where needed, transnational collaboration.

Hence, in this section, we will emphasize and strive to outline the necessity for an advanced, forward-thinking **'next-level' mindset** to establish a **comprehensive (integrated and transnational), cross-sectoral climate and circular Carbon Management Strategy/System**. This includes a discussion of the major issues of such a strategy and how to navigate the technology innovation challenge from a specific European policy perspective. Special attention will be given to the significance of a collaboration between the plastics sector and the cement and waste management sectors. Lastly, we will address the pressing need for effective carbon pricing and accounting regulations within the EU. We finish by proposing a radical simplification of the EU CO_2 accounting framework.

15.3.2 *Leveraging a* 'Next-Level' *Mindset for Circular Carbon Management*

Although increasing energy efficiency and switching to renewables are viable options for many industrial processes, there remains the seemingly unsurmountable barrier of the HTA industrial sectors, such as steel, cement, chemicals/polymers and waste incineration, to achieving climate neutrality. This barrier essentially emerges from CO_2 emissions stemming from chemical processes (e.g., cement production, incineration of plastics waste) and from the material-inherent carbon content (e.g., plastics). They therefore cannot be abated by simple energy technology switches. A more promising approach, advocated here, involves the transformation and integration of the HTA industries into an *all-circular* industrial carbon framework.

15.3.2.1 Deepening the Understanding of the Challenges for HTA Industrial Sectors

Over the past few years, numerous reports and scientific papers addressing the feasibility and challenges of an industrial approach to a cross-sectoral, circular carbon management (cCM) have become available [Friedmann 2020; Bhardwaj 2021; da Mata Costa 2021; Friedmann 2021; Meys 2021; UNECE 2022]. An illustration of a so-called concept-based solution for an industrial-technological circular carbon cluster is shown in Fig. 15.6 (open source at UNECE 2022).

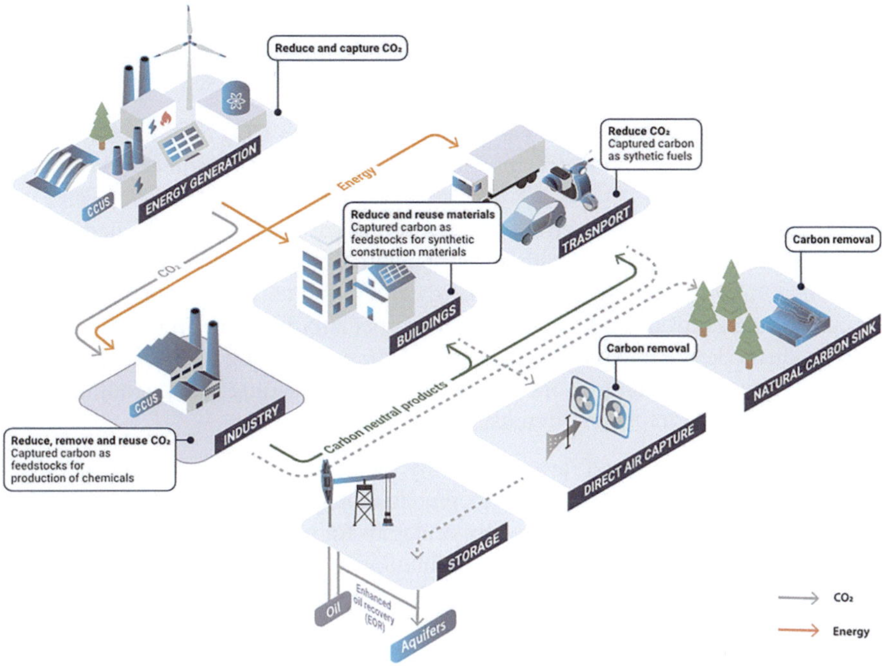

Fig. 15.6 Overall illustration of a circular carbon economy (CCE) with technology-based solutions for carbon capture, utilization and/or storage (CCUS) embedded in other sectoral clusters. (*Source* UNECE 2022)

Also recently launched to address the pressing need for sustainable industrial practices and corresponding carbon management strategies and policies were the *Renewable Carbon Initiative* (RCI) with its "*Manifesto for the next European Commission (2024–2029)*"[18] and "*The Antwerp Declaration for a European Industrial Deal*".[19] Significant new impulses and additional momentum for CCU and cCM in industry and politics are anticipated from these initiatives. This already has become visible in the two recent European Commission Communications.[20,21]

Reflecting on the content of these publications and initiatives, the following evaluation emerges with regard to gaining a better understanding of the necessary innovations and attributes for an *all-circular* carbon and plastics economy.

[18] https://renewable-carbon.eu/news/sign-the-rci-manifesto-for-the-next-european-commission-2024-2029/.

[19] https://antwerp-declaration.eu/.

[20] *"Securing our future - Europe's 2040 climate target and path to climate neutrality by 2050 building a sustainable, just and prosperous society"*, COM(2024) 63 final; https://climate.ec.europa.eu/document/download/2ccd7710-5fc3-420f-aeb8-9a3af271f970_en.

[21] *"Towards an ambitious Industrial Carbon Management for the EU"*, COM(2024) 62 final; https://eur-lex.europa.eu/legal-content/EN/TXT/PDF/?uri=CELEX:52024DC0062.

(1) Establishing a comprehensive carbon management strategy that serves the complex needs of the HTA industrial sectors requires the integration of **three essential attributes**:

- *"cross-sectoral"* refers to the carbon flow, where the CO_2 emissions from one sector (e.g., cement industry) are used as feedstock in another sector (e.g., chemical sector);
- *"circular"* reflects the closed-loop circularity of carbon flows in repeated process and product cycles, ideally with no carbon emissions released to the atmosphere;
- *"transnational"* encompasses any energy and/or matter flows across borders and regions (e.g., Europe and MENA region), especially including renewable energy or products derived thereof. This is crucial for establishing an optimized, scaled and all-encompassing network needed for CCU technologies, depending on the location and geographical circumstances.

(2) For a circular plastics and carbon economy, **vast amounts of green hydrogen are needed**. Hence, existing or under-development strategies for green hydrogen and its infrastructure must be complemented by strategies for other substances such as green methane, methanol or ethanol via CCU technologies and for ammonia, respectively. This is also necessitated to ensure maximizing the use of existing infrastructure for transport and storage.

From a **European perspective**, the most pressing concerns for achieving comprehensive plastics and carbon management within a circular framework are:

- **Availability of "green" electricity and hydrogen:** the enormous quantities of these energy sources and/or feedstock types are unlikely to become available on European or national scales. This underscores the urgent need for the establishment of transcontinental supply chains and for secure, long-term agreements.
- **Reuse of captured carbon:** any strategies for utilizing captured carbon as feedstock for products versus its use for synfuels in transportation and mobility, whether from techno-industrial activities or from biomass, must be carefully aligned and balanced.
- **Market integration of transition actions:** current pricing mechanisms, particularly regarding hydrocarbons derived from CO_2 using green hydrogen, result in increased production and product costs. Addressing this issue requires innovative market mechanisms to mitigate these costs effectively, such as markets for differences and alike.

15.3.2.2 The Pivotal Roles of the Cement Sector and the Waste Incineration Sector

In the transition of the HTA sectors to climate neutrality and a circular carbon economy, pivotal roles must be assumed by (1) the plastics and (petro)chemical industry, (2) the cement industry, and (3) the waste incineration sector. This is based on the following reasons:

- As their reliance on carbon is inherent to their processes and/or products, these sectors cannot be decarbonized by simple energy technology switches.
- In numerous countries, especially in Europe, significant cross-sector value chain activities are already established between these sectors. These include the use of significant amounts of plastics waste as refuse-derived fuels for energy recovery either in cement production or in waste incineration plants (e.g., for heating of buildings).
- **Cement** is an extensively used construction material. The global demand for cement is expected to rise 12–23% by 2050 (UNECE 2022).
- The cement industry is responsible for about 6–7% of total global emissions (ca. 2.5 Gt CO_2 p.a.) [UNECE 2022]. In the EU, the cement industry makes up about 9% of the EU ETS (European Emission Trading System)].[22] Conversely, roughly two-thirds of the emissions in cement production are process reaction related (i.e., limestone calcination) and are therefore essentially un-abatable.
- In comparison to the cement sector, the (municipal) waste incineration sector contributes significantly less to total CO_2 emissions (ca. 1% of total CO_2 emissions).
- Despite this, waste incineration, often coupled with energy recovery, currently plays a decisive role in managing plastics waste in Europe. Referring to data for 2020 for the EU27 + 3, the breakdown of post-consumer plastics waste treatment routes was as follows (PE 2022): 42% incineration with energy recovery, 35% recycling, and 23% landfill. Notably, in all countries with landfill restrictions (9 out of 30), the proportion of incineration with energy recovery exceeds 50%, surpassing even 70% in some countries.
- Simultaneously, as pointed out in Sect. 15.2, there is an ecological cost–benefit limit for mechanical recycling of plastic waste estimated at max. 50% (Pilz 2014).
- Considering the evolving landscape, particularly the anticipated implementation of landfill restrictions in more European and non-European countries, the importance of **plastics waste treatment by incineration is expected to persist in Europe and even rise globally**. Consequently, we project that the significance of solid waste incineration, encompassing all forms of plastics containing waste, will continue to escalate as a viable and attractive waste management solution, especially when further **bolstered by the integration of CCU**.

Clearly, on a regional and global scale, **cross-sectoral collaborations** especially between

[22] While both numbers for the cement industry currently are slightly less than the emissions of the steel industry, different technological approaches exist and are pursued to significantly decarbonize steel production Thus, the steel industry currently seems to shift towards an electrical steel technology in a first transition phase and also intensively explores green hydrogen-based technologies. And yet, since the production of steel also requires a substantial amount of carbon (C), the ultimate technology mix remains open and may also include CCU-based circular carbon management technologies.

the cement industry, the waste incineration and the plastics sector are pivotal
for establishing a ***sustainable, all-circular*** **carbon and plastics economy**.

Signifying an enormous leap in industrial transformation and exemplifying the commitment of the cement industry to cross-sectoral cCM, the **Carbon2ProductAustria (C2PAT)** project stands out. Illustrated in Fig. 15.7, this project was conceived through collaboration among a consortium comprising Lafarge (now Holcim Österreich; cement), Borealis (polyolefins), OMV (fossil fuels, petrochemicals), and VERBUND (renewable energy supply and green hydrogen), supported by several Austrian universities. The remarkable objective of C2PAT is to establish cross-sectoral, circular carbon value chain pathways at full-plant scale (i.e., 750 kt CO_2 annually). Thus, C2PAT aims to capture industrial CO_2 emissions from cement production and convert these with green hydrogen into feedstock for renewable-based chemicals and plastics. By integrating CO_2 emissions as raw material, C2PAT strives to promote climate neutrality, circularity, and innovation within the industry. Despite its groundbreaking potential, the project is currently on hold due to funding constraints from the EU Innovation Fund and as a result of EU regulatory challenges surrounding carbon accounting and pricing of CCU initiatives.

Finally, complementing the efforts by the cement sector, a similar interest in using CCU technologies in waste incineration processes has emerged more recently (see [WE 2024]).

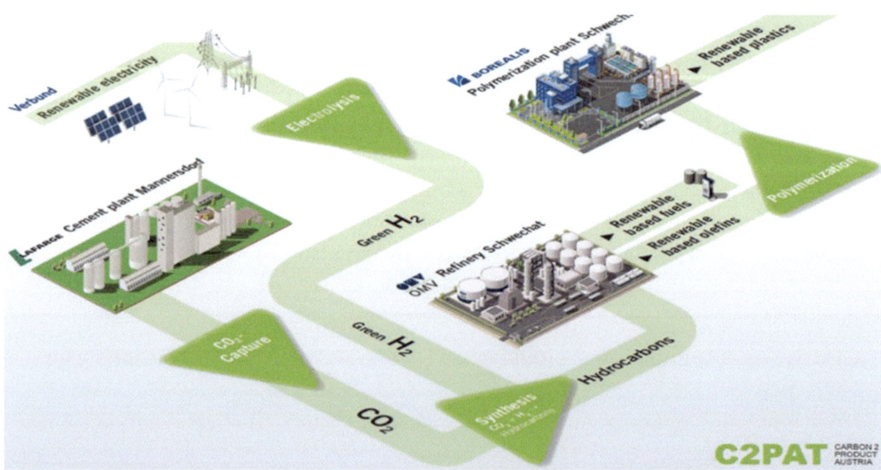

Fig. 15.7 The C2PAT concept for cross-sectoral, circular carbon management. (*Source* C2PAT consortium, June 2021)

15.3.3 Carbon Management Market Initiatives and Carbon Policy Regulations

15.3.3.1 International Carbon Management Market Initiatives and Accounting Schemes

To facilitate and accelerate the deployment of cCM solutions at scale necessarily requires an incentivizing policy and regulatory framework. On the international level (non-EU), two recent carbon market initiatives targeting HTA sectors' CO_2 emissions merit mention.

One is the **CCS+** initiative,[23] formed as an alliance of organizations in June 2021 and dealing primarily with carbon capture and storage (CCS), carbon capture and utilization (CCU), and carbon dioxide removal (CDR) solutions in various forms. CCS+ aims at "*unlocking the potential of certified* CO_2 *capture, utilization, removal and storage solutions around the world*". The focus is on:

- Advancing carbon accounting and monetization schemes for a range of carbon management solutions, including CCS and CCU from point sources as well as other CDR options via direct air capture (DAC) or photosynthesis (biomass).
- Providing a comprehensive modular framework that allows for a life-cycle analysis-based assessment of the environmental impacts of various carbon capture routes.
- Establishing a methodological framework for verification and certification of emission reductions and/or removals from carbon capture technologies and projects.

The activities by CCS + may be used for informed, evidence-based government policies that incentivize investments in environmentally sound and economically viable technologies.

The other recent initiative, dated 2022, is the US Inflation Reduction Act (IRA) [US IRA 2022]. It is a significant milestone in US climate policy and is set to be a game-changer for implementing carbon capture technologies at all scales, particularly attractive for large-scale undertakings. The legislation package of the US IRA provides attractive incentives via tax credits for carbon capture, utilization and/or storage (CCUS). They will be compared to other carbon prices and price estimates for CO_2 in the next section.

15.3.3.2 European Carbon Management Market Initiatives and Accounting Schemes

A cornerstone for incentivizing greenhouse gas (GHG) reductions in the industry and the power sector via carbon pricing is the **EU Emissions Trading System** (EU

---

[23] https://www.ccsplus.org.

ETS),[24] the world's first and biggest carbon market. Established in 2005, the EU ETS operates on the *'cap and trade'* principle, supposed to offer a mechanism for reducing GHGs cost-effectively.[25]

Financed via revenues from the EU ETS, the **EU Innovation Fund** (EUIF)[26] adds a new mechanism to the EU ETS by extending the original 'cap and trade' scheme with a funding instrument for targeted low-carbon technology innovations. Contributions from the fund to the projects are based on calls for competitive bidding.

Overall, compared to the US IRA, the policy mix supporting and incentivizing CCUS technologies in the EU is far more fragmented and complex [Thielgas 2022]. Three main deficits compared to US policy options, such as the US IRA, may be identified:

- While the importance of CCU in contributing to climate and circular economy strategies and targets is acknowledged, CCU is generally considered a medium- to long-term option. As of now, there is a clear lack of effective short-term incentive schemes for CCU, with existing schemes showing a preference for CCS over CCU.
- The crucial role of CCU and circular carbon streams for achieving climate neutrality of the HTA industries is underrated, often not recognized, and even misjudged. While establishing a cross-sectoral carbon economy is considered essential, it is not adequately addressed in current policies.
- Regarding procedural matters, access to existing incentive and funding schemes in the EU, especially for CCU, is overly complex and frequently prohibitive.

15.3.3.3 Carbon Pricing in Context: In Search of an Adequate and Effective CO_2 Price?

Carbon pricing stands out as a pivotal tool in combatting climate change, but its role also expands to policies and regulations to incentivize the adoption of CCU technologies across sectors. Table 15.4 offers a comparison of proposed carbon pricing strategies from academic research, aimed at achieving net-zero climate goals and also accounting for carbon emission damages (using the social cost of carbon (SCC) approach), alongside existing carbon pricing schemes in the US and the EU. While much could be said about each of the data in Table 15.4 (see also comments in footnotes), for the purpose of this section and to maintain brevity, they primarily serve as a basis for comparison to what is assessed by industry experts to be a price (cost) range for CO_2 to warrant a viable business case for cross-sectoral carbon use from

[24] https://climate.ec.europa.eu/eu-action/eu-emissions-trading-system-eu-ets_en.

[25] There are several concerns about the effectiveness and impact of the EU ETS. First, the caps on emissions are not stringent enough to meet the ambitious climate goals set by the EU. Second, the price signals of the carbon market are subject to volatility and can be influenced by external factors such as economic downturns or policy changes, thus leading to unpredictability in carbon pricing. Third, it is therefore difficult for businesses to plan their investments in emission reduction technologies accordingly.

[26] https://climate.ec.europa.eu/eu-action/eu-funding-climate-action/innovation-fund_en.

point-sources via CCU to produce plastics. This cost range is estimated at 150–180 Euro/t of CO_2.

Academic research provides a range of estimates for carbon pricing approaches. For example, the near-term to net-zero (NT2NZ) approach suggests a 2030 price range of 124 $\$_{2018}$/t of CO_2, aligning with the lower end of experts' estimates for CCU at 150 €/t of CO_2. In contrast, the SCC approach, incorporating social and environmental costs, yields a higher estimate of 190 $\$_{2020}$/t of CO_2. Additionally, there is a potential for SCC to affect global climate justice financing schemes like the Adaptation and Mitigation Fund or the Loss and Damage Fund, now hotly debated at UN climate summits. Comparing these estimates to expected CCU costs provides further insight into the financial viability of CCU and circular carbon management (cCM).

Then there is the EU Plastics Waste Levy (PWL) for non-recycled plastic packaging waste of 80 cents/kg (equivalent to 800 €/t). Introduced by the EU from 2021 onwards to fill the Brexit revenue gap, member states have the option to finance these funds either from their budgets (i.e., funded by all taxpayers), or by passing the costs on to manufacturers, retailers, consumers, etc. For packaging plastics waste that is incinerated and perhaps used for CCU (as discussed in Sect. 15.3.2.2), the PWL of 800 €/t can be stoichiometrically translated to a *fictitious* CO_2 price of ca. 255 €/t, the value provided in Table 15.4. This number significantly exceeds even the CO_2 price of the SCC. Hence, this may serve as just one example of how policies of climate relevance may inadvertently create a bias and a non-level playing field, as perhaps overlooked by lawmakers in this context (Lang 2020).

Table 15.4 Range of *'carbon pricing'* according to academic research and US and EU policies

Academic Research[1]	United States (US)	European Union (EU)
To reach net-zero by 2050[2]: NT2NZ price range in 2025: **34-64 \$/t CO_2** NT2NZ price range in 2030: **77-124 \$/t CO_2** **Social cost of carbon (SCC)**[3]: **190 \$/t CO_2**	**US Inflation Reduction Plan (IRA): tax incentives for CCUS**[4] · from point sources (industry): **85 \$/t CO_2** · from the atmosphere (DAC): **180 \$/t CO_2**	**EU ETS system** (direct, rather fluctuating) History 01/2022 to 02/2024[5] **54-105 €/t CO_2** **EU Plastics Waste Levy**[6] (indirect, fictitious but fixed) **255 €/t CO_2**

[1] The support by Simon Lang (PhD Candidate, Yale School of the Environment) in collecting these data is acknowledged
[2] Price ranges for CO_2 in US$\$_{2018}$ for the US in 2025 and in 2030, respectively, based on a near-term to net-zero (NT2NZ) approach, with a net-zero by 2050, as an alternative to the social cost of carbon (SCC) approach [Kaufmann 2020]
[3] Social cost of carbon (SCC) measures the monetized value of the damages to society caused by an incremental metric ton of CO_2 emissions. Data provided here are in US$\$_{2020}$ and represent most recent central estimates by the US Environmental Protection Agency (EPA) based on a near-term discount rate of 2% [EPA 2023]
[4] US Inflation Reduction Act (IRA) [US IRA 2022]
[5] Here lowest and highest value from 01/2022 to 02/2024 [Source: https://tradingeconomics.com/commodity/carbon]
[6] Fictitious CO_2 price derive from EU Plastics Waste Levy when waste is incinerated [Lang 2020]

For comparison also of interest, according to a recent study by the Boston Consulting Group (BCG) on carbon dioxide removal (CDR) technologies (Azarabadi 2023), current direct air capture (DAC) costs range from 900–1,700 \$/t of CO_2. The study suggests DAC costs could decrease by 90% in the coming decades, emphasizing its potential in achieving net-zero goals. However, carbon capture and utilization (CCU) from point sources clearly remains preferable for now.

In conclusion, the widespread adoption of circular carbon technologies at scale requires urgent and radical policy adaptations and reforms, especially within the EU's carbon accounting regulations, such as those governing the EU ETS and innovation funding schemes. These adjustments are vital for the effective implementation of cross-sectoral climate and circular carbon economy strategies making use of CO_2 from point sources as a new feedstock.

Undoubtedly, for **widespread circular carbon technology adoption at scale**, **urgent policy reforms are needed**, particularly within **EU regulations**.

15.3.4 A Framework of CO_2 Accounting: A Conceptual Proposal

Considering the complexities and intricacies of existing CO_2 accounting schemes in the EU, there is a pressing need for a more comprehensive, straightforward, and less bureaucratic approach. Hence, a **radically simplified CO_2 accounting scheme** to address these challenges and encompass all industrial carbon flows is proposed below. It aims to lay the groundwork for a more effective scheme but requires significant revisions and reforms of the current system.

Our proposal for a **comprehensive Carbon Accounting (cCA)** framework aims to integrate all industrial carbon flows, particularly also circular flows as a crucial component of cross-sectoral cCM, which are not adequately dealt with by existing methodologies. The cCA framework accurately represents the structure, activities, and responsibilities inherent in cCM. Figure 15.8 provides an overview of carbon flows in such a system and the corresponding cCA scheme.

Beginning with the primary emissions from the use of fossil fuels and from chemical processes, the proportion of emissions that are captured will progressively increase. These captured emissions will be either temporarily (i.e., intermediate storage, CCiS, for later use by CCU) or permanently stored through CCS techniques. Over time, as industries progressively decarbonize, an increasing portion of captured emissions will be utilized via CCU in the creation of products or in processes involving and using synthetic carbon, such as plastics, but also energy carriers like synthetic fuels. Some of these products, like plastics, will undergo renewed recycling, for instance, in cement or incineration plants equipped with CCU. Furthermore, a

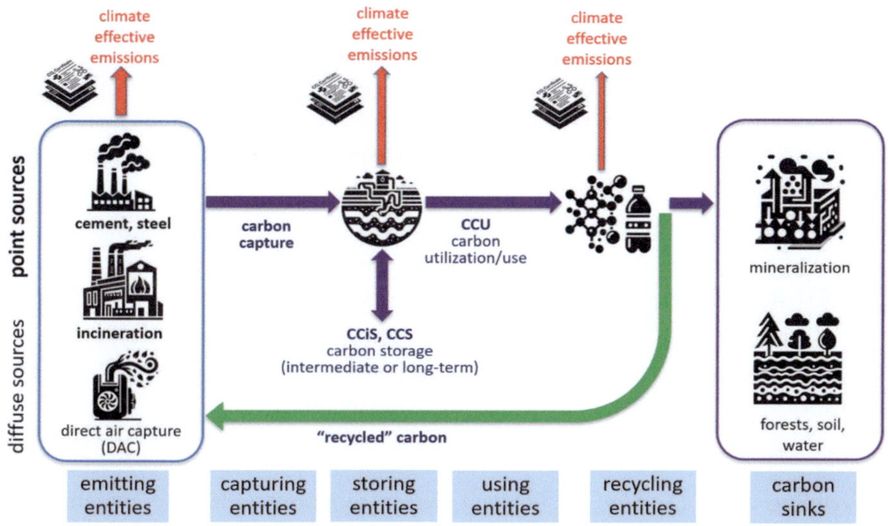

Fig. 15.8 Carbon flows for a comprehensive Carbon Accounting (cCA) including the option for cross-sectoral, circular Carbon Management (cCM); arrow color code: red - climate effective emissions, purple - captured carbon, green - recycled carbon. (*Source* Authors)

segment of the captured emissions may be directed towards carbon sinks, for example, through mineralization in cement structures or to enhance the humus content of soil.

The proposed cCA scheme comprises **three essential components**:

- *Carbon Flows*: the scheme distinguishes between *climate-effective carbon flows* (i.e., emissions into the atmosphere), originating, e.g., from primary emissions and leaks, and *non-climate-effective carbon flows*, which include captured, utilized (CCU) or stored (CCS) emissions.
- *Carbon Registries*: two types of registries are established to mirror the two types of carbon flows; one for total emissions collected in the EU Transaction Log (EUTL), the other for captured emissions documented in the additional registry for captured carbon (EUCC).
- *Responsibilities of Entities*: various entities are responsible for different stages of monitoring, documenting (in the registries) and reporting of carbon flows, such as, e.g., emissions from installations, capturing, storing, using, recycling, or transferring to sinks. Additionally, installations participating in the EU ETS must annually surrender carbon certificates corresponding to their emissions, whether captured or not, aligning with their climate impact.

Thus, the proposed cCA approach not only spans various sectors and activities, it also radically simplifies carbon accounting procedures while remaining flexible and scalable, accommodating factors such as the share of carbon capture, use, recycling, and the duration of carbon storage. Additionally, when combined with appropriate carbon pricing schemes, the system integrates incentives for innovations and the rapid scaling of technologies so urgently needed, including carbon capture, storage, utilization, and fully circular carbon recycling processes. Such pricing mechanisms may encompass subsidies, tax credits, research grants, procurement incentives, and alike. Overall, these features of the proposal enable comprehensive management and tracking of carbon emissions and flows while fostering environmentally beneficial technological advancements.

15.3.5 Key Takeaways and Policy Tasks

Rather than with a traditional summary, we conclude this section with some key takeaways and a 'wish-list' of tasks for policy adaptations.

Our **key takeaway messages** include:

- Enabling an *all-circular* plastics and carbon economy within a comprehensive (integrated and transnational), cross-sectoral climate and circular carbon management strategy and system demands a mindset grounded in understanding the HTA industrial sectors and their intricate policy interactions, both on a regional and global (transnational) scale.
- Cross-sectoral, circular Carbon Management (cCM) is key to enabling an *all-circular* plastics and carbon industry. Especially pivotal are collaborations between the cement industry, the waste incineration, and the plastics sector.

Our **policy task catalogue** ('wish-list') encompasses:

- Facilitating and fostering the *all-circular* plastics and carbon transition necessitates a well-designed, effective, and coherent policy and regulatory framework across multiple levels and dimensions.
- For widescale adoption of circular carbon technologies, urgent policy reforms are necessary, notably in carbon accounting and pricing and especially within the EU.
- The EU Net-Zero Industry Act should also recognize carbon capture and utilization (CCU) and cCM as essential technology fields, along with carbon capture and storage (CCS).
- The current EU ETS Directive lacks rewards for negative emissions; also, technological non-biogenic C/CO2 emissions from HTA industries and alike, used in CCU and in cross-sectoral cCM schemes, are to be treated equal to biogenic C/ CO_2 emissions.
- The EU Circular Economy Action Plan (CEAP) needs to be adapted to recognize CCU and cCM as viable circular economy technology routes especially also for plastics.
- Incineration of plastics (and of organic matter of non-biogenic feedstock) coupled with thermal energy use and CCU, driven by renewable energy, should be recognized as a viable 'chemical' recycling route in EU and national policies. Clearly, starting with the process of waste incineration followed by capture, reduction, and hydrogenation of CO_2 to the synthesis of monomers and new polymers, all

steps along this process chain represent 'chemical' processes to justify such a categorization. This again, would signify a major technology-open step towards a more simplified and progressive legislation for carbon accounting.

Finally, based on the attributes and features of comprehensive industrial carbon management with an aligned carbon accounting scheme described in this section, the following nomenclature is proposed to indicate alignment with climate and circular economy objectives:

For processes and activities: **circular Carbon Management (cCM)**
[or, in terms of strategies: circular Carbon Management Strategy (cCMS)]
For carbon accounting regulations: **comprehensive Carbon Accounting (cCA)**

15.4 Challenges and Perspectives—*'Making a Way out of no Way'*[27]

Reinhold W. Lang and Stefan P. Schleicher

Having addressed the potential CO_2 demand for plastics and explored the technospheric CO_2 availability, and after having examined the policy requirements for enabling an *all-circular* carbon and plastics industry, we now revisit the overarching question posed at the beginning of chapter 15:

Is the establishment of a climate-neutral, 'all-circular' Carbon and Plastics Economy (c^2CPE) Within the Technosphere of the human enterprise both desirable and feasible?

When accepting the principal desirability of such a scenario for reasons related to mitigating climate change and ensuring prosperity and equity-motivated Sustainable Development, the question may perhaps be formulated more simply as: **Can such a c^2CPE scenario be achieved, and if so, how?** In attempting to provide an answer, we first reflect on two studies of particular relevance to this question from a plastics perspective.

First, we refer again to OECD's *Global Plastics Outlook* report (OECD 2022), previously cited in Sect. 15.2. In addition to projections of global and regional plastics use by 2060, this report also addresses forecasts concerning plastics waste and its management, as well as issues of plastics pollution (i.e., mismanaged plastics waste, and plastics leakage to the environment incl. micro-plastics), again on a global and

[27] The present chapter represents a shortened version of a chapter in a report by the authors entitled *"Industrial Carbon Management: Navigating towards a climate-neutral, all-circular industrial Carbon and Plastics Economy"*; available online at: https://carboncyclecircle.at/ and https://www.rethinkconstruction.net/.

regional scale. Three scenarios, outlining a spectrum of policy responses, denoted as the *Baseline* scenario (representing business-as-usual), *Regional Action* policy scenario, and *Global Ambition* policy scenarios are covered. They signify a progression in ambition toward addressing what is widely recognized as a significant environmental challenge with far-reaching implications for ecosystems and human health. Projections indicate that, by 2060, the shares of recycled plastics (excluding CCU-based plastics, which are not explicitly considered) in total plastics consumption for these scenarios are estimated to be 12%, 29%, and 41%, respectively.

Some key results from this OECD report concerning the consumption and waste management of plastics are depicted in Fig. 5.9, which compares global projections for 2060 for these three scenarios with data for 2019. In the *Baseline* scenario, the challenges associated with mismanagement and leakage of plastics waste worsen considerably by 2060 compared to 2019. However, even in the other two scenarios, these issues remain significant as they still continue to accumulate over time.

Second, a significant, comprehensive, and in-depth analysis titled *'Towards circular plastics within planetary boundaries'* was conducted by a team of scientists from RWTH Aachen University, ETH Zurich, and the University of California, Santa Barbara (Bachmann 2023). The study integrates projected plastics production (consumption) data and life cycle analysis (LCA) data and links these to the framework of planetary boundaries, originally proposed by (Rockström 2009; Steffen 2015).

The planetary boundary framework encompasses nine Earth system dimensions, such as for example climate change, biosphere integrity, biogeochemical flows of nitrogen and phosphorus, and aerosol loading, collectively defining the planetary *'safe operating space'* (SOS). Through a sophisticated model accounting for mass and energy flows throughout the plastics supply chain, sustainability thresholds are established.

One result of the model investigations of this research work is depicted in Fig. 15.9. Here we opt to present the findings of this study in the authors' own words (Bachmann 2023):

> "… we show that even a circular, climate-optimal plastics industry combining current recycling technologies with biomass utilization transgresses sustainability thresholds by up to four times. However, **improving recycling technologies and recycling rates up to at least 75 % in combination with biomass and CO_2 utilization in plastics production** can lead to a scenario in which **plastics comply with their assigned safe operating space in 2030**. … **even enhanced recycling cannot cope with the growth in plastics demand predicted until 2050.** Therefore, achieving absolute sustainability of plastics requires a fundamental change in our methods of both producing and using plastics."

Some remarks are warranted here. First, to achieve an overall recycling rate of 75%, as required to remain within the SOS planetary boundaries, the output efficiencies in the three essential process steps - (a) collection, (b) sorting and pre-processing, and (c) conversion & recovery (mechanically and/or chemically) - must now be above 90% (!) for each of these steps.[28] This is clearly not feasible globally by 2030. Second,

[28] The total output efficiency calculates by multiplying the efficiencies of each process step: $0.9^3 = 0.73 = 73\%$.

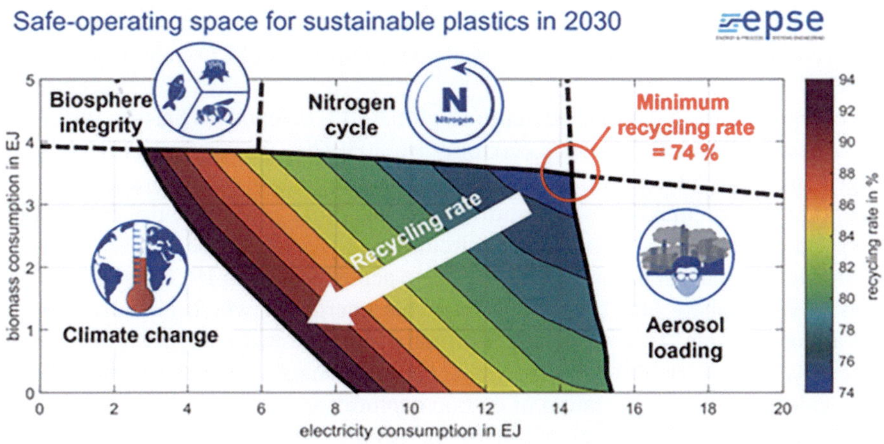

Fig. 15.9 Absolute environmental sustainability of plastics as a function of renewable resource consumption and recycling rates. The planetary footprints are calculated for production volumes in 2030. The plastics industry's assigned share of the SOS is limited to 1.1% for all Earth-system processes. The colored area represents the zone of sustainable plastics in which all planetary footprints are below the assigned share of SOS. Each point in the colored zone represents a plastics industry that consumes wind power (x-axis) and biomass (y-axis). Plastics are considered unsustainable in white areas. Labels of the lines surrounding the solution space refer to the planetary footprint exceeded first when leaving the solution space. Bachmann 2023; this illustration courtesy of A. Bardow, ETH Zurich (Bardow 2024)

apparently, **no scenario could be devised to cope with the increasing amounts of plastics produced and used as of 2050**.

Of course, the purely ecological planetary boundaries approach leaves aside the human dimension inherent in the anthropocentric concept underlying the UN Sustainable Development Agenda 2030 (and beyond). In this context, it is worthwhile to emphasize that the underlying assumptions in our upper-bound assessments of global plastics demand as of 2040 and beyond in Sect. 15.2 are consistent with the equity criterion of Sustainable Development. This criterion aims to enhance global prosperity while simultaneously achieving a more balanced distribution across different world regions. In fact, the pursuit of the UN Sustainable Development Goals (SDGs), with the aspiration of *"a world we want for all, leaving no one behind"* (see footnote [3] in Sect. 15.1) may even be interpreted as necessitating the *"upper-bound"* benchmark of our plastics consumption analysis, thereby providing a more plausible measure for realizing this objective.

And yet, the conclusions drawn from the two studies cited and discussed above paint rather bleak perspectives. This brings us back to the dilemma alluded to in the introductory Sect. 15.1, referencing the African American expression, *'Making a Way Out of No Way'*, also used by the civil rights movement.[29] Faced with this dilemma, we propose drawing once more an analogy to the biosphere, particularly

[29] https://nmaahc.si.edu/explore/exhibitions/making-way-out-no-way.

its natural carbon cycle. Considering the natural planetary carbon cycle of CO_2 as the *'breathing of planet Earth'*, the annual natural turnover of CO_2 amounts to about 350 Gt.[30]

For comparison, the projected fictitious global CO_2 demand for plastics for a fully defossilized, all-circular plastics economy by 2050/2060, estimated in Sect. 15.2, ranged from 2 to 5 Gt CO_2 p.a., roughly representing 1% of the natural CO_2 cycle. The point we are making is straightforward: if natural processes, driven by solar energy and photosynthesis, manage a circular turnover of 350 Gt of CO_2 per year, we should and must continue to seek means to achieve an extra 1% circular turnover in the Technosphere also driven by renewable energy, of which globally there is still a huge surplus. This remains our primary civilizational navigation obligation. Currently, the only viable option is circular carbon management in the Technosphere, with widespread implementation of carbon capture and utilization (CCU).

To conclude these reflections on future perspectives for plastics, we wish to reiterate what is **most urgently needed** to facilitate, encourage, and incentivize a climate-neutral and all-circular Carbon and Plastics Economy (c^2CPE): **comprehensive, consistent, and forward-looking progressive policies and politics**. These policies must address the complexity of the ongoing low-carbon and circular-carbon transition with an **informed, technology-open mindset**.

Finally, to sum it up and to succinctly underscore our *'categorical'* main conclusions and beliefs, here are our **three most important take-away messages**:

Takeaway message (1)

For an **all-circular Carbon and Plastics Economy** in the **Technosphere,** an **amount of CO_2** corresponding to **ca. 1% of the natural CO_2 breathing cycle** is required.

This objective represents a benchmark aim for navigating human enterprise.

Take-away message (2)

While there may **not be a single 'silver bullet'** for establishing a ***sustainable, all-circular*** **Carbon and Plastics Economy**, there could indeed be a bundle of just a few,

with **cross-sectoral carbon capture & utilization (CCU) being one of them!**

Take-away message (3)

[30] https://climate.mit.edu/ask-mit/how-much-carbon-dioxide-does-earth-naturally-absorb.

Without carbon circularity in the Technosphere
[and hence, without circular Carbon Management (cCM) and carbon capture & utilization (CCU)],
there can be **NO climate neutrality** and **NO Sustainable Development!**

15.5 CO_2 as Raw Material for Polymer Synthesis

By Walter Leitner and Guido Schroer

15.5.1 General

The use of carbon dioxide (CO_2) as raw material for chemical reactions has been of general interest for a long time. A major driver was always the scientific challenge resulting from the obvious difference between nature's approach to material synthesis utilizing CO_2 as the only carbon source as compared to industrial organic chemistry where mainly fossil carbon sources are exploited. With the increasing awareness of the need to transform the linear petrochemical "value chain" into a circular anthropogenic carbon cycle, the research field has seen an enormous dynamic (Klankermayer & Leitner 2015; Leitner 2021) and the first examples of industrial implementation are becoming visible also in relevance to the production of polymers.

Fundamentally, the conversion of CO_2 as feedstock faces a combination of thermodynamic and kinetic challenges (cf. Fig. 15.11). The thermodynamic driving force is provided typically by a high-energy reaction partner such as hydrogen or highly reactive substrates (e.g. epoxides, Grignard reagents, etc.). Their generation and availability need to be considered and integrated into the overall process design. Synthetically useful C–C or C-H bond formation or removal of one of the two oxygen atoms at CO_2 encounter usually very high kinetic barriers making the use of highly active, selective, and robust catalysts indispensable (Klankermayer & Leitner 2016). Consequently, catalysis research and process systems engineering are the core scientific areas for successful technology development in CO_2 utilization (Poliakoff 2015).

Initially, the concept of "carbon capture and utilization (CCU)" was developed in the context of CO_2 mitigation attempting to find economic outlets for captured emissions from fossil power plants or industrial processes. It was therefore mainly discussed as an additional or complementary approach to "carbon capture and storage (CCS)" technologies due to the obvious difference in scale between CO_2 emissions from energy and carbon demand in chemistry. Obviously, polymers and plastics offer an attractive option as strategic area for technology options to fulfil those goals.

Incorporating CO_2 into mainly carbonate-based polymer architectures has been at the centre of academic and industrial developments in this context.

With the increasing deployment of "de-carbonized" technologies in the energy sector, however, the paradigm has shifted towards the recognition of CO_2 as the main carbon source in the "de-fossilization" of the chemical sector including chemical and fuel production (Bardow 2021). One possible CO_2 source to be considered would be polymer waste, ultimately resulting in a closed carbon cycle from feedstock to product and back.

In the context of circular plastics, the overall carbon circle comprises thermal recycling by recovering the energy from the waste material through incineration combined with the re-use of the carbon by CO_2 conversion. The additional primary energy required for the necessary chemical reduction and partial or full dexoygenation of CO_2 can be achieved either by electrochemical processes or via low-carbon ("green") hydrogen generated by water electrolysis. Significant reductions in greenhouse gas emissions are thus possible depending on the level of de-carbonization of the energy supply (Leitner & Bardow 2018). As building blocks and base chemicals are being produced from CO_2-based production routes, they will become available for polymer production also resulting in a most direct impact on existing materials and products.

While the perspective to introduce CO_2 as "green carbon" (Han 2022) into existing value chains of plastic production seems very attractive on a business level due to the retention of existing assets, it also comes at a significant price on a systems level. It requires chemical reduction of CO_2 to hydrocarbons or mildly oxide compounds that mark the starting points of the chemical value chain. As inferred from Fig. 15.11, this requires substantial amounts of energy provided in the form of electricity for water electrolysis or co-electrolysis to generate hydrogen or carbon monoxide, respectively. A recent study (VCI 2023) of the Association of the German Chemical Industry (VCI) estimates that the reduction of greenhouse gas emissions from the chemical value chain to net zero could be achieved via the production of the most important base chemicals from a mix of renewable carbon sources comprising chemical recycling (ca 10%), biomass utilization (ca 30%), and CO_2 (ca. 50%), with the remaining 10% still coming mainly from natural gas. The energy input required for this scenario in the form of green electricity would amount to ca 630 TWh, exceeding the total amount of electricity consumed in Germany today (2023: 530 TWh) (Umweltbundesamt 2023). Consequently, the transition to a de-fossilized chemical and plastic production aiming at a closed carbon cycle will be intricated with the transition of the energy-chemistry nexus often referred to as "power-to-X" technologies (Wikipedia 2024).

Two major strategies emerge for the exploitation of CO_2 as a carbon source in the synthesis of polymers (cf. Fig. 15.12). On the one hand, the production of known base chemicals offers immediate entry into existing value chains. As of today, base chemicals with high energy content will be a global commodity rather than produced on a national level. Thus, renewable energy from privileged areas can be harvested as C-sources into established production and processes. On the other hand, the use of CO_2 to construct specific functionalities and to generate new polymer architectures offers opportunities for leapfrog innovations. In line with the principles of

Green Chemistry (Anastas 2020) such new processes and products can ultimately deliver better performance, reduce environmental impact, and be a priori designed for circularity.

15.5.2 Base Chemicals and Monomers from CO_2 for Existing Value Chains

The catalytic conversion of CO_2 to petrochemical base chemicals or known monomers entering directly into existing value chains of the polymer industry is investigated very intensively as discussed also in the previous sections of this Chapter. In the context of circular plastics, the overall carbon circle comprises thermal recycling by recovering the energy from the waste material through incineration combined with the re-use of the carbon by CO_2 conversion. The additional primary energy required for the necessary chemical reduction and partial or full dexoygenation of CO_2 can be achieved either by electrochemical processes or via low-carbon ("green") hydrogen generated by water electrolysis.

Synthesis gas (syngas), the mixture of CO and H_2, can be generated from CO_2 via high-temperature co-electrolysis or alternatively via the reverse water gas shift (rWGS) equilibrium. Through the Fischer–Tropsch process, syngas can be converted to hydrocarbons providing a potential route to synthetic naphtha. The electrochemical conversion allows to control of the CO/H_2 ratio to some extent adjusting it in principle depending on the desired further conversion in various catalytic applications (Eichel 2017). Also, pure CO, an interesting feedstock directly for polymer building blocks, can be obtained by low-temperature co-electrolysis (Schmid 2018).

Formic acid and formate salts or esters are also accessible with thermochemical and electrochemical technology options (Leitner 1995). While formic acid has little direct use in the polymer industry, it has been suggested especially in the form of its methyl ester as a possible liquid and easily transportable CO carrier (Lolli 2013). Methanol is a particularly attractive target to be produced from CO_2 and renewable energy (Schlögl 2022) as it can be further converted into the key olefinic monomers ethene and propene (MTO process) or aromatics (MTA process) (cf. Fig. 15.13). Esters formed with methanol such as methylmethacrylate (MMA) are also widely used as monomers in the polymer industry. Formaldehyde is another important intermediate for plastics and resins, and is available from "green" methanol by already existing technology such as the formox process. The direct production of formaldehyde from CO_2 remains currently challenging, however (Liang & Daasbjerg 2022).

Beyond the mentioned C1 intermediates and building blocks, the production of ethene from CO_2 is studied intensively. A possible route can be envisaged via ethanol as the primary product followed by dehydration. This is operated for bio-based ethanol on a large industrial scale (Braskem 2023). Technologies for CO_2-based ethanol are commercialized using catalytic (e.g. Air Company 2024) and bio-catalytic (e.g. LanzaTech 2024) processes. The latter has been demonstrated within a production route for polyethylene glycol (PEG)-based surfactants (Unilever 2021). Due to the high market demand for "green" ethanol in other areas including fuels, a direct electrochemical route to ethylene from CO_2 would be very promising. Currently, copper catalysts are the focus of numerous research projects in this direction (Jiao 2018).

Oxalic acid has been also described as an attractive C2 intermediate from carbon dioxide formed by dimerization of the radical anion of CO_2 under electrochemical conditions (Schuler 2021). While the current use of oxalic acid in the polymer industry is not very prominent, it can be reduced to ethylene glycol opening thus a potential CO_2-based pathway to PEG-type and PET-type polymers.

Isocyanates are the second component for the production of polyurethanes. The generation of the isocyanate functional group from CO_2 would not only open a possibility for carbon circularity but also replace phosgene as the C1 building block currently in use. The reaction of amines with CO_2 to form the RN = C = O unit suffers from the formation of water as the stoichiometric bi-product, however. Reactive co-reagents such as $POCl_3$ or epoxides have been proposed as "water scavengers" to shift the unfavourable thermodynamic equilibrium and to avoid the undesired hydrolysis of the isocyanate unit (Kroke 2021). The energetic and financial costs associated with the separation or regeneration of the scavenger make these approaches not feasible, however. Achieving the desired N–C bond formation initially in the form of methylcarbamates as precursors to the isocyanates offer a potentially more attractive alternative. Methylformate has been shown to allow for such a pathway using it as a C1 source under conditions of oxidative carbonylation (cf. Fig. 15.14) (Langanke & Leitner 2020).

15.5.3 CO_2 as Building Block for New Processes and Products

CO_2 and its derivates can be indirectly or directly integrated into monomers and building blocks used for polymer synthesis. These approaches have the potential to lead to process and product innovation beyond the reduction of the carbon footprint in existing value chains. While initially considered as "drop-in" alternatives for known fossil-based building blocks, development is pointing towards use in known or even novel applications capitalizing on advantageous properties of enhanced materials. Potential environmental benefits include for example the replacement of hazardous or toxic reagents in the production (e.g. phosgene, CO) or improved end-of-life properties for bio-degradation or chemical recycling. Needless to say that also the actual function of a given application may be improved with new chemical architectures.

Initial work focused on the indirect incorporation of CO_2 as a monomer for polycarbonate and polyurethane synthesis. In particular, the co-polymerization of epoxide monomers with CO_2 has been studied intensively in academia and industry, and highly effective catalysts have been developed allowing control of the chemo-selectivity (cyclic carbonate *vs.* polymer), the CO_2 incorporation (alternating versus statistical), and even the stereo-chemistry of the polymer architecture.

Research for the incorporation of CO_2 into epoxides dates back to the 1970s and initial work from Inoue et al. on the reaction of propylene oxide and carbon dioxide using $ZnEt_2/H_2O$ as a catalyst (Inoue 1969). In the following years and decades intensive research has been conducted on the synthesis of alternating polycarbonates as well as polyethercarbonate polyols from various epoxides. A variety of different catalyst systems have been developed and deployed mostly using Cr, Co, Al, and/or Zn as active metal compounds (Darensbourg 2007; Coates 2004). Focusing on resulting polyethercarbonate polyols, attention was mainly directed to propylene and ethylene oxide as epoxides (cf. Fig. 14a). For polycarbonates, cyclohexene oxide was investigated as a benchmark substrate but also epoxides stemming from natural resources, such as limonene oxide, were used (cf. Fig. 14b, c) (Byrne 2004, Auriemma 2015).

The synthesis of CO_2-based polycarbonates for industrial applications was developed up to the scale where materials were used in actual work items and parts for electronic devices in a longstanding collaboration of BASF and the Technical University of Munich (Klaus 2011). Challenges in mechanical (brittleness) and thermal (glass transition temperatures) properties could be overcome at least partly by blending with other bio-based polymers.

Technologies for CO_2-based polyethercarbonate polyols reached market maturity and have been partly also commercialized. Namely, three examples will be explained in more detail:

- Covestro AG's Cardyon®
- Novomer Inc.'s Converge®
- Econic Technologies Ltd.

A breakthrough in the design and production of polypropylene carbonate polyols was based on the development and deployment of a bimetallic catalyst system, such as double metal cyanides (DMC) or dinuclear Zn/Mg catalysts, under precisely controlled activation and reaction parameters (Chapmann 2015). Covestro AG commercialized materials resulting from corresponding technologies (Langanke 2014) as Cardyon® in 2016. The life cycle assessment (cradle-to-gate analysis) for CO_2-based polyethercarbonate polyols from propylene oxide with 20wt % CO2 demonstrated an 11–19% reduction in greenhouse gas emissions (GHG emissions) and a reduction of fossil resources of 13–16% compared to conventional polyether polyols (Bardow 2014). Savings resulted mainly from the substitution of epoxide with CO_2 per kilogram of material.

Cardyon® was produced in pilot plant scale in the Chempark Dormagen, Germany, with an annual production of up to 5 kt. The resulting polyethercarbonate polyols were used for the production of soft foams, in particular for the production of polyurethane mattresses together with the Belgium company Recticel, and in a number of other applications including textiles, sport floors, insulation materials, and car parts (cf. Fig. 15.16). The development was recognized as top 3 innovations in the German Future Awards in 2019 (Gürtler 2019). Although initial results on the product properties were promising Covestro put the commercial production on hold, shifting focus towards circular economy and closed-looped recycling of conventional propylene oxide polyols and resulting materials.

Novomer Inc. commercialized a technology for the production of CO_2-based polyethercarbonate polyols under the brand name Converge®. Resulting polyols were mainly used for the manufacturing of polyurethanes and application focus was directed towards CASE (coatings, adhesives, sealants, and elastomers) applications. The usage resulted in advanced properties of resulting materials, such as an increased load bearing or improved abrasion resistance. The Converge® technology was sold in 2016 to Saudi Aramco valued at up to US $ 100 million strengthening Saudi Aramco's downstream expansion and diversification of their product portfolio (Aramco 2016).

Econic Technologies Ltd. was founded in 2011 as a university spinoff from the research group of Prof. Charlotte Williams at the time at Imperial College London, UK, with a focus on the development of new catalysts and processes for the synthesis of polyethercarbonate polyols (Econic 2016). Process and technology enable the replacement of up to 30% of fossil-based polyol with CO_2. The major application focus was once again directed towards cost-competitive polyurethane products with comparable performance and a lower carbon footprint. The company signed license contracts with the Chinese and Indian polyol and polyurethane companies Changhua Chemical Technology Company Ltd. and Manali Petrochemicals, respectively, in recent years and collected a Series D funding of € 12.3 million in 2022.

In addition to the direct incorporation of CO_2 into existing polyol value chains, the indirect use of CO_2 via C1-based polyformaldehyde (pFA) units offers the opportunity to generate innovative polyol architectures (cf. Fig. 15.16). As a current example *en route* to potential commercialization, the production of polyoxymethylene polypropylene ether co-polyols will be described.

As part of an industrial-academic consortium led by Covestro AG, a process for the synthesis of novel polyether co-polyols was developed within the Kopernikus programme "P2X" (Kopernikus 2024) with the aim to use C1 building blocks from CO_2/H_2 as non-fossil backbone in the polymer industry. It was demonstrated that a section of the propylene oxide repeating unit – either conventional or also from renewable sources - can be substituted by polyoxymethylene units leading to polyoxymethylene-propylene oxide co-polymers (PME-polyols). Paraformaldehyde (pFA) or formaldehyde is used as substrate for the polyoxymethylene unit, opening the possibility to generate the C1 units ultimately from "green" hydrogen and CO_2 via methanol. The costs of manufacturing and greenhouse gas emissions (cradle-to-gate analysis) of the novel PME-polyols are advantageous to conventional propylene oxide polyols (Ausfelder 2022). The process and material development is continued in the next funding phase of the Kopernikus project with the aim to spin out a start-up company for the commercialization in promising application fields. Core work includes the optimization and scaleup of the co-polyol synthesis, application-based tailoring of material properties and application development in speciality applications, e.g., for novel polyurethanes in CASE applications.

15.6 Summary and Outlook

The use of carbon dioxide as a carbon source for chemical production is recognized as a major pillar of a closed anthropogenic carbon cycle. While it seems highly attractive to recycle "plastic" directly as material or chemically by reversing the polymerization back to the monomers, these approaches are facing significant challenges due to the composite nature of the actual products and/or the mixture formation upon depolymerization. Therefore, energy-intensive separation processes are inevitably part of the resulting concepts of circularity. Converging the carbon content of the recycling streams into gaseous CO_2 evades this situation, especially for waste materials that are particularly difficult to obtain in relatively pure form. On the other hand, the conversion of CO_2 to useful building blocks is also associated with a high (chemical) energy demand (cf. Fig. 15.10). Consequently, the GHG reduction needs to be analyzed on a case-to-case basis to identify the potential for the various recycling strategies.

In a first approximation, the necessary energetic driving force is associated with the level of chemical reduction according to the C1 products carbon monoxide/formic acid (2-electron reduction), formaldehyde (4-electron), methanol (6-electron) and methane (8-electron) (Klankermayer 2016). The production of base chemicals has the advantage to be compatible with the existing value chains, infrastructure, and assets of the polymer industry. It will therefore be an important option to address "de-fossilization" in a short to medium time frame reacting to regulatory frameworks. At the same time, the use of CO_2 for novel processes and products offers the potential to combine the quest for carbon neutrality with product and process innovation. Addressing these challenges will be facilitated by academic-industrial collaborations and networks and offer exciting opportunities for entrepreneurship in start-up initiatives.

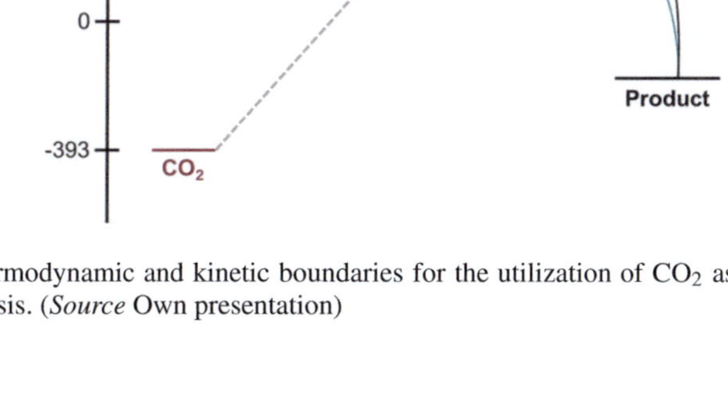

Fig. 15.10 Thermodynamic and kinetic boundaries for the utilization of CO_2 as raw material in chemical synthesis. (*Source* Own presentation)

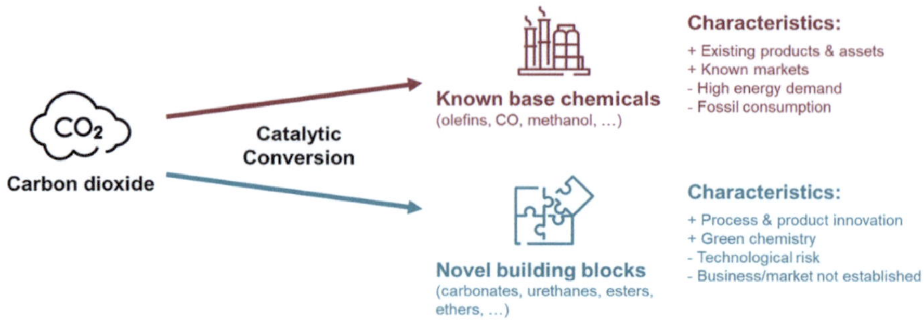

Fig. 15.11 Products resulting from catalytic CO_2 conversion for polymers synthesis and respective characteristics. (*Source* Own representation)

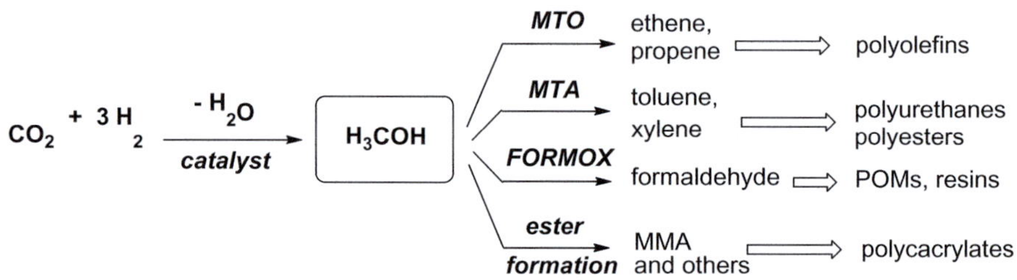

Fig. 15.12 Methanol as a central molecular connection between CO_2 and the polymer value chain. (*Source* Own representation)

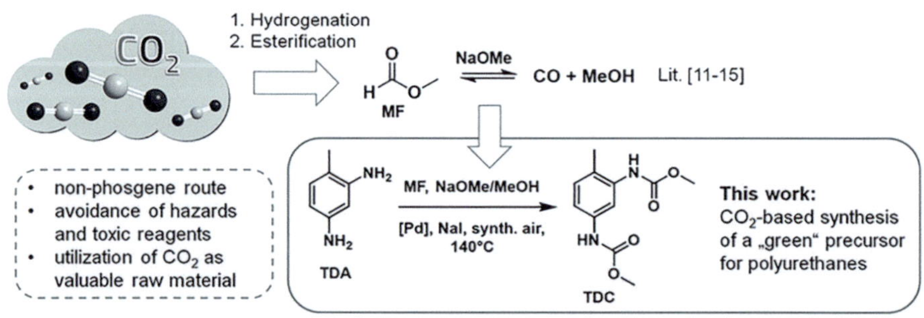

Fig. 15.13 Isocyanates from CO_2 via oxidative carbonylation with methylformate as CO carrier; reproduced from. (*Source* Langanke & Leitner 2020)

Ultimately, carbon dioxide can be envisaged as raw material for the generation of many more functional groups relevant to the polymer industry. The imagination and creativity of chemists can devise new synthetic pathways by retrosynthetic analysis of the target molecular structures back to CO_2 as feedstock. Fundamental research on catalysis is required to tap into this potential as an important contribution to carbon management in the framework of circular plastics.

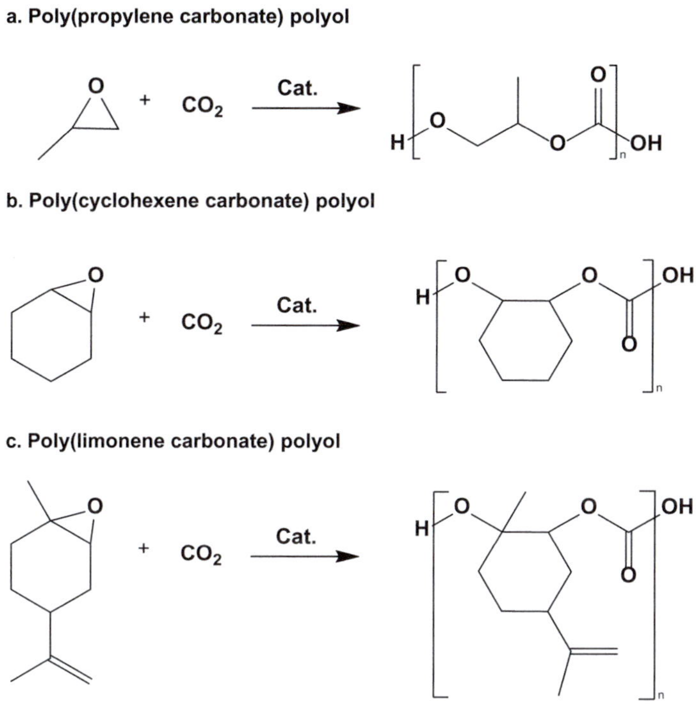

Fig. 15.14 **a-c** Synthesis of: **a** poly(propylene carbonate) polyol from propylene oxide and CO_2; **b** poly(cyclohexene carbonate) polyol from cyclohexene oxide and CO_2; and **c** poly(limonene carbonate) polyol from limonene oxide and CO_2 (simplified, stereochemistry not shown) (*Source* Own representation)

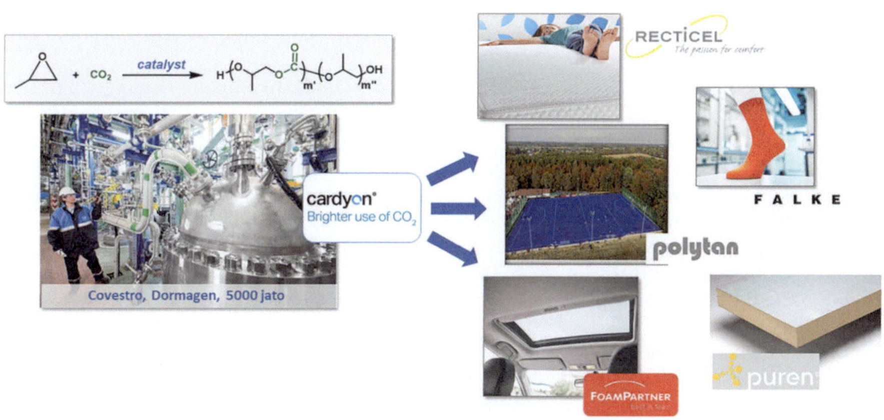

Fig. 15.15 Demonstration plant for the production of CO_2-based polyethercarbonate polyols on 5 kt/a scale and examples for successful material testing of the products (*Source* Own presentation)

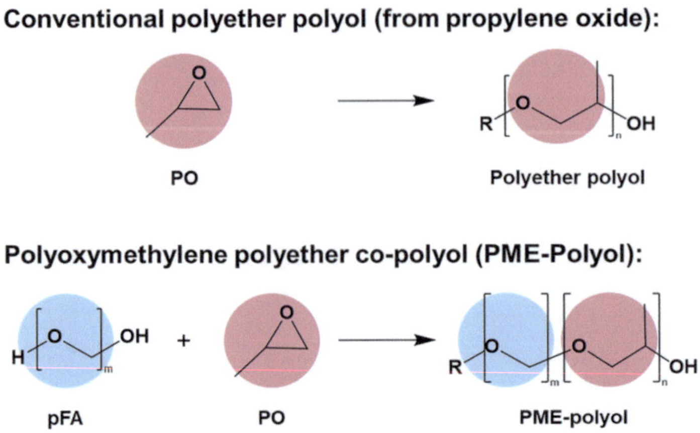

Fig. 15.16 Synthesis of conventional polyether polyol (from propylene oxide, top) and polyoxymethylene polyether co-polyols (from propylene oxide and paraformaldehyde, bottom) (*Source* Own presentation)

References

Air Company 2024: https://www.aircompany.com/

Anastas 2020: J. B. Zimmerman P. T. Anastas, H. C. Erythropel, WL, *Science* **2020**, *367*, 397–400.

Aramco: Aramco's Media & Executive Communications Department, "Saudi Aramco acquires Novomer's polyol business and associated technologies, enhancing its downstream expansion strategy", can be found under https://www.aramco.com/en/news-media/news/2016/acquires-novomers-polyol-business-downstream-expansion, 2016**.**

Auriemma 2015: F. Auriemma, C. De Rosa, M. R. Di Caprio, R. Di Girolamo, W. C. Ellis, G. W. Coates, *Angew. Chem.* 2015, *127*, 1231–1234.

Ausfelder 2022: F. Ausfelder, D. D. Tran, in *Options for a sustainable energy system with power-to-x technologies*, Frankfurt am Main, Germany, 2022.

Azarabadi 2023: Azarabadi, H., et al.; Article *"Shifting the Direct Air Capture Paradigm"*; Boston Consulting Group (BCG), June 2023; https://www.bcg.com/publications/2023/solving-direct-air-carbon-capture-challenge.

Bachmann 2023: Bachmann, M., Zibunas, C., Hartmann, J., Tulus, V., Suh, S. Guillen-Gosalbez, G., Bardow, A.; *"Towards circular plastics within planetary boundaries"*; nature sustainability, published online March 06, 2023; https://doi.org/10.1038/s41893-022-01054-9.

Bardow 2014:N. von der Assen, A. Bardow, *Green Chem.* 2014, *16*, 3272–3280.

Bardow & Leitner 2018: J. Artz, int. al. A. Bardow, W. Leitner, *Chem. Rev.* **2018**, *118*, 434–504.

Bardow 2021: Bardow et.al. "Emerging technologies: general discussion" Faraday Discuss.**,** 2021, 230, 388–412

Bardow 2024: Presentation *"Driving in (carbon) circles: An environmental perspective on sustainable transport"*; presented by A. Bardow (ETH Zurich) at Symposium *"Energiewende im Schwerverkehr"*, Austrian Academy of Sciences (ÖAW), Vienna, 21 Feb. 2024.

Berg 2023a: vom Berg, C. and Carus, M., Report "Making a Case for Carbon Capture and Utilisation (CCU) – It Is much more than just a Carbon Removal Technology"; Renewable Carbon Initiative (ed.), Hürth, July 2023; https://renewable-carbon.eu/publications/product/making-a-case-for-carbon-capture-and-utilisation-ccu-it-is-much-more-than-just-a-carbon-removal-technology/

Berg 2023b: vom Berg, C. and Carus, M., "Renewable Carbon Initiative (RCI) Shapes the Future of the Chemicals and Materiels Sector"; Chemical Industry Digest – Special Annual on Sustainability, June 2023; https://renewable-carbon.eu/publications/product/rci-shapes-the-future-of-the-chemicals-and-materials-sector/

Bhardwaj 2021: Bhardwaj, A., McCormick, C., Friedmann, J.; Report *"Opportunities and Limits of CO2 Recycling in a Circular Carbon Economy: Techno-Economics, Critical Infrastructure Needs, and Policy Priorities"*; Columbia University, Center on Global Energy Policy, May 2021; https://www.energypolicy.columbia.edu/publications/opportunities-and-limits-co2-recycling-circular-carbon-economy-techno-economics-critical/.

Braskem 2023: "Braskem expands its biopolymer production", can be found under https://www.braskem.com

Burke 2023: Burke, M., Zahid, M., Diffenbaugh, N., Hsiang, S. M.; NBER Working Paper Series *"Quantifying Climate Change Loss and Damage consistent with a Social Cost of Greenhouse Gases"*; Working Paper 31658; National Bureau of Economic Research, Cambridge, MA/US; Sept. 2023; https://www.nber.org/papers/w31658.

Byrne 2004: C. M. Byrne, S. D. Allen, E. B. Lobkovsky, G. W. Coates, *J. Am. Chem. Soc.* 2004, *126*, 11404–11405

CEMBUREAU 2023: Position Paper "A strong Business Case for Carbon Capture & Utilization Investments" December 2023.

Chapman 2015: A. Chapman, C. Keyworth, M. Kember, A. Lennox, C. Williams, *ACS Catal.* 2015, *5*, 1581–1588

Coates 2004: G. W. Coates, D. R. Moore, *Angew. Chem. Int. Ed.* 2004, *43*, 6618–6639

da Mata Costa 2021: da Mata Costa, L. P. et al.; *"Capture and Reuse of Carbon Dioxide (CO_2) for a Plastics Circular Economy: A Review"*; Processes 2021, April2021; https://www.mdpi.com/2227-9717/9/5/759.

Darensbourg 2007: D. J. Darensbourg, *Chem. Rev.* 2007, *107*, 2388–2410

EC 2024: European Commission (2024). European Union Transaction Log (EUTL). https://www.eea.europa.eu/en/datahub/datahubitem-view/98f04097-26de-4fca-86c4-63834818c0c0

Econic 2016: Econic Technologies Ltd., "Redeeming CO2 - pioneering the transformation of CO_2 into high-performance materials", can be found under https://www.econic-technologies.com/, 2016.

Eichel 2017: S. R. Voit, int. al., R. Eichel, *Angew. Chem. Int. Ed.* **2017**, *56*, 5402–5411.

EPA 2023: *"Report on the Social Cost of Greenhouse Gases: Estimates incorporating Recent Scientific Advances"*; Docket ID No. EPA-HQ-OAR-2021–0317; U.S. Environmental Protection Agency, Washington DC, Nov. 2023; https://www.epa.gov/system/files/documents/2023-12/epa_scghg_2023_report_final.pdf.

EU 2018a: European Commission – ANNEXES to the Communication from the Commission "A European Strategy for Plastics in a Circular Economy", Brussels, 16 January 2018; https://eur-lex.europa.eu/legal-content/EN/TXT/?qid=1516265440535&uri=COM:2018:28:FIN

EU 2018 : Directive (EU) 2018/852 of the European Parliament and of the Council of 30 May 2018 amending Directive 94/62/EC on packaging and packaging waste, Document 32018L0852; https://eur-lex.europa.eu/eli/dir/2018/852/oj.

EU 2019: European Commission - Press release "The European Green Deal sets out how to make Europe the first climate-neutral continent by 2050, boosting the economy, improving people's health and quality of life, caring for nature, and leaving no one behind", Brussels, 11 December 2019; https://ec.europa.eu/commission/presscorner/detail/en/ip_19_6691

EU 2020: European Commission - Communication "A new Circular Economy Action Plan - For a cleaner and more competitive Europe", Brussels, 11 March 2020; https://eur-lex.europa.eu/resource.html?uri=cellar:9903b325-6388-11ea-b735-01aa75ed71a1.0017.02/DOC_1&format=PDF

Friedmann 2020: Friedmann, J., et al., Report *"Net-Zero and Geospheric Return: Actions Today for 2030 and Beyond"*; Columbia University, Center on Global Energy Policy,

Sept. 2020; https://www.globalccsinstitute.com/resources/publications-reports-research/net-zero-and-geospheric-return-actions-today-for-2030-and-beyond/.

Friedmann 2021: Friedmann, J., Agrawal, M., Bhardwaj, A., Report *"Evaluating Net-Zero Industrial Hubs in the United States: A Case Study of Houston"*; Columbia University, Center on Global Energy Policy, June 2021; https://www.energypolicy.columbia.edu/publications/evaluating-net-zero-industrial-hubs-united-states-case-study-houston.

Gürtler 2019: C. Gürtler, W. Leitner, B. Stange, "CO2 – A Raw Material for Sustainable Plastics", can be found under https://www.deutscher-zukunftspreis.de/en/team-1-2019, 2019.

Han 2022: M. He, Y. Sun, B. Han, *Angew. Chem. Int. Ed.* **2022**, *61*, e202112835.

ICCDU 2024: https://iccdu2024.scimeeting.cn/en/web/index/

Inoue 1969: S. Inoue, H. Koinuma, T. Tsuruta, *J. Polym. Sci., Part B: Polym. Lett.* 1969, *7*, 287–292 IPCC 2023: AR6 Synthesis Report: Climate Change 2023; Intergovernmental Panel on Climate Change, March 2023; https://www.ipcc.ch/report/sixth-assessment-report-cycle/

Jiao 2018: M. Jouny, W. Luc, F. Jiao, *Ind. Eng. Chem. Res.* **57**, 2165–2177.

Kähler 2023: Kähler, F., Porc, O. and Carus, M., RCI Scientific Background Report "Carbon Flows Report-Compilation of supply and demand of fossil and renewable carbon on a global and European level". Editor: Renewable Carbon Initiative, October 2023; available at: www.renewable-carbon-initiative.com

Kaufman 2020: Kaufman, N., Barron, A. R., Krawczyk, W., Marsters, P., McJeon, H.; "*A near-term to net zero alternative to the social cost of carbon for setting carbon prices*"; Nature Climate Change, 2020; https://www.nature.com/articles/s41558-020-0880-3.

Klankermayer & Leitner 2015 : J. Klankermayer, W. Leitner, *Science* **2015**, 350, 629–630.

Klankermayer 2016: J. Klankermayer, S. Wesselbaum, K. Beydoun, W. Leitner, *Angew. Chem. Int. Ed.* **2016,** *55*, 7296 –7343.

Klaus 2011: S. Klaus, M. W. Lehenmeier, C. E. Anderson, B. Rieger, *Coord. Chem. Rev.* 2011, *255*, 1460–1479

Kopernikus 2024: https://www.kopernikus-projekte.de/projekte/p2x

Kroke 2021: F. Gründler int. Al. E. Kroke, *Eur. J. Inorg. Chem.* **2021**, *2021*, 2211–2224

Langanke & Leitner 2020: C. Hussong, J. Langanke, W. Leitner, Chem. Ing. Tech. **2020**, *92*, 1482–1488.

Lang 2010: Lang, R. W., Inaugural Lecture at Johannes Kepler University Linz, Austria; March 2010.

Lang 2017: Lang, R. W., Polymerwerkstoffe für Energieeffizienz und regenerative Energie/Stoff-Technologien, in acatech MATERIALIEN – Werkstoffe und Materialien für die Energiewende (editors Bolt H.,Arzberger I., Berger C.), p. 21ff, acatech – National Academy of Science and Engineering Germany, March 2017; https://www.acatech.de/publikation/werkstoffe-und-materialien-fuer-die-energiewende/.

Lang 2020: Lang, R. W., Guest Commentary *"Die EU Plastikabfallabgabe – Missglückter Versuch einer Umweltsteuer oder Chance?"*; Die Presse (Vienna, Austria); 8 Aug. 2020; Online: https://www.diepresse.com/5854480/die-eu-plastikabfallabgabe-missgluckter-versuch-einer-umweltsteuer-oder-chance?from=rss

Lang and Fischer 2021: Courtesy of Lang, R. W. and Fischer, J., Johannes Kepler University Linz, Austria; 2021.

Langanke 2014: J. Langanke, A. Wolf, J. Hofmann, K. Böhm, M. A. Subhani, T. E. Müller, W. Leitner, C. Gürtler, *Green Chem.* 2014, *16*, 1865–1870.

LanzaTech 2024: https://lanzatech.com/

Leitner 1995: W. Leitner, *Angew. Chem. Int. Ed. Engl.* **1995**, *34*, 2207–2221.

Leitner, C. Gürtler, *Green Chem.* **2014**, *16*, 1865-1870.

Leitner 2021: N. Kinzel, C. Werlé, W. Leitner, *Angew. Chem. Int. Ed.* **2021**, *60*, 11628–11686.

Liang & Daasbjerg 2022: H.Q. Liang, K. Daasbjerg et al. *Angew. Chem. Int. Ed. Engl.* **2022**, *61*, e202204008.

Lolli 2013: G. Lolli et al. *Green Chem.* **2013**, *15*, 1146–1149.

Markowitsch 2023: Markowitsch, C., Lehner, M., Maly, M.; "Comparison and techno-economic evaluation of process routes for lower olefin production via Fischer-Tropsch and methanol synthesis"; Int. J. of Greenhousgas Gas Control 129 (2023).

Meys 2021: Meys, R., et al.; *"Achieving net-zero greenhouse gas emission plastics by a circular carbon economy"*; Science 374, 71–76; Oct. 2021.

nova-Institut 2024: illustrations retrieved March 3, 3024 from https://renewable-carbon.eu/publications/?search=1&publication-type=graphics

OECD 2022: OECD Report - Policy Highlights, "Global Plastics Outlook - Policy Scenarios to 2060", OECD Environment Directorate - Environment and Economy Integration Division", 2022; https://read.oecd-ilibrary.org/view/?ref=1143_1143481-88j1bxuktr&title=Global-Plastics-Outlook-Policy-Scenarios-to-2060-Policy-Highlights.

Paulik 2024: Paulik, C., Johannes Kepler University Linz, Austria; Personal communication on carbon losses in polymerization of polyolefins, March 2024.

PE 2022: Paper "Circular EconomyPLUS: Recommendations for action for a German Circular Economy Strategy", PlasticsEurope Deutschland e.V., 25 November 2022. https://plasticseurope.org/de/knowledge-hub/circular-economy-plus-recommendations-for-action-for-a-german-circular-economy-strategy/.

PE 2023: Plastics – the fast Facts 2023; PlasticsEurope, Brussels/Belgium, 2023; https://plasticseurope.org/knowledge-hub/plastics-the-fast-facts-2023/.

Pilz 2014: Pilz, H., Background Report "Criteria for eco-efficient (sustainable) plastic recycling and waste management: Fact based findings from 20 years of denkstatt studies" (Version 1.2), denkstatt GmbH, Austria; September 2014.

Poliakoff 2015: M. Poliakoff, W. Leitner, E.S. Streng, "The Twelve Principles of CO_2 Chemistry", Faraday Discuss., 2015, 183, 9–17

RCI 2023: Renewable Carbon Initiative (RCI) Manifesto for the next European Commission (2024–2029); Renewable Carbon Initiative, August 2023. https://renewable-carbon.eu/publications/product/renewable-carbon-initiative-rci-manifesto-for-the-next-european-commission-2024-2029-pdf/

Rockström 2009: Rockström, J., et al.; *"Planetary boundaries: exploring the safe operating space for humanity"*; Ecol. Soc. 14, 32 (2009).

Schlögl 2022: R. Schlögl, *Angew. Chem. Int. Ed. Engl.* **2022**, *61*, e202007397.

Schmid 2018: T. Haas int. Al. G. Schmid, *Nature Catal.* **2018**, *1*, 32–29.

Schuler 2021: E. Schuler, M. Demetriou, R. Shiju, G.-J. M. Gruter, *ChemSusChem* **2021**, *14*, 3636–3664.

Steffen 2015: Steffen, W., et al.; *"Planetary boundaries: guiding human development on a changing planet"*; Science 347, 1259855 (2015).

Umweltbundesamt 2023: https://www.umweltbundesamt.de/daten/energie/stromverbrauch

UN 2015a: Adoption of the Paris Agreement -Proposal by the President, United Nations Frame work Convention on Climate Change, UN FCCC/CP/2015/L.9/Rev.1, 12 December 2015; https://unfccc.int/resource/docs/2015/cop21/eng/l09r01.pdf.

UN 2015b: Transforming our world: the 2030 Agenda for Sustainable Development, United Nations General Assembly, Resolution A/RES/70/1, 21 October 2015; https://www.un.org/en/development/desa/population/migration/generalassembly/docs/globalcompact/A_RES_70_1_E.pdf.

UN 2022a: Operationalizing leaving no one behind (LNOB); United Nations Sustainable Development Group (UNSDG); World Commission on Environment and Development, 2022; https://unsdg.un.org/sites/default/files/2022-04/Operationalizing%20LNOB%20-%20final%20with%20Annexes%20090422.pdf.

UN 2022b: Draft Resolution "End plastic pollution: Towards an international legally binding instrument", Environment Assembly of the United Nations Environment Programme, UNEP/EA.5/L.23/Rev.1, 2 March 2022; https://wedocs.unep.org/bitstream/handle/20.500.11822/38522/k2200647_-_unep-ea-5-l-23-rev-1_-_advance.pdf?sequence=1&isAllowed=y.

UN COP28 2023a: Online-Presentation at COP28, Dubai; 5 December 2023

UN COP28 2023b: Outcome of the first global stocktake - Proposal by the President; Draft decision -/ CMA.5; United Nations Framework Convention on Climate Change, UN FCCC/PA/CMA/2023/ L.17, 13 December 2023; https://unfccc.int/sites/default/files/resource/cma2023_L17_adv.pdf.

Unilever 2021: The worlds first laundry capsule made using recycled carbon emissions" found under https://www.unilever.com/news/news-search/2021

VCI 2023: https://www.vci.de/langfassungen/langfassungen-pdf/vci-study-greenhouse-gas-neutrality-in-the-german-chemical-industry.pdf

VCI & VDI 2023: Abschlussbericht "Chemistry4Climte – Wie die Transformation gelingen kann", Verband der Chemischen Industrie e.V. (VCI) und Verein Deutscher Ingenieure e.V. (VDI), April 2023; https://www.vci.de/vci/downloads-vci/publikation/broschueren-und-faltblaetter/final-c4c-broschure-langfassung.pdf.

vito 2023: Report "CO_2 from cement industry needed for decarbonation of EU27 society in time-frame 2030 - 2050", Study accomplished under the authority of CEMBUREAU – The European Cement Association, April 2023; https://cembureau.eu/media/tkuf0c2e/230414-vito-final-2023-sct-r-2946-cembureau.pdf.

Wikipedia 2024: https://en.wikipedia.org/wiki/Power-to-X

Prof. Dr. mont. Reinhold W. Lang is Professor Emeritus at Johannes Kepler University Linz, Austria, where he headed the Institute for Polymeric Materials and Testing (IPMT) and presided over the Polymer Engineering & Plastics Technology study programs until 2022. Since the early 1990ies, he has also strongly been engaged in teaching and research in the broad, transdisciplinary topic "Polymer Technologies for Sustainable Development". On a socio-political level, he has been active as a co-initiator of forumFUTURE, a 'competence forum for sustainable development in politics, business and society' founded in 2019.

Prof. Dr. Stefan P. Schleicher obtained his academic degrees from the University of Technology in Graz and the University of Vienna. He held several positions at academic institutions in Vienna, Bonn, Philadelphia, and Stanford. His research focuses on economic modeling and economic policy issues in the context of energy and climate and the disruptive developments that are unfolding. He has contributed to international research networks such as IPCC for climate and LINK for trade and has coordinated projects in the EU Framework Program. He currently works at the Wegener Center for Climate and Global Change at the University of Graz and is a consultant at the Austrian Institute of Economic Research in Vienna.

Prof. Dr. Walter Leitner is Managing Director at Max Planck Institute for Chemical Energy Conversion (MPI-CEC) and a Professor at RWTH Aachen. He studied chemistry in Regensburg and held various positions in academia at the University of Oxford, the University of Jena, and Max Planck Institut Mülheim an der Ruhr. Since 2007 he established together with Bayer MaterialScience AG a catalytic center at RWTH Aachen that he still leads.

Dr. Guido Schroer is Project Leader of "Power2Polymers", a BMBF-funded spinoff of the Kopernikus programme "P2X", working towards commercialization of novel CO_2 and P2X-based polyols and resulting polymer products. He has worked as a strategy consultant at the Boston Consulting Group focusing on sustainability and transformation projects within the chemical industry before and holds a PhD in polymer and technical chemistry from RWTH Aachen University.

Chapter 16
Mass Balancing for Circular Feedstock

Ulrich Liman and Peter Orth

Contents

Summary

This chapter talks about making plastics in a way that's better for the environment by using materials like recycled plastics, plants, and CO_2 instead of oil and gas. These new feedstocks help protect the climate and ensure we have enough resources for the future. A key part of this change is the mass balance approach. This method keeps track of how much of these new feedstocks are used in making plastics. It makes sure that products stay high-quality while becoming more sustainable.

The chapter also explains how this approach fits with European laws that encourage using recycled materials. Using the mass balance method is important for getting investments in recycling technologies and meeting sustainability goals.

Overall, this chapter gives a clear picture of how we can make plastic production greener. It highlights the need for new ideas and support from laws to achieve a circular economy, where resources are reused and recycled.

U. Liman (✉)
ULC—Ulrich Liman Consulting, Langenfeld, Germany
e-mail: druliman2@gmail.com

P. Orth
OPC—Orth Plastics Consulting, Cologne, Germany

P. Orth et al. (eds.), *Circular Economy of Plastics*,
https://doi.org/10.1007/978-3-031-78829-1_16

16.1 Introduction

The aim of the circular economy with plastics is to decouple the fossil feedstock oil and gas for the production of plastics. Alternative feedstock sources are recyclates or recycled secondary raw materials from chemical recycling (carbon from the technosphere), biomass (carbon from the biosphere) and CO_2 (carbon from the atmosphere). These products can gradually replace fossil feedstock and thus contribute to climate protection and security of feedstock supply. The raw materials transition demanded by politics and society will be based on these three carbon sources.

Although chemical recycling—and pyrolysis in particular—is still in its early stages, the first pilot plants (cf. Chap. 13) have started up and developments over the next few years are foreseeable. Hence, the proportion of secondary raw materials[1] coming from chemical recycling is still small. However, primary raw materials from the refining and processing of biomass are already available to the market on a technical scale (> 1 Mt), while primary raw materials obtained from the industrial reduction of CO_2 are currently only available in small quantities—due to a lack of inexpensive energy for the corresponding large-scale processes. The urea synthesis, in which CO_2 is used on an industrial scale, is an important exception.

Plastic waste in general, including those intended for chemical recycling, is subject to the regulatory waste regime and therefore—in contrast to fossil raw materials—to a large number of national and European legal acts for dealing with them; the same applies to organic waste. Companies that handle waste must be registered or licensed by national authorities to ensure they have the necessary infrastructure, capacity, and management not least for documentation and reporting. Only for CO_2, although it is a gaseous waste, no such regulations apply.

The products obtained from plastic waste using chemical recycling processes, as well as the products from biomass or from the reduction of CO_2 (basic chemicals, monomers and other building blocks) obtained using completely different processes, are chemically identical to the primary raw materials obtained today using petrochemical means. Chemically identical means interchangeable and physically indistinguishable, which enables the use of established procedures and processes as well as existing plants. This is essential for the path to the circular economy and makes it easier and faster.

For all areas of the chemical industry in the broadest sense (petrochemicals, chemistry, plastics production, etc.), it is important to use these alternative raw materials in existing large and efficient production plants and thus continue to produce the entire spectrum of basic chemicals and monomers required in chemistry. During the current transformation phase, raw materials from chemical recycling are used in these plants together with bio-based (in the future also with CO_2-based) raw materials and fossil-based raw materials. The ratio of the volume flows of similar raw materials from different sources depends on the availability of the respective, chemically identical

[1] The terms primary and secondary raw materials used here must be checked and adjusted against this background.

raw material, while none of these circular raw materials can be physically traced in the production process.

In all cases, the goal of the transformation is the long-term abandonment of fossil-based raw materials. In the future, plastics producers can and will use up to 100% circular raw materials of varying origins to produce a product without changing the product properties—ultimately this is the goal of the raw materials transition.

Since an unchanged, specified product quality is generated based on existing and approved processes, the entire value chain can continue to use its established processes unchanged. Due to a lack of measurability, credibility and transparency are essential prerequisites for the success of this important part of the raw materials transition.

16.2 Historical Development

With this approach, transparency must be ensured in an appropriate manner for the regulator (whose focus is, among other things, on compliance with waste regulations) and for the market or customer (whose attention is on the use of circular raw materials and the product's carbon footprint). To enable statements about the content of circular raw materials in a product, the mathematical approach of mass balancing was developed. The amount of circular raw materials used is allocated to a product based on its individual recipe according to mass balance principles, taking into account all yields and losses. This is an accounting allocation of the use of circular raw materials for a specific product (FAPU 2015). The proportion of circular raw materials is determined using the mass balance approach derived from thermodynamics. This approach is helpful if two or more products in a balanced area (tank, plant, site) can no longer be physically separated (see above), but their origin or history should be assigned to a specific product. The mass balance approach is captured in the international standard ISO 22095:2020 (ISO 2020) defining the rules for allocating starting (input) materials to end products.

A transparent, standardized, and auditable allocation mass balance according to the chain of custody principle (EMF 2019) using the credit method of raw materials from recycling (among others) provides a calculated statement about the content of raw materials from the circular input materials shared in the target product. Mass balance approaches are equally and simultaneously suitable for alternative raw materials such as CO_2, biomass, and secondary raw materials from chemical recycling. They are therefore a prerequisite and basis for the raw materials transformation and the resulting changes in chemical production, and are therefore an essential component of companies' current investment decisions.

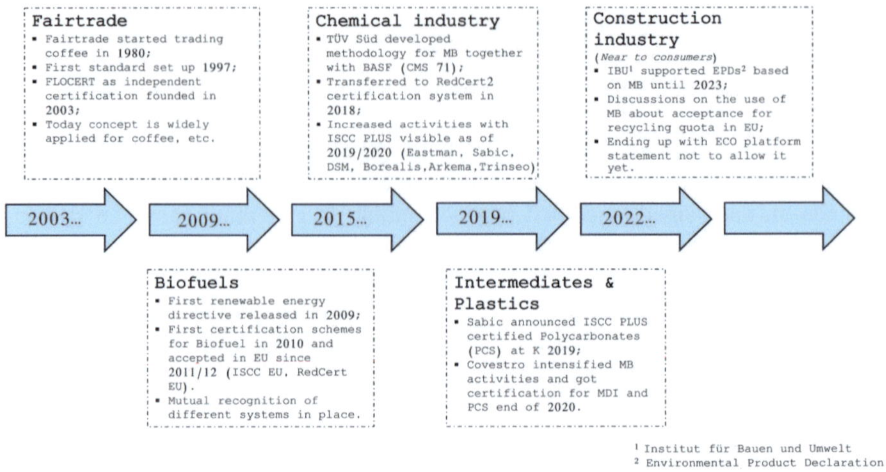

Fig. 16.1 Historical development of the Mass Balance (MB) concept. (*Source* Own representation)

Such or similar approaches were pursued by politicians for many years, took place in various economic sectors and were accepted by the market. Examples include "green electricity" that many energy producers or European railway companies such as DB offer and promote (DB 2024), fair trade products on the market such as coffee or cotton (Fair Trade 2022), wood products such as paper from sustainable forestry (FSC 2022), or fuel additives such as bio-ethanol (Biokraft 2009). European agricultural policy, which is interested in the further development of the bioeconomy, plays a pioneering role here.

The procedure originally developed for the use of renewable raw materials was significantly further developed for use in chemistry and plastics production and operated by BASF (FAPU 2015). It was later further worked on in working groups at (PlasticsEurope 2020) and the Ellen MacArthur Foundation (EMF 2019) to propagate it and develop a standard for the use of circular raw materials. Fig. 16.1 shows the historical development of the Mass Balance (MB) approach in more detail.

16.3 Allocation Aspects

An important component of the mass balance concept is the accompanying verification of traceability embedded in the value chain—the "Chain of Custody" (EMF 2019), which leads from the plastic waste originator (the consumer or user) via the waste manager and recycler to the plastic producer and the processor and finally via the OEM or retailer to the customer (i.e. the consumer again) (cf. Fig. 3.5). The use of digital methods (cf. Chap. 19) may enable and facilitate the work and success of the chain of custody.

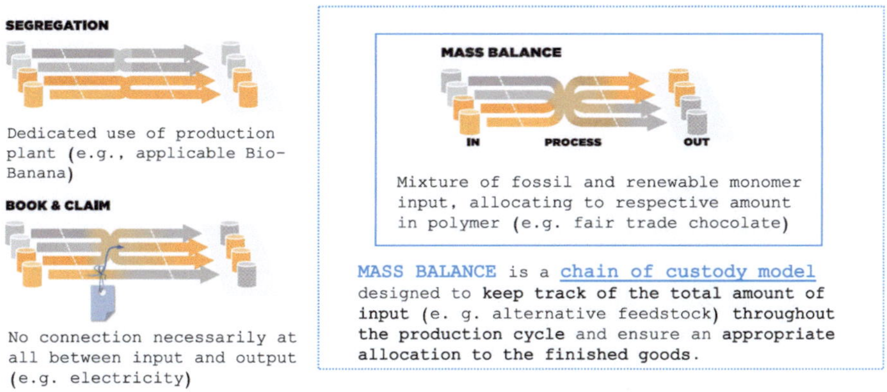

Fig. 16.2 Different kinds of chains of custody. (*Source* EMAF 2019)

Below Fig. 16.2 shows a representation of possible allocation variants (different chain-of-custody models) developed by the Ellen MacArthur Foundation working group (EMF 2019). Different from the Segregation model (e.g. bio-bananas) or the Book & Claim model (electricity) the Mass Balance concept offers a tracking of alternative feedstocks throughout the complete production process to the finished good for the end-user.

However, the situation in a petrochemical or chemical company is much more complex when several chemically identical raw material streams from different sources are brought together for a process or a process chain—in the simple case of a tank, out of which a steam cracker is fed. Figure 16.3 illustrates the allocation of recycled feedstock in the products if the input material is a mixture of 90% fossil and 10% non-fossil raw materials (e.g. pyrolysis oil resulting from the pyrolysis of plastic waste) and the output is used for overall five different purposes (fractions):

When discussing polymers A and B as target products, the four cases shown in Fig. 16.3 arise—the allocation rules. For the sake of simplicity, the proportion of process energy is left out of consideration.

Rolling Average

If there is no allocation, then each fraction mathematically contains 10% shares of circular raw materials, which cannot be identified as such.

Proportional

If the proportion of circular raw material is allocated to a mass unit of polymer A or B, then those contain 100% of it.

Polymers Only

If one only looks at the polymers, two mass units of polymers A and B will each contain 100%.

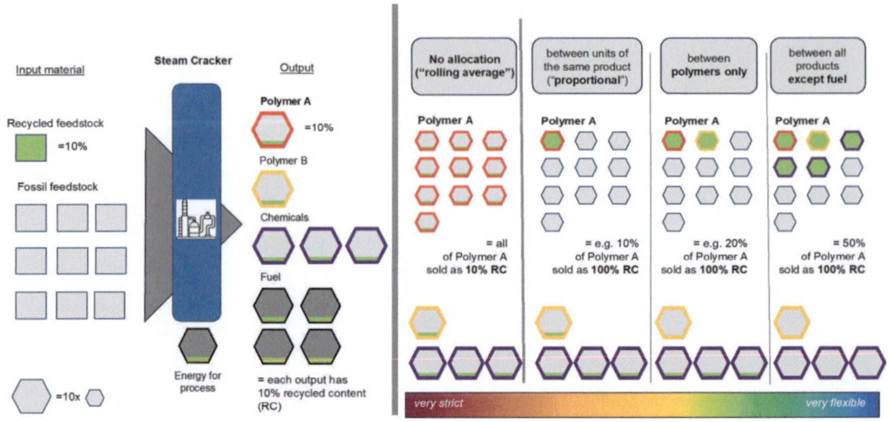

Fig. 16.3 Mass Balance concept applied to polymer production based on the mixture of fossil and recycled feedstock. (*Source* Schmidt 2023)

Fuel Use Exempt

If fuel and fuel components are not taken into account, 50% of polymers A and B will contain 100% circular raw material.

With the "Fuel use exempt" allocation rule, circular raw materials and in particular pyrolysis oils from chemical recycling in large-scale plants such as steam crackers and subsequent process steps can be mixed with other raw materials and still be assigned to a defined target product, for example, a polymer with a mandatory recycled content.

An allocation of recycled secondary raw materials without the possibility of mass balance allocation in selected target products could otherwise only be achieved through separate, small, and less efficient production facilities. The associated economic and ecological disadvantages as well as additional resource consumption would be counterproductive to the pursued complete recycling of plastics (ThinkTank 2023).

Process losses and circular raw materials that flow proportionately into the production of fuels are not counted as recycled content in target products. This allocation is supported by associations of the plastics industry and the chemical industry (VCI and PlasticsEurope 2023) as this appears to be compatible with waste regulation and economical in the industrial circular economy currently being developed. In addition, it enables the detection of recycled content in plastic products, as required by the regulator and the market.

16.4 Auditing and Certification

Such an approach must be audited and certified to create acceptance among customers and credibility among the public and to meet regulatory requirements. In the context of the EU Renewable Energy Directive (RED) (EU 2009), the German Federal Ministry of Agriculture (BMEL) and the German Agency for Renewable Resources (FNR) initiated a consulting project in 2008 to find a solution for proving the sustainability of biofuels (Feige 2016). This resulted, among other things, in a series of certification activities for the mass balance-based use of renewable raw materials. TÜV SÜD developed the certification standard CMS 71 (TÜV SÜD 2021) for the use of renewable raw materials, which has now been transferred to REDCert GmbH and is also used by the International Sustainability and Carbon Certification (ISCC), which emerged directly from the above-mentioned project. Both organizations have since expanded their certification activities to include a larger field of renewable or circular raw materials and energies (cf. Fig. 16.4): "REDcert²—system principles for the certification of sustainable material flows in the chemical industry" (REDCert 2010) and ISCC Plus (ISCC 2010). There are now more than 30 certification organizations working in this field worldwide (EMAF 2019).

Certification	General Information
ISCC (International Sustainability & Carbon Certification) Plus	▪ International Sustainability & Carbon Certification. ▪ Comitted to the **sustainable, social, economic practices** for the **processing and use** of different kinds **of biomass and their products**. ▪ Market applications for food & feed, chemical and energy industries.
REDcert²	▪ Certification of **sustainable material flows in the chemical industry.** ▪ It is a biomass balance approach, a verification method that can show how fossil raw materials have been replaced with sustainable biomass.
RSB Advanced Products Standard	▪ Roundtable on Sustainable **Biomaterials.** ▪ Drives the **development of a bio-based and circular economy** on a global scale through sustainability solutions, certification, and collaborative partnerships in **biomass, biomaterials and bioproducts.**
Certified Sustainable Palm Oil RSPO™ Supply Chain	▪ Roundtable on Sustainable Palm Oil. ▪ Transform markets to make **sustainable palm oil** the norm.

Fig. 16.4 Main mass balance certifications in the chemical industry. (*Source* Covestro 2021)

Plastics producers such as BASF, Covestro, Dow, Ineos Styrolution, Sabic, and Trinseo have now had locations in Europe, Asia and the USA certified according to ISCC-PLUS and supply raw materials and plastics that contain shares of circular raw materials from plants installed there. In this way, they give their customers the opportunity to enter the market with such traceable, certified products.

A rapid, legally secure applicability of suitable mass balance approaches is a key factor for scaling chemical recycling and for the provision of corresponding plastics with a significant proportion of recyclate for the entire value chain. Mass balances make a significant contribution to meeting the demand for products based on recycled raw materials in the market, including meeting the targets for PET beverage bottles (30%) under the Single-Use Plastics Directive (SUPD 2019).

16.5 Mass Balance Approach and European Legal Framework

In its study on "Mass balancing for chemical recycling" (ThinkTank 2023), the Industrial Resource Strategy think tank points out that regulation on mass balance allocation as part of the EU Single-Use Plastics Directive[2] serves as a model for further legal acts. To demonstrate the scope of this guiding principle, Table 16.1 lists important legal acts and others under preparation addressing recyclate content. Without the applicability of mass balance approaches according to the proposed Fuel Use Exempt allocation model, the necessary investments to service the recyclate uses regulated or expected here are unlikely to be made. The Fuel Use Exempt allocation rule should therefore receive comprehensive regulatory recognition beyond the scope of the EU Single-Use Plastics Directive.

[2] Art. 6, paragraph 5 Directive (EU) 2019/904 of the European Parliament and of the Council.

Table 16.1 EU directives and regulations that address recycled content in plastics. (*Source* ThinkTank 2023)

Legal act	Status	Relation to recyclate use
Single-use Plastic Directive (EU) 2019/904	In force	30% recycled content target in plastic beverage bottles by 2030
Sustainable design for mobile phones and tablets—ecodesign	Regulation in preparation	Presentation of requirements for recyclate content to reduce carbon footprint
Packaging Regulation	Draft published	Minimum recycled content for plastic packaging from January 1st, 2030
End-of-life vehicle Directive	In preparation	Targets for recycled content expected from 2030
Ecodesign Regulation for Sustainable Products	Legislative advice	30 delegated acts to be adopted by 2030 for various products, including targets for recycled content
Construction Products Regulation	Legislative advice	Option to set minimum requirements for recycled content in all products in scope
EU Taxonomy Regulation	In force	Criteria-based minimum requirements for recycling to make significant contributions to the transition to a circular economy

References

Biokraft 2009: Biokraftstoff-Nachhaltigkeitsverordnung 2009 zuletzt geändert 08.12.21

BKV 2021: BKV/Dechema/PlasticsEurope/VCI: Forschungspolitische Empfehlungen zum chemischen Kunststoffrecycling – neue Verfahren und Konzepte; Frankfurt, 05.07.2021

https://dechema.de/kunststoffrecycling/_/DiskussionsP%20DECHEMA_PED_VCI%20-%20Fo.polit.%20Empfehungen%20zum%20chem.%20Recycling%20-%20final%20050721.pdf (Last access: Aug. 22, 2024)

DB 2024: https://nachhaltigkeit.deutschebahn.com/de/gruene-transformation/ressourcenschutz (Last access: Aug. 21, 2024)

EMF 2019: Ellen MacArthur Foundation White Paper "Enabling a Circular economy for Chemicals with the Mass Balance Approach". https://emf.thirdlight.com/link/f1phopemqs36-8xgjzx/@/preview/1?oEllen%20MacArthur%20Foundation%20White%20Paper%20"Enabling%20a%20Circular%20economy%20for%20%20Chemicals%20with%20the%20Mass%20Balance%20Approach". (Last access: Aug. 22, 2024)

EU 2009: Directive 2009/28/EC of the European Parliament and of the Council of 23 April 2009 on the promotion of the use of energy from renewable sources (RED Directive)

Fair Trade 2022: https://www.fairtrade-deutschland.de/ (Last access: Jan. 26, 2022)

Feige 2016: Feige A. in Friedl, E., Spindler, E.A., Zertifizierung als Erfolgsfaktor, ISBN 9783658097004 Springer Wiesbaden 2016, S. 189

FSC 2022: https://fsc.org/en/certifed-forests (Last access: Jan 26, 2022)

ISCC 2010: https://www.iscc-system.org/process/overview/ (Last access: Jan. 27, 2022)

ISO 2020: ISO 22095:2020, Chain of custody — General terminology and models
FAPU 2015: Klein, D., Frietsch, S., Gleiche Produkteigenschaften – aber mit nachwachsenden Rohstoffen bei der Herstellung chemischer Produkte über den Massenbilanzansatz, FAPU 87, Jan. 2015, S.3
PlasticsEurope 2020: https://plasticseurope.org/knowledge-hub/mass-balance-approach-to-accele rate-the-use-of-renewable-feedstocks-in-chemical-processes/ (Last access: May 20, 2024)
REDCert 2010: https://www.redcert.org/ (Last access: Jan. 21, 2022)
Schmidt 2023: „Berechnung des Rezyklateinsatzes nach Art. 6(5) SUPD". Presentation by I. Schmidt, Industrievereinigung Kunststoffverpackungen e.V., Nov 2023.
SUPD 2019: Directive (EU) 2019/904 of the European Parliament and of the Council of 5th June 2019 on the reduction of the impact of certain plastic products on the environment
ThinkTank 2023: Thinktank Industrielle Ressourcenstrategie „Massenbilanzierung für das Chemische Recycling", Karlsruhe 2023. https://www.thinktank-irs.de/wp-content/uploads/2023/10/RZ_THINKTANK_Massenbilanzierung-chemisches-Recycling_A4_DE_Web-1.pdf (Last access: Aug. 22, 2024)
TÜV SÜD 2021: https://www.tuvsud.com/de-de/branchen/energie/erneuerbare-energien/erneuer[1]bare-rohstoffe (Last access: Jan 21, 2022)
VCI and PlasticsEurope 2023: https://plasticseurope.org/de/2023/03/03/positionspapier-massenbil anzen/ (Last access: Aug. 22, 2024)

Dr. Ulrich Liman is a chemist by education, he joined Bayer AG in 1989 in marketing of Polyurethane and Polycarbonates and held various managerial positions in Japan, China and Germany. Until 2022 he was responsible for polyurethane R&D and held a seat in the Supervisory Board of Covestro AG.

Dr. Peter Orth is a chemist by education; with OPC—Orth Plastics Consulting he leads a group of senior consultants with decades of professional experience in leading positions in the German and European plastics industry and its associations. His main focus is on innovation and sustainability in cooperation with science, industry and politics. He is currently working in the fields of CO_2 as a feedstock, resource efficiency, and recycling technologies.

Part IV
Cross-Cutting Topics

Part IV summarizes some cross-sectional technologies and methods that are important for the circular economy of plastics: Life Cycle Assessment (LCA) and Product Carbon Footprint (PCF), standardization and digitization. This also includes logistics, which is covered in Chap. 9. These technologies and methods are not limited to a single segment of the cycle but are equally important for all stakeholders involved. They lead to ongoing, step-by-step developments of procedures and processes as well as products and organizations. They are catalysts for product innovations and structural changes in the circular value chain. In the circuit diagram, their place is, therefore, in the hub, in the center.

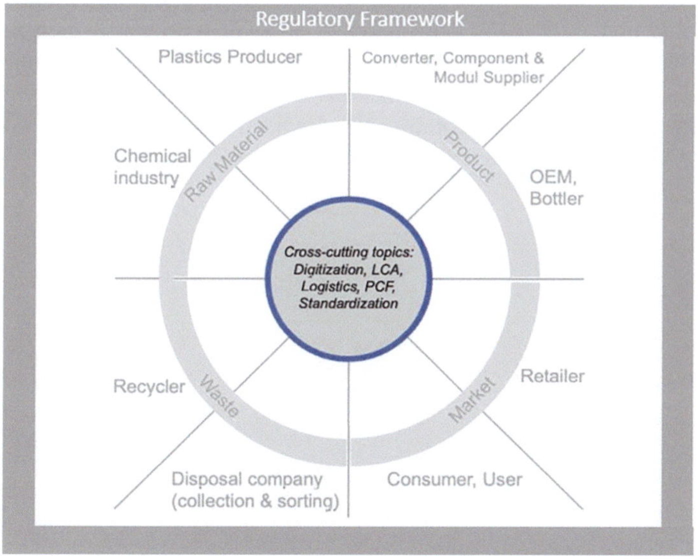

Chapter 17 presents the Life Cycle Assessment method (LCA) and its possibilities. The focus is on using this methodological apparatus to compare plastic recycling processes—mechanical recycling and chemical recycling—with incineration and with energy production. Particular attention is paid to recording and reducing the greenhouse gas emissions caused by these processes as well as to an emissions-neutral plastics economy by moving away from fossil raw materials.

The Product Carbon Footprint (PCF) emerged from atmospheric chemistry; it is actually part of the LCA methodology. Because of its importance in the context of this book, a separate Chap. 18 is dedicated to it. Indicating the greenhouse gases (GHG) attributable to a product and its associated processes based on the GHG Protocol, expressed in CO_2 equivalents, is becoming part of many companies' ESG reporting. The Together for Sustainability (TfS) initiative accompanies the use of PCF for the chemical and plastics industry and, together with specialized service providers and their extensive databases, supports the sophisticated modeling and calculation of the PCF for internal company purposes as well as for public use.

The circular economy relies more heavily on the use of digital methods and solutions for products and processes than the prevailing linear economy. Since all actors in the circular value chain are involved in its function, they also need all the information about its processes, products and product requirements. Concepts for cross-company data rooms and expanded governance structures create transparency and promote business activities. Systematic interfaces between companies to convey information promote the use of knowledge and skills. Digitization is a central enabler for the successful functioning of the circular supply chain, including the evolving recyclate business. This is elaborated in Chap. 19 using concrete examples.

Standards are essential for the circular economy and play an important role in the transformation, as shown in Chap. 20. It explains how and in what context standards apply for the use of plastics in the circular economy, how plastic waste and the recyclable materials made from it are to be dealt with, and which people and committees are involved. Standardization is a process at national, European and global levels, a game of complex interactions with a high level of commitment from the affected economic sectors. Binding standardization is becoming part of circular economy-related regulatory activities in European politics.

Chapter 17
Comparative Life Cycle Assessments of Circular Plastics (LCA)

Ronja Hermanns, Raoul Meys, and Manfred Renner

Contents

Summary

Plastics are mainly based on fossil resources, causing greenhouse gas emissions during current production, processing, and waste treatment. To reduce greenhouse gas emissions, a circular economy is seen as a very promising solution. Therefore, several scientific publications investigate the environmental benefits of a circular economy by comparing waste treatments such as recycling and incineration using life cycle assessments. These studies show that the waste hierarchy for treating plastic waste is also generally valid for reducing greenhouse gas emissions. Sorted plastic fractions should primarily be recycled mechanically, while plastics not suited for mechanical recycling should be kept in the cycle by means of chemical recycling. If plastic waste cannot be recycled mechanically or chemically, waste incineration

R. Hermanns (✉) · R. Meys
CarbonMinds GmbH, Cologne, Germany
e-mail: Ronja.Hermanns@carbon-minds.com

R. Meys
e-mail: raoul.meys@carbon-minds.com

M. Renner
Fraunhofer Institute UMSICHT, Oberhausen, Germany
e-mail: manfred.renner@umsicht.fraunhofer.de

P. Orth et al. (eds.), *Circular Economy of Plastics*,
https://doi.org/10.1007/978-3-031-78829-1_17

with energy recovery is the best option compared to other alternatives. However, there are also exceptions to the waste hierarchy. Co-incineration in cement plants, e.g. can reduce greenhouse gas emissions as compared to recycling, when plastic waste replaces coal or heavy oil. While a circular economy can substantially reduce greenhouse gas emissions, carbon capture and utilization (CCU) and biomass as feedstock must also be taken into account for net-zero greenhouse gas emission plastics.

17.1 Introduction

In addition to the problem of littering, other environmental impacts of plastics have recently gained public attention. A particular problem is the almost exclusive production of base chemicals and plastics from oil to cover the energy and feedstock demands of chemical processes. An estimate by the International Energy Agency (IEA) assumes an increasing demand for chemicals, leading to 20% of global oil demand by 2050 (IEA 2018). The increasing use of fossil resources inevitably leads to a scientifically proven contribution of plastics to man-made climate change (Meys et al. 2021). Some studies even show that plastics could account for almost 15% of the yearly global greenhouse gas emissions budget allowed by the 1.5 °C target in 2050 (Zheng und Suh 2019; Ellen MacArthur Foundation 2016).

To reduce environmental impacts, implementing a circular economy is gaining increasing attention and interest. For this purpose, mechanical and chemical recycling should be used to avoid landfilling and incineration of plastics and, thus, the ongoing use of fossil resources (cf. Fig. 17.1). However, scientific studies show that a well-founded analysis is necessary to ensure that mechanical and chemical recycling effectively reduce environmental impacts (Geyer et al. 2016; Shen et al. 2010; Meys et al. 2020; Jeswani et al. 2021; Patel et al. 2000).

For an overview of the environmental impact analysis of circular plastics, the following chapter summarizes scientific studies on life cycle assessments (LCA) of various processes for recycling and treatment of plastic waste. The summary reflects the state of science and practice and identifies the advantages and disadvantages of the processes. The final section provides a broader outlook on the possibilities of net-zero greenhouse gas emission plastics.

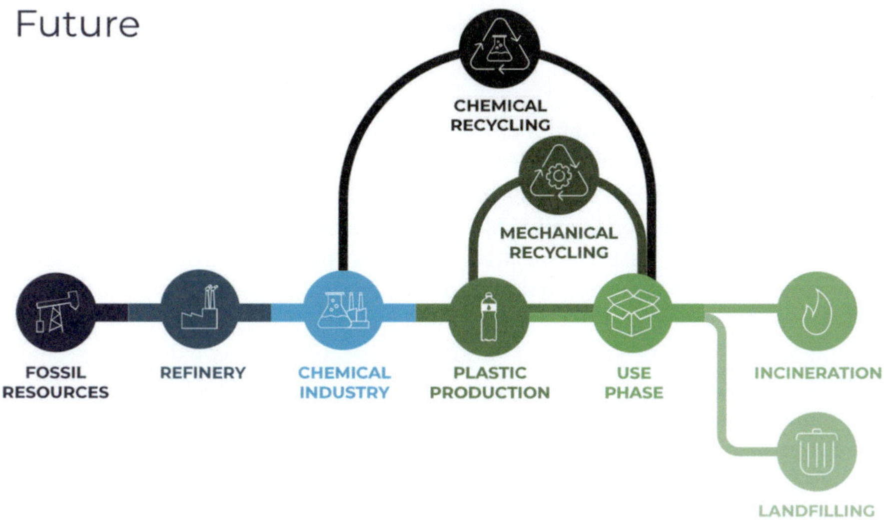

Fig. 17.1 Potential circular plastic cycles through mechanical and chemical recycling (*Source* Own representation)

17.2 Comparative Life Cycle Assessment of Circular Plastics

A *Life Cycle Assessment (LCA*) is used to calculate the environmental impacts of products or processes over their entire life cycle from *cradle to grave* and thus make them visible and comparable (Kloepffer and Grahl 2009). A particular application of LCAs is the impact analysis of circular plastics. This chapter provides an overview of comparative LCAs of different recycling processes (chemical and mechanical recycling) with conventional incineration processes (waste incineration or co-combustion in cement plants). As landfilling is to be systematically reduced in the European Union (less than 10% in 2035) to strengthen the circular economy, landfilling is not discussed further (European Parliament 2023).

Section 17.2.1 first explains the methodological basis for the LCA of circular plastics. The following sub-sections summarise LCAs from the literature. Section 17.2.2 compares waste incineration and recycling, Sect. 17.2.3 compares co-incineration and recycling, and Sect. 17.2.4 compares mechanical and chemical recycling. Finally, Sect. 17.3 summarises the overarching conclusions from a life cycle perspective for the circular economy, with a focus on Europe. The following sections focus on greenhouse gas emissions, as they represent the main environmental impact of current concerns. In some places, relevant differences between greenhouse gas emissions and other environmental impacts are also highlighted and explained.

17.2.1 Methodological Principles

The methodology for LCAs of circular plastics is based on DIN EN ISO 14040 and 14044. The standard allows a fair comparison between the treatment of plastic waste and recycling processes. The basis for a fair comparison is the functional unit (EN ISO 14040 2021; DIN EN ISO 14044 2021; Alhazmi et al. 2021). For the functional unit's definition in an LCA, various perspectives can be chosen according to the literature. The waste perspective is the most frequently used one, whereby a distinction is often made between the use of 1 kg of plastic waste and a certain amount of plastic waste in a region or per year (Antelava et al. 2019; Astrup et al. 2015). In addition to the waste perspective, a product perspective selects individual products as the functional unit. Therefore, when comparing different fuels in cement plants, the production of one ton of cement is also used as a functional unit (Galvez-Martos and Schoenberger 2014). A published study by BASF is an example in which the pyrolysis of plastics is analysed from waste and product perspectives. In this study, the functional unit is defined once as the treatment of 1 ton of mixed plastics waste (waste perspective) and once as the production of 1 ton of polyethylene granulate in virgin quality (product perspective) (Russ et al. 2020).

All LCAs must be based on the same functional unit to holistically compare different recycling and incineration processes. Since all potential incineration and recycling processes are dependent on the available plastic waste, for LCAs of circular plastics, the waste perspective provides a robust basis for various comparisons and simplifies the understanding of the most beneficial technologies (Antelava et al. 2019; Astrup et al. 2015; Davidson et al. 2021; Mayer et al. 2019). While the same composition and quality of plastic waste is always defined as the functional unit for comparison, plastic waste generation can be considered the same for all processes. Therefore, the environmental impacts for the life cycle stages of plastic waste generation are also the same for incineration and recycling. In comparative LCAs, the same environmental impacts from certain life cycle phases, such as the generation of plastic waste, can be neglected, since the result of the comparative LCAs is not affected (Eriksson and Finnveden 2009).

If incineration and a recycling process are compared with each other based on the waste perspective, two effects must be taken into account (IN4climate.NRW (Ed.) 2020). These two effects are (1) all environmental impacts of the processes and (2) all products substituted on the market must be considered. For the products on the market, a so-called "avoided burden" is often used in life cycle assessments (DIN EN ISO 14044 2021; EN ISO 14040 2021).

Figure 17.2 shows the products resulting from incineration and recycling as well as their credits. Material products result from mechanical and chemical recycling, and energetic products from incineration, such as waste incineration or co-combustion in cement plants (Davidson et al. 2021; IN4climate.NRW (Ed.) 2020). This methodological approach can be used for comparative LCAs of all incineration and recycling processes. For this purpose, only the environmental impacts of the respective incineration and recycling processes must be calculated, and the valuable product must be credited. The credit strongly depends on the defined primary product substituted. The resulting credit may vary, e.g. by a factor of 20 for the impact on climate change (coal vs. wind energy) (Ecoinvent 2019). Such different credits can significantly influence the result of the LCA and, thus, the results of comparing recycling and incineration processes (Eriksson und Finnveden 2017).

Regardless of the credit, the results of a technology comparison can also vary greatly between the environmental impacts investigated. For this reason, an LCA always considers various environmental indicators to obtain a holistic picture of the environmental impact of processes (DIN EN ISO 14044 2021; EN ISO 14040 2021).

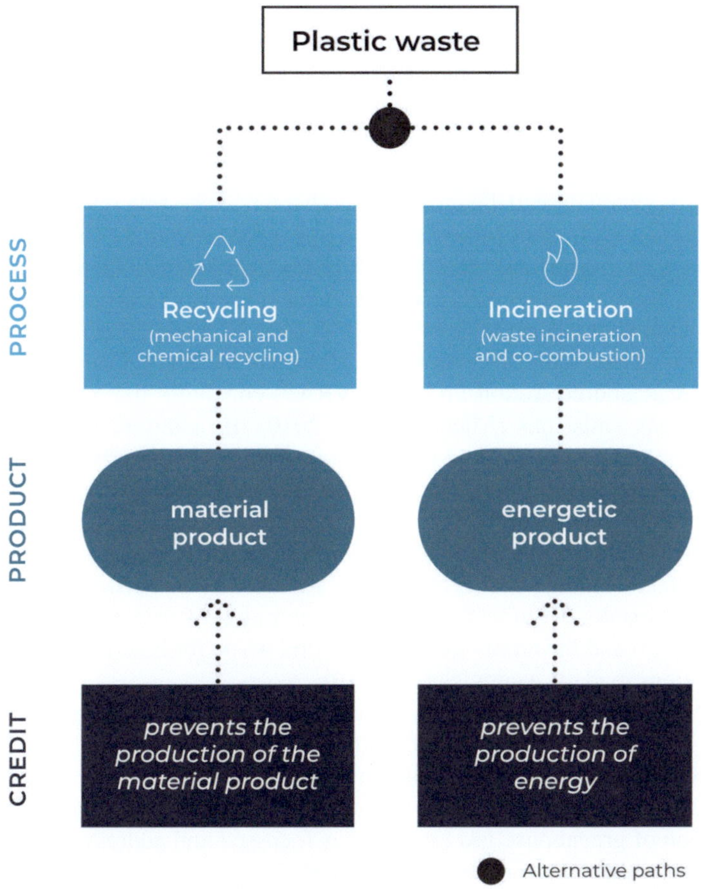

Fig. 17.2 Consideration of the environmental impacts of the additional functions of the recycling and incineration processes (*Source* Own representation)

17.2.2 Waste Incineration and Recycling in Comparison

Many studies investigate the comparison between waste incineration and recycling (IN4climate.NRW (Ed.) 2020; Hermanns et al. 2023; Meys et al. 2020; Alhazmi et al. 2021). In this chapter, recycling refers to mechanical and chemical recycling if no explicit distinction is made between these processes. Most studies evaluate whether a fundamental switch to recycling has a higher environmental benefit than waste incineration with energy recovery. Throughout this chapter, unsorted plastics waste is assumed for a fair comparison, as it may be directly incinerated. In the case of recycling, the comparative LCA includes the sorting step as part of the system boundary.

The discussion of the environmental benefits of recycling compared to incineration has been on the scientific agenda for some time already, with the first detailed review articles summarizing studies from 2000 to 2009 (Cleary 2009; Lazarevic et al. 2010; Laurent et al. 2014; Tyskeng and Finnveden 2010; Michaud et al. 2010). Since a detailed analysis of > 50 independent studies before 2010, each with several LCAs and scenarios would exceed the scope of this chapter, hence only relevant review articles are considered. In their review on recycling, Lazarevic et al. compare 47 LCAs showing clear environmental benefits of recycling over waste incineration (Lazarevic et al. 2010). Mechanical recycling reduces greenhouse gas emissions in 79% of the 39 studies considered (Lazarevic et al. 2010). Chemical recycling leads to a reduction in 100% of the studies considered (Lazarevic et al. 2010). Consistent with this, the Waste & Resource Action Program (WRAP) concludes that recycling reduces greenhouse gas emissions (Michaud et al. 2010). In addition, the WRAP study differentiates between individual polymer types such as polyethylene, polypropylene, polystyrene, polyethylene terephthalate, and polyvinyl chloride. Moreover, in 90% of the studies, there is also a reduction in abiotic resource depletion, energy use, and acidification (Lazarevic et al. 2010; Michaud et al. 2010). However, this clear environmental advantage of recycling is not reflected in all environmental impacts. For example, results of studies for the supersaturation of nutrients in water bodies (eutrophication) and the amount of residual solid waste for landfills vary depending on assumptions made (Lazarevic et al. 2010).

More recent studies and reviews also confirm the environmental advantage of recycling over waste incineration (Davidson et al. 2021; Meys et al. 2020; Dong et al. 2018; Antelava et al. 2019; Helm et al. 2023; Hermanns et al. 2023; Almadhi et al. 2023; Bachmann et al. 2023; Alhazmi et al. 2021; Volk et al. 2021). While the reduction of greenhouse gas emissions is focussed and addressed in all studies, some also show benefits for other environmental impacts, such as abiotic depletion potential, cumulative energy demand, and fossil depletion (Alhazmi et al. 2021; Meys et al. 2020; Hermanns et al. 2023; Volk et al. 2021). However, for environmental impact indicators, such as eutrophication and acidification results are not clear and vary according to the recycling technologies compared (Meys et al. 2020; Hermanns et al. 2023; Alhazmi et al. 2021). The comparison of all studies shows that two technical parameters mainly influence the comparison between recycling and waste incineration: (1) substitution of virgin materials, and (2) the credit for the produced energy from waste incineration.

(1) **Substitution of virgin materials**

Quality of materials recycled (recyclates) is particularly important. The higher the recyclate quality, the higher the number of virgin feedstocks and resulting products that can be substituted in the market. In return, substituting higher-quality products increases the environmental credit of a recycling technology. For mechanical recycling, the ratio between the recycled material product and the initially substituted product is defined by the substitution rate (Ragaert et al. 2017). The substitution rate is 1 if the recycled material product replaces virgin plastics. For chemical recycling, no substitution rate is defined since products are chemically identical or similar to those produced with fossil feedstock like naphtha (Michaud et al. 2010). The following paragraphs, therefore, deal separately with mechanical and chemical recycling.

Mechanical recycling is often defined by the substitution rate, which indicates the quality ratio between the recycled material product and the substituted product (Ragaert et al. 2017).

The lower the substitution rate, the lower the credit and the environmental benefits of mechanical recycling. Lower credits and environmental benefits affect all environmental impacts (Lazarevic et al. 2010; Michaud et al. 2010; Volk et al. 2021). Jeswani et al. (2021) show in their sensitivity analysis for a plastic mix of polyethylene, polypropylene, and polystyrene that reductions in greenhouse gas emissions compared to energy use in the waste-to-energy plant are possible even for poor recyclate qualities unsuitable for its intended use. Complementing this study, Lazarevic et al. and Shonfield et al. compare environmental impacts such as eutrophication, terrestrial acidification, energy use, and fossil depletion in addition to greenhouse gas emissions. They show that below a substitution rate of 0.5, it was no longer possible to make a clear statement on the environmental benefits of mechanical recycling over waste incineration (Lazarevic et al. 2010; Michaud et al. 2010; Shonfield 2008). Thus, defining the substitution rate as realistically as possible for LCAs is crucial to avoid over- or underestimating environmental impacts. Due to mostly unavoidable downcycling, a complete substitution (1:1) with equivalent quality is, in reality, the exception (Geyer et al. 2016; Schyns and Shaver 2021). Various studies have classified a ratio of around 1:0.8 as a more realistic substitution rate (Rigamonti et al. 2009; Turner et al. 2015; Zheng und Suh 2019). While the substitution rate is defined for closed cycles (closed-loop), mechanical recycling in open-loop systems often leads to other recyclates, which can also provide a high ecological added value (Ragaert et al. 2017). One example is recyclates from PET beverage bottles reused as nonwoven textiles (bvse-Fachverband Kunststoffrecycling 2020). Therefore, the discussion on substitution rates leads to the conclusion that mechanical recycling processes should aim to substitute products with high environmental impacts, regardless of whether the materials are in closed or open loops (Geyer et al. 2016). Recycling rates may be used as alternative to substitution rates basically looking at the potential amount of substituted primary products (Demets et al. 2021).

Unlike recyclates (mechanical recycling products), chemical recycling products replace products further up the value chain, such as naphtha, basic chemicals, or monomers (raw materials). These conventional products and their equivalents from chemical recycling are usually chemically very similar or identical. However, further production steps, such as steam cracking and polymerization, are necessary to produce comparable end products. For this reason, products of chemical recycling generally have a lower ecological credit than products of mechanical recycling (Meys et al. 2020). Exceptions, where chemical recycling delivers high credits, may be processes delivering high-value products (i.e. upcycling). In upcycling, polymer waste is converted to new chemicals offering higher value. Examples include the processing of PET to cyclohexanedimethanol and ethylene glycol (Meys et al. 2020). Compared to waste incineration, however, chemical recycling processes such as pyrolysis and gasification with lower-value material products also show environmental benefits (Lazarevic et al. 2010; Jeswani et al. 2021; Russ et al. 2020; Hermanns et al. 2023).

(2) **Credits for energy produced**

This credit represents environmental impacts that would have occurred with conventional energy production, now replaced by waste incineration. In most of the abovementioned studies, energy credits are applied to a country's average electricity and heat production or specific fossil energy sources (coal, natural gas, oil) (Lazarevic et al. 2010; Michaud et al. 2010). For the average energy supply with a high share of fossil energy sources, waste incineration provides high credits for electricity and heat (Rigamonti et al. 2009) (cf. Fig. 17.2). Despite high energy credits for waste incineration, recycling processes usually achieve environmental advantages in these LCAs (Michaud et al. 2010). An exception was shown by Jeswani et al. in their sensitivity analysis: if waste incineration with credit for coal-based electricity and thermal energy from heavy oil is used, waste incineration leads to lower overall greenhouse gas emissions than plastic waste recycling (Jeswani et al. 2021). However, most countries plan to switch energy production to renewable and lower-emission technologies like wind, solar, hydropower, and biomass (European Commission 2016). Using renewable energy decreases the credit for greenhouse gas emissions from the energy product. Decarbonization given, the environmental impact of greenhouse gas emissions from waste incineration will be even worse than recycling (Zheng und Suh 2019; Jeswani et al. 2021; Russ et al. 2020; Hermanns et al. 2023).

Overall, the literature on comparative life cycle assessments shows clear advantages for recycling over waste incineration (Jeswani et al. 2021; Zheng und Suh 2019; Russ et al. 2020; Hermanns et al. 2023). While literature focuses heavily on the environmental impacts of global warming potential and fossil depletion, other environmental impacts, such as acidification and eutrophication, are considered less important. In a holistic view, all environmental impacts should be taken into account to consider all ecological consequences—including local/regional ones—and to take timely measures if necessary. Integrating renewable feedstock and energies might also lead to increased environmental impacts. Large-scale use of biomass e.g. leads to increased acidification of soils and eutrophication (Walker und Rothman 2020).

In the case of renewable electricity from wind farms and solar plants, more metals, water, and land are consumed (World Energy Council 2004; Meys et al. 2021). Therefore, the advantages or disadvantages of recycling processes over waste incineration regarding other environmental impacts have not been fully investigated yet. However, Bachmann et al. 2023 show that net-zero emission plastics can only be produced within the planetary boundaries if recycling technologies are improved and extended (Bachmann et al. 2023).

17.2.3 Co-incineration and Recycling in Comparison

Another form of using plastic waste for producing energy is co-incineration in cement plants, differing from conventional waste incineration in terms of waste utilization and substituted fuel. Plastic waste is used in cement plants for heat generation mainly as a substitute for coal (Cossu et al. 2017; Meys et al. 2020; Russ et al. 2020). Here plastic waste is advantageous having higher calorific value than coal or other alternative fuels, such as used tires (Bourtsalas et al. 2018; Cossu et al. 2017; Rahman et al. 2015).

In their study, Rigamonti et al. analyse different disposal scenarios, including three with 20% co-incineration of plastic waste in cement plants. They show that co-incineration achieves the relatively highest greenhouse gas reduction (Rigamonti et al. 2014). Due to the high efficiency of mixed plastic waste as a fuel and the high credit from substituting coal, it is more beneficial to use mixed plastics in co-incineration than to recycle them (Rigamonti et al. 2014; Lazarevic et al. 2010; Meys et al. 2020; Kosajan et al. 2020; Russ et al. 2020). Russ et al. even show that using mixed plastics in co-incineration in cement plants instead of coal can reduce approx. 1.6 t CO_2 compared to pyrolysis (chemical recycling) (Russ et al. 2020).

However, the advantages of co-combustion in the cement plant only arise if coal is substituted. If other fuels, such as biomass or natural gas, are used, there are clear advantages for mechanical or chemical recycling in terms of greenhouse gas emissions (Russ et al. 2020). Reasons for the different impacts result from alternative fuels leading to lower greenhouse gas emissions than coal, which is why the credit for the energetic product of co-combustion decreases (Meys et al. 2020; Russ et al. 2020).

In addition to global warming potential, other environmental indicators must also be considered for a holistic comparison between recycling and co-incineration. For environmental indicators such as abiotic resource consumption or fossil resource depletion, ecological advantages of recycling over co-combustion in a cement plant are evident, even if coal is substituted by plastic waste (Meys et al. 2020; Lazarevic et al. 2010). The ecological advantages of recycling for the other two environmental indicators and the strong dependence on the fuel used in the cement plant (coal vs. biomass/natural gas) show that no clear statement can be made about the ecological advantages of co-incineration plastic waste in cement plants.

17.2.4 Chemical and Mechanical Recycling in Comparison

The following section compares the two recycling routes mentioned. Only sorted plastic waste is considered, as mechanical recycling without sorting is currently not technically feasible. This section distinguishes between mono and mixed plastic waste from sorting.

Mono-material plastic waste

Large volumes of mono-plastic waste (only one component) result from packaging and consist of polyethylene, polypropylene, PET (polyethylene terephthalate), or polystyrene. Lifecycle assessment analyses exist for this mono-plastic waste that compare mechanical recycling with chemical recycling, such as pyrolysis, gasification, depolymerization, solvolysis, or chemical upcycling.

Compared to mechanical recycling, pyrolysis and gasification result in higher greenhouse gas emissions for mono-plastics (Meys et al. 2020; Shen et al. 2010; Vogt et al. 2021; Volk et al. 2021). For instance, Meys et al. and Schwarz et al. underline these overall results for polyethylene, polypropylene, PET, and polystyrene (Meys et al. 2020; Schwarz et al. 2021). Moreover, for PET waste, a detailed analysis was done by Shen et al. (2010) using different methodological approaches to specifically investigate PET beverage bottles' mechanical and chemical recycling processes into textile fibres. Results over the entire product life cycle show that mechanical recycling reduces 23% more greenhouse gas emissions than pyrolysis or gasification (Shen et al. 2010).

However, during mechanical recycling, degradation mechanisms occur, resulting in a limited number of recycling cycles influenced by different factors such as temperature, additives included in the plastic, and added stabilizers (Schyns and Shaver 2021). After some cycles, products are no longer suited for mechanical recycling but are still suitable for pyrolysis or gasification. Additionally, mono-plastic wastes, such as polystyrene foams, are unsuitable for mechanical recycling, as the blowing agent must be removed beforehand, and additives impair mechanical recycling (Schyns and Shaver 2021).

Besides pyrolysis and gasification, other existing chemical recycling technologies, such as depolymerization and solvolysis could lead to a potential greenhouse gas emission reduction (Biessey et al. 2023; Ragaert et al. 2023). Another depolymerisation process is the methanolysis of polyester to dimethyl terephthalate and ethylene glycol using methanol as a solvent. This process is based on data from Eastman's methanolysis technology, showing impact reductions in more than 90% of evaluated indicators as compared to fossil dimethyl terephthalate. Meys et al. 2020 also confirm a greenhouse gas reduction potential for the methanolysis of PET compared to mechanical recycling. Besides PET, depolymerization shows ecological potential for lower greenhouse gas emissions than mechanical recycling of polyethylene, polypropylene, and polystyrene (Meys et al. 2020; Schwarz et al. 2021). For polystyrene, solvolysis using polar solvents shows additional reduction potential (Schwarz et al. 2021; García et al. 2009). Besides defining other chemical recycling technologies, chemical upcycling is gaining attention, defining the conversion of plastic waste to higher value materials (Meys et al. 2020; Choi et al. 2022). For PET, chemical upcycling to bis(2-hydroxyethyl)terephthalate, 1,4-benzenedimethanol, or 1,4-cyclohexanedimethanol has the potential to reduce greenhouse gas emissions compared to mechanical recycling (Meys et al. 2020). Most of the discussed other chemical recycling technologies and chemical upcycling are still at the research and development stage, hence only preliminary life cycle assessments based on simplified data are currently available. Therefore, it is difficult to assess whether concrete ecological advantages arise in industrial plants, as complete life cycle assessments are lacking (Vogt et al. 2021; Bachmann et al. 2023).

Mixed plastics waste

Sorting a mixture of plastic wastes almost always results in, sorted, mixed plastics (Schyns und Shaver 2021; Ragaert et al. 2017). Although less contaminated than unsorted plastic waste, such mixtures still consist of several types of plastics. Mechanical recycling of such mixed plastics presents particular challenges since different chemical properties and the presence of different additives do not result in homogeneous masses, and decomposition can occur during extrusion (degradation) (Schyns and Shaver 2021; Ragaert et al. 2017). Due to such difficulties, mixed plastics waste can often only be processed into relatively low-value products such as park benches, flower pots, or fillers (Huysman et al. 2015). Comparative LCAs of recycling processes show the environmental benefits of mechanical recycling in terms of greenhouse gas emissions, assuming that both recycling processes fully replace primary products (Michaud et al. 2010; Shonfield 2008). In recent literature, however, the complete substitution of primary products is viewed critically; for this reason, credits for primary products should be critically questioned (Demets et al. 2021). For instance, the subsequent sensitivity analysis of Michaud et al. (2010) confirms the statement using Shonefield's calculations. In his sensitivity scenario, only 20% of primary products (virgin plastics) are substituted. Otherwise, lower value products such as wood (40%) for park benches or concrete (40%) for flower pots are substituted (Shonfield 2008; Huysman et al. 2015). Scenario comparison shows that greenhouse gas emissions for 20% virgin substitution can even switch from reducing to increasing greenhouse gas emissions (Michaud et al. 2010). Based on this large change pyrolysis provides the more environmentally beneficial recycling technology for the sensitivity scenario (Michaud et al. 2010). The example illustrates that comparative life cycle assessments of mixed plastics waste recycling processes strongly depend on the quality of products that can be substituted in the market. For this reason, no general statement can be made on environmentally optimal recycling technology, as separate LCAs would have to be carried out for specific mixed plastics and their specific qualitative substitution possibilities.

For mixed plastic waste, chemical recycling processes such as pyrolysis and gasification may achieve clear environmental benefits, in particular, for mixed plastic waste not suited for mechanical recycling (Arena 2012; Antelava et al. 2019; IN4climate.NRW (Ed.) 2020; Al-Salem et al. 2010; Hermanns et al. 2023). In this regard, several studies state that chemical and mechanical recycling should be considered as complementary technical solutions. Sorted plastic waste should be recycled mechanically, and mixed plastic waste should be recycled mechanically as long as high-quality recyclates can be produced (Michaud et al. 2010). Otherwise, mixed plastic waste should be recycled chemically (Davidson et al. 2021). By following both process routes, the amount of recycled plastic waste can be significantly increased, and incineration minimized. This complementarity is supported by several scientific studies (Davidson et al. 2021; Helm et al. 2023; Bachmann et al. 2023; Meys et al. 2021; Volk et al. 2021).

17.3 Conclusions for the Circular Economy

The previous sections provide a detailed picture of the environmental impacts of various incineration and recycling processes used to treat and recycle plastic waste. At a higher level, individual comparisons clarify that the waste hierarchy regarding greenhouse gas emissions remains true but with some exceptions (BMUV 2017).

Mono and mixed plastics fractions are produced when plastic waste is collected and sorted. Mono-plastic fractions should be recycled mechanically, recyclates should be used instead of conventional plastic products to reduce greenhouse gas emissions (Michaud et al. 2010). This also applies to mixed plastics, provided that high-quality recyclates can be produced from them. Otherwise, mixed plastic waste should be chemically recycled, preferably producing high-value products (Schwarz et al. 2021; Meys et al. 2020, 2021). If this is not possible, pyrolysis techniques should be used to produce chemical feedstock, such as naphtha.

It is also clear from the individual analyses that mechanical and chemical recycling should be used complementary. Combining both process routes offers an excellent opportunity for cascading and optimal circularity of collected and sorted plastic waste while avoiding waste incineration (Davidson et al. 2021).

If plastic waste cannot be recycled mechanically or chemically, waste incineration is the next most advantageous option. In this case, care must be taken to ensure that electricity and heat generation are utmost efficient. However, the literature review also shows that while decarbonization of the electricity and heat sectors is increasing, waste incineration becomes even less attractive (Jeswani et al. 2021; Hermanns et al. 2023).

An exception to the validity of the waste hierarchy is co-incineration in the cement plant. In some cases, fossil fuels, such as coal or heavy oil, are replaced by plastic waste. This can lead to reduced greenhouse gas emissions compared to recycling processes (Russ et al. 2020; Volk et al. 2021). In contrast, recycling processes have a clear advantage when substituting other fuels such as biomass or natural gas (Meys et al. 2020; Russ et al. 2020). The chemical upcycling of polyethylene terephthalate into high-value chemicals can also reduce greenhouse gas emissions compared with mechanical recycling, thus contradicting the classic waste hierarchy (Meys et al. 2020).

While the picture is relatively consistent for greenhouse gas emissions, the ranking of incineration and recycling processes regarding other environmental impacts is less clear. If plastic waste is treated in waste incinerators with the substitution of renewable electricity from wind power or biomass, benefits could result in other environmental impacts. For instance, the large-scale use of biomass increases soil acidification and the large-scale generation of wind and solar power leads to increased consumption of metals, water, and land (Walker und Rothman 2020; World Energy Council 2004).

The holistic ecological assessment of the various incineration and recycling processes reveals clear trends for a circular economy. It becomes clear that plastic waste should increasingly be collected separately and then sorted into pure fractions where possible. This way, the volume of plastic waste for the various recycling routes can be significantly increased. It is also evident that chemical and mechanical recycling are complementary processes and should not be seen in direct competition. Cascading of plastic waste to maximum circularity through both recycling routes should remain the primary focus for a circular economy.

17.4 Net-Zero Greenhouse Gas Emission Plastics through Circular Economy

To achieve net-zero, all greenhouse gas emissions must be set to zero across the entire value chain. Two life cycle phases need special consideration for plastics, unlike other basic materials such as steel or cement. Firstly, plastics production's process- and energy-related greenhouse gas emissions must be reduced to zero. Secondly, emissions at the end of life, particularly from incineration in waste treatment, must be reduced to zero (Zheng und Suh 2019).

Here, mechanical and chemical recycling processes play a crucial role in reducing greenhouse gas emissions from production and waste treatment at the same time. Studies show that implementing either recycling process would reduce global greenhouse gas emissions between 25 and 64% (Zheng und Suh 2019; Meys et al. 2021). While the reduction shows the high potential of recycling, it also illustrates that recycling alone is not enough to achieve net-zero greenhouse gas emission plastics (100% reduction). Even if all plastics worldwide were to be recycled mechanically or chemically, residual waste would inevitably be produced. Such residual wastes, in turn, can only be incinerated (Ragaert et al. 2017). While recycling cannot be 100% efficient, the resulting demand gap must be filled with virgin plastics. For these virgin plastics, renewable resources, such as biomass or CO_2-based feedstocks, need to be used to achieve overall net-zero greenhouse gas emission plastics. In another article by the authors of this chapter (Meys et al. 2021), it was shown that only by combining mechanical and chemical recycling processes with Carbon Capture and Utilization (CCU) and the use of biomass net-zero greenhouse gas emission plastics can be achieved.

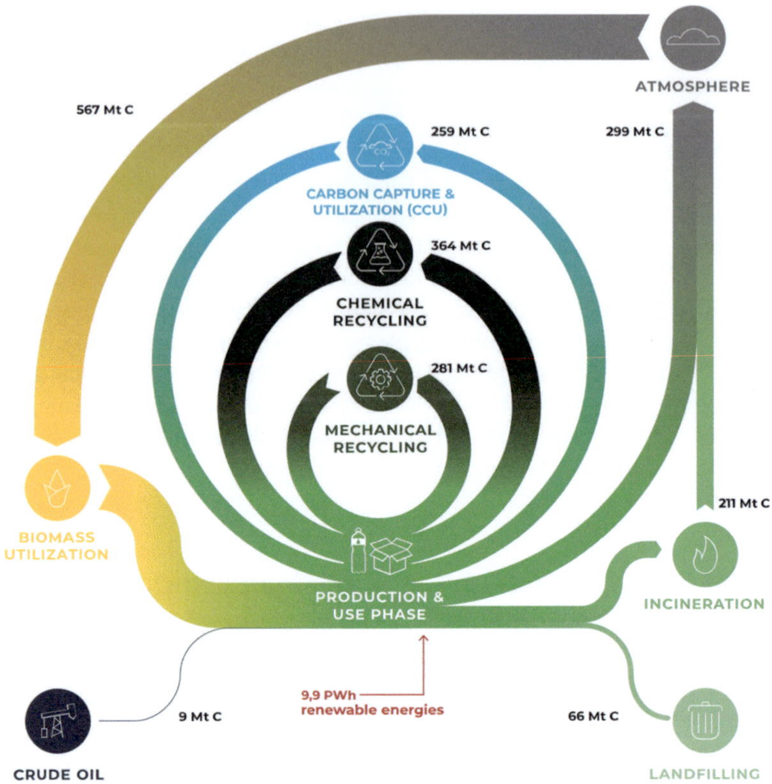

Fig. 17.3 Future carbon fluxes for net-zero greenhouse gas emission plastics (*Source* Meys et al. 2021; reprinted with permission from AAAS)

In such a net-zero greenhouse gas emission plastics industry, residual emissions from incineration of non-recyclable wastes are compensated for by using biomass and the associated uptake of carbon dioxide. Additionally implemented CCU technology enables carbon dioxide from production processes such as fermentation or biomass gasification to be reprocessed into valuable products with renewable hydrogen from renewable energies (cf. Fig. 17.3).

References

Alhazmi, Hatem; Almansour, Faris H.; Aldhafeeri, Zaid (2021): Plastic Waste Management: A Review of Existing Life Cycle Assessment Studies. In: Sustainability 13 (10), S. 5340. https://doi.org/10.3390/su13105340.

Almadhi, Abdulmajeed; Abdelhadi, Abdelhakim; Alyamani, Rakan (2023): Moving from Linear to Circular Economy in Saudi Arabia: Life-Cycle Assessment on Plastic Waste Management. In: Sustainability 15 (13), S. 10450. https://doi.org/10.3390/su151310450.

Al-Salem, S. M.; Lettieri, P.; Baeyens, J. (2010): The valorization of plastic solid waste (PSW) by primary to quaternary routes: From re-use to energy and chemicals. In: Progress in Energy and Combustion Science 36 (1), S. 103–129. https://doi.org/10.1016/j.pecs.2009.09.001.

Antelava, Ana; Damilos, Spyridon; Hafeez, Sanaa; Manos, George; Al-Salem, Sultan M.; Sharma, Brajendra K. et al. (2019): Plastic Solid Waste (PSW) in the Context of Life Cycle Assessment (LCA) and Sustainable Management. In: Environmental management 64 (2), S. 230–244. https://doi.org/10.1007/s00267-019-01178-3.

Arena, Umberto (2012): Process and technological aspects of municipal solid waste gasification. A review. In: Waste management (New York, N.Y.) 32 (4), S. 625–639. https://doi.org/10.1016/j.wasman.2011.09.025.

Astrup, Thomas Fruergaard; Tonini, Davide; Turconi, Roberto; Boldrin, Alessio (2015): Life cycle assessment of thermal waste-to-energy technologies: review and recommendations. In: Waste management (New York, N.Y.) 37, S. 104–115. https://doi.org/10.1016/j.wasman.2014.06.011.

Bachmann, Marvin; Zibunas, Christian; Hartmann, Jan; Tulus, Victor; Suh, Sangwon; Guillén-Gosálbez, Gonzalo; Bardow, André (2023): Towards circular plastics within planetary boundaries. In: Nat Sustain 6 (5), S. 599–610. https://doi.org/10.1038/s41893-022-01054-9.

Biessey, Philip; Vogel, Julia; Seitz, Mathias; Quicker, Peter (2023): Plastic Waste Utilization via Chemical Recycling: Approaches, Limitations, and the Challenges Ahead. In: Chemie Ingenieur Technik 95 (8), S. 1199–1214. https://doi.org/10.1002/cite.202300042.

BMUV (2017): Leitfaden zur Anwendung der Abfallhierarchie nach § 6 Kreislaufwirtschaftsgesetz (KrWG) - Hierarchiestufen Recycling und sonstige Verwertung. Available online under https://www.bmuv.de/fileadmin/Daten_BMU/Download_PDF/Abfallwirtschaft/krwg_leitfaden_abfallhierarchie_bf.pdf, last access in 2017.

Bourtsalas, A. C.; Zhang, Jiao; Castaldi, M. J.; Themelis, N. J.; Karaiskakis, Alexandros N. (2018): Use of non-recycled plastics and paper as alternative fuel in cement production. In: Journal of Cleaner Production 181, S. 8–16. https://doi.org/10.1016/j.jclepro.2018.01.214.

bvse-Fachverband Kunststoffrecycling (2020): PET Recycling Saubere Wiederverwertung. Online verfügbar unter https://www.bvse.de/gut-informiert-kunststoffrecycling/nachrichten-recycling/6441-pet-getraenkeflaschen-hohe-recyclingquote-und-steigender-einsatz-von-recyceltem-material.html.

Choi, Jiho; Yang, Inchan; Kim, Sung-Soo; Cho, Se Youn; Lee, Sungho (2022): Upcycling Plastic Waste into High Value-Added Carbonaceous Materials. In: Macromolecular rapid communications 43 (1), e2100467. https://doi.org/10.1002/marc.202100467.

Cleary, Julian (2009): Life cycle assessments of municipal solid waste management systems: a comparative analysis of selected peer-reviewed literature. In: Environment international 35 (8), S. 1256–1266. https://doi.org/10.1016/j.envint.2009.07.009.

Cossu, R.; Garbo, F.; Girotto, F.; Simion, F.; Pivato, A. (2017): PLASMIX management: LCA of six possible scenarios. In: Waste management (New York, N.Y.) 69, S. 567–576. https://doi.org/10.1016/j.wasman.2017.08.007.

Davidson, Matthew G.; Furlong, Rebecca A.; McManus, Marcelle C. (2021): Developments in the life cycle assessment of chemical recycling of plastic waste—A review. In: Journal of Cleaner Production 293, S. 126163. https://doi.org/10.1016/j.jclepro.2021.126163.

Demets, Ruben; van Kets, Karen; Huysveld, Sophie; Dewulf, Jo; Meester, Steven de; Ragaert, Kim (2021): Addressing the complex challenge of understanding and quantifying substitutability for recycled plastics. In: Resources, Conservation and Recycling 174, S. 105826. https://doi.org/10.1016/j.resconrec.2021.105826.

Dong, Jun; Tang, Yuanjun; Nzihou, Ange; Chi, Yong; Weiss-Hortala, Elsa; Ni, Mingjiang (2018): Life cycle assessment of pyrolysis, gasification and incineration waste-to-energy technologies: Theoretical analysis and case study of commercial plants. In: The Science of the total environment 626, S. 744–753. https://doi.org/10.1016/j.scitotenv.2018.01.151

DIN EN ISO 14044, 02.2021: Umweltmanagement—Ökobilanz—Anforderungen und Anleitungen.

EN ISO 14040, 02.2021: Umweltmanagement—Ökobilanz—Grundsätze und Rahmenbedingungen.

Ecoinvent (2019): ecoinvent Data V. 3.5: Swiss Centre for Life Cycle Inventories.

Ellen MacArthur Foundation (Hg.) (2016): The new plastics economy. Rethinking the future of plastics. World Economic Forum.

Eriksson, Ola; Finnveden, Göran (2009): Plastic waste as a fuel - CO2-neutral or not? In: Energy Environ. Sci. 2 (9), S. 907. https://doi.org/10.1039/b908135f.

Eriksson, Ola; Finnveden, Göran (2017): Energy Recovery from Waste Incineration—The Importance of Technology Data and System Boundaries on CO2 Emissions. In: Energies 10 (4), S. 539. https://doi.org/10.3390/en10040539.

European Commission (2016): Paris Agreement. Available online under https://ec.europa.eu/clima/eu-action/international-action-climate-change/climate-negotiations/paris-agreement_en.

European Parliament (2023): Abfallwirtschaft in der EU: Zahlen und Fakten. Available online under https://www.europarl.europa.eu/news/de/headlines/society/20180328STO00751/abfallwirtschaft-in-der-eu-zahlen-und-fakten?at_campaign=20234Economy&at_medium=Google_Ads&at_platform=Search&at_creation=DSA&at_goal=TR_G&at_audience=&at_topic=Recycling&gclid=CjwKCAjw3dCnBhBCEiwAVvLcu8qmC3Xg2CEuBiWkNLOo_Q5POwmL-zoaL_OVbzp_w8b-prgqFUpTgRoC8SAQAvD_BwE, zuletzt aktualisiert am 22.06.2023, zuletzt geprüft am 05.09.2023.

Galvez-Martos, Jose-Luis; Schoenberger, Harald (2014): An analysis of the use of life cycle assessment for waste co-incineration in cement kilns. In: Resources, Conservation and Recycling 86, S. 118–131. https://doi.org/10.1016/j.resconrec.2014.02.009.

García, María Teresa; Gracia, Ignacio; Duque, Gema; Lucas, Antonio de; Rodríguez, Juan Francisco (2009): Study of the solubility and stability of polystyrene wastes in a dissolution recycling process. In: Waste management (New York, N.Y.) 29 (6), S. 1814–1818. https://doi.org/10.1016/j.wasman.2009.01.001.

Geyer, Roland; Kuczenski, Brandon; Zink, Trevor; Henderson, Ashley (2016): Common Misconceptions about Recycling. In: Journal of Industrial Ecology 20 (5), S. 1010–1017. https://doi.org/10.1111/jiec.12355.

Helm, Levi T.; Murphy, Erin L.; McGivern, Alexis; Borrelle, Stephanie B. (2023): Impacts of plastic waste management strategies. In: Environ. Rev. 31 (1), S. 45–65. https://doi.org/10.1139/er-2021-0117.

Hermanns, Ronja; Kraft, Axel; Hartmann, Philipp; Meys, Raoul (2023): Comparative Life Cycle Assessment of Pyrolysis—Recycling Germany's Sorted Mixed Plastic Waste. In: Chemie Ingenieur Technik 95 (8), S. 1259–1267. https://doi.org/10.1002/cite.202300041.

Huysman, Sofie; Debaveye, Sam; Schaubroeck, Thomas; Meester, Steven de; Ardente, Fulvio; Mathieux, Fabrice; Dewulf, Jo (2015): The recyclability benefit rate of closed-loop and open-loop systems: A case study on plastic recycling in Flanders. In: Resources, Conservation and Recycling 101, S. 53–60. https://doi.org/10.1016/j.resconrec.2015.05.014

IEA (Hg.) (2018): The Future of Petrochemicals: IEA Publications International Energy Agency.

IN4climate.NRW (Ed.) (2020): Chemical Plastics Recycling—Potentials and Development Perspectives. A contribution to defossiling the chemical and plastics processing industry in NRW. A discussion paper by the Circular Economy Working Group.

Jeswani, Harish; Krüger, Christian; Russ, Manfred; Horlacher, Maike; Antony, Florian; Hann, Simon; Azapagic, Adisa (2021): Life cycle environmental impacts of chemical recycling via pyrolysis of mixed plastic waste in comparison with mechanical recycling and energy recovery. In: Science of The Total Environment, S. 144483. https://doi.org/10.1016/j.scitotenv.2020.144483.

Klöpffer, Walter; Grahl, Birgit (2009): Ökobilanz (LCA). Ein Leitfaden für Ausbildung und Beruf. Weinheim: Wiley-VCH. https://onlinelibrary.wiley.com/doi/book/10.1002/9783527627158

Kosajan, Vorada; Wen, Zongguo; Fei, Fan; Doh Dinga, Christian; Wang, Zhaojia; Zhan, Jiayu (2020): The feasibility analysis of cement kiln as an MSW treatment infrastructure: From a life cycle environmental impact perspective. In: Journal of Cleaner Production 267, S. 122113. https://doi.org/10.1016/j.jclepro.2020.122113.

Laurent, Alexis; Bakas, Ioannis; Clavreul, Julie; Bernstad, Anna; Niero, Monia; Gentil, Emmanuel et al. (2014): Review of LCA studies of solid waste management systems--part I: lessons learned and perspectives. In: Waste management (New York, N.Y.) 34 (3), S. 573–588. https://doi.org/10.1016/j.wasman.2013.10.045.

Lazarevic, David; Aoustin, Emmanuelle; Buclet, Nicolas; Brandt, Nils (2010): Plastic waste management in the context of a European recycling society: Comparing results and uncertainties in a life cycle perspective. In: Resources, Conservation and Recycling 55 (2), S. 246–259. https://doi.org/10.1016/j.resconrec.2010.09.014.

Mayer, Felix; Bhandari, Ramchandra; Gäth, Stefan (2019): Critical review on life cycle assessment of conventional and innovative waste-to-energy technologies. In: The Science of the total environment 672, S. 708–721. https://doi.org/10.1016/j.scitotenv.2019.03.449.

Meys, Raoul; Frick, Felicitas; Westhues, Stefan; Sternberg, André; Klankermayer, Jürgen; Bardow, André (2020): Towards a circular economy for plastic packaging wastes—the environmental potential of chemical recycling. In: Resources, Conservation and Recycling 162, S. 105010. https://doi.org/10.1016/j.resconrec.2020.105010.

Meys, Raoul; Kätelhön, Arne; Bachmann, Marvin; Winter, Benedikt; Zibunas, Christian; Suh, Sangwon; Bardow, André (2021): Achieving net-zero greenhouse gas emission plastics by a circular carbon economy. In: Science (New York, N.Y.) 374 (6563), S. 71–76. https://doi.org/10.1126/science.abg9853.

Michaud, Jean-Charles; Farrant, Laura; Jan, Olivier; Kjær, Birgitte; Bakas, Ioannis (2010): Environmental benefits of recycling—2010 update. WRAP Material change for a better enviroment. Available online under www.wrap.org.uk.

Patel, Martin; Thienen, Norbert von; Jochem, Eberhard; Worrell, Ernst (2000): Recycling of plastics in Germany. In: Resources, Conservation and Recycling 29 (1–2), S. 65–90. https://doi.org/10.1016/S0921-3449(99)00058-0.

Ragaert, Kim; Delva, Laurens; van Geem, Kevin (2017): Mechanical and chemical recycling of solid plastic waste. In: Waste management (New York, N.Y.) 69, S. 24–58. https://doi.org/10.1016/j.wasman.2017.07.044.

Ragaert, Kim; Ragot, Cloé; van Geem, Kevin M.; Kersten, Sascha; Shiran, Yoni; Meester, Steven de (2023): Clarifying European terminology in plastics recycling. In: Current Opinion in Green and Sustainable Chemistry 44, S. 100871. https://doi.org/10.1016/j.cogsc.2023.100871.

Rahman, Azad; Rasul, M. G.; Khan, M.M.K.; Sharma, S. (2015): Recent development on the uses of alternative fuels in cement manufacturing process. In: Fuel 145, S. 84–99. https://doi.org/10.1016/j.fuel.2014.12.029.

Rigamonti, L.; Grosso, M.; Giugliano, M. (2009): Life cycle assessment for optimising the level of separated collection in integrated MSW management systems. In: Waste management (New York, N.Y.) 29 (2), S. 934–944. https://doi.org/10.1016/j.wasman.2008.06.005.

Rigamonti, L.; Grosso, M.; Møller, J.; Martinez Sanchez, V.; Magnani, S.; Christensen, T. H. (2014): Environmental evaluation of plastic waste management scenarios. In: Resources, Conservation and Recycling 85, S. 42–53. https://doi.org/10.1016/j.resconrec.2013.12.012.

Russ, Manfred; Gonzalez, Maria; Horlacher, Maike (2020): Evaluation of pyrolysis with LCA—3 case studies. BASF SE, Corporate Sustainability.

Schwarz, A. E.; Ligthart, T. N.; GoBizarro, D.; Wild, P. de; Vreugdenhil, B.; van Harmelen, T. (2021): Plastic recycling in a circular economy; determining environmental performance through an LCA matrix model approach. In: Waste management (New York, N.Y.), S. 331–342. https://doi.org/10.1016/j.wasman.2020.12.020.

Schyns, Zoé O. G.; Shaver, Michael P. (2021): Mechanical Recycling of Packaging Plastics: A Review. In: Macromolecular rapid communications 42 (3), e2000415. https://doi.org/10.1002/marc.202000415.

Shen, Li; Worrell, Ernst; Patel, Martin K. (2010): Open-loop recycling: A LCA case study of PET bottle-to-fibre recycling. In: Resources, Conservation and Recycling 55 (1), S. 34–52. https://doi.org/10.1016/j.resconrec.2010.06.014.

Shonfield, Peter (2008): LCA of Management Options for Mixed Waste Plastics. WRAP Material change for a better enviroment.

Turner, David A.; Williams, Ian D.; Kemp, Simon (2015): Greenhouse gas emission factors for recycling of source-segregated waste materials. In: Resources, Conservation and Recycling 105, S. 186–197. https://doi.org/10.1016/j.resconrec.2015.10.026.

Tyskeng, Sara; Finnveden, Göran (2010): Comparing Energy Use and Environmental Impacts of Recycling and Waste Incineration. In: J. Environ. Eng. 136 (8), S. 744–748. https://doi.org/10.1061/(ASCE)EE.1943-7870.0000206.

Vogt, Bryan D.; Stokes, Kristoffer K.; Kumar, Sanat K. (2021): Why is Recycling of Postconsumer Plastics so Challenging? In: ACS Appl. Polym. Mater. 3 (9), S. 4325–4346. https://doi.org/10.1021/acsapm.1c00648.

Volk, Rebekka; Stallkamp, Christoph; Steins, Justus J.; Yogish, Savina Padumane; Müller, Richard C.; Stapf, Dieter; Schultmann, Frank (2021): Techno-economic assessment and comparison of different plastic recycling pathways: A German case study. In: Environ. Res. Lett. 25 (5), S. 1318–1337. https://doi.org/10.1111/jiec.13145.

Walker, S.; Rothman, R. (2020): Life cycle assessment of bio-based and fossil-based plastic: A review. In: Journal of Cleaner Production 261, S. 121158. https://doi.org/10.1016/j.jclepro.2020.121158.

World Energy Council (2004): Comparison of Energy Systems Using Life Cycle Assessment.

Zheng, Jiajia; Suh, Sangwon (2019): Strategies to reduce the global carbon footprint of plastics. In: Nat. Clim. Chang. 9 (5), S. 374–378. https://doi.org/10.1038/s41558-019-0459-z.

Ronja Hermanns is the Lead Process Engineer for Sustainable Chemical Life Cycle at Carbon Minds in Cologne/Germany. She is responsible for developing and evaluating net-zero strategies for the chemical industry, working with environmental think tanks, chemical associations, and companies. She is pursuing a Ph.D. in life cycle assessments, focusing on environmental, economic, and social impacts of the chemical value chain's transition in cooperation with Ruhr University Bochum.

Dr.-Ing. Raoul Meys strives to make a relevant contribution to society in the transition to climate neutrality. As a Scientist at RWTH Aachen University, his work has had a profound impact, including publications in Science, PNAS, and Nature. This impact is furthered in his role as Managing Director of Carbon Minds of Cologne. He regularly maintains an open exchange with his network partners in the industry, associations, politics, and science.

Prof. Dr.-Ing. Manfred Renner holds a doctorate in mechanical engineering. In 2022 he took over the management of the Fraunhofer Institute for Environmental, Safety and Energy Technology UMSICHT in Oberhausen, Germany, and holds the Chair for Responsible Process Engineering at the Faculty of Mechanical Engineering at Ruhr-Universität Bochum. He also heads the Fraunhofer Cluster of Excellence Circular Plastics Economy CCPE.

Chapter 18
Product Carbon Footprint (PCF)—Scopes and Strategies

Peter Orth

Contents

Summary

The findings of atmospheric chemistry about the accumulation of anthropogenic gases in the atmosphere and their influence on ozone depletion and ultimately on the climate on our planet led to the adoption of the Montreal and Kyoto Protocols on Greenhouse Gases (GHG) and their mitigation and avoidance. As a result, the concept of the Product Carbon Footprint (PCF) evolved to determine the GHG emissions caused by a product and ultimately by an organization. Introduced at the beginning of the 2000s the GHG Protocol developed by WRI and WBCSD became the de facto standard therefor. Emissions are categorized into three scopes to give companies the opportunity to specifically quantify and mitigate their emissions. The Together for Sustainability (TfS) initiative is further developing and substantiating the GHG Protocol for the chemical and plastics industry with its complex value chains. The focus is on methodological approaches for the reliable and comparable calculation of the PCF in the context of a life cycle assessment (LCA) as well as for the verification of the results to make them reliably available and manageable for internal company measures as well as for use in the market and towards policymakers and regulators.

P. Orth (✉)
OPC—Orth Plastics Consulting, Cologne, Germany
e-mail: opc.cologne@t-online.de

P. Orth et al. (eds.), *Circular Economy of Plastics*,
https://doi.org/10.1007/978-3-031-78829-1_18

18.1 Historical Development of the PCF

The concept of the Product Carbon Footprint (PCF) has evolved significantly over the past few decades. In the 1970s and 1980s, the primary focus of environmental chemistry was on reducing local pollution and managing waste. In 1995 Paul J. Crutzen, Mario J. Molina, and Frank S. Rowland were awarded the Nobel Prize in Chemistry for their work in atmospheric chemistry and especially for their efforts in studying the formation and decomposition of atmospheric ozone, which led to the Montreal Protocol on Substances that Deplete the Ozone Layer (UNEP 2024).

A few years later in 1997, the Kyoto Protocol marked a significant milestone by setting binding emission reduction targets for developed countries, bringing GHG emissions into the global spotlight (UNFCCC 1997). With the growing awareness of climate change, the focus shifted towards understanding and mitigating greenhouse gas (GHG) emissions. The early 2000s saw the development of methodologies to quantify GHG emissions at the product level, leading to the concept of the PCF. The launch of the GHG Protocol by the World Resource Institute (WRI) and the World Business Council for Sustainable Development (WBCSD) provided a standardized framework for measuring and managing GHG emissions, it has been instrumental in the development of PCF methodologies (WRI 2002).

Green House Gases (synonymously called Kyoto gases) according to the Kyoto protocol[1] are a group of gases contributing to global warming and climate change as follows: carbon dioxide (CO_2), methane (CH_4), and nitrous oxide (N_2O) (the non-fluorinated gases) together with hydrofluorocarbons (HFCs), perfluorocarbons (PFCs), sulphur hexafluoride (SF_6), and nitrogen trifluoride (NF_3) (the fluorinated gases) (cf. Fig. 18.1).[2]

The different gases have different climate change and global warming impacts, the total impact being described in CO_2 equivalents (CO_2e), a metric measure representing all greenhouse gases by converting them to the equivalent amount of CO_2. The PCF is then calculated by considering the seven gases, measured by mass and converted into CO_2 equivalents using the 100-year global warming potential (GWP) coefficients of the IPCC 6th Assessment Report (IPCC 2021) including the chemical effects of CH_4 and N_2O, and including CO_2 from land use and land use change (BASF 2022).

18.2 The GHG Protocol

The GHG Protocol is a comprehensive global standardized framework for measuring and managing GHG emissions from private and public sector operations, value chains, and mitigation actions. Established by WRI and WBCSD, the GHG Protocol

[1] NF_3 was added to the six GHGs mentioned in the Kyoto Protocol later on.

[2] It should be mentioned that water vapour being an important constituent of the atmosphere acts as a GHG, too, is contributing more to the greenhouse effect than CO_2 or CH_4.

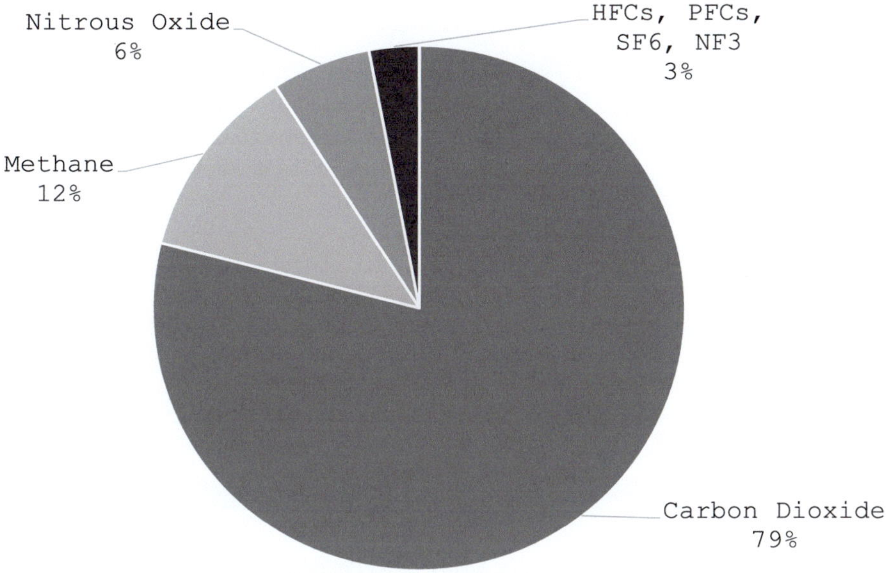

Fig. 18.1 Global GHG emissions 2018. *Source* Crippa et al. (2021)

aims to provide a consistent approach to GHG accounting. It offers various standards, including the Corporate Standard, the Corporate Value Chain (Scope 3) Standard, and the Product Standard, which are widely used for PCF calculations. Today, the GHG Protocol is used by thousands of companies and organizations worldwide, making it the de facto standard for GHG accounting. The "Built on GHG Protocol" mark recognizes calculation tools that are in conformance with GHG Protocol standards (Fig. 18.2).

The GHG Protocol categorizes emissions into three scopes to help organizations comprehensively measure and manage their GHG emissions (cf. Fig. 18.3):

Fig. 18.2 "Built on GHG Protocol" mark. *Source* GHG Protocol (2024)

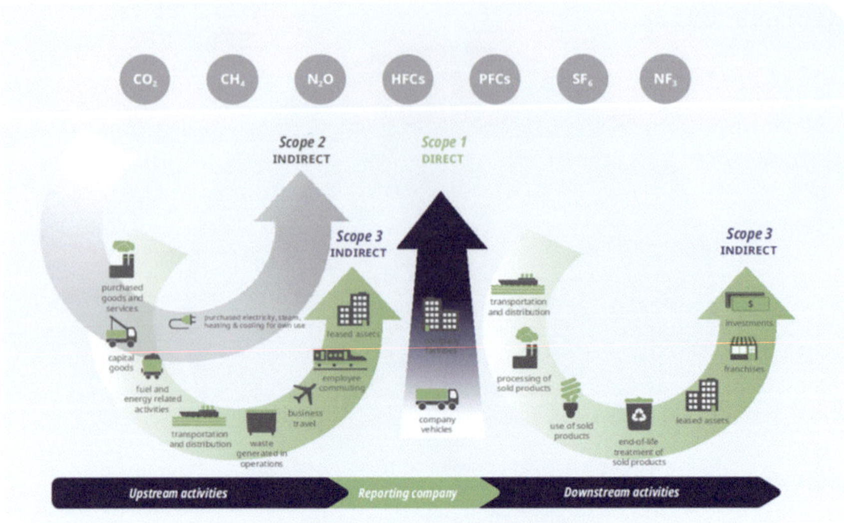

Fig. 18.3 GHG Protocol emissions scopes. *Source* GHG Protocol (2024)[3])

- Scope 1: Direct emissions from owned or controlled sources. For a plastics manufacturing company, these emissions primarily come from the combustion of fossil fuels in production processes and facility operations.
- Scope 2: Indirect emissions from the generation of purchased electricity, steam, heating, and cooling consumed by the reporting company.
- Scope 3: All other indirect emissions that occur in the value chain, both upstream and downstream. This scope normally is the most challenging to measure and manage, covering areas such as raw material extraction, transportation, product use, and end-of-life disposal.

Mitigation strategies

The following entries provide examples of basic GHG avoidance strategies.

[3] With kind permission of GHG Protocol: Technical Guidance for Calculating Scope 3 Emissions, p. 6.

Scope 1 emissions

- Process optimization: Improving energy efficiency in polymerization and extrusion processes.
- Switch to cleaner energy: Transitioning from oil or gas combustion to the use of electricity can significantly reduce direct emissions. Electrification of steam crackers as reported by BASF, Sabic, and Linde (CEN 2024) or LyondellBasell and Technip (TEN 2023) are pathbreaking endeavours.
- Carbon Capture and Storage (CCS): Implementing CCS technologies can capture CO_2 emissions before they reach the atmosphere.

Scope 2 emissions

- Renewable energy procurement: Investing in renewable energy sources such as wind, solar, or hydropower. BASF and RWE have committed to sourcing more electricity from renewable sources (RWE 2024).
- Energy efficiency measures: Upgrading to energy-efficient equipment and optimizing energy use in operations.
- Power Purchase Agreements (PPAs): Entering into PPAs to secure a long-term supply of renewable energy. Covestro has engaged in PPAs to support renewable energy projects (Örsted 2019).

Scope 3 emissions

- **Supplier engagement**: Collaborating with suppliers to reduce emissions in the supply chain. The "Together for Sustainability" (TfS 2024a) initiative of the chemical industry focuses on sustainability in the supply chain.
- **Product design**: Designing products for durability, recyclability, and lower carbon intensity.
- **End-of-Life management**: Enhancing recycling programs and promoting circular economy principles. LyondellBasell's initiatives in mechanical and chemical recycling aim to convert plastic waste back into secondary raw material and/or feedstock (cf. Sect. 10.4).

18.3 The Industry Perspective (TfS)

Together for Sustainability (TfS) is an initiative of the chemical industry (including the plastics manufacturing industry) founded in 2011 that aims to improve sustainability practices within the supply chain (TfS 2024a). Since the chemical industry faces unique challenges in PCF calculation due to its complex and global supply chains and diverse product portfolios TfS provides a platform to collaborate on sustainability assessments, including PCF calculations, offering best practices, tools, and especially methodologies to ensure accurate and consistent PCF Scope 3 calculations. The initiative developed a Product Carbon Footprint Guideline for PCF calculation for chemical materials which enables suppliers and corporations to produce and share high-quality carbon footprint data, "audit-ready", ISO-compliant, and accepted by the GHG Protocol and "open source", useful for global industries using chemical materials.

With the launch of the Product Carbon Footprint (PCF) Guideline in 2022 (TfS 2024b), TfS completed one of its most impactful projects to date. Experts from several chemical companies collaborated to tackle the chemical industry's Scope 3 challenge and formulated the first set of chemical industry specific guidelines for PCF determination. The PCF Guideline is an important enabler for companies to navigate the 20s being a "Decade of Action". During this work, TfS identified some improvement potentials for the Corporate Greenhouse Gas accounting that are addressed in the PCF White Paper (TfS 2024c). There are three points of pivotal importance:

- Biogenic carbon accounting
- Mass balance as a transition mechanism
- Recycled materials and content

Biogenic Carbon

Biogenic carbon refers to carbon derived from biological sources, such as crops, trees, and food waste (cf. Chap.14). It is considered neutral in terms of its impact on the carbon cycle when sustainably managed. As these carbon sources become more important, the development of a framework for their accurate accounting becomes essential. However, their emissions-saving benefits are not appropriately represented in the GHG protocol's Scope 3.1 process "to calculate carbon emissions for purchased goods and services in corporate accounting as the carbon removed from the atmosphere is not considered." Therefore, TfS suggests carbon accounting methodologies using the so-called $-1/+1$ gross flow accounting approach, which acts as a balance sheet for carbon emissions and removals. Through this approach depicted in Fig. 18.4, emitting a ton of carbon is recorded as " $+1$" (negative impact) and removing a ton of carbon is recorded as "-1" (positive impact) to determine the overall effect on the environment (TfS 2024c).

Mass Balance

Since the Mass Balance (MB) approach is not defined in the current GHG protocol and cannot be applied so far, the adoption of an MB approach (cf. Chap. 16) needs to be considered as a market-based mechanism reinforcing the industry's transition towards sustainable practices. TfS therefore proposes a system expansion in the corporate accounting of companies of the GHG protocol to enable accurate reporting, comprising elements like:

- Being site-specific, to allow the sustainable portion of the feedstock to be assigned to a specific product or output;
- Considering other elements of renewable feedstock (such as N, H, O) taken into consideration for the calculation of attributable amounts in addition to carbon;
- Extending the accounting period (for balancing of secondary raw materials going short) to one year;

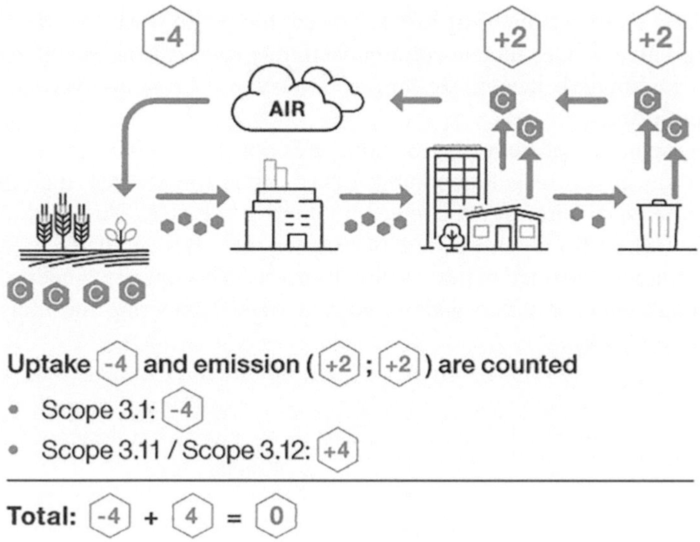

Fig. 18.4 $-1/+1$-approach considering biogenic CO_2. *Source* TfS (2024c)[4]

- Using chemical value-related properties, e. g., the 'lower heating value' (LHV), carbon content, functional units etc. which are preferable as the basis for the calculation (e.g., for mixed plastic waste) or carbon content.

Recycled Content

With the increasing use of recycled materials and content as secondary raw material in the industry, there is a need to harmonize approaches to carbon accounting to recognize the positive effects of a circular economy at both corporate and product-levels. GHG emission reduction using recycled materials should be reflected in corporate GHG accounting for both the users and producers of secondary materials. However, the GHG Protocol's cut-off accounting method falls short in addressing the complex interactions involved. In this regard, TfS has identified four challenges to be dealt with:

- Limitations in the Cut-off approach: The GHG Protocol Corporate Value Chain (Scope 3) Standard employs a cut-off methodology for all materials, which can lead to calculations based on OECD Global Statistics distorting companies' positive contributions to the circular economy to their detriment.
- Complexity of waste incineration: Waste incineration with energy recovery needs to be appropriately accounted and the impact be attributed respectively. Incineration is the ultimate waste treatment method, where energy recovery is a byproduct.

[4] With kind permission from TfS: Improving and Harmonizing Scope 3 Emissions – TfS White Paper, p. 6.

The GHG emissions intensity of this process hinges on waste carbon content and heating value. Under the cut-off approach, this impact is allocated to recovered energy and may disincentivize proactive waste reduction and optimized energy recovery efforts.

- Non-Recognition of Recycling Benefits in Scope 3.1: An issue arises when products produced via recycling exhibit higher cradle-to-gate PCF than their virgin counterparts, despite having a lower cradle-to-grave PCF. This in turn leads to a situation where the climate benefit of recycled materials compared to fossil materials cannot be reflected in purchasing decisions. This opacity hinders companies from confidently embracing recycled materials, hampering the transition to a circular economy.
- Non-Recognition of increasing importance of Circular Contributions: A comprehensive circular economy contribution extends beyond material reuse. The current GHG protocol approach fails to recognize the increasing importance of the circular economy. It does not include accounting approaches for:

(i) recycling-enabled products (e.g., a polymer which is designed for better recycling),
(ii) enhanced recyclability (e.g., design for recycling in the automotive industry),
(iii) empowerment of other products' recyclability (e.g., by the introduction of additives).

An inclusive view of circular economy incentives should encompass these aspects.

18.4 Methodology and Databases

Accurate PCF calculation requires robust frameworks and methodologies that consider the entire lifecycle of a product. Therefore, Lifecycle Assessment (LCA, cf. Chap.17) is a key methodology used in PCF calculations, assessing environmental impacts from raw material extraction to end-of-life disposal. High-quality data with assured quality is crucial for accurate PCF calculations. This includes primary data from operations and suppliers based on partnerships or data-sharing agreements and secondary data from databases. Various software tools, such as GaBi and SimaPro, are available to assist in PCF calculations by providing databases and modeling capabilities. The frameworks and methodologies employed by industry leaders and service providers provide valuable insights into PCF management.

BASF's Methodology for Product Carbon Footprint Calculation (2022) outlining BASF's own methodology to calculate cradle-to-gate product carbon footprints is a technical guidance to help other companies and LCA analysts. The methodology is based on ISO 14067:2018 for the carbon footprint of products, which builds on the principles and requirements of the ISO standards 14,040:2006 and 14,044:2006 for life cycle assessment. The principles and requirements the methodology is structured alongside are as follows:

- Definition of the declared unit—1 kg of product regardless of its state,
- Product system—GHG emissions in CO_2 equivalents,
- System boundary—cradle-to-gate,
- Data requirements—for primary (company-specific) or secondary data (Scope 2 and Scope 3),
- Cut-off criteria—aiming for completeness (closed mass and energy balance),
- Allocation—multi-output and End-of-Life,
- LCA methodology—focussing on GHG emissions and considering fossil removals and biogenic carbon,
- Sensitivity analysis, quality, and reporting.

Carbon Minds of Cologne/Germany has developed the cm.chemicals database (Carbon Minds 2024), a comprehensive LCA and PCF database focusing exclusively on chemical value chains. The database covers the production of over 1000 chemical products and materials including polymers and plastics, modelled in up to 190 regions, e.g., Europe or North America, and countries, e.g., Germany, the United States or China. The database aims to provide data to conduct representative and reliable, quality assured and science[5] based LCA studies of chemical value chains, compliant with ISO standards 14,040 and 14,044 for Life Cycle Assessment, and ISO standard 14,067 for Product Carbon Footprint calculations. The database can be used for compliant PCF calculations according to Together for Sustainability (TfS 2024a) or the Greenhouse Gas Protocol guidelines (GHG 2024), specifically for plants, suppliers, technology, and consumption. The methodology underlying the cm.chemicals database was certified by TÜV Rheinland for compliance with ISO 14040/14044 standards.

Data quality criteria according to Carbon Minds are as follows:

- Technological representativeness—technologies applied throughout the supply chain
- Geographical representativeness—according to regions and countries
- Time-related representativeness—data change over time
- Completeness—covering all relevant flows
- Reliability—respecting different sources
- Methodological appropriateness and consistency

Other databases or platforms like Ecochain (2024), Ecoinvent (2024), GaBi (2024), openLCA (2024), or SimaPro (2024) provide extensive datasets for various materials, processes and applications. To assure transparency and reliability certification plays an important role. Several certification organisations have already specialised in PCF certification. Amongst others these are Veritas (2024), Trust (2024), GL (2024), SGS (2024), TÜV SÜD (2024) and TÜV Rheinland (2024).

ISO Standard ISO 14067:2018 (2024) specifies principles, requirements and guidelines for the quantification and reporting of the Product Carbon Footprint consistent with standards on Life Cycle Assessment (ISO 14040 and ISO 14044). It is

---

[5] Institutes from RWTH Aachen and ETH Zurich.

applicable to PCF studies as discussed above. However, the standard addresses only a single impact category: climate change. It does not assess any social or economic aspects or impacts, or any other environmental aspects and related impacts potentially arising from the life cycle of a product.

18.5 Conclusions

The Product Carbon Footprint approach provides a holistic view of a product's environmental impact, considering all stages of its lifecycle. Frameworks like the GHG Protocol, ISO standards, and the TfS Guidelines ensure consistency and comparability across different products and companies. PCF data can help in better decision-making by identifying hotspots and opportunities for emission reductions (McKinsey 2023). However, accurate PCF calculations require extensive data collection, which can be resource intensive. The methodologies can be complex and require specialized knowledge and tools. Differences in data quality and methodological choices can lead to variability in PCF results.

Companies that effectively manage and reduce their PCF can gain a competitive edge in the market. Proactively managing PCF can help companies stay ahead of regulatory requirements. Transparent PCF reporting can enhance stakeholder trust and engagement.

However, inaccurate or misleading PCF claims can damage a company's reputation. Failure to manage PCF effectively can lead to a loss of market share as consumers and businesses increasingly prefer sustainable products.

By understanding and effectively managing the Product Carbon Footprint, companies in the chemical and plastics industry can not only contribute to global sustainability efforts but also enhance their market position and stakeholder relationships. The management of the Product Carbon Footprint in the plastics industry involves a multifaceted approach that includes understanding and mitigating Scopes 1, 2, and 3 emissions, adopting best practices from industry leaders, and complying with regulatory requirements.

The combined efforts of companies, research institutions, and regulatory bodies are crucial in driving the transition towards a low-carbon, circular economy (TfS 2024).

References

BASF 2022: BASF Methodology for Product Carbon Footprint Calculation. August 2022.pdf.

Bureau Veritas 2024: https://www.bureauveritas.ch/en/our-services/life-cycle-assessment-product-carbon-footprint (Last access: Jun. 30, 2024).

Carbon Minds 2024: Stellner, L., Kalousdian, A., Goßen, J., Ossolińska, A., Kätelhön, A., Vögler, O., Hermanns, R., Suh, S., Bardow, A., Meys, R., Methodology cm.chemicals. Version 2.01,

March 2024. Carbon Minds GmbH, Cologne. https://www.carbon-minds.com/wp-content/uploads/2024/04/cm_chemicals_methodology_V2.01_2024.pdf.

Carbon Trust 2024: https://www.carbontrust.com/what-we-do/product-carbon-footprint-labelling (Last access: Jun. 30, 2024).

CEN 2024: https://cen.acs.org/business/petrochemicals/Sabic-BASF-Linde-start-electric/102/i10 (Last access: Jun. 30, 2024).

Crippa, M., Guizzardi, D., Solazzo, E., Muntean, M., Schaaf, E., Monforti-Ferrario, F., Banja, M., Olivier, J.G.J., Grassi, G., Rossi, S., Vignati, E., GHG emissions of all world countries - 2021 Report, EUR 30831 EN, Publications Office of the European Union, Luxembourg, 2021, ISBN 978-92-76-41547-3, https://doi.org/10.2760/173513, JRC126363

DNV GL 2024: https://www.dnv.com/services/? (Last access: Jun. 30, 2024).

Ecochain 2024: https://ecochain.com/ (Last access: Jun. 30, 2024).

Ecoinvent 2024: https://ecoinvent.org/database/ (Last access: Jun. 30, 2024).

GaBi 2024: https://sphera.com/solutions/product-stewardship/life-cycle-assessment-software-and-data/ (Last access: Jun. 30, 2024).

GHG Protocol 2024: https://ghgprotocol.org/corporate-value-chain-scope-3-standard.

IPCC 2021: IPCC 6th Assessment Report of WG 1. Climate Change 2021: The Physical Science Basis. https://www.ipcc.ch/report/ar6/wg1/.

ISO 14067:2018 (2024): https://www.iso.org/standard/71206.html.

McKinsey (2023) The Scope 3 challenge: Solutions across the materials value chain. https://www.mckinsey.com/industries/metals-and-mining/our-insights/the-scope-three-challenge-solutions-across-the-materials-value-chain

Örsted 2019: https://orsted.com/en/media/news/2019/12/799146415974816 (Last access: Jun. 17, 2024).

Open LCA 2024: https://www.openlca.org/open-source/ (Last access: Jun. 30, 2024).

Rapid Transition 2019: https://rapidtransition.org/stories/back-from-the-brink-how-the-world-rapidly-sealed-a-deal-to-save-the-ozone-layer/ (Last access: Jun. 19, 2024).

RWE 2024: https://www.rwe.com/forschung-und-entwicklung/projektvorhaben/offshore-to-x/ (Last access: Jun. 17, 2024).

SGS 2024: https://www.sgs.com/en-hk/news/2024/06/sgs-launches-environmental-claim-verification-mark-for-product-carbon-footprint (Last access: Jun. 17, 2024).

SimaPro 2024: https://support.simapro.com (Last access: Jun. 30, 2024).

TEN 2023: https://www.ten.com/en/media/news/technip-energies-lyondellbasell-and-chevron-phillips-chemical-sign-mou-electric-cracking (Last access: Jun. 30, 2024).

TfS 2024a: https://www.tfs-initiative.com/ (Last access: Jun. 24, 2024).

TfS 2024b: TfS Product Carbon Footprint Guideline. https://www.tfs-initiative.com/app/uploads/2024/03/TfS_PCF_guidelines_2024_EN_pages-low.pdf.

TfS 2024c: Improving and harmonizing Scope 3 reporting. TfS Product Carbon Footprint White Paper. https://www.tfs-initiative.com/app/uploads/2023/10/TfS-2097-White-Paper-EN_hi_single_page_no_crop_.pdf.

TÜV SÜD 2024: https://www.tuvsud.com/de-de/branchen/energie/erneuerbare-energien/energiezertifizierung/carbon-footprint-verifizierung (Last access: Jun. 19, 2024).

TÜV Rheinland 2024: https://consulting.tuv.com/aktuelles/product-carbon-footprint-und-corporate-carbon-footprint (Last access: Jun. 30, 2024).

UNEP 2024: https://ozone.unep.org/treaties/montreal-protocol.

UNFCCC 1997: Kyoto Protocol to the United Nations Framework Convention on Climate Change. https://unfccc.int/resource/docs/convkp/kpeng.pdf.

WRI and WBCSD 2002: http://pdf.wri.org/ghg_protocol_2004.pdf.

Dr. Peter Orth is a chemist by education; with OPC—Orth Plastics Consulting he leads a group of senior consultants with decades of professional experience in leading positions in the German and European plastics industry and its associations. His main focus is on innovation and sustainability in cooperation with science, industry, and politics. He is currently working in the fields of CO_2 as a feedstock, resource efficiency, and recycling technologies.

Chapter 19
Digitization—Enhancing the Circular Economy

Ulrike Lange, Andreas Bastian, Nico Pastewski, and Wei Min Wang

Contents

U. Lange (✉) · N. Pastewski · W. M. Wang
VDI Technologiezentrum GmbH, Berlin, Germany
e-mail: lange_u@vdi.de

N. Pastewski
e-mail: pastewski@vdi.de

W. M. Wang
e-mail: wang@vdi.de

A. Bastian
Plastship GmbH, Waldems, Germany
e-mail: bastian@plastship.com

P. Orth et al. (eds.), *Circular Economy of Plastics*,
https://doi.org/10.1007/978-3-031-78829-1_19

Summary

The plastics circular economy needs digital methods and solutions for its products and processes. Actors in the value chain need to exchange raw materials and product requirements as well as material composition and availability. Here, concepts for cross-company data rooms and expanded governance structures can create transparency and promote larger business activity. A system of interfaces between companies for the transfer of information enables the efficient use of existing knowledge and available skills. In this way, challenges in the use of recyclates can be addressed through data-based approaches by depicting the status quo of the material, providing security for material information and requirements, supporting the exchange between customers and suppliers, and enabling access to innovations. Cooperations in managed marketplaces promote exchange between companies, minimize risks and generate network effects. Modern approaches such as digital database systems or artificial intelligence provide answers to the challenges of data use. Circular business models become economically possible based on mutual trust. In practice, the digital mapping of product cycles is currently only incomplete. Reasons include a lack of data availability, a lack of interoperability or a poor understanding of the relationships. Companies are also confronted with the simultaneous nature of many technological developments in the context of digitization. An efficient circular economy is based on material flows that are aligned in the upstream stages of the value chain in such a way that plastic recyclates can ultimately be used in high-quality applications. This is only possible reliably if the necessary information is available along the entire value chain.

19.1 Standardization—Prerequisite for the Digitization of the Plastics Cycle

Product-related data and information are an important asset alongside the actual products. In many steps of a product life cycle, large amounts of data and information on material composition and manufacturing processes are recorded, collected, and processed. However, they are often stored in closed company-internal IT systems, so sharing this data is only possible to a limited extent. These IT systems work with heterogeneous and sometimes proprietary data models and formats, which further complicates data exchange. To keep raw materials, materials, products and/or product components in the cycle, a sufficient basis of process and product-related data and information as well as networking along the value chain is required, for which norms and standards are essential (cf. Chap. 20). A common, cross-value-added language creates an understanding for the harmonized design of interfaces and data formats, data qualities and access regulations.

Standardization requirements for the key topics of the circular economy (plastics, ICT & electronics, textiles, batteries, packaging, buildings & and municipalities as well as digitization) were developed by DIN, DKE and VDI together with stakeholders as part of the Circular Economy standardization roadmap (DIN and DKE 2023). 38 standardization needs were identified for digitization, which has a horizontal impact on the key topics. These include the "management of technical and legal interfaces"[1] as a prerequisite for functioning collaboration between actors in the value chain or the "establishment or adaptation of standardized mechanisms to ensure data quality and trustworthy information in the digital product passport".[2] For the key topic of plastics, standardization requirements for "data processing, documentation, conformity assessment and measurement procedures"[3] were defined, as necessary for transparent and consistent data networking along the product life cycle. Despite the diversity of plastic types and products, the targeted sorting and processing of waste streams and the high-quality production of plastic recyclates should be possible.

The transfer of the identified standardization needs into applicable standards is important because legislation in Europe is pushing the transformation towards a circular economy (cf. Chap. 21) and specifies digitization measures—specifically the introduction of a digital product passport—for certain product groups. The first framework standards for the digital exchange of cycle-relevant data and information already exist, such as the product-specific DIN EN 17410 for window and door profiles made of PVC-U (cf. Sect. 11.2.4). DIN SPEC 91446 (2021) specifies a "classification of plastic recyclates by data quality levels for use and (internet-based) trade". This was followed by DIN SPEC 91481 (2024), which focuses on polyamide. Further work is underway on these DIN SPECs to convert them into European and international norms and standards—a general requirement to respond to the reach of value chains beyond national borders. European and international interoperability is promoted by the joint "Digital Product Passport" committee of DIN and DKE, which reflects the work of the standardization organizations CEN and CENELEC, addressed by the European Commission, to harmonize norms and standards for the further development of the digital product pass. This counteracts fragmented and interoperable individual solutions at national and international levels (DIN and DKE 2023).

19.2 Digital Technologies as a Response to Challenges in the Plastics Cycle

Digital technologies are defined as "solution components for the digitally networked company, which consist of networking operational application systems, software and hardware components so that they create data-based added value in an application through data processing" (Stich et al. 2022). The annual study "Digitization

[1] Ibid. p. 45.

[2] Ibid. p. 42.

[3] Ibid. p. 114.

of the Economy in Germany" defines, among other things: the current technologies and trends in digitization in the categories of networking, virtualization, and data processing. For the plastics cycle, the following selected, non-exhaustive digital technologies and trends are listed, which are either in development or already in use (cf. Table 19.1).

The challenge for the plastics industry is to map the complex product structure with a common, standardized vocabulary and to define quality requirements for products and data with the associated testing processes. Digital technologies and trends have the potential to help overcome many challenges in the use of recycled materials (cf. Table 19.2).

Table 19.1 Selected technologies and trends in digitization. (*Source* Stich et al. 2022)

Categories	Technologies	Trends
Networking (Combination and forwarding of data and information)	Wireless networks such as WLAN (5G) and WLAN AX (6G) or Distributed Ledger Technologies (e.g. blockchain)	Cloud Computing "as a Service "
Virtualization (Digital image of reality created using data)	System virtualization to replicate hardware or software	Digital twin or digital shadow
Data processing (Data acquisition, direct processing, derivation of measures)	Artificial intelligence (computer programs, independently solving problems)	Data analytics via automated application of algorithms

Table 19.2 Challenges in the plastics cycle and opportunities of digital approaches. (*Source* Own presentation)

Current challenges in the plastics value chain	Opportunities for digital solutions in the plastics cycle
• Incomplete transfer and lack of information about material characteristics along the chain • Reliable access to price and availability information related to a potentially suitable recyclate quality • Insufficient definition of recyclate quality due to quality fluctuations • Lack of correlation between recyclate type and potential application	• Representation of the status quo • Ensuring security on both the customer and supplier sides • Structuring the exchange of information between customers and suppliers • Access to innovations

19.3 The Digitization of the Plastics Value Chain

Circular economy requires a new level of transparency and trust between the actors involved, as functional circular solutions require a lot of information about products and processes. Transparent and consistent data exchange at the value chain level (see Fig. 19.1) is of central importance for circular management. Digitization helps to reduce hurdles that exist due to information asymmetries and high transaction costs between actors in the product life cycle (Barteková and Börkey 2022). For efficient recycling of plastics and plastic products, data on products, their distribution and use as well as raw materials, semi-finished products, production processes, transport, use and disposal are required. This life cycle data can be used specifically in the individual value creation stages to implement circular concepts or to enable them in subsequent stages. What is important for this is a system of interfaces between companies that enable trustworthy, legally secure, and needs-based transfer and tracking of cycle-relevant data and information.

Digitization based on value creation stages plays an important role, particularly in production technology. Within a value chain stage (cf. Fig. 19.1), digital solutions enable the collection, processing and analysis of relevant data and information about products and their manufacturing processes. In this way, companies can monitor their internal processes using sensors and optimize them in a targeted manner by evaluating the collected data to reduce resource consumption. AI algorithms can also be used to analyze patterns in process flows and resource consumption, enabling real-time interventions and predictive measures. Linking the consumption data of machines and systems with resource management systems allows deviations in the production process to be identified at an early stage and counteracted automatically. Digital solutions at the individual value chain levels, with the data and information generated and stored therein, form the basis for digital networks across the value chain. It is important to take a lifecycle-wide view. However, digitization measures and the infrastructure required for them (data centers, cloud storage, blockchains) must not outweigh the savings achieved to actually save resources (Plattform Industrie 4.0 2021).

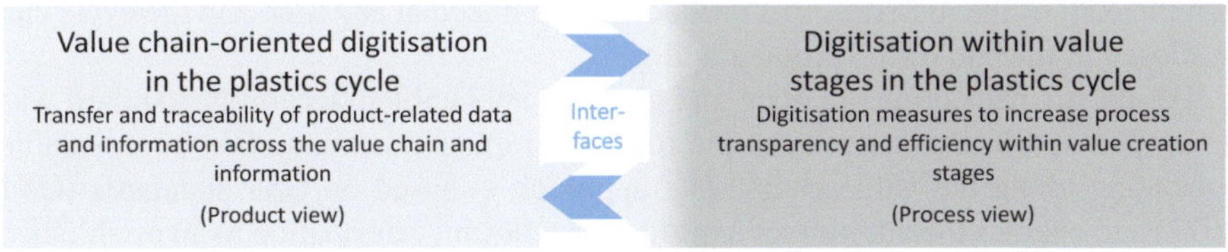

Fig. 19.1 Levels of digitization in the plastics cycle. (*Source* Own presentation)

19.4 Higher Level Digitization Solutions in the Plastics Value Chain

Ideally, every actor contributes to closing the entire information cycle with specific product and process-related data. This connection is the subject of the "AI Application Hub for Plastic Packaging" funding measure, which started in 2022. The AI application hub consists of the innovation laboratories KIOptiPack (design and production) and K3I-Cycling (loop closure), which map the value chain of plastic packaging. In addition to providing practical and innovative AI-supported tools as well as AI-optimized sorting and recycling tools, a central network platform serves to link the individual focus areas (BMBF 2023).

Current research projects such as this AI application hub as well as already established digitization solutions show the development towards a digitally networked plastic industry to leverage ecological and economic resource efficiency potential. Solutions such as the digital product passport, the digital twin, and digital platforms play essential roles here.

19.4.1 Digital Product Passport

As a central instance, the Digital Product Passport (DPP) collects and integrates product-related data along the life cycle phases. The actors involved can access data necessary for their core activities via the DPP. At the same time, they provide data about the DPP to other actors. For the DPP in the plastics industry, the basic information in the form of a material passport—i.e. plastic-type including additives—is crucial. These must be linked to the product in such a way that access is still possible even after years of use and products on the market as well as products returning from the market can be reliably identified and their condition described and evaluated for optimal recycling. Across the value creation stages, the material passport is continuously supplemented with data on materials, recycling specifications or information on the ecological footprint or product carbon footprint (cf. Chap. 18). This means that the path of the materials in a product can be traced at any time and the cycle can be closed at the end of the product's life.

For plastic packaging, an open and globally applicable traceability standard was developed as a best practice example for the complete documentation of recyclable packaging along the value chain. The approach is based on GS1 standards (GS1 2022) and enables the addition of exploitation-relevant information to an established infrastructure and the creation of digital product passports (R-Cycle 2022). The combination of approaches such as HolyGrail 2.0 for household packaging waste enables the transfer of information along the chain as well as the use of digital watermarks for the targeted sorting of waste (AIM 2022). By expanding the DPP with development and usage data, these can be expanded into "Sustainability Twins", which are an important basis for holistic optimization. Based on collected usage data from current products, subsequent products can be optimized in terms of circularity during development and a holistically optimal product variant—a compromise between sustainability, performance, and profitability—can be set using simulation and an LCA (Jacobs et al. 2023).

19.4.2 Digital Twin

Virtual images play a crucial role in digitization (VDI-TZ 2021). A digital twin created by combining a digital master with a digital shadow enables individual product instances to be mapped, monitored, and controlled in real time. The digital master contains the development and production data and reflects the condition of a product at the beginning of the product's life (as-designed and as-built). The digital shadow corresponds to an instantiation of the digital master for a single product; it is continuously expanded during the usage phase by usage and operational data (Lünneman et al. 2019). However, different concepts and data formats of digital twins make their dissemination in practice difficult. The concept of the Asset Administration Shell (AAS) promises to help, presenting the data and information of various products in a standardized format (Plattform Industrie 4.0 2021).

In a good practice example, a digital twin is introduced in the automotive industry as part of a corporate cooperation with the aim of achieving transparency in the value chain. The use of blockchain technology enables selective access to information from this digital twin to store material origin, environmental balance, recycled content, and other information in batches. Using QR codes, the following supply chain actors can update the digital twin that reflects the physical supply chain. The model enables easy creation of CO_2 balances (Product Carbon Footprint (PCF), cf. Chapt. 18) and simplified mass balances without an additional audit (Tiedemann 2020). Nevertheless, automated data collection, its link to the digital material batch, and automated sustainability assessment are still a challenge to be solved for implementation on an industrial scale.

19.4.3 Platform as a Service (PaaS)

Another important development in the design of interfaces is digital platforms. While in the past raw material exchanges were used to sell residual quantities, surpluses, and spot quantities, new generation platforms close the information gap between application and recyclate with new approaches. Managed marketplaces (e.g. plastship 2024, cirplus 2024) with services to promote exchange between companies, minimize risks and create network effects should be highlighted. The overarching goals of the European Plastics Strategy (cf. Chap. 21) to create a stable buyer market for recycled plastics (10 Mt of recyclates in 2025) (COM 2015) underline the importance of digital platform models for the plastic circular economy. They enable supply and demand to be linked for sectors of the economy that were previously only rarely in contact. The existence of recyclate manufacturers results from the existence of buyer markets for the material qualities that have been supplied to date. If new customer markets with new requirements are added, the requirement profiles and the associated opportunities are initially unknown. Customer segments emerge that may be more willing to pay, but also have completely different quality requirements. Future consumer markets are also not sufficiently informed about the connections and possibilities of plastics recycling and formulate requirement profiles that a recyclate manufacturer can hardly meet. Digital platforms can help across markets to minimize the discrepancy between supply and demand by making existing experiences usable for the industry, showing possibilities, and mapping them in a data-technical and product-centric manner. This is done with a focus on previously available recyclate types and requirements of the processing industry and an active role in building a stable customer/supplier relationship. The role of managed marketplaces goes beyond simply providing a user interface. They are also relevant in the plastic circular economy because at every point in the value chain—manufacturers, distributors, waste disposal companies, sorters, recyclers, plastics processors and even manufacturers—the output of the preliminary stage becomes the input of the following value chain stage. This results in higher liquidity in the market compared to marketplaces that only represent one level of value creation (managed marketplaces).

19.5 Digitization Solutions for Individual Value Creation Stages

Specific application scenarios for digital solutions in the life cycle of plastics and plastic products need to be considered for each individual value creation stage (cf. Fig. 19.2). Which digital solution can be used sensibly depends on the core activities of the actors and the resulting data requirements. At the same time, when using digital solutions, subsequent value creation stages must also be considered if product-related data must be collected and shared in sufficient quality. Table 19.3 summarizes possible solutions for the following core issues.

Digital solutions and challenges in the plastics cycle are presented below. The examples are exemplary of the identified problems in the cycle—from plastic production to recycling. The structure of the section is based on the structure of the plastic cycle as shown in Fig. 19.2 being a simplified version of Figs. 1.1 and 6.1.

19.5.1 Chemical and Plastics Manufacturing Industries

The vertically integrated chemical and plastics manufacturing industry has a key position in the plastics cycle. For them, information about the secondary raw materials

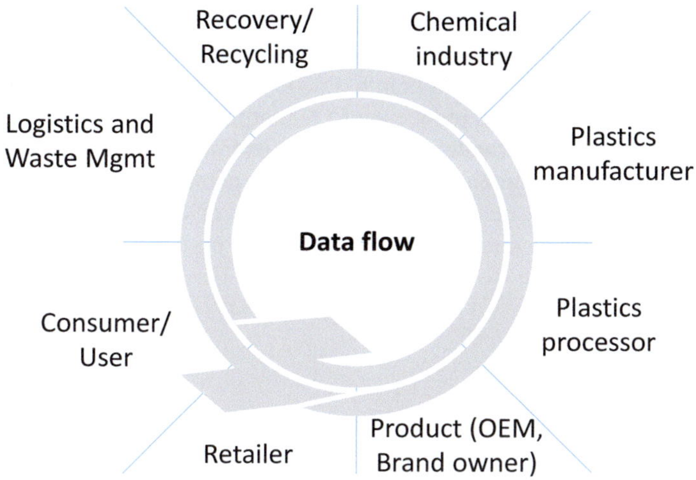

Fig. 19.2 Actors in the plastics cycle. (*Source* Own presentation)

Table 19.3 Key topics for the digitization of the plastic circular economy (*Source*: Own presentation)

Process technology	Material flow management	Validation and evaluation
• Measuring and monitoring the behavior of items in recycling • Optimization of the recipe production of recyclates and hybrids • Optimization of sorting and processing technology • Identification of suitable recyclates	• Internal and inter-company monitoring and optimization of material flows and quantity flows • Internal and inter-company data collection and data storage for the traceability of material flows • Increasing the availability of material flows and supply chain security	• Auditing of the individual value creation stages • Auditing of material qualities • Creation and tracking of mass and environmental balances

coming from plastic waste streams is just as relevant as information about the plastics produced for their customers. The quality of the information provided is crucial for the quality of recyclables returned to the cycle. The aim is, on the one hand, to use optimal technology and processes to bring the properties of the recyclates closer to those of the new material, and on the other hand, to optimize them for the next life cycle using, among other things, appropriate equipment. Compounders optimize the formulation development of recyclates with digital tools to predict the effect of additives or, based on appropriate analytics, to take measures to improve and eliminate errors for the desired product quality. As part of a funding program, a digital twin of the entire production process for producing recyclate is to be created to optimize sub-processes based on data. This also includes automatically creating recipes to produce recyclate (DigiRess 2024). Using digital test protocols, compounding data is recorded in an organized manner, processes are accelerated, and, in the long term, virtual recipe development is made possible via the digital representation of the compounding line (SKZ 2024). Digital methods help to create recipes that are cost-effective and functional, save resources and reduce the CO_2 footprint of the recyclates produced.

Digital services for information management and material purchasing are available as databases with comprehensive data sets that enable the selection of secondary raw materials of defined qualities and the tracking of capacities and availability (etc.). Individual plastics manufacturers offer online portals for order management and information gathering for the recyclate portfolio they offer (Borealis, Covestro, LyondellBasell, etc.).

19.5.2 Plastics Processing

To use recyclates in high-quality applications, the material origin and availability, quality and price are the decisive factors for the plastics processor. The use of recyclate requires knowledge of the material flows and the processing processes; suppliers should be audited, have appropriate quality assurance and an outgoing goods inspection specified for the target application. Fluctuations in material quality must be recorded, analyzed, and documented from batch to batch. Finally, the use of recyclate must be documented by the plastics processor, which requires certification systems. What is important is the significance of the audits, which are offered by several certification bodies. A digital audit platform simplifies the organizational collaboration between audited companies, auditors, certification bodies and certification systems (RecyClass 2024).

Digital methods are also used to increase the proportion of recycled materials in products. For example, one company designs extrusion blow molding systems for hollow plastic bodies that are made from recycled and virgin materials. The spiral distributor is part of the tool head in which recyclate, and virgin material are brought together and must react optimally to different qualities of flakes and granules. With the help of a digital twin, different tool heads with different parameters are created for the spiral distributor. An AI constantly simulates the production process and selects the best variants, which in turn are examined for similarities and adapted again. The process is repeated until the optimal spiral distributor is created. In the example shown, the recycled content could be increased to 85% (VDI ZRE 2021).

19.5.3 Product Manufacturer

For brand owners, the path to the circular economy brings new challenges: control of supply chains, sustainability reporting, minimum standards for green claims in the EU or the use of environmental seals in marketing, to name just a few points. This is where digital approaches can demonstrate their strengths to trace material flows and material origins, carry out certifications, communicate sustainability metrics clearly and securely, and monitor products and packaging on the market (cf. Sect. 8. 3). For the comprehensive management of plastic packaging and its components, for the assessment of country-specific circularity suitability or system participation, maintaining comprehensive information in reliable databases is essential. There are approaches for those placing the packaging on the market to create a comprehensive picture of their packaging, consider its suitability for circular circulation and determine system participation costs. This also makes it possible to optimize the use of packaging across companies (Recyda 2024).

System approaches and digital marketplace offerings will also become available for incentives and mass balancing in the future. Distributors who are not allowed to use recycled materials in their products can purchase credits from companies with proven use of recycled materials to compensate for the lack of recycled materials and thus meet their obligation to use recycled materials (cf. Chap. 21). There is great synergy potential here for distributors, processors using recyclates, and plastic recyclers. The distributor thus fulfills his legal obligations (if his products are recyclable), the processor receives an incentive to use additional recyclates, and the recycler receives a bonus on his marketing price. In addition, reliable data is collected and proven by a certificate (CRC 2024).

19.5.4 Retailer

This paragraph deals exclusively with plastic packaging. Here, circularity is the product property that, with particular attention to collection and sorting, processing, and reuse, indicates the extent to which packaging can be available to the market again as a raw material after the end of the usage phase. Since packaging consists of different components, made from different raw materials and this composition in turn must be evaluated against the current technological status of recycling, there is a need for management and definition solutions for retailers and distributors of packaging (Recyda 2014).

For several years now, the circularity of packaging can be assessed using a digital tool at the European level and audited and certified by external and approved certification bodies. To reflect the current technological status, findings from material testing and experiences from recycling are incorporated into evaluation criteria. Not only distributors use a digital tool, but also certification bodies that create and manage their reports using the system provided, which increases the quality and efficiency of the reporting (RecyClass 2024).

For packaging distributors, solutions are relevant that are not only at the point of sale for the consumer but also include industrial and commercial packaging. Packaging licensing processes for quantity reports to the Central Packaging Register (ZSVR) are being further digitized. Distributors can digitally submit mandatory quantity reports for sales, transport and outer packaging and stay up to date on the licensed quantity and receive evaluations via a portal (RIGK 2024). The points mentioned for branded goods companies are also relevant for retailers, as retailers are an important part of the infrastructure for taking back used packaging.

19.5.5 Private and Industrial Consumers and Users

There are apps for consumers that help find the right way to dispose of packaging waste, create carbon footprints or raise awareness of recycling as such (Jacobs et al. 2023). There are approaches for industrial users to digitally map the disposal route and its subsequent processing. Digital waste management for companies allows the central management of disposal processes from material flow management to invoicing (Recourcify 2024). Digital solutions for registering returns, recording return quantities and creating return protocols are also finding their way into the collection of post-consumer waste such as pesticide containers (RIGK 2024a). These and other approaches can be used to predict quantities and qualities and to better manage material flows.

19.5.6 Logistics and Waste Management

To recycle plastic waste, the waste sources must be logistically linked to specialist disposal companies and the operators of sorting systems. Several companies have already put their business model in this regard on a digital basis (Sennder 2024, Cargoboard 2024). However, many specialist waste disposal companies not only deal with plastics, but also different types of waste. Here it is necessary to decide which waste will be treated within a service regarding available systems and which offers will be submitted in waste tenders. Concepts must be developed, and evidence must be provided. In the simplest case, an order will be processed digitally, and marketing options will be established via apps and/or digital platforms. Digitally captured market movements could be created to support strategic decisions. All of this can be represented digitally, but usually requires consideration of the individual case.

Intelligent material flow management has great potential for the future, as it is a prerequisite for creating and controlling large, uniform waste volume flows that can be planned and recycled cost-effectively. Due to the complexity of the matter, this is an essential area of application for AI in the future and offers the opportunity to optimize material flows or to organize and link completely new ones across different stages of the chain or cycle.

19.5.7 Recovery and Recycling

It is essential for plastic waste recyclers to look at different material flows and their quality and to decide whether the current process design in combination with the material flow composition allows the fulfillment of a customer specification. If the composition of the material stream is known, predictions can be made about the quality of the recyclate. However, given the complex product requirements, the recyclate quality has so far been assessed by the customers themselves and considered in relation to the application. There are solutions for this from the research landscape. The ReVise-Up project, which started in 2023, aims, among other things, to use sensor-based material flow characteristics to adapt sorting, preparation, and plastics processing processes to fluctuating material flow properties (SKZ 2023). Another research group developed a quality model for recycled plastics (QMRP), considering material degradation, degree of mixing and contamination of plastics, cleaning, and reprocessing. The resulting quality assessment, which was successfully confirmed through test runs, helps recyclers classify their recyclate qualitatively (Golkaram et al. 2022).

19.6 Digitization to Close the Carbon Cycle

In a circular economy, plastic recyclates are used again as raw materials. It works optimally when the use of recyclate in products and their suitability for recycling is maximum and the number of resources that are added to the cycle as the primary material and removed as waste is minimal. In other words, the goal is to use carbon in the form of polymers for as long as possible and to keep it in as many applications as possible.

These points concern individual stages of the value chain or the interfaces between them. However, the real strength of digitization in the circular economy is that it enables an overall view of all stages in the cycle. But the reality is that real problems and challenges cannot be overcome with a keyword like "digitization". It is much more the interface between real objects and their representation in data, which will be crucial. If you look at the entire value creation cycle, for a perfect system, all structural, technical, and economic prerequisites must first be created to reuse plastic recyclates in as many applications as possible. It is then necessary to define which information (can) be collected where and how and which information carried through the chain makes sense and can be used when and for what purpose. The price of complete transparency would probably be a loss of intellectual property (IP) and unique selling points (USP) for many parties involved. It would also hardly correspond to the current structure of the plastics industry if there were a digital, completely transparent solution. Individual actors in the value creation cycle will benefit from digital solutions but will also help define and enforce standards and thus make the circular economy tangible and organizable.

References

AIM 2022: Pioneering Digital Watermarks for smart packaging recycling in the EU. Available at:https://www.digitalwatermarks.eu/ https://www.digitalwatermarks.eu/

Barteková und Börkey 2022: Digitalisation for the transition to a resource efficient and circular economy. OECD Environment Working Papers 192, OECD Publishing. Available at: https://www.oecd.org/publications/digitalisation-for-the-transition-to-a-resource-efficient-and-circular-economy-6f6d18e7-en.htm.

BMBF 2023: Project sheet „AI application hub plastic packaging". Bundesministerium für Bildung und Forschung (BMBF). Available at: https://ki-hub-kunststoffverpackungen.de/publikationen

Borealis 2024: MyBorealis: Take the easy way. Available at: https://www.borealisgroup.com/my-borealis (Last access: Aug. 20, 2024)

Cargoboard 2024: Booking transport has never been so easy. https://cargoboard.com/de (Last access: Aug. 20, 2024)

Cirplus 2024: Cirplus - Source and sell recycled plastic. Available at: https://www.cirplus.com/ (Last access: Aug. 20, 2024)

COM 2015: A European Strategy for Plastics in a Circular Economy. Available at: https://www.europarc.org/wp-content/uploads/2018/01/Eu-plastics-strategy-brochure.pdf

Covestro 2024: Order@covestro. Available at: https://order.covestro.com

CRC 2024: Certified Recycled Content. Available at: https://crc.earth/ (Last access: Aug. 20, 2024)

DigiRess 2024: Digitization of the production of recyclates from polyamide waste (DiRePro). Available at: https://www.digiress.de/projekte/DiRePro.php (Last access: Aug. 20, 2024)

DIN und DKE 2023: Deutsche Kommission Elektrotechnik, Elektronik und Informationstechnik und VDI Technologiezentrum GmbH: Deutsche Normungs-Roadmap - Circular Economy, Berlin. Available at: https://www.din.de/resource/blob/892606/06b0b608640aaddd63e5dae105ca77d8/normungsroadmap-circular-economy-data.pdf

DIN EN 17410:2021: Kunststoffe - Geregelter Recyclingkreislauf von Fenster- und Türprofilen aus PVC-U. Beuth Verlag GmbH, Berlin. Available at: https://www.beuth.de/de/norm/din-en-17410/339503681

DIN SPEC 91446:2021–12: Klassifizierung von Kunststoff-Rezyklaten durch Datenqualitätslevels für die Verwendung und den (internetbasierten) Handel. Beuth Verlag GmbH, Berlin. Available at: https://www.beuth.de/de/technische-regel/din-spec-91446/346496956

DIN SPEC 91481:2024–02: Klassifizierung von Kunststoff-Rezyklaten auf Polyamid-Basis durch Datenqualitätslevel für die Verwendung und den (internetbasierten) Handel. Beuth Verlag GmbH, Berlin. Available at: https://www.beuth.de/de/technische-regel/din-spec-91481/375901933

Golkaram 2022: Golkaram, M, Mehta, R, Taveau, M, Schwarz, A, Gankema, H, Urbanus, JH, Simon, L. Cakir-Benthem, S, Harmelen, T: Quality model for recycled plastics (QMRP): An indicator for holistic and consistent quality assessment of recycled plastics using product functionality and material properties. Journal of Cleaner Produktion, Volume 362, 2022. Available at: https://www.sciencedirect.com/science/article/abs/pii/S0959652622019151

GS1 Germany 2022: GS1 Standards. Available at: https://www.gs1-germany.de/gs1-standards/

IMDS 2024: IMDS Information pages. International Material Data System. Available at: https://public.mdsystem.com/de/web/imds-public-pages (Last access: Aug. 20, 2024)

Industry IoT Consortium 2022: The Industrial Internet Reference Architecture Version 1.10. Available at: https://www.iiconsortium.org/iira/

Jacob, A., Hadick, E., Nerb, L.; Amrhein, U. (2023): PolyProblem Report: The Circularity Code. Available at: https://polyproblem.org/reports/report-der-circularity-code/ (Last access: Aug. 20, 2024)

Lünnemann 2019: Lünnemann, P, Wang, W, Lindow, K: Smart Industrial Products – Smarte Produkte und ihr Einfluss auf Geschäftsmodelle, Zusammenarbeit, Portfolios und Infrastrukturen. Available at: https://www.ipk.fraunhofer.de/de/medien/markt-trendstudien/smarte-produkte.html

LyondellBasell 2024: Customerxpress - LyondellBasell Ecommerce Portal. Available at: https://lyb.customerxpress.com (Last access: Aug. 20, 2024)

Plattform Industrie 4.0 2018: RAMI 4.0 - Ein Orientierungsrahmen für die Digitalisierung. Available at: https://www.plattform-i40.de/IP/Redaktion/DE/Downloads/Publikation/rami40-eine-einfuehrung.html

Plattform Industrie 4.0 2021: Industrie 4.0 und Nachhaltigkeit – Zehn Thesen, wie digitale Geschäftsmodelle Nachhaltigkeit in der Industrie 4.0 fördern. Available at: https://www.plattform-i40.de/IP/Redaktion/DE/Downloads/Publikation/Thesen-Nachhaltigkeit-Geschaeftsmodelle.html

Polyglobe 2024: Polyglobe – Die intelligente Art der Markttransparenz. Available at: www.polyglobe.net (Last access: Aug. 20, 2024)

R-Cycle 2022: Circular economy in practice.https://www.r-cycle.org/kreislaufwirtschaft-in-der-praxis-2/ Available at: https://www.r-cycle.org/kreislaufwirtschaft-in-der-praxis-2/ (Last access: Aug. 20, 2024)

RecyClass 2022: Recycled Plastics. Available at: https://recyclass.eu/get-certified/recycled-plastic/#1. (Last access: Aug. 20, 2024)

Recyda 2024: Recyda - Packaging Sustainability Management. Available at: https://www.recyda.com/ (Last access: Aug. 20, 2024)

Resourcify 2024: Mehr Wert. Weniger Abfall. Available at: https://www.resourcify.com/de/ (Last access: Aug. 20, 2024)

RIGK 2024: RIGK launcht Kundenportal für Kunden im RIGK- und RIGK-G- sowie PICK-UP-SYSTEM. Available at: https://www.rigk.de/wissenswert/news/rigk-launcht-kundenportal-fuer-kunden-im-rigk-und-rigk-g-sowie-pick-up-system (Last access: Aug. 20, 2024)

RIGK 2024a: Digitalisierte Rücknahme und Entsorgung von Verpackungen bei PAMIRA– just in time und kontaktlos zu Corona Zeiten. Available at: https://www.rigk.de/wissenswert/presse/digitalisierte-ruecknahme-und-entsorgung-von-verpackungen-bei-pamira-just-in-time-und-kontaktlos-zu-corona-zeiten (Last access: Aug. 20, 2024)

Sennder 2024: Netzwerkeffekte nutzen. Available at: https://www.sennder.com/de/blog/route-optimization (Last access: Aug. 20, 2024)

SKZ 2023: „ReVise-UP“ zur Optimierung der Prozesseffizienz des werkstofflichen Kunststoffrecyclings mittels Sensortechnik gestartet. Available at: https://www.skz.de/presse/revise-up-zur-optimierung-der-prozesseffizienz-des-werkstofflichen-kunststoffrecyclings-mittels-sensortechnik-gestartet.

SKZ 2024: Projekt: Digitales Versuchsprotokoll für die Compoundierung. Available at: https://www.skz.de/forschung/kooperationsprojekte/digitales-versuchsprotokoll-compoundierung (Last access: Aug. 20, 2024)

Stich 2022: Stich, V, Stroh, M, Abbas, M, Frings, K, Kremer, S: Digitalisierung der Wirtschaft in Deutschland—Technologie- und Trendradar 2022. Hrsg. Bundesminisiterium für Wirtschaft und Klimaschutz. November 2022. Available at: https://www.de.digital/DIGITAL/Navigation/DE/Lagebild/Technologie-und-Trendradar/technologie-und-trendradar.html

Tiedemann 2020: Tiedemann, Y: Porsche forciert die Rückverfolgung von Werkstoffen. Available at: https://www.automotiveit.eu/technology/porsche-forciert-die-rueckverfolgung-von-werkstoffen-104.html

UL Prospector: Plastics & Elastomers. Available at: https://www.ulprospector.com/plastics/en (Last access: Aug. 20, 2024)

VDI-TZ 2021: Digitale Zwillinge - Heute schon weit verbreitet, morgen allerorten. Available at: https://www.vditz.de/service/publikationen/details/digitale-zwillinge-heute-schon-weit-verbreitet-morgen-allerorten (Last access: Aug. 20, 2024)

VDI ZRE 2021: Recycling plastics 4.0: Artificial intelligence and digital twin save resources. VDI Center for Resource Efficiency (VDI ZRE). Available at: https://www.youtube.com/watch?v=gOsVjdZZCOg (Last access: Aug. 20, 2024)

Dr.-Ing. Ulrike Lange studied waste management and contaminated sites at TU Dresden, completed her doctorate on the ecological and economic effects of the illegal shipment of waste electrical and electronic equipment and subsequently worked as a research assistant at Intecus GmbH on specialist studies for the Federal Environment Agency and the Federal Ministry for the Environment, among others. Since March 2016, she has been working as a technology consultant at VDI Technologiezentrum (TZ) GmbH in Berlin, focusing on the circular economy.

Andreas Bastian studied business administration at the University of Cologne (B.Sc.) and RWTH Aachen University (M.Sc.). Working student with RIGK GmbH (Gesellschaft zur Rückführung industrieller und gewerblicher Kunststoffverpackungen mbH) in Wiesbaden (development of a recyclability assessment tool, business field analyses). Today he acts as Managing Director of plastship GmbH in Waldems.

Dr.-Ing. Nico Pastewski is Division Manager with VDI Technologiezentrum (TZ) GmbH. After studying engineering, he started his career in 2006 with Fraunhofer Institute for Industrial Engineering (IAO) in Stuttgart and subsequently held various management positions in industry and consulting. His professional focus is on sustainability and digitalization in the context of research and innovation. Since 2019, he has been responsible for the fields of sustainability and health at VDI-Technologiezentrum GmbH (VDI-TZ) and heads their Berlin office.

Wei Min Wang is a technology consultant in the Resource Efficiency and Circular Economy Division of the VDI Technologiezentrum (TZ) GmbH in Berlin. Since October 2021, he is responsible for the technology area of digitalization and Industry 4.0 in the "Centre for Resource Efficiency". He studied industrial engineering focusing on software engineering and was then a research assistant at the Department.

Chapter 20
Standardisation

Stefanie Bierwirth and Francesco Degli Innocenti

Contents

20.1 Summary

Standards are essential for the circular economy; they are a means of communication and help to transfer information and find suitable partners and materials to close the cycle. They thus play an important role in the transformation of the linear economy to a circular economy.

This chapter starts by exploring how standards are created, which other deliverables (specifications, technical reports, and CWAs/IWAs) exist, and who develops

S. Bierwirth (✉)
DIN e.V., Berlin, Germany
e-mail: Stefanie.Bierwirth@din.de

F. Degli Innocenti
Firenze, Italy
e-mail: fdi@novamont.com

P. Orth et al. (eds.), *Circular Economy of Plastics*,
https://doi.org/10.1007/978-3-031-78829-1_20

standards. This leads us to the committees which are partaking in standardisation for plastics in the Circular Economy and the standardisation activities on plastics recycling (including less conventional forms of recycling, such as organic/biological recycling) and recyclates, the relevant committees on the one hand, and the standards and specifications dealt with by them on the other hand. Regarding European standardisation activities this includes the Standardisation Request M/584 developed based on the work of the Circular Plastics Alliance (CPA) initiated by the European Commission. Finally, emerging issues in the field of plastics recyclates, recycling, and bio-based plastics are addressed.

20.2 What is Standardisation?

Standards are technical rules specifying requirements for products, processes, or services. They allow the economy to function smoothly, pave the way for innovations to access markets, identify quality requirements and even relieve governmental regulation of the need to define technical details relevant to laws and regulations. The creation of technical rules, i.e. standardisation, is organised by various national, European, and international organisations and their employees. The technical content of standards, however, is drafted by experts authorised by the relevant stakeholders—such as industry, science, non-governmental organisations, governmental agencies, and environmental protection institutions. Standardisation organisations provide a neutral platform for the development of standards.

In a global circular economy, markets and value chains extend across national borders. For this reason, umbrella organisations for standardisation at the European and international levels are specifically relevant. The three biggest organisations at European level are the "European Committee for Standardization" (CEN), the "European Committee for Electrotechnical Standardization" (CENELEC, for electrotechnical standardisation) and the "European Telecommunications Standards Institute" (ETSI, for telecommunications), corresponding respectively to the "International Organization for Standardization" (ISO), the "International Electrotechnical Commission" (IEC) and the "International Telecommunication Union" (ITU) at international level.

Focussing on CEN and ISO, broad participation, great expertise, and diverse backgrounds of experts involved ensure relevance of topics and representation of the state of the art, leading to high acceptance and thus a broad application of standards. In more detail, each CEN and ISO member country (e.g. Slovenia) appoints one national standardisation body as its representative as CEN and ISO member body (e.g. SIST, Slovenian Institute for Standardization) [1]. Experts are delegated by these CEN and ISO member bodies, and additionally by liaison or partner organisations. These experts from across Europe or the world develop the technical content of standards in working groups (WG), operating in consensus. Superordinated to WGs, technical committees (TC) and subcommittees (SC) act as strategic bodies and collect comments and votes of national members on draft standards via ballots and consultations. The rules on the management of standardisation processes and participation in CEN/CENELEC and ISO/IEC are available to the public (cf. ISO/IEC Directives [2] and CEN/CENELEC Internal Regulations [3]).

The difference between CEN and ISO, however, is not only based on the geographical scope of member bodies and experts. European Standards (EN) are not published as stand-alone documents but must be adopted by CEN members as national standards and can only be purchased in this form (which are, for example, "NF EN"-standards for France, "DIN EN"-standards for Germany, etc.; the "Association française de normalisation " (AFNOR) and the "Deutsche Institut für Normung e. V." (DIN) also generally translate into French and German respectively when adopting European standards). Existing national standards with a scope overlapping with a European standard must be withdrawn. European Standards are therefore consistently valid in all CEN member countries and thus form the foundation for a uniform single European market. This is also why all draft standards issued by CEN are additionally open to public comment through their national members.

ISO Standards, on the other hand, are documents in their own right and can, but need not necessarily be adopted as national standards (e.g. "UNI ISO" for Italy). If identical standardisation interests exist at the European and international levels, standards are ideally developed simultaneously either under the leadership of CEN or ISO, where ISO is preferred due to the participation of member countries from all over the world. The results are "EN ISO" standards, which must be adopted by the CEN member countries as national standards (e.g. "NBN EN ISO" standards for Belgium), as explained above for EN standards. Cooperation between ISO and CEN is governed by the "Vienna Agreement" [4], its equivalent in the electrotechnical standardisation is the "Frankfurt Agreement" [5].

In addition to standards, other standardisation deliverables exist, which may be created without publishing a draft. These deliverables can be developed in a standardisation committee (TC/SC and WG), resulting in, e.g. Technical Specifications (TS) or Technical Reports (TR); the latter must not contain technical requirements, but may summarise the state of the art in, e.g. an informative manner.

Standardisation deliverables may also be developed outside standardisation committees, but still under the roof of CEN and ISO. Interested parties may form a "workshop" for a specific topic and develop a CEN Workshop Agreement (CWA) or an International Workshop Agreement (IWA). A workshop may consist of fewer stakeholders than present in a standardisation committee and does not need to work in consensus; that's why the development time may be shorter than a standardisation process. CWAs and IWAs may contradict other workshop agreements, but not existing standards, thus enabling the compatibility of innovation with existing standards. At the same time several innovations may compete through possibly contradictory workshop agreements. Similar deliverables exist in some member bodies, e.g. "National Technical Agreements" (NTA) at "Nederlandse Normalisatie-instituut" (NEN) and "DIN SPEC" at DIN.

20.3 Link between Standardisation and Governmental Regulation

The application of standards is voluntary, they are technical rules which reflect the state of the art and do not in themselves have any legal force. They may, however, take on a more pronounced status if reference is made to them in administrative regulations at the national or European level.

At European level, the New Approach and the New Legislative Framework stipulate that only essential requirements are included in EU directives and EU regulations. The European Commission instead drafts standardisation requests (formerly called "Mandates") to CEN/CENELEC mandating the development of standards to specify these essential requirements. Such standards are called "harmonized standards". By listing these standards in the Official Journal of the European Union (OJEU), they are officially connected to a certain directive or regulation. When such listed standards are followed, the assumption of conformity applies, i.e. the essential requirements of the associated legal act covered by the standard are considered fulfilled. However, the application of listed standards remains voluntary and compliance with the essential requirements can also be demonstrated by other means. Also, not all standardisation requests by the EC need to be directly linked to a directive or regulation. An example is mentioned below.

20.4 Standardisation for the Plastics Circular Economy

Standards deal with very different topics, encompassing e.g. screw threads, medical devices, quality management, artificial intelligence, or quantum technology. For their development, a wide variety of technical committees, subcommittees, and working groups with relevant experts are needed.

The Circular Economy, even if focussed on plastics solely, comprises a multitude of aspects, which cannot be dealt with by one standardisation committee alone. Relevant committees need to keep each other's work in mind when developing new standards around the Circular Economy, since environmental management plays just as important a role as design for recycling for various products or recyclate characterisation and all aspects ideally align to enable circularity. Due to the need for standards to support the Circular Economy, new standardisation projects arise continuously, and the associated committee landscape continues to change. The full diversity of the area cannot be captured in an exhaustive list and is thus briefly illustrated using three examples. Standardisation work in the plastics committees, mainly focussing on the material itself, is described in Sect. 20.5.

Example 1: Environmental Management and Overarching Aspects

Since the 1990s, environmental management activities have been translated into standards by the international technical committee ISO/TC 207 "Environmental Management" [6]. Among others, standards on life cycle assessment (LCA) and greenhouse gas management are drafted here, and environmental labelling and auditing are considered here, too. The well-known ISO 14000 standards on environmental management systems [7] also originate from here. ISO/TC 323 "Circular Economy" [8] was only established in 2018 and specifies principles for the implementation of a circular economy in organisations, including measurement of circularity. Another high-profile project, currently still under development, is the creation of a "Product Circularity Data Sheet" [9]. The traceability of supply chains [10] also relevant for circularity is being standardized in ISO/TC 308 "Chain of Custody" [11].

Example 2: Design for Recycling and Recyclability in Electrical Engineering

In the field of design for recycling or recyclability European standard EN 45555 [12] needs to be mentioned. It deals with the recyclability and reusability of energy-related products. It was developed in CEN/CLC/JTC 10 "Energy-related products – Material efficiency aspects for eco-design", a Joint Technical Committee (JTC) between CEN and CENELEC. In this example, there is a strong reference to plastics, since plastics are used for housings, insulation, adhesives, etc. The same committee also produced a standard [13] dealing with the use of recycled materials including plastics. The use of recyclates in products is a prerequisite for the circularity of materials.

Example 3: Packaging

In the field of packaging ISO/TC 122 "Packaging" [14] with its subcommittee SC 4 "Packaging and the environment" and the European committee CEN/TC 261/SC 4 "Packaging and Environment" [15] have drawn up several standards relating to a circular economy of packaging. However, these committees work on a cross-material basis, i.e. not only plastics are targeted in their standards. There are standards for reuse [16], for material recycling [17], and for biological recovery [18]. Requirements for recycled plastics are also specified for packaging for the transport of dangerous goods [19].

20.5 Standardisation Activities in the Plastics Committees on CEN and ISO Levels

20.5.1 Relevant Bodies

On the international level, the central body focussing on plastic materials and their environmental aspects, such as recycling, microplastics, or biobased and biodegradable plastics, is ISO subcommittee SC 14 "Environmental aspects" [20], established in 2017 within the Technical Committee ISO/TC 61 "Plastics". The independently operating SC 14 comprises five working groups, listed in Table 20.1.

The equivalent committee at the European level is CEN/TC 249 "Plastics" [21], including working groups WG 9 "Biobased and biodegradable plastics", WG 11 "Plastics recycling", WG 24 "Environmental aspects", and WG 26 "Agricultural plastic products—Design-for-recycling, use, removal, collection, and recycling".

Table 20.1 Working groups of ISO/TC 61/SC 14 "Environmental aspects"

#	Title
WG 1	Terminology, classifications, and general guidance
WG 2	Biodegradability
WG 3	Biobased plastics
WG 4	Microplastics in the environment
WG 5	Mechanical and chemical recycling

Table 20.2 Working groups of CEN/TC 249 "Plastics"

#	Title
WG 4	Decorative laminated sheets based on thermosetting resins
WG 5	Thermoplastic profiles for building applications
WG 7	Thermoplastic films for use in agriculture
WG 9	Bio-based and biodegradable plastics
WG 11	Plastics recycling
WG 13	Wood Plastics Composites (WPC)
WG 16	Welding of thermoplastics
WG 21	Profiles for windows and doors
WG 24	Environmental aspects
WG 25	Static thermoplastic tanks for above-ground storage of fuel
WG 26	Agricultural plastic products—Design-for-recycling, use, removal, collection and recycling

WG 24 monitors the need for standardisation with relevance to environmental aspects and deals with terminology specifying, e.g. terms and definitions related to recycling and biodegradable plastics (EN 17615 [22]). The full list of CEN/TC 249 working groups is shown in Table 20.2.

Since there are numerous junctures between bodies and their standards within and across the mentioned levels, exchange and cooperation among these technical committees and working groups are actively shaped and promoted (cf. 20.4).

20.5.2 Important Standards and Standardisation Deliverables on Plastics Recyclates and Plastics Recycling

Let us now look at projects developed in the committees mentioned in 20.5.1, in particular, ISO/TC 61/SC 14/WG 5 "Mechanical and chemical recycling" and CEN/TC 249/WG 11 "Plastics recycling". In 20.5.3 focus will be shifted towards biodegradable and biobased plastics.

ISO/TC 61/SC 14 "Environmental aspects"

During the first meeting of WG 5 "Mechanical and chemical recycling" in 2019 the experts agreed to develop a Technical Report [23] to help map existing standards and standardisation deliverables and identify future standardisation needs—i.e. new topics to be addressed by WG 5. ISO/TR 23891:2020 [23] also provides an overview of recycling systems, including a brief description of the associated processes. The identified existing documents include standards and standardisation deliverables from other ISO/TCs and from outside ISO (e.g. CEN/CENELEC, ASTM and several national standards bodies). The focus of this Technical Report is on mechanical and chemical recycling, while biodegradation and energy recovery were not considered.

After the publication of ISO/TR 23891 [23], a standardisation project on testing and characterisation of polypropylene and polyethylene recyclates was initiated and published in 2023: ISO 5677:2023 [24], focusing on mechanically recycled PP and PE for further processing e.g. by injection moulding, blow moulding, and extrusion, and specifying test methods needed to determine relevant characteristics.

Having been developed in ISO/TC 61 before the inception of ISO/TC 61/SC 14 and WG 5, ISO 15270:2008 [25] identifies the different recycling and recovery options for post-consumer and pre-consumer plastics waste and assists in the selection of appropriate processes including energy recovery and biodegradation. The standard also provides guidance for the development of new standards and specifications concerning plastic waste recovery. Given that recycling technologies advanced considerably since 2008 a revision of this standard was started in 2022. According to these developments, the standard was split up into five parts, a general part and four parts on mechanical, physical, chemical, and organic/biological recycling, respectively (definitions from ISO documents can be accessed via [26]). The differentiation between the recycling technologies will be included in the revision as well as underlying definitions. To help the relevant experts participating in each part, the work on the different parts is spread over three of SC 14's working groups, WG 1, WG 2, and WG 5. The revision of ISO 15270:2008 is ongoing at the time of finalisation of this text.

In addition to ISO 15270–4 on chemical recycling, WG 5 is working on a standard on gasification [27] intended to provide general principles for chemical recycling by gasification and help the user to choose a suitable gasification technology.

All published and running projects of ISO/TC 61/SC 14 are listed on the ISO website [20].

CEN/TC 249 "Plastics"

While the focus in standardisation on the international level is on recycling technologies—except for ISO 5677:2023 [24], European standardisation activities are currently influenced considerably by a standardisation request by the European Commission.

As part of the European Plastics Strategy, the European Commission launched the Circular Plastics Alliance (CPA) [28] at the end of 2018. The declaration [29] of the CPA was signed by more than 300 stakeholders, its main goal being the use of ten Mt of plastic recyclates per year in products in Europe from 2025 onwards. The CPA's work focuses on five sectors: agriculture, automotive, construction, electrical and electronic equipment (EEE), and packaging. In a matrix structure, four thematic complexes, i.e. design for recycling, collection and sorting, recycled content, and research and development, are addressed and interconnected with each of the five sectors [30]. The European Commission identified standardisation as an important tool for the implementation of this ambitious goal. Thus, corresponding standardisation needs in the sectors named above have been specified covering design for recycling, characterisation of plastics wastes and recyclates, and use of recyclates in products, and the European Commission issued standardisation request M/584 on "Plastics Recycling and Recycled Plastics" to CEN and CENELEC based on those needs. The standardisation request was accepted by CEN and CENELEC in 2022 [31].

Whenever a standardisation request is accepted, CEN and CENELEC guarantee the drafting of new, or revision of existing standards specified therein by a fixed target date. The standards and standardisation deliverables under M/584 must be published by 2 August 2025.

An important prerequisite for the development of high-quality standards within the framework of this standardisation request once again is a constant exchange between the standardisation committees concerned since a consistent definition of interfaces and basic principles is needed.

M/584 involves nine CEN and CENELEC technical committees [32] and has a major impact on the work in CEN/TC 249; more than half of the standards under M/584 are drafted here in four of CEN/TC 249's WGs (WG 7, WG 11, WG 21, and WG 26). "Design for recycling" as one thematic complex is dealt with in all affected TCs for their specific products or sectors, whereas "characterisation of plastics wastes", "characterisation of plastics recyclates", and "use of plastics recyclates in products" are all concentrated in CEN/TC 249/WG 11 "Plastics recycling" comprising 20 standards, 17 of which are under development at the time of finalisation of this text. Among these are standards for characterisation of plastics wastes and recyclates first published in 2007 and now being revised. These standards specify properties of plastics wastes and recyclates to be declared for the transfer from supplier to purchaser, distinguishing between mandatory and optional characteristics. Relevant test methods for the determination of individual properties are also listed but no characteristic values, limit values or ranges are fixed to be met. Each polymer type is dealt with in a separate standard. An immediate impact of the standardisation request is the development of an additional standard on the characterisation of ABS recyclates (EN 18067, see Table 20.3) and the transformation and expansion of EN 15347:2007 [33] on waste characterisation into a standard series.

Future standard EN 18065 (see Table 20.3), based on DIN SPEC 91446:2021–12 [34], also follows the logic of not specifying any limit/characteristic values or ranges, but establishes four lists of properties corresponding to four different "data quality levels" (DQLs) independent of polymer type, where more properties must be specified when moving up from the lowest to the highest DQL. When purchasing

Table 20.3 CEN/TC 249/WG 11 standards developed under M/584 (all standards, except for EN 15346, EN 15347-1 and EN 15348, are under preparation at the time of finalisation of this text)

Standard	Title
EN 15347-series (revision of EN 15347:2007 and new parts 2 to 6)	Plastics—Sorted plastics wastes — Part 1: General Characterization — Part 2: Quality grades of sorted polyethylene (PE) wastes and specific test methods —Part 3: Quality grades of sorted polypropylene (PP) wastes and specific test methods — Part 4: Quality grades of sorted poly(ethylene terephthalate) (PET) wastes and specific test methods — Part 5: Quality grades of sorted poly(vinyl chloride) (PVC) wastes and specific test methods — Part 6: Quality grades of sorted polystyrene (PS) wastes and specific test methods
EN 18065 (new)	Recycled plastics—Classification by data quality levels for use and (digital) trading
EN 15342 (revision)	Plastics—Recycled Plastics—Characterization of polystyrene (PS) recyclates
EN 15344 (revision)	Plastics—Recycled plastics—Characterization of polyethylene (PE) recyclates
EN 15345 (revision)	Plastics—Recycled Plastics—Characterization of polypropylene (PP) recyclates
EN 15346 (revision)	Plastics—Recycled plastics—Characterization of poly(vinyl chloride) (PVC) recyclates
EN 15348 (revision)	Plastics—Recycled plastics—Characterization of poly(ethylene terephthalate) (PET) recyclates
EN 18067 (new)	Plastics—Recycled plastics—Characterization of acrylonitrile–butadiene–styrene (ABS) recyclates
EN 18064-series (new)	Plastics—Quality recommendations and basis for specifications for application of plastic recyclates in products — Part 1: General aspects — Part 2: Polyethylene (PE) — Part 3: Polypropylene (PP) — Part 4: Poly(ethylene terephthalate) (PET) — Part 5: Poly(vinyl chloride) (PVC) — Part 6: Polystyrene (PS) — Part 7: Acrylonitrile- butadiene-styrene (ABS)

recycled plastics, this approach is intended to enable buyers to select the amount of information required for their own purposes and thus to compare recyclate offers more easily. Eventually, a series of seven parts on quality recommendations for the use of plastic recyclates in products (future standard series EN 18064, see Table 20.3) was initiated as a direct result of M/584. In contrast to the revised standards on recyclate characterisation and DQLs, this series will provide typical values or ranges for individual properties and different applications. The series comprises a general part and six polymer-specific parts.

The list of standards described above and planned to be published under the responsibility of WG 11 and under M/584 in August 2025 may be found in Table 20.3.

In addition to the standards listed in Table 20.3, a Technical Report on the classification of plastics recyclates as post-consumer and post-industrial recyclates is underway [35]. Future possible standardisation projects are mentioned in 20.6.

Complementary to the standards under development mentioned above are existing European Technical Specifications (CEN/TS) on sampling of recyclates and plastic waste [36], on sample preparation [37], and on the determination of impurities in recyclates [38].

A comprehensive approach for an entire product life cycle is pursued by European standard EN 17410 [39] describing a controlled loop for PVC-U window and door profiles. It defines the various phases of the life cycle of the profiles, specifies measuring points for quality assurance, and describes design for recycling. The standard links the measuring points and phases with standards for windows and doors and window and door profiles, as well as with the documents on recyclate characterisation described above, thus combining them into a consistent system of standards.

Eventually, the use of uniform terms in standards supports their comprehensibility and establishes an unambiguous language. Terminology standard EN 17615 [22] developed in CEN/TC 249/WG 24 "Environmental aspects" and published in 2022 is intended to ensure exactly that for the environmental aspects of plastics.

20.5.3 Important Standards and Standardisation Deliverables on Biodegradable Plastics and Biobased Plastics

The plastic types that enjoyed enormous commercial success in the second half of the twentieth century were all fossil-based and resistant to biodegradation. More recently, plastics have emerged whose chemistry is not based on carbon of fossil origin but on carbon found in newly formed organic matter. At the same time, research has developed plastics that are susceptible to biodegradation by microorganisms, similar to plant and animal residues. The emergence of these plastics with biological properties (i.e. of biological origin or subject to biodegradation) has caused much confusion in the public and among users. The reality is that synthetic materials may be of fossil origin and be non-biodegradable or of biological origin and be biodegradable. However, they can also have a biological origin but not be biodegradable, or a fossil origin and be biodegradable. This has caused quite some misunderstanding because natural materials are both biological and biodegradable, and we instinctively expect these two properties to be inextricably linked, which is not the case with synthetic materials (cf. Chap. 14). Again, standardisation has brought order to the sector by creating definitions, test methods, and specifications allowing both bio-based and biodegradable products to be clearly and unambiguously defined. In the following passages, the two categories are discussed separately, recognising that they can co-exist in the same products, but conceptually representing different characteristics that should not be confused.

Biodegradable plastics

Historically, interest in biodegradable plastics began in the early 1990s. The value proposition of biodegradable plastic products was recognised early on in their ability to be recovered through organic (biological) recycling in the same way as "natural" bio-waste (food and garden waste).

The real impetus for biodegradable plastics came from the 1994 Packaging Directive.

The Directive introduced this innovation: packaging can also be recovered using "organic recycling" if biodegradable. The definition contained in Article 3 of the original Directive stated that "organic recycling shall mean the aerobic (composting) or anaerobic (biomethanisation) treatment, under controlled conditions and using micro-organisms, of the biodegradable parts of packaging waste, which produces stabilised organic residues and methane. Landfill shall not be considered a form of organic recycling." The term *stabilised organic residues* refers to soil improvers that can be reused in the agricultural and nursery-gardening sector (i.e. compost). It is a genuine form of recycling, although it differs from other methods (such as 'mechanical' recycling) that involve releasing a material on the market that is the same as the input waste (e.g. producing paper from paper waste, glass from glass waste, plastic from plastic waste). In organic recycling, the resulting product differs significantly from the original but has both economic and environmental value, making it an essential component of a genuine cycle. The European Commission mandated CEN (M/200, see [40]) to prepare a series of standards to show a presumption of compliance with the essential requirement of Directive 94/62/EC. In accordance with Mandate M/200, CEN developed a series of standards on prevention, reuse, recycling, energy recovery, and organic recovery of packaging waste. In particular, the harmonised standard (cf. 20.3) EN 13432 [41] pertains to packaging that can be recovered through organic recycling. As such it provided the presumption of conformity to the essential requirements for packaging with regard to 'packaging recoverable in the form of compost' and 'biodegradable packaging' in accordance with the European Directive.

The EN 13432 standard was developed by the CEN Technical Committee CEN/TC 261, which deals with packaging. EN 13432 has become an international reference and the technical basis for the following standards being similar in terms of technical content and requirements:

- EN 14995 [42] prepared by CEN/TC 249/WG 9 "Bio-based and biodegradable plastics"
- ISO 17088 [43] prepared by ISO/TC 61/SC 14/WG 2 "Biodegradability"
- ISO 18606 [18] prepared by ISO/TC 122/SC 4 "Packaging and the environment"

After the resolution of the compostable plastics and packaging issue, standardisation turned its attention to the biodegradability of plastics considered a useful characteristic in various sectors, especially agriculture. Plastics are widely used in agriculture today for various applications such as greenhouses, irrigation, and mulching. For many applications, recovery and recycling is possible and practical (e.g. tunnel and greenhouse film). In other cases, however, recovery and recycling can be difficult due to contamination with soil and crop residues. Thin mulch films, e.g. can break during collection, making recovery problematic. In these cases, the use of biodegradable plastics can be advantageous as the films biodegrade directly in the soil, solving the problems of recovery and recycling. CEN/TC 249/WG 7 "Thermoplastic films for use in agriculture" prepared on this subject the standard EN 17033 [44]. This standard has had the merit of clearly defining the test methods and characteristics of biodegradable mulching films, bringing order to the sector and enabling the development of product certification, an essential factor in a regulated market. Shortly after, the International Standard ISO 23517 [45] with similar technical content was adopted.

Since its establishment in 1992 through the efforts of Dr Hideo Sawada [46], the working group ISO/TC 61/SC 14/WG 2 "Biodegradability" (originally known as WG 22 and placed in ISO/TC 61/SC 5 "Physical–chemical properties") has been very active in developing test methods to measure biodegradation under different conditions (high/low temperature; solid, liquid, heterogeneous state) and using different microbial inocula (compost, soil, seawater, wastewater treatment plants).

During the last decade, interest has turned from biodegradation under controlled conditions to biodegradation in the open environment, i.e. under uncontrolled conditions. In the case of compostable plastics, the end of life is programmed to occur in a composting plant under controlled conditions. Even in the case of plastics used in agriculture, the end of life is localised in space (the agricultural field) and in time (the duration of cultivation), although it is subject to weather conditions. In the case of uncontrolled release, however, the conditions are unpredictable. The interest in this type of event is obviously not related to a planned end-of-life, as no one thinks of littering as a waste management option. On the contrary, these are exceptional conditions that can occur for products and applications that are particularly susceptible to littering. For example, plastics used in fish farming must be recovered after use and disposed of or even recycled if feasible. However, these fishing gears are subject to breakage with the production of fragments. In case of dispersion, biodegradability is a factor that mitigates the environmental risk caused by these fragments. Therefore, the assessment of biodegradability is an important part of the risk assessment of products with a high risk of accidental spillage. In order to assess this factor, the working group "Biodegradability" has already developed several test methods for biodegradation with marine microorganisms (e.g. ISO 19679 [47]), defined specific criteria (e.g. ISO 22403 [48]), and outlined useful parameters for predicting the environmental risk (ISO 5148 [49]).

Biobased Plastics

Until about a century ago, all products, with the obvious exception of mineral products, were "bio-based". During relatively short a time, with the advent of fossil resources and synthetic materials, the scenario of everyday products has been overturned. Today, most things are of fossil origin. In recent years, however, there has been a growing awareness of the need to rebalance this trend, not so much to return to the past, but to initiate a gradual process of decoupling industry and fossil resources, with the aim of reducing dependence on the latter on the one hand and supporting rural development and innovative bio-industries on the other.

In 2006, the European Commission identified bio-based products as an innovative and promising sector to be promoted through support measures, and particularly the development of technical standardisation, as part of the so-called "Lead Market Initiatives". In fact, it was necessary to develop the terminology, test methods, and criteria to unambiguously identify and classify bio-based products, an indispensable prerequisite for any support action.

Technical Committee CEN/TC 411 "Bio-based products", set up in response to Mandate M/492 of the European Commission [50], has produced numerous standards that now allow the bio-based products sector to be supported by specific references.

According to standard EN 16575 [51], bio-based products are derived totally or partially from biomass, i.e. from materials of biological origin (plants, trees, algae, marine organisms, microorganisms, animals, etc.), except for fossil materials, which are also of biological origin but are limited to geological and non-renewable formations. How is it possible to distinguish a bio-based product from a fossil one? The question is not simple, because chemically identical substances can be obtained from both fossil raw materials and biomass. Chemistry cannot help in this case, and it is necessary to rely on physics by measuring the amount of radioactive carbon (^{14}C) present in the material, using the same approach as archaeology to date organic finds. Atmospheric CO_2 is also made up of a very small percentage of molecules containing carbon in radioactive form (^{14}C), constantly replenished in the upper part of the atmosphere. Plants do not distinguish between "normal" CO_2 and radioactive $^{14}CO_2$ but absorb both. Therefore, all biological matter contains ^{14}C in the same proportion as atmospheric $^{14}CO_2$. This characteristic of the atmosphere is faithfully reproduced in all organisms in the food chain. In short, ^{14}C is a marker for biological matter contained in products, too. However, this marker disappears over time as radioactive elements decay. Therefore, the oil that has been deposited in the Earth's crust for millions of years has lost all the ^{14}C that was present in the original biological materials. Fossil materials do not have an atmospheric "signature", they are not radioactive. This difference has been exploited to develop a method to identify and quantify the use of biological raw materials in products by measuring their radioactivity. The methodology for determining the carbon content of biological origin is described in standard EN 16640 [52].

As materials of biological origin contain not only carbon, but also oxygen, hydrogen, nitrogen, etc., another method has been developed to count these components as well. This method is based on the manufacturer's declaration of the formulation/recipe of the bio-based product. Based on the formulation, it is possible to calculate the content of substances of biological origin and the total percentage of the different components (C, H, O). These statements are validated using the "^{14}C method" and elemental analyses applicable to the specific product. This alternative approach is described in EN 16785–1 [53]. Due to its complexity, this method of determining the content of materials of biological origin preferably requires the intervention of a certifier as a guarantor of the truthfulness of the information and the correct application of the validation procedure.

CEN/TC 411 has also addressed issues related to the sustainability of raw materials of biological origin to provide tools for determining the impacts of the cultivation of renewable resources, which obviously affect the sustainability of the final bio-based products. EN 16751 [54] applicable to the biological part of all bio-based products, including all three pillars of sustainability: environmental, social, and economic aspects. This European standard can be used to provide information on the sustainability of biomass production and the supply chain for the bio-based part of the product. This means that if the product is only partly bio-based, the standard can only be used for the bio-based part, as it does not address the non-"bio" (fossil, mineral) parts of the product.

Life Cycle Assessment (LCA) requirements for bio-based products are set out in EN 16760 [55]. This standard provides guidelines and requirements for assessing the life cycle impacts of bio-based products, focusing on the specificities of the biological part of the product. It also includes studies that do not cover the entire life cycle, such as "cradle-to-gate" studies.

Data obtained by applying the different standards must be communicated transparently along the commercial chain, both in business-to-business transactions and, ultimately, in communication with consumers. The core of EN 16848 [56] is a "scheme" to be used for the preparation of an information package accompanying bio-based material to be sent by the supplier to the customer, similar to what happens today with technical and safety data sheets. The bio-based semi-finished product will be accompanied by a card containing the specific information required for eco-design purposes and to establish communication with the next customer along the value chain. Finally, EN 16935 [57] specifies how to inform the final consumer about the characteristics of bio-based products through transparent and non-misleading "business-to-consumer" communication and specific labelling. In a nutshell, the statement "product of biological origin" must be accompanied by the percentage content of substances of biological origin (cf. Chap. 16) and by information on how to dispose of the product ("end of life"), which is essential if the consumer is responsible for sorting waste and needs to understand how to manage the bio-based product. It is also possible to provide information on the sustainability of the biomass.

The definitions, standards, and specifications established by CEN/TC 411 were adopted by CEN/TC 249/WG 9 to draft the standard EN 17228 [58]. This standard is particular to plastics and polymers and fulfils the European Commission Mandate M/430 [59].

20.6 Future Topics

At the time of finalisation of this text, the Standardisation Request M/584 still defines the European standardisation work in CEN/TC 249/WG 11 and about 20 standards will be published until August 2025. However, once these standardisation projects will be sufficiently mature, a revision of the Technical Report ISO/TR 15353 [60] could be initiated. This Technical Report so far focusses on guidelines for the development of new standards for recycled plastics but could be extended or transformed into a document additionally introducing the many newly developed and revised standards of WG 11.

Even if calculation of recycled content and traceability of plastics recyclates are not addressed in M/584, these topics become increasingly important and potentially more complex due to the developments in recycling technologies. Thus, in the near future, EN 15343:2007 [61], dealing with traceability and determination of recycled content will prospectively be revised. European legislation such as the Single-Use Plastics Directive will probably be pathbreaking in its further development.

On the international level, the establishment of definitions and basic principles regarding physical and chemical recycling presumably will be the focus of ISO/TC 61/SC 14/WG 5 in the coming years. However, there is still little direct cooperation between CEN and ISO in the field of recyclate and recycling standardisation (there are hardly any "EN ISO" documents in this area), thus, the question as to whether greater harmonisation is sensible and desirable will have to be answered.

The need for further standardisation in the field of biodegradable plastics has been thoroughly analysed by CEN/TC 249/WG 9 and published in the Technical Report CEN/TR 17910 [45].

There are issues that need to be revised, such as the standard EN 13432 [41] on packaging that can be recovered by organic/biological recycling. Other standards need to be extended in scope, such as EN 17033 [44], which is currently only applicable to mulch films, but could be extended to other tools used in agriculture. Finally, there are areas that are still under development and need to be completed, such as biodegradation in the uncontrolled open environment in the event of littering. In some cases, these objectives are already included in the action plans of the relevant working groups, either because they are scheduled for periodic review or because they are needs that are well known to stakeholders and experts in the field. In any case, the technical report CEN/TR 17910 [62] provides a good starting point for future activities.

For standardisation to support and promote the implementation of the Circular Economy, it is necessary to address emerging topics in good time.

20.7 Outlook

To enable a circular economy, handover between different parties in the cycle must function smoothly. Clear requirements as to which information must be made available to facilitate communication between recyclate producers and buyers, as well as agreement on specifications. Standards can include these requirements and thus promote the circular economy. All stakeholders can participate in the drafting of standards on the neutral platform that the standardisation bodies provide, which is why the resulting standards are considered comprehensible and relevant to the stakeholders. For this to work, interested parties need to participate actively in standardisation or at least contribute their perspective by commenting on published draft standards via their national standards bodies.

The Circular Economy cannot be introduced by making changes at just one point in a complex (system of) value chain(s); adjustments to the linear system must be made at many instances. Standardisation is an ideal tool for these adaptations, as it allows the stakeholders to engage in detailed discussions with each other and to network across and between different sectors. This is true in any sector, and even more so in those where the potential for ambiguity is high due to the difficulty of translating complex properties, which can only be characterised by sophisticated analytical systems, into simple terms. In the field of bio-based and biodegradable plastics, the possibility to rely on unambiguous standards allows for an orderly market, which is a prerequisite for the development of innovations and their commercial growth.

References

1. A list of CEN and ISO members can be found here, CEN: https://standards.cencenelec.eu/dyn/www/f?p=CEN:5, ISO: https://www.iso.org/members.html
2. https://www.iso.org/directives-and-policies.html
3. https://boss.cen.eu/reference-material/RefDocs/Pages/
4. https://www.cencenelec.eu/about-cen/cen-and-iso-cooperation/
5. https://www.cencenelec.eu/about-cenelec/cenelec-and-iec-cooperation/
6. ISO/TC 207: https://www.iso.org/committee/54808.html
7. Examples: ISO 14001:2015, Environmental management systems — Requirements with guidance for use, or ISO 14006:2020, Environmental management systems — Guidelines for incorporating eco-design. Lists of all standards are also available via [6] and the corresponding pages of the subcommittees.
8. ISO/TC 323: https://www.iso.org/committee/7203984.html
9. ISO/AWI 59040, Circular Economy — Product Circularity Data Sheet
10. ISO 22095:2020, Chain of custody — General terminology and models
11. ISO/TC 308: https://www.iso.org/committee/6266669.html
12. EN 45555:2020–04, General procedure for assessing the recyclability and recoverability of energy-related products
13. EN 45557:2020–09, General procedure for the assessment of the percentage of recycled material in energy-related products
14. ISO/TC 122: https://www.iso.org/committee/52040.html
15. CEN/TC 261/SC 4: https://standards.cencenelec.eu/dyn/www/f?p=205:7:0::::FSP_ORG_ID:6473&cs=15F41CC9B62F6E42ED3749377F32EBB14

16. ISO 18603:2013, Packaging and the environment — Reuse
17. ISO 18604:2013, Packaging and the environment — Material recycling
18. ISO 18606:2013, Packaging and the environment — Organic recycling
19. EN ISO 16103:2021–12, Packaging — Packaging for the transport of dangerous goods — Recycled plastics
20. ISO/TC 61/SC 14: https://www.iso.org/committee/6578018.html
21. CEN/TC 249: https://standards.cencenelec.eu/dyn/www/f?p=205:7:0::::FSP_ORG_ID:6230&cs=170C05F24EF08AC268C42E03B8F786259
22. EN 17615:2022, Plastics — Environmental aspects — Vocabulary
23. ISO/TR 23981:2020, Plastics — Recycling and recovery — Necessity of standards
24. ISO 5677:2023, Testing and characterization of mechanically recycled polypropylene (PP) and polyethylene (PE) for intended use in different plastics processing techniques
25. ISO 15270:2008, Plastics — Guidelines for the recovery and recycling of plastics waste
26. ISO Online Browsing Platform: https://www.iso.org/obp
27. ISO 13390, Plastics - Chemical Recycling - Gasification
28. https://ec.europa.eu/growth/industry/strategy/industrial-alliances/circular-plastics-alliance_en
29. Declaration of the CPA: https://ec.europa.eu/docsroom/documents/36361
30. CPA Working rules: https://ec.europa.eu/docsroom/documents/44545
31. M/584: https://ec.europa.eu/growth/tools-databases/enorm/mandate/584_en
32. https://www.cencenelec.eu/news-and-events/news/2022/brief-news/2022-09-07-plastics-in-circular-economy/
33. EN 15347:2007, Plastics — Recycled Plastics — Characterisation of plastics wastes
34. DIN SPEC 91446:2021–12, Classification of recycled plastics by Data Quality Levels for use and (digital) trading
35. CEN Technical Report (under development), Plastics recycling — Classification of plastic recyclates as postconsumer recyclates (PCR) and postindustrial recyclates (PIR)
36. CEN/TS 16010:2020, Plastics — Recycled plastics — Sampling procedures for testing plastics waste and recyclates
37. CEN/TS 16011:2013, Plastics — Recycled plastics — Sample preparation
38. CEN/TS 17627:2021, Plastics — Recycled plastics — Determination of solid contaminants content
39. EN 17410:2021, Plastics — Controlled loop recycling of PVC-U profiles from windows and doors
40. M/200 standardization request to CEN for standardisation and study related to packaging and packaging waste https://law.resource.org/pub/eu/mandates/m200.pdf
41. EN 13432:2000, Packaging. Requirements for packaging recoverable through composting and biodegradation. Test scheme and evaluation criteria for the final acceptance of packaging
42. EN 14995:2006, Plastics — Evaluation of compostability — Test scheme and specification
43. ISO 17088:2021, Plastics — Organic recycling-Specifications for compostable plastics
44. EN 17033:2018, Plastics — Biodegradable mulch films for use in agriculture and horticulture — Requirements and test methods
45. ISO 23517:2021, Plastics — Soil biodegradable materials for mulch films for use in agriculture and horticulture — Requirements and test methods regarding biodegradation, ecotoxicity and control of constituents
46. Sawada H., 2005, Biodegradability of plastics a path to prevent pollution. ISO Focus June 2005:12–15 https://committee.iso.org/files/live/sites/tc61/files/ISO%20Focus%20June%202005_Biodegradability%20of%20plastics.pdf
47. ISO 19679:2020, Plastics — Determination of aerobic biodegradation of non-floating plastic materials in a seawater/sediment interface — Method by analysis of evolved carbon dioxide
48. ISO 22403:2020, Plastics — Assessment of the intrinsic biodegradability of materials exposed to marine inocula under mesophilic aerobic laboratory conditions — Test methods and requirements

49. ISO 5148:2022, Plastics — Determination of specific aerobic biodegradation rate of solid plastic materials and disappearance time (DT50) under mesophilic laboratory test conditions
50. M/492: https://law.resource.org/pub/eu/mandates/m492.pdf
51. EN 16575:2014, Bio-based products — Vocabulary
52. EN 16640:2017, Bio-based products — Bio-based carbon content — Determination of the bio-based carbon content using the radiocarbon method
53. EN 16785–1:2015, Bio-based products — Bio-based content — Part 1: Determination of the bio-based content using the radiocarbon analysis and elemental analysis
54. EN 16751:2016, Bio-based products — Sustainability criteria
55. EN 16760:2015, Bio-based products — Life Cycle Assessment
56. EN 16848:2016, Bio-based products — Requirements for Business to Business communication of characteristics using a Data Sheet
57. EN 16935:2017, Bio-based products — Requirements for Business-to-Consumer communication and claims
58. EN 17228:2019, Plastics — Bio-based polymers, plastics, and plastics products — Terminology, characteristics and communication
59. M/430: https://law.resource.org/pub/eu/mandates/m430.pdf
60. CEN/TR 15353:2007, Plastics — Recycled plastics — Guidelines for the development of standards for recycled plastics
61. EN 15343:2007, Plastics — Recycled Plastics — Plastics recycling traceability and assessment of conformity and recycled content
62. CEN/TR 17910 2022 Biodegradable plastics — Status of standardization and new prospects

Stefanie Bierwirth studied physics at the University of Göttingen. After completing her physics diploma, she joined the Helmholtz Center Berlin for Materials and Energy (HZB). Since 2016 she works at DIN—German Institute for Standardization (DIN)—and is responsible for standardization committees at German, European and international level for the DIN Standards Committee Plastics (FNK) in the areas of plastics recycling, as well as environmental aspects of plastics.

Francesco Degli Innocenti is a biologist by education. He started his professional career at Max Planck Institute for Molecular Genetics in Berlin, Germany, and worked for 3 decades as Director *Ecology of Products and Environmental Communication* (ECOPEC) with Novamont, Novara, Italy, where he dealt with the sustainability bioplastics. Today he is a consultant and coordinates the Scientific and Technical Committee of *Assobioplastiche*, the Italian Association of Bioplastic.

Part V
Regulatory Framework

The change from a linear plastics economy based on fossil feedstock to a circular plastics economy based on non-fossil, i.e. circular, feedstock requires a regulatory framework that must evolve in line with the current state of development of the circular economy. This regulatory framework covers all phases of the plastics cycle. Whereas in the past regulations focused mainly on waste, current and planned regulations increasingly take into account aspects of sustainable product design and sustainable product use. The circuit diagram illustrates this claim with a square frame. With the Green Deal, the Circular Economy Action Plan and previously the Plastics Strategy, the EU Commission has developed a far-reaching circular economy strategy, particularly for plastics and their applications, which is successively being implemented via specific EU regulations, such as the draft Regulation on Packaging and Packaging Waste (PPWR).

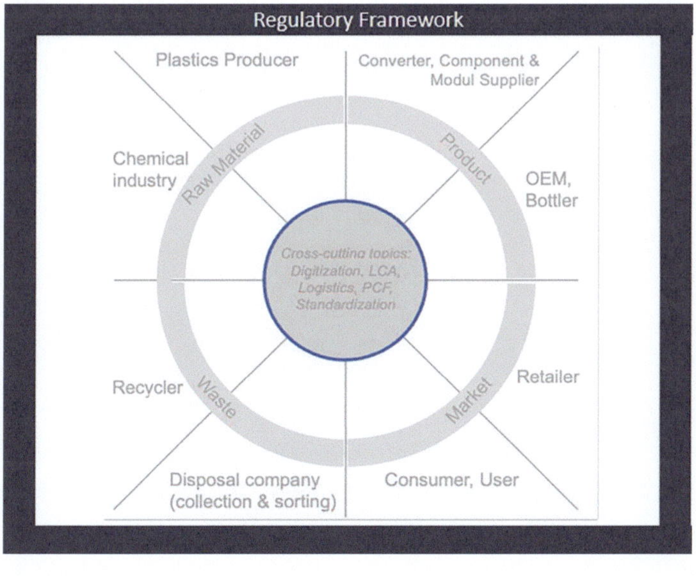

Chapters 21 and 22 present the status of the regulation (until March 2024) for the development of plastic value circles, considering the genesis of these regulatory measures to help understand the prioritization of the regulations by the legislator, the instruments used, and the goals set. The explanations focus on the EU regulatory framework, as this is crucial for a functioning circular economy in the common market. The interaction between national legislation and EU regulation is discussed in a separate section of Chap. 21.

Chapter 21 provides an overview of the most important European legal regulations for the stakeholders of the plastics value circle. In addition to regulations in the key plastic markets (packaging, construction, automotive, and disposables outside the packaging sector), topics such as REACH and food contact regulations are also addressed, as these must be complied with by many players when dealing with recyclates. In addition, a brief introduction is given to the European taxonomy legislation and the European Plastics Tax. Not all regulations aimed at closing the loop are dealt with in Chap. 21. Aspects of the export of plastic waste can be found in Chap. 5; regulations connected with the marine litter problem in Chap. 12. Chapter 22 deals in detail with the new regulatory core element of the recycled content quota applied in the Single Use Plastics Directive (SUPD), the draft Packaging and Packaging Waste Regulation (PPWR) and possibly in other planned regulations on the circular economy such as in the automotive sector. There is no experience yet with the implementation of such quota requirements in practice, but they are already leading to desirable and undesirable effects on the market. Open questions regarding implementation, particularly in the event of recyclate shortage, are discussed constructively and critically. The polymer-specific substitution quota with its potential benefits and risks is discussed as a supplementary or alternative steering instrument for the production of recyclates and for the feedstock transformation in total.

Chapter 21
Regulation in Europe

Silvia Freni-Sterrantino, Antonello Romano, and Jürgen Bruder

Contents

S. Freni-Sterrantino (✉)
EuPC, Brussels, Belgium
e-mail: silvia.freni01@gmail.com

A. Romano
Stora Enso, Brussels, Belgium
e-mail: antonello.romano@storaenso.com

J. Bruder
BPC-Bruder Plastics Consulting, Potsdam, Germany
e-mail: plastics@jbruder.de

P. Orth et al. (eds.), *Circular Economy of Plastics*,
https://doi.org/10.1007/978-3-031-78829-1_21

Summary

This chapter focuses on the regulatory provisions contained in various horizontal as well as product-specific regulations that have an impact on how plastics are manufactured, used, and disposed of. It starts with providing a historical outline of the strategic documents and regulations in Europe since this represents the background to describe how the current regulations affect plastics. Key political documents (such as Green Paper on Plastic Waste, EU Plastics Strategy, Circular Economy Action Plan) are explained that have paved the way for the adoption of the most impactful requirements and obligations that are shaping the EU market for plastics.

The first section explains the regulations and proposals of the European legislator that are intended to specifically promote the development of a circular economy in important plastics application areas. The pioneering roles of the regulations on single-use plastic products and on packaging are highlighted.

A second section is dedicated to selected horizontal regulations to be taken into account by players in a plastics circular economy and often represent important regulatory cornerstones.

In the third section, the links between EU and national activities, including legislation for plastic products and waste, are analysed on examples like national taxes or product bans.

The chapter concludes with a summarised assessment of the EU regulations to promote a circular plastics economy, the outstanding experiences with recently adopted regulations and an outlook on possible future focal points of the EU's plastics policy.

21.1 Historical Outline

Throughout the past decades, the European Union (EU) policymakers have been working towards advancing plastic waste management in Europe as well as efficient new products and manufacturing practices. This was made possible by recognizing plastics and plastic waste as valuable materials and resources, respectively. As such, plastics recycling emerged as an important element of the EU's circular economy and a crucial step in helping to solve the plastic waste crisis.

The EU has always been at the forefront of environmental protection worldwide. It has one of the most ambitious waste legislations in the world, which is the result of decades of dedicated and focused policies and law-making. EU's first Environmental Law goes back to 1973 when the first Environmental Action Programme was adopted. This set general environmental objectives to be achieved by Member States. Today, there are over 300 acts including regulations, directives, decisions, and recommendations that make up the EU environmental legislation.

Following the adoption of the first Waste Framework Directive in 1973, in 1987, environmental protection made its way into the EU Treaties. The Single European Act (SEA 1987) introduced a new 'Environment Title', which provided the first legal basis for a common environmental policy with the aims of preserving the quality of the environment, protecting human health, and ensuring rational use of natural resources.

Switching the focus on plastics, when looking for the origins of the regulations on recycling, it is inevitable to come across the German Packaging Ordinance (GPO 1991), which is associated with the name of the then German Federal Environment Minister Klaus Töpfer. Although the Packaging Ordinance referred to all materials in the packaging sector, it posed challenges for plastic packaging, since, unlike glass or paper packaging, there was virtually no recycling capacity available. However, the Packaging Ordinance included demanding recycling targets with the objective of driving the uptake of recycling infrastructures and recycling activities.

A pioneering requirement of the Packaging Ordinance was the formulation of product responsibility rules in accordance with the polluter-pays principle. The resulting rules and waste take-back obligations, for which the Green Dot of Duales System Deutschland (DSD) became the symbol—were gradually adopted and implemented throughout almost all of Europe.

The Packaging Ordinance was the blueprint for the EU Packaging and Packaging Waste Directive (PPWD 1994). Waste policy for plastics was, therefore, long regarded as regulating plastic packaging waste, and this is sometimes still the case at present. Since approximately 50–60% of plastic waste in the EU is packaging waste (PlasticsEurope 2022), and since packaging, as an everyday and short-lived object of use, passes through the hands of the end consumer during purchase, use (the opening or emptying of goods), and even disposal, such a political prioritisation was understandable and justified.

Further to the above, at the beginning of the 2000s, the EU passed important product and substance-related directives which included rules and requirements on recycling, for example in the area of end-of-life vehicles (ELV 2000), waste electrical and electronic equipment (WEEE 2002, 2012) and for the Restriction of use of hazardous substances in electric and electronic (E&E) equipment (RoHS 2002), which are still highly relevant today for the use of plastics and the recycling of plastic waste in these product segments.

In 2008, after lengthy negotiations, the EU was able to agree on what we can define as an ante-litteram, yet very seminal "circular economy law", referred to as the Waste Framework Directive (WFD 2008). It is important to mention that the WFD created the so-called 'waste hierarchy'. The waste hierarchy lists the priority actions that must be undertaken to tackle waste generation:

1. Prevention;
2. Preparation for re-use;
3. Recycling;
4. Waste recovery, e.g., energy recovery;
5. Disposal.

The waste hierarchy has since developed into a widely known paradigm that guides not only waste disposal, but as is clear from 1. and 2. also how products are conceived and designed.

The role of a regulatory pioneer in the recycling of plastic waste and the introduction of a circular plastics economy in Europe was assigned to the EU Commission in the early 2010s, with the publication of the Green Paper Strategy for Plastics in the Environment (Green Paper 2013) which was then followed by the first Circular Economy Package in 2015 (CEP 2015) and the Plastics Strategy (Plastics Strategy 2018).

In this context, it must be stressed that the adoption of the EU Green Deal in 2019 (Green Deal 2019) further boosted the efforts to transition towards a circular, clean, and resource-efficient economy. The EU Green Deal highlights the circular economy as an important, independent, and overarching pillar of the EU climate protection strategy. Adopted by the EU Commission led by Ursula Von der Leyen, the European Green Deal, is an ambitious plan that aims at transforming the EU's economy into a fair, sustainable, and prosperous model. The European Green Deal makes circularity the principal tool to progressively de-couple resource consumption from economic growth, therefore potentially reducing the EU's dependence on imported and virgin raw materials, and its vulnerability to resource price volatility, while providing new and innovative business opportunities.

Key policy developments: the EU plastics strategy

The idea of moving away from a linear economic model (where plastics were made, used, and disposed of) to a circular economy was cemented by the specific analysis and consultations that led to the adoption of the EU Plastic Strategy. The latter, in fact, identified the need to move away from a resource efficiency-centred approach that only looked into product light-weighting, product shelf-life extension, resource and feedstock saving, to seek to scale up the market for recycling and recyclates and gradually replace the need for virgin raw materials by increasing the use of secondary raw materials.

Under the Plastic Strategy recyclability and the possibility to use secondary raw materials in new products are seen as endemic elements of product design.

The Plastics Strategy presented by the EU Commission in January 2018 is a policy plan that sets a fundamental passage for the plastics industry in the direction of a circular economy. The publication of the EU Plastics Strategy was preceded by intensive work aimed at analysing the plastic material flow. The work was led by the EU Commission with the support of research branches (e.g., the Joint Research Centre) and accompanied by intense public debate and stakeholder consultations. The Plastics Strategy Working Staff Document (Staff Document 2018) issued in support of the Strategy contains an in-depth analysis of, among other things, the role of the plastics industry in the EU, the status of recycling, the extraction and use of recyclates, the collection and littering of plastic waste, the need for innovation and investment, and international cooperation. It thus supports and complements the shorter-text strategy document in terms of scientific analysis and justification of necessary policy measures.

The EU Plastics Strategy focuses mainly on the necessary steps to ensure that the plastic industry can in the future become sustainable and rid itself of externalities such as litter, emissions, and pollution. The document revolves around the following four main axes:

- Making recycling profitable for business (e.g., new rules on packaging to improve the recyclability of plastics and increase the demand for recycled plastic content);
- Curbing plastic waste;
- Driving innovation and investment;
- Spurring global change.

There are three annexes to the strategy document which identify around 40 targeted measures to achieve the above-listed goals. These measures, among others, include:

- Drafting and adoption of the Single Use Plastics Directive (SUPD 2019);
- Implementing of the target of reusing 10 Mt of recycled plastics in 2025 by initiating the *Circular Plastics Alliance* (CPA 2018);
- Review of the EU Packaging and Packaging Waste Directive (PPWD 1994), in particular regarding essential requirements for plastic packaging;
- Inclusion of key points of the Plastics Strategy in the EU Green Deal (Green Deal 2019) and the EU Circular Economy Action Plan (CEAP 2020); and
- Commissioning of a series of studies (e.g., on plastics in agriculture, on bioplastics, on eco-modulation of licence fees, on recyclate use quotas, etc.) by the EU Commission, designed in such a way that they can serve as a basis for further regulatory measures.

The Plastics Strategy remains important for the future of the entire plastics industry in the EU as it is aimed at changing the way the material is conceived and perceived. It recognises the value of plastics as a key enabler for a modern lifestyle whilst acknowledging its externalities. Among other things, the main changes, or better, thought-shifting approach that the Plastics Strategy introduced include the following:

- It looks at the material stream over its entire life cycle, both from an industrial policy perspective and from an environmental and resource point of view;
- It prompted the transformation of the production and use from a linear to a circular model, where waste must be seen as a resource;
- It provided for legislative initiatives and triggered a wide range of voluntary commitments;
- It identified roles and responsibilities for all stakeholders involved in the transformation of the plastics value chain—Including end users and policymakers (!);
- It integrated the fight against littering and in particular marine litter both inside and outside the EU as part of the strategic thinking that must characterise the material stream.

21.2 EU Regulation for a Circular Economy of Plastic Products

21.2.1 Single Use Plastic Products

Planned under the EU Plastic Strategy, the Single Use Plastics Directive (SUPD 2019) was intended very much as a policy response to the marine litter problem. It includes requirements and restrictions aimed at reducing the placing on the market of certain plastic products. The SUPD goes beyond packaging in applying to certain non-packaging products. However, regarding its applicability to packaging, it sets out specific rules that prevail over more general rules contained by the Packaging and Packaging Waste Directive (PPWD 1994), i.e., it acts as a *lex specialis* whereas the PPWD[1] is the *lex generalis.*

From 2021, the SUPD bans plastics product such as cutlery, plates, drinking straws, cotton buds, stirrers, balloon sticks, food packaging made of EPS, all articles made of oxo-degradable plastic. Furthermore, it sets out consumption reduction targets for beverage cups and packaging for ready meals for immediate consumption. It creates labelling obligations for hygiene articles, sanitary towels (pads), tampons and wet wipes, tobacco products with filters, beverage cups.

Also, the SUPD extends the scope of the Extended Producer Responsibility (EPR) scheme. The EPR is a practice and a policy approach in which producers take responsibility for the management of the disposal of the products they place on the market. Under the SUPD the EPR is extended to also cover the clean-up of cigarette filters, balloons, or fishing nets (prevention of ghost nets).

Furthermore, the SUPD sets the obligation for Member States to implement awareness-raising measures aimed at incentivising responsible consumer behaviour (e.g., avoidance of littering) or encouraging the use of reusable alternative products.

Of particular importance is the first EU-wide creation of a product-specific quota for the inclusion of recycled content in plastic beverage bottles.

By creating these provisions, the SUPD created an important precedent: whilst it currently applies to plastic beverage bottles, it will be extended to other packaging types under the rules that will be implemented in the new Packaging and Packaging Waste Regulation (PPWR 2024). Under the SUPD, from 2025, EU Member States must ensure that operators include 25% recycled PET in beverage bottles, and by 2030, 30% of recycled plastics must be used in all plastic beverage bottles made of either PET or any other plastics—e.g., HDPE for milk or dairy drinks.

The SUPD was adopted quite swiftly: it was proposed by the Commission in May 2018, and in less than one year an agreement was reached by the co-legislators (i.e., EU Parliament and Council). The speed of the process, together with the often-imprecise legal text, are creating hindrances to the uniform implementation of the SUPD in the EU member states, as well as to its interpretations by both enforcement

[1] Following the imminent replacement of the PPWD by the Packaging and Packaging Waste Regulation (2024), the PPWR will become lex generalis.

authorities and industrial operators. This is also due to the many delays in the adoption of the implementing provisions by the EU Commission as well as by many diverging national interpretations and exemptions from its obligations that often resulted in de facto barriers to the free movement of goods, or in any case, in diverging market conditions for similar products across the EU. This represents a threat to the integrity of the EU internal market and the free circulation of the products falling under the scope of the Directive.

21.2.2 Packaging

Packaging and Packaging Waste Directive (PPWD)—On the Road to a Packaging and Packaging Waste Regulation (PPWR)

The PPWD can probably be considered the centrepiece of EU legislation on packaging and packaging waste. The objectives of the Directive are the continuous improvement of the environmental performance of packaging and the correct functioning of the EU Internal Market, thereby protecting the free circulation of packaging and packaged goods in all Member States.

The Directive establishes recycling targets for packaging materials, which must be achieved by EU Member States. It also enlists the Essential Requirements for packaging design that must help reduce packaging waste. Packaging that meets these essential requirements must be guaranteed free circulation in the EU.

With the introduction of the PPWD, certain Europe-wide common standards and requirements became applicable for every manufacturer of packaging, every retailer and seller applicable to all packaging across all Member States of the EU.

Regarding packaging material composition, the Directive contains restrictions on the use of certain heavy metals. These rules establish that the amount of lead, cadmium, and hexavalent chromium that must not exceed certain concentration limits (i.e., the sum concentration levels of lead, mercury, cadmium, and hexavalent chromium added together in any sold packaging must not exceed 100 ppm by weight).

Also, the Directive sets out provisions on labels. Packaging labels must not only show information about the product but also include the materials of the packaging, which must be labelled and identified to the same extent.

Under these prescriptions, in general, packaging should be reduced to a minimum and manufactured, where possible with reusable and recyclable materials. Other primary design requirements for all packaging placed on the market require the lowest possible weight and volume of the material, guaranteeing consumer acceptance, hygienic and safety-relevant properties.

Due to their versatility and light weight, the use of plastics increased over the first two decades of the 2000s, and the material became a favorite among retailers and distributors in, among others, the food, but also beverage and pharma sectors. Innovative packaging, ever lighter flexible formats mixed with high performances in terms of shelf-life extension and protection from spoilage made plastics very successful.

In the 2010s, public, media, and political pressure on plastic packaging increased. The main reasons for this were significant leakage of plastic packaging in the environment, and the low rate of recycling and reuse (cf. Staff Document 2018).

The EU Commission already attempted to solve these problems in 2018 by reviewing the PPWD and strengthening some sustainability requirements. The 2018 revision of the Directive introduced higher targets of the overall recycling for packaging (65% in 2025 and 70% in 2030), and higher material-specific targets (such as 55% for plastic by 2030). Furthermore, under the revised PPWD, Member States will have to set up return, collection, and recovery systems. Furthermore, Article 9 paragraph 5 states that, by 31 December 2020, the Commission had to examine the feasibility of reinforcing the essential requirements to improve design for reuse and promoting high-quality recycling, as well as strengthening enforcement, possibly by proposing a legislative proposal.

Following up on this revision clause, in its Circular Economy Action Plan 2.0 for a cleaner and more competitive Europe of March 2020, the EU Commission announced several initiatives to reduce the EU's consumption footprint and increase the EU's circular material use rate by 2030. These would concern the entire life cycle of products, from design and manufacturing to consumption, repair, reuse, recycling, and bringing resources back into the economy. In this context, it also announced a review of the PPWD to ensure that all packaging on the EU market is reusable or recyclable in an economically viable way by 2030.

Both the EU Parliament and the EU Council welcomed the review of the PPWD in a quest for more sustainability and circularity.

From Directive to Regulation

The work to assess the impact of potential new rules on packaging was initiated by the EU Commission in 2019 just shortly after the adoption of the last review to the PPWD.

After an intense debate, in November 2022, the Commission proposed to repeal the original Directive and introduce a new piece of legislation in the form of a regulation (PPWR 2022). Replacing an existing directive with regulation is a trend that can be seen across EU policies from data protection to batteries and e-commerce where the EU opted for regulations to boost harmonisation and the setting of common EU product rules. Whereas a directive is addressed to member states, and must be transposed into national law, a regulation is binding in its entirety and directly applicable to all persons in all Member States. Regulations, therefore, have a direct effect: companies and individuals must directly comply with their rules, and they can also rely on these rules before national courts. Therefore, Regulations are more "harmonising" since they avoid the differing national approaches to transposition.

In fact, the Commission justified the switch to a regulation by citing unilateral packaging policy measures by member states which led to uneven national regulatory frameworks. The move to a regulation was supported by most business stakeholders during the public consultation ahead of the Commission's proposal being published but has faced pushback from some member states that object to this more "harmonising" approach in favour of more decision power for the local authorities. Also, most member states have shown reluctancy regarding the possibility of revising their waste collection, sorting, and discarding systems by means of different and more harmonised extended producer responsibility systems or deposit refund system obligations.

It must be noted, that at the time this chapter is being written a provisional version of the PPWR has been voted by the European Parliament's ENVI Committee and the Council's Committee of the Permanent Representatives of the Governments of the Member States to the European Union (COREPER). The text must still be approved by the European Parliament plenary assembly and the Council of Ministers to become final and officially adopted. It is highly likely that such a current provisional text will be the one that will be published in the EU Official Journal towards the end of 2024. Therefore, we will briefly treat the content of this draft version, which will most likely become the centerpiece of the EU packaging legislation.

The PPWR is wide-ranging but some of the key measures include:

- Targets for packaging waste reduction and mandatory reuse or refill targets in sectors such as retail and catering;
- EU-wide standards for "over-packaging". This includes a maximum allowable empty space in e-commerce packaging and a ban on certain forms of unnecessary packaging (Article 21);
- Design criteria for all packaging to increase recycling rates. From 1 January 2030 packaging will have to comply with the design for recycling criteria and from 1 January 2035 the requirements will be further adjusted to ensure that recyclable packaging is sufficiently and effectively collected, sorted and recycled;
- Targets for minimum recycled content in most forms of secondary plastic packaging from 1 January 2030 of between 10 and 35% depending on the use of the packaging (Article 7—cf. Table 21.1 on proposed targets for recycled content recovered from post-consumer plastic waste);
- A ban on certain formats including single-use plastic packaging for condiments, preserves, sauces, and sugar, and hotel miniature toiletry products (Article 22 and Annex V);
- Mandatory deposit return system (DRS, cf. Sect. 11.2) for single-use plastic drinks bottles and aluminium cans (with a capacity of up to three litres).

21.2.3 Building and Construction Products

The fundamental EU regulation for building and construction products of all used materials is the Construction Products Regulation (CPR 2011). The current and applicable version of the CPR lays down harmonised rules for the marketing of construction products in the EU. The Regulation provides a common technical language to assess the performance of construction products. It ensures that reliable information

Table 21.1 Proposed targets for recycled content recovered from post-consumer plastic waste in the draft PPWR (*Source* PPWR 2024, own presentation)

Type of packaging	From 1 January 2030 (%)	From 1 January 2040 (%)
Contact-sensitive packaging made from PET	30	50
Contact-sensitive packaging made from plastic material other than PET (except single-use plastic beverage bottles)	10	25
Single-use plastic beverage bottles	30	65
Other packaging	35	65

is available to professionals, public authorities, and consumers, so they can compare the performance of products from different manufacturers in different countries.

In March 2022 the EU Commission drafted some new rules on Building and Construction, to amend and review the CPR. The proposed obligations (CPR draft 2022) would

- require manufacturers to design and make products and related packaging in a more environmentally friendly and sustainable way;
- prefer recyclable and recycled materials;
- respect minimum recycling requirements and similar limit values;
- make available, in product databases, instructions for use and repair of the products;
- design products in a way that facilitates re-use, remanufacturing, and recycling.

Legacy substances

It is important to note that the reprocessing of construction products can present additional complexity elements. As construction products are meant to last for long periods of time, it is possible that specific substances, authorized at the time of production, might be subject to restrictions or prohibition once they reach their end of life. This means that when certain products become waste and input for the recycling system, the presence of certain—and no longer authorized—substances risk impairing their subsequent reemployment (these are the so-called "legacy substances").

The provisions on circularity and sustainability are material-neutral and do not create material-specific obligations on, for example, recycling, or uptake of recycled plastic materials. However, future secondary legislation will probably set more specific conditions on the materials that can be used or should be avoided (e.g., blended, mixed and intricate materials, as stated in Art. 22 of the EU Commission draft). This could impact the use of polymers, although it is currently impossible to predict if this will have a detrimental effect on the industry or push new and innovative plastic solutions in the building and construction sector.

21.2.4 Electrical and Electronic Equipment

Waste electrical and electronic equipment is regulated by a dedicated piece of legislation. The dedicated Waste Electrical and Electronical Equipment Directive on this type of waste (WEEE 2002, 2012) has been recast in the context of the 2015 Circular Economy Package.

The E&E sector includes a vast array of products: mobile phones, computers, televisions, fridges, household appliances, lamps but also medical devices. The scope is quite broad to encompass many products made wholly or partially of plastics.

The Directive requires the separate collection and proper treatment of WEEE by setting out targets for its collection, recovery, and recycling:

- Collection rate: The amount of waste electrical and electronic equipment collected must correspond to at least 65% of the average total weight of electrical and electronic equipment placed on the market in the three previous years;
- Recovery rates: Depending on the category of equipment, 75 to 85% of the waste equipment collected each year must be recovered; recovery includes preparation for reuse, recycling, and other (in particular energy) recovery;
- Quotas for recycling and preparation for reuse: Depending on the appliance category, 55 to 80% of the WEEE collected annually must be prepared for reuse or recycled.

The recycling quotas are also mainly achieved by ferrous and non-ferrous metals such as copper, but also precious and special metals. Regardless of this, some processors in the EU recover non-halogenated plastics from WEEE (ABS, PP, PS, PMMA, PC) and send them for mechanical recycling.

Similarly to other waste streams, WEEE plastics might contain a conspicuous list of legacy substances as well as substances of concern, notably brominated flame retardants (BFR) or Persistent Organic Pollutants (POPs). Therefore, WEEE waste, so-called e-waste, is subject to stringent rules regarding its treatment and recycling whether within Europe and beyond. For this reason, the WEEE Directive addresses how to care for shipment of e-waste outside the EU, and sets targets for its collection, recovery, and recycling. The objective of the Directive is to fight illegal waste exports more effectively by making it harder for exporters to disguise illegal shipments of WEEE.

21.2.5 Plastics in Vehicles

One of the so-called waste Directives, applicable to another specific waste stream is the End-of-Life Vehicles Directive (ELV 2000).

Today, the Directive sets clear targets for the reuse, recycling, and recovery of ELVs:

- At least 95% by weight of the unladen car must be reused (in the sense of components for the same purpose) or recovered (recycling or energy recovery);
- At least 85% by weight of the unladen weight must be reused or recycled
- Operators of dismantling facilities must remove large plastic components and prioritise them for reuse or recycling, such as e.g. bumpers, hubcaps, and radiator grilles, if the corresponding materials are not separated during or after shredding in a way that enables recycling" (ELV 2000, Annex 3.2.3.3) before transferring the residual body to a shredder plant or other facility for further treatment.

The regulations provide practically no incentive to separate plastic components to any significant extent and then recycle them, as the recycling rate of 85% can be achieved solely through metal fractions, glass, and liquids from the draining of vehicles. However, the removal of large plastic components stipulated in the text of the ordinance is relativized again, so that no significant quantities from the dismantling of large components have so far been sent for recycling.

The legislation, thus covers all materials, including plastics, for which it prohibits the use of hazardous substances when manufacturing new vehicles (such as lead, mercury, cadmium, and hexavalent chromium) except in defined exemptions when there are no adequate alternatives.

For the future, the ambition is to review the ELV Directive to include more specific rules concerning the design stage of a new vehicle and the choice of materials. In fact, at the time of writing the present chapter, the European Commission has presented a proposal for a new repealing regulation for this type of waste (EU 2023).

The proposed regulation will cover all aspects of a vehicle from its design and placement on the market until its final treatment at the end-of-life. This will entail:

- Rules to improve circular design of vehicles to facilitate removal of materials, parts, and components for reuse and recycling;
- Recover more and better-quality raw materials;
- Establish more stringent rules for the Extended Producer Responsibility;
- Ensure that at least 25–30% of plastic components used to build a vehicle come from recycling (of which 25% from recycled ELVs).

The provision setting out the obligation to include a minimum recycled plastics content is currently being discussed with stakeholders and, most likely, is expected to be set at 25% of the total weight of plastics contained in the vehicles delivered to the waste management operators.

It is important to remember, similarly to the construction sector, that vehicles on average have long lives and contain legacy substances (see above). For this reason, the new regulation will most likely include specific exemptions or derogations for recycled materials to be used for new products.

21.3 Other EU Regulations Impacting the Circular Economy of Plastics

21.3.1 REACH and Polymers

To complete our journey through the legislative requirements on plastics that apply in the EU, it is also necessary to go through some of the key elements of Regulation EC/1907/2006 concerning the Registration, Evaluation, Authorisation, and Restriction of Chemicals and establishing a European Chemicals Agency, the REACH Regulation (REACH 2006).

Polymers are chemicals, and REACH has deeply changed the way chemicals are controlled in the EU. The REACH Regulation entered into force on 1 June 2007. It gave greater responsibility to the industry to manage the risks from chemicals and to provide safety information on the substances to improve the protection of human health and the environment through the better and earlier identification of the intrinsic properties of chemical substances.

The system introduced by the REACH Regulation creates a unified framework for all substances and moves the responsibility to test chemicals from the authorities to producers and also introduces obligations for downstream users.

Under the REACH provisions, all substances manufactured or imported in quantities above 1 metric tonne per year into the EU must be registered. The registration of substances must also take place when they are contained in articles, and it is intended to be released under normal or reasonably foreseeable conditions of use.

The information to be submitted for general registration purposes includes a technical dossier containing data on the properties of chemicals, their uses, and safe ways of handling them.

When a registration dossier is submitted to ECHA, the latter can decide to carry out an evaluation process. A substance evaluation is carried out whenever a given chemical substance might constitute a risk to human health and/or the environment.

The REACH Regulation created two procedures aimed at minimising the risks arising from the placing on the market and use of chemicals—the Restriction and the Authorisation. A Member State or the European Commission can trigger either procedure.

Title VIII of the REACH Regulation regulates the chemical Restriction procedure on the manufacturing, import, use, or placing on the market of substances. A Member State, or ECHA, at the request of the European Commission, can start the Restriction procedure when they are concerned that a certain substance poses an unacceptable risk to human health or the environment. Restrictions are imposed by including a substance or group of substances in Annex XVII to the REACH Regulation. Each entry in such Annex shows restricting conditions on their marketing and use. The Restriction of substances is an ongoing procedure and Member States, and the Commission can propose to further restrict the manufacturing and use of substances not yet included in Annex XVII to REACH.

According to the European Chemicals Agency (ECHA), a Restriction is a preferred option over the Authorisation if:

- there is reasonably good knowledge of the types of uses and on the main use conditions;
- some uses are essential for society, and it should be possible to continue these uses (at least for a defined period of time) but Article 58(2) of the REACH Regulation does not seem to allow an exemption from the Authorisation requirement;
- a use should be phased out if it is reasonably clear that there are no circumstances under which an authorisation for a use is justified.

Once the Restriction has been adopted, industry must comply with the conditions it sets out. That means all operators, including manufacturers, importers, distributors, downstream users and retailers. The Member States are responsible for enforcing the Restriction.

Title VII of the REACH Regulation regulates the Authorisation procedure. An Authorisation is a preferred option if there is little knowledge or a lot of uncertainty on the types of uses or the details of the uses themselves. An Authorisation is required for the substances that get listed in Annex XIV to the REACH Regulation. After the date set in Annex XIV for each substance, they may not be used or placed on the EU market unless they are authorised.

The European Commission and the Member States decide together on the inclusion of substances in Annex XIV. The first step of the authorisation process consists of substances identified by ECHA as being *substances of very high concern* (SVHC). Then they are included in the so-called Candidate List. SVHCs that can be included in the "Authorisation List" include very hazardous substances such as carcinogens, mutagens, substances which are toxic for reproduction (CMR) classified in category 1A or 1B according to the CLP Regulation, etc. (CLP 2008) Also, these can be all substances for which there is scientific evidence of probable serious effects to human health or the environment and which therefore give an equivalent level of concern to those substances listed above (for example, endocrine disruptors).

From the date of inclusion in the Candidate List, suppliers of articles that contain substances on the Candidate List in a concentration above 0.1% (w/w) must comply with duties to inform the recipient of the article pursuant to Article 33 of the REACH Regulation. This obligation applies when the article contains substances of the Candidate List in a concentration above 0.1% weight by weight (w/w) of the entire article and to articles consisting of component articles, when the component retains its shape, surface, or design when integrated into the article, and contains an SVHC in a concentration above 0.1% (w/w) of the article's component.

Information duties must allow the safe use of the article and includes, as a minimum, the name of that substance. There is no tonnage trigger for this obligation.

ECHA then recommends to the European Commission the entries to be included in the "Authorisation List". The European Commission takes the last decision about the inclusion of substances to Annex XIV to the REACH Regulation. For each entry, Annex XIV to the REACH Regulation determines the latest application date for authorisation and the sunset date (i.e., the date from which the substances cannot be placed on the market or used without a valid authorisation for the specific use).

Applicants for Authorisations (which may be manufacturers, importers, downstream users, and only representatives) will have to submit a substitution plan analysing the availability of alternatives and the technical and economic feasibility of substitution.

Polymers and monomers: specific REACH provisions

Polymers are exempt from registration obligations in light of the low concern related to their high molecular weight. While polymers are exempt from registration, monomers as well as reactants and additives which are employed to manufacture polymers must be registered.

It must be noted that in case a polymer is identified as hazardous to human health or the environment, other REACH provisions might find application:

- Safety data sheets;
- Restriction;
- Authorisation.

Monomers used in the manufacture of polymers may have to be registered even though it is only the final polymer that is being imported or used in the EU. As mentioned in relation to registration requirements, manufacturers or importers of polymers must register monomers or any other substance not yet registered by an actor up the supply chain, when both the following conditions are met:

- the polymer consists of 2% weight by weight (w/w) or more of such monomer substance(s) or other substance(s) in the form of monomeric units and chemically bound substance(s);
- the total quantity of such monomer substance(s) or other substance(s) makes up 1 t or more per year.

Polymers are usually supplied as formulated mixtures. According to the REACH legislation, each substance of a mixture must be managed individually. This means that the additives present in a polymer formulation may require separate registration following the rules for registration based on tonnage and scope. This can also apply to residual monomers (those that do not become part of the polymer).

21.3.2 Food Contact Materials—Plastics and Recyclates

21.3.2.1 Food Contact Materials

Since 1976, the European Community has been working on the harmonization of the laws of the European Union's Member States regarding materials and articles intended to come into contact with food. In 1976, the EU adopted its first Directive on this topic (Council Directive 76/893/EEC of 23 November 1976 on the approximation of the laws of the Member States relating to materials and articles intended to come into contact with foodstuffs, (Council Directive 1976) to bring a solution to the problems caused by the fragmented national regulations that affected the functioning of the European Market.

Packaging areas requiring harmonised rules, or at least to be considered for such rules were plastics, including varnishes and coatings, regenerated cellulose, elastomers, and rubber, paper and cardboard, ceramics, glass, metals, and alloys, wood including cork, textile products, and paraffin and microcrystalline waxes.

Since then, most of the implementing rules that aimed at making the framework provisions of the Directive operative have dealt with monomers and adjuvants for plastic products (although, to a limited extent, elastomers and rubber).

Towards the end of the 1980s, there was an additional urge to improve the harmonisation of the rules on food contact packaging. This was necessary to facilitate the increasing intra-EU trades and remove market barriers for food. The so-called Framework Directive (1989) provided the general safety criteria applicable to all food contact materials and called for the adoption of specific directives covering individual categories of food contact materials.

The Framework Directive remained in effect from 1989 to 2004, when the European decision-makers adopted a Framework Regulation covering materials that are intended to come into contact with food (Framework Regulation 2004).

The harmonizing effect of the Regulation was very powerful and immediate as it directly applied in all EU Member States.

The 2004 Regulation, which is still in force, covers all food contact materials and articles. It ensures that food contact materials do not threaten human health, or change the composition, odour, or taste of the food product. As it covers all the materials that come in contact with food, it ends up covering numerous applications spanning from food packaging to manufacturing machinery, conveyor belts to consumer kitchen goods.

The Regulation foresees the adoption of specific implementing regulations to create operative measures for various material groups, including plastics and recycled plastics, regenerated cellulose, and active and intelligent materials.

21.3.2.2 Plastics in Food Contact Materials

Among the material-specific regulations to be adopted under the Framework Regulation, the EU decision-makers adopted requirements on plastic food contact materials.

Regulation (EU) 10/2011 on plastic materials and articles intended to come into contact with food (also known as the Plastic Implementing Measure—PIM (PIM 2011)) is the most wide-ranging EU legislation on a specific category of food-contact material.

It applies to:

- plastics, in monolayers and multilayers, whether or not bound by adhesives;
- plastic layers in multi-materials and articles (i.e., plastics combined with other materials such as paper and/or aluminum); and
- printed plastic articles and those covered by coatings, in addition to plastic layers or coatings forming gaskets in caps and closures, are also covered by the Plastics Regulation.

The Regulation does not cover ion exchange resins, rubber, and silicone fall outside its scope since these materials were meant to be covered by separate harmonizing rules.

The PIM includes a positive list, also known as the "Union List" which consists of permissible monomers and other starting substances. The PIM also covers additives (other than colorants), and some polymer processing aids (PPAs). It is important to mention that the PIM creates exemptions from the possibility of only using the listed substances based on the functional barrier principle. A "functional barrier" is a barrier consisting of one or more layers of any type of material which ensures that the final material or article complies with Article 3 of the Framework Regulation. However, certain conditions must be met to make use of the functional barrier exemption; that is, the non-listed substances must:

- Be separated from food by a barrier;
- Migrate only at a specific level (i.e., less than 10 parts per billion (ppb)—detection limit);
- Must not belong to either of the following categories:
  - carcinogenic, mutagenic, or toxic to reproduction (CMR) pursuant to CLP-Regulation (CLP 2008), or
  - substances in nano form.

In addition, the functional barrier principle allows for the use of listed substances in cases when these do not comply with the Restrictions and/or specifications contained in the Union List.

Under the PIM, the use of colorants and solvents must comply with national laws, subject to compliance with the mutual recognition principle. The latter is a generic principle applying in the field of free movement of goods, i.e. where no harmonised rules exist at the European level, products lawfully marketed in one Member State can be sold in other Member States regardless of complying or not with the national technical rules of these Member States.

Regarding polymer processing aids (PPAs), those listed in the Union list must comply with the restrictions and specifications set therein. However, the PIM does not establish a positive list of PPAs and provides that PPAs that are not listed in the PIM itself are regulated by national laws, subject, again, to the mutual recognition principle.

PPAs used in plastics manufacture and non-intentionally added substances (NIAS) that may be present in food-contact plastics are not subject to positive list requirements whether at the EU or national level.

On NIAS, which are not only impurities that may be present in raw materials used but also reaction and degradation products that can be formed during the manufacture and use of plastics, the PIM demands that any potential risk they trigger must also be assessed based on "internationally recognized scientific principles on risk assessment". However, the EU provisions (Regulations and guidelines) do not provide specific guidance to the industry on how such risk assessment must be carried out.

The PIM establishes an overall migration limit (OML) for all plastic substances that come in contact with food. The OML defines the maximum permitted amount of substances released from a material or article into food simulants. Specific migration limits (SMLs) and/or residual quantity limits (QM) in the material for certain substances are also specified.

The Regulation details the requirements to assess compliance with the applicable SMLs and OML. Foods intended for infants and young children (ages ranging between 0 and 3 years) are subject to more stringent requirements.

A layer in direct contact with the food must always comply with the compositional requirements of the Regulation. However, in a plastic multi-layer material (MMML—means a material or article composed of two or more layers of different types of materials, at least one of them a plastic layer), all the possible plastic layers must be composed of substances listed on the Union List, unless used behind a functional barrier.

Under the PIM, operators that supply substances, intermediate products, or final articles (except at retail stage), must issue a Declaration of Compliance (DoC). The latter must at least include the following information:

- Identity and address of the business operator;
- Identity of the materials and articles;
- Date of declaration;
- Confirmation that the materials and articles comply with the Framework Regulation (2004);
- And, if a functional barrier is used, that it complies with the appropriate requirements.

The PIM significantly changed the market of plastic food contact materials in the EU, although it never achieved a complete market harmonization. To improve both harmonisation and increase the safety and sustainability of food contact materials, the European Commission is currently conducting several studies to support a future amendment of the Framework Regulation as well as of the material implementing provisions. A draft for the review of the current rules could already be issued by the end of 2024.

21.3.2.3 Recycled Plastics in Food Contact Materials

While the Food Contact Materials Framework Regulation and its implementing rules will most likely undergo a thorough revision process, the European Commission has already reviewed the existing rules on recycled plastic materials and articles intended to come into contact with foods. This is Commission Regulation (EU) 2022/1616 (EU 2022) which entered into force on 10 October 2022, repealing Regulation (EC) No 282/2008 (EC 2008).

This Regulation aims at increasing the use of recycled plastic in food contact materials while ensuring their safety. It demands that materials and articles intended to come into contact with food must be both chemically and microbiologically safe.

The reason for overhauling the previous Regulation (EC) No 282/2008 (one of the implementing acts of the Framework Regulation on food contact materials)) and adopting a new set of implementing provisions on the use of recycled plastics was dictated by the enhanced obligations on the uptake of recycled polyethylene terephthalate (PET) in bottles, as well as in other types of food containers that was notably driven by the adoption of the Single Use Plastic Directive.

The 2008 Regulation (EC 2008) never fulfilled the goals that it initially envisaged. In fact, under the 2008 Regulation, the Commission did not authorise any of the more than 200 recycling processes assessed as safe by the EFSA. For this reason, the use of recycled plastics (e.g., in PET beverage bottles) has until now only been authorized under national legislations, in accordance with transitional provisions in force under the 2008 Regulation.

Following the adoption of the new 2022 Regulation (EU 2022) on the use of recycled plastics, the Commission has assured that the pending authorisation requests, as well as the newly received ones, will be swiftly decided upon under the assumption that once compliance with the new rules is ensured, an authorization must be quickly granted.

Among other things, one of the key factors that would simplify and speed up the procedures is that applications for authorizations for recycling procedures will have to be submitted at the company level and no longer for individual facilities (although the problems and impacts that could be created by the significant differences in the waste input materials in each facility must still be concretely verified).

The 2022 Regulation applies to all types of recycled plastic materials and various recycling technologies, e.g., mechanical recycling, recycling from product loops which are in a closed and controlled chain, use of recycled plastic behind a functional barrier, as well as some forms of chemical recycling but excluding complete chemical recycling into starting substances—for these, compliance with the PIM (PIM 2011) must still be assessed.

The Commission must authorise the recycling technologies and, since 10 July 2023 only recycled plastic obtained from authorised existing recycling technologies may remain on the market for food contact material use, except for recycled plastic obtained from the development of a novel—whether mechanical or chemical recycling—technology (i.e., any technology considered different from existing technologies that in turn needs a new authorization).

New technologies must meet stringent safety requirements to be considered suitable technologies, including the monitoring of the contaminants in recycled plastic; the publication of a detailed report and test reports on the safety of recycled plastic produced, and the maintenance of continuous dialogues with recyclers using the new technology.

The final food contact material made of recycled material shall comply with the requirements of Plastics Regulation (EU) 10/2011 (PIM 2011) as regards composition and the restrictions and specifications laid down in the Regulation, including migration limits.

21.3.3 ESPR—Ecodesign for Sustainable Products Regulation

In the context of the 2020 Circular Economy Action Plan (CEAP 2020), in March 2022, the European Commission adopted the Sustainable Products Initiative (SPI). The latter is a non-binding communication meant to *make sustainable products the norm* (COM 2022). The SPI shall be implemented via three means: a new eco-design regulation, the EU Strategy for Sustainable and Circular Textiles and the revision of the Construction Products Regulation (CPR draft 2022).

The new eco-design regulation (Eco-design 2022) was proposed to replace the existing Eco-design Directive 2009/125/EC (Eco-design 2009) and update descending eco-design requirements (including on energy savings and energy labelling) by incorporating many new principles of a circular economy in various families of products.

As the previous Directive, the Eco-design for Sustainable Products Regulation (ESPR) establishes a framework to set eco-design requirements for specific product groups in order to significantly increase their circularity, energy performance, and other environmental sustainability aspects. It foresees the setting of performance and information requirements for almost all categories of physical goods placed on the EU market. For groups of products that share sufficient common characteristics, the framework will also allow to set horizontal rules.

The framework allows for the setting of a wide range of requirements, such as product durability, reusability, upgradability, and reparability; the presence of substances that inhibit circularity; energy and resource efficiency; minimum recycled content; remanufacturing and recycling; carbon and environmental footprints; and information requirements, including a Digital Product Passport.

As it is a framework-setting regulation, it is fair to say that it does not set out specific requirements, these will have to be developed via delegated acts to be adopted for specific categories of products. In addition to the 31 product categories already covered by eco-design requirements under the 2009 Directive, in March 2023 the European Commission, following an assessment of the Joint Research Centre (JRC 2023), has announced that eco-design criteria might be developed for newly identified end-use products and intermediary products including, among others mattresses and furniture, plastics and polymers; also, new criteria will be developed in detail, including post-consumer recycled content.

The Commission is expected to publish a Communication including a list of new products and horizontal measures to be prioritized under the proposal of the ESPR in the first quarter of 2024.

21.3.4 European Taxonomy

Next to the CEAP, the European Commission has adopted in 2021 a parallel package on sustainability in the finance sector, the so-called Sustainable Finance Package.

In the context of this Package, a taxonomy for economic activities has been established to meet the EU's climate and energy targets for 2030 and reach the objectives of the European green deal. The European taxonomy is a classification system that defines criteria for economic activities aligned with a net zero trajectory by 2050 and broader environmental goals other than climate for private investment purposes.

The EU taxonomy allows financial and non-financial companies to share a common definition of economic activities that can be considered environmentally sustainable. In this way, it plays an important role in helping the EU scale-up sustainable investment, by creating security for investors, protecting private investors from greenwashing, helping companies become more climate-friendly, and mitigating market fragmentation.

The Taxonomy Regulation entered into force on 12 July 2020. It establishes the basis for the EU taxonomy by setting out the four overarching conditions that economic activity has to meet in order to qualify as environmentally sustainable:

1. Sustainable use and protection of water and marine resources
2. Transition to a circular economy
3. Pollution prevention and control
4. Protection and restoration of biodiversity and ecosystems.

Under the Taxonomy Regulation, the Commission had to come up with the actual list of environmentally sustainable activities by defining technical screening criteria for each environmental objective through delegated and implementing acts. For the economic activity concerning the manufacturing of plastic packaging and recycling of plastic waste, the EU Taxonomy provides a list of characteristics and technical criteria to be respected for these activities to be recognized as taxonomy-aligned.

It is worth noting that, for example, the manufacturing of plastic packing goods will be considered to contribute to the circular economy and hence be taxonomy-aligned if it meets a new set of criteria related to the use of circular feedstock (i.e. specific recycled content targets), to design for recycling and reuse. Annex B—Technical Screening Criteria annexed to the Taxonomy Regulation foresees that plastic packaging formats must include at least 85% of mechanically recycled post-consumer material, chemically recycled, biobased or CCU (Carbon Capture and Utilization) based material. While designed for reuse the packaging product has to be designed to accomplish with or prove its ability to accomplish with on average a minimum of 10 trips or rotations in a system for reuse, where a certain amount of rotation is ensured, a system for reuse be established and the packaging can be reused/repurposed for the function.

These criteria are not aligned with existing or coming packaging regulations, as the Taxonomy was published prior to the adoption of the PPWR (2024) thus it is not aligned with the new regulation applicable to packaging.

Similar detachment may be found regarding other types of economic activities for which specific ad hoc criteria have been identified and which are detached from the relevant, applicable legislation. To mention just a few, the Taxonomy includes criteria for (non-exhaustive list):

- Manufacture of chemicals
- Manufacture of electrical and electronic equipment
- Manufacture of food products and beverages
- Construction or renovation of buildings
- Transportation
- Waste management.

Therefore, an economic operator wanting to have its economic activity be recognized in line with the Taxonomy will have to first abide by the relevant sector regulations, then ensure the more stringent criteria foreseen in the Taxonomy are respected.

21.3.5 The EU Plastic Tax—A New Circular Economy EU Budget Resource

Among other things, the Plastics Strategy started the discussion around possible fiscal measures that should have discouraged the reliance on and use of fossil-based plastic packaging. The Strategy in fact stated that: 'the Commission will explore the feasibility of introducing measures of a fiscal nature at the EU level (Plastic Strategy 2018, p.11, footnote 41). The footnote detailing this statement specifies that: 'the modalities of such a potential fee would have to be decided based on the assessment of its contribution towards meeting the strategy goals. On top of that, in the context of the preparation of the post-2020 Multiannual Financial Framework, it could be considered as one of the potential options to generate revenues for the EU budget' (ibid.).

Intense discussions were triggered by the planning of this budget measure, which now having become reality, is still widely, yet wrongfully, referred to as a tax.

After these lengthy discussions and despite the fierce opposition of the plastic industry, the Commission presented its proposal for a "plastic tax" (Own Resources 2020). In reality, this took the form of an EU Budget 'own resource' contributing to the 2021–2027 EU budget. Own resources are de facto the EU Commission revenues (the money that the Commission spends to operate, and are covered by the universality principles—i.e. it cannot be gate-fenced to be used for specific policy areas). Other than the plastic tax, the other three main sources of revenue for the 2021–2027 EU budget are customs duties, contributions based on the Value Added Tax (VAT) collected by Member States, and direct contributions by EU countries, based on Gross National Income (GNI).

The plastic 'Own Resource' is therefore a contribution to be paid by Member States to the EU to be calculated based on the amount of non-recycled plastic packaging waste produced by each member state in a given year. All Member States agreed to the contribution (to approve new own resources, the Council needs to reach unanimity) for a foreseen amount of approximately 8 billion euros.

Under the rules creating the new plastic own resource, starting from 1 January 2021, the contribution is calculated by the weight of non-recycled plastic packaging waste with a uniform rate of EUR 0.80/kg.

The contributions are to be calculated based on Eurostat data, which the Member States already collect and provide under existing reporting obligations (e.g. the provisions of the Packaging and Packaging Waste Directive and its Implementing Decision (EU 2019) regarding the reporting on packaging recycling to verify the attainment of the recycling targets).

Given that the exact data is reported to Eurostat in July of the year N + 2, the European Commission for 2021, had to first calculate the contributions based on forecasts. This is a standard practice also applied to other sources of revenue to the EU budget. Once final data is available, the European Commission will adapt the calculations of Member States contributions accordingly.

This means that, in 2021 (after the entry into force of the Own Resources Decision), Member States will pay their contributions based on forecasts. This contribution will be adjusted after July 2023 when the final data will be available.

Intensive work is being done in several member states to create national implementation measures with a view to passing on the burden to economic operators (whatsoever they may be). This is being done irrespective of tax and duty hurdles, enforcement problems, and, above all, the risk of ecologically counterproductive effects through increased use of non-recyclable non-plastic materials (cf. Sect. 21.4.2).

21.4 EU and National Regulation—Selected Issues

When looking at the European regulations applicable to plastics, the influence that many national legislative initiatives had on European legislation and the effect in shaping the environmental requirements for many products across Europe needs to be mentioned.

In areas where there is European legislation in force, the presence of a national rule, that is in contrast with the European one, can negatively affect the free circulation of goods in the Internal market and the certainty of law.

21.4.1 General Principles

The Treaty on the Functioning of the European Union (TFEU 2012) clearly establishes the allocation of competences between Member States and the EU and also the hierarchy of values of the EU.

There are areas of exclusive competence of the EU or the Member States; while a list of other areas—such as Internal market, environment, consumer protection, or safety concerns in selected public health matters—fall into the scope of *shared* competences. *Shared* means that the EU and the Member States are able to legislate and adopt legally binding acts. Member States exercise their own competence where the EU does not exercise, or has decided not to exercise, its own competence.

As a general rule, this means that if the European Union adopts a regulation on ELVs for example, Member States will have to follow this regulation and ensure its enforcement.

However, the legal basis of a legal act can affect how Member States might wish to take discretionary initiatives on a subject disciplined by European law. For example, the proposed new ESPR or ELV Regulation are based on Article 114 TFEU, which aims to ensure the functioning of the single market; while the Single Use Plastics Directive is based on the protection of the environment, Article 192 TFEU.

The choice of one basis rather than another can have repercussions on the discretion some Member States can play in adopting national legislation. The environmental protection principle under Article 192 allows the Member States to use the precautionary principle[2] and hence, adopt more stringent rules in their own territory, thus negatively affecting the free circulation of goods.

According to TFEU, Union policy on the environment "shall aim at a high level of protection considering the diversity of situations in the various regions of the Union. It shall be based on the precautionary principle and on the principles that preventive action should be taken that environmental damage should as a priority be rectified at source and that the polluter should *pay*" (ibid., Article 191). This in turn means that Member States, when justified by the need to protect the environment,

[2] On the precautionary principle, see EP Think Tank (2015).

act swiftly and according to the precautionary principle, can adopt unilateral protective measures: "the protective measures adopted pursuant to Article 192 shall not prevent any Member State from maintaining or introducing more stringent protective measures. Such measures must be compatible with the treaties. They shall be notified to the Commission" (ibid., Article 193).

On the contrary, a legal act based on the freedom of circulation of goods inhibits this discretion. Article 114 TFEU is the legal basis for the adoption of measures "for the approximation of the provisions laid down by law, regulation, or administrative action in Member States which have as their object the establishment and functioning of the internal market". Once the measure is adopted on that basis, Article 114 (5) restricts Member States' discretion to introducing "national provisions based on new scientific evidence relating to the protection of the environment or the working environment on grounds of a problem specific to that Member State arising from the adoption of the harmonisation measure…".

This means, that the EU places more importance on the protection of the environment and allows examples of more stringent environmental requirements (*over-implementation* or gold-plating) of its rules. This discretion by Member States can be still subject to the scrutiny of the European justice on how the precautionary principle is being exercised, if proportional to the desired objective.

21.4.2 National Taxes

Taxes and other fiscal measures fall into the exclusive competences of the Member States. The freedom that Member State enjoy in this field allows for the complete detachment from European legislations or policy rationales.

Thanks to this freedom, some Member States have decided to tax certain plastic products (e.g. Sweden). However, since 2018 approximately, due to the increased discomfort of NGOs and the general public towards plastics coupled with the need to secure funds for national treasuries, some Member States have started to propose taxes on the consumption of plastic products.

During the preparatory work for the Single Use Plastics Directive (SUPD 2019), eventually adopted in the summer of 2019, the urge to act against plastics reached its peak. Eventually, the final text of the SUP Directive allowed Member States to adopt any necessary measure, including economic instruments to achieve a sustained reduction of certain single-use plastic items.

Although almost out of the EU, in July 2021 the United Kingdom (UK Tax 2021) was the first country to propose a plastic packaging tax, followed by Spain the year after (Spanish Tax 2022). Both countries are the only ones having it in force as of 2022 and 2023, respectively.

In 2019, Italy announced its own tax to be enforced as of 2020 (Italian Tax 2019). It is similar in scope to the Spanish one and has been repeatedly postponed; it is now expected to enter into force as of July 2024.

Along the lines of the U.K. and Italy, the German government announced the introduction of a plastic tax based on non-recycled plastic packaging, by the end of 2023. Up to now, the German government, like most of the Member States, has paid the EU plastic tax (i.e., the 2020 new own resource) from their state budget. Following the dispute over the state budget for 2024, the government announced the possible introduction of a plastic packaging tax to help offset the current €60 billion budget deficit. The introduction of the measure however has been postponed to 2025 because of the need for more time to formulate a legislation limiting the bureaucratic burden.

Already in May 2023, Germany adopted a Single-Use Plastics Fund Act (so-called Einwegkunststofffondsgesetz—EWKFondsG) to implement a national single-use plastics levy as a special charge for goods released on the market from 2024 with first payments in 2025 (EWKFondsG 2023). This levy will be mainly used to co-finance cleaning measures by municipal waste management companies.

It is important to mention that these national taxes can clearly have an effect on intra-EU trade of plastics and plastics consumption but they can also incentivize the incorporation of recycled content: the UK plastics tax foresees a discount for products that have at least 30% of recycled content; while the Spanish and the Italian taxes foresee a discount equal to the amount in percentage of recycled material (for example 15% recycled plastics amounts to a 15% discount on the final tax payable).

Although there is no statutory link between these taxes and the payment of the above-mentioned European Plastics Tax (Own resource 2020), since the adoption of the latter in 2020, Spain has indicated that the funds collected via the established plastic packaging tax will be used to pay the European levy, while other Member States have not made this association: some have explicitly decided to not adopt a plastics tax in response to the levy, while others have indicated that the due amount will be withdrawn from their national budgets.

It is important to not confuse taxes (fiscal measures) with charges (levies or contributions) for lightweight plastic carrier bags that cannot be provided free of charge at the point of sale of goods or products, cf. Article 1 of (EU) 2015/720 (Carrier bags 2015) and from the Extended Producer Responsibility fees to be paid by any economic operator introducing packaging in the Single market to pay for the collection, preparation for re-use and recycling, reuse and recycling of packaging items (WFD 2008, Article 8).

21.4.3 National Bans

The Treaty of the Functioning of the European Union encodes environmental protection as a very important value of the Union. Thanks to the discretion allowed for by Articles 191–193, some Member States have taken unliterally initiatives in banning certain products in the past years, thus negatively affecting the freedom of circulation of goods in the Single Market.

Notable examples are:

- In 2016, France has adopted measures to reduce the consumption of single-use cups and plates;
- In 2022, France has adopted the 3R legislation aimed at reducing single-use plastic packaging products by 2040;
- In 2022, France, Spain, and Italy have announced a Restriction of packaging for fruits and vegetables under 1.5 kg.

All these initiatives, although not initiated by any EU-based policy specifically, had the effect to influence future Restriction proposals of the European Commission.

21.4.4 Plastic Pacts

The launch of the Circular Plastics Alliance by the European Commission in 2019 has influenced similar initiatives at the Member States level.

The CPA focusses only on reaching the objective of the Plastics Strategy's pledge campaign, i.e. to incorporate 10 Mt of recyclates into new products by 2025. Hence, for some Member States, the scope of the CPA was not inclusive enough for the various issues related to plastics and its governance. A quick reaction, shortly after the CPA launch, was the creation of the so-called European Plastic Pact (EPP), initiated by France, Netherlands, and Denmark (EPP 2023).

The EPP was open to any stakeholder of the plastics value chain and of the most committed governments and focused on all levels to reduce the release of plastics into the environment:

- by improving the recyclability and reusability of products by design,
- by shifting to a more responsible use of plastics,
- by increasing collection, sorting, and recycling,
- and by incorporating more recycled materials into new products and packaging.

This initiative eventually wound up in 2023 due to the recent developments brought up by the EU legislation and the Global Treaty to End Plastic Pollution (UNEP 2023) which provide a new and wider opportunity to drive forward the plastics agenda in Europe (see closing statement of EPP (2023)) and in the rest of the world. Other countries joined the EPP or created their own Plastics Pacts or ad hoc strategies to take care of plastics from a national point of view and thus foresaw more ambitious measures than the ones stemming from the Plastics Strategy and descending regulations. Eventually, most Member States adopted a plan on plastics (with a major focus on household packaging streams) and their reduction at the national level.

All these pacts are non-binding documents that have strategic, high-level but often short-term objectives and visions on how to best manage plastics products, recycling, and other end-of-waste disposals. Despite great ambitions and full governmental support, most of these initiatives are currently dormant and many public actors have moved to a more influential forum, like the UN-level Global Treaty on Plastics mentioned and the revisions of many pivotal regulations for plastics (and other materials) in the EU.

21.5 Status and Outlook for Circular Economy Related Regulation

Since the publication of the Plastics Strategy (2018), many regulations and proposals have been published at the EU level that relate to plastic products and are intended to contribute to the development of a circular plastics economy. Never before have so many and such detailed regulations and proposals for products made from one material been developed in such a short period of time. The reasons for this are already listed in the Plastics Strategy and are still up to date:

- Low rate of recycling and reuse of plastics
- Significant leakage of plastics into the environment
- High dependence on virgin fossil feedstock (Staff Document 2018).

Future regulations will be derived even more strongly than in the past from the need to protect the climate and therefore promote the substitution of fossil feedstock by circular raw materials (recyclates, biomass, CO_2; cf. Part III).

The regulation of plastic products following up on the publication of the Plastics Strategy to date has focused primarily on single-use products and packaging. It is foreseeable from strategy documents and regulatory proposals that EU legislators will also turn their attention to other plastic applications such as vehicle plastic parts, construction products, or agricultural plastics (cf. Sect. 11.4) in the future, e.g. concerning separate collection or recycled content targets.

The SUPD and PPWR regulations are the first in the EU to stipulate mandatory requirements for the use of recycled plastics for defined groups of packaging. In doing so, the legislator is taking account of the fact that the level of development of a circular economy is not measured by how much waste is put to recycling, but by how much secondary material replaces raw material. As the first compliance date for the recycled content requirement of PET bottles is 2025, it is not yet possible to report on how the implementation of these circularity rules is measuring up. One question that has been intensively discussed in connection with the specification of recycled content is the availability of sufficient amounts of recyclates (cf. Chap. 22). Some unintended scenarios are possible:

- Lack of sufficient polyolefin recyclates to meet the recycled content targets of PPWR;
- Very high share of imported PET recyclates from non-EU countries to meet the targets of SUPD and PPWR.[3]

[3] "As of October 2023, the DG-SANTE Union Register announced around 100 recycling or decontamination installations outside the EU (not including the UK) accredited for import of recycled

Delegated acts from the EU Commission are still pending for the SUPD. The ones triggered by the PPWR will only be adopted in a later stage (e.g., after the formal entry into force of the PPWR itself). This secondary legislation is intended to ensure harmonised implementation by 2030.

A particularly important regulation for implementing the regulations on the proportion of recycled materials in products still is the outstanding act on the use of mass balance methods. A transparent, standardized, and auditable allocation system is required to calculate the use of recycled materials (from mechanical and chemical recycling) in accordance with a mass balance based on the chain-of-custody principle (cf. Chap. 16). However, the EU legislator is not only an initiator of new regulations in terms of the circular economy, but also takes up developments that have already taken place through initiatives by industry or the member states. Examples of this are the planned harmonised regulations on an EU-wide standardised set of design-for-recycling criteria for plastic packaging or the modulation of eco-fees in accordance with design criteria. In Europe are already well-functioning systems implemented such as in Germany for DfR-criteria (ZSVR 2023) or in Sweden and France for eco-modulation of the EPR fee.

In addition to the regulatory measures to implement SUPD and PPWR and the expected measures in other plastics application areas and plastics-relevant product regulations via ESPR, the global stage will become increasingly important for politics, the media, and the public in the coming years by the so-called Global Plastics Treaty (UNEP 2023). The EU and the member states will have submitted corresponding proposals to the negotiations planned until the end of 2024 based on the extensive set of EU and national regulations and existing experience in the plastic circular economy. NGOs and European companies in the plastics supply chain are also taking part in the UN negotiations. It is interesting to note that many actors from these sectors who have previously been involved in the Circular Plastics Alliance[4] or have participated in national Plastic Pact initiatives have moved to a more influential forum, like the UN Global Treaty on Plastics and the revisions of many pivotal regulations for plastics in the EU. If it is possible to incorporate some essential EU components of the plastics policy into the global treaty text, this would be a success not only for marine protection but also for the development of a circular economy of plastics on the global level.

plastic materials and articles intended to come into contact with foods. Over half of these registrations are in non-OECD countries, where the export of plastic waste from the EU will be prohibited. This means that in future any recycled PET imported from non-OECD countries would not be from EU PET waste and so the EU would be using other countries' waste to meet targets" (Systemiq 2024).

[4] The Circular Plastics Alliance (CPA) is an initiative that was founded a few months after the Plastics Strategy was adopted at EU level at the end of 2018. Under the leadership of the EU Commission, around 300 companies, associations, municipal institutions and other stakeholders have set themselves the goal of using ten million tonnes of PCR recyclates in plastic products by 2025 (CPA 2018).

References

Carrier bags 2015: Directive (EU) 2015/720 of the European Parliament and of the Council of 29 April 2015 amending Directive 94/62/EC as regards reducing the consumption of lightweight plastic carrier bags.https://eur-lex.europa.eu/legal-content/EN/TXT/?uri=celex%3A32015L0720 (Last access: May 22, 2024)

CEAP 2020: Communication from the Commission to the European Parliament, the Council, the European Economic and Social Committee and the Committee of the Regions *A new Circular Economy Action Plan For a cleaner and more competitive Europe.* COM/2020/98 final https://eur-lex.europa.eu/legal-content/EN/TXT/?uri=COM%3A2020%3A98%3AFIN (Last access: May 22, 2024)

CEP 2015: Communication from the Commission to the European Parliament, the Council, the European Economic and Social Committee and the Committee of the Regions *Closing the loop - An EU action plan for the Circular Economy*, so-called Circular Economy Package. COM(2015) 614 final. https://eur-lex.europa.eu/legal-content/EN/TXT/?uri=CELEX%3A52015DC0614 (Last access: June 12, 2024)

CLP 2008: Regulation (EC) No 1272/2008 of the European Parliament and of the Council of 16 December 2008 on classification, labelling and packaging of substances and mixtures. https://eur-lex.europa.eu/eli/reg/2008/1272/oj (Last access: May 12, 2024)

COM 2022: Communication from the Commission to the European Parliament, the Council, the European Economic and Social Committee and the Committee of the Regions *On making sustainable products the norm.* https://eur-lex.europa.eu/legal-content/EN/TXT/?uri=CELEX%3A52022DC0140 (Last access: May 22, 2024)

Council Directive 1976: Council Directive 76/211/EEC of 20 January 1976 on the approximation of the laws of the Member States relating to the making-up by weight or by volume of certain prepackaged products. https://eur-lex.europa.eu/legal-content/EN/TXT/?uri=CELEX%3A31976L0211 (Last access: June 12, 2024)

CPA 2018: Circular Plastics Alliance, launched in the Context of the Plastic Strategy in 2018. https://ec.europa.eu/commission/presscorner/detail/en/IP_18_6728 (Last access: May 22, 2024)

CPR 2011: Regulation (EU) No 305/2011 of the European Parliament and of the Council of 9 March 2011 laying down harmonised conditions for the marketing of construction products. https://eur-lex.europa.eu/LexUriServ/LexUriServ.do?uri=OJ:L:2011:088:0005:0043:EN:PDF#:~:text=This%20Regulation%20lays%20down%20conditions,CE%20marking%20on%20those%20products. (Last access: June 12, 2024)

CPR draft 2020: Proposal for a Regulation laying down harmonised conditions for the marketing of construction products, amending Regulation (EU) 2019/1020 and repealing Regulation (EU) 305/2011. https://ec.europa.eu/docsroom/documents/49315 (Last access: June 12, 2024)

EC 2008: Commission Regulation (EC) No 282/2008 of 27 March 2008 on recycled plastic materials and articles intended to come into contact with foods. https://eur-lex.europa.eu/legal-content/EN/TXT/?uri=CELEX:32008R0282 (Last access: May 22, 2024)

Ecodesign 2009: Directive 2009/125/EC of the European Parliament and of the Council of 21 October 2009 establishing a framework for the setting of ecodesign requirements for energy-related products. https://eur-lex.europa.eu/legal-content/EN/ALL/?uri=CELEX:32009L0125 (Last access: May 22, 2024)

Ecodesign 2022: Proposal for a REGULATION OF THE EUROPEAN PARLIAMENT AND OF THE COUNCIL establishing a framework for setting ecodesign requirements for sustainable products. https://eur-lex.europa.eu/legal-content/EN/TXT/?uri=CELEX:52022PC0142 (Last access; June 12, 2024)

ELV 2000: Directive 2000/53/EC of the European Parliament and of the Council of 18 September 2000 on end-of life vehicles. https://eur-lex.europa.eu/legal-content/EN/ALL/?uri=celex%3A32000L0053 (Last access: May 22, 2024)

EPP 2023: https://europeanplasticspact.org

EU 2019: Commission Implementing Decision (EU) 2019/665 of 17 April 2019 amending Decision 2005/270/EC establishing the formats relating to the database system pursuant to European Parliament and Council Directive 94/62/EC on packaging and packaging waste.https://eur-lex.europa.eu/legal-content/EN/TXT/?uri=CELEX:32019D0665 (Last access: Aug. 20, 2024)
EU 2022: Commission Regulation (EU) 2022/1616 of 15 September 2022 on recycled plastic materials and articles intended to come into contact with foods, https://eur-lex.europa.eu/legal-content/EN/TXT/?uri=CELEX%3A32022R1616 (Last access: June 12, 2024)
EU 2023: Proposal for a Regulation of the European Parliament and of the Council on circularity requirements for vehicle design and on management of end-of-life vehicles. https://eur-lex.europa.eu/legal-content/EN/TXT/?uri=CELEX%3A52023PC0451 (Last access: Aug. 20, 2024)
EU 2024: https://single-market-economy.ec.europa.eu/industry/sustainability/buildings-and-construction_en (Last acces: Aug. 20, 2024)
Framework Directive 1989: Council Directive 89/109/EEC of 21 December 1988 on the approximation of the laws of the Member States relating to materials and articles intended to come into contact with foodstuffs. https://eur-lex.europa.eu/legal-content/EN/ALL/?uri=CELEX%3A31989L0109 (Last access: May 22, 2024)
Framework Regulation: Regulation (EC) No 1935/2004 of the European Parliament and of the Council of 27 October 2004 on materials and articles intended to come into contact with food and repealing Directives 80/590/EEC and 89/109/EEC https://eur-lex.europa.eu/legal-content/EN/ALL/?uri=celex%3A32004R1935 (Last access: May 22, 2024)
Green Deal 2019: Communication from the Commission to the European Parliament, the European Council, the Council, the European Economic and Social Committee and the Committee of the Regions *The European Green Deal.* COM/2019/640 final. https://eur-lex.europa.eu/legal-content/EN/TXT/?uri=COM%3A2019%3A640%3AFIN (Last access: June 12, 2024)
Green Paper 2013: Green Paper A 2030 framework for climate and energy policies. COM(2013) 169 final. https://eur-lex.europa.eu/LexUriServ/LexUriServ.do?uri=COM:2013:0169:FIN:en:PDF (Last access: Aug. 20, 2024)
GPO 1991: German Packaging Ordinance 1991. (Verordnung über die Vermeidung von Verpackungsabfällen vom 12.Juni 1991. Bundesgesetzblatt Teil I vom 20.06.1991. pdf available
Italian Tax 2019: Legge 160/2019, art. 1, commi 634 e succ, Gazzetta Ufficiale della Repubblica Italiana
JRC 2023: Joint Research Center: Ecodesign for Sustainable Products Regulation—preliminary study on new product priorities. Technical Report (draft) 2023. https://commission.europa.eu/energy-climate-change-environment/standards-tools-and-labels/products-labelling-rules-and-requirements/sustainable-products/about-sustainable-products_en (Last access: Aug. 20, 2024)
Own Resources 2020: Council Decision 2020/2053 of 14 December 2020 on the system of own resources of the European Union. https://eur-lex.europa.eu/legal-content/EN/TXT/?uri=CELEX%3A32020D2053&qid=1609775612824 (Last access: June 12, 2024)
PIM 2011: Plastic Implementing Measure. Commission Regulation (EU) No 10/2011 of 14 January 2011 on plastic materials and articles intended to come into contact with food. https://eur-lex.europa.eu/legal-content/EN/ALL/?uri=CELEX%3A32011R0010 (Last access: May 22, 2024)
PlasticsEurope 2022: Publication “Plastics—the Facts 2022”.https://plasticseurope.org/knowledge-hub/plastics-the-facts-2022/ (last access: May 22, 2024)
Plastics Strategy 2018: Communication from the Commission to the European Parliament, the Council, the European Economic and Social Committee and the Committee of the Regions *A European Strategy for Plastics in a Circular Economy.* COM/2018/028 final https://eur-lex.europa.eu/legal-content/EN/TXT/?qid=1516265440535&uri=COM:2018:28:FIN (Last access: Aug. 20, 2024)
PPWD 1994: European Parliament and Council Directive 94/62/EC of 20 December 1994 on packaging and packaging waste. https://eur-lex.europa.eu/legal-content/EN/TXT/?uri=celex%3A31994L0062 (Last access: Aug. 20, 2024)
PPWR 2022: Proposal for a Regulation of the European Parliament and of the Council on packaging and packaging waste, amending Regulation (EU) 2019/1020 and Directive (EU) 2019/904, and

repealing Directive 94/62/EC. COM/2022/677 final https://eur-lex.europa.eu/legal-content/EN/TXT/?uri=CELEX%3A52022PC0677 (Last access: June 12, 2024)

PPWR 2024: Comporomise Paper between Council and EU-Parliament, 15 March 2024. https://data.consilium.europa.eu/doc/document/ST-7859-2024-INIT/en/pdf (Last access: March 25, 2024)

REACH 2006: Regulation (EC) No 1907/2006 of the European Parliament and of the Council of 18 December 2006 concerning the Registration, Evaluation, Authorisation and Restriction of Chemicals (REACH), establishing a European Chemicals Agency. https://eur-lex.europa.eu/legal-content/EN/ALL/?uri=celex%3A32006R1907 (Last access: May 22, 2024)

RoHS 2002: Directive 2002/95/EC of the European Parliament and of the Council of 27 January 2003 on the restriction of the use of certain hazardous substances in electrical and electronic equipment. https://eur-lex.europa.eu/legal-content/EN/TXT/?uri=celex%3A32002L0095 (Last access: May 22, 2024)

SEA 1987: The Single European Act, 1987. https://eur-lex.europa.eu/legal-content/EN/TXT/?uri=celex%3A11986U%2FTXT (Last access: May 22, 2024)

Spanish Tax 2022: Ley no. 7/2022 of 8 April 2022 on the management of waste and contaminated soil to promote a circular economy (Plastic Tax Law).

Staff Document 2018: COMMISSION STAFF WORKING DOCUMENT IMPACT ASSESSMENT Reducing Marine Litter: action on single use plastics and fishing gear Accompanying the document Proposal for a Directive of the European Parliament and of the Council on the reduction of the impact of certain plastic products on the environment. SWD/2018/254 final - 2018/0172 (COD). https://eur-lex.europa.eu/legal-content/EN/TXT/?uri=CELEX%3A52018SC0254 (Last access: June 12, 2024)

SUPD 2019: Directive (EU) 2019/904 of the European Parliament and of the Council of 5 June 2019 on the reduction of the impact of certain plastic products on the environment. https://eur-lex.europa.eu/eli/dir/2019/904/oj (Last access: May 12, 2024)

Systemiq 2024: Potential impacts on recycled PET imports on the system vision for a circular PET/polyester economy in Europe. www.systemiq.earth/pet-polyester (Last access: Aug. 20, 2024)

TFEU 2012: Consolidated version of the Treaty on the Functioning of the European Union. https://eur-lex.europa.eu/legal-content/EN/TXT/?uri=celex%3A12012E%2FTXT (Last access: May 22, 2024)

UK Tax 2021: Finance Act 2021, UK Public General Acts 2021 c. 26 PART 2. https://www.legislation.gov.uk/ukpga/2021/26/part/2/enacted (Last access: May 22, 2024)

UNEP 2023: Revised draft text of the international legally binding instrument on plastic pollution, including in the marine environment. https://www.unep.org/inc-plastic-pollution/session-4/documents (Last access: May 22, 2024)

WEEE 2002: Directive 2002/96/EC of the European Parliament and of the Council of 27 January 2003 on waste electrical and electronic equipment (WEEE). https://eur-lex.europa.eu/resource.html?uri=cellar:ac89e64f-a4a5-4c13-8d96-1fd1d6bcaa49.0004.02/DOC_1&format=PDF (Last access: May 22, 2024)

WEEE 2012: Directive 2012/19/EU of the European Parliament and of the Council of 4 July 2012 on waste electrical and electronic equipment (WEEE) (recast) https://eur-lex.europa.eu/legal-content/EN/TXT/?uri=CELEX:02012L0019-20180704 (Last access: May 22, 2024)

WFD 2008: Directive 2008/98/EC of the European Parliament and of the Council of 19 November 2008 on waste and repealing certain Directives. https://eur-lex.europa.eu/legal-content/EN/TXT/?uri=celex%3A32008L0098 (Last access: May 22, 2024)

ZSVR 2023: Minimum standard for determining the recyclability of packaging—edition 2023. https://www.verpackungsregister.org/en/foundation-authority/minimum-standard-pursuant-to-section-21/2023-edition (Last access: Aug. 20, 2024)

Silvia Freni Sterrantino was Senior Legal and Regulatory Affairs Manager at European Plastics Converters Association. Before joining EuPC, she worked as an attorney in leading European law firms and in the Directorate-General for Competition of the European Commission. Admitted to the Italian Bar Association, she studied law at the universities of Siena, Cambridge, and London, where she earned an LL.M. in Competition Law & IP at the London School of Economics and Political Science.

Antonello Romano is a public affairs professional holding a law degree from the University of Siena, Italy, and a master's degree (LLM) in European Law from the University of Lund, Sweden. He currently works as Corporate Affairs Director for Stora Enso, focusing on packaging, waste, and chemical legislation. Prior to that, he worked as Senior Advocacy Manager for PlasticsEurope, focusing on Circular Economy, Packaging, Recycling, and Food Contact Materials.

Dr. Jürgen Bruder is a chemist with an additional degree in business. He worked in higher education and the chemical industry, before becoming Director General of IK, the German plastic packaging association. He held positions in national, European, and global committees including the Global Plastics Alliance. Today he deals with the transformation of the plastics industry to a circular economy focusing on plastic packaging and plastic applications in agriculture.

Chapter 22
Mandatory Recycled Content Quotas—A Paradigm Shift in Regulation

Isabell Schmidt

Contents

Summary

This chapter provides an overview of applied and possible steering instruments to accelerate the transition to a circular economy and to significantly increase the production and use of recyclates.

Particular attention is paid to the mandatory recycled content quota already applied in the SUPD for PET beverage bottles and in the draft PPWR for most plastic packaging applications. From the perspective of plastic packaging manufacturers, the risks of applying this quota in the event of a shortage of recyclate are critically examined and, at the same time, ways of minimizing these risks are identified.

Finally, the polymer-specific substitution quota is investigated as an alternative quantitative steering instrument to promote the production of recyclates. Addressees of such a quota would be the producers and importers of virgin and recycled plastics.

I. Schmidt (✉)
Industrievereinigung Kunststoffverpackungen e.V., Bad Homburg vor der Höhe, Germany
e-mail: i.schmidt@kunststoffverpackungen.de

P. Orth et al. (eds.), *Circular Economy of Plastics*,
https://doi.org/10.1007/978-3-031-78829-1_22

22.1 The European Green Deal: A Game Changer for Plastics Recycling

The European Green Deal (COM 2019) and the EU Commission's new Circular Economy Action Plan (COM 2020) brought forward a paradigm shift in plastics recycling (cf. Chap. 1 and 21). In European and German legislation, recycling was previously viewed from the perspective of waste disposal, classified as a kind of material waste recovery and compared with alternative disposal methods such as energy recovery and mere waste disposal in the form of incineration without energy recovery and landfilling. This classification is expressed in the waste hierarchy of the EU Waste Framework Directive (EU 2008).

As crude oil and other fossil raw materials were still available to the chemical industry in almost unlimited quantities, there was no need to resort to alternative, more expensive raw materials from recycling or other carbon sources. Therefore, the production of virgin plastics is still based on fossil raw materials.[1] Plastic recycling had to find niches in plastics processing to be competitive with virgin plastics production. As mechanically recycled plastics are an imperfect—i.e. not qualitatively equivalent—substitute for virgin plastics, they found limited acceptance in markets where they could meet quality requirements and offer a price advantage. Demanding applications—food contact items in particular—remained inaccessible, with a few exceptions only. Improving plant technology to achieve higher quality levels generally did not pay off due to the competition with virgin plastics.[2] The perspective has changed now with the Green Deal and the subsequent need to phase out fossil feedstock to achieve climate neutrality by 2050. This heralded not only a Europe-wide energy transition but also the transformation of the raw materials market to a circular economy. Recycling is no longer seen primarily as a solution to a waste problem, but as the most important measure to secure the supply of raw materials for the European industry under the conditions of climate neutrality. This paradigm shift is reflected in the introduction of recyclate use quotas for plastics in certain applications such as packaging or vehicles in European legislation to detach markets from fossil-based virgin plastics. The first recyclate use obligations were introduced with the Single-Use Plastics Directive (EU 2019) for beverage bottles. As of 2025, PET beverage bottles must therefore contain an average of 25% recyclates.

---

[1] In 2022, approx. 98% of the approx. 48 Mt of virgin plastic production in Europe was based on fossil raw materials. Less than 2% were circular materials, of which appr. 700 kt were bio-based and appr. 100 kt were chemically (PlasticsEurope 2020, p. 37). The plastics manufacturing industry plans to increase this proportion to 2,800 kt by 2030 by investing in chemical recycling (ibid. p. 16).

[2] One exception to this was the use of recyclate in PET beverage bottles. Due to the high rate of separate collection and the low contamination of this waste stream, which in Germany and other countries is favored by the deposit system for single-use PET bottles (cf. chap. 11.2), a bottle-to-bottle recycling loop was already established in the 2010s.

22.2 Regulatory Approaches for the Transformation to a Circular Economy for Plastics

However, statutory recyclate use quotas for certain product segments are not the only way to bring forward the transformation of the raw materials market to a circular economy. Kirschstein et al. (2023) make a fundamental distinction between price, quantitative and information instruments as well as regulatory and systemic instruments (cf. Table 22.1). Pricing instruments influence directly prices in the virgin and recycled plastic markets. Quantitative instruments, on the other hand, including recycled content quotas, set targets for the absolute or relative product volumes traded on the market. Finally, information instruments regulate the communication of decision-relevant information between the market players involved. Kirschstein et al. (2023) consider regulatory instruments as legal requirements and bans to influence market activity on a selective basis. Finally, systemic instruments include measures to shape self-contained market segments, such as deposit and return systems for specific products (cf. Chap. 11). In principle, the instruments mentioned can be defined and can also be used in combination with each other.

However, regulatory market intervention is fundamentally essential for the transformation to a climate-neutral circular economy, as a market-driven shortage of fossil fuels in the short to medium term resulting in a "natural" adjustment of the economy cannot be assumed. Environmental awareness of consumers and voluntary commitments of companies can at best positively support the transformation but cannot bring it about. In the natural differentiation of markets, price and quality characteristics of products are decisive alongside ecological purchasing arguments. The "eco-niche" cannot expand to become the overall market unless it has price and quality leadership.

Moralizing talks of market failure merely demonstrate a fundamental lack of understanding of the self-regulating mechanisms of a market economy. This needs to be understood to exploit them in line with the transformation goals without stifling the market economy by over-regulation and bureaucracy, thus jeopardizing economic prosperity. The scale of the challenge should not be underestimated, as Europe's

Table 22.1 Overview of instruments for increasing plastics recycling. *Source* Kirschstein et al. (2023) p. 20

Instruments				
Price	Quantitative	Information	Regulatory	Systemic
Plastic tax	Recycling quotas	Labels	Thermal recovery ban	Deposit return
VAT reduction	Recycled content quotas	Marking	Public procurement requirement	Voluntary take back
CO_2 levies	Substitution quotas	Campaigns	Product standards and norms	Branch solutions
License fees				
Subsidies				

economic upturn has been based on the use of fossil resources since the end of the 19th century. A careful scientific impact assessment should therefore be carried out by the legislator. If the transformation fails economically, at least the social acceptance of the Green Deal itself would be at stake.

22.3 Risks of Mandatory Recycled Content Quotas for Specific Products

Schmidt (2020) has formulated six principles for effective and market-oriented regulation to increase the use of recyclates, which can be used to evaluate legal instruments (cf. Table 22.2). On this basis, AGVU and IK (2020) point out the weaknesses and risks of product-related recycled content quotas.

A fundamental difficulty with product-related specifications for the use of recyclates is determining the quota level. If the quota is too high, AGVU and IK (2020) point out the risks to quality and product safety if recyclate available in sufficient quantity and quality for the regulated markets cannot be ensured. However, if it is too low, its stimulating effect on investments in recycling will fizzle out. When setting the quota level, legislators must also take into account the demand for recyclates

Table 22.2 Six principles of effective and market-driven regulation for increasing the use of recyclates. *Source* based on Schmidt (2020), AGVU and IK (2020)

1. Increase recyclate volumes in the overall market Only measures that lead to an increase in the use of recyclate in the overall market and thus result in a significant reduction in fossil-based virgin plastics and climate-damaging emissions make sense. Simply shifting the recyclate to other product applications has no ecological advantage
2. Ensure quality and product safety The quality and product safety of plastic products must not be compromised. Consumer protection is the top priority. With the development of new fields of application, the qualities of recyclates must therefore also be further developed
3. Maintain market economy principles Preserving market economy principles is the most efficient way to achieve the goals of recyclate use in the overall market, minimize the risk of supply bottlenecks and respond to new developments in technology and the market
4. Avoid competitive disadvantages for the European economy Europe has a vital plastics industry in which all stages of the cycle are represented, from raw material production to processing and recycling. The competitiveness of this sector must be ensured, also in order not to jeopardize necessary investments in the circular economy
5. Ensure the free movement of goods in the EU internal market In order not to jeopardize the free movement of goods in the EU internal market, legal requirements for the use of recyclates should apply throughout Europe
6. Enable enforceability Legal regulation must be enforceable with as little bureaucracy as possible on the part of the authorities and industry

from other sectors. The proposal for statutory recyclate use quotas in packaging was soon followed by further proposals for recyclate use quotas in the automotive sector (cf. Sect. 21.2). The automotive industry's demand for PP and PET recyclates from packaging recycling is just as foreseeable as its willingness to pay higher prices than packaging manufacturers. This can lead to a precarious supply situation in the packaging market and increase the risk for packaging manufacturers not to meet their quotas and therefore not be allowed to place their packaging on the market.

Various studies show that a lack of recyclate poses a real threat to supply chains. One bottleneck is the separate collection of plastic waste and its re-sorting for recycling. In the packaging sector alone, the Europe-wide need to achieve the recycled content quotas proposed in the draft EU Packaging Regulation (PPWR 2024) will increase by at least over 4.4 Mt PCR by 2030 compared to the 2021 level (Conversio 2023b). In order to achieve the quotas for 2030 in the area of polyolefin packaging, the production of polyolefin recyclates must be increased fivefold (ibid.). Nevertheless, the EU Commission points out in its early warning report that 19 Member States are at risk of not achieving their plastics-specific recycling targets for 2025 (COM 2023). This includes major economies such as France, Italy, Spain, and Poland. Even the Netherlands, a pioneer in plastics recycling, is at risk. A market study by KPMG warns of supply bottlenecks with plastic waste in the Dutch recycling industry: In 2030, 1.2 Mt, corresponding to 55% of the demand forecast, would be unmet (Hesselink 2023). Even after optimizing waste collection and sorting, such as re-sorting residual waste, the local supply of plastic waste would remain insufficient. A shortage of recyclates would affect small and medium-sized manufacturers in particular, not being able to obtain recyclates in the required qualities on the open market, or only at significantly poorer conditions.

Determining the quota level is also made more difficult by the fact that the use potential for recyclates in different products in an industrial sector can vary greatly. In the packaging sector, for example, the potential for the use of recyclates ranges from currently 0% for food packaging made of polyolefins to 100% for PET beverage bottles. Kirschstein et al. (2023, p. 46) therefore point out that the extension of the recyclate use quotas applicable to PET beverage bottles to other packaging segments or types of plastic cannot be realized in a generalized manner and requires a precise examination of the quality requirements of the packaging concerned and the collection system used. While the conditions for a closed recycling loop for single-use PET beverage bottles are favorable due to the separate collection in the deposit system, other plastic wastes have significantly higher levels of contamination and heterogeneous compositions, which makes the production of high-quality recyclate much more difficult. Recyclate use quotas for plastics in contact-sensitive packaging made of polyolefins and other high-performance applications, where the quality requirements for the raw materials are very high and hardly allow any tolerances, are therefore unrealistic for the time being (Kirschstein et al. 2023, p. 46). If the EU nevertheless makes recyclates mandatory for all food packaging from 2030 as part of the Packaging Ordinance (PPWR 2024), it will do so in order to incentivize investment in chemical recycling in particular and will review the available recycling technologies by 2028.

Another major point of criticism relates to the high bureaucratic effort required for verification and official enforcement (cf. AGVU and IK 2020, Kirschstein et al. 2023). As the recyclate content cannot be quantified using laboratory analysis, complex detection systems are required to trace the material flows throughout the supply chain back to the source of the plastic waste. This is the only way to ensure that the recycled waste meets legal requirements for post-consumer plastic waste. The immense number of manufacturers and products to be monitored, including imports from third countries outside the EU, must also be taken into account. The Central Agency Packaging Register puts the number of registered distributors of packaging in Germany alone at more than 1 million (ZSVR 2024) stemming from 160 countries (more than 400,000 registrations from China alone) and points to existing challenges in enforcing the system participation requirement of the Packaging Act. In comparison, monitoring the recyclate use quotas for packaging would be considerably more complex. In addition to additional raw material costs, packaging manufacturers would also incur monitoring and administrative costs (Kirschstein et al. 2023, p. 46). These contribute to an increase in consumer prices. The risk of unfair competitive conditions due to enforcement deficits should not be underestimated. European packaging manufacturers are already complaining about unfair competition due to imports of semi-finished products for packaging production with dubious information on the recycled content.

As a result of the uncertain supply situation and the difficulty of calculating cost increases, trends towards circumventing the requirements by using other packaging materials can already be observed on the market. In particular, the market for coated paper packaging, which is exempt from the PPWR's recyclate usage quotas if it contains less than 5% plastic, is growing. This packaging is growing rapidly in Germany at around 5–6% per year (GVM 2021), while the consumption of plastic packaging is declining. However, this trend is being criticized even by environmental protection NGOs because the new types of paper composites lead to a shift in environmental pollution and an increase in packaging waste and are also difficult to recycle (GVM and ifeu 2023; GVM 2021; Stravens 2023; NABU n.d.; WWF 2022).

22.4 Reduction of the Risks of Mandatory Recycled Content Quotas

The EU responds to the risks of a recyclate undersupply for the supply chain security of packaged goods within the framework of the EU Packaging Regulation by authorizing the Commission to grant exemptions for certain packaging—except for packaging made of PET and single-use beverage bottles—after appropriate examination (PPWR 2024, Article 7(12)). However, there is no legal entitlement to exemptions for the economic operators concerned. In addition, the review is only to take place two years before the start of the obligation to use recyclates, which, if there are no exemptions, would hardly allow any more time to switch to other packaging materials. In order to create more planning certainty, further instruments to reduce supply risks are therefore being discussed, which are considered below.

22.4.1 *Priority Access*

Under the Packaging Regulation, the EU calls on the member states to provide a comprehensive infrastructure for the collection and sorting of packaging waste in order to ensure the availability of plastic raw materials for recycling and the supply of recyclates (PPWR 2024, Articles 48(2) and 49). In addition, it gives member states the option of granting priority access to recyclates for product applications that allow repeated, high-quality recycling (PPWR 2024, Article 48(2)). The European beverage industry in particular had made such demands, fearing a loss of recycled PET from beverage bottles to other applications (Packaging Europe 2023).

Here, the EU is attempting to solve a problem caused by overly detailed market regulation through further state intervention in the market—an approach which, in the author's opinion, increases the risks and should therefore be examined very carefully. After all, it remains unclear how a priority right of access for certain applications would be structured under market economy principles. If the supply of raw materials in the circular economy is no longer based on the principle of price formation based on supply and demand on the market, but instead on state allocations to companies in certain sectors, the foundations of the market economy are dangerously called into question. What remains unclear with such an approach is how prices would be determined. This approach is therefore likely to be welcomed at best by the economic sectors benefiting from a priority right of access, while the recycling industry would no longer be able to sell its goods on the market at free prices—a serious market intervention that would have a negative impact on the willingness to invest and is sharply criticized by the recycling industry (Sustainable Plastics 2023).

22.4.2 *Higher Flexibility Through Credit Trading or Pooling*

Various economic players are proposing a solution to reduce supply risks aiming to increase the flexibility of recyclate use. For the ecological transformation of the raw materials sector, it is irrelevant in which applications the recyclates are used, as long as there is a continuously advancing substitution of fossil-based raw materials in the plastics industry as a whole. The supply risks of individual manufacturers could, therefore, be mitigated by enabling a balance between manufacturers of different products: Lower recyclate use in products with high hurdles could be offset by higher recyclate use in other plastic products—across the boundaries of individual companies and segments. Trading in credits or certificates along the lines of greenhouse gas emissions trading would be an instrument to make this possible. Such a system is described and propagated by the Federal Association of Plastics Packaging Manufacturers in Germany (IK 2023), among others.

For example, a manufacturer of paint buckets voluntarily using more recycled material in its products than required by law would receive credits that it could sell to manufacturers of food packaging that are not yet able to meet the quota and are dependent on the purchase of credits to compensate for this. Both sides benefit from trading in credits: the paint bucket manufacturer receives a financial incentive to use more recycled materials thanks to the proceeds from the credits. The buyer of the credits, on the other hand, can compensate for its legal obligation to use recyclate and remain on the market with its products in the event of a shortage of the required recyclate qualities. This could mitigate the risks for supply chains and still secure the demand for recyclates. Government intervention, such as the granting of exemptions or priority recyclate allocations, would be unnecessary.

Pooling works in a similar way, whereby several obligated companies can form a pool to jointly achieve the recycled content quotas. Within such pool, it would be possible to achieve a balance across company boundaries. The draft EU Packaging Regulation already provides for such an approach to meet the reusable requirements in the beverage sector (PPWR 2024, Article 29(12)).

From a macroeconomic perspective, there is also a case for flexibilization through credit trading or pooling. Just as greenhouse gas emissions trading ensures that greenhouse gas emissions are first reduced where this is possible at the lowest cost, credit trading for recyclates allows fossil raw materials to be replaced first in products where the economic and legal hurdles are lowest. A credit system thus reduces the economic costs of the transformation.

A criticism of such system is that it promotes the reuse of recyclates in applications with lower quality requirements and therefore downcycling. However, the legislator could set limits to this by excluding applications in which the recyclate does not replace virgin plastic from the credit trade. In addition, all products manufactured with PCR should themselves meet design-for-recycling requirements so that they can be recycled again. With regard to consumer communication, it should be stipulated that only the recycled content actually contained in the product may be shown in order to avoid misleading advertising claims. Finally, in order to reduce the risk of fraud due to falsely issued certificates, the trade-in certificates should be just as strictly controlled as the trade-in recyclates itself. However, the monitoring effort does not differ significantly in both cases, as monitoring the use of PCR also requires a system for tracing documents back to the origin of the waste (IK 2023).

22.4.3 Comply-or-Pay Mechanism

Granting compensation payments for unfulfilled recyclate use quotas represents a further approach to mitigating marketing risks in the event of a supply shortage of recyclates. Manufacturers would have a choice here: either they comply with the requirements for the use of recyclates (comply) or they must make a compensation payment (pay).

Marketing risks due to lack of recyclate can be averted by a comply-or-pay mechanism. The fixed amount of the compensation payment would give investors in recycling and legally obliged recyclate purchasers more financial planning security, as both sides would know the maximum costs per ton of recyclate they can calculate with. However, investments in recycling would only be worthwhile if the production costs for the recyclate were below the specified level of the compensation payment. An appropriate level of compensation is therefore crucial. It must provide an effective incentive for investment in recycling, but should not be prohibitively high for manufacturers of products with recycled content quotas.

A fund organized by the private sector or the state would have to be set up to collect and manage the compensation payment in the form of a special levy. It would also need to be clarified how the money collected can be earmarked and returned to the private sector in order to promote investment in the circular economy. However, the legal requirements for such special levies are extremely complex. They were examined as part of a study by the German Federal Environment Agency for economic instruments based on own resources for non-recycled plastic packaging waste (UBA 2023). However, if the levies were to flow into the general tax budget, the transformation of the plastics sector would be deprived of financial resources, which should be avoided.

The disadvantages of a comply-or-pay mechanism therefore include the bureaucratic effort involved in the central collection, administration, and spending of the special levies in a fund. The instrument also leads to lower demand for recyclates the more companies make use of the possibility of a compensation payment—In contrast to a credit or pooling system, where a constant overall demand for recyclates is always guaranteed. Table 22.3 provides a summarized overview of the instruments discussed for reducing the risks of product-specific recyclate use quotas.

22.5 Substitution Quotas for Plastic Resins as an Alternative Quantitative Instrument

The steering instrument of a polymer-specific substitution quota was brought into discussion in 2020 (AGVU and IK 2020; Schmidt 2020) and has since been deepened, particularly in the VDI Round Table (2022) stakeholder process. The concept of this quota is not based on an individual plastic product or product group, but on the polymeric material and thus on primary plastic manufacturers and recyclers the addressees. In the system according to Kirschstein et al. (2023), substitution quotas, like recyclate use quotas, represent a quantitative instrument (see Table 22.2).

Legislators can use substitution quotas to set targets for the substitution of fossil raw materials in the production of plastics. The obligated parties here are the suppliers of plastics for the processing industry, i.e. manufacturers and importers of virgin

Table 22.3 Comparison of instruments for reducing the risks of product-specific recycled material usage quotas. *Source* Own representation

	Priority Access	Credit trading/Pooling	Comply-or-Pay mechanism
Protecting supply chains from a lack of recyclates	+ Reduction of demand competition, but no protection if technologies for required qualities are lacking (e.g. for polyolefins in food contact)	+ Reduction of risks through greater flexibility in the use of recyclates, but no guaranteed protection, as credits can also become scarce	+ + Guaranteed protection, as financial compensation is possible
Efficiency of enforcement/ bureaucracy	− − Probably requires a state application and allocation system for recyclates	+ Hardly any additional effort compared to monitoring recyclate usage quotas	− Requires the establishment and management of a private or state fund
Safeguarding market economy principles	− − Strong intervention in the market economy through state allocation of (secondary) raw materials	+ + Greater market-economy efficiency through tradable credits	0 Difficult to distribute collected funds in line with the market
Incentives for investments in recycling	− − Free pricing for recyclers not possible on the market	0 Overall demand for PCR remains stable, but declines in particularly high-quality grades (e.g. food contact)	− Overall demand for PCR is falling; investments in recycling are only worthwhile if production costs are below the defined level of the compensation payment

plastics, as well as manufacturers and importers of recycled plastics. The latter must ensure that a certain minimum proportion of the plastics sold are not made from fossil raw materials but from recycled ones. The instrument is aimed at a legally controlled shortage of fossil-based virgin plastics and an increase in the supply of recycled plastics on the market. Substitution quotas can be set for plastic as a material in general or for individual plastic types such as PE, PP, PET or PS (Kirschstein et al. 2023; AGVU and IK 2020; KRU 2019).

In contrast to the product-related recyclate use quotas discussed so far, setting targets for plastics processors in certain sectors such as packaging, construction or automotive, the application does not play a role here. Substitution quotas are applied further up the value chain. Here, the market is even less differentiated, and the number of companies and products is much more manageable.

For example, a substitution quota of 30% for polyethylene (PE) would oblige market suppliers of PE to produce at least 30% of the PE sold using recycled raw materials.[3] Suppliers of virgin plastics would have various options for meeting the legal substitution quota. They would either have to invest in recycling themselves or buy in plastic recyclates or secondary raw materials such as pyrolysis oil from chemical recycling. They could market the recycled plastics separately or offer compounds (mixtures) of virgin material and a certain proportion of recycled material.

A supplementary trading system with certificates can give the economy even more flexibility. In this case, all suppliers of plastics, including compounders and recyclers, would receive certificates that entitle them to sell fossil-based virgin plastics. The quantity of certificates allocated to a company depends on the quantity of recycled plastics it sells. A supplier of polyethylene would therefore receive certificates for 30 tons of PE recyclates sold, which would entitle him to sell 70 tons of fossil-based polyethylene—to stay with the example of a substitution rate of 30%. As recyclers only selling recyclates do not need these certificates themselves, they could sell their certificates on the market. A conventional plastics producer that does not want to invest in recycling could also fulfill its obligation in this case by purchasing such certificates from other companies, e.g. from the recycling industry. Like the already described certificate trading on the part of plastics processors (cf. Sect. 22.4.2), this ensures that investments in recycling take place where they generate the highest added value.

Substitution rates are already used as economic indicators for the progress of the raw materials change. The European statistics authority Eurostat has calculated the circular material use rate annually since 2004 (Eurostat 2024; EEA 2024a). This provides a breakdown by material type of the proportion of material resources used in the EU originating from recycled waste. The Resources Commission at the German Environment Agency also recommends the introduction of a substitution rate at the level of individual materials as an indicator to make the success of the circular economy measurable (KRU 2019). The industry is also increasingly using substitution rates as indicators. For example, the association of plastics producers PlasticsEurope reports the proportion of circular raw materials (recyclates, biomass, CO_2) in plastics production and processing in Europe (PlasticsEurope 2024b). Several discussion papers and studies deal with polymer-specific substitution quotas as a possible steering instrument (AGVU and IK 2020; UBA 2021; Kirschstein et al. 2023; RESAG 2022; VDI 2022). However, substitution quotas have not yet been used as a legal requirement in the sense of a control instrument.

As a significant advantage of substitution quotas over product-related recycled content quotas the legislator can specifically control the share of secondary raw materials in the overall plastics market without having to issue and monitor detailed regulations for specific products. In order to set statutory quota levels for individual

[3] Depending on the definition, the substitution of the fossil raw material could also take place in total through recyclates, bio-based, and CO_2-based raw materials.

Table 22.4 Recycling and substitution quotas for polymers in the EU 27 + 3 for 2021.[4] *Source* Bruder and Lindner (2023)

Polymer	Volume processing [Mt]	Collected waste [Mt]	Mechanical recycling [Mt]	Recycling quota in %	Substitution quota in %	Sources[5]
PE-HD	6.3	5.06	1.97 Input	39	N/a	(1); (2)
PE-LD/LLD	8.5	7.96	2.96 Input	40	N/a	(1); (2)
PP	10.0	5.72	1.49 Input	26	N/a	(1); (2)
PET	4.7 (Incl. recyc.)	N/a	1.32 Output	N/a	28	(4)
PVC	5.2	2.44	0.60 Input	25	N/a	(3)
EPS	1.6	0.51	0.15 Input	30	N/a	(1); (2)
Total	55.6 (Incl. recyc.)	29.5	10.2 Input 5.5 Output	34.6	9.9	(1)

polymers in a long-term, plannable step-by-step process of gradually increasing quotas, the actual degree of substitution must first be determined as a guide. However, the existing indicators do not yet have a sufficient breakdown by polymer; only for PET can a substitution rate of 28% be specified for the EU 27 + 3 market in 2020 (Eunomia 2022). Based on the recycling rates for standard polymers determined in studies to date, which are always higher than the substitution rates depending on the recyclate yields of the recycling processes, the degree of substitution of fossil-based virgin material for the respective types of plastic can therefore only be estimated (cf. Table 22.4).

Substitution quotas mean that the supply of quality-assured recycled plastics for the plastics processing industry is improving, while the price of fossil-based virgin material is rising due to the relative shortage. This makes the use of recyclates in the plastics processing industry more attractive. Due to the self-regulation of the market, the increase in recyclates use will initially take place in those plastics applications where the legal, technical, and economic hurdles are lowest. As regulators gradually increase the substitution quotas, the plastics processing industry has less and less fossil-based virgin plastics available, which is reflected in price increases on the market. As a result, it is worth investing in increasingly sophisticated recycling technologies enabling the use of recyclates in ever more demanding applications. In this way, the transformation of the plastics market is progressing from lower to higher quality requirements until the use of recyclates ultimately becomes unavoidable, even for contact-sensitive applications such as food contact.

However, there is no risk of supply bottlenecks with suitable recyclates and quality defects, as in the case of product-related recycled content quotas, as long as plastics

[4] PET and PVC 2020.

[5] Sources: (1) PlasticsEurope (2022); (2) Conversio (2023a); (3) Conversio (2021); (4) Eunomia (2022).

processors can still switch to virgin material in the event of higher quality requirements. However, the dwindling supply of fossil-based virgin plastics provides a financial incentive for the use of recyclates. The polymer-specific substitution quota also has clear advantages over product-specific quotas in terms of enforcement, as it addresses relatively few players with large market shares and a high leverage effect.

Substitution quotas result in cross-financing of the recycling industry through the proceeds from linear, fossil-based plastics production. The transformation to a circular economy is made the core task of conventional plastics producers, as it were, and can no longer be handed over to a separate recycling industry, which has difficulty getting off the ground due to competition with fossil-based virgin plastics. This addresses a recurring obstacle to the circular economy of plastics, namely the price competition between recyclates and virgin materials and the associated lack of investment power in the recycling sector (cf. Sect. 22.1). Nevertheless, certain market risks associated with this model must be taken into account. For example, imports of plastic pellets must of course be subject to the same substitution quotas in order to avoid competitive disadvantages for the European industry. However, imports of plastic products from third countries could not be regulated if there are no specifications for the use of recycled materials for these products. Supply risks cannot be ruled out either if the recycling sector is not supplied with sufficient quantities of separated waste.

References

AGVU and IK 2020: Gesetzliche Mindestquoten für Rezyklate in Kunststoffverpackungen? Diskussionspapier zur aktuellen Debatte. https://newsroom.kunststoffverpackungen.de/wp-content/uploads/2020/06/Diskussionspapier_AGVU_IK_-Mindesteinsatzquoten-f%C3%BCr-Kunststoff-Rezyklate-FINAL.pdf (Last access. Aug. 20, 2024).

Bruder and Lindner 2023: Bruder, J., Lindner, C.: Entwicklung einer Kunststoffkreislaufwirtschaft – Herausforderungen an alle Kreislaufakteure und an die Politik (Developing a Circular Plastics Economy - Challenges for all Circular Economy players and policymakers). Manuscript submitted for publication by Hanser Verlag.

COM 2019l: The European Green Deal. Communication from the Commission to the European Parliament, the European Council, the European Economic and Social Committee and the Committee of the Regions. COM/2019/640 final.

COM 2020: A new Circular Economy Action Plan. For a cleaner and more competitive Europe. Communication from the Commission to the European Parliament, the European Council, the European Economic and Social Committee and the Committee of the Regions. COM/2020/98 final.

COM 2023: Report from the Commission to the European Parliament, the Council, the European Economic and Social Committee and the Committee of the Regions identifying Member States at risk of not meeting the 2025 preparing for re-use and recycling target for municipal waste, the 2025 recycling target for packaging waste and the 2035 municipal waste landfilling reduction target. Brussels, 8.6.2023; https://eur-lex.europa.eu/legal-content/EN/TXT/PDF/?uri=CELEX:52023DC0304 (Last access: Aug. 20, 2024).

Conversio 2021: Report elaborated for VinylPlus: PVC waste in EU 27+3 countries 2020.

Conversio 2023a: Estimation by Conversio 2023.

Conversio 2023b: Forecast Model „Use of recyclates in Europe 2020 to 2030". https://www.conversio-gmbh.com/res/Forecast_Model_2030_Europe_Sept21_2023.pdf (Last access: Aug. 20, 2024).
EEA 2023: European Environment Agency: Many EU Member States not on track to meet recycling targets for municipal waste and packaging waste; https://www.eea.europa.eu/publications/many-eu-member-states/many-eu-member-states-not (Last access: Aug. 20, 2024).
EEA 2024: European Environment Agency: Circular material use rate in Europe; https://www.eea.europa.eu/en/analysis/indicators/circular-material-use-rate-in-europe?activeAccordion=ecdb3bcf-bbe9-4978-b5cf-0b136399d9f8 (Last access: Aug, 20, 2024).
EU 2008: Directive 2008/98/EC of the European Parliament and of the Council of 19 November 2008 on waste and repealing certain Directives. https://eurlex.europa.eu/eli/dir/2008/98/oj/eng (Last access: August 20, 2024)
EU 2019: Directive (EU) 2019/904 of the European Parliament and of the Council of 5 June 2019 on the reduction of the impact of certain plastic products on the environment. https://eur-lex.europa.eu/eli/dir/2019/904/oj (Last access: Aug. 20, 2024).
Eunomia 2022: https://www.petcore-europe.org/news-events/409-pet-market-in-europe-state-of-play-2022.html (Last access: Aug. 20, 2024).
Eurostat 2024: Circular material use rate; https://ec.europa.eu/eurostat/web/products-datasets/-/sdg_12_41 (Last access: Aug. 20, 2024).
GVM 2021: GVM Gesellschaft für Verpackungsmarktforschung (2021): Substitution von Kunststoffverpackungen durch papierbasierte Verbunde. Endbericht. https://newsroom.kunststoffverpackungen.de/wp-content/uploads/2021/04/2021-05-18-Bericht-Substitution-durch-Papierverbunde.pdf (Last access: Aug. 20, 2024).
GVM and ifeu 2023: GVM Gesellschaft für Verpackungsmarktforschung, ifeu – Institut für Energie- und Umweltforschung Heidelberg gGmbH (2021): Material Efficiency of Packaging in Comparison. https://www.slideshare.net/Newsroom-Kunststoffverpackungen/material-efficiency-of-packaging-in-comparison-pdf (Last access: Aug. 20, 2024).
Hesselink, Tom 2023: Plastic feedstock for recycling in the Netherlands. Market study by KPMG Advisory N.V. for the VNCI, VA, Plastics Europe NL, and Versnellingstafel Chemische Recycling, Den Haag, 13.10.2023.
IK 2023: IK Industrievereinigung Kunststoffverpackungen: Zertifikate für Rezyklate. Lieferketten schützen – Rezyklateinsatz sinnvoll gestalten. https://newsroom.kunststoffverpackungen.de/zertifikate-fuer-rezyklate/ (Last access: Aug. 20, 2024).
Kirschstein et al. 2023: T. Kirschstein, C. Klöppelt, A. Sanllorente, K. Reh, A. Hofmann, F. Auer (2023): Ökonomische Bewertung von Instrumenten zur Förderung des Kunststoffrecyclings (Economic evaluation of instruments to promote plastics recycling). Fraunhofer-Zentrum für Internationales Management und Wissensökonomie IMW, Fraunhofer-Institut für Umwelt-, Sicherheits- und Energietechnik UMSICHT. Halle/Sulzbach-Rosenberg.
KRU 2019: Ressourcenkommission am Umweltbundesamt (KRU): Substitutionsquote - Ein realistischer Erfolgsmaßstab für die Kreislaufwirtschaft! Position der Ressourcenkommission am Umweltbundesamt, Juli 2019; https://www.umweltbundesamt.de/sites/default/files/medien/421/publikationen/190722_uba_kommp_substitutionsquote_bf.pdf (Last access: Aug.20).
NABU n.d.: Papier, Pappe & Karton. Nur scheinbar öko – Papier sparen ist wichtig! https://www.nabu.de/umwelt-und-ressourcen/ressourcenschonung/papier/index.html (Last access: Aug. 20, 2024).
PlasticsEurope 2024a: The Circular Economy for Plastics – A European Analysis. https://plasticseurope.org/knowledge-hub/the-circular-economy-for-plastics-a-european-analysis-2024 (Last access: Aug. 20, 2024).
Packaging Europe 2023: Beverage industry urges EU to include priority access to recyclates in PPWR proposal. 15.11.2023; https://packagingeurope.com/news/beverage-industry-urges-eu-to-include-priority-access-to-recyclates-in-ppwr-proposal/10618.article. (Last access: Aug. 20, 2024).

PPWR 2024: Proposal for a regulation on packaging and packaging waste, amending Regulation (EU) 2019/1020 and Directive (EU) 2019/904, and repealing Directive 94/62/EC, 16.4.2024. https://eur-lex.europa.eu/legal-content/EN/TXT/?uri=CONSIL%3AST_7859_2024_INIT (Last access: Apr. 22, 2024).
PlasticsEurope 2022: https://plasticseurope.org/knowledge-hub/plastics-the-facts-2022 (Last access: Aug. 20, 2024).
PlasticsEurope 2024b: The Circular Economy for Plastics. A European analysis. March 2024. https://plasticseurope.org/knowledge-hub/the-circular-economy-for-plastics-a-european-analysis-2024/ (Last access: Aug. 20, 2024).
RESAG 2022: UMK-Sonderarbeitsgruppe „Rezyklateinsatz stärken" (RESAG): Bericht an die Umweltministerkonferenz (UMK); https://mluk.brandenburg.de/sixcms/media.php/9/RESAG-Bericht%202022-01-28.pdf (Last access: Feb. 12, 2024).
Schmidt 2020: Schmidt, I. Den Rezyklatmarkt effektiv und marktgerecht fördern (Promote the recyclate market effectively and market-driven). Presentation AGVU-Orientierungstag 10.09.2020.
Stravens, M. (2023): Disposable Paper-based Packaging for Food. The false solution to the packaging waste crisis, Amsterdam, The Netherlands: Profundo.
Sustainable Plastics 2023: Plastics Recyclers Europe warns against priority access to recycled PET. 23.10.2023; https://www.sustainableplastics.com/news/plastics-recyclers-europe-warns-against-priority-access-recycled-pet (Last access: Feb.12, 2024).
UBA 2021: Betz, J., Hermann, A., Bulach W., Hermann,C., Dieroff, J., Mehlhart, G. (Öko-Institut e.V.) im Auftrag des Umweltbundesamts, Texte 128/2022; https://www.umweltbundesamt.de/sites/default/files/medien/479/publikationen/texte_128/2022_pruefung_konkreter_massnahmen_zur_steigerung_der_nachfrage_nach_kunststoffrezyklaten_und_rezyklathaltigen_kunststoffprodukten.pdf (Examination of concrete measures to increase the demand for recycled plastics and plastic products containing recycled plastics) Last access: Aug. 20, 2024.
UBA 2023: Umweltbundesasmt (2023): Untersuchung ökonomischer Instrumente auf Basis des EU-Eigenmittels für nicht recycelte Kunststoffverpackungsabfälle. Zwischenbericht. Texte 124/2023. https://www.umweltbundesamt.de/publikationen/untersuchung-oekonomischer-instrumente-auf-basis (Last access: Aug. 20, 2024).
VDI 2022: Circular Economy für Kunststoffe neu denken. Wie die Transformation zur zirkulären Wertschöpfung gelingen kann. https://www.vdi.de/themen/politischer-dialog/dialog-events/vdi-round-table-circular-economy. (Last access: Dec. 12, 2022).
WWF 2022: Beck-O'Brien, M., Egenolf, V., Winter, S., Zahnen, J., Griesshammer, N. (2022): Everything from wood – The resource of the future or the next crisis? How footprints, benchmarks and targets can support a balanced bioeconomy transition. WWF Germany. https://www.wwf.de/fileadmin/fm-wwf/Publikationen-PDF/Wald/WWF-Study-Everything-from-wood.pdf (Last access: Aug. 20, 2024).
ZSVR 2024: https://www.verpackungsregister.org/en/information-orientation/news-press/press-releases-news/detail/news/from-55000-to-1000000-how-the-lucid-packaging-register-is-improving-fairness-and-transparency-in-the-packaging-recycling-market (Last access: Aug. 20, 2024).

Dr. Isabell Schmidt is Managing Director at IK Industrievereinigung Kunststoffverpackungen and responsible for the areas of circular economy, environment and sustainability. The association represents the interests of around 300 predominantly medium-sized member companies in packaging legislation at the German and European levels. Before joining IK, the graduate geo-ecologist completed her doctorate in the field of life cycle assessment at the University of Karlsruhe (TH) in cooperation with BASF and worked at KPMG Sustainability Services in management consultancy in Germany and abroad.

Chapter 23
Outlook and Challenges

Peter Orth and Jürgen Bruder

Contents

23.1 Outlook

Quantitative Path to 2050

The plastics and chemicals industry currently consumes more than 500 Mt of oil per year as feedstock. In order to meet climate change targets, the goal for 2050 is to replace fossil fuel use with renewable energy and to replace fossil feedstocks in plastics and chemicals, aiming for net zero use of circular and renewable feedstocks. The transition milestones are to achieve a 30% reduction by 2030, 60% by 2040 and 100% by 2050.

A growing and more ambitious global population will require around 1,000 Mt of plastic per year by 2050, a demand that industry will have to meet. By 2050, more than 500 Mt of plastic recyclates, more than 200 Mt of biomass and more than 100 Mt of CO_2 captured and converted into useful chemicals through Carbon Capture and Utilisation (CCU) should be used annually.

P. Orth (✉)
OPC - Orth Plastics Consulting, Cologne, Germany
e-mail: opc.cologne@t-online.de

J. Bruder
BPC- Bruder Plastics Consulting, Potsdam, Germany
e-mail: plastics@jbruder.de

P. Orth et al. (eds.), *Circular Economy of Plastics*,
https://doi.org/10.1007/978-3-031-78829-1_23

Technological Advancements

Technological advances are underway to increase the efficiency of mechanical and chemical recycling to recycle 90% of collected plastic waste.

Mechanical recycling technologies are developing rapidly. Return systems for standardized products—involving the consumer—allow waste to be collected by type and recycled almost completely, advanced analytical techniques allow ever better sorting of mixed waste, solvent-based technologies are entering the market, and dedicated waste logistics are creating larger and more homogeneous waste streams. Mechanical recycling is the first priority process, where material streams are bundled, yields are increased, and product qualities are aligned with those of virgin materials. Existing processes and facilities are being scaled up to increase efficiency and cost-effectiveness.

Advanced recycling technologies, such as chemical recycling using methods like pyrolysis or gasification, have made significant progress in pushing the boundaries of what can be chemically recycled. Chemical recycling processes are implemented on a large scale for specific waste streams and are continuously developed to maximise the recovery of waste streams. The technology is being scaled up to meet global needs. However, chemical recycling remains economically challenging and energy intensive.

The logistics of waste collection need to be further developed technically, with the aim of generating as large and uniform material flows as possible from the outset, which can ideally be recycled on a job basis.

The pre-treatment of waste (collection, sorting, processing) involves deciding on the appropriate recycling process for each waste stream to optimise raw material and energy efficiency. Life Cycle Assessments (LCA) provide decision support and guidance in this process.

The sorting and recovery of residues and non-recyclable waste are efficiently used for power generation, the CO_2 generated in the process is captured for CCU processes and not emitted into the atmosphere, slags are recycled, and pollutants are removed from the cycle.

The different recycling technologies are thus integrated into an efficient circular economy system for plastics.

23.2 Challenges

Technical Challenges

The aim of improving the quality of mechanical recyclates is to meet virgin standards and thus become competitive. The development of chemical recycling and biomass conversion technologies must focus on versatility and cost-effectiveness. Carbon Capture and Utilisation (CCU) processes to convert CO_2 into basic chemicals, monomers and building blocks need to be scaled up by the industry.

Economic Challenges

High capital costs are required for new recycling plants, biomass conversion facilities and CO_2 utilisation infrastructure. Solutions must be found to enable competition with the relatively low cost of fossil-based chemicals, especially when oil prices are low. Reliable supply chains need to be established for multiple feedstocks, including post-consumer plastics, biomass and CO_2. Ongoing consolidation in the recycling industry is helping to create larger and better-financed entities with greater investment capacity.

Ecological Challenges

Biomass sourcing must not lead to deforestation or compete with food production. The high energy requirements of recycling and conversion processes must be met by using renewable energy. The overall carbon footprint of new processes must be minimised compared to traditional fossil fuel-based methods.

Challenges for the Industry

The development of circular business models is mandatory for all stakeholders in the plastics value chain. At the corporate level, producer responsibility and product stewardship include participation in circular systems and solutions, control over products and due diligence throughout their life cycle. This includes the standardisation of processes, products and services and product development with a holistic Design for Circularity approach, the provision of transparent product information to customers and consumers, and the digital identification of products as they move through the cycle.

The result at the product level is Design for Sustainability, which, in addition to the functionality, economy and quality demanded by the customer, also includes energy efficiency, durability, repairability and recyclability, as well as safety for consumers or users, society and the environment.

European companies have invested billions of euros in circular economy projects for all materials in recent years. However, there is still a gap between claims and commitments and actual environmental impacts that need to be closed.

Challenges for Consumers and Society

Transparent and comprehensible product information is a prerequisite for sustainable consumption and responsible use of products. The provision of such information increases the competence and awareness of consumers and users. However, it is also in their hands that the disposal of the product and thus the decision on its end-of-life is made. In a circular economy, this includes an appreciation of waste as a resource that needs to be managed appropriately to ensure reuse or recycling, rather than simply getting rid of it.

Many European countries have high levels of consumer participation in recycling programmes. This cultural change is supported by comprehensive recycling schemes and public awareness campaigns. However, consumer behaviour remains inconsistent. While awareness has increased, convenience often trumps sustainability, leading to persistent problems with plastic waste. In addition, engagement and participation in recycling and sustainable consumption practices remain low in most countries and regions of the world.

In parallel, the fight against littering, i.e. the uncontrolled release of waste into the environment as rubbish, must be stepped up. This requires, on the one hand, increased consumer information and education, the operation and optimisation of appropriate waste collection systems and their image, and, on the other hand, the sanctioning of offenders, for which the enforcement authorities must be empowered and enabled.

Role of Policy

Policy hast to provide incentives for innovation either via increased funding for R&D in recycling technologies, biomass conversion, and CO_2 utilization and/or via tax incentives for companies investing in sustainable technologies and infrastructure.

Regulatory support must trigger higher recycling rates, higher circular raw material quotas and the use of recycled content in products. Implementation of carbon pricing mechanisms should make fossil-derived feedstocks less economically attractive.

In national and European legislation, a move away from a focus on waste management towards a focus on resource management is overdue; for the raw material transition, we need a resource hierarchy.

Consistent and long-term policies must ensure stability and predictability in environmental regulations and incentives to encourage sustained investment in defossilization technologies.

Green procurement governmental policies to preferentially purchase products with high recycled and/or renewable content should help create markets. And mandated standards and certifications should trigger recycled and renewable content in plastics and chemicals.

International agreements and collaborations to harmonize standards, share best practices, and support developing countries in their transition efforts need to be pursued.

There are considerable discrepancies in regulations between different regions. While Europe has stringent policies, many countries have yet to develop robust frameworks, leading to a fragmented approach to plastic waste management.

Establishing global standards for plastic products and recycling processes can harmonize efforts across different regions, making international trade and compliance easier and more efficient.

Policy Initiatives

Europe has been at the forefront of regulatory initiatives, with the EU's Circular Economy Action Plan detailing the legislative framework and goals and aiming to ensure all plastic packaging on the EU market is recyclable by 2030. This is driving advancements in packaging design and recycling processes across member states.

Some countries have implemented successful circular economy approaches and models. Germany's dual system for packaging waste collection and recycling serves as a model for other countries. The draft German National Circular Economy Strategy (NKWS[1]) may serve as a blueprint for analysing and establishing consistent CE strategies elsewhere and on the European level.

The NKWS intends to preserve the value of raw materials and products for as long as possible, using materials sparingly and keeping them in the cycle for as long as possible and ultimately strengthening the resilience and innovative power of industry. A life cycle perspective is central to this—from product design to production and consumption, logistics, recycling, reparability and reuse. All stages of value creation are geared towards circularity.

This includes a carbon management strategy to promote defossilization, further development of sorting and recycling technologies, recyclate use quotas for plastics (in particular a polymer-specific substitution quota), limiting the variety of materials, setting standards and norms, further development of circular economy legislation (including at international level), using public procurement as a lever, setting economic incentives, improving financing together with targeted research funding.

Dr. Peter Orth is a chemist by education; with OPC—Orth Plastics Consulting, he leads a group of senior consultants with decades of professional experience in leading positions in the German and European plastics industry and its associations. His main focus is on innovation and sustainability in cooperation with science, industry, and politics. He is currently working in the fields of CO_2 as a feedstock, resource efficiency, and recycling technologies.

Dr. Jürgen Bruder is a chemist with an additional degree in business. He worked in higher education and the chemical industry, before becoming Director General of IK, the German plastic packaging association. He held positions in national, European, and global committees including the Global Plastics Alliance. Today he deals with the transformation of the plastics industry to a circular economy focusing on plastic packaging and plastic applications in agriculture.

[1] German National Circular Economy Strategy (in German: Nationale Kreislaufwirtschaftsstrategie—NKWS) https://www.bmuv.de/themen/kreislaufwirtschaft/kreislaufwirtschaftsstrategie.

Glossary

Bioeconomy

A bioeconomy is a sustainable and biobased economy oriented towards the natural cycle of materials, the supply of which feeds the world sufficiently and healthily and provides people with high-quality products from renewable raw materials. It encompasses the production, development and use of biological resources, processes and systems to provide products, processes and services in all economic sectors as part of a sustainable economic system. It holds the potential for sustainable solutions that conserve resources while creating wealth.

Biogenic Raw Materials

These include both primary raw materials such as wood and products of the first and second processing stages such as cellulose, starch, monomeric carbohydrates, chitin, fats and oils (also of animal origin) and—so far only to a small extent—proteins. The main advantage of biogenic raw materials is their neutrality regarding the global carbon dioxide balance, their high functionality ("nature's synthesis input") and, in certain cases, a particularly favourable eco-profile.

Biomass

is the biodegradable fraction of products, wastes and residues from agriculture of biological origin (animal and vegetable matter), forestry and related industries, including fisheries and aquaculture. The biodegradable fraction of industrial and household waste is also included in biomass according to this definition.

Carbon Capture and Usage (CCU)

CO_2 is captured either from an (industrial) process or directly from the air (*DAC—Direct Air Capture*) and fed into an industrial, material use. The captured CO_2 becomes a carbon source and thus a raw material for chemical or biotechnological processes.

P. Orth et al. (eds.), *Circular Economy of Plastics*,
https://doi.org/10.1007/978-3-031-78829-1

Carbon Cycle

Basically, the path that carbon takes through the Earth system, passing through various stations on land (lithosphere and pedosphere), in water (hydrosphere), in the air (atmosphere) and in the biosphere. Man-made emissions of greenhouse gases such as carbon dioxide and methane profoundly alter this natural cycle. In the specific case of plastics, this refers to the recycling of the carbon bound in them via the material recycling of waste or via CO_2 into circular raw materials for plastics production.

Carbon Dioxide (CO_2)

A naturally occurring gas, also a by-product of the combustion of fossil fuels such as oil, gas and coal, the burning of biomass, and land-use changes and other industrial processes. It is the most important anthropogenic greenhouse gas affecting the Earth's radiation balance. It is the "reference gas" against which the other greenhouse gases are measured and therefore has a Global Warming Potential (GWP).

Carbon Dioxide Equivalent (CO_2 e)

Emissions of greenhouse gases other than carbon dioxide (CO_2) are converted into CO_2 equivalents according to their global warming potential for better comparability ($CO_2 = 1$).

Carbon Footprint

is the total amount of greenhouse gases (including carbon dioxide and methane) produced by our actions (e.g. in the manufacture, use and disposal of products). The carbon footprint is thus a measure of the environmental impact of products. Also applied to people or groups of people.

Chain of Custody

An important component of the mass balance concept is the accompanying verification of traceability embedded in the value chain—the "*chain of custody*"—which leads from the plastic waste producer (the consumer or user) via the disposer and recycler to the plastic producer and processor and finally via the OEM or retailer back to the customer (cf. Fig. 1.1). The use of digital methods enables and facilitates the work and success of the *chain of custody*.

Chemical Recycling

The depolymerisation of polymers using chemical or thermochemical processes, whereby the macromolecules are broken down into monomers or other basic chemical *building blocks*, which are then used again as raw materials in the chemical or plastics-producing industry. The plastic waste thus undergoes a recycling path to materials for the original purpose or for another purpose.

Circular Business Model

In contrast to the linear logic of mining, production, consumption, disposal with isolated optimization and profit maximization by individuals, the circular business model is based on the coordination of all actors active in a value cycle. Here, one actor acts as a central orchestrator to achieve joint activities for circularity at the system level. All loop participants share a circular vision and distribute profits in a way that ensures long-term commitment from all contributing actors. The use of digital technologies plays a crucial role in this.

Circular Economy

The concept of the Circular Economy aims to decouple economic growth from resource consumption. The circular, resource-conserving economy is intended to go far beyond the classic "circular economy" in the sense of recycling. Material and energy cycles are closed—as far as possible and ecologically sensible. Circular Economy encompasses the entire life cycle of a product: The design should enable durability, reprocessing and reparability. The use phase should be intensified and extended. At the end of the life cycle, the various recyclable materials should be separated as far as possible by sorting and dismantling and materially processed for reuse. With these measures, the Circular Economy can pursue both ecological goals such as climate protection and resource conservation, as well as economic goals such as competitiveness and independence from raw materials. In addition, the focus can also be on social goals, such as employment and local value creation.

Circular raw material Raw materials in the circular economy (see also *Renewable Carbon*); in the case of plastics, the carbon compounds (monomers etc.) required for their synthesis, produced from various carbon-containing wastes (plastic waste etc.), from biomass or from CO_2.

Closed-Loop/open-Loop Recycling

Reuse of recyclates in the same application from which the input materials originate. In *open-loop recycling*, on the other hand, the recyclates are also used in other applications. *Open-loop recycling* enables a wider range of applications for the recyclates, which means a larger market and thus possibly higher demand - but also the risk of higher quality losses in recycling. A *closed loop* in the sense of one hundred percent permanent recycling of all materials is desirable, but physically not possible and an approach to this is increasingly suboptimal thermodynamically.

Decarbonisation

The renunciation of fossil fuels based on the element carbon, such as coal, crude oil and natural gas or their derivatives. Their purpose is to use the thermal energy released by their combustion, whereby the final combustion product CO_2 is released into the atmosphere as gaseous waste, where it acts as a greenhouse gas and thus contributes to climate change.

Defossilization

The conversion of an economy with the aim of replacing fossil energy sources with renewable alternatives.

Downcycling

Recycling process that involves the production of materials of lower quality and reduced functionality.

Energy Recovery

Incineration with energy decoupling, usually in a waste incineration plant is currently still the last resort for plastic waste that cannot be recycled in any other way, as at least the chemically stored energy can be (partially) used in this way. Due to their molecular structure, plastics are high-calorific materials that are suitable for the generation of useful energy (district heating, industrial heat, industrial steam, steam for electricity generation). The calorific value is sometimes higher than that of fossil fuels such as coal or even fuel oil. Another possibility is the production of substitute fuels (RDF) to replace fossil fuels such as coal or oil. In its five-stage waste hierarchy, the Closed Substance Cycle Waste Management Act initially favours recycling or material recovery over energy recovery.

Fossil Raw Materials

Fossil raw materials such as crude oil, natural gas and coal are currently the most important primary energy sources in the global economy. One of the most important fractions in the fractional distillation of crude oil is naphtha or crude gasoline, from which the basic chemicals for the chemical industry and thus also the raw materials for plastics production are produced by steam cracking.

Greenhouse Gas (GHG)

Gases in the earth's atmosphere that trigger the so-called greenhouse effect. Greenhouse gases can have a natural, but also an anthropogenic (man-made) origin. Carbon dioxide, methane and water vapour are the most common of these gases https:www.nature.org/en-us/get-involved/how-to-help/carbon-footprint-calculator/.

Life cycle

"The totality of the successive and interrelated phases of existence of a product from the processing of the raw material to its disposal." *End-of-life*: The end-of-life phase in the life cycle begins when the product under consideration and its packaging are disposed of by the user and ends when the product is returned to nature as waste or enters the life cycle of another product (for example as a recycled input). Waste: "A substance or object as defined in Annex I to Directive 2006/12/EC which the holder discards, intends to discard or is required to discard."

Mechanical Recycling

After appropriate preparation and sorting, the dried plastic regrind is either provided in the form of *flakes* or processed into agglomerates or regranulates in the next process step. In agglomeration, the particles of the regrind are bonded together by surface melting to achieve uniformity of particle shape and size and higher bulk densities. To obtain homogeneous and free-flowing regranulates, the ground material is fed through an extruder and, if necessary, compounded (equipped) with additives (carbon black, pigments, reinforcing fibres, etc.). Melt filters ensure the separation of any impurities still contained (wood fibres, paint particles, etc.). The homogenized melt is finally granulated.

Microplastics

Plastic parts and fibres < 5 mm, mostly < < 1 mm, which are produced by corrosion or mechanical abrasion of larger parts or are produced specifically for applications in cosmetics, for example (*microbeads*) and enter the environment because of human activity. The term was defined based on physical properties; it has not yet been refined from an environmental science perspective.

PIR

Post-industrial recyclate (PIR) or pre-consumer recyclate: Recyclate made of post-industrial material which means material that is separated from the waste stream during the manufacturing process. It does not include the reuse of materials from reworking, regrinding or scrap that arise during a technical process and can be reused in the same process.

Post Consumer Material

Material from households, commercial and industrial facilities or institutes (who are end users of the product) that can no longer be used for its intended purpose. This includes returns of material from the distribution chain. (ISO 14021).

Post Consumer Recyclate (PCR)

Recyclate made of post-consumer material which means material from households, commercial and industrial facilities or institutes (who are end users of the product) that can no longer be used for its intended purpose. This includes returns of material from the distribution chain.

Post Industrial Material

Material that is separated from the waste stream during the manufacturing process. It does not include the reuse of materials from reworking, regrinding or scrap that arise during a technical process and can be reused in the same process (ISO 14021).

Recovery Rate

The recovery rate can be defined in the same way as the recycling rate. National Recycling Rate is defined as the quantity of material recovered in the country plus quantities exported for recovery minus material imported intended for recovery out of total waste generated in the country. Recovery of plastic waste includes recycling, energy recovery and organic recovery (aerobic and anaerobic) of biodegradable plastics.

Recyclability

"Recyclability means the compatibility of packaging with the management and processing of waste by design, based on separate collection, sorting in separate streams, recycling at scale, and use of recycled materials to replace primary raw materials" (PPWR 2024, Art 3, (31a)).

Recycled Content Rate or Quota = recyclate Usage Rate

The term recycled content rate is commonly used for the "minimum percentage of recycled material recovered from post-consumer plastic waste" in the draft PPWR, but also used in the SUPD and further strategic documents. The percentage relates to defined "packaging types and formats", e.g. single-use plastic beverage bottles and are "calculated as an average per manufacturing plant and year". (PPWR 2024: Compromise Paper between Council and EU Parliament, 15 March 2024. https://data.consilium.europa.eu/doc/document/ST-7859-2024-INIT/en/pdf (Last access: 25 March 2024).

Recyclate

Material that has been reprocessed through a manufacturing process from recovered (remanufactured) material and made into a final product or component for incorporation into a product. Recovered materials in this report are understood to be *post-consumer materials. The* decision to focus exclusively on *post-consumer recyclates is* justified by the fact that the purpose of closing resource loops in a *circular economy* is thus focused on the loop between *post-consumer* and production.

Recycling Rate

The UN definition is as follows: The National Recycling Rate is defined as the quantity of material recycled in the country plus quantities exported for recycling minus material imported intended for recycling out of total waste generated in the country…. National recycling rate can be presented by type of waste, including e-waste, plastic waste, municipal waste, and others" (UN SDG 12, Target 12.5: https://unstats.un.org/sdgs/metadata/).

Recycling rates for plastic waste therefore represent a ratio of the waste fed into the recycling process and the total waste generated, with the calendar year serving as a temporal reference basis. In the EU as of 2020, the numerator of the quota, i.e. quantity of recycled material, must be determined at measurement points defined for each material. The measurement point is defined as the input amount to the last recycling process through which waste materials are actually further processed into products, materials, or substances. For plastic waste, the measurement point is usually before the extruder for the production of regranulate. This procedure largely avoids the quota-increasing inclusion of impurities, adhesions, foreign substances, or moisture removed before the last recycling step. Determining the denominator of the quotient requires a detailed market analysis, as it is not the (separately) collected quantity but the total quantity of waste generated in the reference year. Depending on the related amount of plastic waste from which the recycling rate is determined, there are recycling rates for plastic packaging, for plastics from the construction sector etc., and the recycling rate for plastic waste in general.

Recycling rates do not tell us anything about the quantity and quality of recyclate that is produced in the recycling process.

Renewable carbon includes all carbon sources that help avoid or replace the use of additional fossil carbon from the geosphere.

Substitution Rate

The substitution rate is defined in the literature as the percentage of secondary material used in relation to the total material input (primary plus secondary material). In the plastics industry, it is used to measure the degree of substitution of fossil-based virgin material by

recyclates or by all circular raw materials (recyclates, biomass, CO2).

The substitution rate can be applied to the volume of all plastics at the national or European level, or to the volume of individual polymers at the European level.

Upcycling

Recycling process with the aim of producing materials of the highest possible quality and functionality. The term suggests a thermodynamically impossible increase in quality.

Recommended Readings

Summary

The following literature collection is intended to give the reader the opportunity to deepen the topics in this book and to place them in larger contexts. It includes studies, reports, policy papers, position papers, scientific articles, etc. from politics, business, and society with a focus on Europe. These deal with different aspects of global warming, species extinction, resource limitations, and environmental pollution, and the subsequent tension between growth and sustainability. The focus of this selection is the challenges for plastic and the plastics industry dealt with in this work: the release of plastic waste into the environment, the development of a circular economy for plastics, the defossilization of the plastics value chain and the solutions and measures developed for this.

This selection compiled by the editors is inherently subjective and incomplete.

International Institutions

Food and Agriculture Organization (2023) The State of Food Security and Nutrition in the World 2022. https://openknowledge.fao.org/server/api/core/bitstreams/67b1e9c7-1a7f-4dc6-a19e-f6472a4ea83a/content

International Energy Agency (2018) The Future of Petrochemicals. https://www.iea.org/reports/the-future-of-petrochemicals

International Energy Agency (2021) Net Zero by 2050: A Roadmap for the Global Energy Sector. https://www.iea.org/reports/net-zero-by-2050

International Energy Agency (2021) World Energy Outlook 2022. https://www.iea.org/reports/world-energy-outlook-2022

Intergovernmental Science-Policy Platform on Biodiversity and Ecosystem Services IPBES (2019): Global assessment report on biodiversity and ecosystem services of the Intergovernmental Science-Policy Platform on Biodiversity and Ecosystem Services. E. S. Brondizio, J. Settele, S. Díaz, and H. T. Ngo (editors). IPBES secretariat, Bonn, Germany. 1148 pages. https://doi.org/10.5281/zenodo.3831673

Intergovernmental Panel on Climate Change (2022) Climate Change 2022: Impacts, Adaptation and Vulnerability. https://www.ipcc.ch/report/ar6/wg2/

International Renewable Energy Agency (2024) Tracking SDG 7: The Energy Progress Report 2024. https://www.irena.org/Publications/2024/Jun/Tracking-SDG-7-The-Energy-Progress-Report-2024

International Standardisation Organisation (2024) ISO policy brief: Combatting plastic pollution with International Standards. https://www.iso.org/files/live/sites/isoorg/files/store/en/PUB100483.pdf

OECD (2018) Business models for the circular economy. https://web-archive.oecd.org/2018-10-22/496906-policy-highlights-business-models-for-the-circular-economy.pdf

OECD (2023) Towards eliminating plastics pollution by 2040. https://www.oecd.org/environment/plastics/Interim-Findings-Towards-Eliminating-Plastic-Pollution-by-2040-Policy-Scenario-Analysis.pdf

OECD (2022) Global Plastics Outlook. https://www.oecd-ilibrary.org/environment/data/global-plastic-outlook_c0821f81-en

OECD (2022) Global Plastics Outlook. Policy Scenarios to 2060. https://www.oecd.org/environment/global-plastics-outlook-aa1edf33-en.htm

UN (2022) World Population Prospects 2022. https://population.un.org/wpp/

UNEP (2022) Global Chemicals Outlook II. https://www.unep.org/topics/chemicals-and-pollution-action/chemicals-management/global-chemicals-outlook

UNEP (2020) Sustainable bioenergy use: a clear path to biodiversity and sustainable economic development. https://www.unep.org/news-and-stories/story/sustainable-bioenergy-use-clear-path-biodiversity-and-sustainable-economic

UNEP (2022) Intergovernmental Negotiating Committee on Plastic Pollution. https://www.unep.org/inc-plastic-pollution

UNEP (2023) The New Plastics Economy Global Commitment. https://www.unep.org/new-plastics-economy-global-commitment

UNEP (2023) Turning Off the Tap: How the World Can End Plastic Pollution. https://www.unep.org/resources/turning-off-tap-end-plastic-pollution-create-circular-economy

United Nations Framework Convention on Climate Change (2023) Race to Zero Campaign. https://climateaction.unfccc.int/Initiatives?id=Race_to_Zero

International Resource Panel (2020) Global Resource Outlook 2019. https://www.resourcepanel.org/reports/global-resources-outlook-2019

World Bank (2020) What a Waste 2.0: A Global Snapshot of Solid Waste Management to 2050. https://datatopics.worldbank.org/what-a-waste/

World Bank (2022) Squaring the Circle. https://thedocs.worldbank.org/en/doc/67912b7a7143fa774f26ce22bed8144e-0080012022/original/EU-CE-REPORT-WEB.pdf

European Institutions

European Environment Agency (2023) Waste Recycling Europe. https://www.eea.europa.eu/en/analysis/indicators/waste-recycling-in-europe

European Environment Agency (2024) The Role of Plastics in Europe's Circular Economy. https://www.eea.europa.eu/publications/the-role-of-plastics-in-europe

European Environment Agency (2023) Economic instruments and separate collection systems — key strategies to increase recycling. https://www.eea.europa.eu/publications/economic-instruments-and-separate-collection

European Investment Bank (2023) Cutting Plastic Pollution. https://www.eib.org/en/publications/20220248-cutting-plastics-pollution

European Research Council (2024) Annual report on the ERC activities and achievements in 2023. https://erc.europa.eu/news-events/erc-publications

European Union (2024) Plastics Strategy. https://environment.ec.europa.eu/strategy/plastics-strategy_en

European Commission (2024) Circular Economy Action Plan. https://environment.ec.europa.eu/strategy/circular-economy-action-plan_en

European Commission (2024) Single-Use Plastics. https://environment.ec.europa.eu/topics/plastics/single-use-plastics_en

Eurostat (2023) EU's circular material use rate slightly up in 2022. https://www.resourcepanel.org/reports/global-resources-outlook-2019

Horizon Europe (2024) https://research-and-innovation.ec.europa.eu/funding/funding-opportunities/funding-programmes-and-open-calls/horizon-europe_en

Governments

German Federal Ministry for the Environment (2024) Draft of a National Circular Economy Strategy (in German). https://www.bmuv.de/download/entwurf-einer-nationalen-kreislaufwirtschaftsstrategie-nkws

National Institutions

Betz, J., Bachmann, H., Huber, I. et al. (2024) Kunststoffabfälle aus Deutschland: Handlungsempfehlungen zu einer umweltgerechten Behandlung im In- und Ausland. https://www.umweltbundesamt.de/sites/default/files/medien/11850/publikationen/47_2024_texte_kunststoffabfall.pdf

DIN, DKE, and VDI (2023) Standardization Road Map Circular Economy. https://www.din.de/resource/blob/906910/0d691bed63405ae85f281336ed71162c/standardization-roadmap-circular-economy-data.pdf

United States Environmental Protection Agency (2024) National Recycling Strategy. https://www.epa.gov/circulareconomy/national-recycling-strategy

Ressourcenkommission am Umweltbundesamt (2023) Opportunities and limits of recycling in the context of the circular economy. https://www.umweltbundesamt.de/en/publikationen/opportunities-limits-of-recycling-in-the-context-of

Associations

BDE, PED, and VCI (2023) Common vision of a circular economy with plastics (in German). https://plasticseurope.org/de/2023/09/06/pressemitteilung/

Cefic (2022) 2021 Sustainability Progress Update. https://cefic.org/app/uploads/2021/10/Sustainability_progress_update_2021.pdf

Chemistry for Climate (2023) How the transformation of chemistry can succeed. (in German). https://www.vci.de/vci/downloads-vci/publikation/broschueren-und-faltblaetter/final-c4c-broschure-langfassung.pdf

GKV, BDE, and bvse (2022) New rules for plastic recyclates in contact with food (in German). https://kunststoffverpackungen.de/wp-content/uploads/2022/11/leitlinien-gkv-bde-bvse_rezyklate_in_kontakt_mit_lebensmittel.pdf

International Solid Waste Association (2024) Global Waste Management Outlook. https://www.iswa.org/wp-content/uploads/2024/05/15-May-IFAT-GWMO-2024-Session-Note-final.pdf

PlasticsEurope (2022) Circular Economy Plus https://plasticseurope.org/de/2022/11/25/circular-economy-plus/

PlasticsEurope (2023) Plastics – The Facts 2022.https://plasticseurope.org/knowledge-hub/plastics-the-facts-2022/

PlasticsEurope (2024) The Circular Economy for Plastics: A European Analysis. https://plasticseurope.org/de/2024/03/19/circular-economy-report/

VCI and PlasticsEurope (2023) Position paper on mass balance assessments (in German). https://www.vci.de/presse/pressemitteilungen/mehr-recycling-durch-mehr-transparenz-massenbilanzen-anerkennen-dringende-investitionen-anstossen.jsp

THINKTANK Industrielle Ressourcenstrategien (2023) Chemical Recycling Entrepreneur Forum (in German). https://www.thinktank-irs.de/wp-content/uploads/2023/07/RZ_THINKTANK_Broschuere_Unternehmerforum-Chemisches-Recycling_DE_Web.pdf

Corporations

BASF (2023) Chemical Recycling – Accelerating Circularity. https://www.basf.com/dam/jcr:2bc62e51-321a-3f2d-a070-c8cf63cfe20c/basf/www/global/documents/en/sustainability/we-drive-sustainable-solutions/circular-economy/chemcycling/FactSheet_ChemCycling_2021.pdf

Coca-Cola Company (2024) World Without Waste Progress Report 2022. https://www.coca-colahellenic.com/en/a-more-sustainable-future/mission-2025/world-without-waste

Eastman Chemical Company (2024) 2023 Sustainability Report. https://www.eastman.com/content/dam/eastman/corporate/en/media-center/resources/eastman-sustainability-report-2023.pdf

LyondellBasell Industries (2024) 2023 Sustainability Report. https://www.lyondellbasell.com/49688c/globalassets/sustainability/2023_lyb_sustainability_report.pdf

Nestlé (2024) Creating Shared Value and Sustainability Report 2023. https://www.nestle.com/sites/default/files/2024-02/creating-shared-value-sustainability-report-2023-en.pdf

Novamont (2024) 2023 Impact Report. https://www.novamont.com/public/Bilancio%20di%20sostenibilit%C3%A0/Novamont%20Relazione%20Impatto%202023_EN.pdf

Unilever (2024) Sustainability Performance Data. https://www.google.com/search?client=firefox-b-d&q=Unilever+2024+Sustainable+Living+Plan%3A+Performance+Report

ThinkTanks and Consultants

Agora Industry (2023): Chemicals in transition. The three pillars for transforming chemical value chains. https://www.agora-industry.org/fileadmin/Projekte/2022/2022-02_IND_Climate_Positive_Chemistry_DE/A-EW_300_Chemicals_in_transition_EN_WEB.pdf

Boston Consulting Group (2021) The Value of the Circular Economy. https://www.bcg.com/videos/unlock-value-of-circular-economy

BKV (2022) Stoffstrombild Kunststoffe in Deutschland 2021: Zahlen und Fakten zum Lebensweg von Kunststoffen. https://www.bkv-gmbh.de/1150-studie-stoffstrombild-kunststoffe-in-deutschland-2021.html

Center for Digital Technology and management (2022) The Future of Waste Management 2021.https://www.cdtm.de/wp-content/uploads/2023/03/Trend-Report-F21-The-Future-of-Waste-Management.pdf

Deloitte (2023) The Future of Materials. https://www2.deloitte.com/us/en/insights/industry/oil-and-gas/the-future-of-materials.html

Fluchs, S., Neligan, A., Schleicher, C. et al. (2022) Circular business models. How circular are companies? (in German) IW-Report, Nr. 27, Köln / Berlin. https://www.iwkoeln.de/studien/sarah-fluchs-adriana-neligan-wie-zirkulaer-sind-unternehmen.html

McKinsey & Company (2022) Advanced Recycling: Opportunities for Growth. https://www.mckinsey.com/industries/chemicals/our-insights/advanced-recycling-opportunities-for-growth

Systemiq (2023) Circular PET and Polyester. https://www.systemiq.earth/pet-polyester

Science

Bachmann, M., Zibunas, C., Hartmann, J. et al. (2023) Towards circular plastics within planetary boundaries. *Nat Sustain* **6**, 599–610 (2023). https://doi.org/10.1038/s41893-022-01054-9

Bertling, J., Brandt, M.-J., Blömer, J., Lange, V., Lehmann, L. (2024) Glossary on Reuse / Mehrweg-Glossar. Terms and Definitions of Reusable Packaging Systems / Begriffe und Definitionen zu Mehrweg-Verpackungssystemen. https://doi.org/10.24406/publica-2478

Caro, D., Albizzati, P.F., Cristobal Garcia, J. et al. (2023) Towards a better definition and calculation of recycling. EUR 31,409 EN, Publications Office of the European Union, Luxembourg, 2023, ISBN 978–92-76–98,958-5. https://doi.org/10.2760/636900,JRC131531

Fraunhofer Gesellschaft (2022), Annual Report 2022 (p. 42). https://www.fraunhofer.de/s/ePaper/Annual-Report/2022/index.html#42

Fraunhofer UMSICHT and TNO (2023) From plastic free to future-proof plastics. https://www.tno.nl/en/newsroom/2023/06/from-plastic-free-future-proof-plastics/

International Institute for Sustainable Development (2020), Doing more with less - sustainable production and consumption report". https://www.iisd.org/articles/deep-dive/doing-more-less-ensuring-sustainable-consumption-and-production

Kähler, F., Porc, O., Carus, M. (2023) RCI carbon flows report – Compilation of supply and demand of fossil and renewable carbon on a global and European level. https://renewable-carbon.eu/publications/product/the-renewable-carbon-initiatives-carbon-flows-report-pdf/

Karali, N., Khanna, N., Shah, N. (2024) Climate Impact of Primary Plastic Production. https://escholarship.org/uc/item/12s624vf

Kerps, A., Dobers, K., Jarmer, J.-P. et al. (2024) Challenges in comparative life cycle assessment of recycling plastics – position paper, Fraunhofer Cluster of Excellence Circular Plastics Economy CCPE (Ed.), Oberhausen/Dortmund/Pfinztal, February 2024. https://www.ccpe.fraunhofer.de/content/dam/ccpe/en/documents/position-paper/fraunhofer-ccpe-position-paper-LCA-web.pdf

Maury, T., Tazi, N., Torres De Matos, C. et al. (2023) Towards recycled plastic content targets in new passenger cars and light commercial vehicles, Publications Office of the European Union, Luxembourg, 2023. https://doi.org/10.2838/834615,JRC129008.

Meng, F., Brandão, M., Cullen, J.M. (2024) Replacing Plastics with Alternatives Is Worse for Greenhouse Gas Emissions in Most Cases. Environ. Sci. Technol. 2024, 58, 2716–2727https://doi.org/10.1021/acs.est.3c05191

Richardson, K., Hardesty, B.D., Vince, J. et al. (2022) Global estimates of fishing gear lost to the ocean each year. https://www.science.org/doi/10.1126/sciadv.abq0135

Vidal, F., van der Marel, E.R., Kerr, R.W.F. et al. (2024) Designing a circular carbon and plastics economy for a sustainable future. Nature **626**, 45–57 (2024). https://doi.org/10.1038/s41586-023-06939-z

NGOs

Circle Economy Foundation (2024) The Circularity Gap Report. https://www.circularity-gap.world/2024l

European Environmental Bureau (2023) Priorities for the Waste Framework Directive (WFD) Revision proposal 2023/0234(COD). https://eeb.org/wp-content/uploads/2023/11/2023_EEB_Position-Paper_Targeted-Revision-WFD.pdf

Ellen MacArthur Foundation (2023) The Global Commitment. Five Years in: Learnings to Accelerate towards a Future without Plastics Waste or Pollution. https://emf.thirdlight.com/file/24/SWAXKg3SWOeL_4OSWZn0SWsy8Dw/The%20Global%20Commitment%20Five%20Years%20In%3A%20Learnings%20to%20Accelerate%20towards%20a%20Future%20without%20Plastic%20Waste%20or%20Pollution.pdf

Greenpeace (2023) Green Wash Danger Zone. https://www.greenpeace.de/publikationen/Greenpeace_Report_Greenwash_Danger_Zone.pdf

World Economic Forum (2024) Circular Transformation of Industries: The Role of Partnerships. https://www3.weforum.org/docs/WEF_Circular_Transformation_of_Industries_2024.pdf

World Wildlife Fund (2022) Living Planet Report. https://www.worldwildlife.org/pages/living-planet-report-2022

MIX
Papier aus verantwortungsvollen Quellen
Paper from responsible sources
FSC® C105338

If you have any concerns about our products,
you can contact us on
ProductSafety@springernature.com

In case Publisher is established outside the EU,
the EU authorized representative is:
Springer Nature Customer Service Center GmbH
Europaplatz 3, 69115 Heidelberg, Germany

Printed by Libri Plureos GmbH
in Hamburg, Germany